# Environmental Science and Engineering
## — *Selected Topics* —

## About the Editors

**Dr. S.N. Kaul** obtained his bachelor and master degrees from IIT, Kanpur and was awarded Ph.D. degree from the University of New Castle Upon Tyne, UK. Dr. Kaul joined CSIR-National Environmental Engineering Research Institute [NEERI] Nagpur in 1970 and worked for over three decades. Dr. Kaul superannuated as Acting Director and Director Grade scientist from NEERI and was also associated with MIT college of Engineering, Pune as its Principal.

Dr. Kaul has published over 350 Papers in National and International journals beside and authorship of over 30 books on Environmental Science and Engineering, 8 Patents and 150 Conference papers. Dr. Kaul has designed and operated several waste treatment plants in India. Dr. Kaul has supervised over a dozen of Ph.D. and 40 M.Tech. students. In addition, Dr. Kaul was responsible for organizing several National and International Conferences and has traveled extensively in India and aboard concerning various R&D programs. Dr. Kaul has received several National and International awards and is expert member of various important National committees on Environmental Science and Engineering.

**Dr. B.N. Rai** is presently working as Professor in the Department of Chemical Engineering & Technology, IT, BHU, Varanasi. Dr. Rai received B.E. (Chem. Eng.) from Bangalore University and M.Tech. (Chem.) and Ph.D. from the Banaras Hindu University. Dr. Rai has more than 20 years of teaching, research and administrative experiences in various capacities and has more than 40 publications in international journals, book chapters, conferences, etc. and delivered invited/key note lectures in various conferences. He has guided more than 20 M.Tech. and 2 Ph.D. students. Dr. Rai was awarded by the Ministry of Non-Conventional Energy, Govt. of India for writing a book in Hindi on Fundamental Topics of Environmental Engineering. His research interests include Environmental Biotechnology and Industrial Pollution and Control. Dr. Rai has also organized international conferences, refresher courses and workshops on waste management. Besides, he has served the University in various capacities such as President, IT Gymkhana, Coordinator, National Youth Festival, etc.

**Dr. D.R. Saini** obtained his B.Tech. and M.Tech. from HBTI, Kanpur and Ph.D. (Chem. Eng.) from IIT-Delhi. Presently working as Principal Scientist at NCL, Pune. Dr. Saini has Coauthored three books, six chapters and published seventy papers in international journals. Dr. Saini has guided 3 Ph.D. and few M.Tech. students.

**Professor Yogesh Chandra Sharma**, born at Aligarh, India obtained Ph.D. in Applied Chemistry 1991 and Doctor of Science (D.Sc.) in 2010. Dr. Sharma has published large number of research articles in international professional journals. Dr. Sharma is referee of nearly four dozens of international professional journals including journals from Royal Society of Chemistry (RSC), American Chemical Society (ACS), American Society of Civil Engineers (ASCE), and journals from other reputed publishers. Dr. Sharma is also Editor in-Chief of the International Journal of Pollution and Solutions (CIP, USA). His current interests are synthesis, characterization and applications of nano-adsorbents, bio-fuel and catalysts. Currently Professor Sharma is teaching at the Institute of Technology, Banaras Hindu University, Varanasi.

# Environmental Science and Engineering

## *— Selected Topics —*

*Editors*

**Prof. (Dr.) S.N. Kaul**

*Principal*

*MIT College of Engineering, Pune*

*Former Acting Director and Director-Grade Scientist, NEERI [CSIR], Nagpur, M.S.*

**Dr. B.N. Rai**

*Professor, Department of Chemical Engineering & Technology, Institute of Technology, Banaras Hindu University, Varanasi, U.P.*

**Dr. D.R. Saini**

*Scientist*

*National Chemical Laboratory [CSIR], Pune, M.S.*

**Prof. (Dr.) Yogesh Chandra Sharma**

*Associate Professor*

*Institute of Technology, Banaras Hindu University, Varanasi, U.P.*

**2013**

**Daya Publishing House®**

*A Division of*

**Astral International Pvt. Ltd.**

**New Delhi – 110 002**

ISBN 978-81-7035-811-4



*Published by* : **Daya Publishing House®**
*A Division of*
**Astral International Pvt. Ltd.**
81, Darya Ganj, Near Hindi Park,
Delhi Medical Association Road,
New Delhi - 110 002
Phone: 011-4354 9197, 2327 8134
Fax: +91-11-2324 3060
E-mail: info@astralint.com
Website: www.astralint.com

*Laser Typesetting* : **Classic Computer Services**
Delhi - 110 035

*Printed at* : **Salasar Imaging Systems**
Delhi - 110 035

PRINTED IN INDIA

*Dedicated to the Sacred Memory*
*of*

**Swami Vivekananda**

# Preface

Environment is being threatened by the anthropomorphic activities carried by human beings. Even nature does play a great role in mitigating some of the man made–adverse activities. Rate of degradation is faster than the rate of re-habilitation of all the components of environment and, therefore, there is a serious need for prevention and control systems to be adopted by all those who are concerned about the environment.

The authors have invited several specialized papers from the experts for inclusion in this book to create awareness among all those who are interested to keep the environment Clean and Green. All the authors are well known in their respective area of specializations. The selected topics on Environmental Science and Engineering will be useful to all practicing engineers, researchers, students, teachers, NGOs and the like. The authors acknowledge the efforts made by all specialists for their contributions in this book and it is hoped that this specialized book will be a useful addition in the field of Environmental Science and Engineering.

*Prof. (Dr.) S.N. Kaul*

*Dr. B.N. Rai*

*Dr. D.R. Saini*

*Prof. (Dr.) Yogesh Chandra Sharma*

# Contents

# Chapter 1

# Sustainable Development: Role of Environmental Impact Assessment and Biotechnology

**S.N. Kaul[1], Lidia Szpyrkowicz[2] and D.R. Saini[3]**
**[1]MIT College of Engineering, Pune, India**
**[2]University of Venice, Venice, Italy**
**[3]CSIR–National Chemical Laboratory, Pune, India**

## ABSTRACT

The emergence of the concept of sustainable development in recent years has brought in the general years has brought in the general realization that societal perceptions must shift towards ecological determination so as to achieve qualitative growth within the limits of eco-system carrying capacity. With this backdrop, the chapter analyses the concepts of sustainable development and carrying capacity; illustrates the shift necessary towards strategies for qualitative growth such as technological innovation, structural change and preventive environmental intervention. As most ecological problems are the cumulative result of environmental and social impacts of human activity in the region, the chapter argues that Environmental Impact Assessment (EIA) could form a major instrument for the assessment of developmental activities in the context of regional carrying capacity. It also delineates implementation measures for designing a viable role for EIA in sustainable development.

The chapter also addresses issues like environmental monitoring, restoration of environmental quality, resource/residue/waste-recovery/utilization/treatment, and substitutions of non-renewable resource base with renewable resources. It includes, the current and prospective applications in these sub-areas of environmental biotechnology, and documents case studies on environmental monitoring (enteric viruses), restoration of environmental quality (oil spill remediation) resource recovery (hydrocarbon recovery from oily sludges, bio-surfactants from distillery spentwash, disulphurization of coal and sour gases), and substitution of non-renewable resources with renewable (conversion of lignocellulosics into value added chemicals)

***Keywords:*** *Sustainable, Environmental impact, Assessment, Carrying capacity, Biotechnology, and Renewable sources.*

## Introduction

The sociocultural roots of our present environmental crisis lie in the paradigms of scientific materialism and economic determinism, which fail to recognize the physical limits imposed by ecological systems on economic activity. The economics must expand within ecosystems, which have limited regenerative capacities. Contrary to the neoclassical theory of continuous material growth, economic activities directly undermine the potential for development through over-exploitation of natural resources, and indirectly compromise future production through over-exploitation of natural resources, and indirectly compromise future production through the discharge of residuals. The entrenchment with qualitative growth as a major instrument of social policy is thus quite paradoxical.

The emergence of the concept of sustainable development in recent years has brought in the general realization that societal perceptions must shift towards ecological determination so as to achieve qualitative growth within the limits of ecosystem carrying capacity. The carrying capacity based planning process, innovative technologies for enhanced material and energy affectivity of production and consumption, structural economic change towards less resource-intensive sectors, and preventive environmental management through increasingly interventionists policies are some of the strategies for reconciling development goals with ecological capabilities.

Environmental Impact Assessment (EIA) could form a major instrument in decision-making and for measurement of sustainability in the context of regional carrying capacity, provided the conceptual framework is extended to the cumulative assessment of developmental policies, plans and projects on a regional basis.

This chapter analyses the concepts of sustainable development and carrying capacity; illustrates the shift necessary towards strategies for qualitative growth such as technological innovation, structural change and preventive environmental intervention; and delineates the role of EIA in achieving the goal of sustainable development.

## Sustainable Development

### Premises and Preconditions

As defined by the United Nations World Commission on Environment and Development, sustainable development must meet the needs of the present generations to meet their own needs and aspirations. It is a process in which the exploitation of resources, the direction of investments, and institutional changes are all made consistent with future as well as present needs.

The concept of sustainable development is based on two underlying premises, *viz.*, symbolic relationship between consumer human race and producer natural systems, and compatibility between ecology and economy.

Clearly, sustainable development must satisfy the enlarged constitutional preconditions of equity and social justice, economic efficiency, ecologic harmony, and endogenous choices.

## The Concept of Carrying Capacity

The concept of sustainable development is closely linked to the carrying capacity of ecosystem. Ecosystem carrying capacity provides the physical limits to economic development and may be defined as the maximum rate of resource consumption and waste discharge that can be sustained indefinitely in a defined planning region without progressively impairing bio-productivity and ecological integrity. While recognizing societal dependence on many ecological resources and functions for its survival, carrying capacity is ultimately determined by the single vital resource of function in least supply. Working within the limits of carrying capacity does not, however, preclude some unavoidable environmental damage in the course of development.

The concept of carrying capacity implies that improvement in the quality of life is possible only when the pattern and levels of production-consumption activities are compatible with the capacities of natural environment as well as social preferences. The carrying capacity-based planning process thus involves the integration of social expectations and ecological capabilities by minimizing future differentials between realized and desired supply/demand patterns, infrastructure/congestion pattern, resource availability/use patterns and assimilative capacity/residual patterns.

The conceptual model for carrying capacity based planning uses a systems approach to estimate the changes in carrying capacity indicators. The differentials between realized and desired carrying capacity are overcome through a combination of institutional, informational and attitudinal strategies aimed at changes in the exogenous driving forces, systems structure, and the aspirations and tolerances that determine the level of desired carrying capacity. An ideal approach to the validation of the concept of regional carrying capacity based developmental planning process will be hierarchical through the stages of a village, district, region, nation and globe in keeping with existing political and administrative boundaries. It is obvious that the success of this approach depends entirely on planner's participation in the process and its acceptance by administrative and political jurisdictions.

## Structural Economic Change

Resource-intensity of economic development has emerged as one of the major determinants to effective environmental management. This situation could be resolved through the strategy of structural economic change which aims at raising the levels of ecologic and economic efficiency by increasing the material and energy effectively of production-consumption activities in order to minimize the expenses on environmental protection while keeping the cost of natural resource exploitation within acceptable limits.

Structural changes involves large scale technological substitution towards environmentally benign technologies such as clean technologies of industrial production that conserve resources, generate less pollution, provide direct economic benefits and stimulate the growth of industry as well as national economy; recycle and reuse technologies for end-of-the-pipe treatment that render the otherwise deemed non-productive activity of waste treatment as profitable proposition through waste utilization; biotechnology for substation of non-renewable with renewable resource base resulting in wide ranging ecologic and economic advantages; and integrated technologies that minimize cross-media transfer of pollutants thus minimizing overall pollution-induced risks in all environmental components rather than addressing single pollutants in specific environmental medium. Structural change also involves replacement of environmentally harmful branches of economy with those which are equally productive and yet environmentally compatible.

A few select examples of structural changes are:

- *Manufacturing Industry: Transition to production processes which save or recycle raw materials and energy*, substation of ecologically harmful products with harmless ones, ecological grouping of industries.
- *Energy Sector*: Rational use of primary energy, greater use of regenerative energy sources, decentralization of supply.
- *Transport Sector*: Reduction in the specific energy consumption of motor vehicles, reduction in total number of motored kilometers, provision of an efficient public transport system.
- *Construction Industry*: Use of renewable and environmentally compatible building materials, saving of land and energy, labour-intensive designs.
- *Agriculture Sector*: Eco-cultivation and biotechnological improvements, use of organic manures and biocides, development of land-use plans that are compatible with species and ecosystem types.

### Preventive Environmental Policy

Implementation of structural economic change warrants increasingly interventionist policies which influence technological considerations preceding investment decision-making in such a way that resource utilization as also the cost of environmental protection and damage are minimized while economic productivity and innovative capacity is maximized.

Preventive environmental policy essentially deals with the conditions that give rise to environmental problems and attempts to readjust these so as to obviate environmental damage. The concept could be implemented by the prevention of all harmful emissions that exceed the assimilative capacity of ecosystems through greater and improved recycling; introduction of low/non emission technologies; pre-emptive substitution of environmentally harmful raw materials and products; conservation of non-renewable resources; use of efficient combustion processes; and active environmental management through greater participation of hereto under-represented public interest in all relevant planning processes thus aiming for strong institutionalization of preventive principle throughout the society.

## Role of Environmental Impact Assessment

EIA is potentially one of the most valuable, interdisciplinary, objective decision-making tools with respect to alternate routes for development, process technologies and project sites. It is an ideal anticipatory mechanism, which establishes quantitative values for parameters indicating the quality of the environment before, during the after the proposed developmental activity, thus allowing measures that ensure environmental compatibility. It presents a clear and concise picture of all benefits and costs associated with alternative courses of action and provides mechanist for merging the concerns for environment and economics in the process of decision-making.

EIA could form a major instrument for the assessment of developmental activities in the context of regions carrying capacity. It its present form, however, EIA suffers from several limitations at the conceptual, methodological and procedural levels.

### Conceptual Limitations

The environmental implications in the development process arise at decisions related to national policies, regional plans and individual projects. However, EIA is presently conceived merely as project level tool and does not address to developmental programs at the policy and planning level. For example, Policy initiative in such areas as international trade, tax incentives, energy conservation, agriculture subsidies, predator control, etc. have never been the subject of environmental review.

Project level decision are may times constrained by existing policies and plans and the range of possible alternatives in project EIA is, therefore, often small or even non-existent. Further, there are no objective screening criteria to decide what type or scale of projects should be subjected into EIA. In India, for example, only those projects which are funded through the Public Investment Bureau (PIB) are subjected to EIA. Project level EIA has, therefore remained a reactive, quasi-regulatory instrument with only marginal effect on project design and implementation.

EIA should ideally be undertaken at the policy and planning levels as the environmental consequences of projects often arise due to higher-level decision. Policy EIA, however, is an extremely complex issue, largely due to the fact that the potential range of alternatives to achieve a desired goal can be almost unlimited. This problem may be resolved to some extent through a hierarchical approach in which the number of alternatives is reduced by defining the problem in terms of a series of choices.

The most appropriate stage for implementing EIA is at the level of district planning, since, at this stage, a reasonable number of alternatives are available to the developer. The assessment of regional supportive and assimilative capacities during formulation of development plans could greatly reduce the requirement project EIA.

Another conceptual limitation of EIA is that it does not incorporate the strategies of preventive environmental intervention. The issues of resource conservation, waste minimization, by-product recovery, improvement in efficiency of equipment, etc. needs to be pursued as explicit goals in EIA.

### Methodological Limitations

Experience over the years has shown that EIAs are always conducted under severe limitations of time, manpower, financial resources and data. India has unique problems in this respect due to its vastness, variety of climates, culture and customs resulting in widely diversified lifestyles, varying nature of terrain, flora and fauna, etc. No reliable comprehensive environmental information base exists and need for extensive data collection makes EIA an extremely cumbersome and time consuming exercise.

The requirement of time, resources and date for EIA could be greatly reduced by focusing the study on a limited number of relevant issues rather than an elaborate listing of values of all environmental parameters. This could be achieved through incorporation of a method for identification of significant issues as a component of EIA.

Scoping is a process which involves determination of significant issues to be considered in an EIA based on the criteria of magnitude, prevalence, frequency and duration for those changes which can be measured/predicted with a reasonable degree of accuracy, such as changes in the physico-chemical-biological environment; and on socioeconomic criteria for those environmental assets which cannot be measured or to which a value cannot be assigned, such as the impact on physical and biotic resources valued by man for recreational or aesthetic purposes.

Other methodological limitations of EIA relate to uncertainties in data which may arise from inadequate design of monitoring network, sampling and measurement errors, or spatial and temporal variations; uncertainties in prediction that may result from approximations made in the development and application of mathematical, physical, experimental or conceptual models; and uncertainties in evaluation that may arise from subjectivity in development of values functions graphs and difficulty in assigning monetary values to environmental intangibilities. Overcoming these limitations requires research and development of improved methodologies.

EIA could also be greatly facilitated by the creation of centralized information base and formulation of environmental quality standards at the regional level. The former may require separate organizational structure.

## Procedural Limitations

Experience with EIA over the years has shown that many projects, due to their scale or nature, do not require an in-depth examination from the environmental point of view. A major limitation of the present environmental process is the lack of objective criteria to decide whether a project requires comprehensive EIA or not.

Screening is a method of selection which allows elimination from the review process of all those projects that do not require detailed EIA, thus avoiding unnecessary expenses and delays in project clearance. Incorporation of screening in the environmental review process requires formulation of project and site related screening criteria for various types of developmental activities.

The public acceptance of impact assessment is another issue of procedural concern. Public hearing for resolving environmental issues is not yet practiced in India and these are taken up, if at all, by a few sensitive groups. Public participation needs to be incorporated in the review process not only to enable consideration of local knowledge and preferences in project planning but also for avoidance of conflict. In this sense, scooping could be a significant step in EIA because it is at this stage in the review process that public involvement could be brought in for identification of issues of public concern.

## Regional EIA a Tool of Sustainable Development

Most ecological problems are the cumulative result of environmental and social impacts of human activity in the region. Planning for sustainable development in the context of ecosystems carrying capacity thus requires systematic identification, quantification and management of cumulative trends in significant environmental variables on a regional basis. Whether these trends are driven wholly by local-regional activity or are the result of larger global factors, the necessary corrective action must be implemented in all jurisdictions on a site-specific basis. Whether these trends are driven wholly by local-regional activity or are the result of larger global factors, the necessary corrective action must be implemented in jurisdictions on a site-specific basis. Functional planning regions need to be identified based on ecological criteria such as climatic and vegetation patterns, soil classification and watershed boundaries rather than political jurisdiction. Within the context of sustainable development, regional EIA could provide the means for estimation of developmental limits imposed by regional carrying capacity.

The concept of regional EIA could be operationalised through regional environmental monitoring and EIA of regional policies and plans. Critics of traditional EIA have long observed that it is impossible to assess the significance of impacts associated with isolated projects in the absence of a wider planning context. Regional EIA could enable individual projects to be designed in the light of preceding development, opportunity costs, and available carrying capacity of biophysical and social systems to cope with unavoidable stress.

## Implementation Measures

Designing a viable role for EIA in sustainable development warrants the following implementation measures:

- ☆ Extension of the scope of EIA to the full range of ecologically and socially relevant developmental activities in a region.

- ✰ Creation of viable infrastructure and delegation of responsibility and functions in respect of EIA to the relevant Central and State Government agencies without ambiguity or redundancy at the two levels.
- ✰ Formulation of explicit sectoral guidelines for environmental review to aid the review authorities, project proponents and consultants undertaking EIA.
- ✰ Incorporation of screening, scooping, and rapid/comprehensive regional EIA concepts in the reviews process for overcoming present limitations to EIA.
- ✰ Incorporation of environmental management plan (EMP), disaster management plan (DMP) and post project monitoring (PPM) in the review process.
- ✰ Delineation of objective environmentally sensitive areas based on the type of regional ecosystem.
- ✰ Introduction appropriate legal measures for implementation of EIA provisions and post-project surveillance.
- ✰ Requirement of special license to act as consultant for undertaking EIA and promotion of recognized groups through appropriate support.
- ✰ Training of decision-makers and senior officials with resource input from international and national agencies; and training of middle level engineers and scientists with support form national institutions.
- ✰ Establishment of a National Environmental Monitoring Organization (NEMO) for creation of a comprehensive centralized information base to aid EIA process.
- ✰ Establishment of an autonomous National Environmental Impact Assessment Agency (NEIAA) for effective implementation of EIA.

## Role of Biotechnology

Environmental biotechnology, in its accepted connotation, refers to waste treatment with some emphasis on resource recovery. The emergence and acceptance of the concept of sustainable development, however, warrants that the scope of environmental biotechnology be enlarged to address issues like environmental monitoring, restoration of environmental quality, resource/residue/waste-recovery/utilization/treatment, and substitution of non-renewable resource base with renewable resources. This enlargement of scope warrants inclusion in R&D of specific applications of biotechnological processes, tools and techniques as also consideration of relevant safety, legal information and institutional issues in the management of environmental and larger socio-economic development issues.

## Environmental Biotechnology Application Areas

### Environmental Monitoring

Several direct and indirect indicators of wastewater and the air characterization are in vogue. These, however, suffer from severe limitations in terms of long test time and infrastructure requirements for analysis.

Urgent need exists for research on the development of sample, rapid and reliable biosensors for viral, microbial and chemical contaminants and pollutants.

## Restoration of Environmental Quality

Inadvertent chemical spills, pesticides present in agricultural runoffs, and leachates from landfills for municipal and industrial sludges result in contamination of sub-surface aquatic and land components of environment.

Application of biotechnology in restoration of contaminated soil and subsurface water offers several advantages compared to conventional withdrawal and treatment procedures such as ability to treat contaminants sorbed to soil or trapped in pore spaces, reduced time requirement and cost-effectiveness.

## Resource/Residue/Waste-Recovery/Utilization/Treatment

Biotechnology could aid waste minimization and resource conservation through application in areas such as improvement of resource quality, low and non-waste technology of production, end-of-pipe treatment on recycle and reuse mode, and utilization of residues/wastes as secondary resources.

Traditional, biotechnological applications in waste treatment have been limited to aerobic treatment of municipal wastewaters and aerobic/anaerobic treatment of sludges and organic wastes from industries. Aerobic and anaerobic fixed-film processes offer cost-effective solution to a host of low/medium/high strength wastewaters with attractive investment payback periods.

Biotechnology could also be instrumental in the detoxification of recalcitrant and hazardous wastes from petroleum refineries; pesticides industry; pulp and paper industry; dye industry; detergent and surfactant manufacturing industry and the chemical industry.

Biotechnology tools and processes such as genetic manipulation, biogumentation and bioleaching could potentially be applied for the detoxification of aromatics, hydrocarbons, surfactants, cyanides and pesticides in various industrial wastewaters.

## Substitution of Non-renewable Resource Base with Renewable Resources

Nearly 600 million tones of crop residues and agro-industrial wastes are generated annual in the country and many of these are amenable to fermentation for production of organic chemicals. Chemicals derived through bioconversion could substitute industrial chemicals in a wide variety of uses.

Shift in energy mix towards renewable is the preferred energy pathway for sustainable development in view of the unprecedented threat posed to the ecosystem if non-renewable resource alone were to be used for meeting the projected demands of energy in agriculture, industry, mining, and needs of galloping population levels.

Highly efficient anaerobic processes have been developed in recent years for recovery of methane from a wide variety of substrates such as biomass, agricultural residues, industrial and municipal wastewaters, and urban solid wastes. Similar research endeavuor is warranted on production of hydrogen and organic chemicals from biomass.

Bio-fertilizers could provide cost-effective solution for ensuring production while maintaining healthy land environment. In situ microbial and fungal cultures improve soil fertility, physico-chemical characteristics of soil, and moisture balance enabling sustained productivity from land mass thereby leading to higher agricultural yields, rapid expansion of green cover, faster rejuvenation of waste lands and such other advantages.

Bio-fertilizers could also be used for restoration of soil and vegetative cover and reconstruction of terrestrial ecology in the aftermath of mining and mineral extraction activities and rehabilitation of

**Table 1.1: Salient Features of Select R&D Achievements in Environmental Biotechnology in India**

| *Sl.No.* | *Salient Features* |
|---|---|
| 1. | **DNA probes for Detection of Enteric Viruses in Drinking Water** |
| | ☆ $10^4$ fold concentration of enteric virus in contaminated water samples achieved using negatively charged membrane filters and urea-argentine phosphate buffer (UAFB) |
| | ☆ Efficacy of the technique demonstrated for raw sewage, and river, lake and drinking water samples. |
| | ☆ Conserved sequence of 154 bp at the 5' end of enteric virus RNA used for detection by RT- PCR. |
| | ☆ Sensitive of the methodology is 40 viral particles. |
| | ☆ Methodology employed in post epidemiological investigations of 1992 gastroenteritis outbreak in Madras |
| | ☆ Water samples collected from affected areas of North Madras revealed the presence of rota virus genome in 7 out of 9 water samples analysed using RT-PCR, conformation to drinking water standards notwithstanding. |
| 2. | **Multiple PCR for Detection of Enteric Viruses in Drinking Water** |
| | ☆ Multiplex PCR allows simultaneous identification of target organisms in a single reaction. |
| | ☆ PCR protocols for detection of enteric viruses, *E. coli, Vibrio Cholera, Salmonella typhimirium* and *Entamoeba histolytica* established. |
| | ☆ Multiplex PCR for E.coli-Vibrio and Salmonella–*E. coli* established to the detection sensitivity of twentyfive cells per reaction. |
| | ☆ Studies underway to delineate a multiplex protocol for *E. coli, Salmonella* and *Vibrio* in a single reaction. |
| 3. | **Rapid Risk Assessment of Chemicals and Wastes for Carcinogenecity** |
| | ☆ An epithelial monkey kidney cell line Ma 104, grown in presence of 10 per cent goat serum was subjected to sublethal dose of 2,4,5 trichlorophenoxy acetic acid. |
| | ☆ Cells observed to be smaller in size, more fibroblastic, and heavily piled up. |
| | ☆ Pleomorphism was noted in the slides prepared with *Giema* and *Papaniculous* stains. |
| | ☆ Cells also showed hyperchomasia and change in nuclear cytoplasmic ratio. |
| | ☆ Treatment of cells with first chlorine wash from pulp and paper mill resulted in occurrence of anisonucleosis and nucleomegaly as well as hyperchromasia. |
| | ☆ Research now directed to detect over-production of cyclins D, and E in the transformed cells using specific monoclonal and polyclonal antibioses, and western blot technique. |
| 4. | **Biotechnological Conversion of Lignocelluloses Substrates to Cellulose, Liquid Glucose and Value-added Chemicals.** |
| | ☆ Nuisance weed parthenium hysterophorus used as s substrate for the production of value added chemicals. |
| | ☆ Process comprises delignification, saccharification, fermentation and recovery. |
| | ☆ Cost-effective production of cellulose achieved through the use of hyper producing (1.5-1.7) Trichloderma reesel Qm9414 with a combination of uv-light and a nitrite. |
| | ☆ Disc gel electrophoresis of cellulose from parent and mutant strains showed an additional protein band in the case of the mutant. |
| | ☆ Supplementation of T. reesel cellulose with enzyme from Aspergilus niger, which is rich in B-glucosidase, enhanced the hydrolytic potential of T. reesel cellulose. |
| | ☆ The weed hydrolysate contained glucose, xylose, arabinose and cellobiose which is comparable to commercial glucose for the production of ethanol and 2, 3–butanediol. |
| | ☆ Production cellulose, liquid glucose and 2, 3 butanediol was 6000 FPU/Kg (wet wt), 300 g/kg. Alkali treated weed (dry wt), and 350 g/kg of alkali treated weed (dry wt), espectively. |

*Contd...*

**Table 1.1–Contd...**

| Sl.No. | Salient Features |
|---|---|

☆ Total recovery of 2,3 butanol from the dilute fermentation broth achieved in multi-stage extraction process using solvent extraction with n-butanol with sodium carbonate as a salting out agent.

☆ Protoplast fusion of T.reesel and A.niger is a progress with a view to producing a hyper B-glucosidase generating strain.

**5. Biotechnological Production of Biosurfactant**

☆ Two bio-surfactant producing cultures, identified as Bacillus licheniformis strain BS1 and Pseudomonas aeruginosa strain BS2, isolated from oil contaminated soil.

☆ Bio-surfactants produced by these cultures with glucose as a substrate reduce the surface tension of distilled water from 27 to 72 dynes/cm. And interfacial tension of water-hexadecane system from 40 to less than 1 dyne/cm

☆ Bio-surfactants produced are lipopeptide and Rhamnolipid (glocolipid) in nature

☆ Fast molecular bombardment mass spectrum of the lipopeptide revealed an approximate molecular weight of lipopeptide in the range of 1050-1150.

☆ After screening of different industrial wastes, distillery waste was selected for bio-surfactant production using *Pseudomonas aeruginosa* strain BS2. Results show that the strain optimally utilizes four-folds diluted distillery waste as a substrate, and produces 1.0-2.0 g/l rhamnolipid.

**6. Preamptive De-sulphurization of Coal**

☆ Removal of pyretic sulphur from coal achieved during froth flotation by using a bacterial preconditioning stage.

☆ Coal (10 per cent slurry) exposed to Thiobacillus ferro-oxidans and Thiobacillus thioxidans

☆ After 2 hours of contact with the organisms, coal is subjected to forth flotation using kerosene as frothing agent at pH of 7.8–8.0.

☆ Pyritic sulphur removal from coal samples with different sulphur concentrations:

| *Raw Coal* | | *Percent Removal* | |
|---|---|---|---|
| *Sulphur (per cent)* | *Ash (per cent)* | *Sulphur (per cent)* | *Ash (per cent)* |
| **Coal from Wardha Valley** | | | |
| 4 | 32 | 80 | 89.45 |
| **Coal Rejects at Washery** | | | |
| 8 | 46 | 70 | 89.00 |
| **Coal Rejects from Mine** | | | |
| 14 | 60 | 55.7 | 86.00 |

☆ Gross calorific value of coal increased from 4436 to 6888 kcal/kg due to ash removal

☆ 60 per cent organic sulphur was removed from coal using this microbial preconditioning process within 96 hours using the mixed culture.

**7. Oil Spill Remediation**

☆ 2 step process

☆ Physical skimming using alkali treated saw dust in Step 1 results in 95 per cent recovery of floating crude oil.

☆ Two bacterial consortia composed of four different organisms, including bio-surfactant producer, results in 65 per cent degradation in 48 hours in step 2.

☆ One of the consortia imparted with pro U operon for osmotolerence by genetic manipulation degrades crude oil under hypersaline conditions.

☆ Bacillus subtillis used for overproduction of acetoin as a de-emulsifying agent for oil recovery from oil emulsions and oily sludges.

**Table 1.2: Cleaner and Cost Effective Technologies for Pollution Prevention and Control**

| *Technology* | *Plant Capacity (Quantity of Wase/Raw Materials)* | *Capital Cost (Rs. Million)* | *Annual Operations and Maintenance Cost (Rs. Million)* | *Annual Benefit (Rs. Million)* | *Investment Payback Period (Months)* | *Product Reuse/ Recycle Recovery* |
|---|---|---|---|---|---|---|
| Production of cellulose, liquid glucose and 2,3 butanediol from lignocelluloses | 25 T/month | 3.2 | 2.9 | 3.6 | 11 | Cellulose, liquid glucose and 2,3 butanediol |
| Wastewater management for sulphidic caustic waste and oily effluents from refinery | 500 $m^3$/hour | 18.5 | 1.25 | 11.5 | 20 | Elemental sulphur, magnesium and talc |
| Production of Bio-surfactants from distillery wastewater | 125 $m^3$/d | 5 | 0.75 | 1.2 | 50 | Bio-surfactants |
| Biomethanation of distillery spentwash | 450 $m^3$/d | 18 | 1.8 | 11.2 | 20 | Biogas |
| Photoproduction of hydrogen from organic waste | 20 $m^3$/d | 4.5 | 1.0 | 0.8 | 68 | Hydrogen gas |
| Microbial de-sulphurization of coal | 12 million T/yr | 3.5 | 2.64 | 2 | 21 | Ferric sulphate and high calorific value coal |
| Microbial de-sulphurization of gases laden with $H_2S$ and recovery of elemental sulphur | 1000 $m^3$/hour fuel gas | 4 | 3.126 | 6.5 | 8 | Elemental sulphur |

wasteland generated through excessive water/Chemical usage and deforestation. Substitution of persistent chemical pesticides with biocides offer the most desirable solution to the problem preservation of terrestrial and aquatic ecology in the face of bio-resistance, bioaccumulation and bio-magnification associated with the use of synthetic organic pesticides.

## R&D at NEERI in Environmental Biotechnology

NEERI is pursuing an active R&D program on Environmental Biotechnology in the thrust areas of Environmental monitoring, restoration of environmental quality, resource/residue/waste-recovery/ utilization/treatment, and substitution of non-renewable resource base with renewable. Some of the notable successes from the A&D are:

- ☆ Bioremediation of contaminated land system.
- ☆ Development of gene probes for rapid detection of water borne pathogens.
- ☆ Technology for efficient remediation of oil spills.
- ☆ Pre-emptive microbial de-sulphurization of coal, oil and sour gases.
- ☆ Production of bio-surfactant, microbial de-emulsifier and biodegradable plastics from industrial wastes.

☆ Production of cellulose and sugars, 2-3 butanediol, and other value added chemicals from nuisance weeds.

☆ Efficient bio-methanation of municipal and industrial wastewaters.

Salient R&D in environmental biotechnology in India are presented in Table 1.1. Cost-benefit analysis of select biotechnologies is presented in Table 1.2. The nature of substrates, initial investments, and investment payback periods for these technologies are attractive to small-scale enterprise in developing countries.

## Conclusion

Policy makers faced with long-term environmental problems often argue that they cannot afford to worry about the remote and abstract when surrounded by the immediate and concrete. The problems which overwhelm us today are precisely those which, through a similar approach, we failed to solve decades ago.

# Chapter 2

# Fly Ash for the Removal of Cr(VI) from Aqueous Solutions and Wastewaters

**Y.C. Sharma[1*], Uma[1], S.N. Upadhyay[2] and S.N. Kaul[3]**

**[1]Department of Applied Chemistry,**

**[2]Department of Chemical Engineering and Technology,**

**Institute of Technology, Banaras Hindu University, Varanasi – 221 005, India**

**[3]MIT College of Engineering, Pune – 411 038, India**

## ABSTRACT

Fly ash, a waste material of thermal power plants, has been used for the removal of chromium from aqueous solutions and wastewaters.The adsorption follows a first order rate kinetics and the value of $K_{ad}$, rate constant for removal was found to be $3.0\times10^{-2}$ $s^{-1}$ at 298 °K, 100 rpm, 2.5 pH and at other optimum conditions. Studies on the transfer of mass from mass transfer. The value of the coefficient of mass transfer, $\beta_1$, $2.15\times10^{-2}cms^{-1}$ at 298 °K, suggested the transfer of Cr(VI) onto the adsorbent surface to be rapid enough. The process of removal is highly dependent on pH of the solutions with maximum removal (89.12 per cent) at pH 2.5. Comparison of the adsorbent used with other non-conventional adsorbents shows that fly ash is a good adsorbent and can be recommended for treatment of metal rich wastewater in general and that of Cr(VI) in particular.

*Keywords: Cr(VI), Removal, PTPS fly ash, pH, Mass transfer.*

---

* Corresponding Author: E-mail: y_sharma_2002@rediffmail.com

## Introduction

Chromium is a very important metal with numerous applications. Its concentration levels on earth's crust are of the order of 100 ppm [1]. It has a wide variety of applications and is extensively used in leather manufacturing, explosives, ceramics, paint, pigments, photography, wood preservation etc.[2]. It is an alloying element of steel and many other alloys as well. Disposal of untreated wastes and effluents from these applications into water resources is one of the major causes of water pollution by chromium [3]. Chromium exists in four valency states namely zero (metallic chromium), bi-, tri- and hexavalent chromium but in nature it occurs in trivalent form as chromite. Tri- and hexavalent forms of chromium are biologically active with later being important from environment view point but its hexavalent form is reported to be 100 times more toxic than its trivalent form [4]. There are no evidences regarding toxicity of Cr (III) but there are several reports regarding toxic and harmful nature of Cr (VI) [5-7]. Exposure to Cr (VI) results in skin allergies like dermatitis, gastro-intestinal ulcers and bronchogenic cancer. It has also been reported to be acutely toxic and a mutagen.

A number of treatment technologies *viz.* reduction, ion-exchange, electro dialysis, electrochemical precipitation, evaporation, solvent extraction, reverse osmosis and chemical precipitation are used for removal of chromium from wastewater. The above methodologies are costly and labour intensive in nature and at the same time require land and site for waste disposal. Adsorption on activated carbon, both in batches as well as in columns has been a popular choice especially for treatment of metal rich industrial effluents [8-11], but its large-scale application is limited by its high cost for developing nations like India. Many scientific workers have reported development of low cost activated carbons from cheaper and easily viable materials [12,13]. A number of waste materials [14-17] have also been used for the removal of metallic species from aqueous solutions and wastewaters.

Millions of tons of fly ash, a by-product of thermal power plants are produced everyday with India producing 100 million tons of it every year. The biggest problem posed by fly ash is its safe disposal. In India alone 4000 acres of land is utilized for disposal of fly ash. In addition to many applications of fly ash, scientific workers have used fly ash as an adsorbent for removal of a variety of metallic pollutants from aqueous solutions [18-22]. Main emphasis of the present work is mainly two fold: first being the application of waste material *viz.* fly ash and second the removal of the Cr (VI) from aqueous solutions and wastewaters. Effect of various important parameters namely concentration, contact time, temperature, pH, ionic strength, adsorbent particle size etc. on process of removal of Cr(VI) has been studied. Kinetic and equilibrium modeling of the process of removal has also been carried out.

## Materials and Methods

### Materials

All the chemicals used in the experiments were procured from B.D.H., Mumbai, India and were of A.R. grade. The adsorbent, fly ash was obtained from a nearby thermal power plant, Patratu Thermal Power Station (PTPS), Jharkhand, India. Fly ash is a cheap, abundant and well known waste material of thermal power plants. Fly ash is basically a mixture of a number of metallic oxides with silica (53.80 per cent by weight), alumina (28.24 per cent by weight) as its major constituents. Other constituents of fly ash are iron oxide, magnesia and titanium dioxide. Composition of fly ash varies and depends on the quality of coal used for combustion. Indian coal is mainly lignite and contains ash up to 30-40 per cent [23]. Combustion of this coal produces more amount of fly ash. In order to remove earthen impurities, fly ash was washed several times with normal water and then with de-ionized water.

## Methods

### Physiochemical Characterization of Fly Ash

The selected adsorbent was sieved mechanically through sieves to maintain a uniform particle size. The average particle size of the fly ash was determined by particle size analyzer, Model HIAC-320 (Royco Instrument Division, USA). Surface area was measured by Lazer zee-meter, Model QS 7 (Quantachrome Corporation, USA) and porosity by mercury porosimeter. Chemical characterization of the adsorbent was carried out by Indian Standard Methods [24].

## Adsorption Experiments

### Kinetic Adsorption Experiments

Batch adsorption experiments were carried out with 1.0 g of fly ash sample with 50 ml aqueous solution of Cr (VI) of desired concentrations at known temperature, pH and ionic strength in polythene bottles on a shaking thermostat with a constant speed of agitation at 100 rpm. Ionic strength of solutions was maintained a $1.0\times10^{-2}$ M $NaClO_4$. The progress of adsorption was assessed by determining residual concentration of Cr(VI) by UV-visible spectrophotometer (Spectronic 20, Bausch and Lomb; USA) at 540 nm [25]. After equilibrium time, the adsorbent was separated by centrifugation from reaction mixture centrifuging at 10,000 rpm for 10 minutes.

## Results and Discussion

### Characterization of the Adsorbent

Fly ash, the adsorbent chosen for the proposed work contains silica (53.80 per cent) and alumina (28.24 per cent) as major constituents. Oxides of Ca, Mg, K, Fe and Mn are also present. Its surface area was found to be 61.82 $m^2g^{-1}$. Other parameters are reported elsewhere [6].

### Effect of Contact Time and Initial Concentration on Removal

Contact time and concentration have pronounced effect on removal of pollutant species from aqueous solutions and industrial effluents [26-29]. A rapid transfer of pollutant species from bulk to the surface of adsorbent cuts short the equilibrium period. Time taken to attain equilibrium is important to know the efficacy and feasibility of the adsorbent for its use in removal of the pollutants. The development of charge on the surface of adsorbent is also governed by the contact time. In the present studies the removal of Cr (VI) decreased from 89.12 to 51.3 per cent by increasing the concentration of Cr (VI) from $2.5\times10^{-5}$ M to $10.0\times10^{-5}$M at 2.5 pH, 150 μm adsorbent particle size, 100 rpm, agitation speed, $1.0\times10^{-2}$M $NaClO_4$ ionic strength and 298°K temperature. The removal is rapid in initial stages, decreases gradually and finally attains equilibrium in 2.4 ksec. An increase in initial concentration of Cr (VI) resulted in increase in the amount adsorbed from 0.12 to 0.27 $mgg^{-1}$ by increasing the concentration of Cr(VI) in solutions from $2.5\times10^{-5}$ M to $10.0\times10^{-5}$ M. this can be attributed to an increase in driving force of the concentration of Cr (VI) in aqueous solutions.

#### Kinetic Studies

In order to decipher the kinetics of removal of Cr (VI) by adsorption on PTPS fly ash, the data was tested with many models and finally the Lagergren's model [30-31] was found to satisfy the experimental data:

$$\log (q_e - q) = \log q_e - (K_{ad}/2.303).t \tag{1}$$

*where,*

$q_e$ and q (both $mgg^{-1}$) are the amounts of Cr (VI) adsorbed at equilibrium and at the time 't' respectively and $K_{ad}$ ($min^{-1}$) is rate constant of adsorption. The straight line graphs of 'log ($q_e - q$) vs t' (Figure 2.1) are single and smooth indicating suitability of proposed model. The values of $K_{ad}$ were determined by slopes of Figure 2.1 and are given in Table 2.1. Values of $K_{ad}$ are exhibiting a decreasing trend with temperature and this proves exothermic nature of the process of Cr (VI) removal.

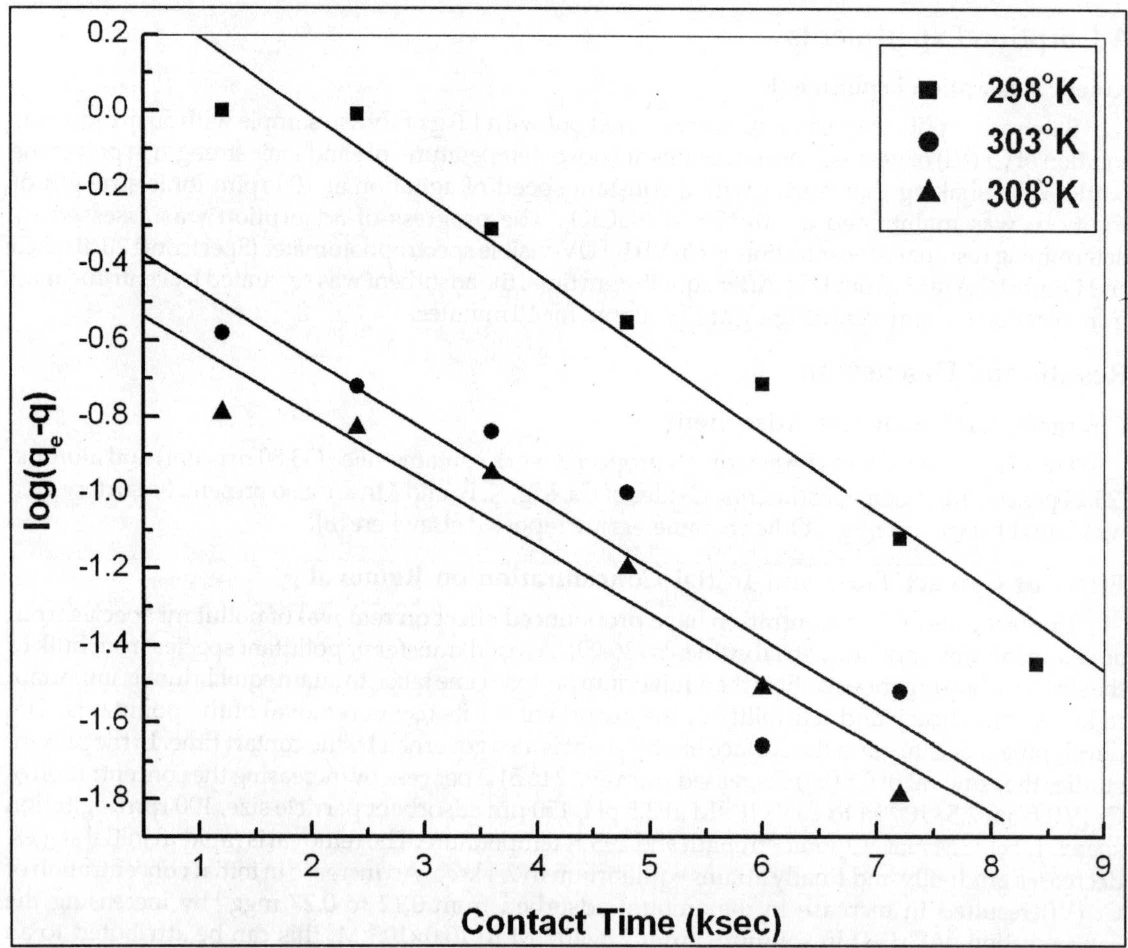


**Figure 2.1: Lagergren's Plot for the Dynamics of Removal of Cr(VI) by Adsorption on Fly Ash**

## Mass Transfer Studies

To understand suitability of adsorbent for any adsorption process, it is necessary to know the transfer of adsorbate to the surface of adsorbent and this possibility was studied by using the following well known model [30]:

**Table 2.1: Values of Rate Constant for Adsorption $K_{ad}$, and Rate Constant for Mass Transfer, $\beta_l$**

| *Temperature (K)* | *$K_{ad}$ ($ml^{-1}$)* | *$\beta_l$ ($\times 10^{-11} cm^2 s^{-1}$)* |
|---|---|---|
| 298 | $3.0\times10^{-2}$ | 2.25 |
| 303 | $2.09\times10^{-2}$ | 1.50 |
| 308 | $1.44\times10^{-2}$ | 0.32 |

$$\ln(C_t/C_0 - 1/(1+mk)) = \ln(mk/(1+mk)) - (1+mk)/mk.\ \beta_l.S_s.t \quad (2)$$

*where,*

$C_t$ and $C_0$ (both $mgl^{-1}$) are the concentrations of adsorbate, Cr (VI), at time 't' and initial concentration, m ($gl^{-1}$) is the mass of the adsorbent, k($lg^{-1}$) is Langmuir's constant, $\beta_l$ ($cms^{-1}$), coefficient of mass transfer and $S_s$ ($cm^{-1}$) is the outer surface of the adsorbent per unit volume of particle free slurry. Value of 'm' was calculated as follows [31, 32]:

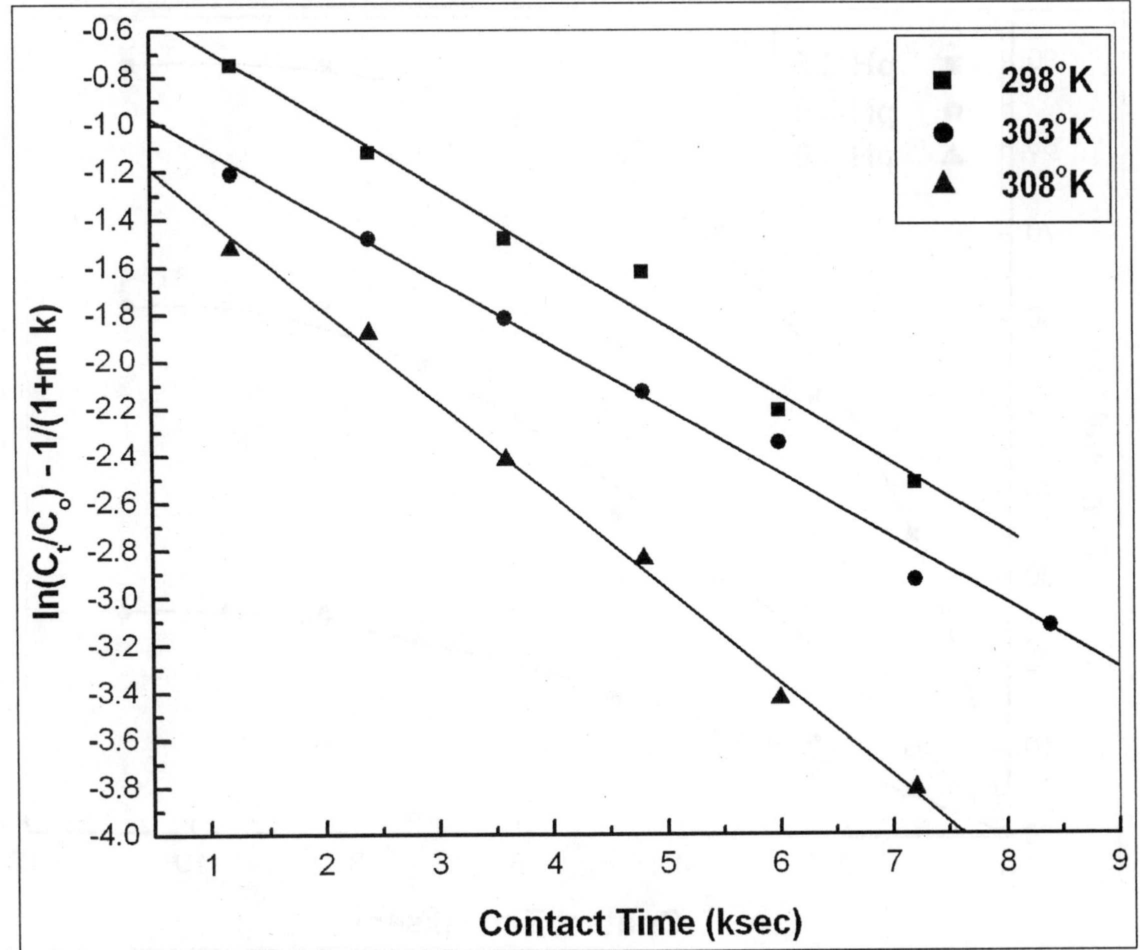


**Figure 2.2: Plot for Transfer of Mass for Removal of Cr(VI) by Adsorption on Fly Ash**

$$m = W/V \tag{3}$$

and $S_s$ was determined by following relation:

$$S_s = 6m/(d_p \rho_p (1 - \varepsilon_p)) \tag{4}$$

*where,*

W(g) is weight of the adsorbent, V(l), volume of the adsorbent sample, $d_p$ (cm) is diameter of adsorbent particle, $\rho_p$ ($gcm^{-3}$) density of adsorbent, and '$\varepsilon_p$', a unit less parameter is the porosity of the adsorbent.

The value of the coefficient of mass transfer, $\beta_1$, is measure of knowing rapidness of the transfer of adsorbate onto adsorbent surface. Its values were calculated at different temperatures by plotting 'ln $(C_t/C_0 - 1/(1+mk)$ vs t' (Figure 2.2). It is clear from perusal of this figure that the graphs are linear at all the values of temperature. This validates applicability of proposed mass transfer model for removal of Cr (VI) by adsorption on fly ash. The deviation of plots near saturation can be attributed to the difference in extent of mass transfer at the initial and final stages of the process of removal. Values of

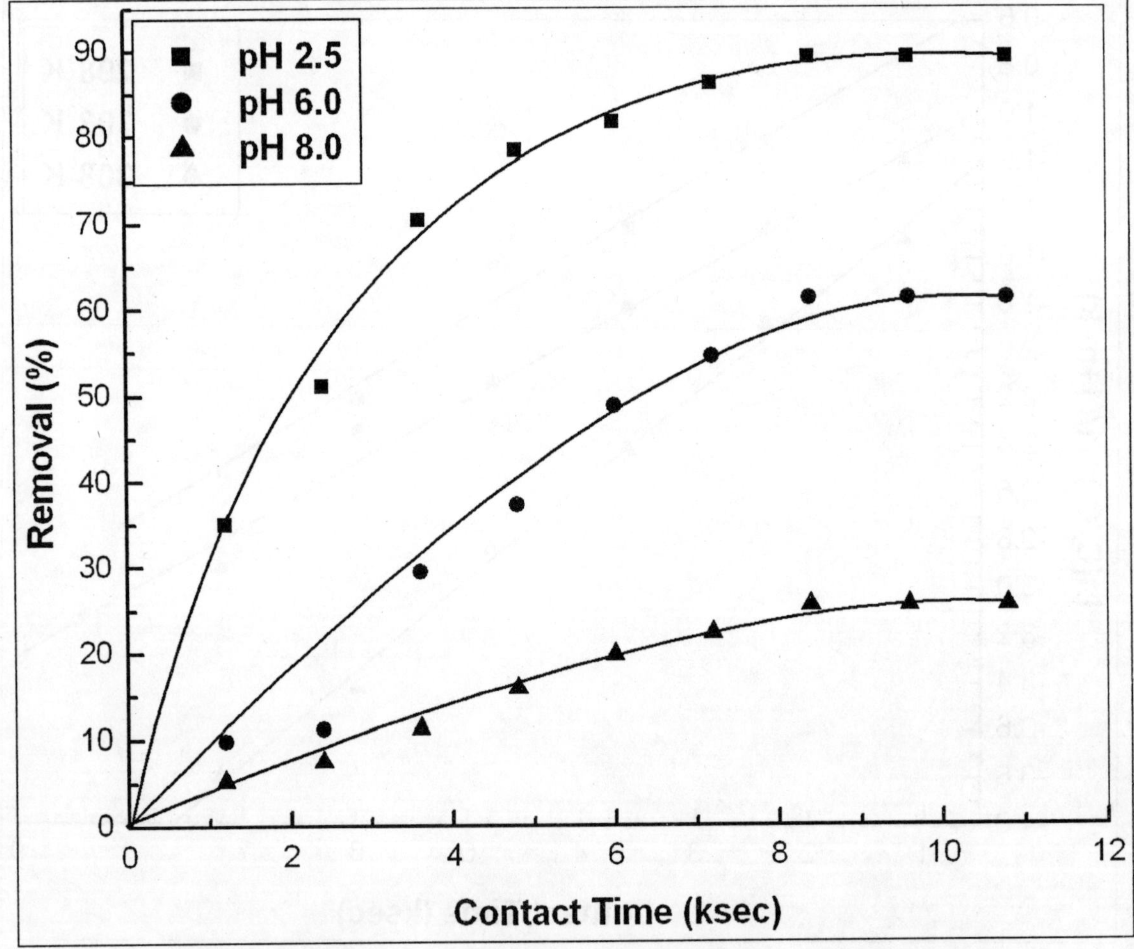


**Figure 2.3: Effect of pH on the Removal of Cr(VI) by Adsorption on Fly Ash**

the coefficient of mass transfer, $\beta_1$, were calculated from slopes and intercepts of the graphs of Figure 2.2 and are given in Table 2.1. The values of '$\beta_1$' was however, found to be $\beta_1$, $2.15\times10^{-2}$cms$^{-1}$ at 298 °K. The values $\beta_1$ indicate a rapid rate of mass transfer for the process undertaken [30].

## Effect of pH on Removal of Cr (VI) by Adsorption on Fly Ash

The pH of solution is reported to affect metallic species by adsorption [33,34] significantly as it has pronounced impact on surface charge of the adsorbent phase. As a result of its prominent effect on removal of pollutants from aqueous solutions, it has been termed as 'master variable'. pH also controls the degree of ionization and speciation of the adsorbate. In the present studies the removal of Cr (VI) by PTPS fly ash was studied at pH values of 2.5, 6.0 and 8.0 respectively and the data has been plotted in Figure 2.3. It is clear from this Figure that removal was maximum (89.12 per cent) at pH 2.5 and it decreased with increase in the pH value. At pH 6.0 maximum removal was 61.2 per cent and it drastically decreased (25.4 per cent) at pH 8.0 (Figure 2.3). The $pH_{zpc}$ of fly ash is 2.8 and this indicates that the surface is positively charged at 2.5 and will be quite favorable for removal $HCrO_4^-$ species. Significant adsorption at neutral and negatively charged surface cannot be explained on the basis of electrostatic attraction only. Specific chemical interactions and surface complexation have been suggested to describe the adsorption of Cr(VI). The adsorption beyond pH 4.5 will necessarily include $CrO_4^{-2}$ ions and the following complexation scheme has been suggested [35]:

$$(SOH)_2 + 2H^+ + CrO_4^{-2} \leftrightarrow CrO_4^{-2}\,(SOH_2^+) \tag{5}$$

Such type of model has also been reported by other scientific workers as well [36] and thus support our findings.

## Conclusions

Fly ashes are by-products and are understood to be nuisance to the environment. Because of their regular generation especially from coal burning, scientists are trying their other applications. Removal of metals by adsorption on fly ash is one such important application.

On the basis of the studies carried out, the following conclusions can be drawn:

1. Fly ash has been successfully used for removal of Cr(VI) from aqueous solutions.
2. Time of equilibrium was found to be 8.4 ksec and higher removal was obtained at lower concentrations.
3. The removal is governed by first order rate kinetics.
4. Mass transfer studies approve suitability of application of fly ash for removal of Cr(VI).
5. Process of Cr(VI) removal is exothermic in nature with maximum removal (89.12 per cent) at 298 °K.
6. Maximum removal is obtained at pH 2.5 and removal decreases with increasing pH values.

The data obtained can be successfully used for designing treatment plants for the treatment of Cr(VI) rich water and wastewaters.

## Acknowledgements

The authors are thankful to AICTE, Govt. of India, to provide financial assistance to one of the authors, Uma.

## References

1. U. Forstner U and. Wittman G.T.W., 1979 Metal Pollution in Aquatic Environment, Springer and Verleg, N.Y.
2. Kannan K. 1995, Fundamentals of Environment Pollution, S Chand and Co. Ltd., N. Delhi,
3. Cho H, Oh D and Kin K, 2005, J. Hazardous Materials, 127, 187.
4. J.O Nriagu, 1988. Environment Pollution, 50, 139.
5. M. Kobya, 2004, Adsorption Sci. Tech, 22, 51.
6. Sharma Y.C, Uma, Srivastava and M. Mahto, 2007, Chemical Engineering Journal, 127, 151.
7. Sharma Y.C and Weng C.H, 2007 J. Hazardous Material, 142, 449.
8. McKay G., 1982 J Chem Tech Biotech, 32, 59.
9. Hernandez G and R. Rodriguez, 1999 J. Non-Cryst. Solids, 246, 209.
10. Seco A., Marzal P., Gabaldon C and Ferrer J., 1999a Journal of Chemical Technology and Biotechnology, 68, 23.
11. Seco A., Marzal P and C. Gabaldon, J. 1999b Chem. Tech. Biotech, 66, 279.
12. Teker M and Imamoglu M., 1999, Turk. J. Chem, 23, 185.
13. Chang P, Wang X, Yu S and W. Wu.2007, Colloids and Surfaces A, 302, 75.
14. Chubar N., Carvalho J.R and Goreia M.J.N., 2003, Colloids and Surfaces, 230, 57.
15. Shukla S.R., Pai R.S and A.D. Shendarkar., 2006 Separation Purif Tech, 47, 141.
16. Shukla S.R. and Pai R.S., 2005, J. Chem Tech. Biotech, 80, 176.
17. Mittal A., Lisha K and J. Mittal, 2007 J Hazardous Mat, (online).
18. Sell N.J., Norman J.C and M.B. Vandenbasch, 1994 Resource Conser. Recl, 10, 279.
19. El-Eswed B and Khalili F., 2006, J. Colloid Interface Sci, 299, 497.
20. Lin C.J., J.E. Chang J.E., Lu M.C., 1999, Presented at the R'99 congress. Environment-expert.com/events/2000/2000.htm.
21. Sharma Y.C., Srivastava J and Pandey J.L., 2005 International synopsia on advances in Mining technology and management IIT-Kharagpur, India.
22. Gitari M.W., Petrok P.W., Etcheberg O., Key D.L, Wuoha E., and C. Okujeni., 2006 J. Environ Sci. Health, 41, 1729.
23. Annual report, Central mining research institute. Dhanbad, India, 2005.
24. Indian Standard Methods for Analysis of Fire Clay and Silica Refractory Materials (1960).
25. APHA, AWWA, WPCF Standard Methods for Examination of Water and Wastewater. $18^{th}$ vol., Washington D.C 1992.
26. Erden E., Karapinar R and R. Donat, J.2004, Colloid Interface Science, 28, 309.
27. Xu Y and L. Axe, 2004, J. Colloid Interface Sci, 28, 373.
28. Apte A.D., Verma S., Tare V and P. Bose, 2005 J. Hazardous Material, 121, 215.
29. Cespedes F.F., Perez M.F., Sanchez M.V. and E.G. Pradas, 2006, Environ. Pollut, 142, 449.

30. Mckay G., Otterburn M.S and Sweeny A.G., 1980, Water Res, 14, 15.
31. Papini M.P., Khaie Y.D., Troia B and M. Majone., 1999, Environ. Sci. Tech, 33, 4457.
32. Sharma Y.C., 2001, J Colloid Interface Sci, 233, 265.
33. Weng C.H and Huang C.P., 1994 J Environ. Eng. Div. ASE, 120, 1470.
34. Nollet H., Rods M., Lutagen P., Meeren P and W. Verstraete., 2003 Chemosphere, 53, 655.
35. Tiwari P.H., Campbell A.B and Lee W., 2002 Can J. Chem, 50, 1642.
36. Stumm W., 1981, Aquatic surface chemistry, Wiley- Interscience, N.Y.

# Chapter 3

# Implementation of the CETPs in India: A Boon for the Small-scale Textile and Tannery Industrial Sectors

**G.R. Pophali[1]*, T. Nandy[1] and S.N. Kaul[2]**
**[1]National Environmental Engineering Research Institute (NEERI), Nehru Marg, Nagpur – 440 020, India**
**[2]Maharashtra Institute of Technology (MIT), Pune – 411 038, India**

## ABSTRACT

This chapter reports the benefits of implementation of Common Effluent Treatment Plant (CETPs) in India. CETPs in Rajasthan and Tamil Nadu have been implemented for the treatment of combined wastewater from small scale textile dyeing and printing units and tannery industries. The concept of CETPs has emerged as an environmentally and economically sustainable method of treatment for industrial wastewaters. This chapter also reports the major priorities and the strategies for implementation of CETPs. The concept of a CETP clearly assigns the roles and responsibilities of individuals, industrialists, local Govt. bodies, Govt. regulatory authorities, experts and consultants and environmentalists for the efficient, transparent and accountable management. It makes effective use of human, technical, and financial and natural resources and financially sustainable *i.e.* being a collective responsibility it does not fail due to financial constraints. The implementation of CETPs suggests that, the CETPs can be effectively used for treatment of various industrial wastewaters emanating from small-scale clusters.

***Keywords:*** *CETPs, Textile wastewater, Tannery wastewater, Small-scale clusters, Benefits of CETPs*

---

* Corresponding Author: E-mail: gr_pophali@neeri.res.in, girishrpophali@yahoo.com

## Introduction

The Indian textile and tannery industries have been gaining substantial socioeconomic significance due to its contribution to social equity and export contributions. Environmental concerns associated with textile industries in Rajasthan and textile industries and tanneries in Tamil Nadu had placed the industries under serious threat of survival due to non-compliance of the effluent discharge standards prescribed by state statutory boards. Rajasthan houses a large number of small scale cotton and synthetic textile, dyeing and printing processing industries in Pali, Balotra, Jodhpur, Bhilwada and Jaipur. Similarly, Tamil Nadu has a large number of textile industries located in Erode and Tirupur and tanneries in Ambur, Vaniambadi, Ranipet and Erode.

Both the States were considered to be one of the most polluted States in the country. The wastewater discharged from these industries used to pollute the rivers flowing adjacent to the cities. This had considerably affected the ground water quality in the region. Detailed field investigations of the area revealed that small-scale industries did not have wastewater treatment facilities due to one or more of the the following reasons:

- Large capital investment required for installation of ETPs
- Cost involved in operation and maintenance, including manpower, energy, chemicals and laboratory etc.
- Land availability constraint.
- Lack of awareness and understanding the seriousness of problem of environmental pollution.

## Implementation of CETPs

In view of the above, cost effective CETPs implemented in Rajasthan and Tamil Nadu for treatment of combined wastewater from small scale textile and tannery industries. The major priorities of the initiative for implementation of CETPs were:

- Prevention of surface and ground water pollution and subsequently aquatic life.
- Prevention of land pollution.
- Conservation of natural resources such as surface and ground water, land and flora and fauna.
- Providing, end-of-the pipe treatment for effluents emanating from various industrial processes, which is environmentally, economically and ecologically sustainable.
- To prevent small-scale industrial sectors from closure and enable them to flourish thereby saving the loss of foreign exchange earnings from export.

The priorities of initiatives were established based on the objective of achieving sustainable development (*i.e.* keeping the steady growth of industries as well as protecting the environment).The major objectives of the initiatives were:

- Implementation of CETPs through latest state-of-the-art treatment technology for small scale textile industrial sectors
- Reuse and recycle of treated effluent

The strategies of the initiatives for the above objectives were executed by performing:

### Inventory of Industries

- ☆ Assessment of wastewater generation and detailed characterization
- ☆ Classification of industries based on quantum of wastewater generated
- ☆ Classification of wastewater based on biodegradability

### Evaluation of Existing Treatment Technologies

- ☆ Assessment of existing technologies for design of CETP to develop treatment packages and optimization based on spatial distribution of CETP

### Treatibility Studies

- ☆ Bench scale physico-chemical and biological R&D studies and development of scale-up factors for full scale plant
- ☆ Ranking of technology (treatment scheme) options
- ☆ Cost estimation of the technology (treatment scheme) options
- ☆ Reuse/recycling and resource recovery
- ☆ Disposal of treated effluent
- ☆ Recommendation of the best treatment option
- ☆ Cost benefit analysis
- ☆ Sharing of financial burden.
- ☆ Solid waste Management

### Mobilization of Resources

#### Financial

- ☆ Various Governmental as well as Non-Govt. Organizations, and Industrial Associations rendered financial resources.
- ☆ Half of the total capital cost was shared equally by the State and Central Govt. and the remaining was shared the small scale industries depending upon the capacity and magnitude of industrial activity.
- ☆ The cost of operation and maintenance of CETPs was borne by the Industrial Associations and Trusts.
- ☆ Human
- ☆ Manpower from local areas and State were mobilized. Human resources for operation control, maintenance, and management of CETP were mobilized by the Trusts and industrial associations which in turn deployed local bodies and individuals.

## The Process Package

### Textile Clusters

The process *(1, 2, 3)* for combined treatment of textile effluent from clusters of small-scale industries as designed and implemented in three industrial clusters is described hereunder (Figure 3.1). The CETP scheme comprises of the following unit operations and processes:

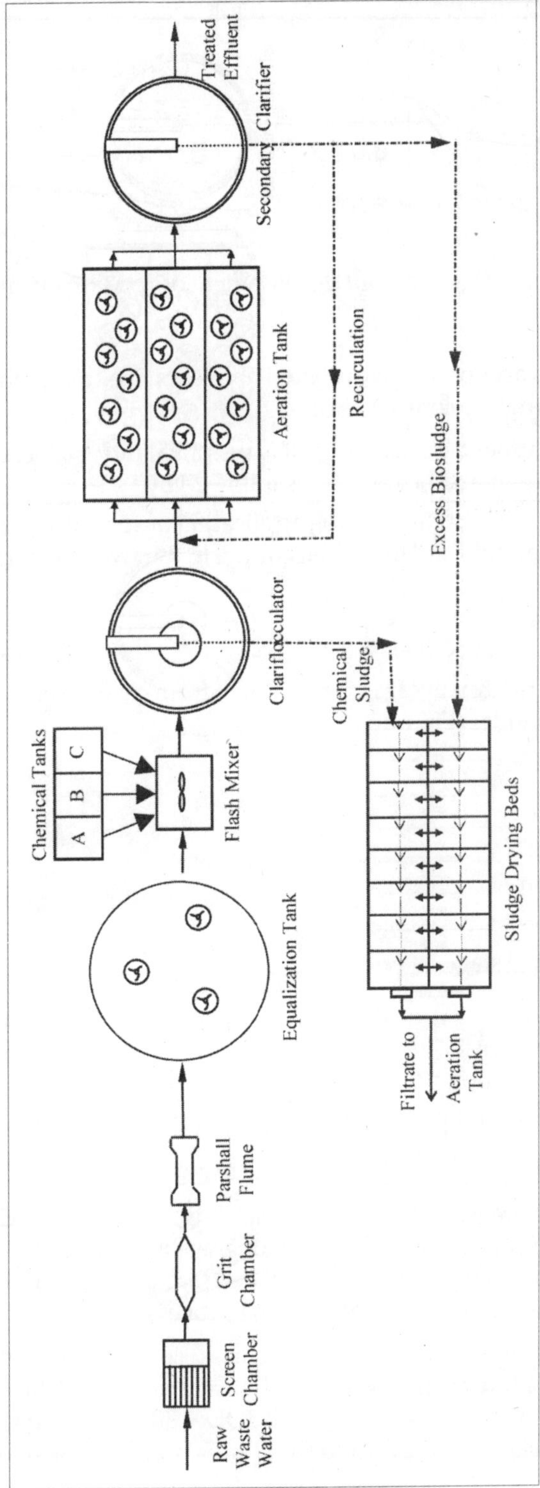


**Figure 3.1: Schematics of CETPs, Designed and Implemented for Textile Sectors in Rajasthan**

- ☆ Mechanical screening
- ☆ Grit removal
- ☆ Equalization
- ☆ Flash (Rapid) mixing
- ☆ Flocculation (Slow mixing) and clarification
- ☆ Biological oxidation (extended aeration)
- ☆ Secondary clarification
- ☆ Reuse of treated effluent for green belt development or recycling in industries.
- ☆ Sludge dewatering.

The CETP unit operations and processes as listed above for the treatment of wastewater from small scale industrial clusters mainly comprises of:

- ☆ Physico-chemical treatment for reduction of excess alkalinity, suspended solids, suspended colloidal material, colour and heavy metals followed by
- ☆ Aerobic biological treatment through the application of specially designed and developed high speed floating aerators and microbial culture to remove biodegradable organic matter substantially.

## Tannery Clusters

The process for combined treatment of tannery effluent from clusters of small-scale industries as implemented is described hereunder (Figure 3.2):

- ☆ Screen removal
- ☆ Rotary screen
- ☆ Equalization
- ☆ Chemical precipitaion
- ☆ Activated sludge process, two stage
- ☆ Secondary settling, two stage
- ☆ Chemical precipitation
- ☆ Pressure sand filtration
- ☆ Activated carbon adsorption
- ☆ Sludge drying and thickening
- ☆ Centrifuge

The design is based on pretreatment comprising settling, screening, and solar evaporation of soak and pickle liquor at individual tannery units. The tanneries connected to CETP involve processing from raw to finished leather, and semi finished to finished leather based on vegetable and chrome tanning process. A very few units have vegetable tanning process.

The suspended solids (SS), biochemical oxygen demand (BOD) and chemical oxygen demand (COD) of the treated effluent conform to the prescribed limits of the State Pollution Control Boards for discharge into Inland Surface Waters (SS < 100 mg/L, BOD < 30 mg/L and COD < 250 mg/L) and onto Land (SS < 100 mg/L, BOD < 100 mg/L and COD < 250 mg/L). The treated effluent from the

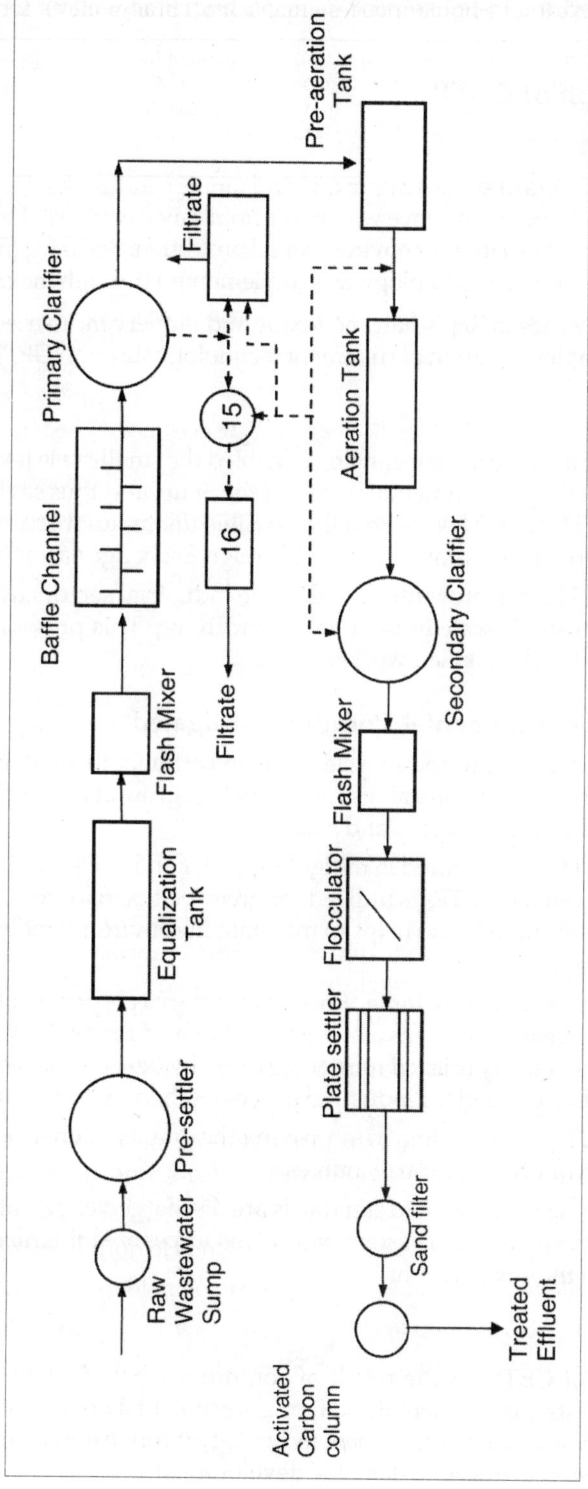


**Figure 3.2: Schematics of CETP Implemented for Tannery Sectors in Tamil Nadu**

CETPs is proposed to be utilized for in-house process applications as well as for the green belt development.

## Benefits of Implementation of CETPs

### Industrial/Economic Impact

Effluents emanating from clusters of small-scale textile and tannery industries were being directly discharged either onto land or into inland surface waters without any treatment. Hence it was high time to provide a suitable treatment alternative for environmental protection. Realizing the fact, treatment scheme based on the latest state-of-the-art technology was implemented through the concept of CETP.

- ☆ Small-scale textile industries in Rajasthan and textile and tannery industries in Tamil Nadu are surviving due to implementation of treatment technology through CETP concept.

### Employment Catalyzed

- ☆ The implementation of the treatment technology enabled the small-scale textile and tannery sectors to comply with the environmental norms of the State, and thus saved thousands of small-scale units from closure. This averted the possible unemployment of the vulnerable sections of the work force and loss of substantial Foreign Exchange earnings from Exports.
- ☆ Implementation of CETPs has prevented small-scale industrial sectors from closure, and therefore thousands of small-scale industries are surviving. This provided employment opportunities to skilled and unskilled workers.

### Quality of Life Improved/Environmental Pollution Mitigated

- ☆ Improved waste treatment, reuse and recycle of treated effluent through implementing of CETPs enabled conservation of natural resources such as ground and surface water. This helps in extension of safe water supply and sanitation.
- ☆ Implementation of CETPs has resulted in mitigating water and soil pollution problem. The final treated effluent from the CETPs is utilized for green belt development, which helps to reduce noise and air pollution. It also helps to maintain the environment hygienic and also improve the aesthetic appearance.
- ☆ The CETPs are located in arid zone of India, where water scarcity is prevalent almost through out the year; hence the treated effluents are effectively reused for green belt development. Public place has been effectively utilized for a noble cause by development of green belt. The treated effluents are also recycled for industrial processes thereby conserving freshwater.
- ☆ Implementation of CETPs has resulted in improving the quality of environment as well as protection and conservation of natural resources.
- ☆ Treated effluents meeting the stipulated standards are discharged either into inland surface waters or onto land, hence pollution of such natural resources are minimized which in turn results in protection of the environment.

### Societal Impact

- ☆ The implementation of CETPs is the result of community based planning wherein the individuals, industrialists, governmental and non-governmental organizations participated in decision-making. The execution of treatment technology through implementation of CETPs is a true example of a step towards sustainable development.

- ☆ The state-of-the-art technology was transferred to the non-Govt. Organizations and Industrial Associations after full one year of training and subsequence capacity building. The CETPs were operated and controlled for a period one year and handed over to the Trusts. The CETPs are being effectively used for industrial effluent treatment in the textile clusters.

## Strategic Impact

- ☆ The concept of CETP implemented through latest state-of-the-art technology has emerged as a most techno-economically viable option for wastewater management from clusters of small-scale industrial sectors.
- ☆ The system clearly assigns the roles and responsibilities of individuals, industrialists, local Govt. bodies, Govt. regulatory authorities, experts and consultants and environmentalists for the efficient, transparent and accountable management.
- ☆ The concept of CETP makes effective use of human, technical, and financial and natural resources and financially sustainable *i.e.* being a collective responsibility it does not fail due to financial constraints.
- ☆ The system is economically viable, as small-scale industries cannot have their own ETPs and socially and culturally accepted, since it offers employment and human settlement in the local areas.
- ☆ The concept of CETP is well suited for any industrial process changes, resulting in changes in the effluent characteristics. The system is accommodative to changing production and consumption pattern, keeping the pollution under control.

# Conclusions

The implementation of the state-of-the-art CETPs in the arid zones of India has proved to be a boon for the small-scale textile and tannery industrial sectors, as it offered economically and environmentally sustainable treatment. The concept of CETP makes effective use of human, technical and financial sources and conserves natural resources. The CETP technology can be effectively used for treatment of various industrial wastewaters emanating from small-scale clusters.

It is further recommended that the following points shall also be considered while implementing the CETPs:

### Wastewater Collection System

- ☆ Wastewater collection and conveyance system shall also be considered as part of the treatment scheme for a CETP as it defines the limits of parameters such as wastewater quantum and characteristics, which give the basis for CETP treatment system and design units.
- ☆ Effluents emanating from industrial processes shall be collected and conveyed to CETP, through a scientifically designed drainage network, which will ensure safe, economical and sophisticated mode of effluent conveyance system.

### Future Industrial Growth

- ☆ Prediction of future trends in industrial growth of the area, which brings about the changes in effluent quantity and characteristics, shall also be considered, while designing the CETP.

## References

1. "Basic engineering package for CETP for Textile Units at Pali, Rajasthan", Feasibility Report, pp. 1, NEERI (1993).
2. "Basic engineering package for CETP for Textile Units at Balotra, Rajasthan", Feasibility Report, pp 2, NEERI (1993).
3. S.N. Kaul, Girish R. Pophali and Tapas Nandy, "An improved Treatment Plant for Textile Wastewaters and an Improved Process Thereof", Patent File No. NF/149/01dt. 14.03.2002, India.

# Chapter 4

# Wastewater Management in India

**S.K. Gupta**
**Centre for Environmental Energy Science, I.I.T., Mumbai, India**

## ABSTRACT

In India the last one decade has witnessed growing water stress, both in terms of water scarcity and quality deterioration, which has prompted many municipalities to look for a more efficient use of water resources, including a more widespread acceptance of water management practices. Wastewater management is a part of integrated waste management which would help in effective management of wastewater. An integrated approach involving wastewater treatment, source reduction, reuse of process water, recycling of treated effluent and waste-minimization would lead to clean and green environment. Three technical areas are dealt with: domestic wastewater, industrial wastewater management, and wastewater re-use. Each approach is reviewed from a technical perspective with a further commentary on economic and social factors that underpin the different techniques that each approach has to offer. The preliminary evaluation indicates that for the management of domestic and industrial wastewater and increased utilisation of reclaimed wastewater, clearer institutional arrangements, more dedicated economic instruments and the set-up of water reuse guidelines are needed. Technological innovation and the establishment of a best practice framework will help, but even more, a change is needed in the underlying stakeholders' perception of the water cycle.

## Introduction

Water consumption is increasing in the world by leaps and bounds. The world's six billion inhabitants are already appropriating 54 per cent of all accessible freshwater reserves. It is predicted that by 2025 human kind's share will be 70 per cent. This estimate reflects only the impact of population growth. The result of water consumption is the generation of wastewater which adds to the woes of already acute problem of water pollution and scarcity.

In 1955, India had more than 5,000 cubic meters, about 177,000 cubic feet, of freshwater per person. By 2013, according to the United Nations children's fund, Unicef, the figure is expected to fall below 1,700 cubic meters, an internationally recognized definition of "water stress."(Bhushan, 2007) The water stress index – the ratio of a country's total water withdrawal to its total renewable freshwater resources – serves as a rough indicator for the pressure exerted on water resources (note, however, that not all water uses are causing comparable stress). By 2025, it could fall below 1,000 cubic meters, the cut-off line for "water scarcity." Due to improper planning of water conservation, today, India is ranked 122 out of 130 nations for its water quality and 132 out of 180 nations for its water availability. The European countries too face the same problem. Approximately half of the European countries, representing almost 70 per cent of the population, are facing water stress issues today. A water stress index above 20 per cent is supposed to necessitate comprehensive management efforts to balance supply and demand, and actions to resolve conflicts among competing uses.

According to UNDP's World Water Development Report, 2003 (WWDR, 2003), the domestic use of water account's for only eight per cent of global water consumption. The agricultural sector is the largest user of water globally and accounts for about 70 per cent of the total freshwater abstraction. However, it is predicted that both these users will be outdone by industry: Water consumption by industries is increasing. In fact, in high income countries, industrial water use already accounts for as much as 59 per cent of the total freshwater consumption; almost twice the amount used in agriculture. It is likely, then, that this will become a global trend even as more and more nations begin to choose industry over agriculture, as a key to economic growth.

Presently, industry accounts for 22 per cent of the global freshwater consumption. It is expected that the figure will double over the next two decades. According to forecasts published in WWDR, 2003, the volume of water consumed per year by industry will rise from 752 $km^3$/year in 1995 to an estimated 1,170 $km^3$/year by 2025.

In India, the general concept is, water is a natural gift and so the citizens need not to pay for its use. In Delhi, only Rs 0.35 are charged for 1 $m^3$ of water supply up to 10 kL/month (Gupta, 2004) water consumption. In other states water charges vary between Rs 20 to Rs 60 per month for any quantity of water consumed (Gupta, 2004). In addition, the consumer is generally not aware about paying for wastewater collection and its treatment. Potable water from different treatment plants is brought to the under ground reservoir through water mains before it is supplied to the people. There are around 1200 such reservoirs/boosters pumping station spread all over to Delhi for smooth and regular water supply. The cost of treatment and distribution as noted in DJB report is about Rs 2.75/kL (Gupta, 2004). It was also noted that this cost does not include water transport charges to the treatment stations.

In India, for example, in a family of 5 persons with daily water requirement of 200 litres per person, per day requirement is 1 kL/day, monthly consumption is about 30 kL. The minimum tax should be at least Rs 80 per month; where as maximum of Rs 45 per month is normally collected. If the collection procedure is correctly applied, it is possible to collect more money, ultimately resulting in good quality of water to the consumer. The water taxes in various countries are presented in Figure 4.1. It can be seen that average cost of water is about 1 US $ per cubic meter. The charges for water supply to consumer in India are far below the average cost.

The sewage and wastewater from industry is a very big water resource if it is discharged in a water source (surface or underground) after proper treatment. Domestic wastewater normally includes

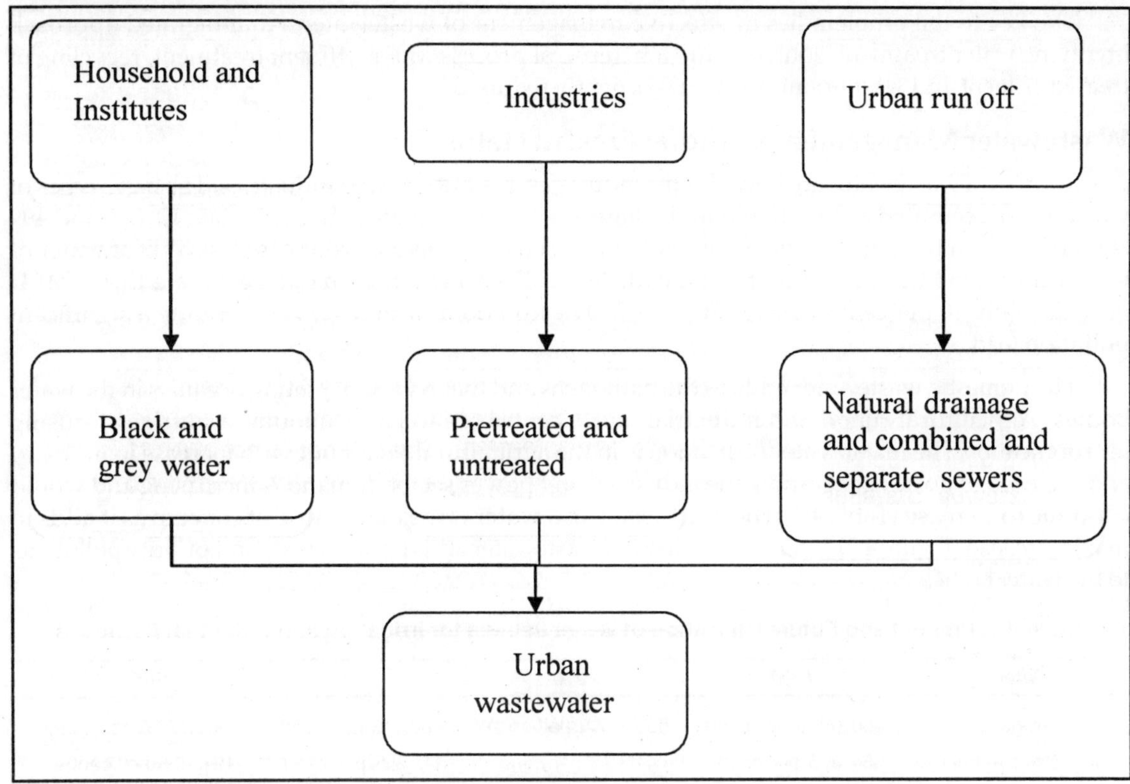


**Figure 4.1: Types of Wastewaters Contributing to Urban Wastewater**

discharge from kitchens, bathrooms, laboratories, hospitals, schools etc., has 99.9 per cent water and 0.1 per cent impurities in the form of suspended and dissolved pollutants. Sewage collection and treatment is one of the most essential components amongst the various infrastructure requirements to improve the quality of life. Existing sewerage system covers only 55 per cent of the total population in class I city. There are nearly 115 sewage pumping stations working all over Delhi to collect the sewage from the peripheral sewer. To lift this sewage to the sewage treatment plant, pumping stations run around 12 to 15 hrs daily. The cost of collection and treatment is about Rs 1.41/kL (Gupta, 2004) but no provision has been made to tax the people for this type of facility.

The cost of industrial wastewater treatment can be in the range of Rs 2 to 5/kL of wastewater and it will depend on the quality of effluent discharged by the industry (Gupta, 2004). Therefore, the total cost of water to consumer should include; raw water cost, water treatment cost, water distribution cost, wastewater collection cost and wastewater treatment cost. If the water supply agencies charge less than the minimum cost required for water treatment and supply, the water quality problems will be present (Gupta, 2004).

If the wastewater generated from the aforesaid uses is not managed in a proper manner, it may result into depletion of water sources because of the continuous utilization of the water.

The key to the problem lies in effective management of wastewater. An integrated approach involving water treatment, source reduction, reuse of process water, effluent treatment, recycling of treated effluent and waste-minimization is urgently required.

## Wastewater Management in India: Present status

Sources of wastewater are households, industries, institutions and industries. The basic types of wastewater generated are as shown in the Figure 4.1. It is estimated that 423 Class I cities and 499 class II towns in India harboring a population of 20 crores generate around 26254 MLD of water of which only 6955 MLD (25 per cent) is treated. About 57000 industries in India generate 13,468 MLD of which only 60 per cent is treated (CPCB, 2004). Non point sources also contribute a significant pollution load.

The domestic wastewater adds to the pathogens and the oxygen depleting organics in the water bodies. Agricultural runoff and industrial waste contribute to large amount of nutrients causing eutrophication.The major water utilization is in the agricultural sector but with increase in industry and the need of power generation, the industrial and power sector demand is increasing and would continue to increase (Table 4.1).The industrial wastewater is the main cause of concern as it adds to maximum load (Figure 4.2). Untreated industrial wastewater adds harmful toxic and priority pollutants to the water bodies.

**Table 4.1: Present and Future Utilization of Water in India for Irrigation and Industrial Purposes**

| *Category* | *1990* | *2010* | *2025* | *2050* |
|---|---|---|---|---|
| Irrigation | 460 (88.6 per cent) | 536 (77.3 per cent) | 688 (73 per cent) | 1008 (70.9 per cent) |
| Industries + Energy | 34 (6.6 per cent) | 41.4 (6 per cent) | 80 (8.5 per cent) | 121 (8.5 per cent)+ 143 (10.1 per cent) |
| Total (including others) | 519 | 693 | 942 | 1422 |

*Source*: National commission for integrated water resources development plan, Ministry of water resources, 1999

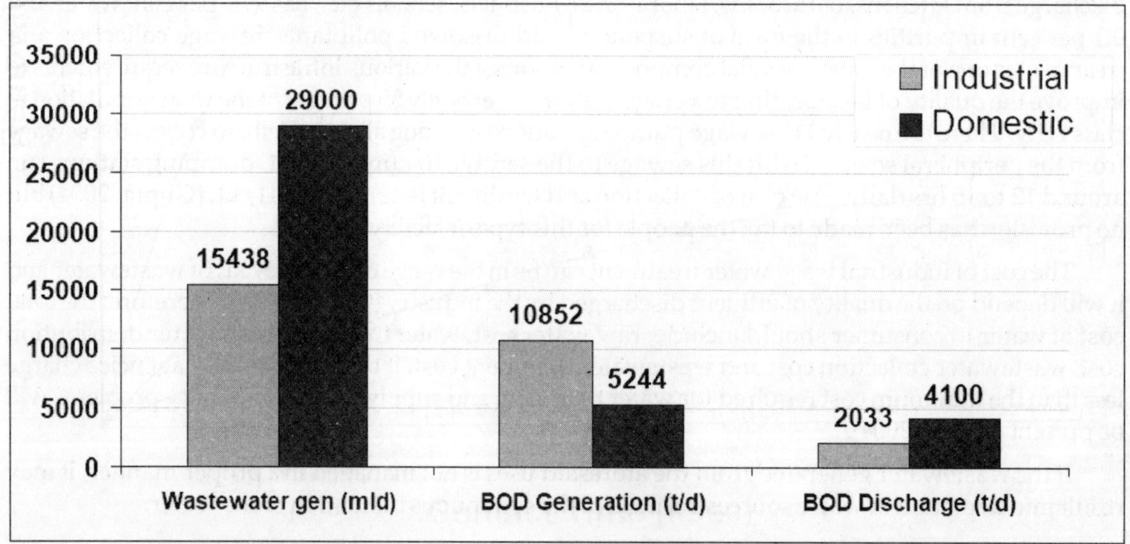


**Figure 4.2: Comparison of Pollution Loads from Domestic and Industrial Sector**

Due to fast urbanisation, industrialisation and agricultural activities demand for water is increasing. Continuing discharge of wastewater in water bodies even after treatment (to whatever stringent level beyond MINAS) may not help maintaining 'wholesomeness' of our aquatic resources steeply. Thus it becomes inevitable to discontinue discharge of wastewater into our dwindling water resources.

The wastewater generated can be broadly divided into two categories lesser toxic domestic wastewater and toxic industrial wastewater. Management of the two types of water is in progress but has not been exploited completely. The management of each and every type of wastewater requires different approach but the common steps are as follows:

**Table 4.2: The Wastewater Management Hierarchy**

| *Elements of Wastewater Management* | *Associated Tasks* |
|---|---|
| Source of generation | Quantification of wastewater, evaluation of techniques of wastewater reduction and determination of wastewater characteristics |
| Reduce wastewater generation at source | Proper education to individual citizens, use and management of proper plumbings |
| Source control | Design of onsite systems to provide partial treatment of the wastewater |
| Recycle wastewater in its current form | Design of recycle piplines, characterization for a specific use |
| Collection | Design of sewers used to remove wastewater from the various sources of generation |
| Transmission and pumping | Design of large sewers used to transport wastewater to treatment facilities |
| Treatment before disposal | Selection, analysis and design of treatment operations and processes to meet specified treatment objectives related to the removal of wastewater contaminants of concern |
| Reuse of wastewater after tertiary treatment | Design of advanced wastewater treatment for a specific reuse |
| Disposal | Design of facilities used for the disposal and reuse of treated effluent in the aquatic and land environment, and the disposal and reuse of sludge on land |

## Domestic Wastewater

As per the latest estimate of Central Pollution Control Board, about 29,000 million litre/day of wastewater generated from class-I cities and class-II towns out of which about 45 per cent (about 13000 mld) is generated from 35 metro-cities alone. The collection system exists for only about 30 per cent of the wastewater through sewer line and treatment capacity exists for about 7000 million litre/ day. Thus there is a large gap between generation, collection and treatment of wastewater. The solution definitely is judicious use of available water resources along with recycling.

The management of domestic wastewater in present scenario is limited to the treatment of wastewater and reuse with treatment in metro cities. The treatment of domestic wastewater consists of preliminary, primary and secondary treatments whereas in addition to these three tertiary treatment is given to the wastewater intended. As a result of the fast-growing urban population, service infrastructure is insufficient to ensure public health. In fact, about 15 per cent of the urban population does not have access to safe drinking water and about 50 per cent is not serviced by sanitary sewers. In 1997, the total volume of wastewater generated in India was 17 $Mm^3/d$ (4,500 mgd), of which 72 percent was collected and only 24 per cent was ever treated. These conditions cause a high number of waterborne diseases in the country (more than 30 million life years according to the World Bank).

The capital city of Delhi is one illustration of failing service infrastructure and deteriorating environment. The growing population in Delhi has led to an increase in the volume of wastewater, yet the current treatment capacity is only about 1.3 $Mm^3/d$ (3,400 MGD)–which is only 73 per cent of the wastewater generated. Another example is Mumbai, where 2.3 $Mm^3/d$ (608 MGD) of raw sewage is discharged into the Arabian Sea. However, there have been some attempts at rectifying these situations. The large, $300 million, Bombay Sewage Disposal Project was approved in 1995 with the financial support of the World Bank. Other efforts have been made in the Calcutta metropolitan area, where 13 sewage treatment plants have been constructed with a total capacity of 386,000 $m^3/d$ (102 MGD) using either activated sludge processes, trickling filters, or oxidation ponds. In addition, the Ganges River program is to include treatment facilities for 6 cities in Uttar Pradesh that will incorporate reuse for agriculture and forestry.

In 1985, over 73,000 hectares (180,000 acres) of land were irrigated with wastewater on at least 200 sewage farms. There has been a dramatic increase in wastewater volumes discharged and used for agricultural irrigation in India. With its current population, cities like Hyderabad can supply wastewater to irrigate an estimated 40,000 hectares (99,000 acres). The law prohibits irrigation of salad vegetables with wastewater, yet the prohibited practice is widespread and government agencies reportedly do not actively enforce regulations governing reuse. Furthermore, in states there is no microbiological standard and hence no parameter to control the level of treatment. Enteric diseases, anemia, and gastrointestinal illnesses are high among sewage farm workers. Consumers of salad and vegetable crops are also at risk.

The domestic wastewater includes sewage from residences, offices, commercial complexes, institutes etc. The basic physical and chemical properties of sewage are listed in the Table 4.3.

**Table 4.3: Characteristics of Domestic Wastewater**

| *Constituent* | *Unit* | *Range* | *Typical* |
|---|---|---|---|
| Total Solids | mg/L | 300-1200 | 700 |
| Dissolved | mg/L | 250-850 | 500 |
| Suspended | mg/L | 100-400 | 220 |
| $BOD_5$ | mg/L | 100-400 | 250 |
| TOC | mg/L | 100-400 | 250 |
| COD | mg/L | 200-1,000 | 500 |
| Total Nitrogen | mg/L | 15-90 | 40 |
| Organic | mg/L | 5-40 | 25 |
| Ammonia | mg/L | 10-50 | 25 |
| Total Phosphorous | mg/L | 5-20 | 12 |
| Chloride | mg/L | 30-85 | 50 |
| Sulfate | mg/L | 20-60 | 15 |
| Alkalinity | mg/L | 50-200 | 100 |
| Grease | mg/L | 50-150 | 100 |
| Total Coliform | colonies/100 mL | $10^6$-$10^8$ | $10^7$ |
| VOCs | μg/L | 100-400 | 250 |

At present in India the traditional approach to pollution abatement due to aforesaid pollutants is of treating those pollutants for safe disposal of sewage according to the effluent standards as prescribed by IS standards. The treatment methods followed for the domestic wastewater include preliminary, primary and secondary treatment as shown in Figure 4.3

Wastewater treatment processes can be categorized into the following three:

- *Physical process*: impurities are removed physically by screening, sedimentation, filtration, flotation, absorption or adsorption or both, and centrifugation;
- *Chemical process*: impurities are removed chemically through coagulation, absorption, oxidation-reduction, disinfection, and ion-exchange;
- *Biological process*: pollutants are removed using biological mechanisms, such as aerobic treatment, anaerobic treatment and photosynthetic process (oxidation pond).

Conventional wastewater treatment consists of the following stages: preliminary, primary, secondary, and disinfection. Municipal wastewater treatment facilities use combinations of physical, biological and chemical treatment technologies. Preliminary and primary treatments are usually physical processes, such as screening for the removal of debris and large solids, and sedimentation. A secondary treatment may utilize biological processes, such as stabilization ponds, trickling filter,

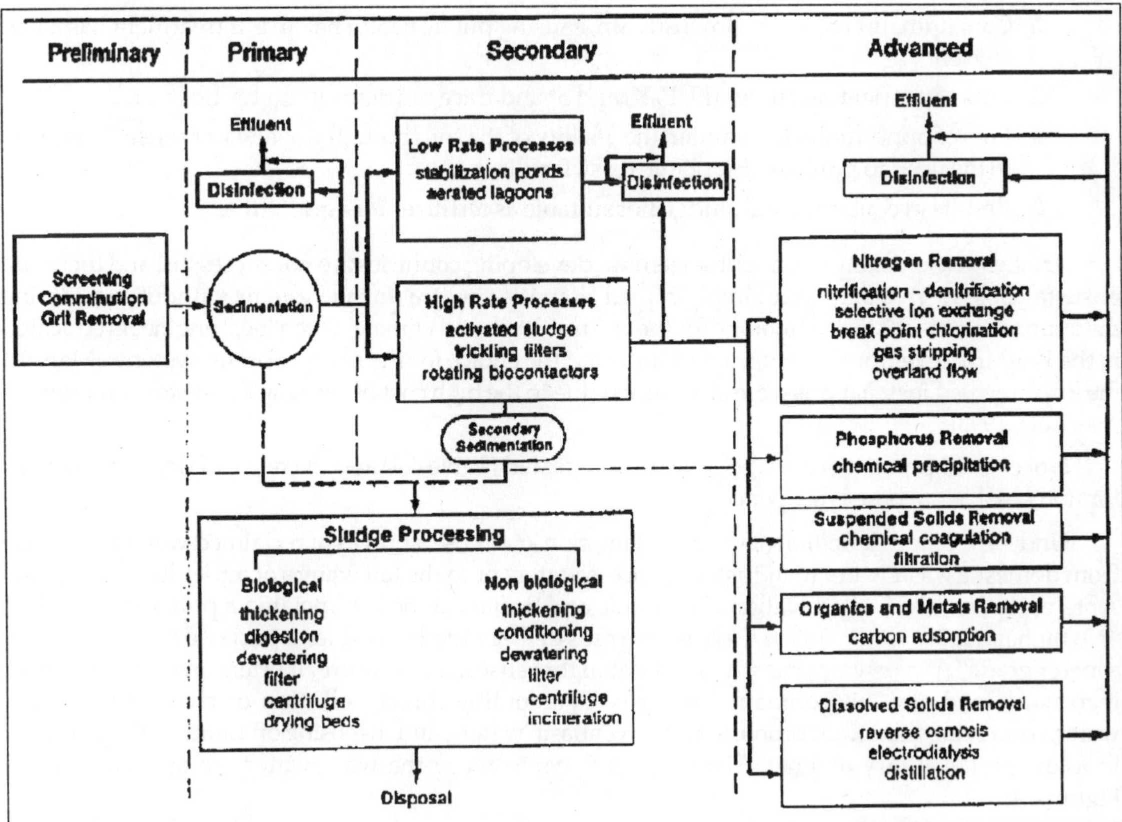


**Figure 4.3: Conventional Domestic Wastewater Treatment Process**

oxidation ditch, and activated sludge, which is then followed by sedimentation of biomass (sludge). Tertiary and advanced treatment is an additional treatment for higher-level removal of specific pollutants, such as nitrogen or phosphorus, which cannot be removed by conventional secondary treatments.

The fundamental precondition for water reuse is that applications will not cause unacceptable public health risks. Untreated wastewater poses a serious risk of water-borne diseases, such as cholera, typhoid, dysentery, plague and elephantiasis. In the 19$^{th}$ century, large-scale applications of untreated wastewater for agriculture triggered epidemics of such water-borne diseases in Europe. With medical and public health advancements, links between untreated wastewater and diseases have become better understood, and measures to minimize exposure to such pathogens have been introduced.

The conventional centralized system flushes pathogenic bacteria out of the residential area, using large amounts of water and often combines the domestic wastewater with rainwater, causing the flow of large volumes of pathogenic wastewater. In fact, the conventional sanitary system transfers a concentrated domestic health problem into a diffuse health problem for the entire settlement and/or region. In turn, the wastewater must be treated where the cost of treatment increases as the flow increases. The abuse of water use for diluting human excreta and transporting them out of the settlement is increasingly questioned and being considered unsustainable (van Leir, 1998). The negative effects of centralized treatment are summarized as:

- ☆ Contamination of water downstream, causing public health hazards if treatment has a low efficiency.
- ☆ Loss of nutrient resources (N, P, K, and S) and trace nutrients in domestic waste.
- ☆ Loss of opportunity to maintain the fertility of the soil through wastewater reuse. This leads to the need to purchase inorganic fossil fertilizer.
- ☆ Results in contaminated sludge not suitable as fertilizer for agriculture.

Another reason many treatment systems in developing countries are not successful and therefore unsustainable are that they were simply copied from Western treatment systems without considering the appropriateness of the technology for the culture, land, and climate. Often local engineers educated in the Western development programs supported the choice for the inappropriate systems. Many of the implemented installations were abandoned due to the high cost of running the system and repairs (van Leir, 1998).

Not even 25 per cent of sewage generated is treated (Table 4.4) and the reuse of domestic sewage for non potable purpose is 0.3 per cent.

Under the broad definition of water reclamation and reuse, sources of reclaimed water may range from domestic wastewater to industrial process waters or to the tail waters of agricultural irrigation systems. In fact, the United Nations Economic and Social Council enunciated a policy in 1958 that, "No higher quality water, unless there is a surplus of it, should be used for a purpose that can tolerate a lower grade". Wastewater has a high potential for reuse in agriculture. It offers an opportunity for increasing food and environmental security by avoiding direct pollution of rivers and surface water, conserving significant proportions of river basin waters, and disposing off municipal wastewater in a low-cost, sanitary manner. The various ways to reuse the wastewater are as shown in the Figure 4.4.

**Table 4.4: Sewage Generation and Treatment in Class I Cities and Class II Towns (2001 population basis)**

| *City Category and Population* | *Number of Cities* | *Sewage Generation, MLD* | *Installed Sewage Treatment Capacity, MLD* | *Capacity Gap in Cities having STPs, MLD (A)* | *Sewage Generation in Cities having no STPs, MLD (B)* | *Total Capacity Gap, MLD (A+B)* | *Planned Treatment Capacity, MLD* |
|---|---|---|---|---|---|---|---|
| Class I cities having more than 10 lac population | 39 | 13503 | 4472 (In 29 cities) | 6135 | 2896 | 9031 | 1549 |
| Class I cities having 5 to 10 lac population | 32 | 3836 | 485 (In 13 cities) | 1293 | 2058 | 3351 | 123 |
| Class I cities having 2 to 5 lac population | 119 | 4807 | 768 (In 34 cities) | 804 | 3235 | 4039 | 4 |
| Class I cities having 1 to 2 lac population | 224 | 4018 | 322 (In 36 cities) | 373 | 3323 | 3696 | 32.5 |
| All the above Class I cities together | 414** | 26164 (100 per cent) | 6047 (23.1 per cent) (In 112 cities) | 8605 (32.9 per cent) | 11512 (44 per cent) | 20117 (76.9 per cent) | 1708.5 (6.5 per cent) |
| Class II towns having 0.5 to 1 lac population | 489** | 2965 (100 per cent) | 200 (>143*) (4.8 per cent) (In 22 towns) | Nil | 2822 (95.2 per cent) | 2822 (95.2 per cent) | 34.1 (1.15 per cent) |
| All Class I cities and Class II towns | 893** | 29129 (100 per cent) | 6190 (21.3 per cent) | 8605 (29.5 per cent) | 14334 (49.2 per cent) | 22939 (78.7 per cent) | 1742.6 (6.0 per cent) |

*: Estimated sewage of the cities having STPs.

**: Delhi being taken as one city.

*Source*: CPCB 2004.

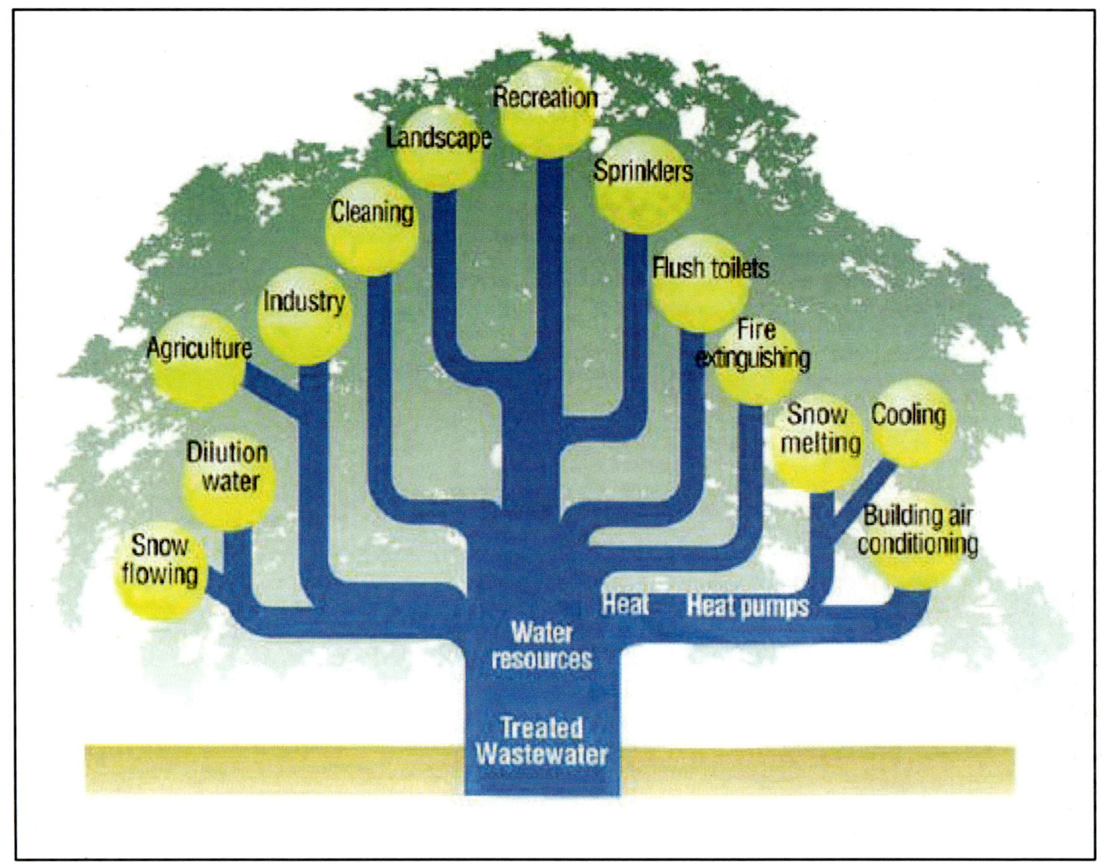


**Figure 4.4: Tree of Water Resource Recycling**
***Source*: MLIT, 2001**

Most of the reclamation and reuse of domestic wastewater in India is done is agriculture and aquaculture as compared to Europe the areas of application are split into four categories: (1) agriculture; (2) industry; (3) urban, recreational and environmental uses, including aquifer recharge; and (4) combinations of the above (mixed uses). The scale of the projects is also split into four classes: very small (<0.1 GL/y), small (0.1–0.5 GL/y), medium (0.5–5 GL/y) and large (>5 GL/y). Reclaimed wastewater is reused predominantly for agricultural irrigation (44 per cent of the projects) and for urban or environmental applications (37 per cent of the projects); in northern Europe, the uses are mainly for urban or environmental applications (51 per cent of the projects) or industrial (33 per cent of the projects).

In the four major project cities of Ahmedabad, Delhi, Kanpur and Kolkatta, wastewater was being used for agriculture (growing vegetables, food grain and horticulture crops) and aquaculture. There are other places too as enlisted in Table 4.5 where the use of domestic wastewater in agriculture and aquaculture is going on.

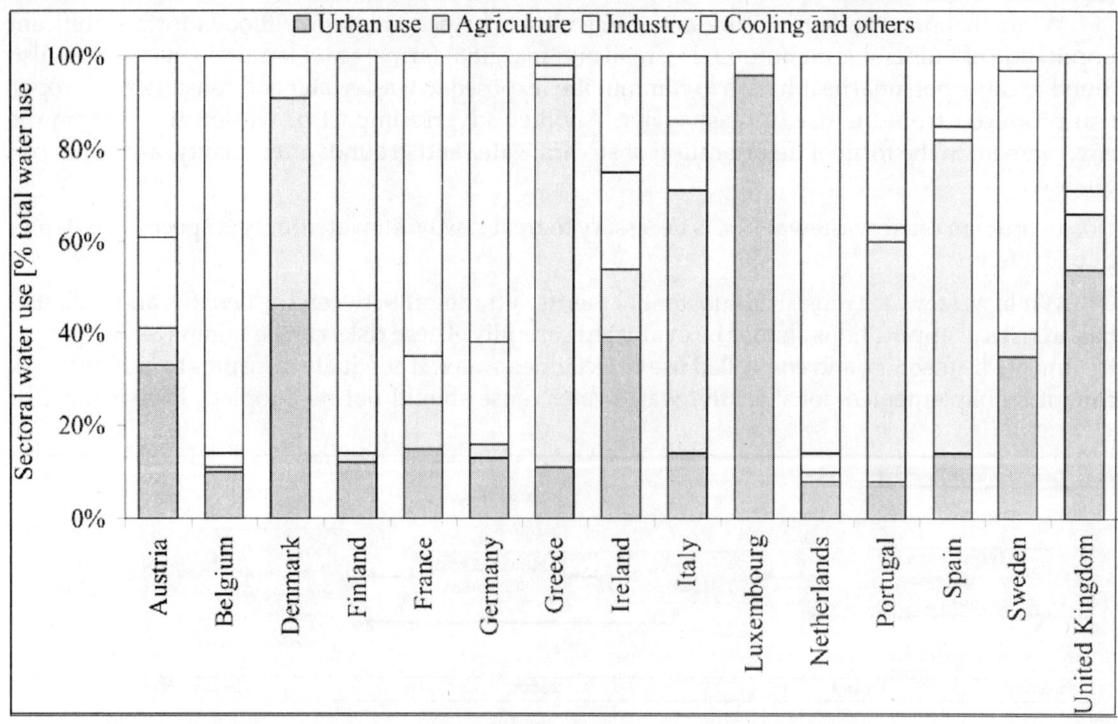


**Figure 4.5: Sectoral Use of Percentage of Water Use in Europe**

**Table 4.5: Use of Wastewater in Agriculture and Aquaculture in India**

| *City* | *Direct Use ha* | *Indirect Use ha* | *Reference* |
|---|---|---|---|
| Ahmedabad | 890 | – | Juwarkar *et al.*, 1988 |
| Amritsar | 1.124 | – | Juwarkar *et al.*, 1988 |
| Bhilai | 607 | – | Juwarkar *et al.*, 1988 |
| Bikaner | 40 | – | Juwarkar *et al.*, 1988 |
| Calcutta | | 12.9 | Mara and Cairncross,1989; Edwards 2001 |
| Delhi | 1.214 | – | Juwarkar *et al.*, 1988; Farooqui 2002 |
| Gwalior | 202 | – | Juwarkar *et al.*, 1988 |
| Hyderabad | 110 | 40,500 | Bucchler and Devi, 2002 |
| Jamshedpur | 113 | – | Juwarkar *et al.*, 1988 |
| Kanpur | 1.3 | – | Strauss and Blumonth, 1990 |
| Lucknow | 150 | – | Juwarkar *et al.*, 1988 |
| Madras | 133 | – | Juwarkar *et al.*, 1988 |
| Madurai | 77 | – | Juwarkar *et al.*, 1988 |
| Nagpur | 1.5 | – | Juwarkar *et al.*, 1988 |
| Trivandrum | 37 | – | Juwarkar *et al.*, 1988 |
| Vadodara | —— | 14.567 | IWMI (Bhamoriya, 2002) |

While this practice (agriculture and aquaculture) helped create livelihoods for a significant population of vulnerable communities (small and marginal farmers and landless), its use was also found to cause potential health risks to communities exposed to wastewater and consumers of crops/ fish produced from the use of wastewater. Another adverse impact of wastewater was on the environment in the form of deterioration of stream water and groundwater quality, as well as soil quality.

In order to reuse wastewater, it is necessary to treat raw wastewater to meet specific needs and public safety.

While wastewater reuse has substantial merits, a trade-off between the benefits and potential health risks of applications should be evaluated carefully. These risks can be minimized by proper treatment, disinfection, and controlled use of reclaimed water. If adequate measures to minimize risk cannot be implemented consistently, wastewater reuse should not be adopted. Proper reuse of

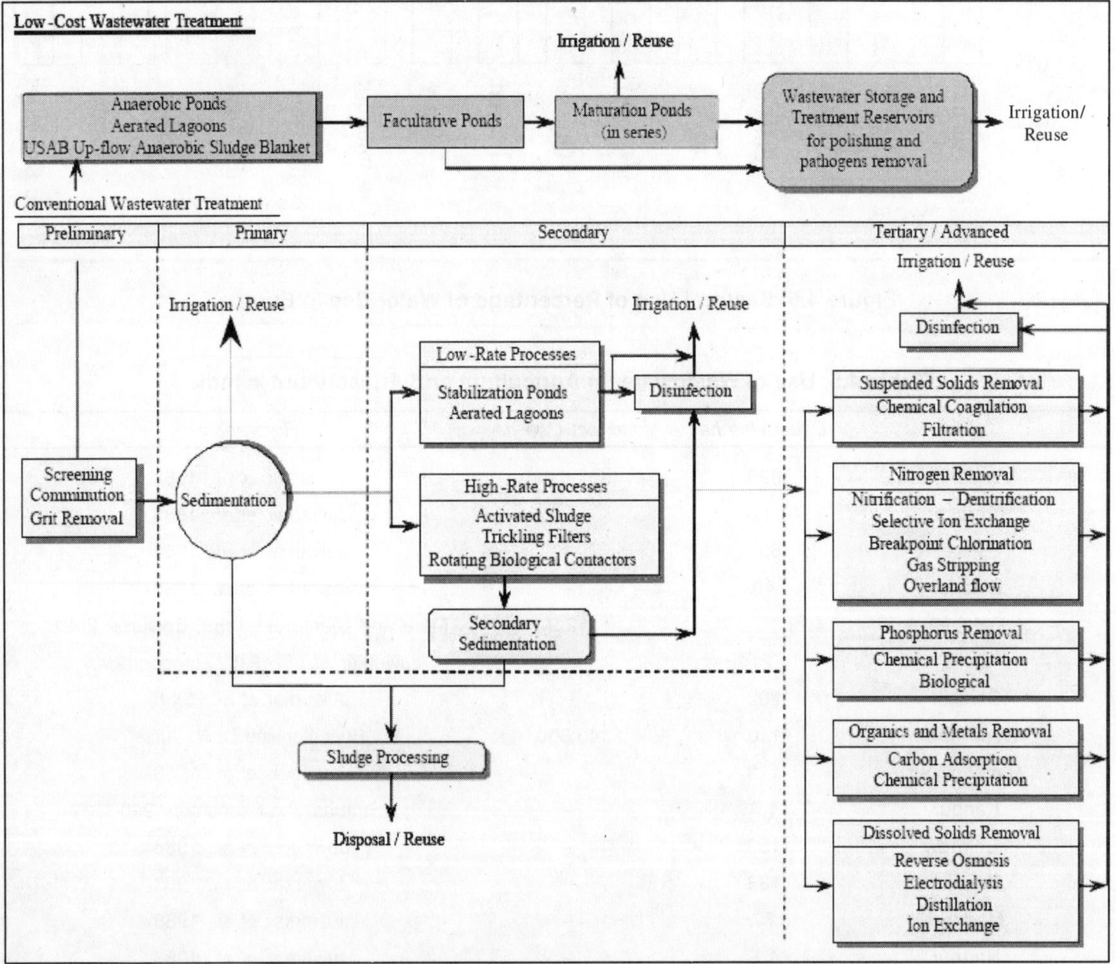


**Figure 4.6: Generalised Municipal Wastewater Treatment Scheme and Points of Reuse Adapted from Asano, Smith, and Tchobanoglous, 1984; R. Tsuchihashi, 2005; Jimenez 2005)**

wastewater should be encouraged also, because in some places it allows the production of crops that can be exported to other countries that have strict regulations on health. A generalized municipal wastewater treatment scheme and points of reuse is given in Figure 4.6.

## Appropriate Treatment Technology

Based on experience from past mistakes in sewage treatment technology, the definition of what is sustainable is clearer. Developers should base the selection of technology upon specific site conditions and financial resources of individual communities.

Although site-specific properties must be taken into account, there are core parts of sustainable treatment that should be met in each case. The criteria for sustainable technology are summarized as:

- ☆ No dilution of high strength wastes with clean water.
- ☆ Maximum of recovery and re-use of treated water and by-products obtained from the pollution substances. (*i.e.* irrigation, fertilization)
- ☆ Application of efficient, robust and reliable treatment/conversion technologies, which are low cost (in construction, operation, and maintenance), which have a long life-time and are plain in operation and maintenance.
- ☆ Applicable at any scale, very small and very big as well.
- ☆ Leading to a high self-sufficiency in all respects.
- ☆ Acceptable for the local population.

One approach to sustainability is through decentralization of the wastewater management system. This system consists of several smaller units serving individual houses, clusters of houses or small communities. Black and gray water can be treated or reused separately from the hygienically, more dangerous excreta. Non-centralized systems are more flexible and can adapt easily to the local conditions of the urban area as well as grow with the community as its population increases (Schertenlieb, 2000). This approach leads to treatment and reuse of water, nutrients, and byproducts of the technology (*i.e.* energy, sludge, and mineralized nutrients) in the direct location of the settlement.

Communities must take great care when reusing wastewater, both chemical substances and biological pathogens threaten public health as well as accumulate in the food chain when used to irrigate crops or in aquaculture. In most cases, industrial pollution poses greater risk to public health than pathogenic organisms. Therefore, more emphasis is being placed on the need to separate domestic and industrial waste and to treat them individually to make recovery and reuse more sustainable. The system must be able to isolate industrial toxins, pathogens, carbon, and nutrients (Rose, 1999).

## Industrial Wastewater

The amount of industrial wastewater generated is almost half as compared to the domestic wastewater however the BOD load exerted is almost double. Only 5 per cent of wastewater is generated by small scale industries and remaining 95 per cent from the large scale industries. The water an industry discharges is always of lower quality than the feed water and this wastewater is discharged downstream. At best the wastewater discharged represents a quality that can be recycled for lower grade of industrial use and at worst represents water quality which is unsuitable for every use other than navigational purposes. Water consumption depends on the type of industry. Whereas thermal power, textiles, pulp and paper and iron and steel are highly water intensive sectors, industrial sectors like chlor-alkali, cement, copper and zinc and plastics require little water.

Data on actual water consumption in India is absent. However, the data on wastewater discharge by various industrial sectors in the country is available. The data on wastewater discharge has been complied by CPCB. According to CPCB, the total wastewater discharged by all major industrial sources is 83,048 million litres per day (mld). This includes 66,700 mld of cooling water discharged by thermal power plants (TPPs). Out of the remaining 16,348 mld of wastewater, TPPs generates another 7,275 mld as boiler blowdown water and overflow from ash pond. In other words, this water is unfit for usage, is seldom used by the industries and usually used for certain agricultural purposes or by villagers to meet their daily needs. According to CPCB the annual water consumption in Indian industry is 40 billion cubic meters and the annual wastewater discharge is about 30.7 billion cubic meter. Therefore, the overall ratio of wastewater discharged to freshwater consumption in Indian industry works out to be about 0.77. That is, for every cubic meter of water consumed by Indian industry, 0.77 cubic meters of wastewater is discharged. India ranks 3rd in inefficient water use by industry *i.e.* it generates a high amount of wastewater or the wastage of water consumed. The inefficiency of proper water use has caused water scarcity which has led to industry community conflicts. The examples are many such as SIV industries (Coimbatore), Sinar Mas Pulp and paper mills (Near Pune), Grasim Industries (Mavoor unit).

**Table 4.6: Inefficient Water Use by Industry**

**Inefficent water use by industry**

The ratio of water consumption and economic value creation in Indian industry is poor. For every cubic metre of water that Indian industry uses, it generates merely US $7.5 economic productivity

| *Country* | *Industrial Water Use (billion cubic metres)* | *Industrial Productivity (million US $)* | *Industrial Water Productivity (US $/cubic metre)* |
|---|---|---|---|
| Argentina | 2.6 | 77171.0 | 30.0 |
| Brazil | 9.9 | 231442.0 | 23.4 |
| India | 15.0 | 113041.0 | 7.5 |
| Korea, Rep. | 2.6 | 249268.0 | 95.6 |
| Norway | 1.4 | 47599.0 | 35.0 |
| Sweden | 0.8 | 74703.0 | 92.2 |
| Thailand | 1.3 | 64800.0 | 48.9 |
| United Kingdom | 0.7 | 330097.0 | 443.7 |

*Source*: World Bank, 2001.

There are industrial estates in India treat the wastewater in CETPs (Common effluent treatment plants). Major 78 common effluent treatment plants were taken as the objective of the study and the performance status is indicated in Table 4.7. Only 6.4 per cent of the CETPs complied to the effluent standards.

Monitoring of water use in industries is gaining importance due to increasing competition and stringent environment norms. Water is no longer perceived as a free commodity; further, innovative technologies, along with modification of existing technologies are commercially available to reduce process water consumption. In general, reduction in industrial wastewater can be achieved through one or a combination of the following measures:

☆ Process modification or change in raw materials to reduce water consumption;
☆ Direct reuse of wastewater;
☆ In-plant reuse of reclaimed wastewater;
☆ Use of treated wastewater for non-industrial purposes.

**Table 4.7: Performance Status of 78 CETPs Studied by CPCB during 2002–2005**

| State | Number of CETPs Studied by CPCB | CETPs Complying pH, BOD, COD, TSS and TDS Standards | | CETPs Complying pH, BOD, COD and TSS but not Complying TDS Standard | |
|---|---|---|---|---|---|
| | | Number | Name | Number | Name |
| AP | 2 | 0 | | 0 | |
| Delhi | 10 | 3 | Mayapuri, GTK, Badli | 6 | Wazirpur, Mangolpuri, Jhilmil, SMA, Nangaloi, Okhla I.A |
| Gujrat | 15 | 0 | | 2 | Ankleshwar, Sachin (0.5MLD) |
| Haryana | 1 | 1* | Kundli-I | 0 | |
| Karnataka | 2 | 0 | | 1 | Pai&Pai |
| Maharashtra | 9 | 0 | | 3 | Thane-Belapur, Ambernath, Patalganga |
| MP | 1 | 0 | | 0 | |
| Punjab | 1 | 1 | Phillore | 0 | |
| Rajasthan | 5 | 0 | | 1 | Jodhpur ** |
| Tamilnadu | 29 | 0 | | 2 | Thiruvai Karur***, TALCO Ambur Thuthipet |
| UP | 3 | 0 | | 0 | |
| Total | 78 | 5 (6.4 per cent) | | 15 (19.2 per cent) | |

*: TDS not determined but likely to be within limits.

**: CETP was under trial; TDS not determined but Cl exceeded.

***: TSS not determined.

The final choice of the measures adopted are dictated by the availability of substitute raw material and technology, nature of pollutants, purity requirement of water for reuse, competing water demands and overall treatment cost. Table 4.7 summarizes some key initiatives implemented for improved water use in different industrial sectors in India and outside India.

The problem with the industrial wastewater is not the quantity but quality too. The pollution load is almost double the domestic wastewater. Of the 17 most polluting industries in 2006 Petrochemicals, pulp and paper, textile, dairy, distillery and tanneries are the most polluting (Table 4.8). The present Indian scenario for some polluting industries and changes to be adopted is discussed in this section.

## Pulp and Paper

The pulp and paper industries can be classified into three-category *i.e.* large-scale pulp and paper mills, agro based pulp and paper mills and waste paper based mills having installed capacity of 4.25 million tonnes. Environmental problems associated with pulp and paper industries vary with

**Table 4.8: Water Conservation Possibilities and Technological Availability for Different Industries**

| *Industry* | *Interventions* | *Technology Adopted* | *Remarks* |
|---|---|---|---|
| Distillery | Spentwash recycling | Modification in existing process | Successful trials with 40 per cent spentwash recycling for molasses dilution and fermentation at Fermpro, Malaysia (Yeoh *et al.*, 2001) 25 per cent spentwash recycle for fermentation being used at SSK Distillery, Akluj, India (Waste Minimization Circle) Semi-continuous process reduced water consumption up to 70 per cent in Sarina distillery, Australia (EnviroNet Australia) |
| Sugar | Recycling of process water from the barometric column<br>Recycling of boiler condenser water<br>Recycling wash water used in chimney cleaning | Developed system for cooling and collecting water | 93 per cent water savings, payback just above 24 months in San Francisco Ameca Plant Jalisco, Mexico (EMCENTRE, 1999; UNEP-WBCSD, 1998) |
| Dairy | Reduced water for washing<br>Recirculation of water with cleaning solution in cleaning area | Improved hose with shut-off nozzle<br>High pressure washing | Reduced water use by 30,000 m$^3$/year in a dairy in Tartu, Estonia (UNEP-WBCSD, 1998) |
| Pulp and paper | Change in feedstock<br>Regeneration of water from the wastewater stream<br>Recycle of paper machine white water, decker and screen white water, condenser cooling water | Screening<br>Clarification<br>Evaporation<br>Close monitoring of operation conditions | Trials in a mill at Jämsänkoski, Finland displayed (UNEP-WBCSD, 1998)<br>75 per cent reduction in freshwater consumption by changing feedstock<br>Zero effluent pulp and paper mill |
| Board | Regeneration of process water from secondary treated effluent | Combined ultrafiltration and reverse osmosis | Successful pilot scale study in Turkey (Yalcin *et al.*, 1999) showed high quality of treated water |
| Tannery | Reuse of main soak for dirt soak<br>Drum soaking instead of pit soaking<br>Reuse of limed water<br>Deliming of water for subsequent use soaking | Chemical treatment<br>Little modification in existing facilities | Recommended by Central Pollution Control Board, India expected to result in significant reduction in water consumption (CPCB, 1999) |
| Textiles | Regeneration, use of dyeing and fulling wastewater | Combined ultrafiltration and reverse osmosis | Tested on pilot scale in Italy (Ciardelli *et al.*, 2000) with estimated treatment cost 1/m$^3$ |
| Semi-conductor | Use of rejected water from deionizing unit as cooling water make-up<br>Recycling of process water | Chemical treatment<br>Microfiltration<br>Ion-exchange resin treatment | Practical experience in SGS-Thomson Micro-electronics, Malta (UNEP-WBCSD, 1998)<br>Case study in Chungli, Taiwan (You *et al.*, 2001) |

*Contd...*

**Table 4.8–Contd...**

| *Industry* | *Interventions* | *Technology Adopted* | *Remarks* |
|---|---|---|---|
| Metal industries | Treatment of heavy metal polluted acidic water | Biological treatment with alkali producing algae (*Spirulina* sp.) followed by anaerobic polishing | Research experience in South Africa (Van Hille *et al.*, 1999) shows significant reduction in metal ion concentration when tested in a bench scale reactor<br>Tannery or saline water can be used as nutrient for algae |
| Petroleum refining | Use of tertiary treated municipal wastewater as process water | Ultrafiltration and reverse osmosis | Significant savings in freshwater consumption through combined ultrafiltration and reverse osmosis of secondary treated municipal sewage in Chennai Petroleum Company Limited, Chennai, India Treatment cost is estimated to be Rs. 28/kL and is economically feasible (Viswanathan and Sethi, 2002) |

**Table 4.9: Estimated Water Pollution Load (in tons) by Industry**

| *Industry* | *Estimates Using Output Intensities* | *Ranking* | *Estimates Using Employment Intensities* | *Ranking* |
|---|---|---|---|---|
| Aluminium | 47469 | 3 | 0 | 16 |
| Copper | 16035 | 6 | 44495 | 9 |
| Zinc | 7737 | 8 | 22923 | 12 |
| Iron and Steel | 1639368 | 1 | 8093409 | 1 |
| Cement | 5168 | 11 | 28000 | 11 |
| Oil Refinery | 4340 | 12 | 16805 | 13 |
| Drugs | 5889 | 10 | 44736 | 8 |
| Petrochemicals | 1818 | 13 | 3805 | 14 |
| Fertilisers | 31480 | 4 | 106644 | 7 |
| Pesticides | 7366 | 9 | 37927 | 10 |
| Caustic Soda | 836 | 15 | 135691 | 5 |
| Pulp and Paper | 86245 | 2 | 801764 | 3 |
| Leather | 894 | 14 | 5316058 | 2 |
| Dyes | 0 | 16 | 1198 | 15 |
| Distillery | 7740 | 7 | 110334 | 6 |
| Sugar | 16747 | 5 | 217639 | 4 |

*Source*: Pandey and Ghosh.

the size and category of the mill. Although all the large-scale pulp and paper mills have adequate treatment systems but still some of the problems like colour in the effluent and solid waste disposal problem still persists. The colour problem is basically due to spillage of black liquor during its handling which ultimately joins the effluent stream. The effluent that emanates from the bleaching section of the mill also contributes the colour. The toxicity is another problem, which is mainly due to use of elemental chlorine during the bleaching.

In small scale agro-based pulp and paper mills, major cause of pollution is discharge of black liquor which is otherwise taken to chemical recovery plant by large scale pulp and paper mills. The absence of chemical recovery plant in small-scale pulp and paper mills is due to their smaller size and high cost involved in its installation.

In waste paper based mills, zero discharge is possible through recycling of wastewater after suitable treatment. But due to poor waste treatment and old fibre recovery technologies, the industries are unable to recycle the effluent.

In the Table 4.10, the present practices adopted by the pulp and paper mills alongwith the technology requirements are outlined.

**Table 4.10**

| *Technologies/Current Practises* | *Requirements* |
|---|---|
| Use of elemental chlorine in bleaching of pulp which is resulting in generation of toxic effluent containing chloro compounds of lignin. It results in generation of AOX, which is highly toxic and carcinogen | Bleaching techniques, which avoid the use of elemental chlorine, should be adopted. Use of chlorine dioxide and use of oxygen at alkali extraction stage should be preferred. It reduces generation of AOX by about 70 per cent. The use of oxygen/ozone as bleaching agents makes the effluent recyclable from this section with no colour |
| Use of kraft pulping process for delignification of wood/bamboo raw material in large-scale mills. | Use of modern pulping process like RDH pulping, Oxygen delignification should be practised. These processes produce pulp of low kappa number and high brightness, which require low bleach chemicals. Low steam requirement with high pulping yield are the attempted benefits |
| Land disposal of lime sludge by the mills having chemical recovery plant | Most of the large-scale mills dispose lime sludge on land. Lime kiln should be used for recalcination of lime sludge so that it can be reused in the process. |
| Discharge of black liquor by the small scale agro-based pulp and paper mills | Chemical Recovery Plant (CRP) is required to be installed by such mills. The black liquor can be taken to CRP for recovery of pulping chemicals. Some agro-based pulp and paper mills have installed CRP and running it successfully. The installation of CRP by all units is necessary to control pollution and colour |
| Removal of colour through lignin precipitation by adding poly-electrolyte | For the industry where chemical recovery is not feasible economically and problem of colour persists, the removal of lignin through precipitation and its conversion into useful and marketable product *e.g.* binders should be practised. |
| Use of obsolete technologies like conical save-all etc. as fibre recovery system | Dissolved air floatation based fibre recovery systems and micro-filters. Use of micro-filters improves the recycling of wastewater significantly. |

## Textile Industry

The textile industry encompasses a range of industrial units, which use a wide variety of natural and synthetic fibres to produce fabrics. The textile industry can be broadly classified in two groups, namely cotton industry and woollen industry.

The effluent generated from textile industry contains various dyes, chemicals, auxiliary chemicals, sizing materials etc. The effluent is usually treated by physico-chemical treatment followed by biological treatment process. However, such treatment systems are not effective for removal of colour, dissolved solids, trace metals etc. In-plant control measures such as process change, recovery/reuse of chemicals, chemical substitutes can significantly reduce pollution from this sector. There are described:

**Table 4.11**

**(a) Process Change/Recovery**

| Area | Technology/Current Practises | Requirements |
|---|---|---|
| Sizing | Starch is most commonly used as sizing material for cotton textiles. During sizing, the starch eliminates the possibility of its recovery. It also contributes high BOD. | Recovery of sizing material: The other types of sizing material, such as carboxymethyl cellulose (CWC), polyvinyl acetate (PVA) are comparatively more expensive but are recoverable. |
| Mercerising | Mercerising waste contains about 4 per cent caustic. Few textile industries have provision for its recovery. | Recovery of caustic: Spent caustic from mercerising as well as other units can be recovered and reused either by membrane separation technology or by most commonly used evaporation method. |
| Dyeing | The textile industry uses various types of dyes to impart the desired quality in the fabrics, which generates coloured effluent. No specific treatment is given before it is discharged | Reuse of dye bath: Instead of discharging the exhausted dyebath, it can be reconstituted by adding appropriate amount of make-up dyes and auxiliary chemicals. The reconstituted dyebath can be reused for dyeing successive batches |
| Printing | The conventional printing involves colour paste application to fabric and subsequent dye fixation. After this, all chemicals other than the dye need to be removed from fabric. This increases effluent generation. | Transfer Printing: The transfer printing transfers dyestuff, previously printed on paper, on the fabric. Only the dyestuff and other chemicals are transferred on the fabric thus eliminating the after-washing. |
| Wastewater Treatment | The wastewater is treated by physico-chemical and/or biological treatment process. Some industry uses filtration alongwith physico-chemical treatment to reuse the specific effluent streams. | Recycling of Treated Wastewater: The effluent can be further treated by activated carbon adsorption process or other advanced treatment process so that the treated effluent can be recycled/reused |

**(b) Chemical substitute**

| Area | Chemical in Use | Substitute required |
|---|---|---|
| Sizing/Process | Conventional Starch based size | Synthetic wrap sizes (PVA) Acrylates) |
| Desizing | Enzymes | Acids |
| Soaping | Conventional Soap | Synthetic Detergents |
| Good Scouring | Soda ash | Sodium acetate |
| Disperse Dyeing and Pigment Printing | Acetic acid | Ammonium sulphate |
| Printing | Gum | Emulsion |
| Oxidation of vat dye | Acetic Acid | Solium bicarbonate |
| Screen Printing machines | Conventional gums | Permanent adhesives |
| Finishing starch based | Temporary finishes | Durable finishes |
| Dyeing | Two stages dye (Disperse, vat. etc.) | Single stage dyes (Tindigosol) |
| Dyeing | Solvent Pthalogen blue | All aqueous Pthalogen blue |
| Dacron Dyeing | Conventional carriers | Monochloro benzene |
| Dye bath | Acetic acid | Formic acid |
| Lubricants used in textile machinery | Carding oils anti-state lube | Non-ionic emulsifiers |

## Tanneries

Tanning is an integral part of leather making process, which converts the raw hides and skins to finished leather, which is used for manufacturing various leather goods.

Tanning industries can be classified into two categories with respect to the kind of tanning practised: vegetable or EI tanning and chrome tanning. Further these processes can be classified into different processes each depending upon the level of tanning a particular industry adopts.

**For Vegetable Tanning**

1. Raw hide to vegetable tanned semi-finished leather.
2. Vegetable tanned semi-finished leather into finished leather.

**For Chrome Tanning**

1. Raw hide to chrome tanned semifinished leather
2. Raw hide to finished leather by chrome tanning
3. Chrome tanned leather into finished leather.

In India, mostly chrome tanning is practised.

The effluent generated from tanning process contains very high TDS, chromium (where chrome tanning is practised), high BOD and COD. So far, the practice to reduce TDS in effluent has been to segregate soak liquor and to minimise the total volume by solar evaporation. The chromium can be reduced by two ways *i.e.* chemical precipitation or chrome recovery. The biological treatment (anaerobic followed by aerobic) is practised to reduce BOD and COD.

The technologies available and the requirements are described in Table 4.12.

**Table 4.12**

| *Technologies/Current Practices* | *Requirements* |
|---|---|
| **(*a*) CHROME RECOVERY** | |
| Adoption of chrome recovery in case of raw hide to finished processes especially is practised by some of the medium and large-scale tanneries. It comprises collection of spent chrome liquor after basification and recovery of chrome from the same. This results in substantial saving besides reducing the concentration of chrome in the trade effluent. The supernatant liquor after recovery of chrome can be used for soaking. The recovered chrome can be used alongwith regular basic chrome sulphate for chrome tanning. | Large and medium units should provide individual chrome recovery system barring those which are practising semi-chrome finished to finished leather. The small-scale units need to adopt the chrome recovery system in a group of 4 to 5 units. |
| **(*b*) SOAKING** | |
| *(a) Reuse of main soak for dirt soak* | *(a) Addition of soaking enzymes* |
| Soaking consists of dirt soak and main soak. The main soak is retained and used for dirt soak for the next batch. | Soaking enzymes @ 0.4 per cent of w/s wt. can be added to achieve uniform and thorough soaking. |
| *(b) Drum soaking instead of pit soaking* | *(b) Reuse of dirt soak* |
| Drum soaking instead of pits soaking which reduces the water consumption besides bringing down the soaking time from 12 hrs. to 3 hrs. | Possibilities of collection of dirt soak liquor and adding polyelectrolyte to flocculate and settle the suspended solids need to be explored. Soak liquor after treatment and filtration can be reused partially in liming/deliming washes and picking. Modern plants are providing these facilities. |

*Contd...*

**Table 4.12–Contd...**

| *Technologies/Current Practices* | *Requirements* |
|---|---|
| **(*c*) LIMING** | |
| *(a) Substitution of paste lime by 85 per cent pure calcium oxide (CaO)* | *(a) Use of Liming Enzymes* |
| This can bring down the quantity of consumption of powered lime to one third of its original quantity. It can also reduce the frequency of cleaning the primary settling containing the lime sludge. | Use of liming enzymes @ 0.4 per cent can reduce Sodium Sulphide ($Na_2S$) Consumption by 40 per cent. Use of soaking and liming enzymes can improve liming quality |
| *(b) Provision of slight slope in the pasting area* | resulting in cleaner and flatter pelts. The marginal increase in |
| By providing a slight slope in the pasting area, the excess liming paste can be effectively collected and used, which is otherwise washed away in the drain by lime yard workers. | cost is more than offset by increase in yield achieved in Wet Blue as well as reduction in $Na_2S$ consumption. At higher enzyme levels, hair can be shaved off without burning which can be screened and used as a fertiliser. |
| *(c) Reuse of relime liquor* | |
| 50 per cent of relime liquor can be retained and reused for liming of subsequent batches. This also reduces water consumption. | Some tanneries are practising; however, others should be encouraged to do so. |
| Other than optimisation of chemical consumption in this section, fleshing can be used for production of high-grade protein. This, if achieved, will solve the problem of solid waste disposal from the liming section. | |
| **(*d*) DELIMING** | |
| Efforts can be made to reduce water consumption in this section by implementing the following measures:<br>a. Use of deliming agents.<br>b. Use of first delimes wash for liming. | Proper deliming agents need to be identified. |
| **(*e*) PICKLING** | |
| *Use of drained float for next batch* | |
| Normally after desired pH is obtained in pickling, 1/3rd float is drained to ensure better chrome exhaustion at lower floats. This liquor can be collected and used for next batch to reduce consumption of salt and acids. | Such facilities need to be installed in selected units for demonstration purposes. |

## Technologies for Waste Minimization in Tanneries

- ☆ Manual/mechanical desalting prior to soaking
- ☆ Reducing hydraulic load in soaking
- ☆ Counter current recycling of soak liquors
- ☆ Lime-free beam house operations
- ☆ Enzyme assisted less-sulfide unhairing
- ☆ Recycling of lime liquor/washings
- ☆ Ammonia-free deliming process
- ☆ Salt-less picking

- ✰ Pickle-free chrome tanning
- ✰ Closed loop pickle-chrome recycling methods involving high exhaust chrome tanning
- ✰ Less chrome and chrome-free tanning methods
- ✰ Full chrome tanning methods
- ✰ Use of Eco-benign post tanning auxiliaries
- ✰ Optimization of float volumes in processing
- ✰ Formaldehyde-free finishing techniques
- ✰ Use of newer finish application systems
- ✰ Use of Roller Coater for improved transfer efficiency
- ✰ Split finishing

## Distillery

The main source of pollution from distilleries is the spentwash, which is produced as a result of fermentation and distillation of the molasses. The spentwash is acidic in character having very high BOD (40,000–50,000 mg/l) and COD (10,000-125,000 mg/l). The different treatment technologies for treatment of distillery effluent have been adopted depending upon the availability of land; filler material for compost and mode of disposal of treated effluent. The existing technologies and technologies required are given in Table 4.13.

**Table 4.13**

| *Technologies/Current Practices* | *Requirements* |
|---|---|
| *Bio-methanation followed by Secondary Biological treatment system* | |
| Large number of industries have provided the system wherein methane gas produced can be used into boilers which saves the fuel to the tune of 60 per cent, besides removing BOD in the order of 80-90 per cent. The secondary biological treatment such as activated sludge process or extended aeration system after bio-methanation brings down the BOD in the range of 300-500 mg/l. The final effluent after mixing with the other wastewater can be used on land for irrigation. | Two stage secondary biological treatment system is needed preferably using diffused aeration system for better efficiency and saving in power. The treated effluent when utilised on land for irrigation should meet the requirement of TDS by mixing with the irrigation water from other sources. |
| *Composting* | |
| In this process, press mud generated from sugar mill is utilised to produce compost by mixing distillery effluent. Both anaerobic and aerobic composting systems are practised. In some plants composting with treated effluent treated through bio-methanation plant is also practised. This system can achieve zero effluent if the press mud quantity matches with the effluent generated. | Adequate capacity of composting facility is needed and the compost quality is required to be checked periodically. There should be adequate holding capacity of effluent so that at no time effluent finds way into the drain or in the ground water. Ground water monitoring is also necessary to ensure that there is no seepage from the composting site. |
| *Incineration* | |
| Following drying of spentwash and using dried material in the boiler for steam generation. | Use of the spent wash alongwith baggasse/biomass for generation of power. |

## Sugar Industry

India is the largest producer of sugar in the World and per capita consumption of sugar in the country is 13,4 kilograms per annum. There are about 400 operating sugar mills, located mainly in the states of Uttar Pradesh, Maharastra, Andhra Pradesh, Karnataka and Tamil Nadu. During 1993-94, these industries have produced 98 Iakh tonnes of sugar. It is a seasonal industry and operates for about 6 months in a year, normally from October to March.

## Pollution Problems

Sugar mills consume around 1,500-2,000 litres of water and generate about 1,000 litres of wastewater for per tonne of cane crushed, the effluent is mainly floor washing wastewater and condensate water. Leakage in valves and glands of the pipeline add sugarcane juice, syrup and molasses in the effluent. The sugar mill effluent has a BOD of 1,000-1,500 mg/litre, but appears relatively clean initially. However, after stagnating for sometime, it turns black and start emitting foul odour. If untreated effluent is discharged, in water courses. It depletes dissolved oxygen in water and makes the environment unfit for aquatic life. If untreated effluent is discharged on land, decaying organic solids and oil and grease clog the soil pores.

Most of the use bagasse as a Sugar mills fuel in boilers, which produces particulate matter, oxides of nitrogen, carbon, sulphur and water the vapours. The particulate matter, usually referred to as fly ash, consists of ash, unburnt bagasse and carbon practicles. Fly ash is very light therefore, if pollution control equipments are not installed, it escapes in the atmosphere through chimney and travels long distances. In such conditions, nearby population suffer from dizziness and irritation in eyes, nose, throat and lungs. The heavier particles, if settle on vegetation then it damages them. Among the solid waste generated by sugar mills, lime sludge and press mud are important. For purifying the sugarcane juice from organic matter, dirt and other impurities, milk of lime is in generates lime sludge. The impurities from the sugarcane juice is either vacuum filtered or press have filtered and removed as press mud. In addition to these, solid wastes are also generated from the pollution control facilities, like ETP sludge and flyash collected from the dusting devices.

## Pollution Standards

The pollution standards stipulate that BOD of effluent should be less than 30 mg/litre for disposal into inland surface waters and less than 100 mg/litre for disposal on land. BOD can be 500 mg/litre, in case land application effluent is envisaged as a secondary treatment system for further removal of BOD.

Regarding water consumption and effluent generation, specified standards are 1,000 litres and 400 litres respectively for per tonne of cane crushed.

As per general emission standards, particulate matter is required to be within 150 mg/Normal cubic metre. In case of horse shoe/pulsating grate and spreader stroker bagasse fired boilers, the particulate matter emission is required to be within 500 mg/Normal cubic metre and 800 mg/Normal cubic metre respectively.

## Pollution Control

There is scope of recycling and reuse of water in sugar mills thereby minimising water consumption and ultimately effluent quantity. The recycling and reuse of hot condensate water can reduce the water consumption to as low as 100-200 litres, as against 1,500-2,000 litres per tonne of cane crushed.

Proper housekeeping, periodic checking and maintenance of pipe joints, valves and glands further reduces the water consumption and effluent quantity.The effluents from the sugar industry can be treated added. The preparation of milk of the lime by conventional biological treatment systems, general, anaerobic biological processes (oxidation ponds and biomethanation) several advantages over aerobic processes (aerated lagoons, activated sludge process). Anaerobic processes are easier to control and operate, produce a lower quantity of sludge and their costs are lower. Anaerobic processes decompose the organic compounds in an atmosphere free of oxygen and consequently require significantly less energy as compared to aerobic processes.

Double Sulphitation Process, already adopted by most of the sugar industries, reduces the quantity of lime sludge and press mud to a I great extent. The lime sludge is usually dumped in low lying areas, whereas press mud is sold to farmers as it can be used as manure. Bagasse is either used as fuel or sold to pulp and paper industry which use them as raw materials. Molasses produced in sugar industry is raw materials for fermentation industries.

Cleaner production (CP) assessment is an analytical method of particular relevance for evaluating options for industrial water reuse. CP has been promoted by UNEP since the late 1980s, and is defined as the 'continuous application of an integrated, preventive environmental strategy to processes, products, and services to increase overall efficiency, and reduce risks to humans and the environment'. CP can be implemented to improve industrial processes, product performance and various services provided in society. The CP assessment methodology is used for systematic identification and evaluation of CP opportunities, thereby facilitating their implementation. The five areas where CP opportunities to be identified through CP assessment: changes in raw materials, technological change, good operating practices, product changes, and on-site reuse and recycling (UNEP, 2004). Water reuse and recycling in the process can be achieved by implementing good operating practices, changing to technologies that require less process water, and on-site reuse and recycling. In addition, CP assessment can be applied to reformulate industrial products to use less water. Additional information on the assessment methodology and examples of applications can be obtained from UNEP.

Potential concerns for industrial water reuse include scaling, corrosion, biological growth, and fouling, which may impact industrial process integrity and efficacy, as well as product quality. These concerns are often interrelated, and may be addressed by the options summarized in Table 4.14.

**Table 4.14: Industrial Water Reuse: Concerns, Causes, and Treatment Options**

| *Concerns* | *Causes* | *Treatment Options* |
|---|---|---|
| Scaling | Inorganic compounds, salts | Scaling inhibitor, carbon adsorption, filtration, ion exchange, blowdown rate control |
| Corrosion | Dissolved and suspended solids pH imbalance | Corrosion inhibitor, reverse osmosis |
| Biological growth | Residual organics, ammonia, phosphorous | Biocides, dispersants, filtration |
| Fouling | Microbial growth, phosphates, dissolved and suspended solids | Control of scaling, corrosion, microbial growth, filtration chemical and physical dispersants |

The water recycle and reuse is practiced in a few industries at present using membrane technologies to reuse the wastewater or recycling the less polluted wastewater in successive operation using "save alls". There have been several projects which do give the optimistic view of present day crisis as mentioned in Table 4.7.

## Wastewater Management – Challenges

The wastewater should be considered as the source of water and properly managed to avail the maximum benefits. India as a country is reusing only 0.3 per cent of wastewater.

In order to successfully implement the municipal wastewater management systems an 8 point criteria is prerequisite for any society which cover policy issues, management approaches, technology selection and financing mechanisms:

1. Secure political commitment and domestic financial resources.
2. Create an enabling environment at national and local levels.
3. Develop integrated urban water supply and sanitation management systems also addressing environmental impacts.
4. Use well-defined time-lines, and time-bound targets and indicators.
5. Select appropriate technologies for efficient and cost-effective use of water resources and consider eco-technology alternatives.
6. Apply demand-driven approaches.
7. Involve all stakeholders from the beginning and ensure transparency in management and decision-making processes.
8. Ensure financial stability and sustainability.
   - ☆ Link the wastewater sector to other economic sectors.
   - ☆ Introduce innovative financial mechanisms, including private sector involvement and public-public partnerships.
   - ☆ Consider social equity and solidarity to reach cost-recovery.

The proper reuse of municipal wastewater could be selected based on following selection criteria:

- ☆ Inventory of potential sources of wastewater for reuse.
- ☆ Identification of the widest possible selection of potential uses.
- ☆ Identification of legal requirements and responsible institutions.
- ☆ Detailed analysis of reuse alternatives.
- ☆ Economic evaluation of alternatives.
- ☆ Financial feasibility check of most promising alternatives.

Despite the fact that water reuse is already becoming an essential and reliable water supply option for many municipalities, there is still significant potential for an increased utilisation of reclaimed wastewater. Several issues are to be talked taking into consideration Indian scenario. The steps will include:

### Re-orientation of the Water Governance Towards Integrated Water Management

The water resource management is still at its infancy. In developing the appropriate means at local level there is a need to go wider in thinking and to gain a good balance between disciplinary expertise and interdisciplinary understanding. Too often in stakeholders consultations water reuse is excluded from the possible integrated water management scenarios and often regardless whether water reuse is or is not a realistic alternative. The challenge for the water reuse specialists here is to educate and re-orient their own institutions to more conscious and sustainable practices by bridging the tight but artificial compartments of water supply and sanitation.

## Need to Strengthen Cooperation among Stakeholders

The tight compartmentalization of water supply and sanitation resulted in poor institutional arrangements on the water cycle management in general, and water reuse in particular. This is a factor that produced a considerable time lag between the feasibility study of water reuse and putting the results into practice, especially (but not only) for those regions where water and sanitation services are run by different entities.

There is much discussion on how water reuse projects should be managed, in particular who should take the leadership and how the responsibilities/liabilities should be divided. This was investigated through an international survey that covered four types of ownership structures: (1) where water and sewerage management is the responsibility of a single corporate entity, where (2) the water or (3) the wastewater company managed the water reclamation project, or (4) where *ad hoc* project-related structures were set up. The survey indicates that the adoption of the suitable ownership structure for the project is all a question of local circumstances, political will, legislation, institutional structure and regulation. On the other hand, successful case studies hint that ownership is not the real issue, rather is that of tied liability, and above all access to financing and cost allocation. Another preliminary conclusion of the survey is that the communication/collaboration between the water and the wastewater sector is always desirable.

## Establishment of Guidelines or Criteria for Wastewater Reclamation and Reuse

Once convinced of the need of water reuse, it is not always easy to obtain a permit for the reuse of reclaimed water, and this despite Indian Government's wide encouragement to reuse wastewater treatment effluent. For several states one of the major problems is the lack of clear criteria on when to reuse and on quality standards. Due to the lack of water reuse criteria, the public administration bodies had to rely on conservative assumptions. This led to various types of misunderstandings and misjudgments.

An extreme example is an agricultural reuse project where the wastewater treatment plant effluent complied with strict standards for unrestricted agricultural irrigation, but the public administration released a permit basically referring to the WHO's recommendations on irrigation with raw wastewater. Although this is an extreme case, it illustrates quite well how urgent the need is for the establishment of water reuse guidelines.

## Economic Instruments

Financing is perhaps the major barrier to wider use of treated wastewater. For the demand and supply prices to match, targeted, time-bound subsidies are important and necessary. The subsidy is generally aimed at allowing the project to operate on a commercial basis while reaching a certain public programme objective. One of the reasons why water supply benefits alone cannot cover the project costs is that there still exist distortions of the water supply market. However, even when the cost recovery principle is applied, externalities such as for instance the scarcity of water and the marginal cost of new sustainable sources of water, *e.g.*, where existing sources are at – or beyond – their sustainable limit, are rarely accounted for. Similarly, the financial, social and environmental burdens of effluent disposal to the environment are rarely considered in the economic analysis. Subsidies cover a number of areas, predominantly planning, technical assistance and research (pilot studies, etc.), construction costs, actions contributing to regional objectives which are not locally cost-effective and pay-for-performance incentives.

## Building Trust, Credibility and Confidence

Even if the authorities favour the application of all the sustainability principles, no rules and no incentives will work without a general acceptance of the stakeholders, *i.e.*, the water and sanitation companies, the community and the consumers alike. Otherwise even basic sustainability principles may be disregarded. It is important to note that the perception revealed by a survey is that, in the view of some public administrations and of the population, treated wastewater still remains basically wastewater. As a basis for building the trust between stakeholders, there is a need to convey simple, clear and reliable information. The establishment of a best management practice framework to provide a basis for structure and transparency in the management and decision processes is very much needed. A sub-optimally managed project may result in adverse health, environmental or financial outcomes that may quickly reduce enthusiasm for water reclamation, hindering its development in the region. In case of failure, one might not get a second chance! For example, in the Netherlands dual-reticulation systems are banned altogether because of one negative experience of cross-connections with the drinking water supply.

Of particular importance are the management practices to reduce and communicate the risk of human exposure. Management practices of quality control and failure management vary considerably from region to region and even from project to project. A common trait in process operation and risk management of the surveyed projects is the adoption of extensive quality control practices and, in particular, the widespread use of instrumentation, control and automation. On the other hand, despite the fact that procedures such as Hazard Analysis and Critical Control Points are increasingly used to direct efforts in process control and monitoring to guarantee hygienically safe reclaimed water, very few surveyed projects have used them. Another interesting point is that very few projects seem concerned about emerging issues such as trace organic contamination.

## Conclusions

The wastewater management sector in India is in an infant phase with unique opportunities for water reuse to be implemented on a larger scale as a sustainable practice within a framework of wastewater management. Success of wastewater management policy depends on individual, local communities and companies as much as on centralized rules and regulations. The changes in the infrastructure are to be brought at the grass root level. In order to fully tap the significant potential for water reuse, clearer institutional arrangements, economic instruments and water reuse guidelines are very much needed (top-down approach), technological innovation and the establishment of a best practice framework will help, but there can be few more pressing and critical goals than to produce a change in the underlying stakeholders' perception of the water cycle (bottom-up approach). As the Danish poet Esther Gress expressed it:

*"If you want to change the world*

*You must change man*

*If you want to change man*

*You must make him want to change".*

## References


1. AK Sengupta, "Safe reuse of wastewater in agriculture and aquaculture- WHO perspective", International Conference on Health and Environment, March 2006.
2. A.N. Angelakisa, B. Durhamb, "Water recycling and reuse in EUREAU countries:Trends and challenges", Desalination, pp.3-12., 2008
3. Chandra Bhushan, "Water use in industry", Down to Earth, November 2007.
4. D. Bixio, C. Thoeye, J. De Koning, D. Joksimovic, D. Savicc, T. Wintgens, T. Melin, "Wastewater reuse in Europe", Desalination 187, pp 89-101, 2006.
5. Gupta B B, "Water Related Problems in India: Conventional and Emerging Technologies", Conference on Market Development of Water and Waste Technologies through Environmental Economics, May 2004, Paris.
6. Luigi Petta, Annika Kramer, Ismail Al Baz, " The EM Water project – promoting efficient wastewater management and reuse in Mediterranean countries", Desalination 215, pp. 56–63, 2007.
7. M. Ernsta, A. Sperlicha, X. Zheng, Y. Gan, J. Hub, X. Zhaoc, J. Wang, M. Jekela, "An integrated wastewater treatment and reuse concept for the Olympic Park 2008, Beijing", Desalination 202, pp. 293–301, 2007.
8. N.K. Saha, M. Balakrishnan, V.S. Batra, "Improving industrial water use: case study for an Indian distillery", Resources, Conservation and Recycling, 43, pp 163–174, 2005.
9. Planning commission India, " Water supply and sanitation", India assessment, 2002.
10. Pandey R., Ghosh S. "Estimating Industrial Pollution in India: Implications for an Effluent Charge." NIPFP, 2002.
11. Parivesh, CPCB, 2002.
12. RC Trivedi, " Sustainable technology options for safe reuse of wastewater in India", International Conference on Health and Environment, March 2006.
13. Rita Hochstrat, Thomas Wintgens, Thomas Melin, " Development of integrated water reuse strategies", Desalination, 218, pp.201-217, 2008.
15. Sam Godfrey, " Global overview of standard setting processes and challenges and way ahead for developing countries", International Conference on Health and Environment, March 2006.
16. SHH Oelofse, P Viljoen, S Taljaard, WAM Botes, "Discharge of water containing waste emanating from land to the marine environment: a water quality management perspective", Water SA. vol. 30(5), pp.56-60., 2004.
17. Thomas Rutkowski, Liqa Raschid-Sally, "Stephanie Buechler, Wastewater irrigation in the developing world–Two case studies from the Kathmandu Valley in Nepal", Agricultural Water Mgmt., Vol.88, pp.83-91, 2007.
18. USEPA, " Guidelines for water Reuse", EPA/625/R-04/10, September 2008.
19. World Bank Report "India Water Supply and Sanitation- Bridging The Gap Between Infrastructure and Service", January 2006.

# Chapter 5

# Investigation of Metal Ions Exchange Between Solids and Solutions in Sediments and Soils by Using Dynamic Electroanalytical Techniques

**Salvatore Daniele*, M. Antonietta Baldo, Carlo Bragato**

**Department of Physical Chemistry, University of Venice, Calle Larga S. Marta, 2137. 30123 Venice, Italy**

## ABSTRACT

This chapter reports on microanalytical methodologies based on mercury microelectrodes to investigate on metal ions concentration profiles that could be established at soil/water interfaces as a consequence of changes in pH or sulphur and oxygen species in the media. In particular, two strategies are considered. The first is based on a combination of anodic stripping voltammetry (ASV) and scanning electrochemical microscopy (SECM) as a simple procedure to monitor metal ions concentration profiles at soil/water interfaces. The second relies on the development of a proper sediment sampling corer that allows the simultaneous monitoring of sulphur, oxygen and metal ions. For the quantification of the electroactive species either external calibration plots or calibrationless procedures were employed.

## Introduction

A class of chemical reactions of environmental interest involves the interaction of metal ions with the solid phase in aquatic systems [1-6]. These reactions take place at interfacial regions and strongly

* Corresponding Author: E-mail: sig@unive.it

depend on local chemical reactivity. Solid/liquid interfaces of real systems include for instance: sediment/water, sewage sludge or, more in general, soil/water interfaces. Investigation on chemical equilibria involving heavy metal ions at the above interfaces advances understanding on several important phenomena such as the dynamic mechanism of immobilisation/remobilisation, transport and fate of metal ions in the aquatic media. There is a large reservoir of trace metals in soils of aquatic systems and remobilization of these metals is important in determining porewater concentrations and fluxes to overlying waters [3]. The chemical factor most closely associated with the solubility and mobility of trace elements is soil pH, with most elements increasing in mobility at lower pH [5,6]. Under areas where rainfall exceeds evaporation, soil acidification is an ongoing natural process, which car. be accelerated by the activity of plants, animals, humans and acid deposition. The latter phenomenon, in particular, has become one of the most important factors, resulting in environment acidification [6].

When metal ions are leached from the soil, steep concentration gradients may result either in the porewater or in the homogeneous water phase [3]. In situ investigation of these phenomena therefore requires microanalytical systems that possess spatial resolution, other than sensitivity, selectivity and accuracy.

This chapter reports on microanalytical methodologies based on mercury microelectrodes [7] to investigate on metal ions concentration profiles that could be established at soil/water interfaces as a consequence of pH changes in the media as well as of changes of specific chemical species such as sulphur and oxygen. In particular, two strategies are considered. The first is based on a combination of anodic stripping voltammetry (ASV) [8] and scanning electrochemical microscopy (SECM) [9] as a simple procedure to monitor metal ions concentration profiles at soil/water interfaces. The second relies on the development of a proper sediment sampling corer that allows the simultaneous monitoring of sulphur, oxygen and metal ions.

The analytical methodology proposed here may be useful for predicting to what extent mobilization of trace elements from soil occurs, in relation, for instance, to salt marshes or ponds acidification caused by acidic rainfall, as well as through weathering processes [10,11]. Here, examples of applications to real samples of sediments and contaminated bauxitic soils are reported.

## Experimental

The soils examined were a freeze dried sediment sample collected in the Lagoon of Venice and a bauxitic soil sampled at about 4 m depth in Marghera Industrial zone (Venice, Italy). The samples investigated were formed by more than 90 per cent of fine particles with dimension < 70 mm diameter. The concentration gradients of metal ions at the water/soil interface was performed by ASV-SECM measurements. In these experiments, the soil samples were equilibrated with aqueous solutions at different pH following two procedures, which will be denoted as A and B. In procedure A, Teflon pots, 5 cm base diameter, were filled with 20 g of the soil sample examined. Soil slurries were then prepared by addition (without stirring) 50 mL Milli-Q water, containing 0.5 M NaCl alone or 0.5 M NaCl and acetate buffer or $HClO_4$ at different concentrations. The concentration of the acid or buffer in the latter mixtures was adjusted so that the aqueous solutions above the "solid" phase achieved different final pH values over the range 3-5. The sample equilibrated with NaCl alone provided pH values over the range 7.4–8.3. Prior to ASV-SECM measurements, the two-phases system obtained upon mixing the soil and the aqueous solution was allowed to equilibrate for a given time. In procedure B, 20 g of the soil sample was mixed with 50 mL Milli-Q water, containing 0.5 M NaCl alone or NaCl and concentrated acetate buffer or $HClO_4$ solution under magnetic stirring and left equilibrating at least for

24 h, up to achieve an uniform pH in both homogeneous aqueous phase and the soil slurry. All ASV measurements in the water phase were performed in the aerated solutions and using linear sweep (LS) voltammetry in the stripping step. Unless otherwise stated, the scan rate employed was 50 mV $s^{-1}$.

Depth profiles of sulfide, molecular oxygen and metal ions in sediments were performed directly in a Plexiglas corer 30 cm long and 5 cm wide. Along the Plexiglas tube, holes 0.5 cm diameter were produced and taped with small stoppers, which were screwed in. From these holes the microelectrode was positioned in the cores. Stoppers were also placed in the bottom of the tube and then taped immediately after the samples were taken in order to minimize sample exposure to air. Samples were brought to the laboratory and stored at 4°C until the day of analysis.

The working microelectrode was a sphere-cap mercury coated platinum microdisk. It was prepared by *ex situ* deposition of mercury, under potentiostatic control, onto the platinum microdisk [7]. The sphere-cap height $h$ was calculated from the sphere-cap volume ($V$) given by: $V = \pi h(3a^2 + h^2)/6$ itself determined from the charge $Q$ ($Q = V\rho F/M$), (*where*, $\rho$ and $M$ are the density and molar mass of mercury, respectively) passed during the electrodeposition. The mercury microelectrodes employed in this chapter were characterized by $h$ values of $10 \pm 0.2$ µm.

The steady-state diffusion limiting current at a platinum disk or at a sphere-cap mercury microelectrode can be written as [7]

$$I_d = k n F D C^b a \quad (1)$$

*where,*

$k$ is a parameter that depends on the geometry of the microelectrode [12], $C^b$ is the bulk concentration of the electroactive species and $a$ is the radius of the microdisk. For the sphere caps employed here, a $k$ value of $6.28 \pm 0.21$ was calculated as reported in detail in ref 7.

The effective radius $a$ of the microdisk (before mercury deposition) was determined by recording the steady-state diffusion limiting current from a 1 mM $Ru(NH_3)_6^{3+}$ + 0.1 M phosphate buffer solution ($D = 5.8 \times 10^{-6}$ $cm^2$ $s^{-1}$), and using equation (1). This equation was also employed for checking the value of $k$ for the mercury microelectrode either before or after its use in ASV measurements.

The reference electrode was a homemade Ag/AgCl electrode. It was constructed by using a silver wire coated electrochemically with 100 mm thick silver chloride, onto which a film of the anionic polymer exchanger Tosflex® resin was deposited. This membrane was then allowed to saturate with $Cl^-$ by prolonged immersion in a KCl saturated solution.

General voltammetric experiments were carried out by using an EG&G Mod. 283 potentiostat/galvanostat (PAR, Princeton, N.J. USA), and the EG&G PAR 270 electrochemical analysis software. A CH workstation 700B was employed for SECM measurements, along with a micropositioning device, which consisted of a set of three step motor stages with a 0.1 µm resolution An optical microscope enabled a rough check of the microtip position above (or within) the solid matter.

## Results and Discussion

### Determination of Metal Ion Fluxes by ASV-SECM

The ion mobility was investigated by performing LS-ASV measurements in combination with SECM to establish the metal ion profiles in the soil-water systems prepared as reported in the experimental section by procedure A. In particular, LS-ASV measurements were carried out at different distances either in the solution above the soil slurry or within it, in a downward direction as is shown in the scheme of Figure 5.1. The measurements were started by positioning the microtip in the bulk solution, *i.e.*, 3–5 mm above the soil surface, where a check for the initial concentration level of metal ions was performed. Usually, three ASV replicates were carried out after the soil/water system was allowed to equilibrate for 10 min. In these measurements the deposition potential was –1.2 V and deposition time was 300 s. Under these conditions, in all cases investigated here, no stripping peak discernible from the background was observed.

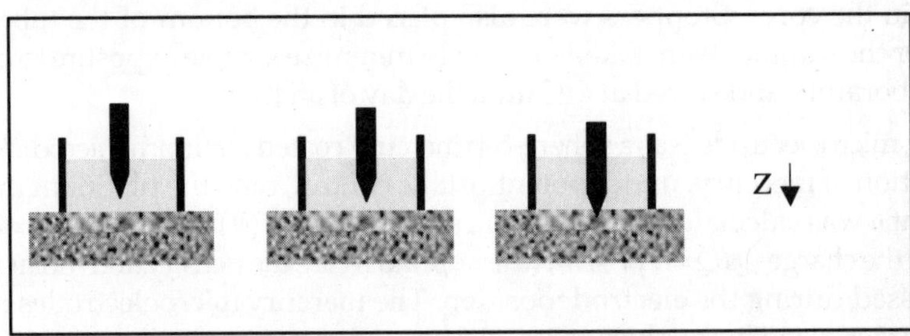


**Figure 5.1: Scheme of LS-ASV Measurements on Water/Soil Systems in the SECM Approach**

After this initial test and a waiting period, which varied from about 5 to 90 min, LS-ASVs were recorded at defined *z* positions, while moving the mercury microtip in increments of 50, 100 or 300 mm (depending on local variation of ASV peak current) along the *z* axis up to a total traveled distance of about 5–6 mm (the end of which represents the arbitrary zero point). Figures 5.2 and 5.3 shows a series of ASVs recorded with the above procedure with the mercury microtip positioned (as checked by the optical microscope) either in the water phase, or within the soil slurry, at different distances. The

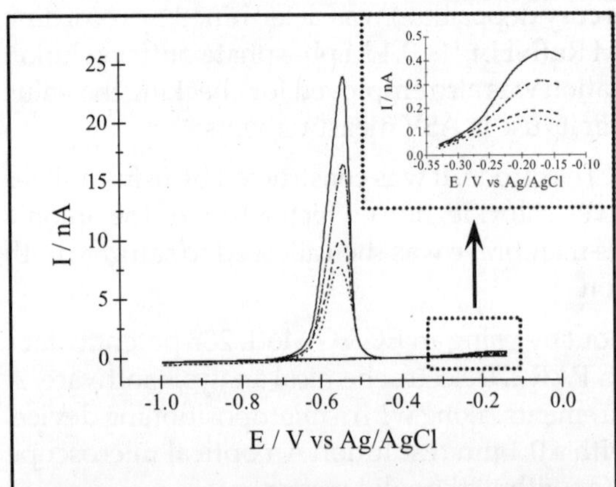


**Figure 5.2: ASV Responses Recorded with a Mercury Microelectrode (*h/a* = 0.99) at Various Distances in a Sediment/Water System**

**Aqueous phase 0.5 M NaCl + acetate buffer, pH = 4.75, equilibration time 70 min, deposition potential –1.0 V, deposition time 1 min, scan rate 50 mVs$^{-1}$; z position (· · · · · ·) 3500 mm, (– · – · –) 1500 mm, (——) 500 mm, (- - - -) 300 mm.**

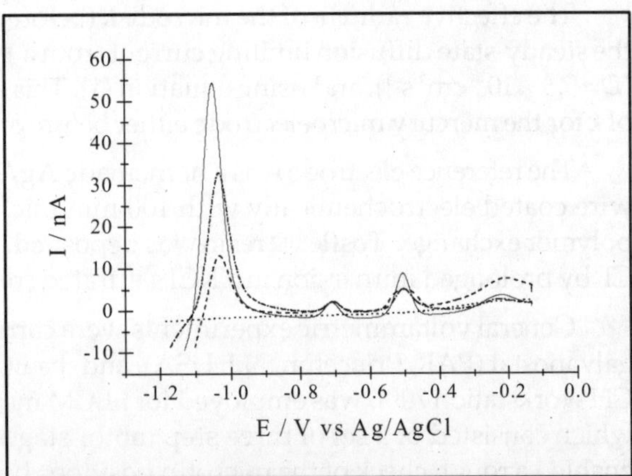


**Figure 5.3: ASV Responses Recorded with a Mercury Microelectrode (*h/a* = 1.02) at Various Distances in a Bauxitic-Soil/Water System**

**Aqueous phase 0.5 M NaCl + acetate buffer, pH = 4, equilibration time 45 min, deposition potential –1.2 V, deposition time 1 min, scan rate 50 mVs$^{-1}$; z position: (· · · · · ·) 5000 mm, (- - - -) 3000 mm, (——) 2300 mm, (– · – · –) 1500 mm.**

sediment sample (Figure 5.2) provided overall a well-shaped anodic peak located at –0.45 V, typical for lead, and a smaller and rather broad one around –0.2 V typical for copper (see also inset in Figure 5.2). At lower pH, a small peak due to cadmium was also evident (see later). The bauxitic soil sample displayed four peaks (Figure 5.3), which, on the basis of the stripping peak potentials, were assigned to Zn (–1.05 V), Cd (–0.67 V), Pb (–0.49 V) and Cu (–0.20 V), with zinc being the overwhelming contaminant, as expected for this type of soil.

Figure 5.4 summarizes typical stripping peak current against distance plots obtained for the above ions in the two samples investigated under different conditions. From this figure it is evident

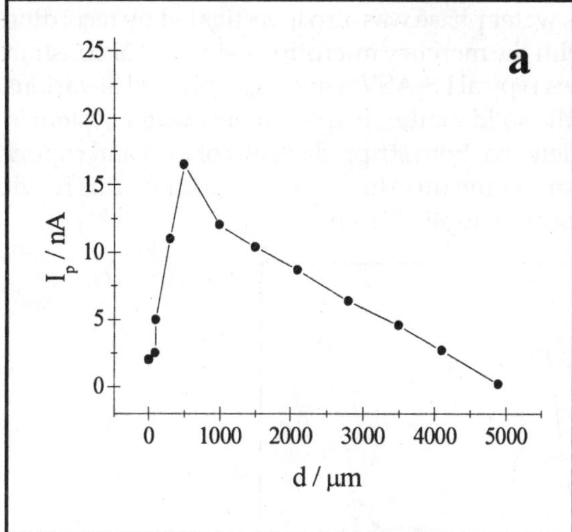


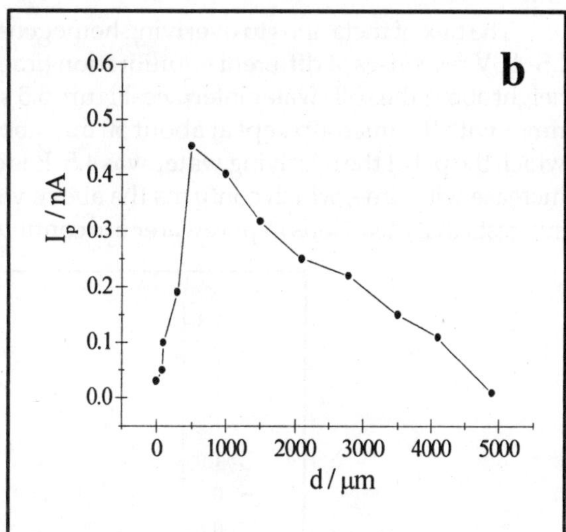


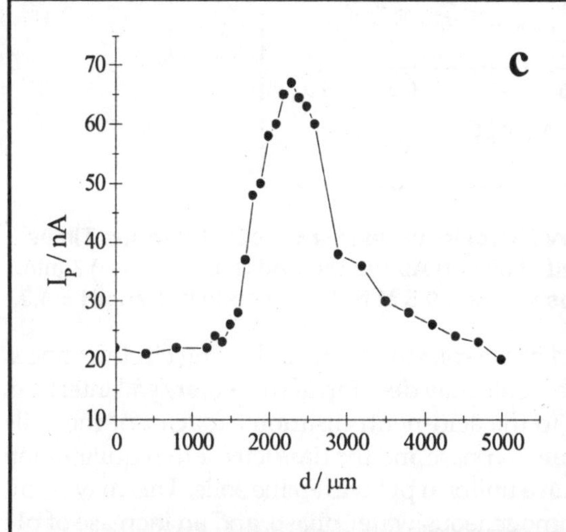


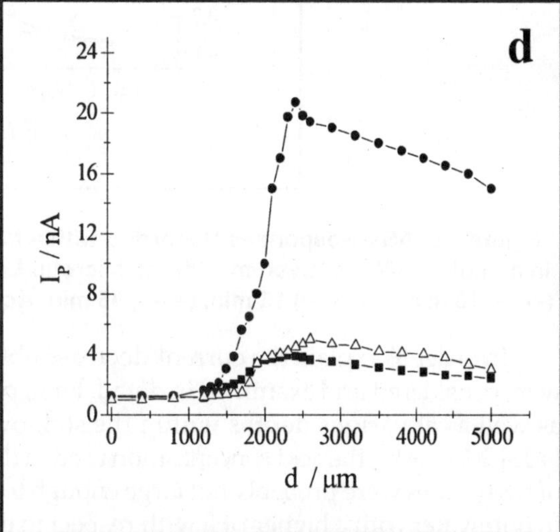


**Figure 5.4: Stripping Peak Currents Against Distance Plots for: (a) Pb, (b) Cu in a sediment/water system equilibrated with aqueous solution 0.5 M NaCl + acetate buffer, pH = 4.80, equilibration time 60 min; (c) Zn, (d) (•) Pb, (■) Cd, (Δ) Cu in a bauxitic-soil/water system; aqueous solution 0.5 M NaCl + acetate buffer, pH = 4.0, equilibration time 90 min.**

that in all cases, the current initially rises, then reaches a maximum and falls to an almost constant lower value. Simultaneous observation of the tip position under the optical microscope indicated that the maximum current occurred when the microtip was positioned very close to the water/soil interface, and that the current decreased as the microelectrode was moved further downward along *z* within the soil slurry. Since the aqueous phase, initially, was free of metal ions, the above current against distance trends suggest that, fluxes of metal ions to overlying homogeneous aqueous phases occur. However, while the initial current increase, as the tip is approaching the soil, is consistent with soil being the source of the electroactive species, the subsequent current decrease as the microtip is moved further is, apparently, not consistent with the above view of the metal ion flux direction.

The flux of metal ions to overlying homogeneous water phase was also investigated by recording LS-ASV responses at different equilibration times with the mercury microtip positioned at constant height above the soil/water interface. Figure 5.5 shows typical LS-ASV responses obtained at various times with the microtip kept at about 50 mm above the solid matter, in a sediment/water system in which the pH of the overlying water was 4.5. It is evident that both stripping peaks of lead and copper increase with time, which confirms the above view about the flux direction. In addition, this result suggests that metal ions in porewater are continuously re-supplied from soil.

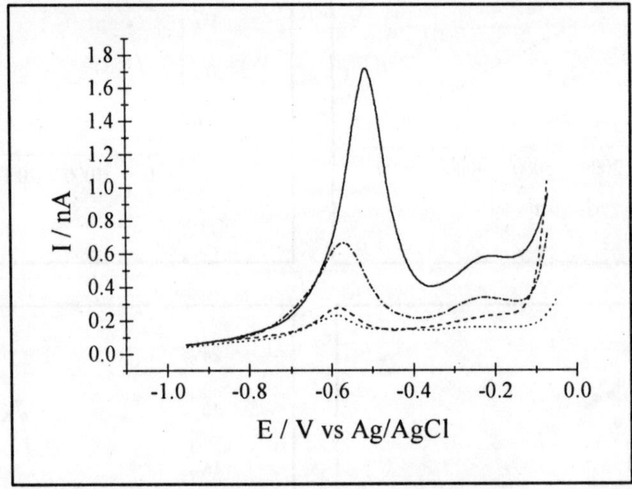


**Figure 5.5: ASV Responses Recorded with a Mercury Microelectrode (*h/a* = 1.02) at Various Times in a Sediment/Water System with the Microtip Located at 50 mm Above the Soil Bed: (· · · · · · ·) 7 min, (- - - -) 10 min, (– · – · –) 15 min, (——) 45 min. Aqueous solution 0.5 M NaCl + acetate buffer, pH = 4.0.**

In order to explain the current decrease obtained in the soil slurry, the following circumstances were considered and examined in detail. First, pH gradients may develop at the water/soil interface, as well as at various depths within the soil, owing to the acid-neutralisation capacity of the soils [4,13]. Moreover, the acid concentration used in the aqueous phase and the time left for the equilibration of two phases were probably not large enough to obtain a uniform pH within the soils. This may result in porewater with a higher pH, with respect to the homogeneous water phase, and an increase of pH with depth. This in turn implies a different availability of free (or labile) metal ions during the preconcentration step in different locations of the soil slurry [14]. Second, when the microtip penetrates the soil slurry, it is likely that a small pool of interstitial water is formed around the active electrode surface. If the volume of the pool is small, as compared with the depletion zone, which develops during the preconcentration step, hindered and restricted diffusion may occur [15], both because of the soil particles packing around the microelectrode, and soil physical characteristics such as porosity and tortuosity of pores [16,17]. These facts would cause a sensible decrease of the effective diffusion coefficients of the metal ions within the soil, ultimately leading to a decrease of the stripping current, even if the metal ion concentration were effectively larger [18].

The influence of pH and physical characteristics of soil on the ASV responses was investigated further. In particular, to verify whether a homogeneous pH within the soil slurry would produce constant current/distance profiles, a series of LS-ASV measurements were performed in samples prepared by procedure B. As mentioned in the experimental section, with the latter procedure no pH gradient should exist between the homogeneous phase and the soil slurry. Thus, in absence of other effects, a constant ASV current against distance plot would be expected. Figure 5.6 shows typical current/distance plots obtained for zinc and lead ions in a bauxitic-soil/water sample equilibrated at pH = 3. As is evident, under these conditions, sigmoidal current/distance plots, with a steep current decrease when the microtip is pushed visibly into the soil slurry are obtained. On average, the current recorded in the soils was about 0.34 (± 0.3) times that of the aqueous phase. Since, under the latter pH condition, complexation reactions of metal ions with soil particles or organic matter, should be negligible [18,19], the current decrease should reflect, essentially, the physical characteristics of the soil or hindered diffusion in the mass transport during the preconcentration step.

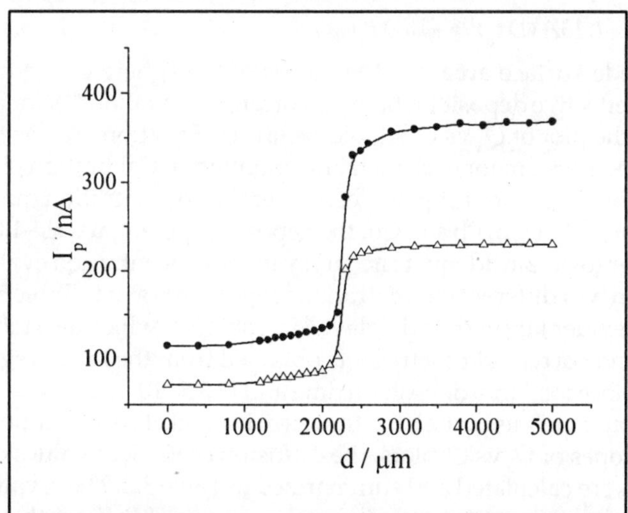


**Figure 5.6: Stripping Peak Currents Against Distance Plots for (a) (•) Zn, (Δ) Pb in a Bauxitic-Soil/Water System Prepared by Procedure B**

To further investigate on the latter aspects, a series of measurements was performed in both sediment and bauxitic soil systems equilibrated with aerated 0.5 M NaCl solution solution for 24-48 h (according to procedure B), and using oxygen as the molecule probes. Cyclic voltammograms at low scan rates (10 mV $s^{-1}$) were recorded with the microelectrode visibly located in the water phase or in the soil slurry, over a potential range where the reduction process of oxygen to hydrogen peroxid occurred. In both cases, sigmoidal waves (not shown) typical for voltammetry with microelectrodes operating under steady-state conditions were obtained [7]. However, the wave height recorded in the homogeneous aqueous phase was about three times larger than that recorded in the soil slurry. Replicate measurements performed in different locations, either in the aqueous phase or in the soil slurry, provided similar results. On average, it was found that the steady-state limiting current in the soil was 0.31 (± 0.2) times (12 replicates) that obtained in the homogeneous solution. This value is in agreement with that found above from ASV measurements. Thus, because no difference in probe molecule concentration should exist between the two phase systems in equilibrium, the wave height decrease in the soil slurry can conceivably be associated with a smaller apparent diffusion coefficient of the molecule, due to restricted diffusion effects.

Overall, the above results confirm that the trends of the plots displayed in Figure 5.4 reflect both the occurrence of pH gradients within soil and the physical characteristics of the soil slurries. The soil characteristics effect can be quantified in terms of formation resistivity factor $F = D_w/D_s$ [16,17], where $D_w$ and $D_s$ are the diffusion coefficients of species in the water and soil phase, respectively. This parameter is related to both porosity and tortuosity of pores in soil [16,17], and therefore, it was taken into consideration to account for the diffusion coefficient dependence of metal ions in the soil slurry, as is illustrated below.

The concentrations of the metal ions were determined from the charge involved in the ASV peaks $Q_a$, assuming that metal accumulated in the mercury film during the deposition step is oxidized back during the stripping step. The relationship between $Q_a$, and $C^b$, derived by using a procedure reported elsewhere [20], is given by:

$$Q_a = nFC^b\,[1.13A\,(D\,t_d)^{1/2} + kaD\,t_d] \tag{2}$$

*where,* A is the electrode surface area, $r$ is the radius of the sphere cap (equal to $a$ in the present situation) and $t_d$ is the effective deposition time. According to equation (2), for a given microelectrode and deposition time, the plot of $Q_a$ vs $C^b$ should be linear. Equation (2) was therefore employed to standardise the electrode system for each metal ion monitored. Calibration plots were performed in 0.5 M NaCl solution for all ions investigated here over the concentration range 0.1 to 500 mM, and using $t_d$ equal to 1 or 2 min. For Cd, Pb and Cu, the deposition potential was –1.0 V, while for Zn it was fixed at –1.2 V. In order to obtain adequate accuracy in the concentration evaluation, the calibration plots were examined in two different concentration ranges as shown in Table 5.1. In any case $Q_a$ vs $C^b$ plots were linear as is evident from data displayed in Table 5.1, which also includes the equations of the calibration plots and correlation coefficients obtained from the linear regression analysis of the experimental points. Since for 2 min deposition time and $D > 5\,´\,10^{-6}$ cm$^2$ s$^{-1}$ (as is the case for the metal ions in the homogeneous aqueous phase), the first term in brackets of equation (2) is negligible within 5 per cent, from the slopes of $Q_a$ vs $C^b$ plots, the diffusion coefficient values of the metal ions in the homogeneous phase were calculated and summarized in Table 5.2. These values compare well with literature data obtained under similar experimental conditions [7]. From these values and using the formation resistivty factor $F = 0.31$, the apparent diffusion coefficients of the metal ions in the soil slurries were evaluated and also included in Table 5.2.

For the quantification of the metal ions in the homogeneous aqueous phase, the experimental ASV charges obtained using $t_d$ equal to 1 or 2 min, along with the calibration equations shown in Table 5.1 were employed. In all other cases, *i.e.*, ASVs obtained in aqueous phase for $t_d$ different from 1 or 2 min, or in the soil slurry (in any case), the concentrations of metal ions were evaluated by using the inverse of equation (2). In the latter situations, $k$ and $a$ values were obtained as described in the experimental section, while the diffusion coefficients are those shown in Table 5.2. For the quantification of metal ions in the soil slurry, it was assumed that the soil physical properties were isotropic.

**Table 5.1: Linear Regression Equations ($Q_a$ vs $t_d$) Calculated for the Metal Ions Investigated at Two different Deposition Times, and Relative Dynamic Ranges**

| *Metal Ion* | *$t_d$ (Min)* | *Dynamic Rang (μM)* | *Linear Regression Equation* $Q_a(nC) = A + B\,C^b$ (μM) | | *$R^2$* |
|---|---|---|---|---|---|
| | | | *A* | *B* | |
| $Cu^{2+}$ | 60 | 0.1–5.0 | 0.02 | 0.27 | 0.998 |
| | 60 | 5.0–500 | 0.05 | 0.22 | 0.999 |
| | 120 | 0.1–5.0 | 0.07 | 0.52 | 0.998 |
| | 120 | 5.0–500 | 0.10 | 0.46 | 0.999 |
| $Pb^{2+}$ | 60 | 0.1–5.0 | -0.07 | 0.76 | 0.998 |
| | 60 | 5.0–500 | 0.11 | 0.62 | 0.999 |
| | 120 | 0.1–5.0 | 0.09 | 1.47 | 0.999 |
| | 120 | 5.0–500 | 0.15 | 1.31 | 0.999 |
| $Cd^{2+}$ | 60 | 0.1–5.0 | 0.03 | 0.36 | 0.998 |
| | 60 | 5.0–500 | 0.07 | 0.31 | 0.999 |
| | 120 | 0.1–5.0 | 0.10 | 0.71 | 0.999 |
| | 120 | 5.0–500 | 0.12 | 0.63 | 0.999 |
| $Zn^{2+}$ | 60 | 0.1–5.0 | -0.03 | 0.52 | 0.999 |
| | 60 | 5.0–500 | 0.08 | 0.43 | 0.999 |
| | 120 | 0.1–5.0 | 0.07 | 0.99 | 0.998 |
| | 120 | 5.0–500 | 0.11 | 0.89 | 0.999 |

**Table 5.2: Diffusion Coefficient Values Determined for the Metal Ions Investigated in Aqueous Solution Containing 0.5 M NaCl ($D_w$), and in the Soil Slurry ($D_s$)**

| *Metal Ion* | *$D_w$ ($cm^2\,s^{-1}$)* | *$D_s$ ($cm^2\,s^{-1}$)* |
|---|---|---|
| $Cu^{2+}$ | $6.2\times10^{-6}$ | $2.0\times10^{-6}$ |
| $Pb^{2+}$ | $8.6\times10^{-6}$ | $2.8\times10^{-6}$ |
| $Cd^{2+}$ | $7.0\times10^{-6}$ | $2.2\times10^{-6}$ |
| $Zn^{2+}$ | $6.6\times10^{-6}$ | $2.1\times10^{-6}$ |

Figures 5.7 and 5.8 show typical concentration profiles determined as described above at different distances both in the aqueous phase and in the soil slurry for samples obtained by procedure A. These profiles actually reflect the current trends observed in Figures 5.2 and 5.3, and suggest that, under those conditions, the flux of metal ions supplied by the sediment is higher or equal to that imposed by the transfer in the homogeneous aqueous phase and diffusion towards the bulk solution. Obviously, soil acts as an infinite source of metal only if the solubilization and/or desorption kinetics is fast enough to allow the re-supply of dissolved metal in the pore water. Under these conditions, soil behaves like a buffer system with respect to dissolved metals.

In order to establish the real availability of metal ions from soil, under homogeneous pH conditions, LS-ASV measurements were performed in the soil slurries that were prepared by procedure B, at different pH. Figure 5.9 summarizes the metal ion concentrations found in both sediment and soil samples.

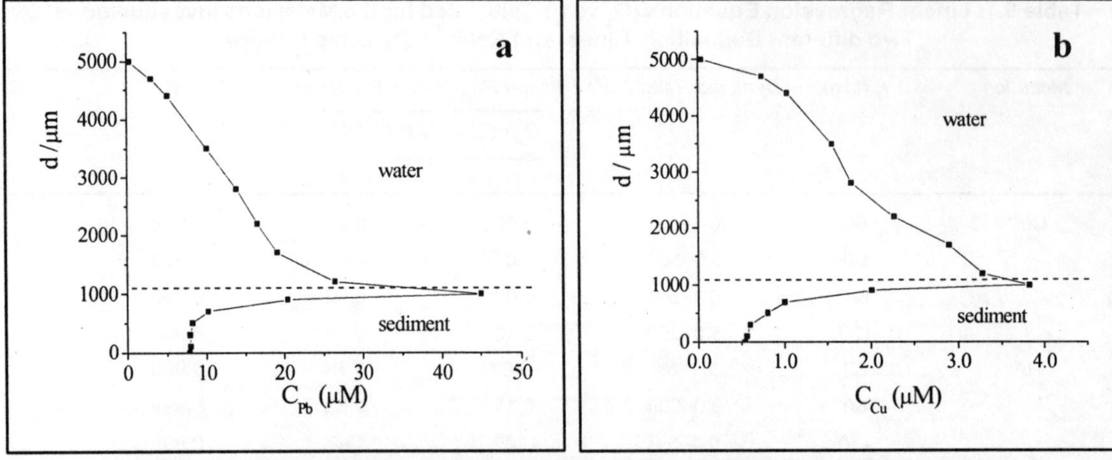


**Figure 5.7: Concentration Profiles for (a) Pb, (b) Cu in a Sediment/Water System Prepared by Procedure A and Equilibrated at pH 4.0. Relative standard deviations (RSD) for concentration data are within 7 per cent.**

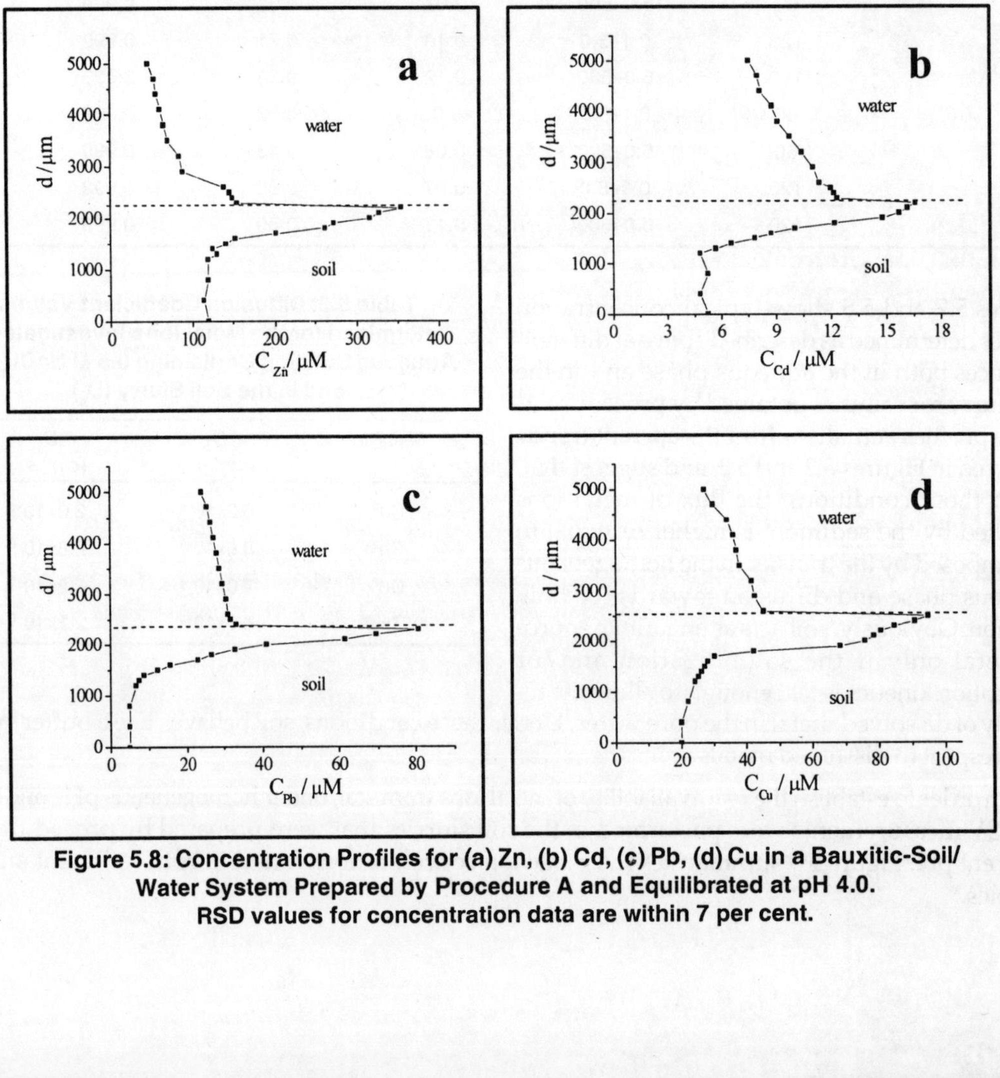

**Figure 5.8: Concentration Profiles for (a) Zn, (b) Cd, (c) Pb, (d) Cu in a Bauxitic-Soil/ Water System Prepared by Procedure A and Equilibrated at pH 4.0. RSD values for concentration data are within 7 per cent.**

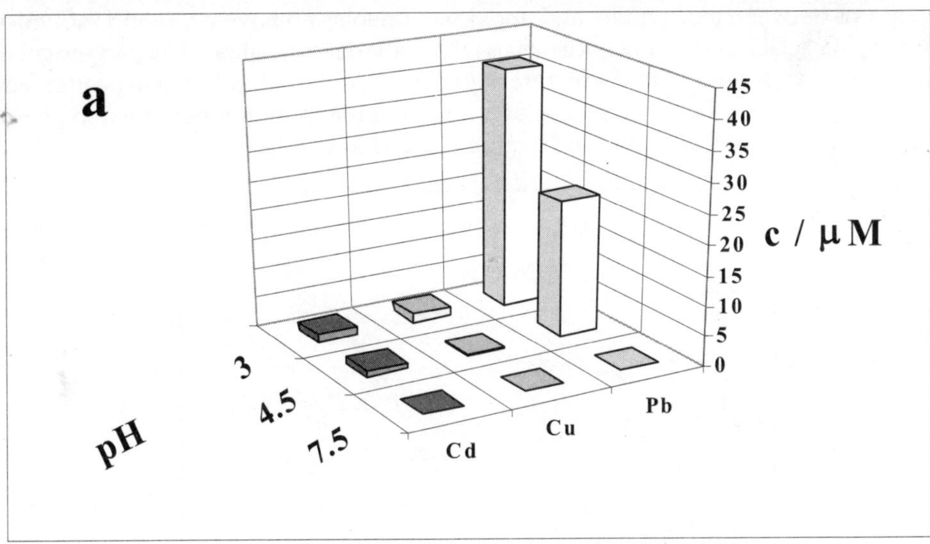


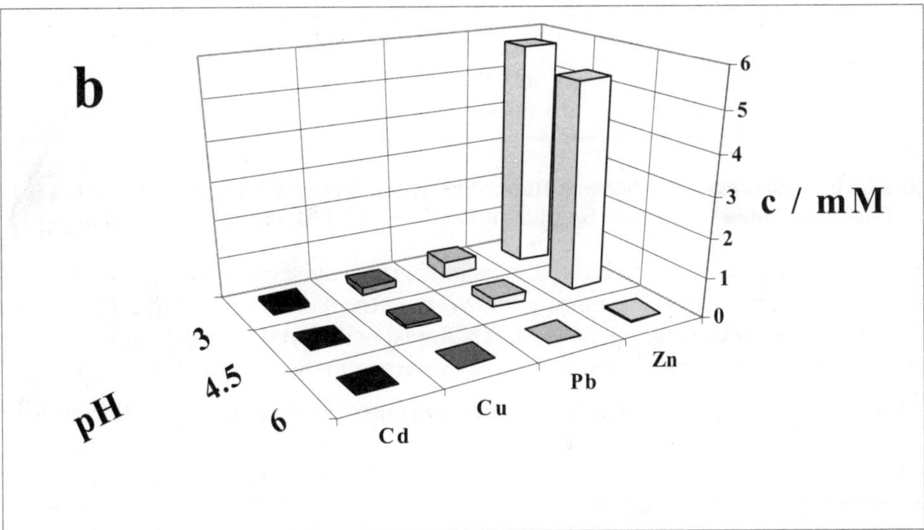


**Figure 5.9: Concentration Histograms of Metal Ions found in Soils Prepared with Procedure B at Different pH Values: (a) Sediments, (b) Bauxitic-soil samples. RSD values for concentration data are within 7 per cent.**

The trends observed are consistent with pH effect on ion mobilization, which increases as pH decreases [14,19]. It is also interesting to note that at lower pH, the concentrations of metal ions in porewater are much larger than those found in the samples prepared by the acidification procedure A. This ultimately justifies the observed metal ion fluxes to overlying acidic water.

## Monitoring of Depth Profiles of Sulphide, Oxygen and Metal Ions

The oxygen content in the water phase of a soil is important parameter, as it affects the oxidation processes that may occur at the water-solid interface [21]. For instance, the major changes that occur in redox conditions between oxic waters and anoxic sediments can have profound influences on the speciation and bioavailability of many trace metals [2]. For the determination of the oxygen concentration in soil, cyclic voltammograms at low scan rate were recorded with the electrode positioned either in the soil or in the overlying aqueous phase, over the potential region where the oxygen reduction process occurs. Figure 5.10 shows typical voltammograms thus obtained.

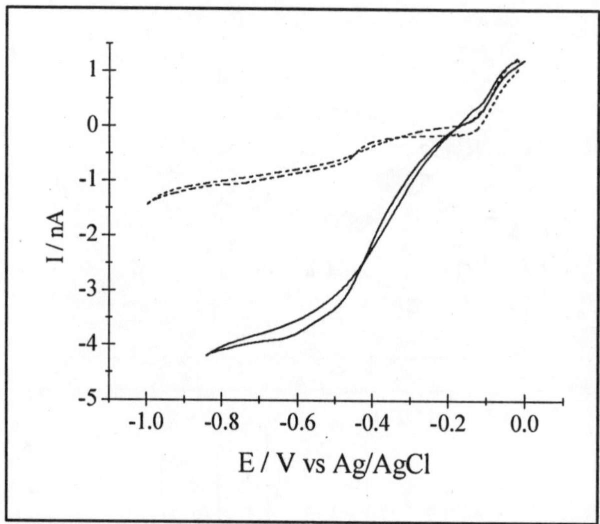


**Figure 5.10: Cyclic Voltammograms Recorded with a Mercury Microelectrode (*h/a* = 1.03) Positioned in (———) Bulk Solution and (- - - - -) Soil Slurry. Scan rate 10 mVs$^{-1}$.**

The concentration was evaluated from the height of the wave (at the plateau) along with equation (1). In the calculation, the following parameters were used: $n = 2$, $D$ equal to $2.12\times 10^{-5}$ cm$^2$ s$^{-1}$ and $6.57\times 10^{-6}$ cm$^2$ s$^{-1}$ in the homogeneous phase and soil, respectively; the $k$ value was that evaluated for the specific mercury microelectrode employed, as described in the experimental section.

Characterisation and monitoring of inorganic sulphur species is also critical for understanding of sulphur and metal cycling in suboxic sedimentary environments. For the detection of sulphide ions in aqueous solution, the cathodic stripping process of HgS accumulated onto the electrode surface through reaction (3) was exploited [20]:

$$HS^- + Hg^\circ \rightleftharpoons HgS + H^+ + 2e \qquad (3)$$

Figure 5.11 shows typical cathodic stripping peaks obtained in the samples investigated.

The deposition potential was set at –0.4 V vs Ag/AgCl, and accumulation times over the range 0-60 s were employed. Since the process involving sulphide is a surface process, a conditioning step was necessary to restore the electrode surface between measurements, and to avoid signal degradation during measurements. This was achieved by maintaining the potential at –1.6 V for 20-30 s. At this potential both reduction of HgS and hydrogen evolution occur, which allowed to restore the electrode surface. For the quantification of sulphide in the sediment samples the calibrantionless procedure based on equation (2) was employed. The diffusion coefficient of sulphide ion was evaluated in surface sea water, spiked with known concentrations of sulphide as described in detail elsewhere

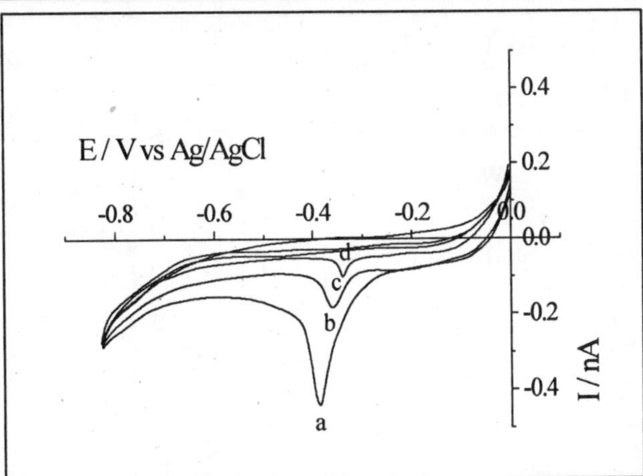


**Figure 5.11: Cathodic Stripping Voltammograms Recorded by Linear Sweep Voltammetry at 50 mV s$^{-1}$ with a Mercury Microelectrode in a Sediment Sample at Different Deposition Times: (a) 60, (b) 30, (c) 15, (d) 10 s. Deposition potential E = –0.4 V vs Ag/AgCl.**

[20], and a value equal to $1.39 \times 10^{-5} cm^2 s^{-1}$ was calculated. This value was corrected for the formation resistivity factor (*i.e.*, $F = 0.31$) and then used for the detection of the sulphide species in the porewater of the soil.

For the metal ions the ASV procedure described in the previous section was employed. However, under natural conditions very low concentrations of metal ions were involved. This required a deposition time of 300 s to obtain valuable and reproducible stripping peaks.

The measurement *in situ* of oxygen, sulphur, and metal ions which are strongly interrelated, provides a way to overcome some undesirable variation of the chemical equilibria involved in these matrices. Since, the content of the latter species depend on the depth at which the sediments are sampled, the depth profiles of sulphur, oxygen and the metal ions $Cd^{2+}$, $Pb^{2+}$ and $Cu^{2+}$ were evaluated by using the same mercury microelectrode placed in the sample core through the holes of the corer at different distances. Figure 5.12 shows depth profiles thus obtained in the sample under its natural pH = 8.3.

These profiles are similar to others reported in the literature in similar environments [22,23]. In fact, the concentration of sulphide in sediments increases with depth, as a result of sulphate reduction during the decomposition of organic matter [23]; oxygen is depleted quickly below the sediment-water interface [23]; relatively small concentrations of $Cd^{2+}$, $Pb^{2+}$ and $Cu^{2+}$ are found in the water phase, they being released to overlying water at the sediment/water surface; even smaller or undetectable amounts of metal ions were found within the sediment, as at pH = 8.3 they are probably associated to particulate matter either inorganic or organic in nature forming very insoluble precipitates. In particular, the excess of sulphide can give rise to rather insoluble metal sulphide compounds, consistently with previous findings [2].

## Conclusion(s)

The analytical methodologies presented in this chapter may be very useful for monitoring trace elements in complex media such as soils and pore water of sediments. In particular, it has been demonstrated that mercury microelectrodes in combination with SECM can be used for monitoring

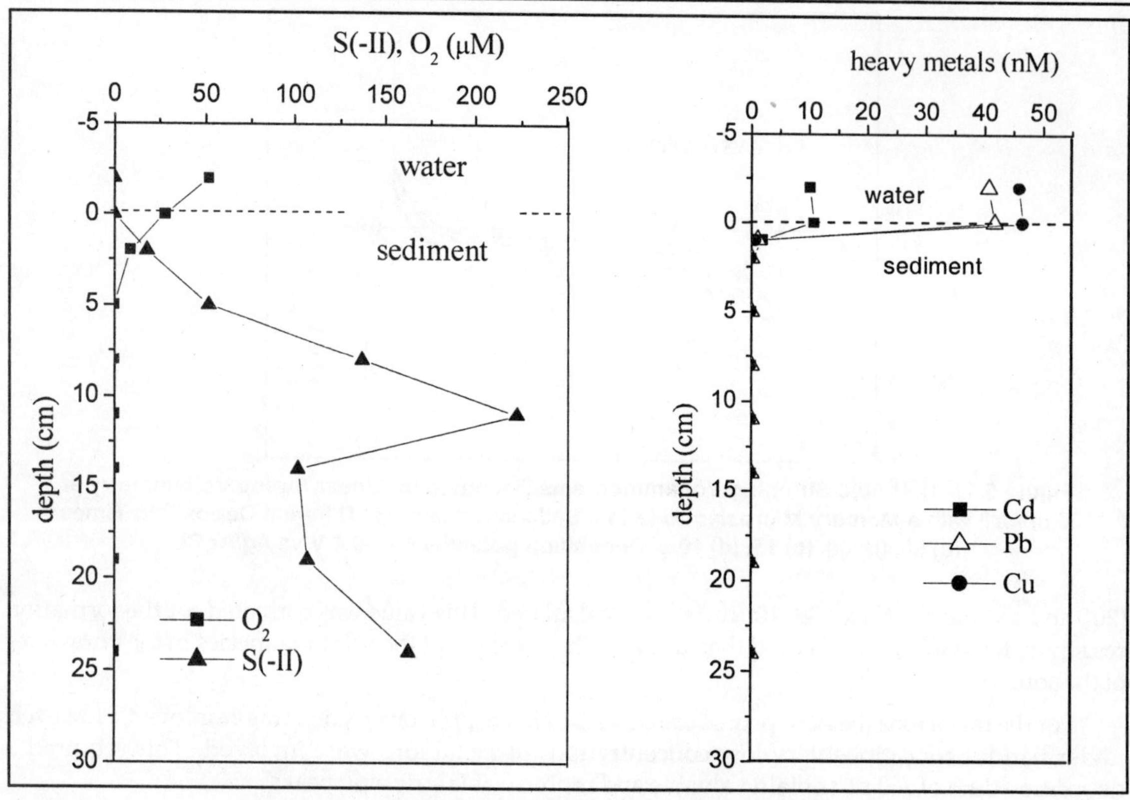


**Figure 5.12: Depth Profiles for Oxygen, Sulphur, Cadmium, Lead and Copper on a Sediment Core Sample**

metal ion concentration profiles at soil/water interface with high spatial resolution. This measurement concept could be applied for the *in situ* investigation on the impact of acid rain on soils or groundwaters, in relation to the mobility of trace metal ions. On the other hand, the design of the sampling sediment corer here presented is useful for monitoring depth profiles of key redox species of environmental interests under the natural conditions of the samples, avoiding any manipulation which could produce some artefacts.

## Acknowledgements

Financial support of the Ministry of University and Scientific Research (IUR) is gratefully acknowledged.

## References

1. W. Stumm, *Chemistry of the Solid-Water Interface*, J. Wiley, New York, 1992.
2. J.W. Morse, *Mar. Chem.* 1994, *46*, 1.
3. H. Ernstberger, W. Davison, H. Zhang, A. Tye, S. Young, *Environ. Sci. Technol.* 2002, *36*, 349.
4. S. Qureshi, B. K. Richards, T. S. Steenhuis, M. B. McBride, P. Baveye, S. Dousset, *Environmental Pollution*, 2004, *132*, 61.

5. C. Carlon, M. Dalla Valle, A. Marcomini, *Environmental Pollution*, 2004, *127*, 109.
6. F.C. Menz, H.M. Seip, *Environmental Science and Policy*, 2004, *7*, 253.
7. M.A. Baldo, S. Daniele, M. Corbetta, G.A. Mazzocchin, *Electroanalysis* 1995, 7, 980.
8. J. Wang, *Stripping Analysis: Principles, Instrumentation and Applications*, VCH, Deerfield Beach, FL 1985.
9. A.J. Bard, M.V. Mirkin (eds.), *Scanning Electrochemical Microscopy*, Marcel Dekker, New York, 2001.
10. T. Wallstedt, H. Borg, *Environmental Pollution*, 2003, *126*, 381.
11. C. Kjoller, D. Postma, F. Larsen, *Environ. Sci. Technol.*, 2004, *38*, 2829.
12. S. Daniele, M.A. Baldo, C. Bragato, G.A. Mazzocchin, *Ann. Chim. (Rome)* 2002, *92*, 203.
13. J.O. Reuss, B.J. Cosby, R.F. Wright, *Nature* 1987, *329*, 27.
14. J. Buffle, In *Complexation Reactions in Aquatic Systems. An Analytical Approach*, Ellis Horwood, Chichester, 1988.
15. C. Amatore, R. S. Kelly, E. Kristensen, W. G. Khur, R.M. Wigthman, *J. Electroanal. Chem.* 1986, 213, 31.
16. R.N. Glud, K. Jensen, N.P. Revsbech, *Geochim. Cosmochim. Acta* 1995, *59*, 231.
17. N. Iversen, B.B. Jorgensen, *Geochim. Cosmochim. Acta* 1993, *75*, 571.
18. J. Pei, M.L Tercier-Waeber, J. Buffle, G. C. Fiaccabrino, M. Koudelka-Hep, *Anal. Chem.* 2001, *73*, 2273.
19. J.Pei, M-L Terrier-Waeber, J. Buffle, *Anal. Chem.* 2000, *72*, 161.
20. M.A. Baldo, S. Daniele, C. Bragato, G.A. Mazzocchin, *Anal. Chim. Acta*, 2002, *464*, 217.
21. N.P. Revsbech, B.B. Jorgensen, T.H. Blackburn, Y. Cohen, *Limnol. Oceanogr.* 1983, *28*, 1062.
22. E. Henneke, G.W. Luther III, G.J. de Lange, G.J., *Mar. Geol.* 1991, *100 (1-4)*,115.
23. P.J. Brendel, G.W. Luther III, *Environ. Sci. Technol.* 1995, 29, 751.

# Chapter 6

# Formaldehyde, Methanol Biodegradation from the Mixture in Anaerobic Digester

**S. Sandhya* and K.Swaminathan**
**National Environmental Engineering Research Institute (NEERI), CSIR Complex, Chennai – 600 113, India**

## ABSTRACT

The treatment of formaldehyde methanol containing synthetic wastewater in an anaerobic digester was studied. Formaldehyde and methanol were oxidized with the formation of volatile acids by mixed microbial culture from the synthetic medium. The biodegradation of methanol and volatile acid begins after exhaustion of formaldehyde in medium. First volatile acids are utilized followed by methanol. Formaldehyde and methanol mixture also follows the same pattern. HPLC analysis of volatile acids showed formation of formic, acetic and propionic acids.

*Keywords: Anaerobic digester, Formaldehyde/methanol biodegradation.*

## Introduction

Formaldehyde and methanol are common chemical solvents widely used in the chemical, textile, paper and wood processing industries. Besides this they are important industrial solvents in the manufacturing of resins, plastics, adhesive and many other products. At higher concentrations, both the pollutants are known to produce adverse effect on humans such as vomiting, gastrointestinal problems [1]. Formaldehyde is also used as an active ingredient in preservative and disinfectants. Its

* Corresponding Author: E-mail: sswami_in@yahoo.com

disposal directly to sewage treatment plant after use may imbalance the performance, due to its antimicrobial effects. The carcinogenicity of formaldehyde to animal and human is well known [2]. Nevertheless, formaldehyde at lower concentration is known to be biodegradable in both aerobic [3,4] and anaerobic systems [5,6]. Hidago *et al.* [4] reported the removal of formaldehyde by suspended cells of heterotrophic bacterium *Rhodococcus erythropolis* UPV1 in defined mineral medium supplemented with phenol, formaldehyde. Gerite and Gode [7] reported formaldehyde degradation at 2-5 mg/l in activated sludge. Bonastre *et al.* [8] observed the partial degradation of formaldehyde at 2100 mg/l in activated sludge. Garnido *et al.* [9] worked with a multiple up flow filter under anoxic conditions using formaldehyde as carbon source. They could eliminate 0.5 kg formaldehyde $m^{-3}$/d efficiently at 1 to 0.3 d. hydraulic retention time. Anaerobic degradation is an alternative to aerobic degradation for formaldehyde removal from wastewater. Formaldehyde concentrations of 20-250 mg/l resulted in 50 per cent reduction in methanogenesis activity [10,11]. The main purpose of the study was to biodegrade formaldehyde, methanol, individually and in mixture by anaerobic/facultative mixed microorganisms in anaerobic digester. The degradation pattern in presence of co-substrates and identification of volatile acids formed during biodegradation using HPLC for predicting the probable metabolic pathway of degradation.

## Experimental

### Organism and Culture Isolation

The formaldehyde degrading bacteria were isolated by successive enrichments of sewage containing 0.05 per cent formaldehyde. The enriched microorganisms were always maintained in mineral salts medium containing formaldehyde. The mineral salt medium used contains mg/l; ammonium chloride 280; potassium di-hydrogen phosphate 250; magnesium sulfate 110; calcium chloride 20; yeast extract 20; trace element solution.

### Reagents and Chemicals

Methanol and Formaldehyde used in study are commercially available solvents and other chemicals used were all reagent grade.

### Analytical Determination

Formaldehyde was determined by Nash procedure using UV- Visible spectrophotometer [12]. Methanol was measured using a Perkin Elmer gas chromatograph equipped with Nukol column (30×0.25id) and flame ionization detector. Nitrogen (1.5 $cm^3$/min) was utilized as carrier gas. The injector and detector temperature were 250°C and 270°C respectively. Volatile acids were determined using Hawlett Packard 1100 Liquid Chromatogram equipped with a C18 ODS column (25 cms×4m id) and a diode-array UV detector. The mobile phase was acetonitrile-phosphoric acid (80:20) at a flow rate of 1. Detection was performed at 210 nm.

### Biodegradation Studies

The biodegradation studies were conducted in anaerobic digester of one liter capacity with 500 ml working volume with outlet to remove samples periodically. Microorganisms were enriched from municipal sewage in presence of formaldehyde and methanol.

## Results and Discussion

### Biodegradation of Formaldehyde

The growth of microorganism on formaldehyde (420 mg/l) as sole carbon source has been studied in the laboratory. The degradation of formaldehyde led to the simultaneous appearance of methanol and volatile acid as shown in Figure 6.1. As soon as volatile acids are formed, there is decrease in pH from 7 to 4.2 followed by increase to 6.5, indicating utilization of formed volatile acid. Methanol is also formed as an intermediate after three days and remains unchanged till nineteen days and followed by removal after 21 days. Formaldehyde was first totally converted to methanol. Omil *et al.* [6] have also reported formation of methanol from formaldehyde during biodegradation of formaldehyde under anaerobic condition. Eiora *et al.* [13] have also reported the formation of methanol and formic acid during formaldehyde degradation. This is also in agreement with results obtained by Kato *et al.* [14] studying formaldehyde biodegradation. They found out an enzyme, which catalyzed dismutation of formaldehyde to form methanol and formic acid. Adroer *et al.* [15] have also reported the formaldehyde biodegradation is initiated by a dismutation reaction, and the biodegradation led to the simultaneous appearance of methanol and formic acid. The degradation of methanol and volatile acid begins after exhaustion of formaldehyde in medium. Table 6.1 shows the results of formaldehyde degradation achieved during 21 days in anaerobic digester operation under identical conditions at 70-988 mg/l of

**Table 6.1: Influent and Effluent Formaldehyde and COD Concentration Observed in Anaerobic Sludge Digester**

| *Formaldehyde, mg/l* | | | *COD, mg/l* | | |
|---|---|---|---|---|---|
| *Influent* | *Effluent* | *Removal, per cent* | *Influent* | *Effluent* | *Removal, per cent* |
| 70 | 10 | 85.71 | 78 | 14 | 82.0 |
| 228 | 40 | 82.45 | 165 | 31 | 81.0 |
| 492 | 30 | 93.90 | 340 | 19 | 94.0 |
| 719 | 60 | 91.65 | 458 | 36 | 92.0 |
| 988 | 135 | 86.33 | 618 | 80 | 87.0 |

The average values are based on a minimum of five observations

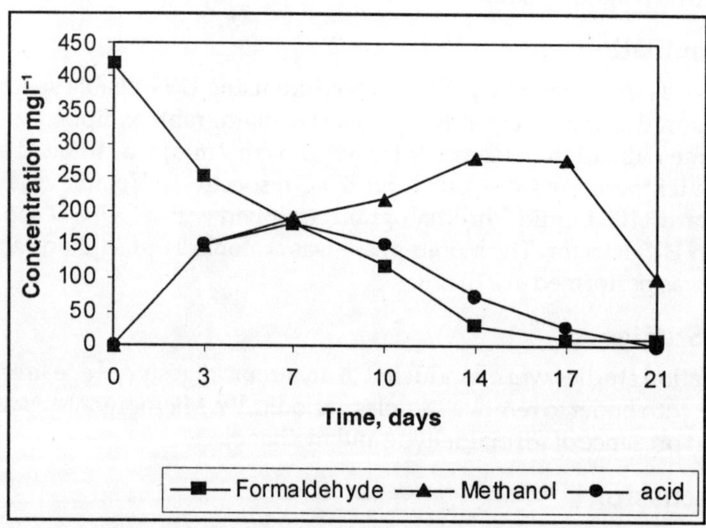


**Figure 6.1: Biodegradation of Formaldehyde and Product Formation**

initial formaldehyde concentration. The influent COD was 78–618 mg/l with COD removal efficiency of 81-94 per cent. The ratio between the COD and formaldehyde concentration in the influent was 0.62-1.11, but this ratio was found to vary for effluent samples, reaching the new value of 0.59–1.4. The maximum influent concentration of 988 mg/l was treated in anaerobic digester. Olivera *et al.* [16] had studied formaldehyde degradation in an anaerobic packed bed reactor and reported that 1158 mg HCHO/l/d was applied as formaldehyde loading rate resulting in 99.7 per cent formaldehyde and 92 per cent COD removal efficiencies, where as Gurrido *et al.* [17] worked with a modified up flow filter under anoxic conditions using formaldehyde as carbon source and obtained 65 per cent removal.

## Biodegradation of Methanol

Batch growth of microorganisms on 536 mg/l of methanol as sole source of carbon has shown the formation of volatile acid (452 mg/l), formaldehyde (240 mg/l) within two days of incubation (Figure 6.2). This was followed by decrease in formaldehyde to nondeductible limit with the concentration of volatile acid remaining constant after 7th days. Methanol concentration decreased gradually till 10th day and reduced to almost nil after 14th day. Table 6.2 shows the results of methanol degradation achieved during 21 days in anaerobic digester operation under identical conditions at 96-972 mg/l of initial methanol concentration. The influent COD was 170–1721 mg/l with COD removal efficiency of was 67–91 per cent. The ratio between the COD and methanol concentration in the influent was 1.77 at all the concentrations and remains constant for effluent samples also. The maximum influent methanol concentration of 972 mg/l was treated in anaerobic digester. Eiora *et al.* [13] have also observed slow methanol removal and formation of formaldehyde and formic acid as product of degradation. Methanol as single carbon source was removed in few days, but in presence of formaldehyde its removal took longer time. This could be due to the inhibition of microorganisms caused by formaldehyde. However, this inhibition could also be caused by the presence of formic acid, since inhibition remained after formaldehyde had been completely removed.

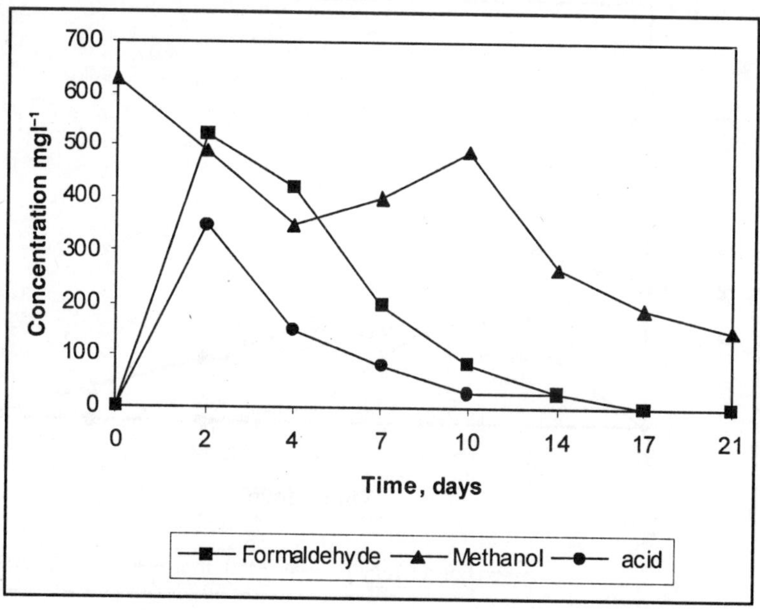


**Figure 6.2: Biodegradation of Methano and Product Formation**

**Table 6.2: Influent and Effluent Methanol and COD Concentration Observed in Anaerobic Sludge Digester**

| Methanol, mg/l | | | COD, mg/l | | |
|---|---|---|---|---|---|
| *Influent* | *Effluent* | *Removal, per cent* | *Influent* | *Effluent* | *Removal, per cent* |
| 96 | 15 | 84.37 | 170 | 17 | 90.0 |
| 232 | 32 | 86.20 | 410 | 36 | 91.0 |
| 462 | 127 | 72.51 | 818 | 144 | 82.4 |
| 689 | 325 | 52.83 | 1220 | 367 | 70.0 |
| 972 | 502 | 48.35 | 1721 | 567 | 67.0 |

The average values are based on a minimum of five observations.

The removal of formaldehyde or methanol by adsorption in the digester can not be ruled out. But it seems to be negligible. Qu and Bhattacharya [5] and Omil *et al.*[6] have reported only 10-11 per cent of a biotic removal in anaerobic reactor.

## Biodegradation of Formaldehyde Methanol Mixture

When formaldehyde (250 mg/l) and methanol (100 mg/l) were added together, formaldehyde was preferentially utilized and exhausted within 10 days. The degradation of formaldehyde and methanol in mixture is shown in Figure 6.3. But there was formation of volatile acid with decrease in pH and it was utilized first followed by methanol. Biological conversion of formaldehyde to methanol is evidence by increase in methanol concentration from 100 mg/l to 134 mg/l till 3 days, but started decreasing after words. An inhibitory effect of methanol on volatile acids degradation was observed.

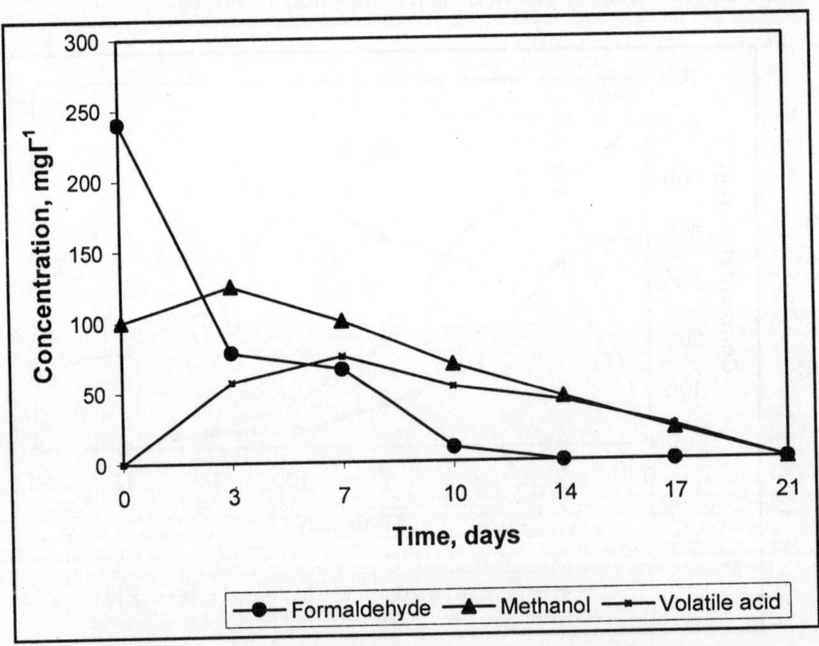


**Figure 6.3: Biodegradation of Formaldehyde and Methanol in Mixture (Formaldehyde 250 mgl$^{-1}$ and Methanol 100 mgl$^{-1}$)**

There exist a possibility that volatile acid formation might be inhibitory to methanol dehydrogenase enzyme and therefore only after its removal it is been expressed. The enzymatic assay has not been investigated, but the results obtained are in agreement with basic characteristics of growth on multiple substrates.

When formaldehyde (500 mg/l) and methanol (500 mg/l) were added to gather, formaldehyde was preferentially utilized and exhausted within 14 days. The degradation of formaldehyde and methanol in mixture is shown in Figure 6.4. At higher concentration of formaldehyde methanol mixture the degradation pattern was different. Volatile acids formation started after 3 days of incubation and decreased after 14 days. The quantity of the formation of volatile acids was less as compared to single compound degradation. The decrease in pH was observed and formaldehyde was utilized first within 17 days followed by methanol. Only 79 per cent of initial methanol was degraded within 21 days. COD removal was gradual and maximum during 21 days of incubation. Methanol concentration has increased to 520 mg/l within 7 days again indicating biological conversion of formaldehyde to methanol.

The degradation of formaldehyde via methanol is thermodynamically favored as it generates

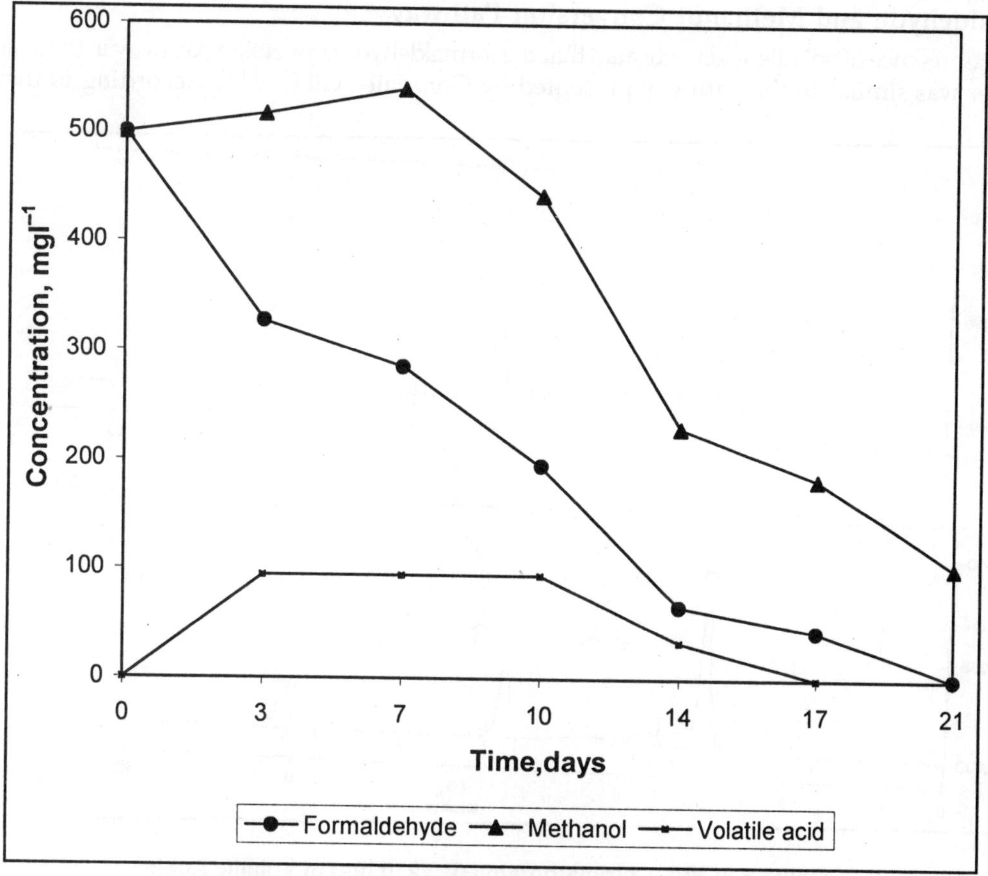


**Figure 6.4: Biodegradation of Formaldehyde and Methanol in Mixture (Formaldehyde 500 mgl$^{-1}$ and Methanol 500 mgl$^{-1}$)**

more energy. At low concentrations of volatile acids and methanol, direct conversion to methane is favored. But at higher concentration, methanol degradation is slow and effluent shows presence of methanol and organic COD. Florencio *et al.* [18] showed that the affinity of methalotrophic methanogens for methanol is 60 times higher than acetogens. Therefore, a low methanol concentration, methalotrophic methanogens are survived. When methanol concentration in effluent is high, the production of volatile acids was strongly stimulated and acetogens survival rate is higher. Methanogens cannot compete over acetogens.

## Monitoring of Volatile Acids

The HPLC analysis of different volatile acids at the end of formaldehyde and methanol degradation is given in Figure 6.5 A for volatile acids formed after biodegradation of formaldehyde. At 500 mg/l of initial formaldehyde concentration, when biologically treated effluent was analyzed, it showed presence of formic acid (1.0 mg/l), acetic acid (2.2 mg/l), propionic acid (3.6 mg/l). The HPLC analysis is shown in Figure 6.5B for volatile acids formed after biodegradation of methanol. Traces of formic acid (0.49 mg/l), acetic acid (1.8 mg/l) and propionic acid (4.2 mg/l) were detected as by product of methanol degradation.

## Formaldehyde and Methanol Conversion Pathway

The presence of volatile acids indicated that the formaldehyde conversion pathway in the anaerobic digester was similar to the pathway presented by Gonzeatee Gil *et al.*[19]. According to them the

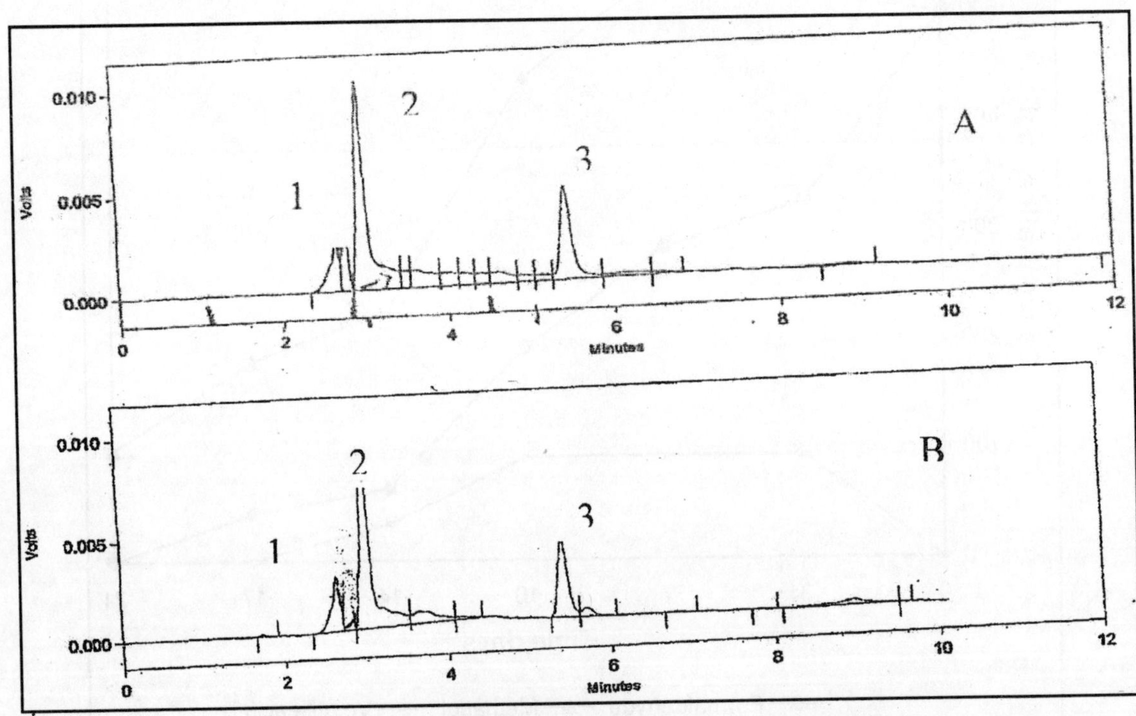


**Figure 6.5: HPL Chromatograms ($l_{max}$ 210 nm) of Volatile Acids**

**A: From formaldehyde and B: From methanol biodegradation.**

**Peaks correspond to 1: Formic acid, 2: Acetic acid, 3: Propionic acid**

formaldehyde was transformed to formic acid and methanol. The HPLC analysis suggests that oxidation of formaldehyde to formic and acetic acids and methanol. Oliveira *et al.*[16] have observed immediate consumption of formic acid formed. The proposed mechanism suggests the oxidation of formaldehyde into formic acid and its rapid reduction to methanol. Eiroa *et al.*[13] also detected formic acid and methanol as formaldehyde degradation product. Omil *et al.*[6]. have suggested that formaldehyde consumption under anaerobic condition is either by direct methane conversion or by methanol generation. Oliveirs *et al.*[5] have also detected low quantities of acetic acid in effluent samples of anaerobic packed bed reactor treating formaldehyde. He has further suggested that consumption of formed formic acid leads to generation of acetic acid. He has suggested that biological or chemical synthesis was occurring in the reactor.

$$HCHO + H_2O \longrightarrow H_2 + HCOOH$$

$$HCHO + H_2 \longrightarrow CH_3OH$$

$$2HCHO + H_2O \longrightarrow CH_3OH + HCOOH$$

The possible biodegradation mechanism seems to be similar to "Cannizzaro Reaction'[16]. Biodegradation of methanol shows generation of formaldehyde and formic acid as intermediate product of degradation. Acetic acid was not detected in the degradation product indicating formic acid to methanol conversion is reversible reaction. Lu and Hegemann[20] reported that there exist a possibility that formaldehyde resistant bacteria may be converting formaldehyde to methanol and formic acid using formaldehyde dismutase

## Acknowledgement

Authors are thankful to Director, NEERI, Nagpur, India for providing support for this research. The authors wish to thank the Department of Biotechnology, New Delhi for financial support.

## References

1. Prado, O. J., Veiga, M. C. and Kennes, C. Biofilteration of waste gases containing a mixture of formaldehyde and methanol. Appl. Microbiol. Biotechnol. 2004, *65*, 235-242.
2. Swenberg, J. A., Kerns, W. D., Mitchell, R. L., Gralla, E. J., Pakov, K. L. Introduction of squamous cell carcinomas of the rat nasal cavity by inhalation exposure to formaldehyde vapor. Cancer Res. 1980, *40*, 3398-3402.
3. Yamazaki,T., Tsugawa, W. and Sode, K. Biodegradation of formaldehyde by a formaldehyde resistant bacterium isolated from seawater. Appl. Biochem. Biotechnol. 2001, *91-93*, 213-217
4. Hidalgo, A., Lopategi, A., Prieto, M., Serra, J.L. and Llama, M.J. Formaldehyde removal in synthetic and industrial wastewater by Rhodococcus erythropolis UPV1. Appl. Microbiol. Biotechnol. 2002, *58*, 260-263.
5. Qu, M. and Bhattacharya, S. K. Toxicity and biodegradation of formaldehyde in anaerobic methanogenic culture. Biotechnol. Bioengi. 1997, *55*, 727-736.
6. Omil, F., Mendez, D., Vidal, G., Mendez, R. and Lema, J. M. Biodegradation of formaldehyde under anaerobic conditions. Enzyme Microb. Technol. 1999, *24*, 255-262.
7. Gerike, K. and Gode, P. The biodegradability and inhibitory threshold concentration of some disinfectants. Chemosphere. 1990, *21*, 799-812.

8. Bonastre, N., deMas. C. and Sola, C. Vanillin equation in kinetic modeling of formaldehyde biodegradation. Biotechnol. Bioeng. 1986, *28*, 616-619.

9. Garnido, J. M., Mendez, R. and Lema, J. M. Simultaneous urea hydrolysis, formaldehyde removal and denitrification in a multifed up flow filter under anoxic and anaerobic conditions. Water Res. 2001,*35*, 691-698.

10. Todini, O. and Pol, H. L. Anerobic degradation of benzaldehyde in methanogenic granular sludge: The influence of additional substrates. Appl. Microbiol. Biotechnol. 1992, *38*, 417-420.

11. Sharma, S., Ramakrishna, C., Desai, J. D. and Bhatt, N. M. Anerobic biodegradation of a petrochemical wastewater using biomass support particles. App.Microbiol. Biotechnol. 1994, *40*, 768-771.

12. Nash, T. The colorimetric estimation of formaldehyde by means of the Huntzch reaction. Bioche. J. 1953, *5*, 416-421.

13. Eiroa, M., Vilar, A., Kennes, C. and Veiga, M. C. Formaldehyde biodegradation in the presence of methanol under denitrifying conditions. *J.Chemical Technol. Biotechnol. 2006*, 81, *312-317*.

14. Kato, N., Shirakawa, K., Kobayashi, H. and Sakazawa, C. The dismutation of aldehyde by a bacterial enzyme. Agric. Biol. Chem. 1983, *47*, 39-46.

15. Adroer, N., Casas, C., deMas, C. and Sola, C. Mechanism of formaldehyde biodegradation by *Pseudomonas putida*. Appli. Microbiol. Biotechnol. 1990, *33*, 217-220.

16. Oliveira, S.V W. B., Moraes, E. M., Adorno, M. A. T., Varesche, M. B. A., Foresti. E. and Zaiat, M. Formaldehyde degradation in an anaerobic packed bed bioreactor. Wat. Res. 2004, *38*, 1685-1694.

17. Gurado, J.M., Garrido, R., Mendez, J. and Lema, M. Simultaneous urea hydrolysis, formaldehyde removal and denitrification in a multifed upflow filter under anoxic and anaerobic conditions, Water Res. 2001, *35*, 691- 698.

18. Florencio, L., Field, J. A. and Lettinga, G. The importance of cobalt for the individual trophic groups in an anaerobic methanol degrading consortium. Appl. Environ. Microbiol. 1994, *60*, 227-234

19. Gonzalez, Gil. G., Kleerebezem, R., van Aelst A., Zoutberg, G. R., versprille, A.L. and Lettinga, G. Toxicity effect of formaldehyde on methanol degrading sludge and its anaerobic conversion in biobed expanded granular sludge bed reactors. Water Sci. Technol. 1999, *40*, 195-202.

20. Lu, Z. and Hegemann, W. Anaerobic toxicity and biodegradation of formaldehyde in batch cultures. Wat.Res. 1998, *32*, 209-215.

# Chapter 7

# Hydrodynamic Assessment of Riverbed for Landscape Planning and Development: A Case Study of River Yamuna in Delhi

**Ritesh Vijay and Apurba Gupta***
**Environmental Systems Design and Modelling Division,**
**National Environmental Engineering Research Institute,**
**Nehru Marg, Nagpur – 440 020, India**

## ABSTRACT

Most of the critical flood problems occur in urban areas where properties at risk are higher and damages tend to be heavier. In the present study, hydrodynamic simulation of the river Yamuna has been assessed in the riverbed under different designated flood flows for existing and modified riverbed geometry. Flood flows for various return periods namely once in 10, 25, 50 and 100 years are estimated based on the past 41 years recorded data using flood frequency analysis. The model is applied successfully over the stretch of 23 km riverbed from Wazirabad barrage in upstream to Okhla barrage in downstream, including manmade structures like bridges and barrages. The results of model simulations for various return periods of the flood under existing riverbed condition are compared with recorded flow and water surface levels at various downstream gauge stations. Calibrated model is used for generating submergence scenarios and land availability in the riverbed under different flood flows with natural and proposed modified riverbed geometry. The additional land free from inundation due to modified riverbed

* Corresponding Author: E-mail: a_gupta@neeri.res.in

is estimated to be 175 ha, 200.1 ha, 198.7 ha and 185.2 ha, under once in 10, 25, 50 and 100 years return flood respectively. The simulation results of the proposed riverbed geometry in the stretch study can be used for riverbed planning and development purposes.

***Keywords:*** *Hydrodynamic, Riverbed, Channelisation, Flood, Submergence*

## Introduction

Flooding is an important natural process. It plays a vital role in the maintenance of floodplain and riverbed ecosystems but also put human life and property at risk (UNDHA, 1997; IFRCRCS, 1998). Researchers investigated changes in floodplain occupancy was a major cause of increasing flood risk and rising flood losses (White *et al.*, 1958, De 2000) even in cites that had experienced population decline. Recent major floods in India are a reminder that vulnerability to floods remains high (Mirza *et al.*, 2002; John, 1996; Uitto and Biswas, 2001). Flood provides an opportunity to draw together and to assess findings from a wide range of research and disasters (Parker, 2000).

The impact of man on rivers and river channels has been widespread throughout the period of habitation of the planet (Brookes, 1988, Vijay *et al.*, 2007). River channelisation encompasses methods of engineering which modify river channels for the purpose of flood control, drainage improvement, maintenance of navigation, reduction of bank erosion etc. (Funk and Ruhr, 1971; Keller and Brookes, 1984; Brookes, 1885). The history of channelisation throughout the world is very old, where canals and ditches were built to carry water and sewage including diversions of streams fro irrigation. It has been estimated that more than 25, 000 km of river had been modified in the USA since the Flood Control Act of 1936 (Leopold, 1977; Oliver, 2005). Similarly history of channelization in India is also old and presently various river channelization projects are under consideration. A number of development options for the land resource of riverbed Yamuna have been suggested by various development agencies. One of the developmental proposals for utilization of the land resource of the river corridor is the channelization of the river Yamuna in Delhi stretch (Prasad *et al.*, 2002). Channelization entails construction of artificial channel for the river that is bounded by dykes or bunds on both sides of the channel. It will stop meandering and over-topping of the river that will lead to reclaiming of vast stretches of land which is otherwise periodically inundated when the river floods. The reclaimed land will be subsequently used for the development of civic infrastructure.

The present study simulates the hydrodynamics of the riverbed of a 23 km stretch of river Yamuna in National Capital Territory (NCT) Delhi, India. A one-dimensional model for simulation of water surface profiles for steady flow conditions has been used with due consideration to channel geometry, structural controls, bed roughness and peak discharge conditions as well as channel modification. The model simulations of water surface profiles have been used for delineating inundation areas and land availability in the riverbed for both the conditions of natural and channel modification using Surfer (Golden software 1996).

## Model Application and Data Requirements

The BOSS RiverCAD which is based on optimized version of HEC-RAS (1997) has been used in the present study. The model simulates water surface profiles for steady gradually varied flow with subcritical flow regime conditions as well as the effects of control structures. Input data for simulation includes starting water surface elevation, discharge, cross-sectional geometry and reach lengths,

Manning's *n* values and other energy loss coefficients, and characteristics of bridges, weirs, and barrages. The model computes water surface profiles in channels with irregularly shaped cross sections, channel modification, velocity, depth etc for a full network of channels (Chow *et al.*, 1988; Chow, 1959). The program uses an iterative procedure called the 'standard step method' to solve energy and energy loss equations in order to compute changes in water surface elevation between adjacent cross sections (BOSS, 2000).

## Study Area

The river stretch considered in the study begins at the upstream station, Wazirabad barrage and ends at Okhla around 23 km downstream in NCT Delhi, India shown in Figure 7.1. It lies between 28° 28′ 05″–28° 54′ 36″ N and 77° 09′–77° 24′ E which falls in the Toposheet No. 53H (SOI). The floodplain is 2 to 3 km wide in the urban stretch of National Capital Territory (NCT), Delhi. The river waterway in the stretch is confined to a width of 450-800 m during the non-monsoon period and the river

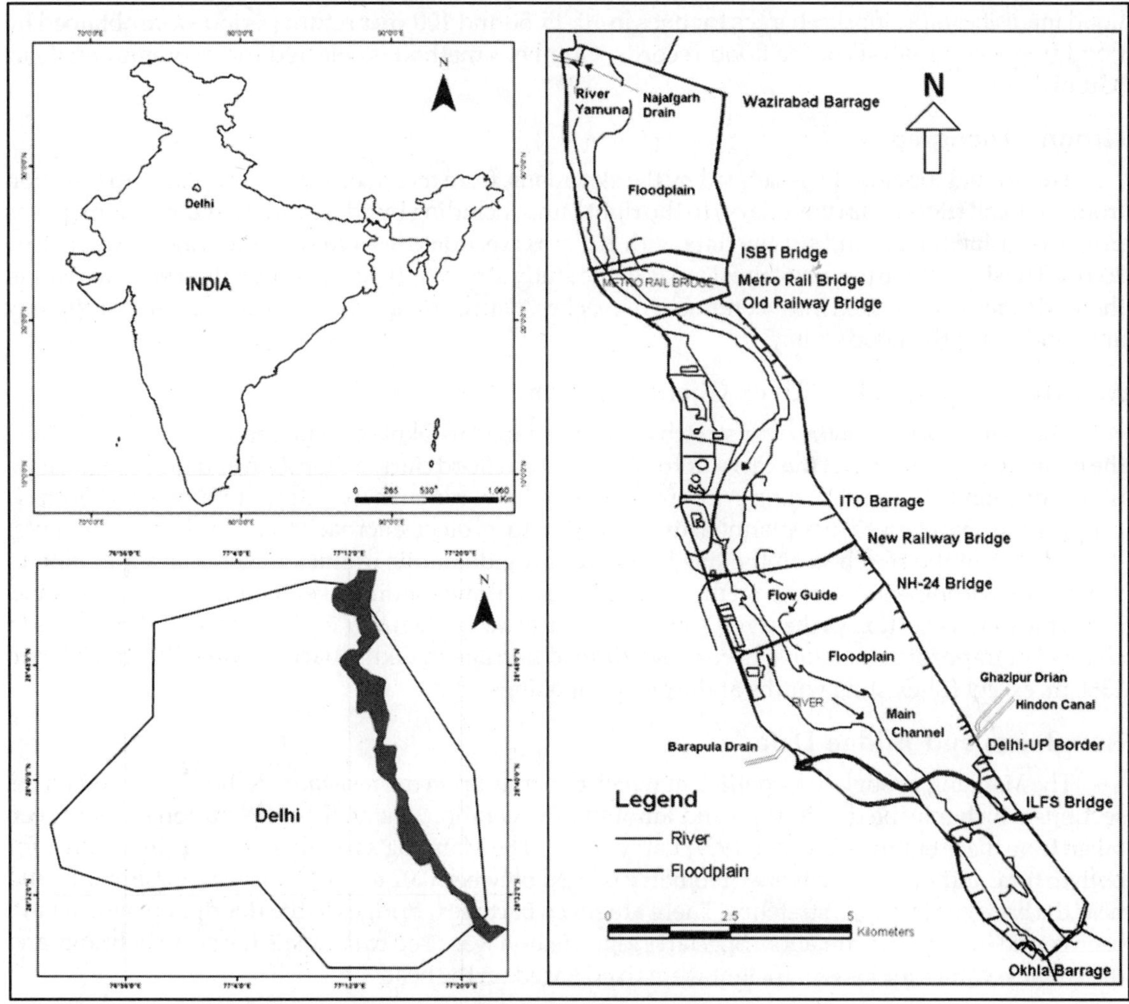


**Figure 7.1: Detailed Location Map of the Riverbed Yamuna in Delhi**

meanders at places, have large areas of floodplain lying vacant. The floodplain area in the downstream of Wazirabad Barrage is 4480 ha up to Okhla, predominant land use being agricultural activities. The other activities in the riverbed leading to pollution problem and environmental damages are encroachments, slums, power station, fly ash dump, cremation ground, bathing ghats, sanitary landfills, water works, sewage treatment plants (Pandit, 1998) and many other unorganized activities. Such type of activities is leading to riverbed degradation. Excess storm water from areas in Haryana and the Sahibi river flow into the Najafgarh drain, which in turn flows to the river Yamuna, downstream of Wazirabad Barrage (CBPCWP, 1982). Najafgarh, supplementary and Barapula drains contribute more than 90 per cent of the wastewater in to river Yamuna without any treatment.

## Flood Frequency Analysis

The peak discharges were estimated based on flood frequency analysis (Punmia and Pande, 1990; Subramanya, 1997) of historic stream flows at a specified river location recorded by Central Water Commission, India. In order to estimate water surface profiles and inundation for different flood intensities, specific discharges for once in 10, 25, 50 and 100 year return period were obtained by flood frequency analysis of the flood records. Gumbel's method is selected for frequency analysis (Gumbel, 1958).

## Ground Topography

The ground topography is defined by the elevations, from mean sea level, along each cross section from left bund (floodplain boundary) to the right bund including bund elevation. A contour map was prepared using the ground survey data with 101 cross sections measured at intervals between 200–300 m. The slope of the terrain of most areas of the study stretch is 0-0.5 per cent only. The total river in the study area was divided into six number of reaches, with due consideration to channel morphology and land use in the floodplain.

## Modified Topography: River Channelization

The channelization of the river stretch from Wazirabad to Okhla considered river dredging along the main river course which can carry upto 10 year return flood during monsoon and flows for all non-monsoon months. The remaining part of the riverbed that remains dry for non-monsoon months is proposed to be dressed for planned use in order to protect encroachment and environmental degradation in the riverbed on the one hand and provide public utilities to general population to minimize environmental stresses on the other. The natural and modified geometry of the river channel at a typical cross-section in the river Yamuna is illustrated in Figure 7.2. The width of the proposed channel of trapezoidal section between Wazirabad barrage to Okhla barrage varies from 750 m to 1350 m, except for existing widths at the bridge openings.

## Roughness and Bridge Data

The Manning's roughness coefficient which depends upon the resistance to flow at various cross sections which may be due to type and amount of vegetation, channel configuration etc. has been taken from the literature (Danniel, 1997; Larry, 2001). The Manning's roughness coefficient values for both natural and modified channel geometry ranged between 0.02 to 0.12 for river course, banks and near the bridges at various stretches. There are seven barrages/bridges in the floodplain, which have been considered in the simulation. Special bridge method was used to simulate the effect of bridge and the suitable values for loss coefficient were fixed based on the head.

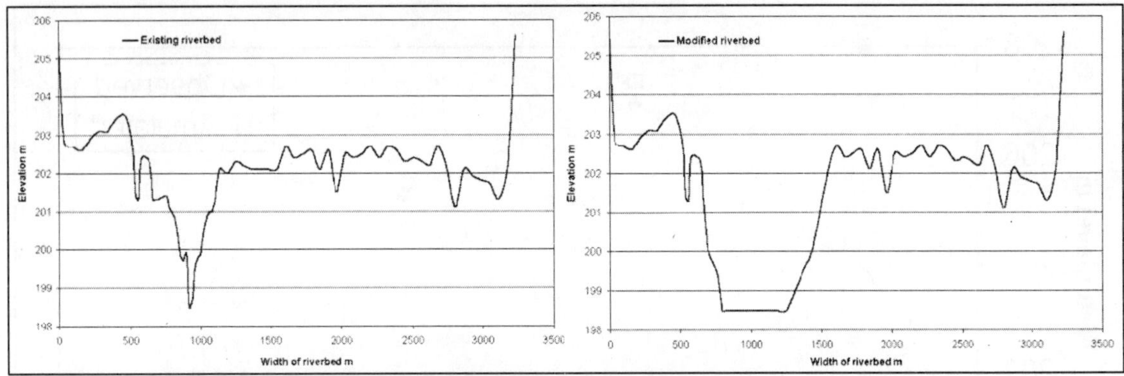


**Figure 7.2: Existing and Modified (Channelisation) Geometry of the Riverbed**

## Hydrodynamic Simulation

Simulations have been carried out for various designated flood flows under existing and proposed channel dredging condition of the riverbed. The starting water surface elevation at most downstream cross section *i.e.* at Okhla barrage required for the computation of water surface profile has been taken from the gauge records. The submergence scenarios of the riverbed for natural and channelized river conditions under different flows were estimated by overlaying the water surface profile data on the DEMs. A comparison was made to estimate the additional land obtained due to proposed channelization in the river course.

## Results and Discussion

Model simulations for water surface elevations along the river stretch from Wazirabad to Okhla at various cross sections separated by 200-300 m interval were used to visualize the impact on land

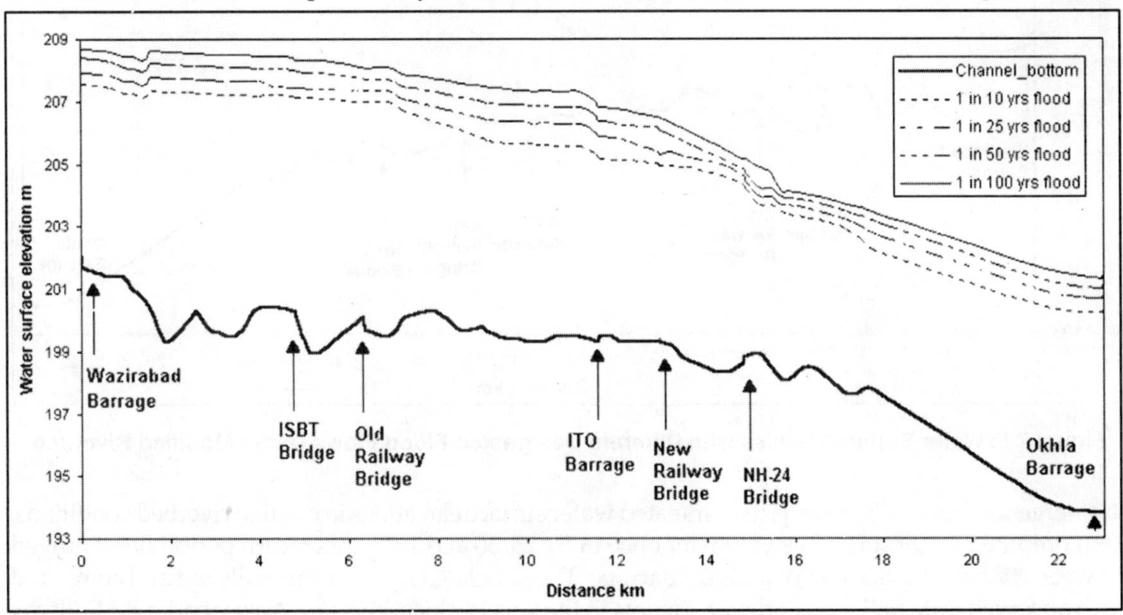


**Figure 7.3: Water Surface Profiles with Different Designated Flood Flows Under Existing Riverbed**

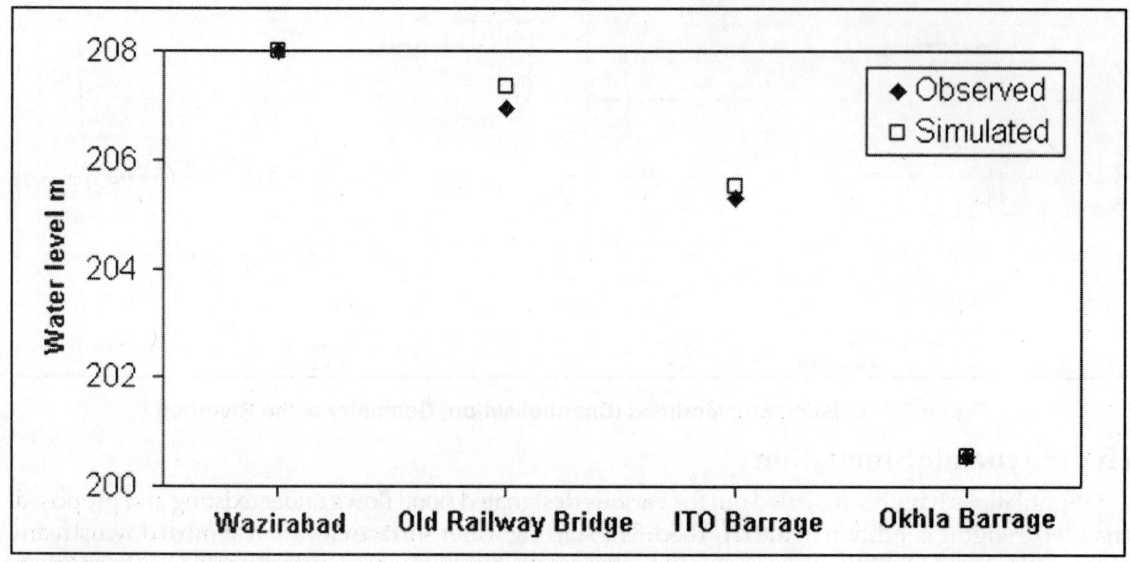


**Figure 7.4: Comparison Between Observed and Simulated Water Levels at Observed Flow**

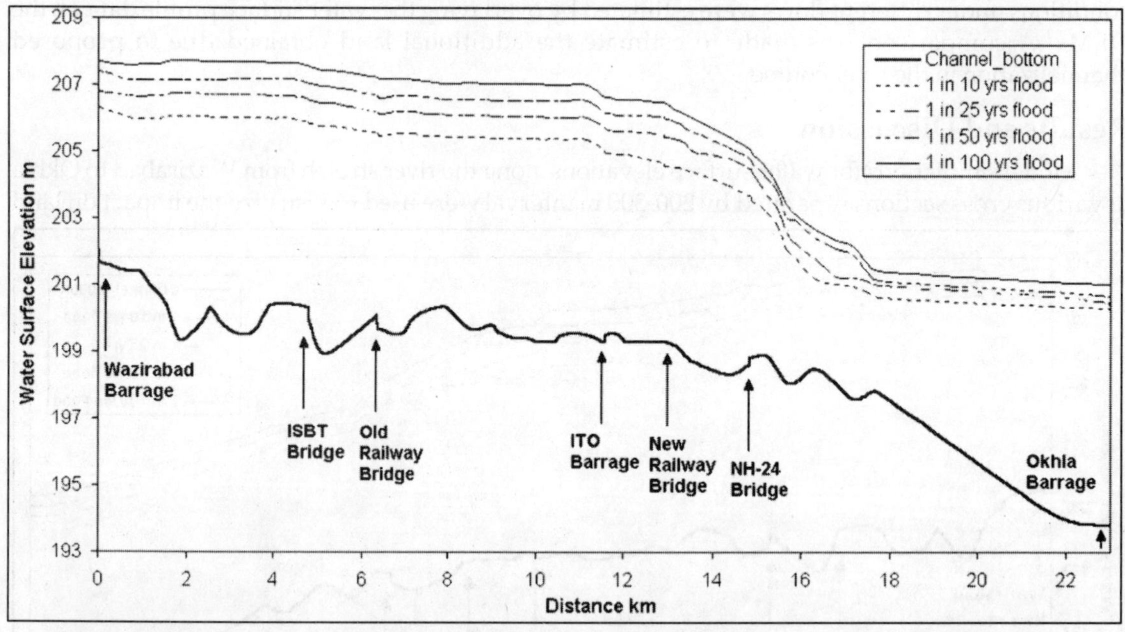


**Figure 7.5: Water Surface Profiles with Different Designated Flood Flows Under Modified Riverbed**

submergence. Figure 7.3 presents the simulated water surface elevations in existing riverbed conditions. The computed water surface elevation for once in 10, 25, 50 and 100 years return period flood ranged between 207.5 m to 208.5 m at Wazirabad Barrage. The simulated results were analyzed and compared with observed data of floods and water levels in the specific locations *viz.* Wazirabad, Old Railway

Bridge, ITO Barrage and Okhla Barrage as described in Figure 7.4. The correlation coefficient between observed and simulated water levels was found to be 0.748. Similarly, Figure 7.5 illustrates the computed water surface elevations for designated flood flows under modified riverbed geometry by channel dredging of the main river channel. It may be clearly visualized from the water surface profiles of existing and modified riverbed geometry that channel dredging reduces the water surface elevation significantly.

The calibrated model was used to delineate the inundation and land availability in combination with Surfer for different flood flow conditions. The inundation due to flood at various locations was considered as the difference between ground and water surface elevation. The inundation obtained was termed as submergence scenarios. Thus the submergence and land area were demarcated by values below zero and above zero, respectively. Figures 7.6 (a) and (b) show the submergence scenarios and land available above the water surface for 10 and 100 year return flood flows under existing and modified riverbed (channel dredging), respectively. The entire submergence shows the non-uniform spread of water over the riverbed due to lesser depth of river at various sections. The inundated area

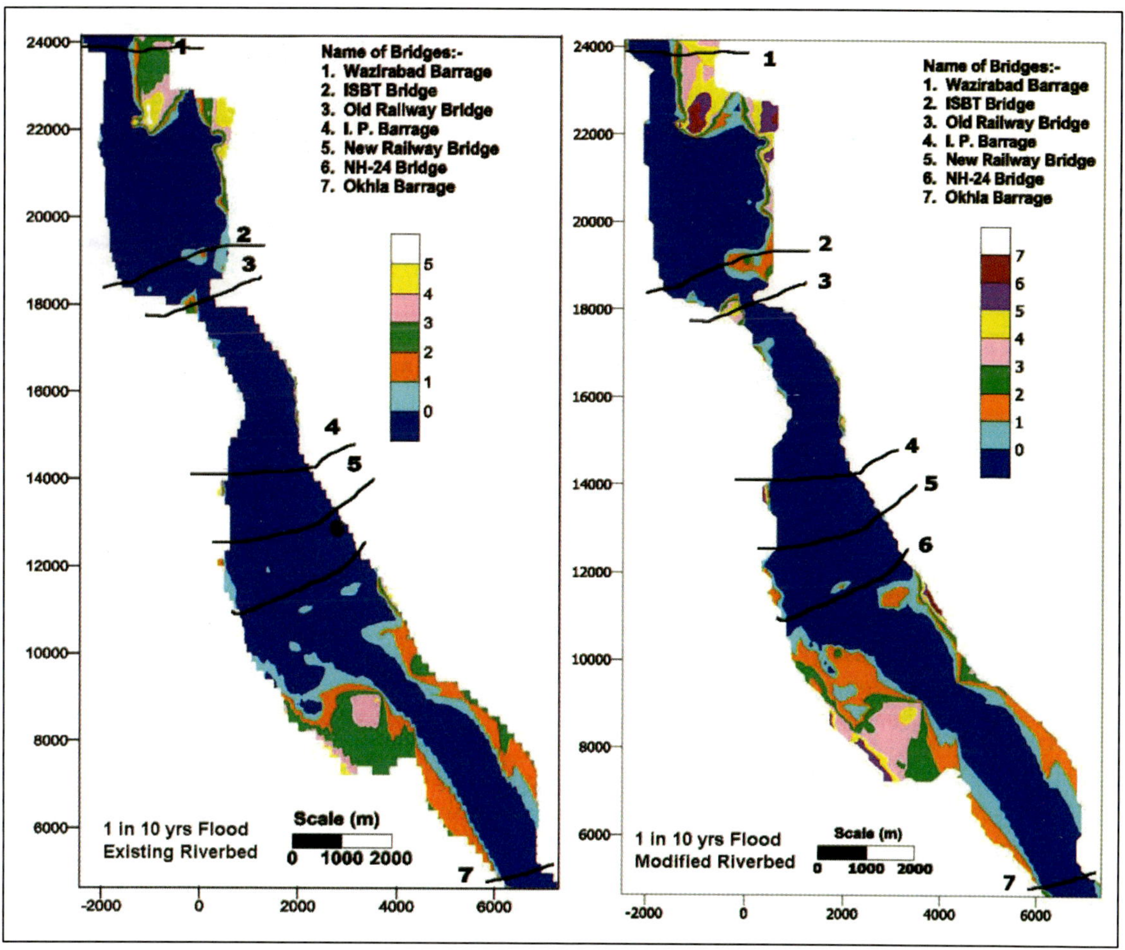


**Figure 7.6(a): Submergence Due to 10 Yrs Return Flood in Existing and Modified Riverbed**

15. Interagency Advisory Committee on Water Data (IACWD), 1982. Guidelines for determining flood flow frequency, Bulletin #17B, US Geological Survey, Office of Water Data Coordination, Reston, VA.

16. John P Cole. 1996. *Geography of the World's Major Regions*, Routledge, Taylor and Francis Group (UK).

17. Keller EA and Brookes A. 1984. Consideration of meandering in channelization projects: selected observations and judgments. In River Meandering, Proceeding of Conference Rivers, American Society of Civil Engineers, Vicksburg, MS, pp. 384-398.

18. Larry W. Mays 2001. *Water Resources Engineering*, John Wiley and Sons, Inc. New York.

19. Mirza M, Dixit A, Nishat, A. 2002. *Flood Problem and Management in South Asia*, Springer.

20. Oliver P. Harmar, Nicholas J. Clifford, Colin R. Thorne, David S. Biedenharn, 2005, Morphological changes of the Lower Mississippi River: geomorphological response to engineering intervention, River Research and Applications, vol 21, issue 10, pp: 1107-1131.

21. Pandit M K. 1998. Yamuna River Rejuvenation Plan-Riparian Ecosystem, Phase-I Report, Centre for Inter-Disciplinary Studies of Mountains and Hill Environment (CISMHE), Delhi, November 1998.

22. Parker DJ. 2000. *Floods*, Vol-1, Routledge Taylor and Francis group, London.

23. Prasad L, Kumar P, Babu CR. 2002. Valuation of Ecological Functions and Benefits: A Case Study of Wetland Ecosystems along the Yamuna River Corridors of Delhi Region, funded by the World Bank through Environmental Economics Research Committee [EERC], IGIDR, Mumbai under "India: Environmental Management Capacity Building Technical Assistance Project" of the Ministry of Environment and Forests, Government of India.

24. Punmia BC and Pande B, Lal B. 1990. *Irrigation and Water Power Engineering*, Standard Publishers Distributors, New Delhi, pp. 123-126.

25. Subramanya K. 1997, *Engineering Hydrology*, 2nd edition, Tata McGraw-Hill, New Delhi.

26. Survey of India (SOI), Toposheet No. 56.

27. Uitto JI, Biswas AK. 2001. *Sustainable Development of the Ganges-Brahmaputra-Meghna Basins*, United Nations University Press, Tokyo.

28. United Nations, Department of Humanitarian Affairs (UNDHA), 1997, Floods: People at Risk, Strategies for Prevention, Geneva: United Nations.

29. Vijay R, Sargoankar A, Gupta A. 2007, Hydrodynamic Simulation of River Yamuna for Riverbed Assessment: A Case Study of Delhi Region, *Environmental Monitoring and Assessment*, 130: 381-387.

30. White GF, Calef WC, Hudson JW, Mayer HM, Sheafer JR and Volk DJ. 1958. Changes in urban occupance of floodplains in the United States; Research paper 57, Department of Geography, University of Chicago.

# Chapter 8

# Treatment of Synthetic Wastewater Containing Phenol and Ammonia-Nitrogen in a Three-Stage Attached Growth System

**Bhaskar Jyoti Deka, Biju Prabha Sahariah and Saswati Chakraborty***
**Department of Civil Engineering,**
**Indian Institute of Technology, Guwahati – 781 039, Assam, India**

## ABSTRACT

The objective of the present research is to study biodegradation of a synthetic wastewater containing high concentration of phenol and ammonia-nitrogen. Single-stage activated sludge process though the most common process for biodegradation is inadequate to achieve nitrification due to presence of phenolic compounds. In the present study a three-stage anaerobic-anoxic-aerobic continuous reactor assembly was used for biodegradation of phenol and ammonia-nitrogen containing wastewater. Cultures were acclimatized in separate batch reactors under anaerobic, anoxic and aerobic environments with sponge cubes as media for biomass immobilization. The acclimatized cultures grown on sponge surface were added in continuous reactors operated under anaerobic (R1) anoxic (R2) and aerobic (R3) environments. Phenol was completely removed in three-stage system form influent concentration of 1000 mg/L. In anaerobic reactor (R1), phenol removal efficiency was almost 29 per cent and average biogas production

* Corresponding Author: E-mail: saswati@iitg.ernet.in

was 326.3 mL/day. In anoxic reactor (R2), residual phenol from R1 was completely removed along with 90 per cent removal of recycled nitrate-nitrogen from aerobic reactor. Ammonia-nitrogen removal efficiency in three-stage system was more than 90 per cent.

***Keywords:*** *Phenol, Ammonia-nitrogen, Fixed-film reactor, Sponge media, Three-stage system.*

## Introduction

Biological treatment generally a cost-effective viable method for treating various industrial wastewaters, using microorganisms for degrading pollutants. Process effluents generated from chemical industries, oil refineries, coal gasification, coke and steel plant, synthetic fuel processing operations, coal carbonization, etc. contain high levels of phenolic compounds and ammonia-nitrogen along with other toxic substances. Due to the presence of these toxic and inhibiting pollutants, biodegradation is difficult and challenging. Activated sludge process (ASP) is the most common feature in most of the studies along with various pre-treatment options. However, single- and two-stage ASPs are reported to be inefficient with inadequate nitrification (Pandey *et al.*, 1991; Luthy, 1981). Two- and three-stage anoxic–aerobic/anaerobic–anoxic–aerobic sequential treatment enhanced process performance (Zhang *et al.*, 1998). The anaerobic stage was used as a pre-treatment step and anoxic–aerobic reactors were used for denitrification with organic removal and nitrification (Chakraborty and Veeramani, 2006; Melcer and Nutt, 1988; Liu *et al.*, 1996). A three-stage anaerobic–anoxic–aerobic attached growth continuous system is selected for the present study with synthetic feed containing phenol and ammonia-nitrogen. The aim of the study was to collect data for design the three-stage continuous growth system for biodegradation of phenol and ammonia-nitrogen. Fixed film reactors are known for their higher efficiency to remove pollutants due to retention of higher amount of biomass. In the present study sponge was used in biological reactors as the inert media for immobilization of biomass.

## Materials and Methods

### Materials

The concentrated acid was made to 1 N/0.1 N by dilution and used for all experimental work. Dextrose anhydrous purified, phenol, potassium nitrate, ammonium chloride purified (99 per cent), $K_2HPO_4$, $KH_2PO_4$ (98 per cent), $NiCl_2.6H_2O$ (97 per cent), $MgSO_4.7H_2O$, $CaCl_2.2H_2O$, $FeCl_3.6H_2O$, $CuCl_2$, $ZnCl_2$, $NiCl_2.6H_2O$ and $CoCl_2$ used for synthetic feed for biological reactors were manufactured by E.Merck (India) Ltd., Mumbai. Yeast Extract was procured from Nice Chemicals Pvt. Ltd., Cochin. Single distilled water prepared in the laboratory was used for preparation of all the reagents and all experimental work, except feed of biological reactor, for which tap water was used.

### Experimental Methods

#### Culture Development

##### *Seed Sludge*

Sludge collected from an anaerobic biogas plant at IIT Guwahati was used as seed for start-up of anaerobic and anoxic reactors. Four batch reactors were started separately in glass container (working volume 5 L each) under anaerobic (R1A and R2A) and anoxic (R3A and R4A) environments. Sewage effluent collected from sewage treatment plant of IIT Guwahati was screened through clean cloth and used as the seed for two aerobic reactors (R5A and R6A).

***Characteristics of Media***

Present reactors were operated as attached growth system. Sponge sheets were procured from local market. Cubes (dimensions: 1 cm×1 cm×1 cm) were made manually from sponge sheet. These sponge cubes were used as inert media for the present work. In each 5 L of reactor (R1A-R6A) 60 g of sponge cubes were added.

***Acclimatization with Synthetic Feed***

In anaerobic reactors R1A and R2A, 500 mg/L of dextrose was added daily for initial 21 days. Along with dextrose, the synthetic feed also included ammonium chloride (50 mg $NH_4Cl$/L) as nitrogen source; phosphate buffer (dissolving 104.5 g/L $K_2HPO_4$ and 72.3 g/L $KH_2PO_4$) of pH 6.8; trace metals solution (1mL/L) and yeast extract (10 mg/L) as nutrients. Phosphate buffer of 5 mL/L of feed was added as phosphorus source and to maintain reactor pH. Composition of trace metal solution was: $MgSO_4.7H_2O$: 10,000 mg/L, $CaCl_2 2H_2O$: 10,000 mg/L, $FeCl_3.6H_2O$: 5000 mg/L, $CuCl_2$: 1000 mg/L, $ZnCl_2$: 1000 mg/L $NiCl_2 6H_2O$: 500 mg/L and $CoCl_2$: 500 mg/L (Chakraborty and Veeramani, 2005). Concentration of phosphate buffer, yeast extract and trace metal solution remained unchanged through out the study. Synthetic feed was dissolved in tap water and added in reactors. For the first 63 days R1A and R2A were operated only with dextrose (500-1000 mg/L/day). Thereafter, on day-64 onwards phenol was added in R1A and R2A along with 1000 mg/L dextrose. In R1A and R2A, phenol concentration was increased gradually to overcome the toxic effect of phenol on microorganisms. Feed phenol concentration reached 1000 mg/L on day-161. As phenol concentration was increased in feed, dextrose concentration was gradually decreased. On day-77, feed dextrose concentration was decreased to 800 mg/L, decreased further to 400 mg/L on day-105 and 100 mg/L on day-181 and became nil on day-221 onwards and R1A and R2A were operated only with phenol (1000 mg/L), ammonia (10 mg/L) and other nutrients.

In anoxic reactors (R3A and R4A) for the initial 21 days 500 mg/L of dextrose and 70 mg/L of nitrate-nitrogen (using potassium nitrate, $KNO_3$) were added (at COD: $NO_3^-$-N ratio of 7.57). Same amount of phosphate buffer, yeast extract and trace metal solutions were added in the feed of anoxic reactors like anaerobic reactors. On day-22 onwards, feed dextrose and nitrate ($NO_3^-$-N) concentrations were increased to 1000 mg/L and 250 mg/L respectively (COD: $NO_3^-$-N ratio of 4.24). On day-31, phenol of 5 mg/L was added in anoxic reactors along with 1000 mg/L of dextrose and 250 mg/L of $NO_3^-$-N. Feed dextrose concentration was decreased in R3A and R4A along with increase in phenol concentration in the feed. On day-79, feed phenol concentration became 500 mg/L and dextrose concentration became 400 mg/L. On day-97 onwards anoxic reactors were operated with 500 mg/L phenol, 450 mg/L of $NO_3^-$-N without any dextrose in the feed (COD: $NO_3^-$-N ratio of 2.64). In three-stage system, effluent of anaerobic reactor is expected to be diluted with nitrified effluent of aerobic reactor so that denitrification can take place in anoxic reactor. Due to this, anoxic reactor was acclimatized up to maximum phenol concentration of 500 mg/L.

In aerobic reactors (R5A and R6A) sponge of 60 g in each reactor was added in 5L of screened sewage. Compressed air was added in R5A and R6A for mixing of effluent sewage and sponge. Synthetic feed consisting of ammonia-nitrogen (using ammonium chloride) was added in aerobic reactors. Concentration of ammonia-nitrogen was increased gradually from 10 mg/L to 250 mg/L in 90 days.

**Continuous Study**

Continuous study is conduced in three PVC (polyvinyl chloride) columns purchased from local market. Diameter of each column is 15 cm and height 118 cm. All three columns were closed at the bottom with a plastic cap of diameter nearly 15 cm and height of 5 cm. These caps were attached with the reactors using M-seal. An iron frame was fabricated to support PVC columns. PVC columns were connected to the iron frame using clamps. The scheme of reactor system used for continuous runs are shown in Figure 8.1. An inlet and outlet were constructed in all three columns as shown in Figure 8.1. Acclimatized cultures from two batch anaerobic reactors R1A and R2A immobilized on sponge cubes (total 120 g in two reactors) were added in one PVC column and renamed as anaerobic reactor (R1). Similarly contents of R3A and R4A were added in the next PVC column and started as anoxic reactor, R2. Reactor R3 was started with contents of R5A and R6A. Working volume (liquid and biomass) of each reactor is 15 L. Anaerobic reactor, R1 was closed at the top with a similar PVC cap like bottom. In the upper cap of R1, one PVC pipe was inserted through a hole and the joining of the cap and pipes was sealed using M-seal. Other end of the PVC pipe was inserted in an inverted cylinder kept in a tub of water for collecting biogas by water replacement system. An air compressor was used for supply of air in aerobic reactor, R3. No external unit was used for mixing of liquid and biomass in anaerobic (R1) and anoxic (R2) reactors. Anoxic reactor R2 was maintained as an open reactor. However, due to narrow mouth and higher depth of the reactor (height: diameter ratio of 7.86) it was expected very little oxygen penetration inside anoxic reactor. It was observed during reactor operation that that color of sludge in R2 was of dark black colour. A blower was used to maintain the temperature of the continuous unit at 30-32°C throughout the study.

Continuous study was started in a three-stage attached growth system consisting of anaerobic (R1)-anoxic (R2) and aerobic (R3) reactors. Initially study was started with feed flow rate of 5 mL/min (7.2 L/day) to anaerobic reactor. All three reactors were operated in upflow mode. One plastic settling tank (each volume 500 mL) was connected to each reactor for liquid-solid separation (Figure 8.1). Supernatants from settling tanks were allowed to flow to a plastic bucket by gravity. Effluent of aerobic

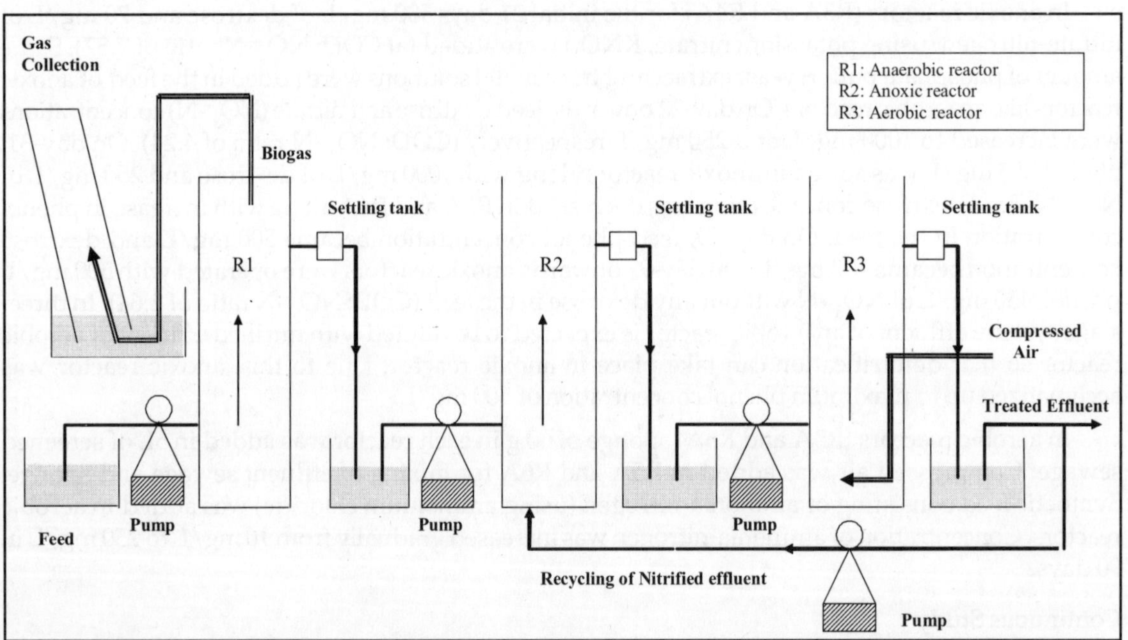


**Figure 8.1: Anaerobic-Anoxic-Aerobic Sequential Biological Attached Growth Reactor System**

reactor R3, was recycled (recycled flow rate 7.2 L/day) form the plastic bucket attached to R3 to anoxic reactor R2. The effluent from R2 was collected similarly in another plastic bucket attached to R2 and sent directly to R3 (flow rate 15.4 L/day). One peristaltic pump (Model: PP-20-4C) was used for supply of feed to R1 and effluent of R1 and recycled effluent of R3 to R2. Another peristaltic pump (Model: ISM-817) was used to deliver the effluent of R2 to R3. Synthetic feed used for continuous study consisted of phenol (1000 mg/L) and ammonia-nitrogen 250 mg/L. Phosphate buffer, yeast extract and trace metal solution (1 mL/L feed) were added as nutrients in synthetic feed in a similar way used during acclimatization. Feed pH was adjusted to 7.0-7.5 by adding 0.1 N HCl/$NaHCO_3$. Feed was prepared daily. Effluents were collected from the supernatant of settling tank attached to each reactor, centrifuged at 1000 rpm for one minute and then used for chemical analysis. In R2 and R3, pH was maintained within 7.0-7.5 by adding 0.1 N HCl/$NaHCO_3$. In R1, $NaHCO_3$ was added after opening the clip of the PVC pipe end.

## Analytical Protocol

All analyses were carried out according to APHA (1998). High Speed Centrifuge (Model: R- 24, Remi India Ltd. Bombay) was used for separation of solids from liquid samples, which was used for all chemical analyses. Solution pH was measured by a digital pH meter (Model: Systronics µpH system 361, India). Phenol was estimated by direct photometric method using 4-aminoantipyrene and measuring the absorbance at 500 nm. Ammonia concentration was measured by phenate method (APHA, 1998). Chemical Oxygen Demand (COD) was estimated by closed reflux titrimetric method. Nitrate-nitrogen concentration was determined measuring absorbance of samples and standards at 210 nm.

## Results and Discussion

## Performance during Acclimatization

### Performance of Anaerobic Reactor

The average performance of anaerobic reactors (R1A and R2A) during acclimatization phase is shown in Figure 3.2(a). These two reactors were operated for initial sixty-three days with dextrose 500-1000 mg/L along with trace metal, phosphate buffer and yeast extract. The objective of dextrose feeding for sixty-three days was to develop anaerobic culture capable of utilizing organic carbon (dextrose) and generating biogas. During this period (initial sixty-three days) almost 70-80 per cent of COD was removed in R1A and R2A. Figure 3.2(a) shows that on day-64, 5 mg/L of phenol was added in R1A and R2A. On day-67, effluent phenol in R1A was 2 mg/L (60 per cent removal). In R2A, corresponding phenol removal was 68 per cent. In R1A and R2A, phenol concentration was increased to 10 mg/L on day-67 and on day-69, almost 50-58 per cent phenol removal was achieved. Along with gradual increase in feed phenol concentration, influent dextrose concentration was gradually decreased in R1A and R2A. R1A and R2A required almost 217 days for acclimatization up to phenol concentration of 1000 mg/L. Initially phenol removal efficiency in R1A and R2A was 50-60 per cent. With increase in influent phenol concentration to 1000 mg/L, removal efficiency increased to 78-97 per cent. It can be seen that after phenol addition COD removal efficiency decreased in R1A and R2A. On day-106, influent phenol and dextrose concentrations were 500 and 400 mg/L respectively and COD removal efficiency decreased to 48 per cent. When reactor was operated only with phenol of 1000 mg/L, COD removal efficiency was nearly 45 per cent.

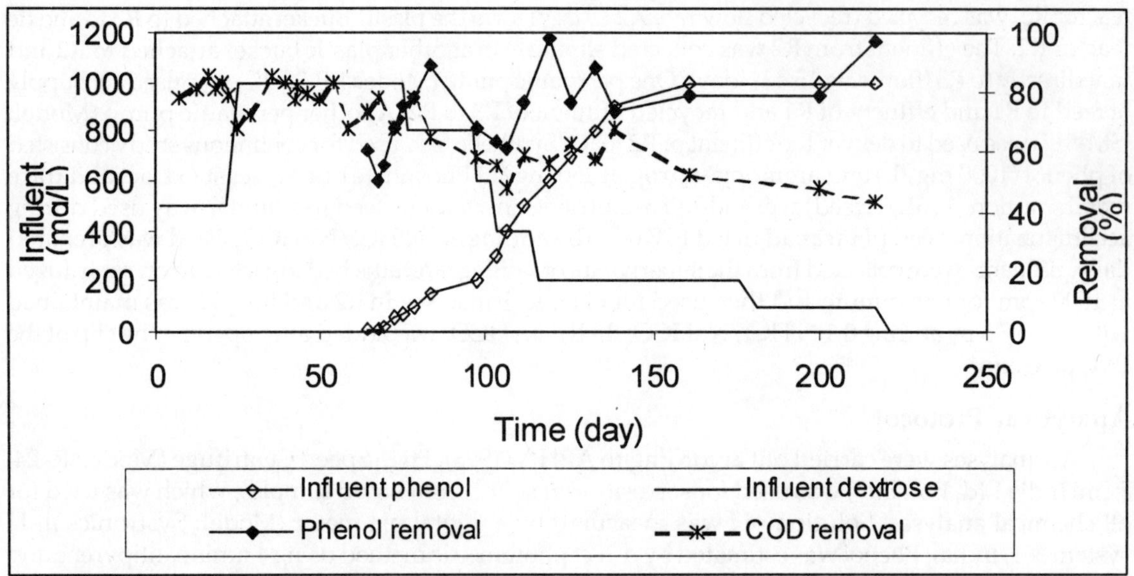


**Figure 8.2a: Average Performance of Anaerobic Reactor (R1A and R2A) during Acclimatization**

**Performance of Anoxic Reactor**

The anoxic units (R3A and R4A) were fed with dextrose (500 mg/L) and $NO_3^-$-N (70 mg/L) up to day-21 (COD: $NO_3^-$-N 7.57). This was calculated based on 1.06 g of COD for 1 g of dextrose. During this period concentration of $NO_3^-$-N was monitored in anoxic reactor and almost complete denitrification was observed [Figure 8.2(b)]. From day-22, influent dextrose was increased to 1000 mg/L and influent

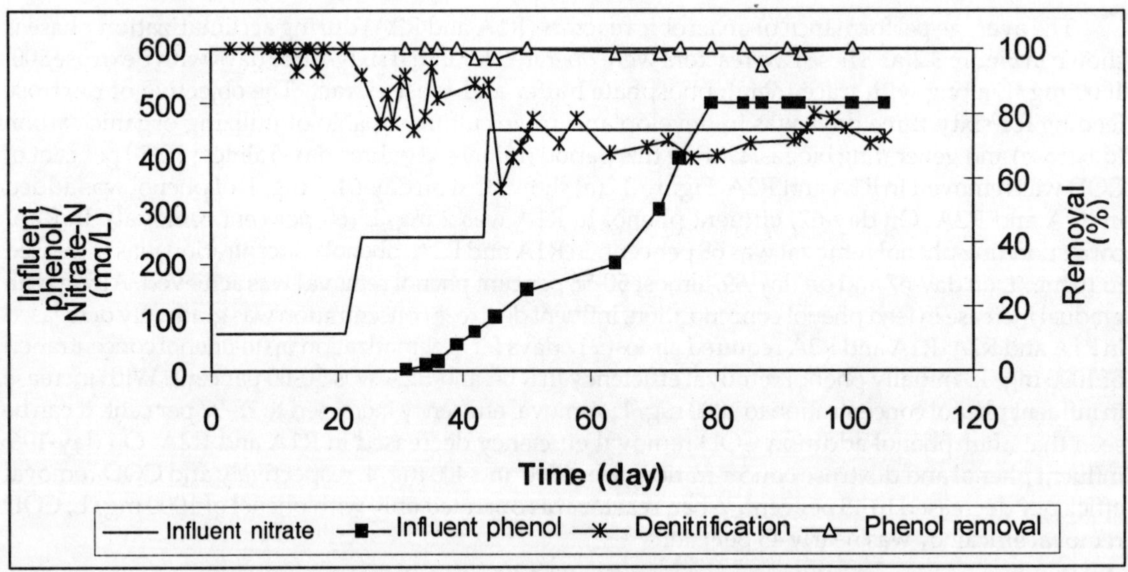


**Figure 8.2b: Average Phenol and Nitrate-Nitrogen Removal Efficiency in Anoxic Reactors (R3A and R4A) during Acclimatization**

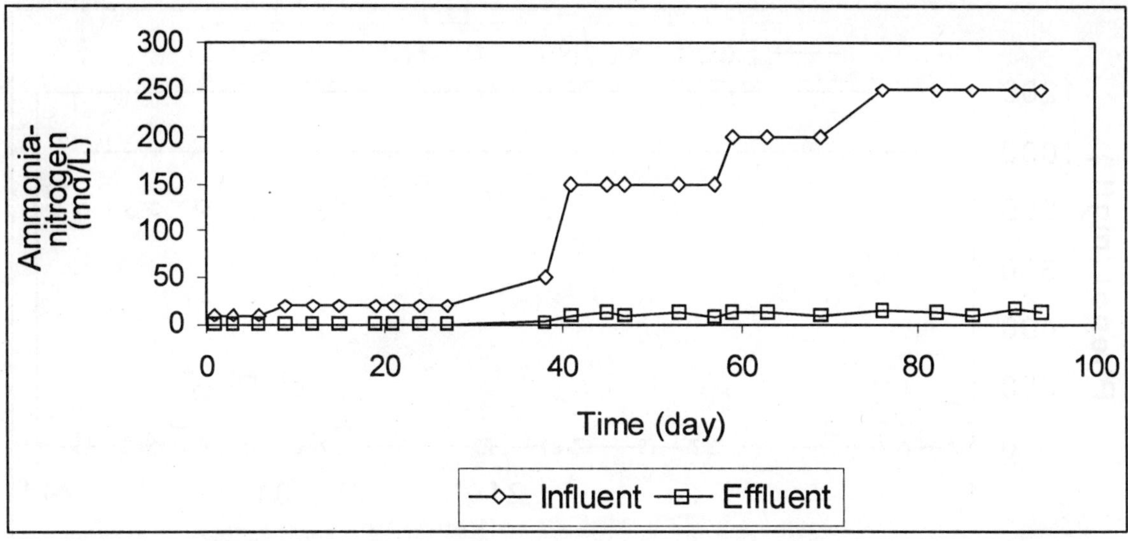


**Figure 8.2c: Average Performance of Aerobic Reactors (R5A and R6A) during Acclimatization**

$NO_3^-$-N was also increased to 250 mg/L (COD: $NO_3^-$-N 4.24). Denitrification was still higher (75-90 per cent). On day-97, feed phenol concentration was 500 mg/L and $NO_3^-$-N 70 mg/L (COD: $NO_3^-$-N 2.64, 1 g phenol gives 2.38 g COD). Denitrification was almost 70-80 per cent. Throughout the acclimation period more than 95 per cent phenol removal was achieved in both the anoxic bioreactors up to phenol concentration of 500 mg/L Figure 8.2(b). These results indicate that both phenol and $NO_3^-$-N were used simultaneously by anoxic microorganisms.

### Performance of Aerobic Reactor

Aerobic bioreactors (R3A and R9A) were acclimated with $NH_4^+$-N (10-250). Effluent $NH_4^+$-N concentration was monitored during acclimation period in aerobic bioreactors and shown in Figure 8.2(c). Up to feed $NH_4^+$-N concentration of 20 mg/L, complete removal of ammonia-nitrogen was achieved. When influent $NH_4^+$-N concentration was increased to 250 mg/L, effluent $NH_4^+$-N concentration was 9-15 mg/L with removal efficiency 94-96 per cent.

**Table 8.1: Average Performance of Three-Stage Continuous System**

| *Sl.No.* | *Parameter* | *Anaerobic (R1)* | | *Anoxic (R2)* | | *Aerobic (R3)* | | *Total Removal (%)* |
|---|---|---|---|---|---|---|---|---|
| | | *Influent (mg/L)* | *Effluent (mg/L)* | *Influent (mg/L)* | *Effluent (mg/L)* | *Influent (mg/L)* | *Effluent (mg/L)* | |
| 1. | Phenol | 1000 | 710 | 355 | 74 | 74 | 0 | 100 |
| 2(a). | Ammonia- | 200 | 159 | 88 | 71 | 71 | 17 | 91.5 |
| 2(b). | nitrogen | 250 | 227.4 | 116.4 | 99 | 99 | 5.4 | 97.8 |
| 3. | COD | 2500 | 2112 | 1124 | 676 | 676 | 136 | 94.5 |
| 4. | pH | 7.2 | 7.06 | - | 7.9 | 7.9 | 6.9 | - |
| 5. | Biogas (mL/day) | - | 326 | - | - | - | - | - |
| 6. | Nitrate-nitrogen | 0 | 0 | 24 | 2 | 2 | 48 | - |

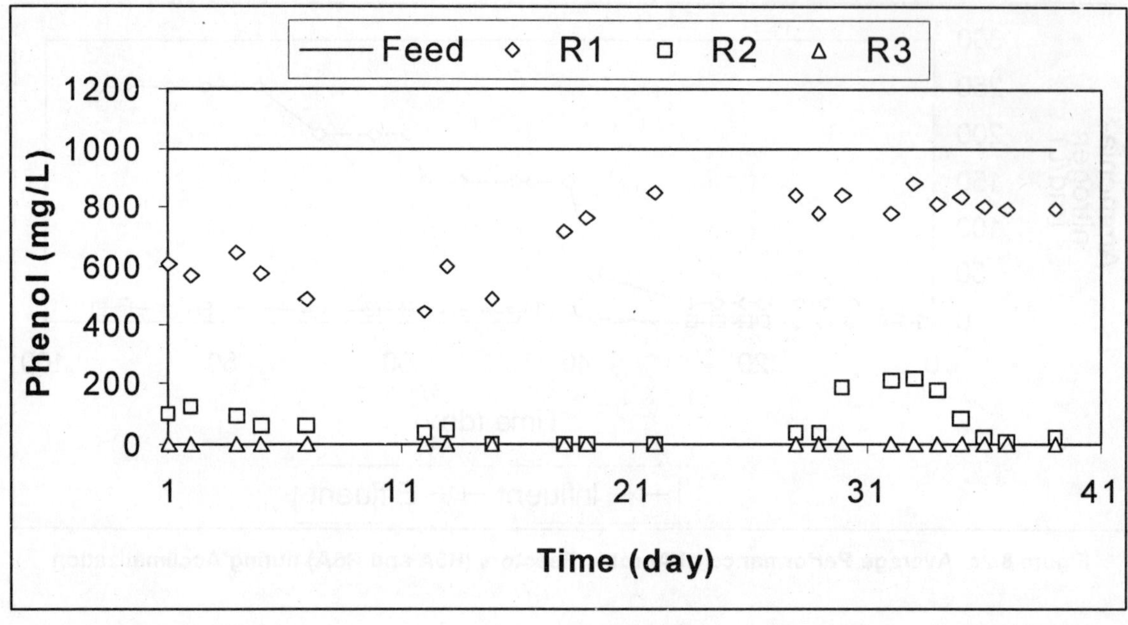


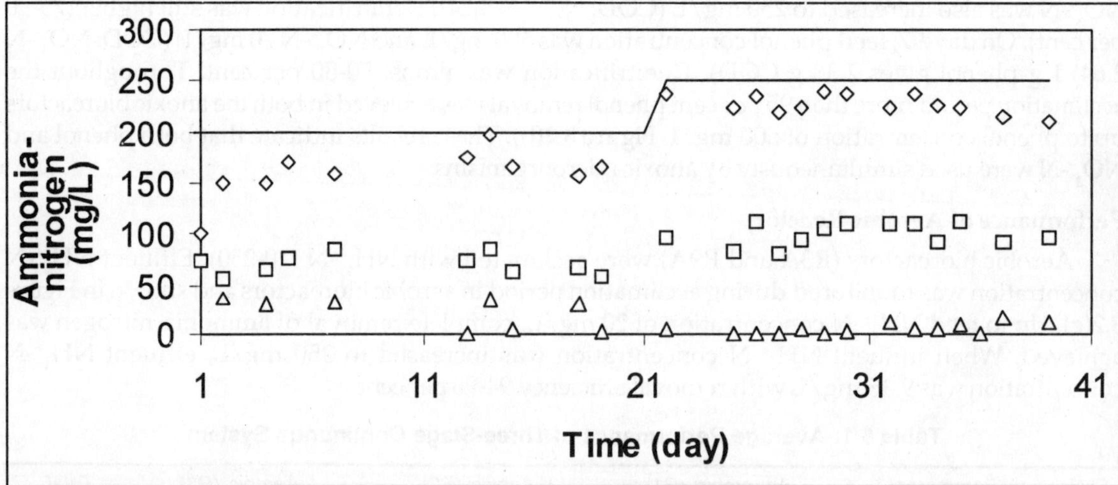


**Figure 8.3: Phenol and Ammonia-Nitrogen Removal in Three-Stage Continuous System**

## Performance of Three-Stage System

Three-stage system is fabricated and continuous operation is started with the acclimatized cultures of anaerobic, anoxic and aerobic reactors. It is operated with feed phenol concentration of 1000 mg/L and $NH_4^+$-N of 250 mg/L. Average performance of the three-stage system is shown in Table 8.1. Performances of the continuous system on phenol and ammonia-nitrogen removal are shown in Figure 8.3.

Table 8.1 shows that phenol was completely removed in three-stage system. Influent phenol loading in R1 was 0.5 kg/m³.day and phenol removal efficiency was almost 29 per cent. In R2, influent phenol loading was 0.17 kg/m$^3$ day of phenol and nearly 79 per cent of phenol was removed. In aerobic reactor residual phenol was removed completely. Ammonia-nitrogen removal efficiency was almost 91-97 per cent (Table 8.1). In R2, 91 per cent of nitrate-nitrogen was removed. In R1, average biogas production was 326.3 mL/day. Biogas production from phenol at 30° is 0.92 L/g of phenol removed. Therefore, 0.38g/day of phenol was consumed for biogas generation. Total phenol removed in R1 was 2.08 g [=7.2 L/day (1000-710 mg/L)]. Hence, 18 per cent of total phenol removed in R1 was used for biogas generation and remaining 82 per cent was used for biomass synthesis.

## Conclusion

In the present work biodegradation of synthetic wastewater containing phenol and ammonia-nitrogen was studied in an attached growth three-stage system. Sponge surface was used as the inert media for growth of biomass. Cultures were acclimatized in separate batch reactors under anaerobic, anoxic and aerobic environments on sponge media. A three-stage attached-growth continuous reactor assembly was constructed using PVC column and reactors were operated at total HRT of 4 days. Phenol was completely removed in three-stage system from influent concentration of 1000 mg/L. Influent phenol loading in R1 was 0.5 kg/m$^3$.day and phenol removal efficiency was almost 29 per cent. In R2, influent phenol loading was 0.17 kg/m$^3$.day of phenol and nearly 79 per cent of phenol was removed. In aerobic reactor residual phenol was removed completely. Ammonia-nitrogen removal efficiency was almost 91-97 per cent. It was calculated that 18 per cent of total phenol removed in R1 was used for biogas generation and remaining 82 per cent was used for biomass synthesis.

## Acknowledgement

Dr. Saswati Chakraborty would like to thank Ministry of Environment and Forest (MoEF), Govt. of India (No. 19-18/2003-RE) for financial support to this work.

## References

1. APHA, WEF, AWWA, (1998), "Standard Methods for the Examination of Water and Wastewater", 20th ed., APHA, Washington, DC, USA.
2. Chakraborty, S. and Veeramani, H. (2005)," Response of pulse phenol injection on an anaerobic–anoxic–aerobic system". Bioresource Technol., 96, 761–767.
3. Chakraborty, S. and Veeramani, H. (2006), "Effect of HRT and recycle ratio on removal of cyanide, phenol, thiocyanate and ammonia in an anaerobic–anoxic–aerobic continuous system", Process Biochem., 41, 96–105.
4. Liu, J., Baozhen, W., Li, W., Chengji, J., Cao, X. and Wang, L. (1996), "Removal of nitrogen from coal gasification and coke plant wastewater in anoxic-aerobic submerged biofilm- activated sludge (SBF-AS) hybrid system", Water Sci. Technol., 34, 17-24.
5. Luthy, R.G. (1981), "Treatment of coal coking and coal gasification wastewaters", J. Water Pollut. Control Fed., 53, 325-340.
6. Melcer, H. and Nutt, S. G. (1988), "Nitrogen control of complex industrial wastewater", J. Environ. Engrg., 114, 166-178.
7. Pandey, R. A., Parhad, N. M. and Kumaran, P. (1991), "Biological treatment of low temperature carbonization wastewater by activated sludge-process-a case study", Water Res. 25, 1555-1564.
8. Zhang M, Tay J.H., Qian Y. and Gu X.S. (1998), "Coke plant wastewater treatment by fixed biofilm system for COD and $NH_3$-N removal", Water Res., 32, 519-527.

# Chapter 9

# Anodic Oxidation of Recalcitrant Organics in Tannery Wastewater on Carbon Anode

**N.N. Rao[1*], Priyanka Joshi[2], Rakesh Lakhena[1], Rohit Misra[1], T. Nandy[1] and S.N. Kaul[2]**

**[1]Wastewater Technology Division, National Environmental Engineering Research Institute, Nagpur – 440 020, India**

**[2]MAEER'S MIT College of Engineering, Kothrud, Pune – 411 038, India**

## ABSTRACT

Leather industry produces large quantities of wastewater with high pollution levels, which has to be treated before disposal. In this study, an electrochemical treatment method that involved oxidation of tannery wastewater at 3D electrodes was explored. The treatment of the wastewater by the electrochemical method using 3D carbon electrodes in an electrochemical reactor comprising of a typical 2 electrode assembly was investigated with the aim of satisfactory elimination of colour and COD removal from tannery wastewater. The effects of pH and current were studied in detail. 3D carbon anodes were proved to have the electro catalytic properties for

* Corresponding Author: E-mail: nn_rao@neeri.res.in

colour and COD removal. Pre-adjustment of pH of tannery wastewater to 4 and applying a current of 0.5A is found adequate to achieve 70-80 per cent colour removal and 50-60 per cent COD removal. The results suggest that the method is a powerful application tool for reducing colour as well as COD from tannery wastewater.

***Keywords:*** *Anodic oxidation, Carbon anode, Recalcitrant organic compounds, Tannery wastewater.*

## Introduction

The tanning industry, one of the leading economic sectors in many countries generates large quantities of heavily polluted wastewater. Tannery wastewater owes its pollution load to a massive presence of various organic substances. However, the characteristics of tannery wastewater vary widely depending on the nature of the adopted tannin process, the amount of water used, the process of hide preservation, the hide processing capacity and the in plant measures followed to reduce pollution. Treatment of this wastewater by conventional biological methods is often inadequate to remove pollutants completely, especially ammonia and tannins (Ramaswamy and Rao, 1991), the latter being characterized by low biodegradability, common in polyphenolic compounds. In contrast, electrochemical techniques are becoming more reliable and are gaining popularity for the treatment of industrial wastewater. In recent years, there has been increasing interest in the use of electrochemical methods for treatment of wastewater. The electrochemical treatment of organic compounds in wastewater is potentially a powerful method of pollution control. Electrochemical methods have been successfully applied in the purification of wastewater from tannery industries. The reduction in colour content and COD in tannery wastewater can be achieved by employing electrochemical oxidation method.

Electrochemical techniques when applied for wastewater are generally being operated at high cell potentials, and thus the anodic process occurs in the potential region of water discharge. In the anodic discharge of water, adsorbed hydroxyl radicals (.OH) can be generated which are strong oxidizing agents of the majority of the organic pollutants (Simond *et al.*, 1997).

The electrochemical systems mainly comprise of 2 electrodes immersed in a suitable electrolyte. One of the electrodes is anode and the other is the cathode, depending upon the polarity connections given to D.C power supply. When requisite current and potential is applied to the system, redox reactions take place at the electrode.

Electrochemical reactors have been successfully used since the early 60's for the recovery of pollutants from metal recovery (Szpyrkowicz *et al.*, 1994, Szpyrkowicz *et al.*, 1995). Among various applications of electrochemical processes, those in which oxidation of pollutants can be achieved is particularly important, as it can offer an alternative to traditional chemical oxidation.

## Material and Methods

The tannery effluent used in the study was collected from a common effluent treatment plant (CETP) located in Tamil Nadu. The wastewater was collected from a number of tanneries that process rawhide to semi-finish stage as well as finish stage which was being treated at the CETP. The tannery wastewater sample was collected from secondary clarifier effluent outlet. Major fraction of biodegradable organic compounds have been treated in activated sludge process at CETP leaving behind non-biodegrable organic compounds.

## Electrochemical Assembly

The electrochemical assembly used for the treatment of tannery wastewater comprised of a direct current power supply, a magnetic stirrer, a 3D carbon anode and cathode of the size 10 cm × 10 cm and 1.3 cm thick and an undivided electrochemical reactor (1500 ml).

## Experimental Studies

Experiments were performed for the treatment of tannery wastewater using an electrochemical assembly wherein 1lit of tannery wastewater was taken in an electrochemical vessel. 3D carbon electrodes (anode and cathode were dipped in it. The electrochemical reaction assembly was connected to a direct current supply and the reaction medium was continuously stirred using a magnetic stirrer. A current of 0.5A was supplied for 1hr for the electrochemical reaction to take place. Samples in the interval of 10 min were collected and analyzed for pH, colour and COD. Similarly investigations were carried out to study effects by varying the current supply to 1.5A. Similar procedures were carried out to study the effects by varying the pH to 4, 7.5 (natural pH of tannery wastewater) and 10.0.

## Analyses

The pH of the test samples was measured using a pH meter (720A, Orion). The colour removal of the samples was measured using a Data Logging Spectrophotometer (Hach). For the measurement to be reproducible within ± 5 per cent Pt-Co units at 455nm, the spectrophotometer was standardized using Pt-Co standard consisting of potassium hexachloroplatinate and cobalt nitrate. The chemical oxygen demand (COD) of the sample was determined using a COD reactor (Hach) and Hach Portable spectrophotometer according to closed reflux method. An attempt was made to identify organic compounds in secondary clarifier effluent using GC-MS.

## Results and Discussion

The physicochemical characteristics of tannery wastewater collected from secondary clarifier outlet are given in Table 9.1. The effluent has pH 7.5-7.8. The concentration of suspended solids is high (100 ± 10 $mgl^{-1}$). It has high TDS in the range 10-12 $gl^{-1}$ that also gives rise to higher electrical conductivity obviating the need for addition of supporting electrolyte. It also contained chloride in sufficiently higher concentration (5000 ± 100 $mgl^{-1}$) thereby providing means of indirect oxidation. The concentration of COD and BOD are within the Standards set forth for treated effluent disposal in India. The filtered effluent still possess brown colour (300 ± 50 Pt-Co Units).

**Table 9.1: Physico-chemical Characteristics of Secondary Clarifier Effluent (CETP, Tannery Wastewater)**

| Sl.No. | Parameter | Magnitude |
|---|---|---|
| 1. | pH | 7.5-7.8 |
| 2. | Colour, Pt-Co units (PCU) | 300 ± 50 (Brown) |
| 3. | Electrical Conductivity (mS $cm^{-1}$) | 11-12.5 |
| 4. | Suspended Solids, ($mgl^{-1}$) | 100 ± 10 |
| 5. | Total Dissolved Solids ($gl^{-1}$) | 10-12 |
| 6. | Chloride ($gl^{-1}$) | 5.0-5.5 |
| 6. | Chemical Oxygen Demand ($mgl^{-1}$) | 200 ± 25 |
| 7. | Biochemical Oxygen Demand ($mgl^{-1}$) | 30 ± 10 |

The treatment of the tannery wastewater using an electrochemical system was evident through the loss of colour in the test solutions. The reaction accompanied only slight change in the pH *i.e.* 6.8 vis-à-vis a pH of 7.5 of the untreated effluent. The observations which were common to all the experiments conducted in the study were water electrolysis leading to gas evolution at anode and cathode, smell of chlorine gas, foam (froth) collection above the liquid level in the reactor and colour removal. The treated effluents became slightly acidic compared to their initial pH (DpH = 1.0-1.5) and conductivity also increased by 3-4 $mscm^{-1}$. There was accumulation of dirt on the carbon anode surface and off-white salty deposit on cathode at the end of electrolytic treatment.

Tannery wastewater contained several organic compounds as identified by GC-MS. The various major compounds and corresponding retention times are given in Table 9.2. Several of these compounds were absent in the electrochemically treated wastewater, notably the low molecular weight compounds. Some higher molecular weight compounds appear to resist oxidation and some new compounds form under the reaction conditions.

**Table 9.2: GC-MS Identification of Organic Compounds Present in Secondary Clarifier Effluent Before and After Electrochemical Treatment**

| Sl.No. | Before | | After | |
|---|---|---|---|---|
| | RT | Compound | RT | Compound |
| 1. | 2.11s | Propinoic acid 3-(butylthio) | – | – |
| 2. | 3.01s | 1H-Imidazole | – | – |
| 3. | 6.01m | 2-hexanol dimethyl | – | – |
| 4. | 7.50m | 1,3-propane diol-2,2-diethyl | – | – |
| 5. | 5.56m | Phenol ($C_6H_6O$) | – | – |
| 6. | 16.60s<br>18.85s | Dioctyl Phenol ($C_{22}H_{38}O$) | – | – |
| 7. | 17.05s | Dibutyl Phthalate ($C_{16}H_{38}O$) | 17.05 | Dibutyl Phthalate ($C_{16}H_{38}O$) |
| 8. | – | – | 19.02 | Octadecenoic acid |
| 9. | – | – | 19.82 | Piperidinol 1,6-dimethyl |

## Effect of pH

The pH of the electrolyte is known to influence the rate of electro-oxidation. Moreover the pH adjustment can also lead to competitive reactions such as enhanced water electrolysis, enhanced deposition of metals at cathode and other reactions. Leaving apart changes in the electrolyte, pH also manifests in changing the electrode potential, *e.g.* 3D anode/electrolyte interface. Therefore, the effect of pH on the treatment of tannery wastewater was examined for different pH *i.e.* 4.0, 7.5 (natural pH of secondary clarifier effluent) and 10.0.

## Effect of pH on Colour and COD Removal

The influence of colour removal is depicted in the Figure 9.1(a-d). The $-\log[(\text{colour})_{t}/(\text{colour})_0]$ vs time plots are shown in the Figure 9.1(b). The reduction in colour was found to be maximum at pH 4.0 (>80 per cent removal). With respect to the colour removal, pH 10 was only slightly better than pH 7.5. However, the colour reduction was very slow at pH = 7.5. The Figure 9.1(c) illustrates the influence of pH on the trend of COD removal from tannery wastewater. It is clear that the acidic pH (4.0) is more effective than the natural pH of the secondary clarifier effluent. The COD reduction was not significantly influenced at pH 10.0. It was observed that the COD removal was initially fast, then reached a plateau and then started to increase as a function of treatment time. This could be attributed to anode fouling due to the deposition of polymeric material as the formation of stable intermediates such as polyphenols.

The kinetic plots displayed in the Figure 9.1(b) indicate that the colour removal reaction can be approximately described as pseudo first order type with respect to colour concentration. This may be explained by the following equation;

$$-d[\text{colour}]/dt = k[\text{colour}]$$

At higher pH (10.0), the reaction does not obey expected linearity. The corresponding k values are shown in the Table 9.3.

**Table 9.3: Kinetic Constants for Colour and COD Removal at Different pH**

| *pH* | $k_{colour}$, *(min⁻¹)* | $k_{COD}$, *(min⁻¹)* |
|---|---|---|
| 4.0 | 0.0125 | 0.0068 |
| 7.5 | 0.0066 | 0.0037 |
| 10.0 | 0.0048 | 0.0029 |

The kinetic constants for the electro-oxidative removal of COD are depicted in the Table 9.3. Th $-\log[(COD)_t/(COD)_o]$ vs time plots as shown in the Figure 9.1(d) are fairly linear ($R^2 = 0.98$) at pH 7.5, whereas the linearity is unsatisfactory at pH 4 ($R^2 = 0.674$) and pH 10.0 ($R^2 = 0.77$). The deviation of the data from the expected trend may be attributed to the simultaneous occurrence of more than one reactions facilitated by the $H^+$ or the $OH^-$. The pseudo first order rate constants (k) are as shown in the Table 9.3. The rate expression can be explained as;

$$-d[COD]/dt = k[COD]$$

## Effect of Current

Anode current is yet another parameter that strongly controls the rate of the reaction in any electrochemical reactor. In the present study, experiments at current 0.5A and 1.5A were conducted and the changes in colour and COD were measured as a function of reaction time. The study was carried out at pH 4.0 as well as 7.5.

## Effect of Current on Colour Removal

The colour removal obtained is given in the Figure 9.2 (a-b). There is a slight increase in colour removal percentage at higher current. However the overall colour removal within 40 min. is similar at both the currents. The corresponding rate constants are shown in Table 9.4. Similarly, at pH 4 also the rate of colour removal is initially the same within 30 min. However, the overall percentage of colour removal was observed to be 74.8 per cent at 1.5A as against 64.7 per cent at 0.5A. In general, the rate constants were higher at high current and at low pH.

## Effect of Current on COD Removal

The effect of current on COD removal is shown in the Figure 9.2(c-d). It is observed that applying higher current is not particularly useful at least initially. Only at 50-60 min, did the COD removal exceed the value at a lower current. Thus, 71.5 per cent COD removal was obtained at 1.5A in a period of 1h as compared to 50 per cent at 0.5A current. The corresponding k values are shown in the Table 9.4. It can be inferred that the COD removal was not significantly improved at higher current.

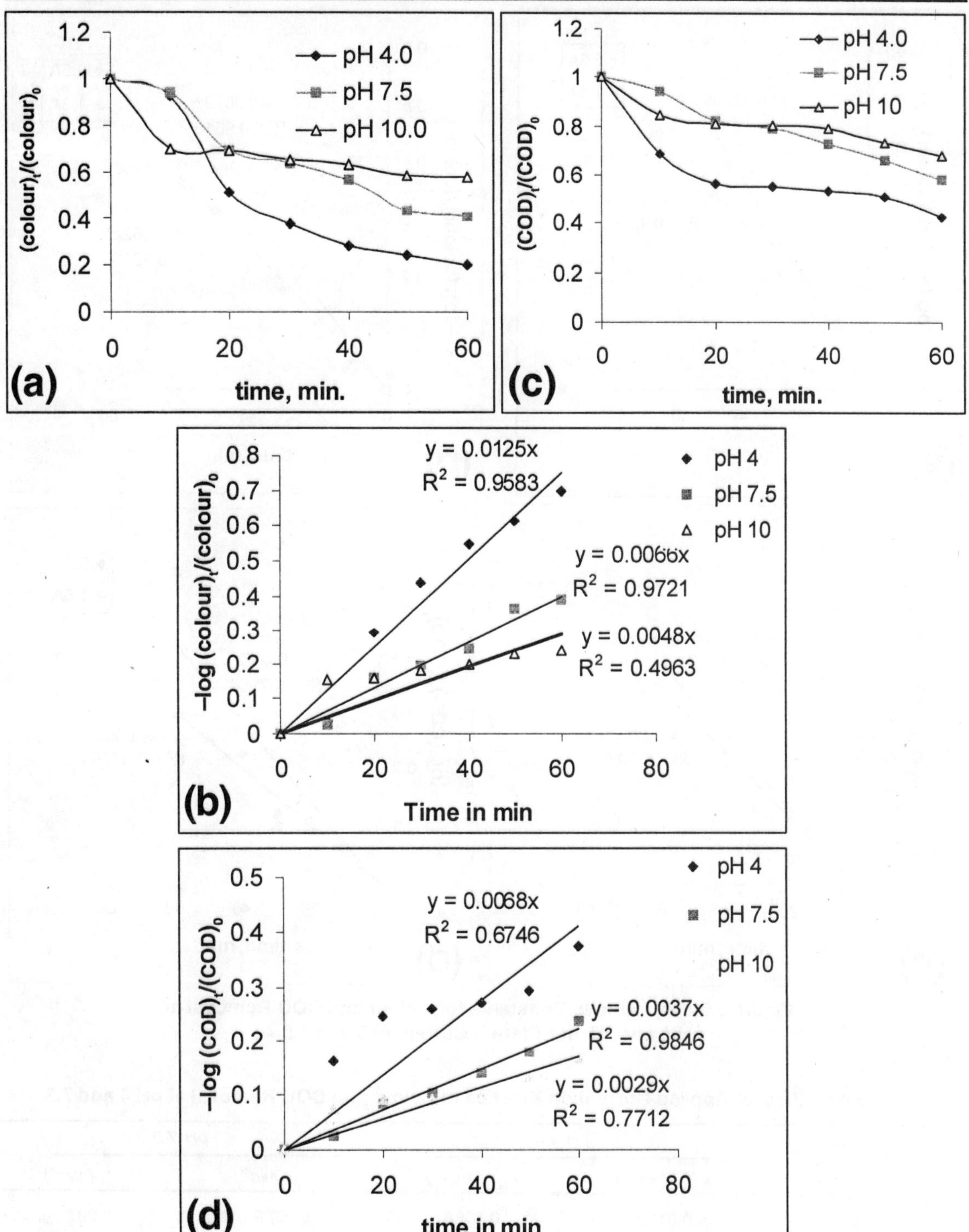


**Figure 9.1(a–d): Trend of Colour and COD Removal and Pseudo First Order Plots at Different pH**

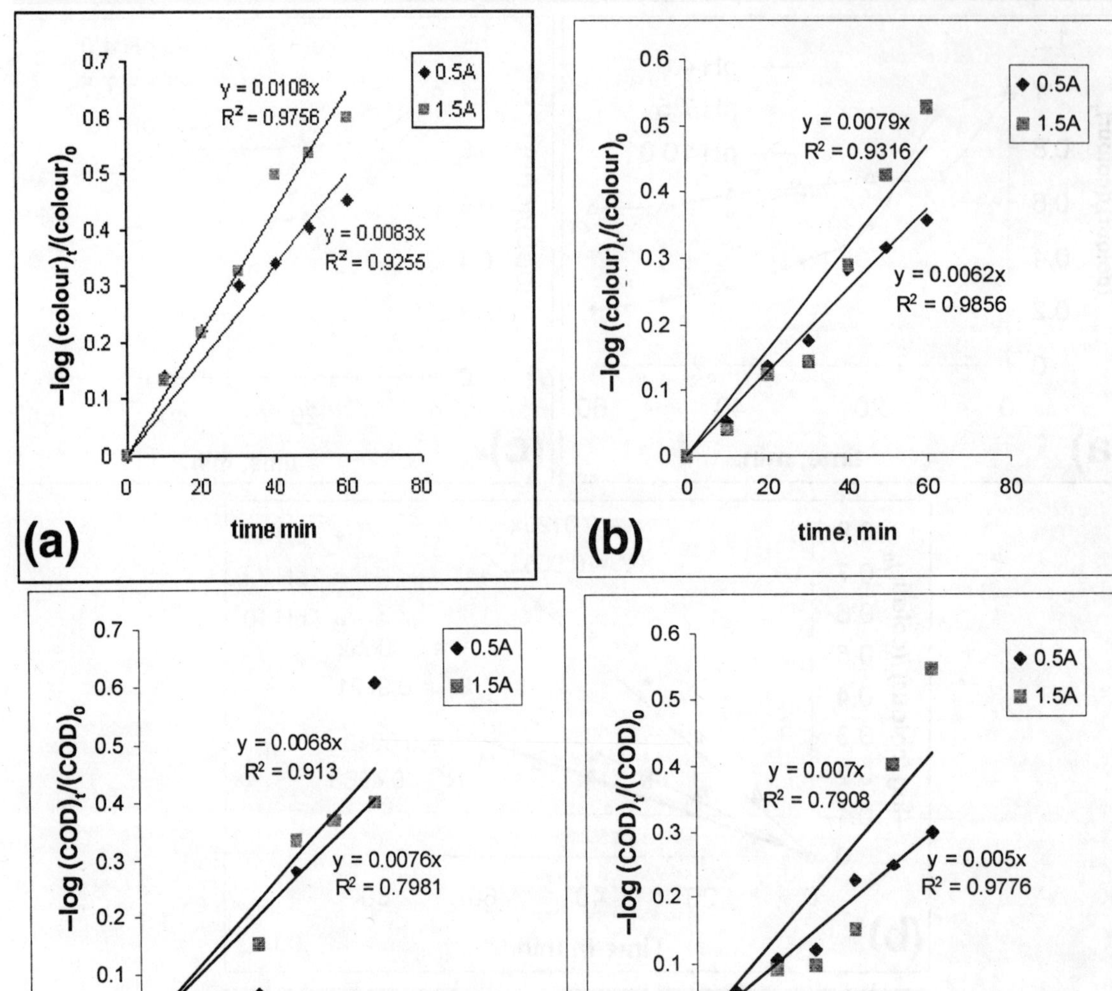


**Figure 9.2(a–d): Kinetic Constants for Colour and COD Removal at pH 4 and 7.5 at Different Current (0.5 and 1.5 A)**

**Table 9.4: Effect of Applied Current on Kinetics of Colour and COD Removal at pH 4 and 7.5**

| *Current* | *pH 4.0* | | *pH 7.5* | |
|---|---|---|---|---|
| | $k_{colour}$, *(min⁻¹)* | $k_{COD}$, *(min⁻¹)* | $k_{colour}$, *(min⁻¹)* | $k_{COD}$, *(min⁻¹)* |
| 0.5 A | 0.0108 | 0.0068 | 0.0079 | 0.007 |
| 1.5 A | 0.0083 | 0.0076 | 0.0062 | 0.005 |

## Conclusions

Tannery wastewater is a complex wastewater with several pollutants present in it. The electrochemical method as described in the study can be efficiently used to treat the wastewater for reducing colour and COD. As a general approach, maintaining a lower pH, say pH 4 and lower current (0.5A) can ensure 70-75 per cent colour removal and 50-60 per cent COD removal. The reactions of colour reduction and COD removal follow first order kinetics. The method is applicable with moderate energy input and medium to high capital cost.

## Acknowledgements

The authors wish to thank DST, New Delhi for financial support and Dr. S. Devotta, Director, NEERI for encouragement.

## References

1. Ramasami, T., Rao, P.G., 1991. International Consultation Meeting on Technology and Environmental Upgradation in leather sector. New Delhi.
2. Simond O, Schaller V, Comninellis Ch. Theoritical model for the anodic oxidation of organics on metal electrodes. Electrochimicia Acta 1997; 42, 2009-2012.
3. Rajeshwar K. Ibanez J.G and Swain G.M. Electrochemistry and environment, J.Appl. Electrochemistry., 24 1077-1091 (1994).
4. Abdo M.S.E., Rasheed S and Al-Ameeri (1987). Anodic Oxidation of a direct dye in an electrochemical reactor. J. Enivron. Sci. Heath A22(1), 27-45.
5. Adams, R.N., Electrochemistry at Solid Electrodes, Marcel Dekker, New York, 1969.
6. Burstein GT, Barnett Cj, Kucernak AR, Williams KR. Aspect of the anodic oxidation of methanol. Catalysis Today 1997; 38(4): 425-437.
7. Carriere J., Jones and Broadbent A.D. (1993) Decolourization of textile dye solutions. Ozone Sci. Eng. 15, 189-200.
8. Comninellis Ch. Electrochemical treatment of wastewater containing phenol. IchemE 70 part B 1992; 219-224.
9. Fleszar B, Ploszynska J., An attempt to define benzene and phenol electrochemical oxidation mechanism. Electrochimicia Acta 1985; 30(1): 31-42.
10. Hwang, J.-Y., Wang, Y.-Y and Wan, C-C (1987). Electrolytic oxidation cuprocyanide electroplating wastewaters under different pH conditions, Journal of Applied Electrochemistry, 17, 684-694.
11. Krasnobrodko IG. Destruction treatment of wastewater from dyes Ed. Chimia, Leningrad 1988): 142-4.
12. Lin S.H and Chen M.L (1997) Treatment of textile wastewaters by electrochemical method, Wat. Res., 868-876.
13. Rao N.N., Szpyrkowicz., Satyanarayan S.,Influence of Anode Material on Electrochemical Oxidation for the treatment of Tannery Wastewater.

14. Rao N.N., Somakeshwar K.M., Kaul S.N., and Szpyrkowicz L., Electrochemical oxidation of tannery wastewater, J. Chem. Technol. Biotechnol., 76, 1124-1131, 2001.

15. Simond O, Schaller V, Comninellis Ch. Theoretical model for the anodic oxidation of organics on metal electrodes. Electrochimica Acta 1997; 42(13-14): 2009-2012.

16. Szpyrkowicz L, Naumczyk J, Zilio-Grandi F. Application of electrochemical processes for tannery wastewater treatment. Toxicological and Environmental Chemistry 1994; 44: 189-202.

# Chapter 10

# A Review on Aqueous Phase Catalytic Oxidation for Environmental Application

**B.N. Rai*, Kiran Singh and R.S. Singh**
**Department of Chemical Engineering and Technology,**
**Centre of Advanced Study, Institute of Technology, BHU,**
**Varanasi – 221 005, India**

## ABSTRACT

From environmental application point of view, catalytic water phase processes are relatively new area with tremendous potential in the near future. This article presents state of art review of applications of heterogeneous water phase catalyst for oxidation and hydrogenation of mainly organic pollutants within past few years. In the present article oxidation along with $TiO_2$ assisted photocatalysis processes have been included. The literature survey reveals that heterogeneous water phase catalytic oxidation has potential over conventional wastewater treatment technologies. New bimetallic catalysts and supports with higher surface area have improved catalytic efficiency in oxidation processes. It was also observed that some waste materials such as red mud as catalyst is a very progressive field.

## Introduction

A large portion of the waste generated by chemical and related industries is in the form of aqueous streams. In many cases, these aqueous streams contain organic compound in too small concentration so that economic recovery is not feasible, yet the concentrations are high enough to pollute the

* Corresponding Author: E-mail: bnrai.che@itbhu.ac.in

environment. During past few years, interest of catalytic water phase processes received importance among the researchers due to its potential to replace the conventional processes for waste treatment. Till date very limited applications of the catalytic water phase processes for waste treatment have been reported. However, ever tightening regulations in treatment of aqueous waste force to search novel methods in the near future. Many industrial waste streams are not suitable for biological processes due to their inherent toxicity, but their treatment by traditional non-catalytic chemical processes or by incineration may be too energy intensive. With adaptable catalyst, energy consumption of various well-known oxidation procedures *e.g.* supercritical water oxidation, wet peroxide oxidation, wet oxidation, and wet air oxidation of stable organics may be decreased (Larachi *et al.*, 2001). One interesting observation is advanced removal of nitrogen containing compounds in wet air oxidation when metal oxide catalyst is present (Deiber *et al.*, 1997). Quality problems of groundwater and freshwater along with better understanding of importance of drinking water quality have already generated several studies dealing with heterogeneous catalytic raw water treatment processes including oxidation and hydrogenating processes.

If catalyst in water phase process is dissolved then it can not possibly be utilized for oxidation and hydrogenation of wastewater because the separation of catalyst will be difficult. In addition, many active homogeneous catalysts, such as some metal salts are a potential environmental problem. Thus, heterogeneous catalytic procedures where the catalyst is in different phase and therefore more likely easy to separate. However, heterogeneous processes are usually more complicated to control. According to Turner (1981), the following five steps are needed to enable heterogeneous reaction: (i) diffusion of reactants to the surface, (ii) adsorption of reactants onto the surface, (iii) reaction on the surface, (iv) desorption of products off the surface, and (v) diffusion of products from the surface.

In adsorption and desorption, surface plays an important role therefore, appropriate support selection may have a remarkable effect on reaction rate. This is possibly a field of the most rapid development in heterogeneous catalysis during past few years. This article consists of publication concerning environmental heterogeneous catalytic water phase applications. All the most important processes, catalysts, and supports are listed.

In the gas phase, catalytic processes have established their strong position among all exhaust gas treatment. However, in water phase, the development has not been as rapid partly due to the difficulties in finding active and stable catalysts. In water phase, the economical facts typically force to use lower reaction temperatures than in gas phase often limits the use of catalysts.

Metal oxides (Cu, Mn, Co, Cr, V, Ti, Bi, and Zn) and various noble metals (Ru, Pt, Rh, Ir, and Pd) have traditionally been used as heterogeneous catalysts. Very recently, immobiization of active homogeneous catalyst to support has reached increasing interest in environmental liquid phase applications. This group of catalysts includes various organometallic compounds, such as "biomimetic" porphyrines and its derivatives, but also some metal salts (Pirkanniemi and Sillanpaa, 2002). For industrial application there are several characteristics needed for catalysts (i) high activity, (ii) resistance for poisoning and stability in prolonged use at elevated temperatures, (iii) mechanical stability and resistance to attrition, (iv) non-selectivity in most cases, and (v) physical and chemical stability in various conditions (Pirkanniemi and Sillanpaa, 2002).

It is noteworthy that due to leaching, a reaction believed heterogeneous might after all be homogeneous. According to Sheldon *et al.* (1998), many common heterogeneous liquid phase catalysts are not stable toward leaching under oxidizing conditions. Rigorous proof of heterogeneity can be obtained only by filtering the catalyst at the reaction temperature before completion of the reaction and testing the filtrate for activity (Sheldon *et al.*, 1998).

## Metal Catalysts

Metal catalysts are almost with no exception noble metals and were supported on metal oxide surface ($TiO_2$, $Al_2O_3$, $ZrO_2$ and $CeO_2$) or on activated carbon. Supporting noble metal catalyst on metal oxide surface has been reported to improve activity and stability of the catalyst (Matatov-Meytal and Sheintuch, 1998). Also catalyst pretreatment such as calcination, sulfurization or reduction is an important factor, which affects remarkably final catalytic activity and/or stability of the catalyst (Karpel Vel Leitner *et al.*, 2000). Qin *et al.* (2001) found activity in total organic carbon (TOC) reduction of most common noble metals in catalytic wet oxidation of chlorophenols is in following order: Pt > Pd > Ru whether the support is activated carbon or alumina.

## Metal Oxide Catalysts

Stability is one of the important physico-chemical properties of the metal oxides. Metals with unstable high oxidation state oxides ($\Delta H^o_{298}$ < 9.5 kJ/mol of O), such as Pt, Pd, Ru, Au, and Ag do not yield stable bulk oxides at moderate temperatures. Most of the commonly used metal oxide catalysts (Ti, V, Cr, Mn, Zn, and Al) have stable high oxidation state ($\Delta H^o_{298}$ > 15.5 kJ/mol of O). Fe, Co, Ni, and Pb belong to group with intermediate stability of high oxidation state oxides ($\Delta H^o_{298}$ = 9.5 – 15.5 kJ/mol of O). Electrical conductivity is another possible classification criterion for example, insulators have low electron mobility, and therefore their catalytic activity is generally poor. In addition, with some exceptions the n-type metal oxides are not catalytically active as oxidation catalysts. However, some of the non-conducting metal oxides are used as support in catalytic processes. According to Kochetkova *et al.* (1992), catalytic activity of metal oxide catalysts during oxidation of phenol shows the following order:

$$CuO > CoO > Cr_2O_3 > NiO > MnO_2 > Fe_3O_3 > YO_2 > Cd_2O_3 > ZnO > TiO_2 > Bi_2O_3$$

It is a well-known fact that metal oxides are usually less active catalysts than noble metals. However, in majority of applications metal oxides are more suitable since they are more resistant to poisoning. In addition, combining two or more metal oxide catalysts may improve non-selectivity and catalytic activity.

## Organometal Catalysts

Organometals, such as metalloporphyrins and phthalocyanines are very novel catalysts in environmental applications. Immobilized organometals have been studied as catalysts in degradation of chlorinated phenols (Meunier and Sorokin, 1997; Sanchez *et al.*, 1999; Pirkanniemi and Sillanpaa, 2002), carbon tetrachloride (Ukrainczyk *et al.*, 1995), aqueous sulfide solutions (Fischer *et al.*, 1997; Iliev *et al.*, 2000), and atrazine (Hequet *et al.*, 2000), organometals may have several advantages over other catalysts which include lower toxicity if leached to effluent during the catalytic treatment process and higher activity in some cases. The uses of immobilized organometals in environmental processes are not well studied to predict the importance in the future.

## Deactivation of the Catalyst

Rapid deactivation of the catalyst in economically reasonable operating conditions is one of the major drawbacks in utilization of heterogeneous liquid phase catalysts. Poisoning, sintering or leaching are the major cause of deactivation of heterogeneous catalyst. Deactivation is mainly dependent on the catalyst, support, and treated water. If halogen-, sulfur- or phosphorus-containing compounds are present in reaction media, possibility of poisoning of noble metals are very high. With proper metal oxide support, poisoning of noble metal catalyst might be reduced. For example, alkali- and alkaline-earth-metals are known suitable for that purpose. Hot and acidic medium promotes solubility of some metal oxides and therefore causes deactivation of heterogeneous catalysts by leaching of metal or metal oxide catalysts (Pirkanniemi and Sillanpaa, 2002).

## Catalyst Support

Usually supports are classified by their chemical nature to organic and inorganic supports. Whatever the support is, it plays an important role in immobilizing active catalyst. Principally, the support has three main functions: (i) to increase the surface area of catalytic material, (ii) to decrease sintering and to improve hydrophobicity and thermal, hydrolytic, and chemical stability of the catalytic material, and (iii) to govern the useful lifetime of the catalyst (Matatov-Meytal and Sheintuch, 1998). In the role of co-catalyst, support may also improve the activity of the catalyst. Presence of heterogeneous surface improves transfer of gaseous oxidant into the solution, which could also be understood as a function of the support (Cooper and Burch, 1999).

Catalytic activity may increase by increasing the surface area which can be achieved by reducing particle size, increasing porosity and applying appropriate support to the catalyst. Particle size cannot be reduced unlimitedly, since pressure loss in industrial size reactors would increase remarkably. By increasing the porosity the surface area of many common supports may be increased upto 700 $m^2/g$ (BET). However, controlling reaction conditions inside the particles is difficult and porosity may decrease selectivity in some cases (Matatov-Meytal *et al.*, 2001). With fibrous or woven support the active surface area of catalyst may be relatively high without any significant pressure loss (Andrews *et al.*, 1996; Matatov-Meytal *et al.*, 2000).

## Metal Oxide Supports

Metal oxide supports have been discovered to improve the stability of noble metal catalysts. The most commonly used metal oxide supports in environmental water phase applications are $TiO_2$ and $Al_2O_3$. However $CeO_2$ was also used by few researchers (Pak *et al.*, 1997).

## Silica

Silica is not widely used material to support metal catalysts as compared to metal oxides. On the other hand, some widely used commercial supports, such as ZSM-5 zeolite are silicate minerals. In immobilization of organometal catalysts silica was used by Fischer *et al.*, 1997 and Sanchez *et al.*, 1999.

## Zeolite

Zeolites are hydrous aluminum silicate minerals, whose molecules enclose cations of sodium, potassium, calcium, strontium, or barium, or a corresponding synthetic compound. Commercial metal-containing ZSM-5 zeolites were used in environmental liquid phase catalysis studies in oxidation as well as in hydrogenation of halogenated organics (Lin and Wang, 2000).

## Fibers

In groundwater denitrification and photocatalytic oxidation of water contaminants, glass- and carbon fibers, and woven fiber cloths have been studied by various researchers as supports. Advantages of fibrous supports are low pressure loss and pore diffusion resistance as well as transparency of UV light in photocatalytic applications as compared to pellet shaped catalysts (Matatov-Meytal *et al.*, 2000, 2001; Piscopo *et al.*, 2000). Matatov-Meytal *et al.* (2000) studied Pd impregnated glass fiber catalysts in denitrification of groundwater. Silica content of the fiber catalyst material was up to 93.5 wt. per cent and Pd content 0.05–0.2 wt. per cent. Surface area of glass fibers was up to 400 $m^2/g$.

Material was reported stable within pH interval of 4.5–8.5. However, changes in pH during denitrification destroyed the catalyst rapidly if pH was not adjusted. Using Pd–Cu instead of Pd supported on glass fiber improved nitrate reduction activity significantly (Matatov-Meytal *et al.*, 2001). Robert *et al.* (1999) and Piscopo *et al.* (2000) impregnated 2–10 wt. per cent of $TiO_2$ particles onto glass-fibers. The material was used for photocatalytic degradation of benzamide. The efficiency was found to be comparable with suspended $TiO_2$ photocatalyst. However, they expected that the optimal efficiency would occur under conditions different from the suspension case.

## Ceramic Materials

Ceramic materials and cement have been studied as longevity improving supports for metal oxide catalysts. Besides an improvement of mechanical strength also chemical and hydrolytic stability and hydrophobicity of oxidation catalyst are improved (Matatov-Meytal and Sheintuch, 1998). Chou *et al.* (2001) used brick grain as a support to a heterogeneous γ-FeOOH catalyst. γ-FeOOH was precipitated onto the surface of brick grain and used as heterogeneous Fenton's catalyst in hydrogen peroxide oxidation of benzoic acid. Because of reductive dissolution of γ-FeOOH below pH 4.4, the heterogeneous catalysis contributes primary to the oxidation at pH 4.4–7.0.

## Pillared Clay

Intercalation of polymeric species with natural clays converts it into a catalytically active material which other wise considered inactive. Abdellaoui *et al.* (1999) studied aluminum and Al–Cu pillared clays in catalytic hydrogen peroxide oxidation of phenol. Aluminum oxyhydroxide pillared clays were not catalytically active, but when the clay was pillared with mixed Al–Cu species the phenol conversion was reported to increase strongly (Abdellaoui *et al.*, 1999). Barrault *et al.* (2000a) found also [Fe–Al] pillared clays active in catalytic hydrogen peroxide oxidation of phenol with TOC reduction upto 80 per cent at 70 °C. It was also observed that leaching of active catalyst was only 0.2 per cent within 4 h of reaction time.

## Red Mud

Red mud is a solid residue of the caustic leaching of bauxite generated in Bayer process which mainly contain iron, titanium, and aluminum oxides (20 per cent, >10 per cent, and <10 per cent, respectively) with small but remarkable amount of silicon, calcium, and sodium oxides. Sulfided red mud was proved catalytically active in hydrogenation of pure organic compounds and hydrodechlorination of tetrachloroethylene, but it has also been studied as a support in catalytic wet oxidation of organic substances present in industrial wastewaters. The active phase of red mud is pyrrhotite, a non-stoichiometric sulfide, which is thermodynamically stable at temperatures above 200 °C. Since red mud is produced as a residue in large amounts and its storage is an environmental problem, finding use for it as a catalyst would be environmentally desirable (Pirkanniemi and Sillanpaa, 2002).

## Polymers

SDB co-polymer (Styrene divinyl benzene) was used by few researchers including Lee and Kim, 2000 and Huang *et al.*, 2001. Huang *et al.* (2001) used SDB as support for Pt catalyst in catalytic oxidative nitrate removal from wastewater. Since plain SDB is water repellent, it is accessible for gaseous reactants. Lee and Kim (2000) used sulfonated SDB as catalyst in wet air oxidation of carboxylic acids.

### Carbon Black

Superior adsorptivity of various (halogenated) organics is a great advantage of carbon black as a support over other support materials. Since most heterogeneous catalytic reactions are diffusion limited, high adsorptivity of support favours high catalytic activity of supported catalyst. However, organic support is not suitable for all purposes.

The regeneration temperature is lower in metal support activated carbon. In that case, wastewater treatment can be performed at low temperature, and the degradation of adsorbed compounds occurs when activated carbon is regenerated. Presence of an active catalyst enables use of relatively low regeneration temperature (200–350°C) (Sheintuch and Matatov-Meytal, 1999). Major advantage in this procedure is energy savings, since there is no need to heat large volumes of wastewater to temperature high enough to activate metal or metal oxide catalysts.

## Oxidation Processes

For conventional non-catalytic water phase oxidation processes long reaction time, relatively high temperature (200 °C) and pressure (70–250 atm) are essential (Matatov-Meytal and Sheintuch, 1998). When heterogeneous catalysts utilized, reaction conditions in many cases can be turned milder. Air or oxygen are generally used as an oxidant in environmental catalytic water phase processes. However, hydrogen peroxide and ozone are more common. The low solubility of oxygen in water and the slow rate of air-to-liquid mass transfer limit use of compressed air. Increase of pressure enhances solubility of oxygen, but it also increases running costs (Matatov-Meytal and Sheintuch, 1998). It must be noted that when expensive reactants such as $H_2O_2$ or $O_3$ are used the running costs are higher and therefore such processes should not replace more economic treatments as the biological degradation if the toxicity of the wastewater does not prevent using them. However, many industrial waste streams are too toxic to be treated biologically and this is where catalytic oxidation (or hydrogenation as well) is economically most favourable.

In drinking water treatment, heterogeneous catalytic ozonation was reported remarkably more efficient in TOC removal than plain ozonation with otherwise similar conditions. In the presence of heterogeneous catalyst ($CuO/Al_2O_3$), TOC removal (humic substances) was approximately 50 per cent within 10 min at 20 °C, without catalyst only about 15 per cent (Legube and Karpel Vel Leitner, 1999). Cooper and Burch (1999) studied heterogeneous catalytic ozonation of halogenated hydrocarbons from groundwater. Conversion of chlorinated phenols was within 30 min of contact time complete in the presence of alumina or transition metal alumina catalyst. Carbonate is known to act as radical scavenger and though cause an inhibition (Andreozzi *et al.*, 1999). The equations are as under:

$$HCO_3^- + HO^\bullet \rightarrow CO_3^{\bullet -} + H_2O \tag{1}$$

$$CO_3^{2-} + HO^\bullet \rightarrow CO_3^{\bullet -} + OH^- \tag{2}$$

Tables 10.1 and 10.2 list heterogeneous catalytic oxidation of phenol and phenolic compounds and non-phenolic compounds respectively.

### Heterogeneous Fenton's Reaction

It is justified to approach Fenton's reaction separately from other metal or metal oxide catalyzed oxidation reactions due to the differences in catalytic method. Iron or copper ions catalyze dissociation of hydrogen peroxide to hydroxyl radical and hydroxide ion [Equation (3) or to hydrogen ion and hydroperoxy radical (Equation (4)] (Centi *et al.*, 2000):

**Table 10.1: Heterogeneous Catalytic Oxidation Studies of Phenol and Phenolic Compounds**

| *Compounds* | *Catalysts* | *Supports* | *Oxidant* | *Reaction Temperature (°C)* | *References* |
|---|---|---|---|---|---|
| Phenol | Al-Cu | Pillared clay | $H_2O_2$ | 20 | Abdellaoui *et al.*, 1999 |
| Phenol | Al-Fe | Pillared clay | $H_2O_2$ | 25-70 | Barrault *et al.*, 2000a,b |
| Phenol | $MnO_2/CeO_2$ | – | Air | 80 | Larachi *et al.*, 2001 |
| Phenol | EX-1144.8 (Cu, Zn, Co, and Al oxides) | Cement | $O_2$ | 130-160 | Pintar *et al.*, 1997 |
| Phenol | CuO, $Al_2O_3$, ZnO, $Ni_2O_3$ | – | $H_2O_2$ | 29.5 | Drijvers *et al.*, 1999 |
| Phenol | $MnO_2$, $TiO_2$, $CuO/Al_2O_3$ | – | $O_2$ | 400 | Yu and Savage, 2000, 2001 |
| Phenol | Fe-ZSM-5 | – | $H_2O_2$ | 100 | Fajerwerg *et al.*, 2000 |
| Phenol | CuO | $\gamma$-$Al_2O_2$ | Air | 120-160 | Fortuny *et al.*, 1999 |
| Phenol | Various Cu-catalysts | Various *e.g.* $\gamma$-$Al_2O_3$-silicate | $O_2$ | 140 | Santos *et al.*, 2001a |
| Phenol | Cu (Engelhard Cu-0203) | – | $O_2$ | 127-180 | Santos *et al.*, 2001b |
| Nitrophenol | Mn-Ce composite oxide precipitates | – | Air | 200-325 | Deiber *et al.*, 1997 |
| Phenolic wastewaters | Ni-Oxide | – | Air | 15-65 | Christoskova and Stoyanova 2001 |
| Drinking water: Chlorophenols | $Fe_2O_3$, $TiO_2$ | $Al_2O_3$ | $O_3$ | 20 | Cooper and Burch, 1999 |
| Phenol, *p*-halogenated phenols | $Fe_2O_3$, CuO, $Cr_2O_3$, CoO, NiO | Activated carbon | Air | 240-300 | Matatov-Meytal and Sheintuch, 1997 |
| *p*-Chlorophenol | Mn, Ce, Pt, Pd, Ru, Pt-Pd, Pt-Pd-Ce | $\gamma$-$Al_2O_3$, C | $O_2$ | 180 | Qin *et al.*, 2001 |
| 2-Chlorophenol | $\alpha$-FeOOH | $SiO_2$ | $H_2O_2$ | 20 | Huling *et al.*, 2000 |
| 2,4,6-Trichlorophenol | Metallophthalocyanines | Amberlite | $H_2O_2$, $KHSO_5$ | 20 | Sorokin and Meunier, 1994 |

**Table 10.2: Heterogeneous Catalytic Oxidation of Non-Phenolic Compounds**

| *Compounds* | *Catalyst* | *Support* | *Oxidizer* | *Temperature (°C)* | *References* |
|---|---|---|---|---|---|
| Acetonitrile, DMF, Luther wastewater | Ru | $\gamma$-$Al_2O_3$ | $O_2$ | 160-260, 130-200 | Dobrynkin *et al.*, 1999 |
| Acetonitrile, DMF, carbamide | $Co_3O_4$, $Fe_2O_3$, $Mn_2O_3$, Zn-Fe-Mn-Al-O, Pt, Ru | $\gamma$-$Al_2O_3$, $CeO_2$ graphite-like carbon, activated carbon | $O_2$ | 160-250 | Dobrynkin *et al.*, 1998 |
| Aniline, $\beta$-alanine, ammonia, nitrobenzene, pyridine, acrylonitrile | Mn-Ce composite oxide precipitates | - | Air | 200-325 | Deiber *et al.*, 1997 |
| Drinking water: halogenated hydracarbons | $Fe_2O_3$, $TiO_2$ | $Al_2O_3$ | $O_3$ | 20 | Cooper and Burch, 1999 |
| Carboxylic acid | $Fe^{3+}/Fe^{2+}$ | ZSM-5 | $H_2O_2$ | 25-70 | Centi *et al.*, 2000 |
| Dye wastewater | Fe | MgO | $H_2O_2$ | - | Pak and Chang, 1999 |
| Acetic acid | Pt | $Al_2O_3$ | $O_2$ | 200-250 | Klinghoffer *et al.*, 1998 |
| Pyridine | Pt, $MnO_2$, ($\alpha$-$Al_2O_3$ $\gamma$-$Al_2O_3$ | $\gamma$-$Al_2O_3$, $CeO_2$ | $O_2$ | 370 | Aki and Abraham, 1998, 1999a,b |
| Pulp bleaching wastewater | Pd-Pt-Ce | $\gamma$-$Al_2O_3$ | $O_2$ | 130-170 | Zhang and Chuang, 1999 |
| Rinse waters (*e.g.* NTA, thiourea) | Metal oxide grains | $TiO_2$ | $O_3$ | 20 | Luck *et al.*, 1997 |
| Rinse waters (*e.g.* NTA, thiourea) | Fe, Mn, Ni, Co, Zn, Ag, Cr, Cu | $Al_2O_3$, $TiO_2$, clay | $O_3$ | 20-40 | Legube and Karpel Vel Leitner, 1999 |
| Succinic acid | Ru | $CeO_2$ | $O_3$ | 25 | Delanoe *et al.*, 2001 |
| Succinic acid | Ru | $CeO_2$ | $O_3$ | 25 | Karpel Vel Leitner *et al.*, 2000 |
| Wastewater, benzoic acid | Fenton $\gamma$-FeOOH | Brick grain | $H_2O_2$ | 24 | Chou *et al.*, 2001 |
| Kraft bleaching wastewater | Ru | $TiO_2$, $ZrO_2$ | Air | 190 | Pintar *et al.*, 2001b,c |
| Various chlorinated hydrocarbons | | Ambersorb | Air | ≤250 | Drago *et al.*, 1994 |
| Trichloroethylene | CuO, $Al_2O_3$, ZnO, $Ni_2O_3$ | - | $H_2O_2$ | 29.5 | Drijvers *et al.*, 1999 |
| Carboxylic acids | Pt, SDB | $Al_2O_2$ | Air | 80 | Lee and Kim, 2000 |
| Nitrogen from wastewater | Pt | SDB | $O_2$ | 110-130 | Huang *et al.*, 2001 |

$$Fe^{2+} + H_2O_2 \rightarrow Fe^{3+} + OH^{\bullet} + HO^{\bullet} \tag{3}$$

$$Fe^{3+} + H_2O_2 \rightarrow Fe^{2+} + H^{+} + HOO^{\bullet} \tag{4}$$

Homogeneous Fenton's reaction is already in use in industrial wastewater purification processes (Centi *et al.*, 2000). Several studies have been published concerning supporting Fenton's catalyst to various carriers, such as brick grain (Chou *et al.*, 2001), MgO (Pak and Chang, 1999), $SiO_2$ (Huling *et al.*, 2000), and zeolite (Centi *et al.*, 2000). Also copper or iron containing pillared clays were used as Fenton's catalysts (Abdellaoui *et al.*, 1999; Barrault *et al.*, 2000 a,b).

## Semiconductor Photocatalysis

Electron hole pairs develop when aqueous solution of semiconductor photocatalyst, such as $TiO_2$ is excited with ultraviolet light ($\lambda$<380 nm) as shown in equation (5). These electron hole pairs have an oxidizing potential of 2.9 V versus NHE, which is enough to oxidize most pollutants present in aqueous waste streams (Hofstadler *et al.*, 1994):

$$TiO_2 + h\nu \rightarrow TiO_2\,(e^- + h^+) \tag{5}$$

The hole produced by irradiation reacts with water or surface-bound hydroxyl ion producing hydroxyl radical (Equation (6) and (7)) (Baird, 1997):

$$TiO_2\,(h^+) + H_2O \rightarrow HO^{\bullet} + H^+ \tag{6}$$

$$TiO_2\,(h^+) + OH^- \rightarrow HO^{\bullet} \tag{7}$$

Electron released by irradiation of photocatalyst combines with dissolved molecular oxygen, producing superoxide radical, $O_2^{\cdot}$ (Baird, 1997). Hydrogen peroxide possibly added, acts as an oxidant, but also as an $e^-$ scavenger instead of dissolved molecular oxygen (Equation (8)). $H_2O_2^{\cdot}$ dissociates to hydroxyl radical and hydroxide ion even easier than $H_2O_2$, due to an extra electron in it (Hofstadler *et al.*, 1994; Baird, 1997):

$$TiO_2(e^-) + H_2O_2 \rightarrow H_2O_2^{\cdot} \rightarrow OH^- + HO^{\bullet} \tag{8}$$

Other semiconductors that could be used as photocatalyst are $WO_3$, $SrTiO_3$, $\alpha$-$Fe_2O_3$, ZnO, and ZnS but generally not as attractive for environmental applications as $TiO_2$. $TiO_2$ is biologically and chemically inert, inexpensive and resistant to photocorrosion and chemical corrosion. For these reasons use of other semiconductors in environmental studies is rather limited (Hoffmann *et al.*, 1995). The major drawbacks in photocatalytic degradation of wastewaters are the need of transparency of treated wastewater at the spectral region semiconductor absorbs and the slow complete mineralization in some cases (Herrmann *et al.*, 1993). Table 10.3 present a review of photocatalytic processes.

## Heterogeneous Catalytic Sonolysis

The combination of sonolysis at 520 kHz and oxidation with hydrogen peroxide using heterogeneous metal oxide catalysis was studied by Drijvers *et al.* (1999). According to them, probably no $HO^{\bullet}$, but only $HO_2^{\bullet}/O_2^{\bullet-}$ radicals were formed from ultrasound excited $CuO^-$ catalyzed decomposition of $H_2O_2$, which would explain similar degradation rates of sonicated TCE solutions with and without $H_2O_2$/CuO addition (Drijvers *et al.*, 1999).

**Table 10.3: Review of Photocatalytic Processes**

| *Compounds* | | *Catalyst* | *Spectral Range and Oxidant* | *References* |
|---|---|---|---|---|
| Dyes | Dodecylbenzene sulfonate, azymphos-methyl, dimethoate | $TiO_2$ and $FeCl_3$ | Solar light | Dominguez *et al.*, 1998 |
| | Special brillant blue | $TiO_2$ on hollow glass tubes | UV | Ray and Beenackers, 1998 |
| | Alizarin red | $TiO_2$ | Visible light | Liu *et al.*, 1999 |
| | Squarylum cyanine | $TiO_2$ | Visible light and $H_2O_2$ | Wu *et al.*, 2000 |
| | Rhodamine B, Orange II, Sulfo-rhodamine B, Flurescein, Alizarin red, Squarylum cyanine, Eosin | $TiO_2$ | Visible light | Wu *et al.*, 2000 |
| | Squarylum cyanine | $TiO_2$ | Visible light | Wu *et al.*, 1999a |
| | Textile wastewater | $TiO_2$ and $Fe^{3+}$ | UV and $H_2O_2$ | Balcioglu and Arslan, 1997 |
| | Methylene blue | $TiO_2$ | UV | Houas *et al.*, 2001 |
| Chlorinated organics and phenolic compounds | 4-Chlorophenol | $TiO_2$ and $SiO_2$ | UV and $H_2O_2$ | Hofstadler *et al.*, 1993 |
| | Tetrachlorethene, trichloroethene, *cis*-dichloroethene, (and toluene, ethylbenzene, xylene) | Pt-$TiO_2$ on ambersorb | UV | Suri *et al.*, 1999 |
| | 2,4-Dichlorphenol | Fenton on Nafton membrane | Visible light and $H_2O_2$ | Sabhi and Kiwi, 2001 |
| | Pentachlorophenol | $TiO_2$ (sol-gel) | UV | Pecchi *et al.*, 2001 |
| | 4-Chlorophenol | $TiO_2$ on silica giberglass | UV | Hofstadler *et al.*, 1994 |
| | Phenol and *ortho*-substituted phenolic compounds: 2-chlorophenol, guaiacol, catechol | $TiO_2$ | UV | Peiro *et al.* 2001 |
| Others | Salicylic acid, (formic acid) | $TiO_2$ | UV and sonolysis | Davydov *et al.*, 2001 |
| | Nitrotoluenes | $TiO_2$ | UV and $O_2$ | Kumar and Davis, 1997 |
| | Formic acid | Fe and $TiO_2$ | UV | Arana *et al.*, 2001 |
| | Benzamide | $TiO_2$ on fiberglass | UV | Robert *et al.*, 1999 |
| | Benzamide | $TiO_2$ on fiberglass | UV | Piscopo *et al.*, 2000 |
| | Dithiocarbamate (Vapam) | $TiO_2$ | UV | Vidal and Luengo, 2001 |

## Conclusion

The key issue of the effective catalyst oxidation of organics in wastewater is the catalyst. In water pollution abatement, solid catalysts offer a novel approach in liquid phase. These processes are capable to replace other application with higher energy consumption. However, in order to develop effective catalyst and process, more kinetics and mechanist studies in different systems in aqueous solutions are needed. Especially in photocatalytic processes research should be focused related to engineering aspects, such as specific reactor design and novel supports. Development of new bimetallic catalysts and high-porosity supports has promoted this field of science.

## References

1. M. Abdellaoui, J. Barrault, C. Bouchoule, N.F. Srasra and F. Bergaya, Catalytic oxidation of phenol by hydrogen peroxide in the presence of [Al–Cu] pillared clays. J. Chim. Phys. Phys.-Chim. Biol. 96 3 (1999), pp. 419–429.

2. R. Andreozzi, V. Caprio, A. Insola and R. Marotta, Advanced oxidation processes (AOP) for water purification and recovery. Catal. Today 53 1 (1999), pp. 51–59.

3. R. Andrews, E. Grulke and G. Kimber, Mechanical properties of porous carbon composites for environmental applications. Fluid/Part. Sep. J. 9 3 (1996), pp. 207–217.

4. S. Aki and M.A. Abraham, An economic evaluation of catalytic supercritical water oxidation: comparison with alternative waste treatment technologies. Environ. Prog. 17 4 (1998), pp. 246–255.

5. S. Aki and M.A. Abraham, Catalytic supercritical water oxidation of pyridine: comparison of catalysts. Ind. Eng. Chem. Res. 38 2 (1999a), pp. 358–367.

6. S. Aki and M.A. Abraham, Catalytic supercritical water oxidation of pyridine: kinetics and mass transfer. Chem. Eng. Sci. 54 15/16 (1999b), pp. 3533–3542.

7. J. Arana, O.G. Diaz, M.M. Saracho, J.M.D. Rodrnguez, J.A.H. Melián and J.P. Pena, Photocatalytic degradation of formic acid using $Fe/TiO_2$ catalysts: the role of $Fe_3^+/Fe_2^+$ ions in the degradation mechanism. Appl. Catal. B 32 1/2 (2001), pp. 49–61.

8. N.C. Baird, Free radical reactions in aqueous solutions: examples from advanced oxidation processes for wastewater and from the chemistry in airborne water droplets. J. Chem. Educ. 74 7 (1997), pp. 817–819.

9. I.A. Balcioglu and I. Arslan, Treatment of textile wastewater by heterogeneous photocatalytic oxidation processes. Environ. Technol. 18 10 (1997), pp. 1053–1059.

10. J. Barrault, M. Abdellaoui, C. Bouchoule, A. Majesté, J.M. Tatibouet, A. Louloudi, N. Papayannakos and N.H. Gangas, Catalytic wet peroxide oxidation over mixed (Al–Fe) pillared clays. Appl. Catal. B 27 4 (2000a), pp. L225–L230.

11. J. Barrault, C. Bouchoule, J.-M. Tatibouët, M. Abdellaoui, A. Majeste, I. Louloudi, N. Papayannakos and N.H. Gangas, Catalytic wet peroxide oxidation over mixed (Al–Fe) pillared clays. Stud. Surf. Sci. Catal. 130A (2000b), pp. 749–754.

12. G. Centi, S. Perathoner, T. Torre and M.G. Verduna, Catalytic wet oxidation with $H_2O_2$ of carboxylic acids on homogeneous and heterogeneous Fenton-type catalysts. Catal. Today 55 1/2 (2000), pp. 61–69.

13. S. Chou, C. Huang and Y.-H. Huang, Heterogeneous and homogeneous catalytic oxidation by supported γ-FeOOH in a fluidized-bed reactor: kinetic approach. Environ. Sci. Technol. 35 6 (2001), pp. 1247–1251.

14. S. Christoskova and M. Stoyanova, Degradation of phenolic wastewaters over Ni-oxide. Water Res. 35 8 (2001), pp. 2073–2077.

15. C. Cooper and R. Burch, An investigation of catalytic ozonation for the oxidation of halocarbons in drinking water preparation. Water Res. 33 18 (1999), pp. 3695–3700.

16. L. Davydov, E.P. Reddy, P. France and P.G. Smirniotis, Sonophotocatalytic destruction of organic contaminants in aqueous systems on $TiO_2$ powders. Appl. Catal. B 32 (2001), pp. 95–105.

17. G. Deiber, J.N. Foussard and H. Debellefontaine, Removal of nitrogeneous compounds by catalytic wet air oxidation. Kinetic study. Environ. Pollut. 96 3 (1997), pp. 311–319.

18. F. Delanoe, B. Acedo, N. Karpel Vel Leitner and B. Legube, Relationship between the structure of $Ru/CeO_2$ catalysts and their activity in the catalytic ozonation of succinic acid aqueous solutions. Appl. Catal. B 29 4 (2001), pp. 315–325.

19. Dobrynkin *et al.*, 1998. N.M. Dobrynkin, M.V. Batygina and A.S. Noskov, Solid catalysts for wet oxidation of nitrogen-containing organic compounds. Catal. Today 45 1-4 (1998), pp. 257–260.

20. Dobrynkin, N.M., Batygina, M.V., Noskov, A.S., 1999. Solid catalysts for wastewater treatment through liquid phase oxidation. In: Korus'99, Proc. Russ.–Korean Int. Symp. Sci. Technol., 3rd 2, pp. 487–490.

21. C. Dominguez, J. García, M.A. Pedraz, A. Torres and M.A. Galan, Photocatalytic oxidation of organic pollutants in water. Catal. Today 40 1 (1998), pp. 85–101.

22. R.S. Drago, S.C. Petrosius, G.C. Grunewald and W.H. Brendley, Jr., Deep oxidation of chlorinated hydrocarbons. ACS Symp. Ser. 552 (1994), pp. 340–352.

23. D. Drijvers, H. Van Langenhove and M. Beckers, Decomposition of phenol and trichloroethylene by the ultrasound/$H_2O_2$/CuO process. Water Res. 33 5 (1999), pp. 1187–1194.

24. K. Fajerwerg, T. Castan, J.-N. Foussard, A. Ferrard and H. Debellefontaine, Dependency on some operating parameters during wet oxidation of phenol by hydrogen peroxide with Fe–ZSM-5 zeolite. Environ. Technol. 21 3 (2000), pp. 337–344.

25. H. Fischer, G. Schulz-Ekloff and D. Wohrle, Oxidation of aqueous sulfide solutions by dioxygen part II: catalysis by soluble and immobilized cobalt (ii) phthalocyanines. Chem. Eng. Technol. 20 9 (1997), pp. 624–632.

26. A. Fortuny, C. Bengoa, J. Font, F. Castells and A. Fabregat, Water pollution abatement by catalytic wet air oxidation in a trickle bed reactor. Catal. Today 53 1 (1999), pp. 107–114.

27. V. Hequet, P. Le Cloirec, C. Gonzalez and B. Meunier, Photocatalytic degradation of atrazine by porphyrin and phthalocyanine complexes. Chemosphere 41 3 (2000), pp. 379–386.

28. J.M. Herrmann, C. Guillard and P. Pichat, Heterogeneous photocatalysis: an emerging technology for water treatment. Catal. Today 17 1/2 (1993), pp. 7–20.

29. M.R. Hoffmann, S.T. Martin, W. Choi and D.W. Bahnemann, Environmental applications of semiconductor photocatalysis. Chem. Rev. 95 1 (1995), pp. 69–96.

30. K. Hofstadler, G. Ruppert, R. Bauer, G. Heisler and S. Novalic, Photocatalyzed treatment of 4-chlorophenol with immobilized titanium dioxide. Trace Met. Environ. 3 (1993), pp. 777–782.

31. K. Hofstadler, Bauer, Rupert, S. Novalic and G. Heisler, New reactor design for photocatalytic wastewater treatment with $TiO_2$ immobilized on fused-silica glass fibers: photomineralization of 4-chlorophenol. Environ. Sci. Technol. 28 4 (1994), pp. 670–674.

32. A. Houas, H. Lachheb, M. Ksibi, E. Elaloui, C. Guillard and J.-M. Herrmann, Photocatalytic degradation pathway of methylene blue in water. Appl. Catal. B 31 2 (2001), pp. 145–157.

33. T.-L. Huang, J.M. MacInnes and K.R. Cliffe, Nitrogen removal from wastewater by a catalytic oxidation method. Water Res. 35 9 (2001), pp. 2113–2120.

34. S.G. Huling, R.G. Arnold, P.K. Jones and R.A. Sierka, Predicting Fenton-driven degradation using contaminant analog. J. Environ. Eng. 126 4 (2000), pp. 348–353.

35. V. Iliev, L. Prahov, L. Bilyarska, H. Fischer, G. Schulz-Ekloff, D. Wohrle and L. Petrov, Oxidation and photooxidation of sulfide and thiosulfate ions catalyzed by transition metal chalcogenides and phthalocyanine complexes. J. Mol. Catal. A Chem. 151 1/2 (2000), pp. 161–169.

36. N. Karpel Vel Leitner, F. Delanoe, B. Acedo and B. Legube, Reactivity of various Ru/$CeO_2$ catalysts during ozonation of succinic acid aqueous solutions. New J. Chem. 24 4 (2000), pp. 229–233.

37. A.A. Klinghoffer, R.L. Cerro and M.A. Abraham, Catalytic wet oxidation of acetic acid using platinum on alumina monolith catalyst. Catal. Today 40 1 (1998), pp. 59–71.

38. Kochetkova, R.P., Babikov, A.F., Shpilevskaya, L.I., Shiverskaya, I.P., Eppel, S.A., Shmidt, F.K., 1992. Liquid-phase catalytic oxidation of phenol. Khim. Tekhnol. Topl. Masel, (4), 31-4: Ref. Matatov-Meytal, Y.I., Sheintuch, M., 1998. Catalytic abatement of water pollutants. Ind. Eng. Chem. Res. 37(2), 309–326.

39. S. Kumar and A.P. Davis, Heterogeneous photocatalytic oxidation of nitrotoluenes. Water Environ. Res. 69 7 (1997), pp. 1238–1245.

40. F. Larachi, I. Iliuta and K. Belkacemi, Catalytic wet air oxidation with a deactivating catalyst analysis of fixed and sparged three-phase reactors. Catal. Today 64 3/4 (2001), pp. 309–320.

41. D.-K. Lee and D.-S. Kim, Catalytic wet air oxidation of carboxylic acids at atmospheric pressure. Catal. Today 63 2–4 (2000), pp. 249–255.

42. B. Legube and N. Karpel Vel Leitner, Catalytic ozonation: a promising advanced oxidation technology for water treatment. Catal. Today 53 1 (1999), pp. 61–72.

43. K.-S. Lin and H.P. Wang, Supercritical water oxidation of 2-chlorophenol catalyzed by $Cu_2^+$ cations and copper oxide clusters. Environ. Sci. Technol. 34 22 (2000), pp. 4849–4854.

44. G. Liu, T. Wu, J. Zhao, H. Hidaka and N. Serpone, Photoassisted degradation of dye pollutants. 8. Irreversible degradation of alizarin red under visible light radiation in air-equilibrated aqueous $TiO_2$ dispersions. Environ. Sci. Technol. 33 12 (1999), pp. 2081–2087.

45. F. Luck, M. Djafer, N. Karpel Vel Leitner, B. Gombert and B. Legube, Destruction of pollutants in industrial rinse waters by advanced oxidation processes. Water Sci. Technol. 35 4 (1997), pp. 287–292.

46. Y.I. Matatov-Meytal and M. Sheintuch, Abatement of pollutants by adsorption and oxidative catalytic regeneration. Ind. Eng. Chem. Res. 36 10 (1997), pp. 4374–4380.

47. Y.I. Matatov-Meytal and M. Sheintuch, Catalytic abatement of water pollutants. Ind. Eng. Chem. Res. 37 2 (1998), pp. 309–326.

48. Y. Matatov-Meytal, V. Barelko, I. Yuranov and M. Sheintuch, Cloth catalysts in water denitrification. I. Pd on glass fibers. Appl. Catal. B 27 2 (2000), pp. 127–135.

49. Y. Matatov-Meytal, V. Barelko, I. Yuranov, L. Kiwi-Minsker, A. Renken and M. Sheintuch, Cloth catalysts for water denitrification II. Removal of nitrates using Pd–Cu supported on glass fibers. Appl. Catal. B 31 4 (2001), pp. 233–240.

50. B. Meunier and A. Sorokin, Oxidation of pollutants catalyzed by metallophthalocyanines. Acc. Chem. Res. 30 11 (1997), pp. 470–476.

51. Pak, D.W., Chang, W.S., Uhm, S.J., Han, S.H., Jung, K.D., Cho, S.H., 1997. Wastewater treatment by catalytic oxidation. Pct Int. Pat. WO 97/08101.

52. D.W. Pak and W.S. Chang, Decolorizing dye wastewater with low temperature catalytic oxidation. Water Sci. Technol. 40 4/5 (1999), pp. 115–121.

53. G. Pecchi, P. Reyes, P. Sanhueza and J. Villasenor, Photocatalytic degradation of pentachlorophenol on $TiO_2$ sol–gel catalysts. Chemosphere 43 2 (2001), pp. 141–146. SummaryPlus | Full Text + Links | PDF (107 K) | View Record in Scopus | Cited By in Scopus (38)

54. A.M. Peiro, J.A. Ayllon, J. Peral and X. Domenech, $TiO_2$-photocatalyzed degradation of phenol and ortho-substituted phenolic compounds. Appl. Catal. B 30 3/4 (2001), pp. 357–373.

55. A. Pintar, G. Beri and J. Levec, Catalytic liquid-phase oxidation of aqueous phenol solutions in a trickle-bed reactor. Chem. Eng. Sci. 52 21/22 (1997), pp. 4143–4153.

56. A. Pintar and J. Batista, Catalytic hydrogenation of aqueous nitrate solutions in fixed-bed reactors. Catal. Today 53 1 (1999), pp. 35–50.

57. A. Pintar, J. Batista and J. Levec, Catalytic denitrification: direct and indirect removal of nitrates from potable water. Catal. Today 66 2–4 (2001a), pp. 503–510.

58. A. Pintar, M. Besson and P. Gallezot, Catalytic wet air oxidation of Kraft bleaching plant effluents in the presence of titania and zirconia supported ruthenium. Appl. Catal. B 30 1/2 (2001b), pp. 123–139.

59. A. Pintar, M. Besson and P. Gallezot, Catalytic wet air oxidation of Kraft bleach plant effluents in a trickle-bed reactor over a Ru/$TiO_2$ catalyst. Appl. Catal. B 31 4 (2001c), pp. 275–290.

60. K. Pirkanniemi and M. Sillanpaa, Heterogeneous water phase catalysis as an environmental application: a review. Chemosphere. 48 (2002), pp. 1047-1060.

61. A. Piscopo, D. Robert, C. Marzolin and J.V. Weber, $TiO_2$ supported on glass fiber for the photocatalytic degradation of benzamide. J. Mater. Sci. Lett. 19 8 (2000), pp. 683–684.

62. J. Qin, Q. Zhang and K.T. Chuang, Catalytic wet oxidation of p-chlorophenol over supported noble metal catalysts. Appl. Catal. B 29 2 (2001), pp. 115–123.

63. A.K. Ray and A.A.C.M. Beenackers, Development of a new photocatalytic reactor for water purification. Catal. Today 40 1 (1998), pp. 73–83.

64. K.J. Reddy and J. Lin, Nitrate removal from groundwater using catalytic reduction. Water Res. 34 3 (2000), pp. 995–1001.

65. D. Robert, A. Piscopo, O. Heintz and J.V. Weber, Photocatalytic detoxification with $TiO_2$ supported on glass-fibre by using artificial and natural light. Catal. Today 54 2–3 (1999), pp. 291–296.

66. S. Sabhi and J. Kiwi, Degradation of 2,4-dichlorophenol by immobilized iron catalysts. Water Res. 35 8 (2001), pp. 1994–2002.

67. M. Sanchez, A. Hadasch, A. Rabion and B. Meunier, Oxidation of 2,4,6-trichlorophenol catalyzed by iron phthalocyanines covalently bound to silica. C.R. Acad. Sci. Ser. IIc: Chim. 2 4 (1999), pp. 241–250.

68. Santos, P. Yustos, B. Durbán and F. García-Ochoa, Catalytic wet oxidation of phenol: kinetics of phenol uptake. Environ. Sci. Technol. 35 13 (2001a), pp. 2828–2835.

69. A. Santos, P. Yustos, B. Durbán and F. Garcia-Ochoa, Oxidation of phenol in aqueous solution with copper catalysts. Catal. Today 66 2–4 (2001b), pp. 511–517.

70. C. Schuth, S. Kummer and S. Kraft, Zeolite based catalysts for the hydrodehalogenation of chlorinated compounds in groundwater. Ufz-Ber. 23 (2000), pp. 58–61.

71. M. Sheintuch and Y.I. Matatov-Meytal, Comparison of catalytic processes with other regeneration methods of activated carbon. Catal. Today 53 1 (1999), pp. 73–80.

72. R.A. Sheldon, M. Wallau, I.W.C.E. Arends and U. Schuchardt, Heterogeneous catalysts for liquid-phase oxidations: philosophers' stones or Trojan horses?. Acc. Chem. Res. 31 8 (1998), pp. 485–493.

73. Sorokin and B. Meunier, Efficient $H_2O_2$ oxidation of chlorinated phenols catalysed by supported iron phthalocyanines. J. Chem. Soc. Chem. Commun. 15 (1994), pp. 1799–1800.

74. R.P.S. Suri, J. Liu, J.C. Crittenden and D.W. Hand, Removal and destruction of organic contaminants in water using adsorption, steam regeneration, and photocatalytic oxidation: a pilot-scale study. J. Air Waste Manage. Assoc. 49 8 (1999), pp. 951–958.

75. J.C.R. Turner, An introduction to the theory of catalytic reactors. In: J.R. Anderson and M. Boudart, Editors, Catalysis Science and Technology vol. 1, Springer-Verlag, Berlin, Germany (1981), pp. 43–86.

76. L. Ukrainczyk, M. Chibwe, T.J. Pinnavaia and S.A. Boyd, Reductive dechlorination of carbon tetrachloride in water catalyzed by mineral-supported biomimetic cobalt macrocycles. Environ. Sci. Technol. 29 2 (1995), pp. 439–445.

77. Vidal and M.M.A. Luengo, Inactivation of titanium dioxide by sulphur: photocatalytic degradation of Vapam®. Appl. Catal. B 32 1/2 (2001), pp. 1–9.

78. T. Wu, T. Lin, J. Zhao, H. Hidaka and N. Serpone, $TiO_2$-assisted photodegradation of dyes. 9. photooxidation of a squarylium cyanine dye in aqueous dispersions under visible light irradiation. Environ. Sci. Technol. 33 9 (1999a), pp. 1379–1387.

79. T. Wu, G. Liu, J. Zhao, H. Hidaka and N. Serpone, Evidence for $H_2O_2$ generation during the $TiO_2$-assisted photodegradation of dyes in aqueous dispersions under visible light illumination. J. Phys. Chem. B 103 23 (1999b), pp. 4862–4867.

80. T. Wu, G. Liu, J. Zhao, H. Hidaka and N. Serpone, Mechanistic study of the $TiO_2$-assisted photodegradation of squarylium cyanine dye in methanolic suspensions exposed to visible light. New J. Chem. 24 2 (2000), pp. 93–98.

81. J. Yu and P.E. Savage, Catalyst activity, stability, and transformations during oxidation in supercritical water. Appl. Catal. B 31 2 (2001), pp. 123–132.

82. Q. Zhang and K.T. Chuang, Treatment of combined bleach plant effluents via wet oxidation over a Pd–Pt–Ce/alumina catalyst. Environ. Sci. Technol. 33 20 (1999), pp. 3641–3644.

# Chapter 11

# Green Environmental Acceptable Synthesis of some New Biologically Active 2–[(4"-substituted Phenyl-6"-Pyrimidin–2"-yl)- Diazenyl-8'-1'/2'-Napthal] Pyrrole

**M.N. Narule[1], J.S. Meshram[2*], P.A. More[3]**
**[1]JSPM, Imperial College of Engineering and Research, Wagholi, Pune, India**
**[2]Department of Chemistry, Rashtrasant Tukadoji Maharaj Nagpur University, Nagpur – 440 033, India**
**[3]Department of Chemistry, Sinhgad College of Engineering, Pune, India**

## ABSTRACT

An efficient synthesis of 2-[(4"-substituted phenyl-6"-pyrimidin-2"-yl)-diazenyl-8'-1'/2'-napthal] pyrrole IV (a-j) has been achieved by reacting corresponding benzylidene acetyl pyrrole II(a-m) with guanidine carbonate followed by coupling the 2-[(6'-substituted phenyl-2'-amino) pyrimidine]pyrrole III(a-m) with 1/2-Naphthal. These compounds were screened for their antibacterial activity against *S. aureus, B. substillis, E. coli* as well as for their antifungal activity against *C. albicans* and *A. niger*. The synthesized compounds are characterized by IR, [1]H NMR, elemental and chemical properties.

***Keywords:*** *Pyrrole, Pyrimidine, Azo compound.*

* Corresponding Author: E-mail: meghasham_n@rediffmail.com

## Introduction

In recent years the pollution free organic reaction and assisted by microwaves in particular have been gained special attention[1,2]. The use of solid supports as well as microwaves[3] has been well established as a pollution free technique, which is currently under much investigation by synthetic organic chemists[4]. As Microwave induced Organic Reaction Enhancement (MORE) chemistry[5,6] allows reaction to occur on a preparative scale in open vessels under solvent free conditions which avoids the risk of high pressures and explosions[7,8] microwave activation rather than conventional heating is preferred as solid supports are rather poor thermal conductors but strong microwave absorbents[9]. One reaction is that the use of microwave activation in organic synthesis can increase the purity of the resulting products, enhance the chemical yield and shorten the reaction time. The other reason is that solvent free reaction avoids organic solvents during the reaction in organic synthesis, lead to a clean, efficient and economical technology[10]. It has many advantages such as high efficiency and selectivity, easy separation and purification, mild reaction conditions and environmental acceptability[11], further the reactions are generally faster and the products obtained are of high purity[12], for increase the efficiency of microwave reaction number of inorganic solid supports like, alumina, silica gel, montmorillonite K10 clay, etc are used as catalyst and as energy transfer, medium that devoid the hazards of solution phase reactions. These solid sup portal microwave organic reaction[13] show various uses such as better selectivity, remarkable reaction rate enhancement, improved yield and associated ease of manipulation[14–16].

Simple and fused pyrrole derivatives are known to possess various biological activities[17–21]. Pyrimidines being an integral part of DNA and RNA possess pharmacological properties such as anti-inflammatory and analgesic[22], blood platelets disaggregating property[23] and antimicrobial activity[24]. Azo derivatives have been reported to possess antibacterial[25], antimicrobial[26], diagnostic aid[27], antineoplastic[28], urinary antiseptic[29], topical dermatological activities[30] and also show pesticidal activity[31]. Most of the azo compounds are widely used as dyes, pigments and also used as indicators in analytical chemistry.

In view of the above mentioned pharmacological activities of pyrrole, pyrimidine and azo compounds a number of the 2-[(4"-substituted phenyl-6"-pyrimidin-2"-yl)-diazenyl-8'-1'/2'-napthal] pyrrole derivatives have been synthesized containing these moieties. 2-[(4'-substituted phenyl-2'-amino) pyrimidine] pyrrole III (a-m) were prepared by reacting benzylidene acetyl pyrrole II(a-m) with guanidine carbonate compound III(a-m) which was coupled with 1/2-naphthal in presence of $NaNO_2$ and HCl at 0-5°C to give 2-[(4"-substituted phenyl-6"-pyrimidin-2"-yl)-diazenyl-8'-1'/2'-napthal] pyrrole IV(a-j).

The UV-Vis-spectra of the azo dyes IV (a-j) were recorded and the values of absorptions ($\lambda_{max}$) and fastness properties are shown in Table 11.1. It is apparent that the wavelength of maximum absorptions azo compound was observed at 200-500nm in EtOH solutions. Variation in $\lambda_{max}$ is being attributed to structural variation of electron–rich aromatic compounds with N=N linkage used for the preparation of these azo compounds.

Structure proof for the synthesized compounds IV a-j was illusidated by IR and $^1$HNMR studies. IR spectrum shows the presence of NH-pyrrole group at 3200-3290cm$^{-1}$, N =N group at 1500-1560-cm$^{-1}$, OH group at 3400-3470cm$^{-1}$, C-N group at 1561-1590cm$^{-1}$, C–Cl group at 750-780cm$^{-1}$, C-$NO_2$ group at 730-757. $^1$HNMR spectrum showed be presence 8.1-8.6 (s, 1H, NH-pyrrole), 6.5–7.0(Ar-H), 5.1-5.7 (s, 1H, OH).

I

II(a-m)

Ethanol/Guanidine carbonate

III(a-m)

ArH $NaNO_2$/HCl

IV(a-j)

| | R | Ar |
|---|---|---|
| 1) | H | 1-$C_{10}H_7O$ |
| 2) | 2-Cl | 1-$C_{10}H_7O$ |
| 3) | 3-Cl | 1-$C_{10}H_7O$ |
| 4) | 4-Cl | 1-$C_{10}H_7O$ |
| 5) | 2-$NO_2$ | 1-$C_{10}H_7O$ |
| 6) | H | 2-$C_{10}H_7O$ |
| 7) | 2-Cl | 2-$C_{10}H_7O$ |
| 8) | 3-Cl | 2-$C_{10}H_7O$ |
| 9) | 4-Cl | 2-$C_{10}H_7O$ |
| 10) | 2-$NO_2$ | 2-$C_{10}H_7O$ |

Scheme-I

**Table 11.1: (UV-VIS Section of Azo compound IV (a–j) fastness properties.**

| Code | Colour | $\lambda_{max}$ | Fastness Properties | | | |
|---|---|---|---|---|---|---|
| | | | Silk | | Wool | |
| | | | Light[a] | Wash[b] | Light[a] | Wash[b] |
| IV a | Red | 474 | 3 | 3–4 | 2–3 | 3–4 |
| IV b | Brown | 466 | 3–4 | 2–3 | 3–4 | 2 |
| IV c | Red | 432 | 2 | 4 | 3–4 | 3–4 |
| IV d | Red | 414 | 3–4 | 3–4 | 2–3 | 2–3 |
| IV e | Orange | 425 | 4 | 2–3 | 3 | 3–4 |
| IV f | Orange | 455 | 2–3 | 3–4 | 3–4 | 2–3 |
| IV g | Red | 460 | 3–4 | 2–3 | 3–4 | 2 |
| IV h | Orange | 476 | 3–3 | 2–3 | 2 | 2–3 |
| IV i | Red | 475 | 3 | 2–3 | 3–4 | 3 |
| IV j | Orange | 454 | 3–4 | 3 | 2–3 | 3–4 |
| IN | EtOH | solution | (4a–j) | | | |

a: Light-fastness: 1: Minimum; 2: Poor; 3: Moderate; 4: Fairly good; 5: Good; 6: Very good.

b: Wash-fastness: 1: Poor; 2: Fair; 3: Good; 4: Very good and 5: Excellent

## Biological Activities

The compounds IV a-j were screened for their antibacterial activity against *Bacillus subtilis, Staphylococcus aureus* and *Escherichia coli* and antifungal activity against *Candida albicans* and *Aspergillus nigar* by filter paper disc technique[13]. Standard antibacterial Streptomycin and antifungal Griscofulvin were also tested under similar conditions for comparison. Results are presented in Table 11.2.

**Table 11.2: Antibacterial and Antifungal Activities of Compounds IV a-j.**

| Compound | Antibacterial Activity | | | Antifungal Activity | |
|---|---|---|---|---|---|
| | S. aureus | B. substillis | E. coli | C. albicans | A. niger |
| IV a | ++ | ++ | + | ++ | ++ |
| IV b | + | +++ | +++ | +++ | +++ |
| IV c | – | ++ | +++ | +++ | ++ |
| IV d | +++ | ++ | ++ | ++ | ++ |
| IV e | + | +++ | +++ | ++ | + |
| IV f | ++ | +++ | ++ | +++ | ++ |
| IV g | +++ | ++ | – | ++ | ++ |
| IV h | ++ | – | + | – | +++ |
| IV i | +++ | ++ | +++ | +++ | – |
| IV j | ++ | + | ++ | + | +++ |
| SM | +++ | +++ | ++++ | | |
| GF | | | | ++++ | +++ |

SM: Streptomycin; GF: Griesofulvin. The inhibition diameter in Mm: (-)<6, (+)7-9, (++)10-15,(+++)16-22, (++++)23-28.

## Experimental Section

All the melting points were determined in open capillaries and are uncorrected. The IR spectra were run in KBr on a Perkin–Elmer infrared spectrophotometer. $^1$H NMR spectra on Brucker AC–300F (300 Hz) NMR spectrometer using DMSO as a solvent using tetramethyl silane as internal standard compounds II (a-m) and III (a-m) are synthesized.[15]

**Table 11.3: Characterization data of compounds IV a–j**

| *Comp.* | *R* | *Ar* | *Mol. Formula* | *M. Pt (°C)* | *RF Value Eluent** | *Yield (%)* | *Analysis Found (Calcd) %* | | |
|---|---|---|---|---|---|---|---|---|---|
| | | | | | | | *C* | *H* | *N* |
| IV a | H | α-$C_{10}H_7O$ | $C_{24}H_{17}N_5$ | 175° | 0.12 | 66 | 73.6 (73.5) | 4.3 (4.3) | 17.9 (17.7) |
| IV b | 2-Cl | α-$C_{10}H_7O$ | $C_{24}H_{16}ON_5Cl$ | 183° | 0.45 | 82 | 67.7 (67.6) | 3.7 (3.6) | 16.4 (16.3) |
| IV c | 3-Cl | α-$C_{10}H_7O$ | $C_{24}H_{16}ON_5Cl$ | 163° | 0.67 | 87 | 67.7 (67.6) | 3.7 (3.6) | 16.4 (16.3) |
| IV d | 4-Cl | α-$C_{10}H_7O$ | $C_{24}H_{16}ON_5Cl$ | 178° | 0.62 | 83 | 67.7 (67.6) | 3.7 (3.6) | 16.4 (16.3) |
| IV e | 2-$NO_2$ | α-$C_{10}H_7O$ | $C_{24}H_{16}O_3N_6$ | 184° | 0.49 | 78 | 66.0 (66.0) | 3.6 (3.4) | 19.2 (19.1) |
| IV f | H | β-$C_{10}H_7O$ | $C_{24}H_{17}N_5$ | 170° | 0.64 | 52 | 73.6 (73.5) | 4.3 (4.3) | 17.9 (17.7) |
| IV g | 2-Cl | β-$C_{10}H_7O$ | $C_{24}H_{16}ON_5Cl$ | 158° | 0.75 | 68 | 67.7 (67.6) | 3.7 (3.6) | 16.4 (16.3) |
| IV h | 3-Cl | β-$C_{10}H_7O$ | $C_{24}H_{16}ON_5Cl$ | 165° | 0.84 | 58 | 67.7 (67.6) | 3.7 (3.6) | 16.4 (16.3) |
| IV i | 4-Cl | β-$C_{10}H_7O$ | $C_{24}H_{16}ON_5Cl$ | 169° | 0.51 | 73 | 67.7 (67.6) | 3.7 (3.6) | 16.4 (16.3) |
| IV j | 2-$NO_2$ | β-$C_{10}H_7O$ | $C_{24}H_{16}O_3N_6$ | 163° | 0.48 | 49 | 66.0 (66.0) | 3.6 (3.4) | 19.2 (19.1) |

Eluents for TLC: Ethyl acetate–Chloroform (6: 4) for IV (a-j).

* Appropriate solvent for crystallization IV(a-j).

IV (a-j): General procedure for the Preparation of 2-[(4"-substituted phenyl-6"-pyrimidin-2"-yl)-diazenyl-8'-1'/2'-napthal] pyrrole:- 0.1 mol of compound III was dissolved in (20ml) 3 per cent HCl was cooled. The solution to 0-5°C and saturated solution of sodium nitrite was added drop wise maintaining ice-cold temperature. The completion of reaction was checked by starch-iodine test. To this solution 0.1gm of α-naphthol/β-naphthol in 20ml of 7 per cent NaOH in a period of 10 min. was added. The coloured solution obtained was stirred for 30min. and then neutralized to pH 7 by adding 10 per cent HCl the solid was separated out, filter dried and crystallized from aqueous ethanol.

IV a: Yield 66 per cent; M.P. 175°C: IR (KBr) 3430(-OH), 3200(NH-pyrrole), 1672(C=0), 1530(N=N), 1567cm$^{-1}$ (C-N), 3010(CH-pyrrole); $^1$HNMR (DMSO-$d_6$), 8.5 (1H, s, NH-pyrrole), 5.7 (s, 1H, OH), 6.3 (Ar- H). 5.1-5.7 (s, 1H, OH).

**Table 11.4: Comparative Study Data of Compounds 3a-m**

| Compd | M.P. (°C) | Reaction Time | | | Yield (per cent) | | |
|---|---|---|---|---|---|---|---|
| | | Microwave | | Classical (hr) | Microwave | | |
| | | Solid Phase (min) | Solvent Phase (min) | | Solid Phase | Sovent Phase | Classical |
| IV a | 175° | 2 | 2 | 8 | 77 | 72 | 61 |
| IV b | 183° | 3 | 3 | 8 | 82 | 80 | 55 |
| IV c | 163° | 2.5 | 5 | 7 | 78 | 73 | 57 |
| IV d | 178° | 3 | 4 | 8 | 85 | 83 | 62 |
| IV e | 184° | 1 | 3 | 7 | 76 | 73 | 57 |
| IV f | 170° | 2 | 3 | 8 | 80 | 78 | 62 |
| IV g | 158° | 3 | 2 | 7 | 85 | 76 | 52 |
| IV h | 165° | 2.5 | 4 | 8 | 76 | 71 | 64 |
| IV i | 169° | 3 | 4 | 7 | 68 | 62 | 59 |
| IV j | 163° | 2 | 4 | 7 | 69 | 64 | 52 |

IVb: Yield 82 per cent; M.P. 183°C: IR (KBr) 3440(OH), 3210(NH-pyrrole), 1672(C=0), 1536(N=N), 1561cm$^{-1}$ (C-N), 3122(CH-pyrrole), 758(C-Cl); $^1$HNMR (DMSO-$d_6$), 8.2 (1H, s, NH-pyrrole), 5.5 (s, 1H, OH), 6.2 (Ar-H).

IV c: Yield 87 per cent; M.P. 163°C: IR (KBr) 3436(OH), 3222(NH-pyrrole), 1672(C=0), 1560(N=N), 1569cm$^{-1}$ (C-N), 3011(CH-pyrrole) 759(C-Cl); $^1$HNMR (DMSO-$d_6$), 8.4 (1H, s, NH-pyrrole), 5.5(s, 1H, OH), 6.9 (Ar-H).

IVd: Yield 83 per cent; M.P. 178°C: IR (KBr) 3450(OH), 3250(NH-pyrrole), 1670(C=0), 1541(N=N), 1587cm$^{-1}$ (C-N), 3020(CH-pyrrole) 755(C-Cl); $^1$HNMR (DMSO-$d_6$), 8.5 (1H, s, NH-pyrrole), 5.3 (s, 1H, OH), 7.0 (Ar-H).

IVe: Yield 78 per cent; M.P. 184°C: IR (KBr) 3456(OH), 3254(NH-pyrrole), 1670(C=0), 1555(N=N), 1575cm$^{-1}$ (C-N), 3020(CH-pyrrole), 732(C-$NO_2$); $^1$HNMR (DMSO-$d_6$), 8.4(1H, s, NH-pyrrole), 5.4 (s, 1H, OH), 6.1 (Ar-H).

IVf: Yield 52 per cent; M.P. 170°C: IR (KBr) 3443(OH), 3244(NH-pyrrole), 1670(C=0), 1557(N=N), 1569cm$^{-1}$ (C-N), 3020(CH-pyrrole); $^1$HNMR (DMSO-$d_6$), 8.3 (1H, s, NH-pyrrole), 5.5 (s, 1H, OH), 6.1 (Ar-H).

IVg: Yield 68 per cent; M.P. 158°C: IR (KBr) 3457(OH), 3258(NH-pyrrole), 1670(C=0), 1544(N=N), 1572cm$^{-1}$ (C-N), 3020(CH-pyrrole) 770(C-Cl); $^1$HNMR (DMSO-$d_6$), 8.6 (1H, s, NH-pyrrole), 5.1 (s, 1H, OH), 6.5 (Ar-H).

IVh: Yield 58 per cent; M.P. 165°C: IR (KBr) 3459(OH), 3230(NH-pyrrole), 1670(C=0), 1538(N=N), 1577cm$^{-1}$ (C-N), 3020(CH-pyrrole) 778(C-Cl);; $^1$HNMR (DMSO-$d_6$), 8.6 (1H, s, NH-pyrrole), 6.8 (s, 1H, OH), 6.2 (Ar-H).

IV i: Yield 73 per cent; M.P. 169°C: IR (KBr) 3447(OH), 3287(NH-pyrrole), 1670(C=0), 1539(N=N), 1589cm$^{-1}$ (C-N), 3020(CH-pyrrole) 780(C-Cl); $^1$HNMR (DMSO-d$_6$), 8.2 (1H, s, NH-pyrrole), 5.3 (s, 1H, OH), 6.7 (Ar-H).

IVj: Yield 49 per cent; M.P. 163°C: IR (KBr) 3439(OH), 3287(NH-pyrrole), 1670(C=0), 1532(N=N), 1590cm$^{-1}$ (C-N), 3020(CH-pyrrole), 757(C-$NO_2$); $^1$HNMR (DMSO-$d_6$), 8.4 (1H, s, NH-pyrrole), 5.3 (S, 1H, OH), 6.8 (Ar-H).

## Conclusion

Researchers across the globe have developed green revolution to design synthesis in organic chemistry. The microwave assisted greener chemical transformations affords excellent product yield, reduced reaction time and minimization or elimination of by product. The result obtained confirms superiority of microwave irradiation over classical heating method (Table 11.4).

## Acknowledgements

The authors are thankful to UGC, New Delhi for R& D project. One of the authors M.N.Narule is thankful to UGC for research fellowship.

## References

1. Kidwai M, Dave B J and Venkataramanan R, Indian *J. Chem.*, 41B, 2002, 2414.
2. Kidwai M, Rastogi S and Vekantaramanan R, *Bull Chem. Soc* Jph, 76, 2003, 203.
3. Kidwai M and Sapra P, Bhushan K R and Misra P, *Synthesis*, 10, 2001, 1509.
4. Kidwai M and Sapra P, *Synth. Commun*. 32, 2002, 1639.
5. Solid supports and catalysis in organic synthesis, edited by K. Smith (Prentice Hall), 1992.
6. Clark J. H, Gullen S R, Barlow S J and Bastock T W. *J. Chem. S*, perkin trans 2, 1994, 1117,
7. Kidwai M and Kumar P, J. *Chem. Research*, 1996, 254.
8. Kidwai M and Kohali P, *Indian J Chem.*, 36, 1997, 1071.
9. Bose A K, Banic Lavinskaia N, Jayaraman M and Manhas M S, *Chemtech*, 27, 1997, 18.
10. Mitra A K, De A and Karchaudhuri N, synth commun, 29, 1997, 573: *Indian J Chem.*, 39B 2000, 311,387.
11. Loupy A, Petit A, Hamclin J, Texier–Boullet F, Jacquault P and Mathe D, *Synthesis*, 1998, 1213.
12. Texier–Boullet F, Latouche R and Hamclin J, *Tetrahedron Lett*, 34,1997, 2123.
13. Loupy A, *Spectra Analyses*, 175, 1993, 33.
14. Caddick S, *Tetrahedron*, 51, 1995, 10403.
15. Oza H B, Joshi D G and Parikh H H, *Heterocyclic Comm.*, 3, 1997, 3.
16. Kumar R S and Jaju B P, *J.Indian Chem. Soc*, 67,1990, 920.
17. M. Hondershausen, *Pestic Sci*, 48(1996), 269-278.
18. B. S. M. Fahmy, M. H. Elnagdi, *J. Chem. Tech*, 15(1980), 30-37.
19. A. Omar, M. E.12MohSen, W. O. M. Aboul, *J.Heterocycl Chem*, 21(1984), 1415.
20. F. L. Sot, T. L. Lambe and R. N. Butter, *J. Chem Soc*, Perkin Trans, 1(1972), 269.
21. R. J. Bundburg, *Comprehensive Heterocyclic chemistry*, Ed. A. R. Katritzky and C. W. Rees, Pergaman Press, Oxford, 1984, 4, 313.
22. S. M. Sondhi, R. P. Varma, *Indian Drugs*, 23(B), 1999, 50.
23. S. K. Nandeeshaiah and S. V. Ambekar, *Indian J. Chem.*, 37(B), 1998, 995.
24. J. M. Parmar, *et al.*, *Indian J. Chem.*, 38(B), 1999, 440.

25. Mietzsch, Klarer, *Ger Pat*, 638, (1936), 701.
26. Gley, Girard, Compt, *Rend. Soc. Biol*, 125, (1936), 1027.
27. Taylor, Thorp, *Bit. Heart J*, 21(1959), 492.
28. Carter, Friedmabn, *Europ, J. Cancer*, 8(1972), 853.
29. Dohrten, Diedrich, *U. S. Pat*, 1(1932), 862, 361.
30. Tirasek *et al., Cersk. Dermatol*, 38(1966), 41.
31. M. N. Narule, J.S. Meshram, *J. Oriental Chemistry* (In press).

# Chapter 12

# Wastewater Management in an Industrial Complex: An Indian Case Study

**K.L. Rao, P. Sridhar and M. Ratna Kumar**

**Jeedimetla Effluent Treatment Ltd., Plot No. 267, Phase I, Industrial Development Area, Jeedimetla, Hyderabad – 500 055, India**

## ABSTRACT

Industrial development is vital to the economic prosperity of a nation. Industries while playing significant role in the production and use of goods and services are also posing substantial threat to environment. Therefore, it is imperative that the industries re-orient their activities such that sustainable development is ensured. The activities related to the prevention of pollution and treatment of pollutants can be an individual effort or a joint effort. The chapter discusses the efforts of the industries, particularly chemical and allied industries, located in the industrial estates in and around Hyderabad, India, in the treatment and containment of pollutants, with emphasis on the handling of wastewaters with heterogeneous organic substances and high concentrations of total dissolved inorganic solids. The experiences gained in the operation of a common facility for the treatment of heterogeneous industrial wastewaters is presented. The applicability of these systems and their economics, under different situations, are discussed and case studies are cited.

***Keywords:*** *Wastewater treatment, Total dissolved inorganic solids, Combined wastewater treatment, Cleaner production, Evaporation, Sustainable development.*

## Introduction

Industrial development is vital to the economic prosperity of a nation. Industrial estates provide benefits to communities, by serving as effective instruments for industrial development-reducing costs of infrastructure and stimulating regional economic activities, but they can also lead to environmental, health and safety problems. Such a situation warrants the initiation of activities that aim at industrial development without environmental degradation. Industries that are aware of the fragile relationship between development and environmental conservation have started adopting methods to reduce environmental degradation. The methods adopted are based on the scientific principles of sustainable development. The activities related to the prevention of pollution and treatment of pollutants can be either an individual effort or a joint effort. The chapter discusses the efforts of the industries, particularly chemical and allied industries, located in the industrial estates in and around Hyderabad, India–in the treatment and containment of pollutants, with emphasis on the handling of wastewaters with heterogeneous organic substances and high concentrations of Total Dissolved Inorganic Solids (TDIS).

There are about ten industrial estates in and around Hyderabad. Bulk drug manufacturing units are mainly concentrated around Hyderabad city in the districts of Ranga Reddy and Medak. At present 120 Bulk drug and 8 dye intermediate manufacturing units are in operation in Andhra Pradesh. 90 per cent of the bulk drug and dye intermediate manufacturing units are in small or medium scale sector. Most drugs and dye intermediates are manufactured by chemical synthesis. Hence their manufacturing processes and unit operations are similar, though the products in each industry are different. Typical waste streams include mother liquors, spent solvents, equipment wash waters, spilled materials, off-specification products and used processing aids. Such wastewaters are generally characterized by high concentrations of organics and dissolved inorganic solids. Wastewaters having a concentration of dissolved inorganic solids more than or equal to the concentration present in sea water are termed 'hypersaline'wastewaters. These wastewaters are difficult to treat using conventional microorganisms commonly found in wastewater treatment plants because of the osmotic stress of the dissolved inorganic solids on the microorganisms[1]. Therefore, the concentration of dissolved inorganic solids in the wastewater should be either reduced or contained to a level conducive for biological treatment systems, for the effective treatment of organic pollutants.

## Treatment Systems Available for the Reduction of TDIS in Industrial Wastewaters

The various alternative techniques available for the reduction and containment of TDIS include-adoption of cleaner technology of production, evaporation and reverse osmosis. Combined wastewater treatment systems are also one of the available processes.

Most of the above mentioned treatment systems are based on the 'end-of-pipe' approach to treat industrial wastewaters. However, this approach first allows, in some cases, the production of waste and wasteful use of resources in manufacturing. It then consumes further resources to convert the wastes generated into different potentially usable/recoverable material or otherwise to other forms without reducing the net contributions to the environment. As experienced in industrialized countries, such an approach is a short-term remedy and cannot lead to sustainable industrial development in the long term.

## Cleaner Production

In 1989 The United Nations Environment Program (UNEP) defined clean production as " the continuous application of an integrated preventive environmental strategy applied to processes, products and services to increase Eco-efficiency and reduce risks to humans and the environment". Cleaner production activities are also captured by such terms as pollution prevention, source reduction, waste minimization or Eco-efficiency[4].

Cleaner production includes initiatives such as better management and housekeeping, substitutions for toxic and hazardous materials, process and product modifications and internal reuse of waste products. At its heart, the concept is about the prevention, not control of pollution. As such, cleaner production is based on environmental solutions that take economics into consideration. Cleaner production technologies are often more cost-effective than control technologies. The costs of dealing with wastes are reduced and there is potential for new markets to be discovered through innovations or the sale of by-products. Cleaner production can reduce environmental risks and liabilities and result in enhanced cost competitiveness. Companies can also improve their public image and gain consumer trust, by showing a commitment to cleaner production. Thus it brings together the powerful combination of economic savings and environmental improvements.

## Case Studies

### Cleaner Technology of Production

#### Case Study of an Industry Manufacturing Dye-Intermediates[5]

VASANT Chemicals Ltd. and Vasant Organics Ltd., are two companies located in the Jeedimetla Industrial Estate of Hyderabad manufacturing dye intermediates. They produce Di-amino stilbene di-sulphonic acid starting from para nitro toluene alongwith other chemicals such as oleum, sulphuric acid, caustic, chlorine and iron powder as major raw materials. Through in-house R&D efforts the industry modified the manufacturing technology as follows.

Adoption of forced air oxidation resulting in the total elimination of chlorine used earlier.

Installation of a filtration system to recover the residual final product from the spent iron sludge and also to reduce the moisture content of the iron sludge for recycling in the cement industry.

The modified process technology of air oxidation includes a triple effect evaporator as a unit operation for concentration of the solution before crystallization and recovery of the condensate water for reuse in the oxidation process. The wastewater generated after dinitro stilbene disulphonic acid recovery contains sodium sulphate with trace quantities of organic impurities. To recover the sodium sulphate from the wastewater as solid and recycle the condensate water, the modified process also incorporated a quadruple effect evaporator and a fluidized bed incinerator. The process modification which includes installation of vacuum nutch filter for the iron sludge has been carried out successfully achieving the benefits mentioned below.

Use of chlorine, which is about 90 TPM, is completely eliminated.

Use of caustic lye is reduced by 60 per cent.

Wastewater generation is reduced by about 73 per cent.

The total dissolved solids load in wastewater is reduced by 64 per cent and 16 TPD of 98 per cent sodium sulphate is recovered as a by product which has been found useful in industries where sodium sulphate is used as a raw material.

The yield of the final product diamino stilbene disulphonic acid, has improved by 20 per cent.

**Economic Viability**

- ✰ Operating costs = Rs. 10.2 Million/annum
- ✰ Savings on recovery of sodium = Rs. 6.6 Million
- ✰ Sulphate @Rs. 1.25/Kg.
- ✰ Savings on recovery of water = Rs. 0.866 Million
- ✰ Savings on treatment of wastewater = Rs. 6.732 Million
- ✰ Net savings = Rs. 4.0 Million
- ✰ Capital investment = Rs. 20 Million
- ✰ Pay back period = 5 years

The A.P. Pollution Control Board under the Hyderabad Waste management project, a joint initiative by the Government of India and Government of Australia has awarded a grant of Rs.1.7 Million to Vasant Chemicals Ltd., for adopting a cleaner production technology.

## Reverse Osmosis

### Case Study of an Industry Manufacturing Bulk-Drugs

One of the industries located in an industrial estate near Hyderabad, has adopted the use of reverse osmosis system for reducing of TDIS in the biologically-treated wastewater to render it fit for recycle in the industry. The details of the system are as follows.

**Characteristics of Treated Wastewater for R.O. System**

*Flow rate, cum./day = 250*

| *Parameter* | *Influent to R.O. System* | *Effluent of R.O* |
|---|---|---|
| TDS, mg/L | 15000 | <500 |
| COD, mg/L | 3000 | <200 |
| BOD, mg/L | 1000 | Absent |

The permeate from R.O. is totally recycled back in the industry. Reject from R.O. with higher concentration of Dissolved inorganic solids, is sent to a quadruple effect evaporator. The condensate of the evaporator is also recycled in the industry. The solids separated by the evaporator are sent to the Treatment, Storage and Disposal Facility (TSDF) for hazardous wastes, located at Dundigal, Hyderabad.

## Evaporation Systems

Many of the bulk drug manufacturing units located in the industrial estates of Hyderabad have adopted the use of Evaporation systems for the separation of total dissolved inorganic solids (TDIS) from wastewater and the recovered salts are either being re-used or disposed off at the TSDF, depending upon the economic value of the recovered solids. The wastewater, after the separation of TDIS, is either treated on-site or at a Common Effluent Treatment plant, for the removal of organic pollutants. However, published data related to the efficiency of these systems are readily not available.

## Efforts of Individual Industries for the Management of Wastewaters

Many of the industries present in and around the city of Hyderabad, that generate more than 40 cum. of wastewater per day, have installed wastewater treatment systems in their industrial premises.

The following table indicates the efforts made by individual industries in treating their wastewaters (Source of information: Zonal Office, R.C. Puram, A.P. Pollution Control Board).

**Table 12.1**

| *Sll.No.* | *Name of the Industry* | *Brief Description of ETP Facility* | *Disposal of Treated Wastewater at* |
|---|---|---|---|
| 1. | Aurobindo Pharma Ltd. | MEE system, Primary and Secondary treatment systems and R.O. system | CETP-PETL |
| 2. | Dr.Reddy's Labs. | Primary and Secondary treatment systems | CETP-PETL |
| 3. | Natco Pharma Ltd. | Primary and secondary treatment systems | CETP-PETL |
| 4. | Ralchem Industries Ltd. | MEE system, Primary and Secondary treatment systems, gaseous and liquid incinerator | CETP-PETL |
| 5. | Siris India Ltd. | Primary and secondary treatment systems | CETP-PETL |
| 6. | Vorin Labs. Ltd. | Primary, Secondary and Tertiary treatment systems | CETP-PETL, JETL |

Almost all the chemical and allied product manufacturing industries that are members of the Common Effluent Treatment Plants (CETPs), have adequate infrastructure to collect, store and neutralize their wastewaters before they are sent to CETPs and comply with the inlet standards stipulated for the member industries for treatment at CETP.

## Combined Wastewater Treatment

### Case Study of a Common Effluent Treatment Plants[6]

Pattancheru Enviro-Tech Ltd. (PETL) is a Common Effluent Treatment Plant (CETP), operating at the Patancheru Industrial area for the treatment of industrial wastewaters generated mainly from bulk drug manufacturing industries. During the course of operations, the plant had the following problems.

1. Variation in the concentration of TDIS, COD and BOD in the influent owing to the change in the bulk-drug market scenario
2. Increase in organic load to be handled by the treatment plant
3. Limited space availability for adding additional units
4. Failure of anaerobic treatment system leading to increase in BOD load onto activated sludge process
5. Compliance to meet the regulatory standards for discharge into inland- surface water

Under the above circumstances, PETL has adopted the use of Pure-Oxygen systems for enhancing the treatment capacity of the plant and meeting the regulatory standards. The adopted system works as follows. Liquid oxygen is converted to the gaseous form and injected into the aeration tanks along with the influent to the aeration tanks. The presence of oxygen in the levels required for the effective stabilization of BOD, is maintained by the mixflo-oxygen systems. The various equipment installed for the supply of oxygen to the aeration tanks, are as follows.

1. Liquid oxygen storage tank (stored at –186°C)
2. Liquid Oxygen vaporizers (for converting liquid oxygen to gaseous form)
3. Oxygen control panel (to control flow rate of oxygen and maintain desired levels)

4. Mixflo pumps (to inject and mix oxygen into aeration tanks)
5. Piping loops (to dissolve oxygen)
6. Ejectors (to create mixing conditions in aeration tanks)
7. Online- D.O.Meters (to monitor and control the flow rates of Oxygen)

**Table 12.2: Advantages and Disadvantages of Using Mixflo- Liquid Oxygen Systems**

| Sl.No. | Parameter | Advantage | Disadvantage |
|---|---|---|---|
| 1. | Capital cost | Low | ——- |
| 2. | Maintenance/running cost | ——- | High |
| 3. | Area required to install Oxygen system | Less, compared to other systems | ——- |
| 4. | Time required for implementation in an existing system | Low | ——- |
| 5. | Oxygen supplies | ——- | To be transported on road by tankers, hence uncertainty in supplies |
| 6. | Power requirement | ——- | High |
| 7. | Controlling Dissolved Oxygen levels | Oxygen flow rates can be increased/ decreased according to fluctuations in the inlet BOD load to maintain desired D.O. level in the system. | |

## Case Study of a Combined Wastewater Treatment Plant in the Jeedimetla Industrial Estate[7]

In 1989, a Common Effluent Treatment Plant (CETP) started functioning under Jeedimetla Effluent Treatment Ltd. (JETL), to address the wastewater management problem in the Jeedimetla Industrial Estate. JETL started operations for treating heterogeneous industrial wastewater. Since then the CETP expanded its treatment facilities to meet the increasing demand for treatment and to improve the quality of treated wastewater to meet discharge standards into sewers. After scientifically establishing the issues to be considered for improving the performance of CETP, JETL decided to use domestic sewage for controlling Total Dissolved Inorganic Solids (TDIS) level in the influent wastewaters and operate the CETP as a Combined Wastewater Treatment Plant (CWWTP).

JETL-CWWTP was in operation since September 1998 and currently, 1000 cum./day of industrial and 1500 cum./day of domestic wastewaters are being treated together and discharged into municipal sewer. Summary data of Treated effluent is presented as Table 12.3.

However, CWWTP is a biological treatment system and is not designed to reduce the concentration of TDIS in the treated wastewater. At present, the total TDIS discharged into municipal sewer is around 35 tons/day. In its continuous endeavour to improve the quality of treated wastewater, JETL decided to establish treatment systems to reduce the total quantity of TDIS discharged through treated effluent into municipal sewer. The approach adopted to achieve the above regulatory compliance is to segregate all the wastewater having TDIS more than or equal to 50,000 mg/L and subject it to a common treatment to separate out the salts from the high TDIS wastewaters.

**Table 12.3: Analytical Data on the Treated Wastewater Quality of JETL-CWWTP**

| Sl.No. | Parameter | Analysis Results as per JETL Lab | | | | Regulatory Standard for Disposal into Public Sewer |
|---|---|---|---|---|---|---|
| | | Min. | Max. | Avg. | S.D. | |
| 1. | Colour and Odour | – | – | – | – | – |
| 2. | Suspended Solids | 10 | 122 | 51 | 32 | 600 |
| 3. | Particle size of Suspended Solids | – | – | – | – | – |
| 4. | pH Value | 7.8 | 8.5 | | | 5.5–9.0 |
| 5. | Temperature | – | – | – | – | – |
| 6. | Oil and Grease | 1.8 | 12.6 | 8.8 | 4.7 | 20 |
| 7. | Total Residual Chlorine | – | – | – | – | – |
| 8. | Ammonical Nitrogen (as N) | 8 | 40 | 22 | 5 | 50 |
| 9. | Total Kjeldahl Nitrogen (as N) | – | – | – | – | – |
| 10. | Free Ammonia (as NH) | – | – | – | – | – |
| 11. | Chemical Oxygen Demand | 810 | 1000 | 920 | 85 | – |
| 12. | Biochemical Oxygen Demand (3 days at 27° C) | 20 | 57 | 40 | 12 | 350 |
| 13. | Arsenic (as As) | Below detectable limit of 0.05 | | | | 0.2 |
| 14. | Mercury (as Hg) | Absent | | | | 0.01 |
| 15. | Lead (as Pb) | Below detectable limit of 0.05 | | | | 1.0 |
| 16. | Cadmium (as Cd) | 0.02 | 0.12 | 0.06 | 0.04 | 1.0 |
| 17. | Hexavalent Chromium (as $Cr^{+6}$) | Below detectable limit of 0.1 | | | | 2.0 |
| 18. | Total Chromium (as Cr) | 0.11 | 0.50 | 0.27 | 0.20 | 2.0 |
| 19. | Copper (as Cu) | 0.05 | 0.51 | 0.27 | 0.16 | 3.0 |
| 20. | Zinc (as Zn) | 0.22 | 1.60 | 0.70 | 0.45 | 15.0 |
| 21. | Selenium (as Se) | Below detectable limit of 0.05 | | | | 0.05 |
| 22. | Nickel (as Ni) | 0.40 | 1.16 | 0.69 | 0.23 | 3.0 |
| 23. | Cyanide (as CN) | 0.15 | 0.40 | 0.24 | 0.09 | 2.0 |
| 24. | Fluoride (as F) | 8.0 | 10.0 | 9.6 | 3.2 | 15.0 |
| 25. | Dissolved Phosphates (as P) | – | – | – | – | – |
| 26. | Sulphide (as S) | Absent | | | | – |
| 27. | Phenolic Compounds (as $C_6H_5OH$) | 0.08 | 2.15 | 0.50 | 0.66 | 5.0 |
| 28. | Radioactive materials: | | | | | |
| | a) Alpha emitters micro cure, mg/l max. | Not Applicable | | | | $10^{-7}$ |
| | b) Beta emitters micro cure, mg/l max. | Not Applicable | | | | $10^{-6}$ |
| 29. | Bio–assay test | Toxicity Factor 4 | | | | 90 per cent survival of fish after 96 hours in 100 per cent effluent |
| 30. | Manganese (as Mn) | 0.90 | 1.20 | 1.31 | 0.26 | 2.0 |
| 31. | Iron (as Fe) | 1.40 | 1.55 | 1.52 | 0.08 | 3.0 |
| 32. | Vanadium (as V) | 0.08 | 0.1 | 0.09 | 0.01 | 0.2 |

Except 1,3,4 and 5 all are expressed in mg/L

## Establishment of MEE system

JETL quantified the volume and concentration of TDIS from those wastewaters having greater than 50,000 mg/L of TDIS concentration. After considering the various technologies available at source or collectively at one place and the economic feasibility, JETL has decided to establish a common MEE system for the wastewaters having TDIS greater than 50,000 mg/L. Representative samples obtained by mixing the individual wastewaters in proportion to flow rate, were subjected to Pilot studies and the feasibility for adopting MEE systems has been established. The characteristics of the wastewater subjected to MEE process and a brief description of MEE are as follows.

### Design Criteria for MEE

Characteristics of wastewater for MEE system:

- ✰ pH: 5.6
- ✰ TDS: 274.13 g/L
- ✰ TDIS: 223.18 g/L
- ✰ COD: 62.4g/L
- ✰ Density: 1182 Kg/Cum.
- ✰ Volume of wastewater: 160 cum./d
- ✰ No. of effects: Four-Quadruple effect Evaporator
- ✰ Type: Falling film followed by raising film

Mass balance for the MEE system is presented as Figure 12.1.

Industrial wastewaters received for treatment in MEE facility are thoroughly homogenized and sent to MEE for evaporation. A portion of the concentrate is recycled back into the system to maintain the concentration of solids at requisite levels. The salt is separated by a centrifuge. Condensate is sent to biological treatment system for treatment and disposal. The centrate from MEE is sent to spray dryer where the wastewater is sprayed into a drying chamber. The salt is being sent for disposal in the landfill of TSD Facility presently operating at Dundigal, Ranga Reddy District. However, the feasibility of recovering individual salts for reuse by industry is also under investigation.

### Capital Costs of MEE and Spray Dryer Systems

The cost of project is around Rs. 40 Million

The integration of MEE and spray dryer system with CWWTP is presented as Figure 12.2.

## Conclusion(s)

End-of-the pipe treatment technologies have been adopted by chemical and allied industries for several years. At present, the industry is adopting cleaner technology of production considering the techno-economic feasibility of adoption. This approach is being regarded as the best–owing to the environmental and economic benefits that accrue to the industry. Adoption of treatment systems like evaporation and reverse osmosis to recover and reuse salts and water is also in vogue. Efforts to treat industrial wastewaters can be an individual or collective effort. Case studies of common facilities for the treatment of industrial wastewaters indicate that such facilities significantly contribute to the growth and sustenance of environmental quality and industry. The successful operation of the CWWTP–JETL has resulted in the optimal utilization of resources like land, water and energy in the industrial

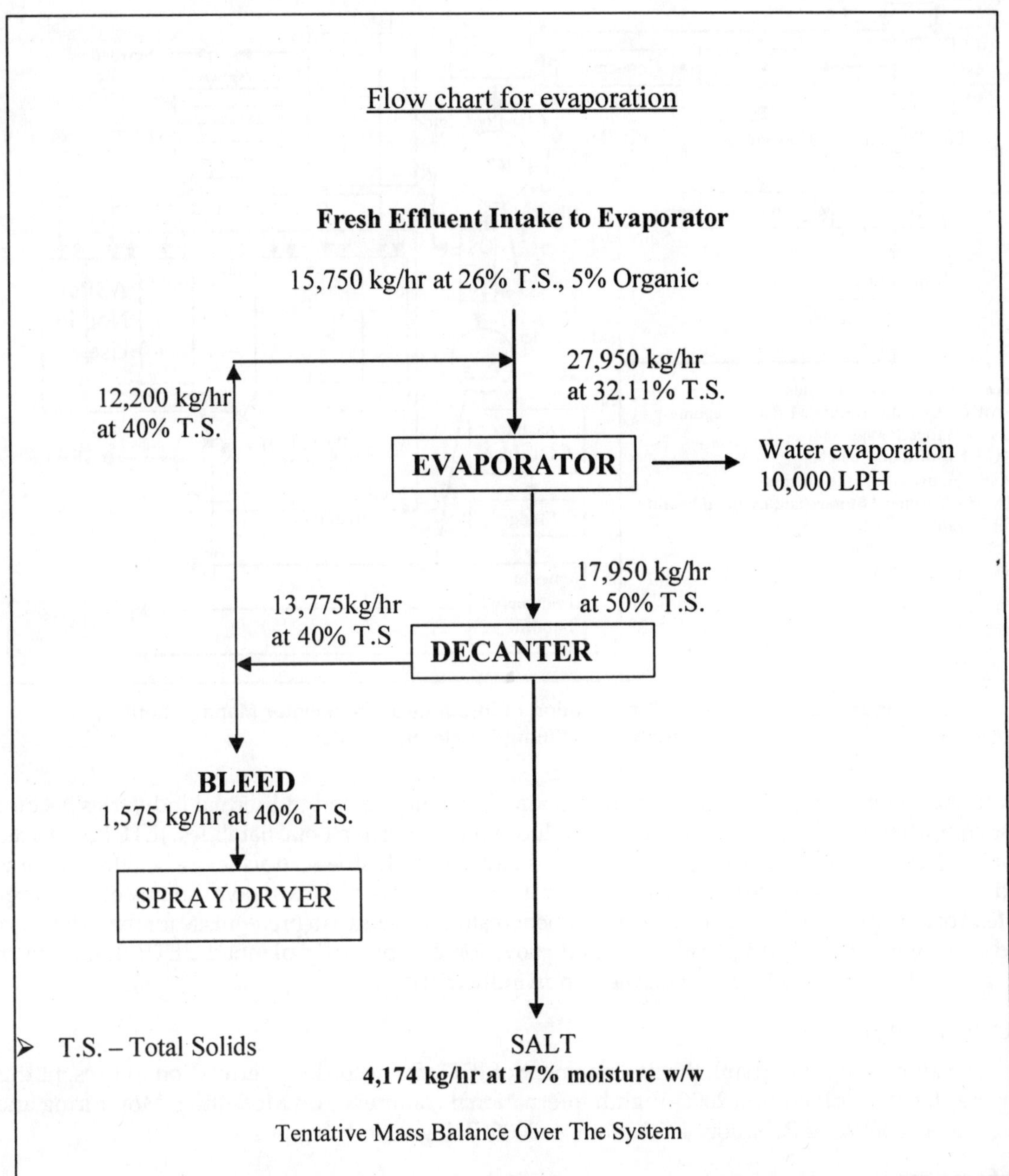
Flow chart for evaporation
Fresh Effluent Intake to Evaporator
15,750 kg/hr at 26% T.S., 5% Organic
27,950 kg/hr
at 32.11% T.S.
12,200 kg/hr
at 40% T.S.
EVAPORATOR
Water evaporation
10,000 LPH
17,950 kg/hr
at 50% T.S.
13,775kg/hr
at 40% T.S
DECANTER
BLEED
1,575 kg/hr at 40% T.S.
SPRAY DRYER
T.S. – Total Solids
SALT
4,174 kg/hr at 17% moisture w/w
Tentative Mass Balance Over The System

**Figure 12.1**

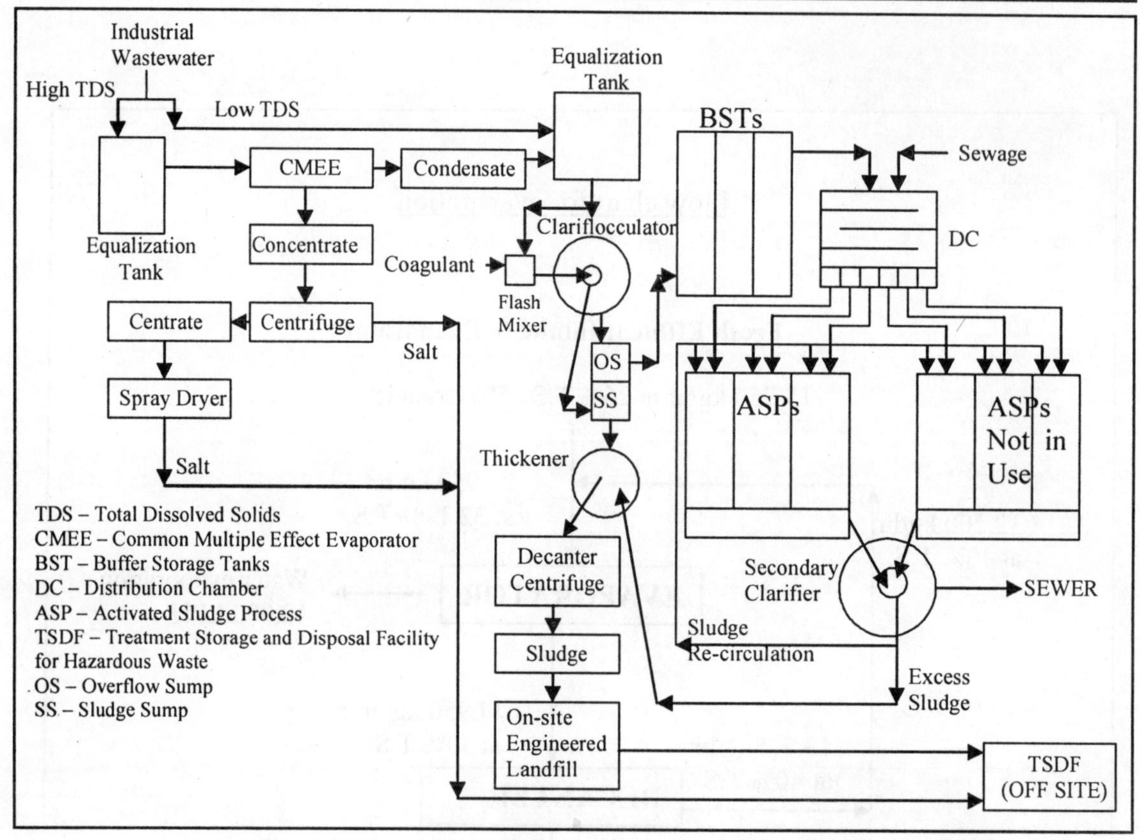


**Figure 12.2: Schematic Presentation of Integrated Wastewater Management at Jeedmetla Effluent Treatment Limited**

estate. Adoption of a common wastewater treatment facility helped individual industries become cost-competitive in global scenario and also reduced their environmental liabilities. JETL has proved that a collective effort to protect environment by small scale industries can achieve a healthy economy and environment besides facilitated regulatory monitoring to confine to a point source of discharge instead of multiple sources. A performing common treatment facility is a prerequisite for the sustenance and development of an industrial estate and provision of a properly planned CETP should be an integral part of infrastructure development in an industrial estate.

## Acknowledgement

The authors wish to thank the management of JETL for according permission to present this paper at the Water Pollution 2006–Eighth International conference on Modelling, Monitoring and Management of Water Pollution, 4-7 September 2006, Bologna, Italy.

## References

1. Subrahmanyam,P.V.R. *Biological Treatment of High Saline Wastewaters*. Paper presented at the 'International Seminar on Combined Wastewater Treatment', organized by JETL, Hyderabad, March 2000.

2. Rao,K.L., Venkayya,S. and Sridhar.P, *Operation of a Combined wastewater Treatment plant vis-a-vis an approach to handle high salt bearing heterogeneous industrial wastewaters*, paper presented at the second International Conference on industrial pollution and control technologies, JNTU, Hyderabad, December, 2001.
3. Brian O'Gallagher, *Waste Management Technologies*, Dept. of Industry, Technology and Commerce, Government of Australia, August 1990.
4. Editorial, Green Business Opportunities, CII Quarterly, Vol. 8, Issue 1, Jan–March 2002.
5. APPCB- *Information bulletin*, Publication No.110, November 2000.
6. Venkata Rao,P.,Suryanarayana Raju,V. and Akella,V.R. *Application of Pure Oxygen for Treatment of Industrial Wastewaters at CETP- A Case Study*, Paper presented at the 'International Seminar on Combined Wastewater Treatment', organized by JETL, Hyderabad, March 2000.
7. Rao,K.L. *Approach adopted by CWWTP-JETL in reducing discharge of TDIS of treated effluent into municipal sewer*, paper presented at the workshop organized during the Annual General Meeting of the Dye-Manufacturers Association of India, Hyderabad, January, 2002.

# Chapter 13

# Reactive Extraction: A Clean Technology for Recovery of Itaconic Acid

**Suresh Kumar[1], Kailas L. Wasewar[2*] and Amit Keshav[3]**
**[1]MTech Student; [2]Assistant Professor; [3]Research Scholar**
**Department of Chemical Engineering, Indian Institute of Technology (IIT)**
**Roorkee, Uttarakhand – 247 667, India**

## ABSTRACT

Itaconic acid also known as methylene succinic acid is an unsaturated acid with total market quoted being 10,000 to 15,000 metric tones per year worldwide. The demand for itaconic acid is growing, especially for its use in the production of various polymers. It is also a valuable material for production of resins, cosmetics, lubricant, thickener, herbicides wool modifier etc. The current methods of producing itaconic acid are both unsustainable and expensive. It is necessary to implement an efficient cleaner technology for the production of itaconic acid. The itaconic acid is prepared by fungal fermentation at a relatively high cost. The fungal fermentation is expensive because it requires dilute feed solutions (approximately 10 wt. per cent glucose) gives a yield of only 50–60 per cent. This is due to the cost and lower efficiency of the current methods for separating the itaconic acid from the fermentation media. The organism most often used for production of itaconic acid is *Aspergillus terreus.* The industrial production of itaconic acid was from catalytic condensation of succinic acid derivatives with formaldehyde. The conventional method for the recovery of itaconic acid consumes more amounts of chemicals, energy and discharges waste. This method is unfriendly to the environment. Thus there is a need to look at other methods for producing itaconic acid. One of the promising technologies for recovery of itaconic acid from fermentation broth is reactive extraction. Reactive extraction is one of the cleaner and energy efficient technology. In present chapter reactive extraction of itaconic acid

* Corresponding Author: E-mail: k_wasewar@rediffmail.com, klw73fch@iitr.ernet.in

was studied using Aliquat 336 in toluene and heptane. The study was to highlight the effect of aromatics (toluene) and paraffins (Heptane). The overall distribution coefficient ($K_D$), loading factor and equilibrium complexation constants $K_E$ were evaluated. The values of $K_E$ in these diluents confirm that extractant is a function of diluent's chosen. Out of these diluents, heptane gives the higher distribution coefficient.

***Keywords:*** *Itaconic acid, Fermentation, Reactive extraction, Aliquat 336, Diluents, Clean technology.*

## Introduction

Itaconic acid (methylene succinic acid) is a dicarboxylic acid obtained by the distillation of citric acid. It is a white crystalline powder that is soluble in water, acetone, and ethanol, moderately soluble in acetic acid, slightly soluble in organic solvents, very slightly soluble in benzene, chloroform, ether, carbon disulfide and petroleum ether. Itaconic acid is a functionalized analog of acrylic acid, the simplest conjugated-alkenoic acid [1].

Itaconic acid finds its place in various industrial applications. Itaconic acid and its polymers can be used to manufacture effective deodorants as they can react with alkaline or acidic odor such as ammonium, amine, and hydrogen sulfide; in paper coatings, metal and concrete paint as it copolymerize with styrene and butadiene to prepare latex. In paints to improve quality and used as fiber carpet sizing agent to make carpet more endurable. Itaconic acid can react with acrylic and methacrylic acid or their esters to prepare resins which can be widely used in emulsion coating, leather coating, coatings for car, refrigerator and other electrical appliances to improve adhesion, colour and weather resistance. With chloro alkyl dimethyl-benzyl-ammonium chloride added, they can be used to prepare water soluble coating for food packaging to reduce bacteria contamination. Esters of itaconic acid can be used in paint, ion-exchange resin, lubricant, binder, plasticizer, and sealant and molding plastics. Itaconic acid derivatives are used in medicine, cosmetics, lubricant, thickener, herbicide and wool modifier. Polymer which contains itaconic acid has special luster and transparency; it is fit for making synthetic cut stone and special lens.

Itaconic acid is manufactured by the catalytic condensation of succinic acid derivatives with formaldehyde [2]. Though, the process is simple it suffers from disadvantages of lot of chemicals consumption, high energy utilization and waste byproducts generation. Fermentation is a relatively clean method to manufacture itaconic acid. *Aspergilluss terreus* species, grown under phosphate-limited conditions, are known to produce itaconic acid during fermentation. The main problem with fermentation technology is that as the acid is generated the pH of system falls [3]. Lowering of pH destroys the bacterial species responsible for growth. This leads to low acid product yield and concentration. Recovery of acid can be a method to gain high yields. Various mass transfer operations methods are available for recovering the acid like electrodialysis, liquid surfactant membrane extraction (LEM), distillitation, ion-exchange, supported membranes, calcium hydroxide precipitation. Each one of them have their advantages and disadvantages [4]. Calcium hydroxide precipitation is good and reliable process used for recovery industrially. Its drawbacks are consumption of large quantities of reagents ($H_2SO_4$ and lime); huge amount of waste generation per ton of acid produced; disposal problem of waste and very poor sustainability. Dialysis have good potential but has drawbacks of frequent cleaning required, membrane fouling and large dialysis unit as compared to fermenter is required. Liquid surfactant membrane extraction (LEM) has large interfacial area for mass transfers in a compact unit. Its drawbacks are complexity of operation and instability. Electrodialysis allow simultaneous separation and concentration of the acid. But requires higher power consumption. Ion-exchange is a reliable technology. Its drawbacks are requires a large amount of chemicals; waste stream generated creates treatment/disposal problem. Distillitation method is well-established technology. Its drawbacks are formation of high-boiling internal esters, dimers and consume more power. Reactive extraction with proper selection of diluents and extractants, can provide high selectivity and extraction, but suffers from toxicity problems of solvents towards microbial strains.

The reactive extraction is one of the cleaner and energy efficient technology widely used for carboxylic acids extraction. With proper selection of extractant and diluents reactive extraction can proved to be an efficient and reliable process with higher concentration and recovery of the acid. In present chapter reactive extraction of itaconic acid using Aliquat 336 in ethyl acetate and toluene was presented. Physical and chemical extraction experiments were conducted. Values of distribution coefficient, degree of extraction using physical and chemical extraction were compared. Chemical extraction proves out to be successful than physical extraction. Equilibrium complexation constant values were found to be higher for Aliquat 336 in toluene than in ethyl acetate, proving it to be better extractant-diluent system.

## Materials and Process

### Materials

Aliquat 336 (Methyltricaprylammonium chloride), is a quaternary amine a mixture of $C_8$-$C_{10}$ with a minimum assay of 80 per cent. The molecular weight is 404.17 and density of 0.888 g/cm$^3$. Itaconic acid purchased from Himedia, india and the diluents ethyl acetate and toluene were of analytical grade and were used without further purification. For titrating NaOH, was obtained from S.D. Fine-Chem Ltd, India. Phenolphthalein solution (pH range 8.2–10.0) was used as an indicator for titrating. The initial aqueous acid concentration range of 0.05 to 0.2 kmolm$^{-3}$ was used.

### Experiments

The experiments were carried out in conical flasks (100 ml). Equal volumes of aqueous and organic phases were added and were made to stirred for 12 hours in a constant temperature (305 K) bath shaker. The solutions were than allowed to settle in separating funnels for 2 hours. Lower phase was subjected to analysis by titration for determination of left acid in aqueous phase. Acid extracted in organic phase concentration was calculated from mass balance.

## Results and Discussion

### Physical extraction

For physical extraction, the distribution coefficient represents the ratio between concentration of extracted species in the volume of corresponding solvent and the total concentration of acid in aqueous phase. Thus,

$$K_{D(diluent)} = \frac{[HA]_{org}}{[HA]_{aq} + [A^-]_{aq}} \quad (1)$$

Figure 13.1 shows the extraction of itaconic acid (0.05–0.2 kmol/m$^3$) by toulene, ethyl acetate. The distribution coefficients were found to be high for ethyl acetate than toulene. The removal of acid by physical extraction is primarily by the solvation of the acid molecules by donor bonds and equilibrium

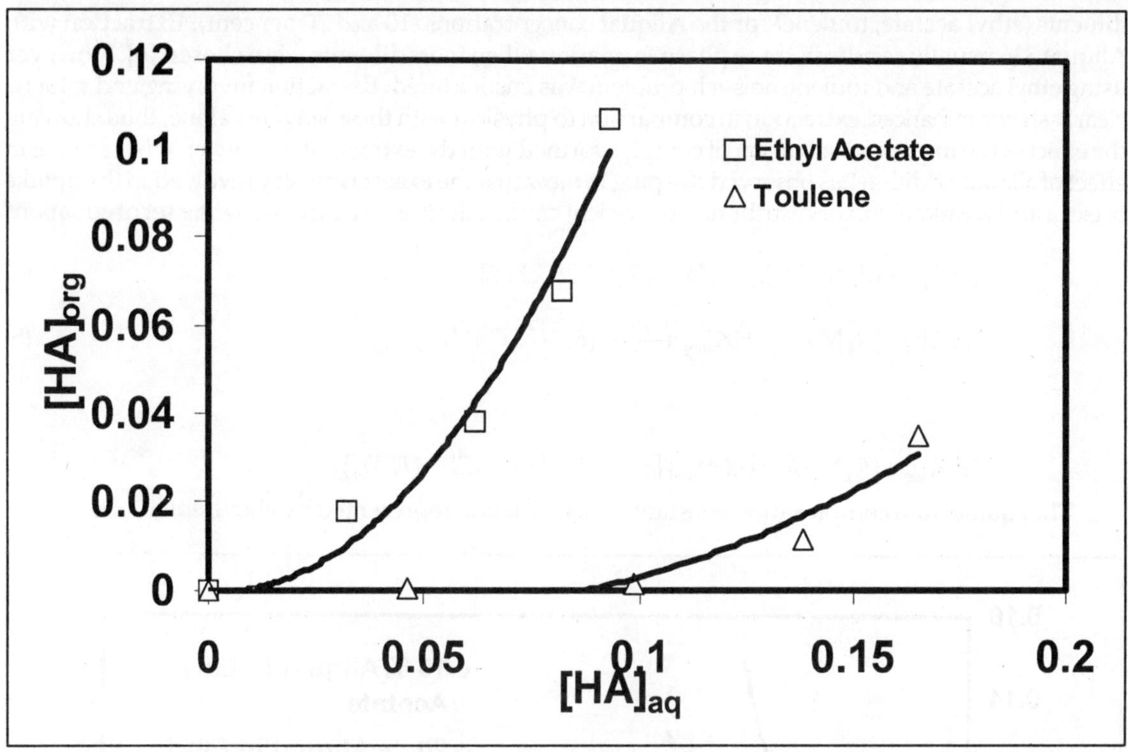


**Figure 13.1: Physical Extraction: Extraction of Itaconic Acid by Ethyl Acetate and Toluene**

solubility is the criteria of determining the maximum loading of the solvent. Ethyl acetate shows higher extraction constant than toluene, yet degree of extraction was found to be less than 50 per cent for ethyl acetate and less than 18 per cent for toluene over whole concentration ranges.

## Chemical Extraction

The low distribution coefficient for pure diluents, suggested the use of an extractant to improve the extraction. Aliquat 336 (10 and 20 per cent), a quaternary amine was used as extractant. Aliquat 336 extracts both the dissociated and undissociated forms of acids [5]. Itaconic acid dissociates in aqueous solution and dimerizes in a non polar solvent. Therefore only acid–extractant complex and undissociated acid are presumed to exist in the organic and aqueous solutions respectively. For the extraction of itaconic acid by Aliquat 336 with chemical interaction, the distribution coefficient of the undissociated molecules is

$$K_{D(HA)} = \frac{[R_4N^+Cl^- : HA]_{org}}{[HA]_{aq} + [A^-]_{aq}} \tag{2}$$

$$\text{Degree of extraction E} = \frac{100 \times K_{D(HA)}}{1 + K_{D(HA)}} \tag{3}$$

The extraction involves a chemical reaction between the quaternary amine and acid. Figure 13.2 shows the plot of organic versus aqueous acid phase equilibrium concentrations for the various diluents (ethyl acetate, toulene) for the Aliquat concentrations (10 and 20 per cent). Extraction with Aliquat 336 usually results in three phase formation when inert diluents were chosen [6]. However using ethyl acetate and toulene no such problem was encountered. Extraction involving and toluene clearly shows enhanced extraction in comparison to physical with these solvents alone, thus showing the effect of the improved solvation of complex formed with the extractant. However, ethyl acetate in effect of aliquat addition has observed Keeping in view that the exact chemistry involved in the uptake of extra acid is unknown, the distribution coefficient can be interpreted by the following set of equations

$$[HA]_{aq} + [R_4N^+Cl^-]_{org} \xleftrightarrow{K_{11}} [R_4N^+Cl^- : HA]_{org} \quad (4)$$

$$[HA]_{aq} + [R_4N^+Cl^- : HA]_{org} \xleftrightarrow{K_{11}} [R_4N^+Cl^- : (HA)_2]_{org} \quad (5)$$

$$\vdots$$

$$[HA]_{aq} + [R_4N^+Cl^- : (HA)_{n-1}]_{org} \xleftrightarrow{K_{11}} [R_4N^+Cl^- : (HA)_n]_{org} \quad (6)$$

The equilibrium complexation constant for the reaction represented by above equations is:

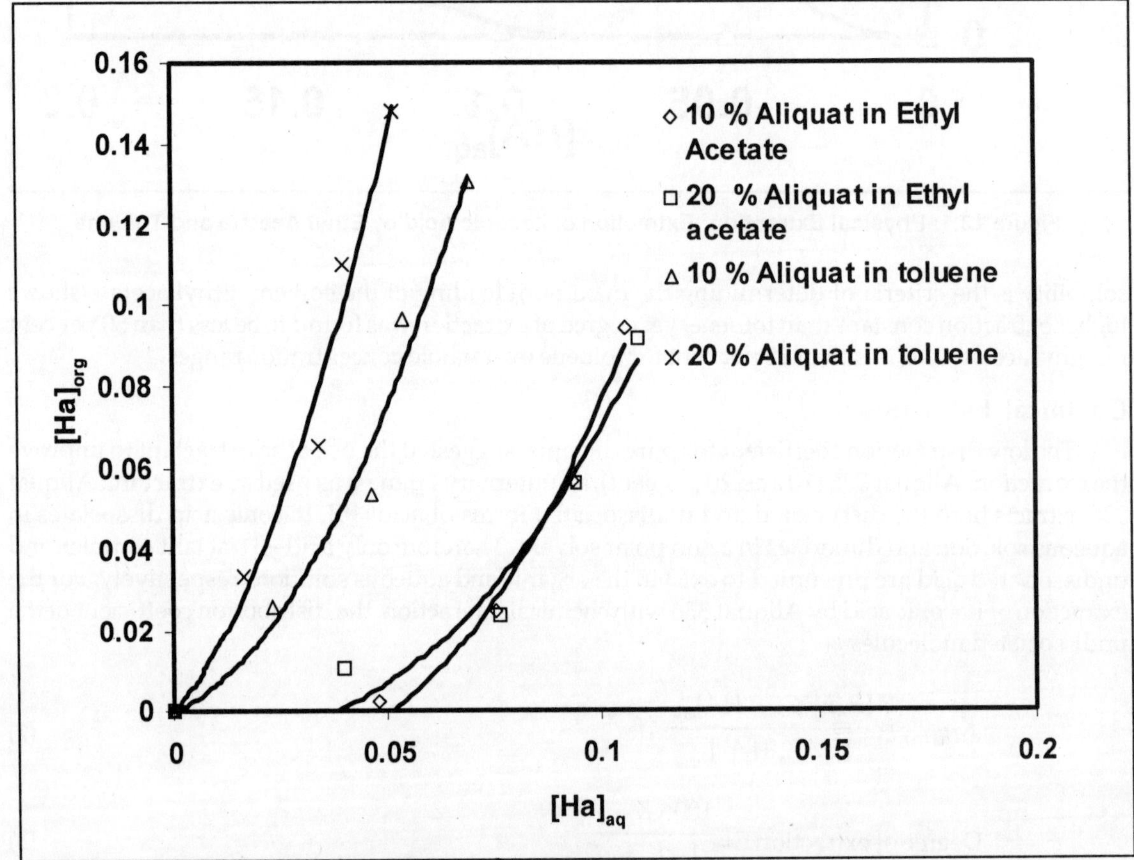


**Figure 13.2: Chemical Extraction of Itaconic Acid Using Various Concentration of Aliquat 336 in Ethyl Acetate and Toluene**

$$K_{En} = \frac{\left[R_4N^+Cl^- : (HA)_n\right]_{org}}{\left[R_4N^+Cl^-\right]_{org}\left[HA\right]_{aq}^{\,n}} \tag{7}$$

The extent to which the organic phase (amine +diluent) can be loaded with carboxylic acid is expressed as the loading ratio, $z$;

$$z = \frac{[HA]_{org}}{[R_4N^+Cl^-]_{i,org}} \tag{6}$$

The value of $z$ depends on the extractability of the acid strength of the acid-base interaction and its aqueous concentration. The stoichiometry of the overall extraction reaction depends on the loading ratio in the organic phase, $z$. Since in the extraction of itaconic acid using Aliquat 336 in various diluents, organic phase is not highly concentrated *i.e.* loading ratios is less than 0.5, the (1:1) complex was formed and the following equation holds

$$\frac{z}{1-z} = K_{E1}[HA]_{aq} \tag{7}$$

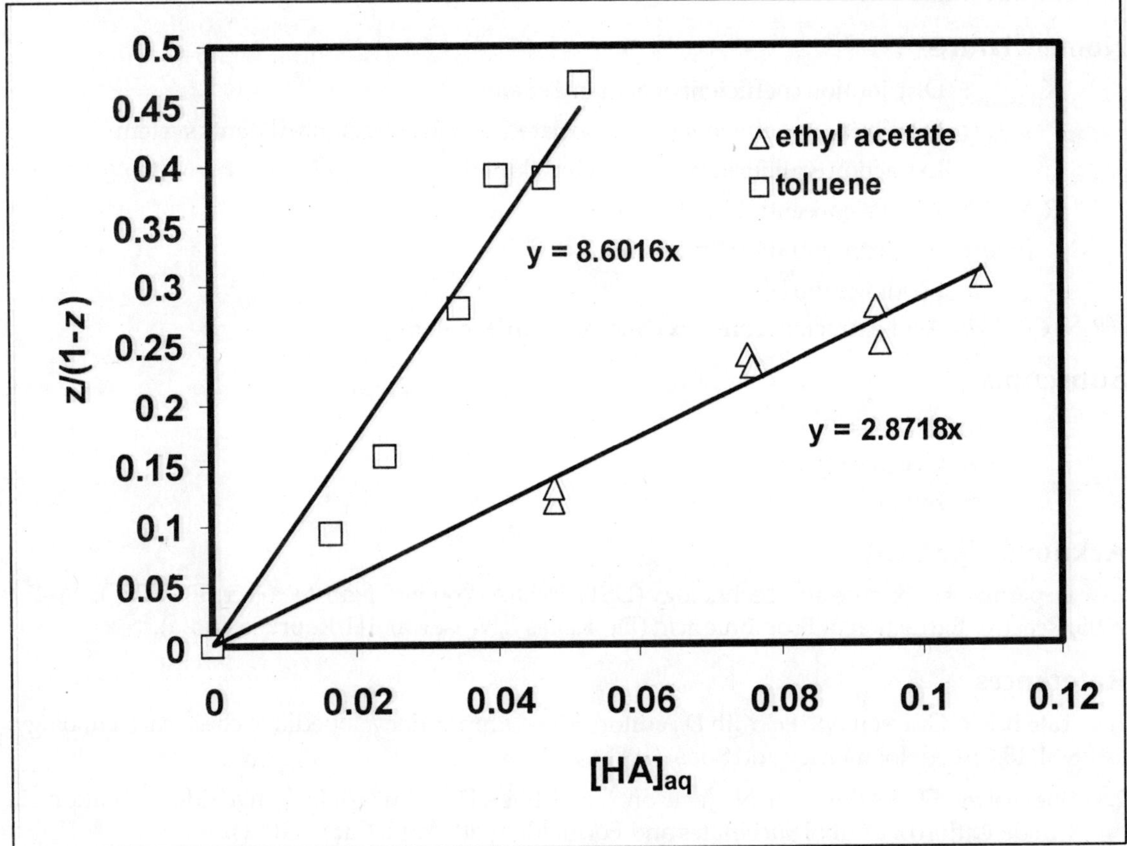


**Figure 13.3: Chemical Extraction: Extraction of Itaconic Acid by Ethyl Acetate and Toluene**

Figure 13.3 shows the plot of *z/1-z* vs $[HA]_{aq}$ for the extraction of itaconic acid using Aliquat 336 in ethyl acetate and toulene. A straight line passing through the origin was obtained. The values of $K_{E1}$ increase in the following trend Aliquat + toluene (8.60) was higher than Aliquat + ethyl acetate (2.87).The large difference among $K_{E1}$ values for the same acid in different diluents indicate that solvation by the diluent is a critical factor in the extraction of acid. Aromatic diluents produce slightly higher $K_{E1}$ values, which has been rationalized as solvation due to interaction of the aromatic electrons with the complexes.

## Conclusion

Amines are effective extractants used for extraction of carboxylic acids. They are generally viscous, so they are dissolved in diluents, to enhance the physical properties. In present chapter Aliquat 336, a quaternary amine was used to extractact itaconic acid form its dilute solutions. Toluene and ethyl acetate were chosen as diluents. Physical extraction give low extractions with degree of extraction less than 50 per cent. Chemical extraction with Aliquat 336 in these diluents provides higher extraction particularly for toluene. Degree of extraction increases from less than 20 per cent for physical to greater than 74 per cent for Aliquat + toluene. Equilibrium complexation constant was also higher for Aliquat + toluene system then for Aliquat + ethyl acetate. However loading ratio was less that 0.5 in all cases so only (1:1) acid: amine complex were assumed to be formed. Higher distribution coefficient and equilibrium complexation constant values for toluene confirm it to be a better diluent for Aliquat 336.

## Nomenclature

$K_{D,(diluent)}$: Distribution coefficient of acid in diluents

$K_{D(HA)}$: Distribution coefficient of undisociated acid in extractant-diluents system

$K_E$: Extraction equilibrium constant [m³/kmol]

$[R_4N^+Cl^-]$: Aliquat concentration [kmol/m³]

$[HA]_{aq}$: Acid concentration [kmol/m³]

z: Loading ratio [-]

$[R_4N^+ Cl^-:HA]$: Acid-extractant complex concentration [kmol/m³]

## Subscripts

aq: Aqueous phase

org: Organic phase

o: Initial

## Acknowledgement

Department of Science and Technology (DST), INDIA, Young Scientist Project, SR/FTP/ETA-43/2005, Reactive Extraction of Propionic acid (Dr. Kailas L. Wasewar, IIT Roorkee, India).

## References

1. Tate B. E.Y, Grayson, M., Eckroth D., editors Kirk-Othmer, Encyclopedia of chemical technology, vol. 13, 3rd ed. John Wiley and Sons, (1981).

2. Shekhawat, D., Kirthivasan, N., Jackson, J. E., Miller, D. J. Citraconic Anhydride Formation via Condensation of Dialkyl Succinates and Formaldehyde. Appl. Catal. (2001).

3. Willke T., Vorlop, K.D., Biotechnological production of itaconic acid. Appl. Microbiol. Biotechnol.; 56, (2001), 289–295.

4. Wasewar, K. L., Pangarkar, V. G., Yawalkar, A. A., Moulijn, J. A., Fermentation of Glucose to Lactic Acid Coupled with Reactive Extraction: A Review, Ind. Chem. Eng. Res., 43, (2004), 5969-5982.

5. Kyuchoukov, G., Yankov, D., Marinova, M., Molinier, J., Albet, J., Mechanism of Lactic Acid Extraction with Quaternary Ammonium Chloride (Aliquat 336), Inst. of Chem. Engg., Bulgaria, Inst. National Polytechnique de Toulouse France, Ind. Eng. Chem. Res., 44, (2005), 5733-5739.

6. Keshav, A., Chand, S., Wasewar, K. L., Proceeding of International Symposium and 60th Annual Session of IIChE in association with International Partners (CHEMCON-2007), 27-30th December, Calcutta, India (2007).

# Chapter 14

# Biofilms in Environment: A Review

**Sharvari V. Gaidhani[1], Dhara P. Sachdev[1,2], Karishma R. Pardesi[2], Smita S. Bhuyan[2] and Balu A. Chopade[1,2]***

**[1]Institute of Bioinformatics and Biotechnology (IBB), University of Pune, Pune, India**

**2–Department of Microbiology, University of Pune, Pune, India**

## ABSTRACT

Microorganisms have a unique ability of adapting and surviving in different environments. One such hostile environment is the rhizosphere a nutrient rich region around plant roots. Plant growth is the consequence of profound activity in this area. Soil microorganism combat adverse conditions such as nutrient deficiency, competing microorganisms, harsh environments, etc. In order to maintain rhizosphere competence, overcome unfavourable conditions, survive and proliferate microorganims acquire their niche by forming biofilms.

Biofilm formation is controlled and regulated process which is carried out in different steps *viz* adherence, formation of microcolonies, biofilm maturation and detachment of cells. Biofilms provide the organism with various advantages including protection, nutrient availability, metabolic co–operativity, acquisition of new genetic traits, defense and acquiring favourable niche. Biofilm formation is a dynamic process depending upon motility, chemotaxis, surface adhesins, matrix components and quorum sensing. Biofilms have acquired importance in clinical fields and is fast gaining importance in agricultural field. The present review gives a comprehensive account of need, process and factors affecting biofilm formation.

***Keywords:*** *Biofilm, Plant Growth Promoting Rhizobacteria (PGPR), Clinical biofilms, PGPR biofilms*

---

* Corresponding Author: E-mail: directoribb@unipune.ernet.in; chopade@unipune.ernet.in

## Introduction

Bacteria are the most successful life forms on earth with respect to the variety niches colonized by them. The important reason behind being the adaptability of bacteria towards any kind of environment (Brown, 1985). They can survive in remarkably diverse habitats. (Pace, 1997, Whitman, 1998). Prokaryotes can inhabit the environment where higher life forms are present or even in the inhospitable environments (Madigan *et al.*, 1997) One such hostile environment is provided by plants and soil under the influence of plant roots. Plants influence their environment by synthesizing complex organic compounds using inorganic material and sunlight. In this process the plants fix carbon, amongst which 40 per cent of it is released in soil in the form of root exudates. They also include compounds which are rich in nitrogen and phosphorus (Lynch, 1990).

The root exudates mainly include chemicals such as indole acetic acid (IAA), indole butyric acid (IBA), flavonides, sucrose, etc (Vancura, 1980, Lugtenberg *et al.*, 1999, Whipps, 1990). The soil which is under immediate influence of these root exudates is known as 'Rhizosphere' whereas the soil which is not under influence of the root exudates is known as 'Bulk soil'. Microbial diversity in rhizosphere is very high as compared to bulk soil due to root exudates secreted by plants (Molina *et al.*, 2000).

Bacteria can colonize on leaves, roots, seeds and internal vasculature of plants. The flora differs significantly due to difference in the type of tissue having unique physical and chemical properties. Until now, perception of bacteria as unicellular life form was deeply rooted since; single cell could be practically obtained by diluting the samples. Various studies have also been carried out by culturing the bacteria in solid and liquid medium, which has enlightened us about the amazing facets of microbial physiology (Davey, 2000).

Since bacteria can colonize and grow practically in any environment, the review herewith deals biofilm formation in rhizosphere. Plant growth promoting bacteria form biofilms by utilizing root exudates as nutrients. It will also take into account various activities being carried out and ecological significance of biofilm formation in plant associated bacteria. Along with this biofilms in reference to the clinical aspects has also been discussed.

## Rhizosphere

Rhizopshere is part of soil in immediate vicinity of plant roots and under the influence of the root exudates. Due to these root exudates, rhizosphere is relatively rich in nutrients as compared to the bulk soil. In response to these exudates the variety of organisms differs. The viability in such an environment strongly depends upon water content of the soil (Normander *et al.*, 1999). There are many factors which determine the rhizosphere competence of bacteria. These mainly include genotype, strain, soil type, pH, minerals, nutrient and water content, species, genotype and physiological state of the plant and the presence of other microbial species (Albareda *et al.*, 2006).

There is a special group of bacteria which can utilize these root exudates in the form of nutrients. They not only utilize the root exudates and acquire shelter on roots of the plants but also help in plant growth promotion hence they are termed as 'Plant Growth Promoting Rhizobacteria' (PGPR). Functions of PGPR include nitrogen fixation, increased supply of nutrients, production of plant growth promoting hormones, solubilization of organic phosphate and control of bacterial, fungal and insect diseases. Presence of nutrient rich environment in rhizosphere, allows many organisms to colonize rhizosphere. Importantly two groups are found they are commnesals (beneficial) and pathogens (disease causing) (Watt *et al.*, 2006). Survival of plant in the rhizosphere primarily depends upon the ability of the plant to get adapted to the local environment. The local environment is constituted by the plants growing in the vicinity and the bacterial population thriving on them. The changes in the environment mainly include growth and development of the neighbouring plants species and microorganisms surviving on the plant. In response to changes in the environment plant roots start producing small molecules and gets engaged in positive and negative communication.

With a wide variety of organisms and root systems around there are three main types of interactions can be seen in rhizosphere they are (Travis *et al.*, 2003).

- ✰ Root–root communication
- ✰ Root–microbe communication
- ✰ Root–insect communication

## Root–Root Communication

Roots of the plant system are constantly in contact with the root systems of the other plants in the vicinity. Roots recognize the invading roots and restrict them with help of chemical messengers which include small molecules and proteins (Stintzi and Browse, 2000; Stotz *et al.*, 2000). Allelopathy is induced in plants by secretion of secondary metabolites which play an important role in establishing plant life on earth. Bais *et al.* (2002c) has reported (±)–catechin as a phytotoxic chemical responsible for invasive behaviour of knapweed in the rhizosphere. (-)–catechin showed the allelopathic effect whereas (+)–catechin exhibited inhibition towards soil borne bacteria (Bais *et al.*, 2002c). From the above example it can be very well understood that the plants utilize the secondary metabolites to regulate the rhizosphere from the detrimental activities of the neighbouring plants. Some parasitic plants belonging to family *Scrophulariaceae* invades the roots of economically important plants such as maize (*Zea mays*), sorghum (*Sorghum bicolor*), millet (*Panicum milaceum*), rice (*Oryza sativa*) and competes with the plant roots to deprive them from water, minerals and essential nutrients (Yoder, 2001).

## Root–Microbe Communication

Root–microbe communication is another important process that characterizes the underground zone. Compounds such as flavonides identified in root exudates play an important role in root–microbe interactions. The root exudates in legumes activate *Rhiozbium melitoli* genes responsible for the nodulation process (Peters *et al.*, 1986). It was observed by Bais *et al.* (2002b) that the hairy root cultures of sweet basil (*Ocimum basilicum*) could produce the rosameric acid (RA) in response to the fungal cell wall extracts from *Phytophthora cinnamoni.* RA showed a potent antimicrobial activity on *Phytophthora cinnamoni, Pythium ultimum* and soil borne pathogens such as *Pseudomonas aeruginosa* (Bais *et al.*, 2002b). Along with this the metabolites produced by the bacteria also have the anti–microbial activity that helps the plant in being protecıed form the pathogens. The cell–cell communication i.e quorum sensing also plays an important role in this process (Walker *et al.*, 2003). Thus, it is understood that roots as well as organism in the rhizosphere have an ability of influencing each other.

## Root–Insect Communication

The root insect interactions are mainly confined to the leaves and the stem. It has not been much explored in terms of the root–insect communication. Few attempts that were made for studying communication include development of a coculture of the hairy roots and aphids (Wu *et al.*, 1999). It was found that the yield of the plant was decreased significantly and the production of the polyacetylenes increased. (Wu *et al.*, 1999). Poly acetylenes are produces as one of the factors of phytoalexin response (Flores *et al.*, 1988). Thus, more attempts can be made so as to find more about the root insect communication.

## Plant Growth Promoting Rhizobacteria (PGPR)

Organisms such as *Pseudomonas* spp. and filamentous bacteria were examined and quantified by Watt *et al.* (2006) using Fluorescence in situ Hybridization (FISH). These organisms were found to colonize 40 per cent of root area on the visible areas such as root hairs, axial grooves, cap cells or soil particles. *Pseudomonas* spp. were distributed evenly along the length of the root; and the filamentous bacteria occurred mostly in the area of the root cap and the mature zone of root (Watt *et al.*, 2006). *Pseudomonas* spp, such as *P. putida*, *P. fluorescence* and various other natural isolates which are commonly found on leaves and in soil also form biofilms (Ude *et al.*, 2006). *Pseudomonas fluorescence* CHA0 acts as a biocontrol agent against pathogenic fungi (Bianciotto *et al.*, 2001).

Amongst the root colonizers, nitrogen fixing symbionts of legumes, *Rhizobium* spp. are very important. The invasion, colonization and nodulation has been reported earlier by Gage, D. J (2004). This process requires attachment of soil particles and plant roots. Organisms such as *Rhizobium leguminosarum* bv *viciae* and others such as *Sinorhizbium melitoli* form biofilms. Clusters of these have been observed in the root hair zone during initial nodulation step (Fujishige, 2006).

PGPR not only includes the root colonizing bacteria, but also free living diazotrophes. *Azospirillum brasilense* releases a phytohormone which is more beneficial in plant growth rather than its ability to fix nitrogen (Burdman 2000, Assmus,1995). Organisms such as *Bacillus subtilis* 60SI is also an important PGPR. It colonizes the roots of *Arabidopsis thaliana* and protects the plant against *Pseudomonas syringae* pv. *tomato*. The protection being provided by the production of a lipopeptide known as surfactin. Surfactin is cidal for pathogen (Bias, 2004). A broad host range rhizobacteria *Paenibacillus polymyxa*, induces drought tolerance in *Arabidopsis thaliana*. It has a range starting from grasses to trees. It not only provides tolerance, but also protection from pathogens by synthesis of several antibiotics (Timmusk, 2004). Few species of saprophytic, non pathogenic *Agrobacterium* harbour a conjugative virulence plasmid which can cause disease such as crown gall, hairy root and root mat (Escobar, 2003).

Thus, it appears that PGPR are the organisms which can colonize plant roots, promote plant growth and reduce disease incidences. When seeds are inoculated with specific organisms it helps the new developing plant in utilizing various nutrients. The three important nutrients required by plants from the soil, include Nitrogen (N), Phosphorus (P) and Potassium (K), some also require sulphur (S) in addition. Though sulphur is available in the soil in various mineral forms then also the plants are not able to utilize it directly, it is first transformed by bacteria to sulphates.

PGPR improves the nutrient uptake of plant by altering plant hormone levels. This changes the rate of root growth and shape by increasing root branching, root mass, root length and the amount of root hair. This helps in increasing the water uptake capacity of plant thus increasing yield of the plant. PGPR promotes plant growth in following ways:

- Increasing nitrogen fixation in legume.
- Increasing the supply of other nutrients, such as phosphorus, sulphur, iron and copper.
- Producing plant hormones such IAA, IBA, flavonides, cytokines, etc.
- Enhancing growth of other beneficial bacteria or fungi.
- Controlling the fungal and bacterial diseases.
- Controlling the insect pests.

These properties of the PGPR have been exploited to large extent and various commercial products are available in the market, which increase the plant productivity (Table 14.1).

**Table 14.1: Commercial Product Used in Agricultural Field**

| Sl.No. | Product | Usage | Effect |
|---|---|---|---|
| 1. | BioBoost | Canola | Increases the yield on an average to eight bushels per acre |
| 2. | Mycostop | Wide range of soil borne seed and root diseases | Control of growth. |
| 3. | Subtilex | Fungal pathogens | Increases nitrogen fixation and controls the fungal pathogens. |

*Source*: Stuart McMillan, Cnadian Organic Grower (http://www.cog.ca/documents/075U_PGPR.pdf).

## Biofilm

### Definition

The microorganisms that are found in the rhizosphere or natural environment exist in form of multicellular aggregates. These multicellular aggregates are known as biofilms. Biofilms are associated with solid surfaces and in intimate contact with other cells (Webb, 2003, Parsek, 2004, Stoodley, 2002). Costerton *et al*. (1995) defined biofilm as "a structural community of bacterial cells enclosed in a self produced polymeric matrix, adherent to an inert or living surface". "A functional consortium of microorganisms organized within an extensive exopolymer matrix", is the description given by Elder and colleagues (Elder *et al.*, 1995). This concept was further simplified as "a community of microbes embedded in the organic polymer matrix, adhering to a surface" by Carpentier and Cerf (Carpentier and Cerf, 1993).

### Ubiquitous Nature of Biofilms

Reports and reviews explain the ubiquity of the biofilms. Ubiquity refers to ability of a bacterium to adhere and form biofilm practically on any surface *i.e.* biotic or abiotic. The choice of substrate ranges from mineral, vegetables, contact lenses, ship hulls, animals, dairy and petroleum pipelines, rocks in streams, historical monuments, buildings, biochemical implants, transcutaneous devices, nutrient sources such as cellulose in paper industry (Carpentier and Cerf, 1993; Costerton *et al.*, 1985; Costerton *et al.*, 1995; Elder *et al.*, 1995). Evidences of presence of biofilms in the early fossil record are available (Westall *et al.*, 2001). They were mainly formed in the hydrothermal, environments, putative biofilm microcolonies have also been identified by morphology in the 3.3–3.4 billion year old South African Kornberg formation (Westall *et al.*, 2001). Filamentous biofilms have also been identified in the 3.2 billion year old deep sea hydrothermal rocks of the Pilbara Craton, Australia (Rasmussen *et al.*, 2001). Such biofilms can also be found in the modern hydrothermal environments such as hot springs (Reysenbach *et al.*, 2001) and deep sea vents (Taylor *et al.*, 1999).

### Steps in Involved in the Biofilm formation

Biofilms are dynamic in nature. When the formation of biofilm is discussed mainly abiotic surfaces are taken into consideration. But biofilm formation also takes place on the biotic surfaces also. The best available biotic surface for the bacteria to form biofilm is various parts of plants such as leaves, seeds, roots, vessels, etc. According to the current model of *Pseudomonas* spp. biofilm formation is a linear process and starts when the free floating bacterial cells getting attached to surfaces. The attachment is followed by growth and finally into a matured, structurally complex biofilms. The process stops when the dispersal of attached cells in the surrounding fluid starts (Costerton and Stewart, 2001, Sauer, 2002). The process of biofilm formation can be elaborated very largely depending upon the need to understanding the process. Recently the proteomic studies carried out on *Pseudomonas aeruginosa* indicated that the biofilm formation of the organism proceeds as a regulated development sequence. It has been proposed to be carried out in five stages, (Sauer *et al.*, 2002; Stoodley *et al.*, 2002). First and second stages involve loose or transient association with the surface, followed by firm adhesion, third and fourth stage is identified by aggregation of cells into microcolonies and subsequent growth and maturation, stage five is characterized by regaining of motility where cells are sloughed or shed off from biofilm. The five main steps can be diagrammatically represented in Figure 14.1.

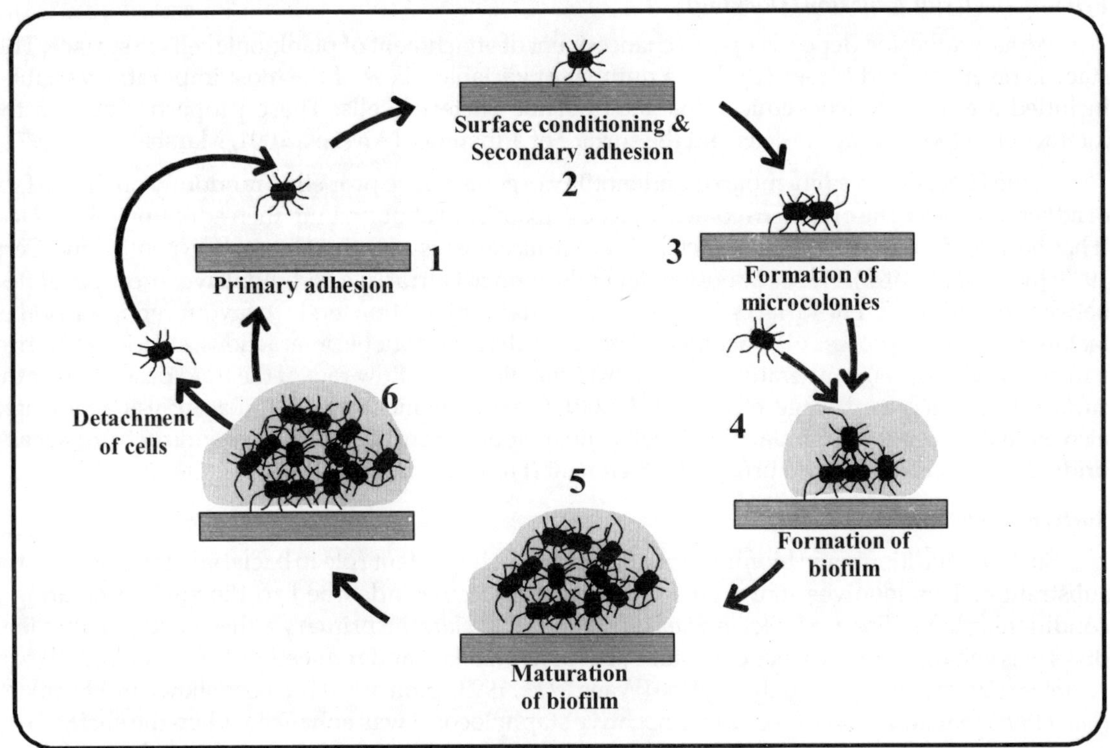


**Figure 14.1: Diagrammatic Representation of Various Steps Involved in the Process of Biofilm Formation**

**1: Primary adhesion of bacteria to the surface; 2: Surface conditioning and Secondary adhesion; 3: Formation of microcolonies; 4: Formation of Biofilm; 5: Biofilm maturation; 6: Detachment of cells.**

## Adhesion

The process can be explained in a very simple manner. The process is regulated or dictated by many variables such as species of the bacterium, surface composition, environmental factors and essential gene products. The mechanism followed for bacterial surface adhesion depends upon the fact, whether the substrate is biotic or abiotic. Attachment of the bacterium to the abiotic surface is mediated through non specific interaction such as hydrophobic interactions. Similarly the adherence to live or devitalized surfaces is achieved through specific molecules such as lectin, ligands or adhesions. (Carpentier and Cerf, 1993). The attachment of bacteria to the roots of plants depends upon its growth in the nutrient rich environment or in response to the specific nutritional conditions (O'Toole and Kolter, 1998)

Bacterial adherence or attachment can be divided into three main steps they are (An *et al.*, 2000, Pearce *et al.*, 1995; Marshall 1985, Boland *et al.*, 2000, Gristina, 1987)

- ✰ Primary Bacterial Adhesion (Docking)
- ✰ Surface conditioning
- ✰ Secondary Bacterial Adherence (Locking)

***Primary Bacterial Adhesion (Docking)***

Primary adhesion depends upon a chance event of attachment of planktonic cells to surface. This stage is reversible and is regulated by a number of variables. Few of the most imporatne variables included are the physicochemical characters of the bacterial cells. These properties govern the interaction between bacteria cells and the surface of adherence (An *et al.*, 2001, Marshall *et al.*, 1971).

Using factors such as chemotaxis and motility organisms are propelled randomly on the surface of adherence. Once the organism comes in close proximity of about < 1 nm then adhesion takes place. It has been reported earlier that he bacterial cell surface posses negative charge (Carpentier and Cerf, 1993; Jucker *et al.*, 1993). Hence adhesion depends upon net attractive and repulsive forces generated between the two different surfaces. Thus the electrostatic interaction tends to favour repulsion. Other factors also govern process of attachment. They include, hydrophobic interactions, stearic hindrance, van der Waal's forces, temperature, pH, hydrodynamic forces, flow rate of the fluid passing over the surface, bacterial growth stage, etc (An, *et al.*, 2001; Carpentier and Cerf, 1993). Bacterial attachment id also mediated through fimbrae, pili, flagella and exopolysacchrides (EPS). These structures act as a bridge between bacteria and surface of attachment (Flint 1997, Kumar 1998).

***Surface Conditioning***

Surface conditioning in biofilm formation plays an important role in bacterial attachment to the substratum. This involves inorganic and organic molecules adsorbed to the surface creating a conditioning layer. For example, in *Staphylococcus epidermidis* the primary adhesion to polyethylene disks was enhanced in presence of surface activated platelate and reduced due to adsorbed plasma proteins relative to uncoated polyethylene (Wang, *et al.*, 1993). Similarly it has been shown by Herrmann *et al.* (1998) that adhesion of coagulase negative staphylococci was enhanced when the surface was coated by various types of plasma proteins which included fibronectin. A dose dependent reduction was observed by Dunne and Burd (1993) due to fibronectin and proteolytic fragments of fibronectin in the adhesion of *S. epidermidis* to a coated plastic surface. Meat or milk products and whey proteins help not only in increasing bacterial adhesion of several milk associated organisms (Kumar 1998).

Thus, studies have shown that the environments where there are plenty of nutrients available nutrients act as a primer to increase the ability for surface attachment (Watrickand Kolter, 2000; Carpentier and Cerf, 1993; Costerton *et al.*, 1995; ZoBell, 1937). The repulsion, if present in between the two surfaces is overcome by specific molecular interactions mediated by adhesins present on the cell surface, such as pili (An, Y. R *et al.*, 2000; Boland *et al.*, 2000).

The efficiency of primary adhesion depends on all the factors mentioned above. But in general the surface chemistry of the cell pushes the equilibrium in favour of adhesion considering that the organic substances tend to aggregate near surface making it nutrient rich and the organisms congregate near such areas (Carpentier *et al.*, 1993; Costerton *et al.*, 1995; ZoBell, 1937).

### *Secondary Bacterial Adhesion (Locking)*

The process of irreversible attachment starts after surface conditioning is completed. As attachment process is irreversible, it is also known as locking. Specificity between adhesions and surface is used for the process of the process (An *et al.*, 2000). During this step cells are found in different forms such as free floating (planktonic) and aggregates. The free floating cells, start synthesizing exopolysacchrides (EPS) and join the process of adhesion. The EPS synthesized, then forms a complex with surface material or receptor or specific ligands such as pili or fimbrae of bacterial cell. This allows organisms to bind firmly to the substratum.

The type of adhesion differs form organism to organism. This can be very explained on the basis of the reports of *Vibiro cholerae* $E_1$Tor by Watnick *et al.* (1999). It was observed that organism uses toxin coregualted pilus during the process of human infection, but when it has to adhere to abiotic surface it uses mannose sensitive haemagglutinins. There have been few reports which say that though EPS plays an important role, then also it is required in later stages and not in the primary process of adherence. *Flavobacterium psychrophilum* is known to cause a bacterial cold water disease known as rainbow trout syndrome (Bernardet *et al.*, 1996; Lorenzen *et al.*, 1997). It possesses a polysaccharide rich capsular structure and presence of loosely attached outer layer. (Crump, I *et al.*, 2001). Moller *et al.* (2003) reported that the adherence of the organism increases due to presence of sialic acid binding lectin. There are several reports that indicate EPS is not necessarily required for the process of adhesion, but is required for the later stages such as development of biofilm matrix (Allison, 1987; Sutherland, 1999).

### Formation of the Microcolonies and Biofilms

Once organisms are attached to the surface then they start accumulating leading to formation of small colonies these colonies. These colonies are known as microcolonies. The basic requirements for the formation of the biofilm are presence of flagella or pili for the purpose of motility preferably twitching motility. It is a type of motility carried out with the help of pili and not flagella. It is a kind of surface translocation produced by an organism which varies micromorphologically from swarming growth. The cells move predominantly singly but aggregates can also be found. The movement is intermittent and jerky and does not regularly follow the long axis of the cell (Lautrop, 1961).

A number of surface attachment defective (sad) mutants were studied. It was found that the organisms which were defective in the flagellar motility showed blockage in the initial attachment of the organism to the surface. Similarly there was another set of sad mutants were also defective in type IV pili. These organisms were able to attach to surface as the wild type were able to, but were unable to form microcolonies. This occurred due to inability of the organism to show twitching motility due to the defective type IV pili. This proved that twitching motility is required for the assembly of a monolayer and formation of the microcolonies in *Pseudomonas aeruginosa*. (O' Toole *et al.*, 1998).

Pratt and Kolter (1998) screened strains of *E. coli* and reported that type I pili are required for initial attachment of organism. After screening it was found that the strain defective in production of type I pili were also defective in biofilm formation. One protein of *E. coli* Ag43 is also required in the biofilm production. This is a phase variation protein required for the cell–surface and cell–cell contact. (Danese *et al.*, 2000) The role of surface structures in *Vibrio cholerae* is similar to those in *E. coli*. Here also flagella are required for bringing organism in close proximity of surface for bacterial spread.

Along with factors such as presence of surface adhesions, flagella, pili, fimbrae another important factor required for the formation of microcolony and biofilm is quorum sensing. These signals are produced by organisms to modulate and coordinate their interactions with plants. Quorum sensing influences formation of biofilms to a great extent (Fuqua and Greenberg 2002).

### Biofilm Maturation

After the process of initial attachment bacterial cells undergo initial adaptation to life in the biofilms. Enhancement of two properties has been observed after adaptation. They include increased synthesis of EPS and development of antibiotic resistance. Irrespective of enhanced ability of these two events there are various other properties that are developed within biofilm by bacterial cell. These properties include increased resistance to UV, increased rate of genetic exchange, altered biodegradation capability and increased secondary metabolite production (O' Toole *et al.*, 2000, Goodman *et al.*, 1994).

In biofilms produced by *Pseudomonas* production of EPS known as alginate is enhanced. (Govan *et al.*, 1996). Though the regulation of alginate production is a complex process, involving a number of regulators two of the few studies have reported that contact of bacteria on an abiotic surface leads to an up regulation of the alginate biosynthesis genes (Darzins *et al.*, 1997, Davies *et al.*, 1993). Along with the alginate biosynthesis the quorum sensing molecules known as acyl homoserine lactones (acyl–HSLs) are also required for biofilm maturation (Davies *et al.*, 1998). *Las I* gene codes for an enzyme required for the synthesis of the N–(3–oxododecanoyl)–L–homoserine lactones. Davies *et al.*, showed that a mutant in the *Las I* possesses the capability of interacting with the surface, but cannot produce the similar kind of biofilm architecture of a *Pseudomonas* biofilm (Davies *et al.*, 1998).

EPS defective mutants of *V. cholerae* are also defective in early stages of biofilm development (Watnick *et al.*, 1999; Yildiz *et al.*, 1999) and are unable to produce biofilm architecture as observed in wild type (Watnick and Kolter, 1999). As alginate is required for the development of biofilms in *Pseudomonas*, similarly colonic acid required in the *E. coli*. Colanic acid the major EPS produced by the *E. coli*. It is also required for the development of characteristic architecture of biofilm of *E. coli* but is not involved in the initial events of the attachment. When a screening for the regulation of the genes involved in biofilm formation, it was found that genes OmpC porin and *wca* locus (synthesis of the colonic acid) are up regulated whereas *fliC* (flagellar synthesis) gene is down regulated (Prignet–Combaret *et al.*, 1999).

### Detachment

The last stage in the development of biofilm is the detachment of the cells from the surface. Detachment of the cells from biofilms returns them back to the planktonic cell life. The probable reason behind returning back to the planktonic cell life is till not clear, but it probably can be starvation of the organism in the biofilm. Alginate degrading enzyme alginate lyase may also play major role in detachment of the cells in *Pesudomonas aeruginosa* (Boyd and Chakrabarty, 1995). Both of them showed that the over expression of enzyme causes speedy detachment of cells from biofilm. Reports of *Pseudomonas fluorescens* suggests that biofilm decreased as incubation time increased (Allison *et al.*, 1998). Not much work has been carried out in this field and needs to be explored to a larger extent.

## Biofilm Formation in Rhizosphere

### Initiation of Colonization

Studies on *Pseudomonas putida* showed that root colonization of the organism was stable. It was observed that bacteria population reached its maximum after 1 -2 days of seed inoculation (Espinosa–Urgel *et al.*, 2002). O'Toole and Kolter (1998 a, 1998 b) reported that colonization of roots is initiated either in nutrient rich environment or in response to specific nutritional conditions.

It has been observed that flagellar motility is required by pathogenic *E. coli* for colonizing host tissue (Giron *et al.*, 2002). But still it remains a controversy whether flagellar motility is required for attachment to plant roots. Repots by de Weger *et al.* (1987) and Turnbull *et al.* (2001a, b) demonstrated that flagellar motility is essential in root colonization of tomato. It also played an important role in, in vitro root colonization of wheat by *Pseudomonas fluorescence*. In similar conditions when studies were carried out on mutants of *P. fluorescens* with impaired motility, it was observed that mutants showed decreased ability of root colonization (Turn bull *et al.*, 2001b).

Bacterial motility towards the roots is dependent upon a factor known as chemotaxis. Movement of *Azospirillum brasilense* towards root exudates has been demonstrated by Van Bastelaere and his co-worker (1999). Not only in *Azospirillum*, but *Pseudomonas fluorescens* shows that activity of chemotaxis. It also plays an important role in root colonization (Vande Brock and Vanderleyden, 1995, Vande Brock *et al.*, 1998).In both organisms chemotaxis plays an important role in the process of root colonization (Vande Brock and Vanderleyden, 1995, Vande Brock *et al.*, 1998, Van Bastelaere *et al.*, 1999).

When initiation events are compared for pathogenic organisms and plant associated bacteria it can be observed that, in pathogenic organisms presence of flagella, pili or fimbrae is essential for initiation. Whereas in plant associated bacteria along with flagella, pili and fimbrae an additional factor known as chemotaxis is also required.

### Attachment to Roots and Rhizosphere

Attachment of the organisms is not only mediated through flagella as adhesin but they can use also use type IV pili and lipopolysaccharide (LPS) (Dorr *et al.*, 1998, Lugtenberg *et al.*, 2001). A protein mediating cell–cell aggregation, Agg A also participates in adhesion and colonization of roots (Bull and Anderson, 1992). The efficiency of root colonizer not only depends on many factor. Root colonization is also regulated by two component system *colR/cols* and a site specific recombinase (Deklers *et al.*, 1998a, 1998 b). *Pseudomonas fluorescence* has been isolated in form of phenotypic variants from roots. This phenotypic variability occurs due to site specific recombinases which participate in DNA rearrangements (Lugtenberg *et al.*, 2001). The variability was not only observed in planktonic cells but also in biofilm formation pattern (Drenkad and Ausubel, 2002) in form of increased antibiotic resistance.

There are other regulatory processes which includes cell–cell communication through quorum sensing signals. Quorum sensing signals are mediated through acyl homoserine lactones (AHLs) in microcolonies associated to plant roots (Steidle *et al.*, 2001).

Also when a comparison is carried out between the adhesion and microcolony formation steps of pathogenic and plant associated bacteria it is found that the quorum sensing signals are required for both. Also the quorum sensing signal producing molecule are same *i.e.* acyl homoserine lactones (AHLs). Thus, it can be said that process of quorum sensing may be same in both clinical and PGPR biofilms. Also, it may be carried out for the same purpose.

## Advantages of Formation Biofilms

Survival of the fittest is a well known theory. Similarly the organisms try to survive by forming biofilms rather than remaining in the planktonic cell form. In laboratory conditions the organisms acquire nutrients from nutritionally rich medium, but in environment, it always needs to compete with other organisms for food. Biofilm formation by the organism helps in overcoming this difficulty. Various advantages to organisms forming biofilms:

- ✫ Protection from environment
- ✫ Nutrient availability and metabolic cooperativity
- ✫ Acquisition of new genetic traits
- ✫ Defense against stress
- ✫ Acquiring a favourable niche

**Protection from Environment**

Organisms forming a biofilm obtain a certain degree of shelter and homeostatic conditions due to production of EPS. This matrix consists of various components such as EPS, proteins, nucleic acids, etc (Davey and O'Toole, 2000).

Depending upon the experiments carried out by Veiga *et al.* (1997) and Wu *et al.* (1996) it was found that EPS produced by methanogenes contributed to the polymer matrix of the flocs. Due to this EPS the organism produced aggregates, which finally became a place for other organisms to reside in. Thus EPS acts as a backbone of the matrix. It also helps in adsorbing dissolved organic compounds such as diclofop methyl and other xenobiotics from the bulk fluids. This helps the community of organisms to concentrate on essential nutrients and growth components (Wolfaardt *et al.*, 1998). Presence of EPS restricts the entry of any antimicrobial agent to enter inside the matrix. EPS performs this function by acting as an ion exchanger (Gilbert *et al.*, 1997). This can be mainly observed in case of hydrophilic and positively charged antibiotics such as aminoglycosides (Nicols *et al.*, 1988, Nicols *et al.*, 1989, Nickel *et al.*, 1985). EPS also plays an important role in sequestering the metals, cations and toxins (Decho, 1990, Flemming, 1993). Hence form the reports it can be concluded that EPS is plays an important role in protecting the organism form environment by binding it in a matrix.

**Nutrient Availability and Metabolic Cooperativity**

The most important advantage that an organism gains in being a part of biofilm is the availability of the nutrients form environment. Biofilms provide an effective means of exchanging nutrients and metabolites with the bulk fluid in the surrounding. This increases the chances of making the nutrients available to organism as well as in removing toxic metabolic wastes from community (Costerton *et al.*, 1995). The metabolic characteristics of bacteria changes as it forms biofilms. This occurs since the architecture of the biofilm makes the availability of the nutrients very easy with metabolic cooperativity. As a result of this organism gains a niche in biofilm. This allows the organism to form a miniconsortia in the environment and hence experience a range of signals within biofilm (Hirsch, 1984).

Environment consists of various types of genetically and nutritionally diverse organisms. The biofilms form an important place where a different kind of association can be seen. This association is observed in the metabolically distinct organisms. The association between the metabolically distinct organisms is called as the syntrophism. Syntrophism refers to a special type association between the metabolically distinct organisms which are dependent upon each other to utilize certain substrates, typically for energy production. The best example of this can be given of methanogenic bacteria. Methanogenic bacteria can convert the complex organic substrates into the simple forms such as methane and carbon dioxide. The entire process is carried out by three interdependent set of organisms. That is the fermentative organisms catabolise the complex organic matter into the acids, alcohols. This is utilized by the acetogenic bacteria to produce acetic acid in larger amounts with few other acids in minor amounts. Finally the organic acids formed are degraded by the methanogenic bacteria to form methane, carbon di oxide and hydrogen (Schink, 1997).

Since biofilms form a microconsortia *i.e.* the multimember biofilms, it results in the metabolic cooperativity in the organisms and the close proximity facilitates the interspecies substrate exchange, removal or distribution of the metabolic products (MacLeod *et al.*, 1990; Horsekott *et al.*, 1998).

**Acquisition of New Genetic Trait**

In biofilms the organism are held together in close proximity with the help of the EPS. Due to this the organism gains the genetic material by the process known as horizontal gene transfer. The horizontal gene is important in the evolution and in maintaining the genetic diversity. It has been well established that the plasmids are found in the wide range of the organisms and conjugation is the process by which the plasmids are transferred to other organisms (Davey and O'Toole, 2000).

By exogenous isolation methods (Fry and Day, 1990) novel plasmids have been isolated from the marine environment (Dalberg *et al.*, 1997). In this study it was found that the plasmids conferred the resistance towards mercury. They were isolated from bacteria in the biofilms combining them with the nutrient deprived recipient cells. Another set of experiments showed that horizontal gene transfer takes place in *Streptococcus* biofilm formed microcosm dental plaque. Also the *Bacillus subtilis* strain harbouring a conjugative transposon which confers resistance to tetracycline was introduced to the system, and the resistance profile of the biofilm bacteria was assessed (Roberts, 1999). Lebaron *et al.* (1997) also showed that the horizontal gene transfer also occurred in the *E. coli* strains in single species biofilms formed on glass beads within reactor vessel.

Another mode of gene transfer is by the process of transduction. Very little has been known regarding the gene transfer in biofilms. A recent report shows that the biofilm of *E. coli* formed on the polyvinyl chloride was susceptible to T4 phage attack (Doolittle *et al.*, 1995).

**Defense against Stress**

Biofilms being a microconsortia of the organisms are resistant to physical forces such as shear forces produced by fluids. Organisms present in biofilm have an ability to withstand nutrient deficiency, pH fluctuations, oxygen radicals, disinfectants, antibiotics, etc as compared to planktonic cells. In some cases the biofilms are also resistant to the process of phagocytosis (Donlan and Costerton, 2002). If the cells try to phagocytise the biofilms then they cause more harm to the tissue surrounding the biofilm. This invulnerability of the biofilms is not understood completely but it may be conferred by number of biofilm specific characteristics. These mainly include slow growth and physiological heterogeneity of the inhabitants. The most important character that induces the property of resistance in the cells is the ability to form EPS.

In addition to the role played by EPS in aggregation and biofilm structure, it also plays a part in defense enabling biofilms to resist shear forces and phagocytosis by inflammatory cells (Donlan and Costerton, 2002). Lewis proposed that tolerance of biofilms is conferred by antimicrobial agents but a solid evidence proving this is still not available (Lewis, 2001). The organisms form biofilms humans so that they can resist the immune response mounted against them. But in nature what poses them to form biofilms? The probable explanation to this can be given as certain bacteria may have evolved in such a way that they can form biofilms in response to the events mimicked by nature as in human body (Shirttiff, 2002). This can be very well explained by taking an example of *Staphylococci* and *Enterococci* which synthesizes the EPS in response to the iron deprivation and osmotic stress which are the conditions mimicked from the human body (Deighton *et al.*, 1993; Baldassarri *et al.*, 2001).

### Acquiring a Favourable Niche

In the course of evolution human beings have evolved system known as immune system which helps combat organisms trying to attack human body. But along with that the organism have also evolved in a such a way that they can with stand immune responses by producing some of the defense substances. One way of achieving this is formation of biofilms.

The adherence of the bacteria to the body surface is mediated by various proteins known as Microbial Surface Component Recognizing Adhesive Matrix Molecules (MSCRAMM). These proteins include various host molecules such as fibronectin, fibrinogen, vitronectin and elastin. They play a very important role in adherence of the organism to the solid surface of the human body (Patti, 1994). The presence of MSCRAMMs can be found abundantly in the *S. aureus* and *S. epidermidis* (Heilmann, 2003). Similar to this the oral streptococci have evolved to bind to the conditioning film on the tooth surfaces which consists of the salivary glycoproteins and lipids. They produce salivary amylase binding proteins and salivary agglutinin glycoprotein for getting attached to the oral pellicle (Loo, 2003; Burne, 1998). Organisms like *Vibrio cholerae, E. coli, P. aeruginosa, Salmonella enterica* have developed various mechanisms such as development of flagella, pili, and fimbrae. (Lejeune, 2003)

The expression of the adhesins is decreased once the adherence is established completely. Thus, it can be said that role of adhesins and motility factors such as flagella, pili and fimbrae is in the initial stage of the attachment and after the formation of the biofilm these proteins are not required. (Jefferson, 2004).

## Factors Required for Biofilm Formation

Formation of biofilm is under strict regulatory control. The process requires presence of the certain factors which are prerequisites for the formation of biofilms. Though the organism responds to the environmental conditions separately, then also the plant growth associated bacteria acts through conserved regulatory systems. The factors required as listed below:

- ✰ Motility and Chemotaxis
- ✰ Surface adhesions
- ✰ Biofilm matrix components
- ✰ Quorum sensing

### Motility and Chemotaxis

Motility involves the movement of the organism using various structures such as flagella, pili, etc. Type of motility differs from the bacterium to bacterium, depending upon the appendage possessed by cell. The different types of motility shown by organism are flagellar swimming, swarming, twitching, and gliding motility (Henrichsen, 1972). The flagellar motility not only helps the organism in the initial attachment of the organism but also in the biofilm formation. Twitching motility is used by *Pseudomonas* species so as to spread across the surface and form the mushroom shaped structures on the surface. Twitching motility is possible due to the retractable Type IV pili (Klausen *et al.*, 2003). Type IV pili mutant of *R. solanacearum* reduced the virulence on tomato, autoaggregation and biofilm formation (Kang *et al.*, 2002). Motility is not only required in the initiation and development of biofilms, but also in their dispersal, an essential mechanism for spreading and colonizing new habitats (Sauer *et al.*, 2002).

Chemotaxis is an event of migration of the cell towards a certain chemical. Plant releases various types of exudates in the soil from specific sites. Many bacteria preferentially move towards these exudates and utilize them as nutrients. This helps them to colonize the surfaces of the plants which are nutritionally rich. *A. brasilense* prefers aerotaxis to check the oxygen tension which allows the process of respiration but at the same time does not degrade the nitrogenase enzyme (Greer–Phillips, 2004). The rate of chemotaxis by different type of bacteria differs with respect to the nutrient (attractant) and the species. Usually a strong reaction is exhibited towards the host plants by the organisms and it is observed that endophytes are attracted more than Rhizobacteria (Bacilio–Jimezo *et al.*, 2003). *R. solanacarum* chemotaxes efficiently towards various organic acids, amino acids and galactose. It also was observed that different isolates of *R. solanacarum* showed different specificity for the host plants rather than the non–host plants (Yao and Allen, 2006).

**Surface Adhesins**

Surface adhesins are the molecules that are required by the bacteria to adhere to surface of substrate. These are the bacterial surface associated molecules which mainly include polysaccharides and proteinaceous pili and fimbrae. These are used for adherence to surface of inert materials, plant structures and to other bacterial cells. Various adhesins studied in plant growth promoting bacteria include richadhesin (Dardanelli *et al.*, 2003), Rhizobium adhesion proteins (Raps) (Ausmees *et al.*, 2001), glucomannan (Laus *et al.*, 2006), arabinaose containing polysacchrides (Burdman *et al.*, 2000), haemagglutinins (Guilhabert and Kirkpatrick, 2005), Large adhesion protein (Lap A) (Espinosa–Urgel *et al.*, 2000), Outer memebrane porin (OmpA) (Torrens *et al.*, 2005).

Members of the family *Rhizobeaceae*, which also includes *A. tumefaciens*, shows production of the small molecules known as the richadhesins. Richadhesin is a calcium binding protein. The function of this protein in *A. tumefaciens* is still unknown but in rhizobial cells it is actively involved in adhesion to the plant roots (Dardanelli *et al.*, 2003, Smit *et al.*, 1989; Swart *et al.*, 1993). The attachment of rhizobial cells is also mediated through a protein characterized by Ausmees, *et al.* (2001). These are larger than the richadhesins and agglutinate the cells by getting attached to the $Ca^{+2}$. These are the unique molecules exhibited by the *R. leguminosarum* and *R. etli*. It has been proposed that the richadhesin is used by the *R. leguminosarum* in binding to the roots at the pH value of greater than 6.5, but if the pH is greater than 6.5 it uses another type of adhesin known as glucomannan under acidic condition. Glucomannan is a high molecular weight adhesions and has been found to be used by *R. leguminosarum* for binding to the pea and vetch lectins (Laus *et al.*, 2006).

*A. brasilense* makes use of the extracellular arabinose–containing polysaccharide in aggregation of the cells. (Burdman *et al.*, 2000). When the cells of *Azospirillum* were preincubated with arabinose it was observed that the agglutination of these cells using the bovine serum albumin was hampered. This reveals that *A. brasilense* cell possesses some lectin molecules, which facilitate the auto aggregation of the molecules and even in the attachment of the cells to the plant roots through the arabinose residues. Similarly the outer membrane proteins are also involved in the process of the aggregation. (Burdman *et al.*, 1999; Castellanos *et al.*, 1998). *X. fastidious* also produces haemagglutinins that mediate the contact between the cells. *Lap A* (Large adhesion protein) is an important adhesin in certain *Pseudomonas* species. *P. putida* makes use of the Lap A for adhering to the abiotic surfaces as well as the seed surfaces (Espinosa–Urgel *et al.*, 2000; Hinsa *et al.*, 2003).

**Biofilm Matrix Components**

The biofilm matrix components are those that help the organism in the adherence to substrate, stabilizing the biofilms, and protect cells from desiccation and other stress conditions formed by them. The main component included in the biofilm matrix components is polysaccharides but may also contain proteins and DNA (Branda *et al.*, 2005). According to the changing environmental conditions, organism has an ability of changing the matrix and the matrix components. It also varies according to the strain and species under study.

Hypermucoid *P. fluorescens* CHA0 has an increased ability to get attached to the roots and the as well as to the arbuscular myccorhizal fungi. This helps us understand that the overproduction of polysaccharides leads to increased ability of the organism to form biofilms. (Bianciotto *et al.*, 2001). In the biofilms formed by wild type *Pseudomonas aeruginosa* exopolysaccharide alginate could not be detected but over production of the same influences the architecture of the biofilm. (Wozniak *et al.*, 2003). *R. leguminosarum* bv. *viciae* synthesizes an acidic exopolysaccharide and a mutant which produces a exopolysaccharide in lesser amounts shows reduced biofilm formation. The formation of biofilm by this organism is also affected by the mutations in the PrsDE type I protein secretion system, to smaller extent due to inability to produce glycanases which modulates the size of the exopolysaccharide chains and hence affecting the biofilm maturation (Russo *et al.*, 2006).

*Xyllela campestris* produces exopolysaccharide viz xanthan gum which also acts as the important virulence factor and is required for the formation of aggregates, but is not involved in the process of initial attachment. Also cellulose has been used by wide variety of the organisms for formation of the pellicle and biofilm. (Romling, 2002). Acetylated cellulose is also an essential component of pellicles. It is formed by the wrinkly spreader mutant of *P. fluorescens* strain SBW25 (Spiers *et al.*, 2005).

**Quorum Sensing**

Quorum sensing is the cell–cell signaling between the organisms. Biofilms often form the site of quorum sensing, since it influences the formation of the biofilm largely. (Fuqua and Greenberg, 2002). The plant associated bacteria use mechanism of quorum sensing so as to coordinate interactions of organism and plants. This regulation includes control of tissue maceration, antibiotic production, toxin release and HGT (von Bodman *et al.*, 2003). Other mechanism such as acyl homoserine lactones (AHLs) in proteobacteria, gamma butyrolactones in *Streptomyces* species, cis–11–methyl–2–dodecanoic acid (DSF) in *Xanthomonas* and *Xylella* species and Oligopeptides in the Gram positive microbes is used for same purpose. Quorum sensing mainly links the cell density to the gene expression. Along with this the organisms produce few diffusible signal molecules which accumulate in the extracellular environment, attain a critical threshold concentration and trigger a response which leads to gene expression. (Sharma *et al.*, 2003). In a survey of the soil it was found that the AHL producing bacteria are more associated to the roots of the plants rather than the bulk soil. (Elastri *et al.*, 2001).

In quorum sensing organisms, few systems have been identified and studied on a large scale. They include *Pseudomonas chlorophis, P. fluorescens, Rhizobium elti, R. leguminosarum, Sinorhizobium melitoli*, etc. Simialrly microbially derived signaling molecules are placed in two main categories viz; amino acids and short peptide pheromones commonly utilized by Gram positive bacteria and fatty acid derivatives such as AHLs utilized by the Gram negative organisms (Kleerebezem *et al.*, 1997; Lazazzera *et al.*, 1998; Shapiro, 1998). Quorum sensing regulates various cellular processes mainly including development of the genetic competence, virulence, biofilm formation and bioluminescence. It also regulates the transition into stationary phase which represents the period of no growth (quiescent stage). The major quorum sensing molecules that have gained major attention include N–butnoyl–L–homoserine lactone, N–(3–Hydroxybutanoyl)–L–homoserine lactone, N–hexanoyl–L–homoserine lactone, N–(3–oxohexanoyl–L–homoserine lactone, N–octanoyl–L–homoserine lactone, N–(3–oxooctanoyl)–L–homoserine lactone, N–(3–oxododecanoyl)–L–homoserine lactone, Cyclo (L–Pro–L–Tyr), cyclo–(Δ Ala–L–Val), 2–Heptyl–3–hydroxyl–4–quinolone, 4–bromo–5- (bromomethylene)–3–(1′–hydroxybutyl)–2 (5H)–furanose, Butyrolactone and 3–Hydroxypalmitic acid methyl ester (Sharma, *et al.*, 2003) (structures shown in Figure 14.2).

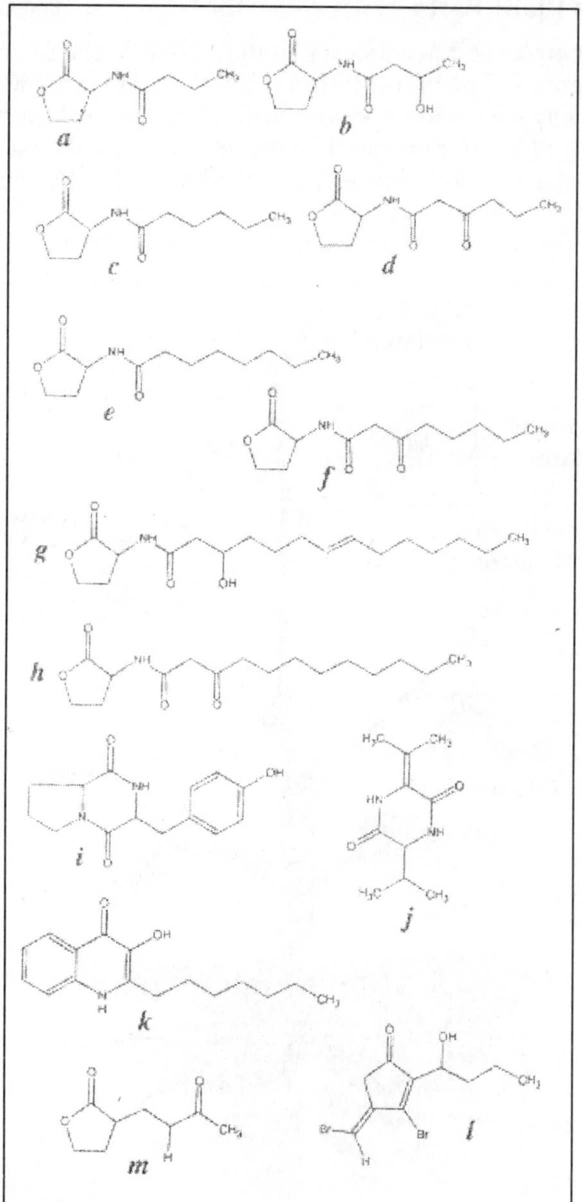

**Figure 14.1: Some Common Microbial QS Molecules**

**a: *N*-butnoyl-L-homoserine lactone; b: *N*-(3-Hydroxybutanoyl)-L-homoserine lactone; c: *N*-hexanoyl-L-homoserine lactone; d: *N*-(3-oxohexanoyl-L-homoserine lactone, N-octanoyl-L-homoserine lactone; e: *N*-(3-octanoyl)-L-homoserine lactone; f: *N*-(3-oxooctanoyl-L-homoserine lactone, g: *N*-(3-hydroxy-7-*cis*-tetradecenoyl)-L-homoserine lactone; h: *N*-(3-oxododecanoyl)-L-Pro-L-homoserine lactone; i: Cyclo-(L-Pro-L-Tyr); j: (ΔAla-L-Val); k: 2-Heptyl-3-hydroxyl-4-quinolone; l: 4-bromo-5-(bromomethylene)-3-(1'-hydroxybutyl)-2(5H)-furanose; m: Butyrolactone and n: 3-Hydroxypalmitic acid methyl ester.**

## Biofilms in Various Plant Parts

Plant surfaces are complex and dynamic environment that can be colonized by the bacteria on a large scale. As seen in Figure 14.3 plants can be colonized at various parts such as leaves, roots, seeds, and internal vasculature. Organisms are deposited on the plant surfaces by active and passive methods. Deposition with the help of wind, rain splash and water flow are the passive methods, where as chemotaxis and motility and included in the active methods (Pedgley, D. E, 1991; Chesnokova, O *et al.*, 1997; Vande Broek, A., 1995). Presence of various organisms on various plant parts are summarized below with the activity being carried out by them (Bronwyn *et al.*, 2004).

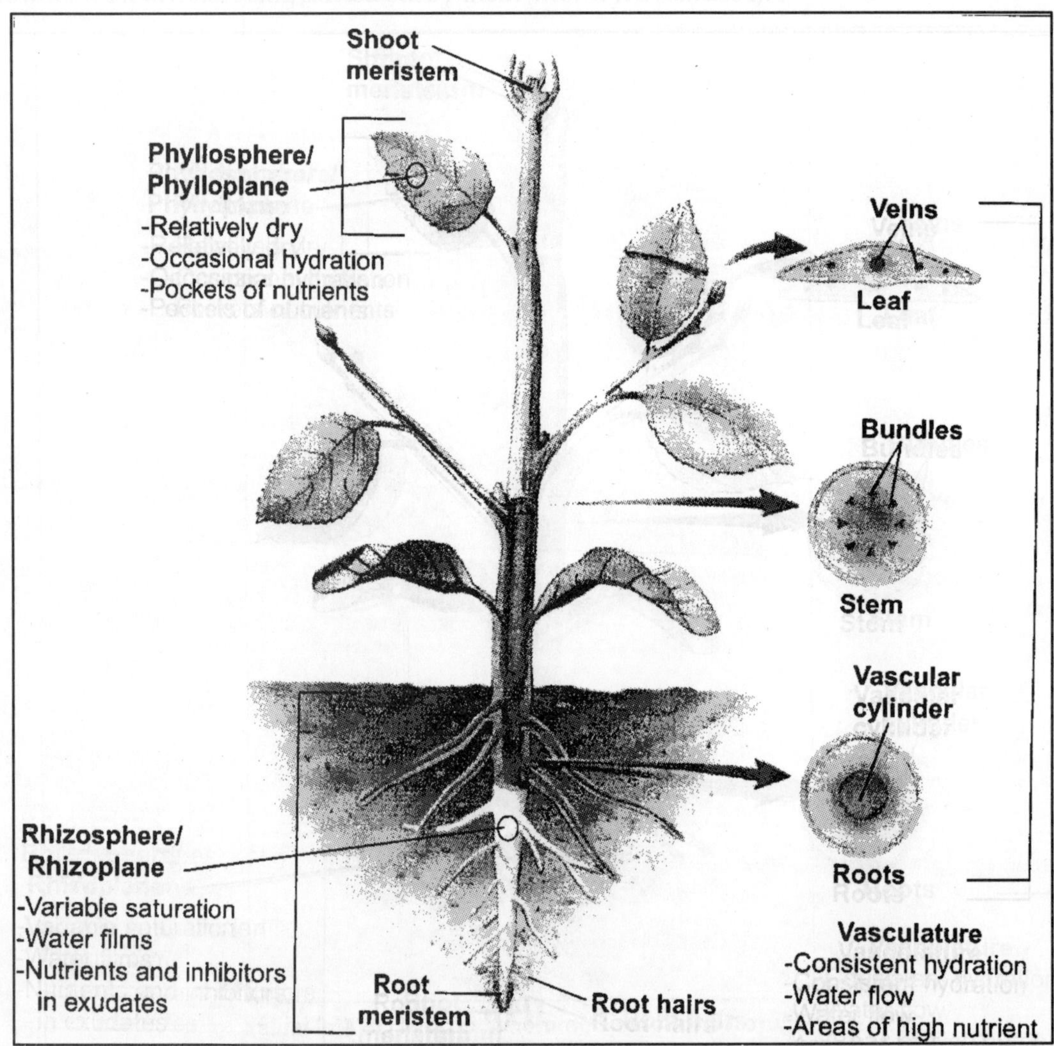


**Figure 14.2: Sites of Microbial Colonization on Terrestrial Plants. Depictionof dicotyledonous plant indicates Phyllosphere/phylloplane on leaves, rhizosphere/rhizoplanar on roots and internal vascular system.**

***Source*: Madar S, Inquiry into Life, 6th edition, 1991; Chapter 7, p 142, McGraw–Hill.**

**Table 14.2: Organisms Found on Various Plant Parts and the Activities being Performed by them**

| *Sl.No.* | *Plant Part* | *Organism* | *Activity* | *References* |
|---|---|---|---|---|
| **1.** | **Roots** | | | |
| | Root surface | *Pseudomonas fluorescens** | Produces the plant growth promoting substances | Turnbull *et al.*, 2001b |
| | | *Agrobacterium tumefaciens** | Neoplasia, non symbiotic nitrogen fixation | Ramey *et al.*, 2004<br>Matthsse *et al.*, 2001 |
| | | Rhizobia** | Nodulation for nitrogen fixation and neoplasia, curling of root hair | Smit *et al.*, 1992 |
| | | *Bacillus cereus**** | Biocontrol agent | Bais *et al.*, 2004 |
| | Root elongation zone and root hair | *Azospirillum brasilense** | Stimulates the root proliferation | Burdman *et al.*, 2000<br>Assmus *et al.*, 1995 |
| | | *Rhizobium* spp.** | Nodulation for nitrogen fixation and neoplasia, curling of root hair | Smit *et al.*, 1992 |
| **2.** | **Vasculature** | | | |
| | Xylem | *Xyllela fastidiosa***** | Pierce's disease of grape vine and citrus variegated chlorosis | Purcell and Hopkins, 1996 |
| | | *Xanthomonas campestris***** | Black rot on cruciferous plants | Dow *et al.*, 2003 |
| | | *Pantoea stewartii* sub sp. *stewartii***** | Stewart's wilt in maize | Leigh and Coplin, 1992 |
| | | *Ralstonia solacearum***** | Lethal wilt | Kang *et al.*, 2002 |
| | | *Calvibacter michiganensis* sub sp. *sepedonicus***** | bacterial ring rot in potato | Marques *et al.*, 2003 |
| **3.** | **Leaf** | *Pseudomonas syringae* pv. *syringae***** | Brown spot disease in bean | Monier and Lindow, 2003 |
| | | *Erwinia chrysanthemi***** | Soft rot disease through rapid maceration of the host tissue | Rojas *et al.*, 2002 |

*: Non pathogenic, Non symbiotic; **: Non pathogenic, symbiotic; ***: Beneficial organism; ****: Pathogenic.

## Conclusion(s)

Present review gives the state of art information on biofilm formation with respect to plant growth promoting rhizobacteria. Literature studies on biofilm and plant associated bacteria reveal that work has been carried out on ubiquity of biofilm in rhizosphere. It appears that there is a dearth of knowledge on steps involved in biofilm formation which can be a prospective field of study. Enormous amount of work has been carried out on genetic basis of biofilm formation from clinical isolates, however similar work on PGPR bacteria is in the stage of infancy. Existing knowledge of properties of PGPR, will form a base line for these studies. Studies can be focused on PGPR as biofilm formers and biocontrol agents as has been reported in *Pseudomonas fluorescens* CHA0. Genetic mechanisms controlling this can be elucidated.

## References

1. Albareda M, Dardanelli M. S, Sousa C, Megias M, Temprano F *et al.* (2006) Factors affecting the attachment of rhizospheric bacteria to bean and soybean roots. FEMS Microiol Lett. 259: 67–73.

2. Allison D. G and Sutherlan I. W, (1987) Role of exopolysaccharides in adhesion of freshwater bacteria. J. Gen. Microbiol. 133: 1319–1327.

3. Allison D. G, Ruiz B, San Jose C, Jaspe A, Gilbert P, (1998) Extracellular products as mediators of the formation and detachment of *Pseudomonas fluorescens* biofilms. FEMS Microbiol. Lett. 167: 179–84.

4. Alok Sharma, Manvika Sahgal and Bhavdish N. Johri, (2003) Microbial communication in the rhizosphere: Operation of Quorum sensing. 85 (8): 1164–1172.

5. An Y. H, R. B Dickinson and R. J Doyle. (2000) Mechanisms of bacterial adhesion and pathogenesis of implant and tissue infection p; 1–27. In Principles, methods and applications. Humana Press, Totowa, N. J.

6. Assmus B, Hutzler P, Kirchhof G, Amann R, Lawerence J. R, Hartmann A (1995) *In situ* localization of *Azospirillium brasilence* in the rhizosphere of wheat with fluorescently labeled rRNA–targeted oligonucleotide probes and scanning confocal laser microscopy. Appl. Environ. Microbiol. 61: 1013–1019.

7. Ausmees N, Jacobson K, Lindberg M, (2001) A unipolarly located, cell surface associated agglutinin, RapA, belongs to a family Rhizobium–adhering proteins (Rap) in *Rhizobium leguminosarum* bv. *trifolii*. Microbiology, 147: 549–59.

8. Bacilio–Jimezo M, Aguilar–Flores S, Ventura–Zapata E, Perez–Campos E, Bouquelet S *et al.* (2003) Chemical characterization of root exudates from rice (*Oryza sativa*) and their effects on the chemotactic response of endophytic bacteria. Plant. Soil.249: 271–77.

9. Bais H. P, Fall P, Vivanco J. M. (2004). Biocontrol of *Bacillus subtilis* against infection of *Arabidopsis* roots by *Pseudomonas syringae* is facilitated by biofilm formation and surfactin production. Plant Physiol. 134: 307–19.

10. Bais H. P, Walker T. S, Schweizer H. P, Vivanco J. M (2000b) Root specific elicitation and antimicrobial activity of rosameric acid from *in hariy root cultures of sweet basil* (*Ocimum basilicum L.*) Plant.Physiol. Biochem. 40: 983–995.

11. Bais H. P, Walker T. S, Stermitz F. R, Hufbauer R. A, Vivanco J. M (2000c) Ennatiomeric dependent phytotocix and antimicrobial activity of (±)–catechin; a rhizo–arbuscualr mycorrhizal fungi in presence of $CO_2$ and flavonols. Appl. Environ. Microbiol. 58: 821–825.

12. Baldassarri L, Cecchini R, Bertuccini L, Ammendolia M.G, Iosi F, Arciola C. R, Montanaro L Di,Rosa R, Ghrerardi G, Dicuonzo G, Orefici G and Creti R, (2001) *Enterococcus* spp produces slime and survives in rat peritoneal marophages. Med. Microbiol. Immunol. 190: 113–120.

13. Bashan Y (1986) Migration of the rhizosphere bacteria *Azospirillum brasilense* and *Pseudomonas flurorescence* towards wheat root s in soil. J. Gen. microbial. 132: 3407–3414.

14. Beatson S. A, Whitchurch C. B, Annalese B and Semmler and Mathick J.S, (2002) Quorum sensing is not required for twitching motility in *Pseudomonas aeruginosa* 184 (13): 3598–3604

15. Bernardet J. F, P Segers, M. Vancanneyt, F. Berthe, K. Kersters and P. Vandamme, (1996) Cutting a Gordian knot: emended description of the family *Flavobacterieaceae* and proposal of *Flavobacterium hydatis* mon–nov (basonym, *Cytophaga aquatilis*, Strohl and tait, 1978). Int. J. Syst.Microbiol, 46: 128–148.

16. Bianciotto V, Andreotti S, Balestri R, Bonfante P, Perotto S (2001) Mucoid mutants of the biocontrol strain *Pseudomonas fluorescens* CHA0 show increased ability in biofilm formation on, mycorrhizal and nonmycorrhizal carrot roots. Mol. Plant–Microbe Interact. 14: 255–60.

17. Bloemerg G.V, Wijfjes A. H. M, Lmers G. E. M, Stuurman N and Lugtenberg B. J. j (2000) Simultaneous imaging of iPseudomonas fluorescens WCS365 populations expressing three different autofluorescent proteins in the rhizosphere: New perspectives for studying microbial communities. Mol. Plant–Microbe. Interact. 13: 1170–1176.

18. Boland T, R. A Latour and F. J Sutzenberger. (2000) Molecular basis of bacterial adhesion, p–29–41. In Y. H An and R. J Frieman (ed). Handbook of bacterial adhesion principles, methods and applications, 1st edition. Humana Press, Totowa, N. J.

19. Boyd A and Chakrabarthy A. M, (1995) Role of alginate lyase in cell detachment of *Pseudomonas aeruginosa.* Appl. Environ. Microbiol. 60: 2355–2359.

20. Branda S. S, Vik S, Freidman L, Kolter R, (2005) Biofilms the matrix revisited. Trends. Microbiol. 13: 20–26.

21. Bronwyn E. R, Maria K, Susanne B von Bodman and Fuqua C, (2004) Biofilm formation in plant–microbe associations, Curr. Opin. Microbiol. 7: 602–609.

22. Brown MRW, Williams P. (1985). The influence of the environment in the envelope properties affecting survival of bacteria in infections. Annu. Rev. Microbiol. 39: 527–56.

23. Burdman S, Okon Y, Jurkevitch E (2000) Surface characteristics of *Azospirillum brasilense* in relation to cell aggregation and attachment to plant roots. Crit. Rev. Microbiol. 26: 91–110.

24. Burdman S, Jurkevitch E, Schwartsburd B, Okon Y (1999) Involvement of the outermembrane proteins in aggregation of *Azospirillum brasilense*. Microbiol. 145: 1145–52.

25. Burne R. A, (1998) Oral streptococci products of their environment. J. Dent. Res. 77: 445–452.

26. Carpentier B and O. Cerf, (1993) Biofilms and their consequences with particular reference to hygiene in the food industry. J. Appl. Bacteriol. 75: 499–511.

27. Castellanos T, Ascencio F, Bashan Y (1998) Cell surface lectins of *Azospirillum* spp. Curr. Microbiol. 36: 241–244.

28. Chesnokova O, Coutinho J. B, Khan I. H, Mikhail M. S, Kado C. I (1997) Characterization of flagella genes of *Agobacterium tumefaciens*, and effect of a bald strain on virulence Mol. Microbiol. 232: 579–590.

29. Costerton J. W, K. J Cheng, G. G. Geesey, J. I Ladd, J. C Nickel, M. Dasgupta and T. J. Marrie (1987) Bacterial biofilms in nature and disease. Annu. Rev. Microbiol. 41: 435–464.

30. Costerton J. W and H. M Lappin Scott (1995) Introduction to microbial biofilms, p 1–11. In H. M Lappin Scott and J. W Costerton (ed) Microbial biofilms, 1st edition. Cambridge University Press. New York, N. Y.

31. Costerton J. W, Lewandowski D. E, Caldwell D. R, Korber and H. M Lappin Scott (1995) Microbial biofilms. Annu, Rev. Microbiol. 49: 711–745.

32. Costerton J. W, P. S. Stewart and E. P. Greenberg (1999) Bacterial biofilms: a common cause of persistent infections. Science, 284: 1318–1322.

33. Costerton J. W, Stewart P. S (2001) Battling biofilms, Sci. Am, 285: 74–81.

34. Crump E. M, M. B Perry, S. C Clouthier and W. W Kay (2001) Antigenic characterization of the fish pathogen. *Flavobacterium psychrophilum*. Appl. Environ. Microbiol. 67: 750–759.

35. Dalberg C, C. Liaberg, V. C Torsvile and M. Hermansson (1997) Conjugative plasmids isolated from marine environments show various degrees of homology to each other and are not closely related to well characterized plasmids, Appl. Environ. Microbiol, 63: 4692–4697.

36. Danese P. N, L. A Pratt, S. Dove and R. Kolter (2000) The outermembrane protein Ag43 mediates cell to cell interactions in *E. coli* biofilms. Mol. Microbiol (37): 424–432.

37. Dardnelli M, Angelini. J, Fabra A (2003) A calcium dependent bacterial surface protein is involved in the attachment of rhizobia to peanut roots. Can. J. Microbiol. 49: 399–405.

38. Darzins A. L, Russell M.A, (1997) Molecular genetic analysis of type iv pili biogenesis and twitching motility using *Pseudomonas aeruginosa* as a model system a review. Gene. 192:109–15.

39. Davey M. E and O'Toole G. A (2000) Microbial Biofilms. From Ecology to Molecular Genetics, Microbiol. Mol. Bio. Reviews. 64: 847–867.

40. Davies D. G, Chakrabarty A. M, Geesey G. G (1993) Exopolysacchride production in biofilms: substances activation of alginate gene expression by *Pseudomonas aeruginosa*, Appl. Environ. Microbiol. 59 (4) 1181–86.

41. Davies D. G, Parsek M. P, Pearson J. P, Iglwski B. H, Costerton J. W, Greenberg E. P (1998) The involvement of cell to cell signals in the development of a bacterial biofilm. Science 280 (5361): 295–98.

42. Decho A. W (1990) Microbial exopolymer secretions inn ocean environment their role in food webs and marine processes. Oceanory. Mar. Biol. Annu. Rev, 28: 73–153.

43. Deighton M and Borland R (1993) Regulation of slime production in *Staphylococcus epidermidis* by iron limitation. Infect. Immun, 61: 4473–4479.

44. de Weger L. A, van der Vlug C. I, Wijfjes A. H, Bakker P. A, Schippers B and Lugtenberg BJJ (1987) Flagella of plant growth stimulating *Pseudomonas fluorescens* strain are required for colonization of potato roots. J. Bacteriol. 169: 2769–2773.

45. Donhan, R. M and Costerton. J. W (2002), Biofilm: survival mechanisms of clinically relevant microorganisms. Clin. Microbiol. Rev. 15:167–193.

46. Doolittle M. M, J. J Cooney and D. E Cladwell (1995) Lytic infection of *E. coli* biofilms by bacteriophage T4. Can. J. Microbiol.41: 12–18.

47. Dow J. M, Crossman L, Findlay K, He Y–Q, Feng J–X, Tang J–L (2003) Biofilm dispersal in *xanthomonas campestris* iscontrolled by cell–cell signaling and is required for full virulence to plants. Natl. Acad. Sci. USA. 100: 10995–11000.

48. Dunne W. M and E. M Burd (1993). Fibronectin interfere with the adhesion of *Staphylococcus epidermidis* biofilm. Antimicrob. Agents. Chemother. 37: 2522–2526.

49. Elastrie M, Delorme S, Lemanceau P, Stewart G, Laue B *et al.* (2001) Acyl homoserine lactone production is more common among plant associated *Pseudomonas* spp. Than among soil borne *Pseudomonas* spp. Appl. Environ. Microbiol. 67: 1198–209.

50. Elder M. J, F. Stapleton, E. Evans and J. K. G Dart (1995) Biofilms related infections in ophthalmology. Eye, 9: 102–109.

51. Escobar M. A, Dandekar A. M, (2003) *Agrobacterium tumefaciens* as an agent of disease. Trends. Plant. Sci. 8: 380–86.

52. Espinosa–Urgel M, Kolter R, Ramos J–L (2000) Genetic analysis of function involved in adhesion of *Pseudomonas putida* to seeds. J. Bacteriol, 182: 2363–69.

53. Espinosa–Urgel M, Kolter R, Ramos J–L (2002) Root colonization by *Pseudomonas putida*. Love at first sight. Microbiology 148: 341–343.

54. Flemming H. C (1993), Biofilms and Environment protection. Water Science Technol. 27: 1–10.

55. Flint S. H (1997), Biofilms in dairy Manufacturing Plant Description, Current Concerns and methods of Contrōl Biofouling, 11: 81–97.

56. Flores H. E, Pickard J. J, Hoy M. W (1988) Production of polyacetylenes and thiophenes in heterotrophic and photosynthetic root cultures of Asteraceae. Biol. Mol. 7: 233–254.

57. Fry J. C and M. J Day (1990) Bacterial genetics in natural environments, 1st ed. Chapman and Hall. Ltd, London, United Kingdom.

58. Fujishije N. A, Kapadia N. N, De Hoff P. L, Hirsch A. M. (2006). Investigations of Rhizobium biofilm formation. FEMS Microbial. Ecol. 56: 195–206.

59. Fuqua C, Greenberg E. P (2002) Listening in on bacteria: acyl–homoserine lactone signaling. Nat. Rev. Mol. Cell Biol. 3: 685–95.

60. Gage D. J (2004) Infection and invasion of roots by symbiotic, nitrogen fixing rhizobia during nodulation of temperate legumes. Microbial. Mol. Biol. Rev 68: 280–300.

61. Gilbert P, J. Das and I. Foley (1997) Biofilms susceptibility to antimicrobials. Adv. Dent. Res. 11: 160–167.

62. Giron J.A, Torres A. G, Freer. E, kaper J. B (2002) the flagella of enteropathogenic *Escherchia coli* mediate adherence to epithelial cells. Mol. Microbiol. 44: 361–379,

63. Goodman A. E, Marshall K. C, Hemansson M (1994) Gene transfer among bacteria under conditions of nutrient depletion in stimulated and natural aquatic environments. FEMS. Microbiol. Ecol. 15: 55–60.

64. Govan J.R.W and V. Derretic (1996) Microbial pathogenesis in cystic fibrosis: mucoid *Pseudomonas aeruginosa* and *Burkholderia cepacia*. Microbiol. Rev. 60: 539–574.

65. Greer–Phillips S. E, Stephens B. B, Alexander G (2004) An energy taxis transducer promotes root colonization by *Azospirillum brasilense*. J. Bacteriol. 186: 6595–604.

66. Gristina A. G (1987) Biomaterial cultured infection microbial Vs tissue integration. Science; 237: 1588–1597.

67. Heilmann. C (2003) Molecular basis of biofilm formation by *Staphylococcus epidermidis*. In Medical Implications of Biofilms (Wilson, M and Devine, D; Eds) vol 1, p- 110–135.

68. Herrmann M, P. E Baudaux, D. Pillet, R. Auckenthaler, P. D. Lew, F. Schumacher, Perdreau, G. Peters and F. A Walvogel (1998) Fibronectin, fibrinogen and Lanin act as mediators of adherence of clinical staphylococcal isolates to foreign material. J. Infect. Dis. 158: 693–701.

69. Hinsa S. M, Espinosa–Urgel M, Ramos J. L, O'Toole G. A (2003) Transition from reversible to irreversible biofilm formation by *Pseudomonas fluorescens* WCS365 requires an ABC transporter and a large secreted protein. Mol. Microbiol. 49: 905–18.

70. Hirsch. P (1984) Microcolony formation and consortia. p–33–393. In K. C. Marshall (ed). Microbial adhesion and aggregation. Springer, Berlin, Germany.

71. Horsekott D, K. Weistt and H Ruden (1998) Better understanding of the pathogenesis of prosthetic valve endocarditis–recent perspectives for prevention strategies, J. Heart. Valve. Div. 7: 313–315.

72. Jefferson K. K (2004) What drives bacteria to produce a biofilm? Minireview, FEMS Microbio. Lett 1–11.

73. Jucker B. A, H. Hanna and A. J. B. Zehnder (1996) Adhesion of the positively charged bacterium *Stenotrophomonas (Xanthomonas) maltophila* 70401 to glass and Teflon. J. Bacteriol. 178: 5472–5479.

74. Kang Y, Liu H, Genin S, Schell M. A, Denny T. P (2002) *Ralstonia solanacearum* requires type iv pili to adhere to multiple surfaces and for natural transformation and virulecnce. Mol. Microbiol. 46: 427–37

75. Klausen M, Heydorn A, Ragas P, Lambertsen L, Aaes–Jorgensen A I *et al.* (2003) Biofilm formation in *Pseudomonas aeruginosa* wild type, flagella and type iv pili mutants, Mol. Microbiol. 48: 1511–24.

76. Kumar C. Ganesh (1998) Significance of microbial biofilms in food industry: a review. Inter. J. Food. Microbiol, 42: 9–27.

77. Laus M.C, Logman T. J, Lamers G. E, Van Brusse A. A, Carlson R. W *et al.* (2006) A novel polar surface polysaccharide from *Rhizobium leguminosarum* binds host plant lectin. Mol. Microbiol. 59: 1704–13.

78. Lebaron P, P. Bauda, M. C Lett, Y. Duvatiflah, P. Simonet, E. Jacq, N. Frank, B. Roux, B. Baleux, G. Faurie, J. C. Hubert, P. Mormand, P. Prieur, S. Schmitt and J. C Block (1997) Recombinant plasmid mobilization between *E. coli* strains in seven sterile microcosms. Can. J. Microbiol, 43: 534–540.

79. Leigh J. A, Coplin D. 1 (1992) Exopolysacchrides in plant–bacterial interactions. Annu. Rev. Microbiol. 46: 307–346.

80. Lejeune P, (2003) Contamination of abiotic surfaces: what a colonizing bacteria sees and how to blur it. Trends. Microbiol, 11: 179–184.

81. Lewis K, (2001) Riddle of biofilm resistance. Antimicrob. Agents. Chemother. 45: 999–1007.

82. Loo C. Y (2003) Oral streptococcal genes that encode biofilm formation. In. Medical Implications of Bioiflms (Wilson, M and Devine D, Eds) Vol 1, p- 212–227.

83. Lorenzen E, I. Dalsgaard and J. Bernadert (1997) Characterization of isolates of *Flavobacterium psychrophilum*, association with cold water disease or rainbow trout fry syndrome I: phenotypic and genomic studies. Dis. Aquat. Org. 31: 197–208.

84. Lugtenberg B. J. J, Dekkers. L Bloemerg G. V (2001) Molecular determinants of rhizosphere colonization by *Pseudomonas.* Annu. Rev. Phytop[athol. 39: 461–490.

85. Lugtenberg B. J. J, Kravchenko L. V and Simons M (1999) Tomato seed and root exudates sugars: Composition, utilization *Pseudomonas* biocontrol strains and role in rhizosphere colonization. Environ Microbiol. 1: 439–446.

86. Lynch J.M, Whipps J.M (1990) Substrate flow in rhizosphere. Plant Soil 129: 1–10.

87. Mac Leod. F. A, S. R Guiot and J. W Costerton (1990) Layered structure of bacterial aggregates produced in an upflow anaerobic sludge bed and filter reactor. App. Environ. Microbiol. 56: 1598–1607.

88. Madigan, M. T, J. M. Mantinko and J. Parker. (1997) Microbial ecology p 532–605. In Brock biology of microorganisms, 8th ed. Prentice Hall, Inc; Upper Saddle River, N.J.

89. Marshall, K.C (1985) Mechanisms of bacterial adhesion at solid water interfaces, p: 133–141. In D, C savage and M. Fletcher (ed). Bacterial adhesion: mechanisms and physiological significance. Plenum Press, New York, N. Y.

90. Marshall K. C, R. Stout and R. Mitchell (1971) Mechanisms of initial events in the sorption of marine bacteria to surfaces. J. Gen. Microbiol, 68: 337–348.

91. Marques L. L. R, De Boe S H, Ceri H, Olsen M. E (2003) Evaluation of biofilms formes by *Clavibacter michiganensis* subsp. *sepedonicus* Phytopathology 93: S57.

92. Matthysse A. G, McMahan S (2001) The effet of the *Agrobacterium* colonization differs between legumes and other dicots. Appl. Environ. Microbiol. 67: 1070–1075.

93. Molina L, Ramos C, Duque E, Ronchel M. C, Garcia J. M, Wyke L and Ramos J. L (2000) Survival of *Pseudomonas putida* KT2440 in soil and in the rhizosphere of plants under greenhouse and environmental conditions. Soil. Biol. Biochem. 32: 315–321.

94. Monier J. M and Lindow S. E (2003) Differential survival of solitary and aggregated bacterial cells promotes aggregate formation on leaf surfaces. Proc. Natl Acad. Sci. USA 100: 15977–15982.

95. Moller J. D, Madsen L, Dalsgaard I, Larsen J. L, (2003) Involvement of a sialic acid binding lectin with haemaggulitination and hydrophobocity of *Flavobacterium psychrophilum.*

96. Nichols W. W, S. M Dorington, M. P. E. Slack and H. L Walmsley (1989) Inhibition of tobramycin diffusion by binding to alginate. Antimicrobiol. Chemother, 32: 518–523.

97. Nichols W. W, M. J Evans, M. P. E. Slack and H. L Walmsley (1989) The penetration of antibiotics into aggregates of mucoid and non mucoid *Pseudomonas aeruginosa*. J. Gen. Microbiol. 135: 1291–1303.

98. Nickel J. C, I. Ruseka, J. B Wright and J. W Costerton (1985) Tobramycin resistance of *Pseudomonas aeruginosa* cells growing in biofilm on urinary tract catheter. Antimicrob. Agents. Chemother. 27: 619–624.

99. Normader B, Hendrinksen N. B, Nybroe O (1999) Green fluorescent protein marked Pseudomonas *fluorescens*: localization, viability and activity in the natural barley rhizosphere. Appl. Environ Microbiol. 65: 4646–4651.

100. O'Toole, G. A and R. Kolter (1998) Initiation of biofilm formation in *Pseudomonas aeruginosa* WCS365 proceeds via multiple, convergent signaling pathways: A genetic analysis. Mol. Mirobiol. 28: 449–461.

101. O'Toole G. A and R. Kolter (1998) Flagellar and twitching motility are necessary for *Pseudomonas aeruginosa* biofilm development. Mol. Microbiol. 30: 295–304.

102. Pace, N.R 1997. A molecular view of microbial diversity and the biosphere. Science 276: 734–740.

103. Pasek M. P, Fuqua C (2004) Biofilms 2003 Emerging themes and challenges in studies of surface–associated microbial life. J. Bacteirol., 186: 4427–4440.

104. Paul E. A and F. E Clark, (1989). Soil Microbiology and Biochemistry. Academic Press. San Diego. Calif.

105. Patti J. M, Allen B. L, Mc Gavin M. J and Hook M (1994) MSCRAMM mediated adherence of microorganisms to host tissues. Annu. Rev. Microbiol. 48: 585–617.

106. Pearce D. M, J Bazin and J. M Lynch (1995) The rhizosphere as a biofilm, p: 207–220. In H. M Lappin Scott and J. W Costerton (ed). Microbial biofilms 1st edition. Cambridge University Press. New York, N. Y.

107. Peters N. K, Frost J.W, Long S. R (1986) A plant flavone, luteolin, induces expression of *Rhizobium melitoli* nodulation genes. Science. 233: 977–980.

108. Pedgley D. E (1991) Aerobiology: The atmosphere as a source and sink for microbes, In Microbial Ecology of Leaves. Edited by Andrews. J. H, Hinrao S. S, Springer–Verlag.

109. Pratt L. A, Kolter R (1998) Genetic analysis of *Escherchia coli* biofilm formation; defining the roles of flagella, motility, chemotaxis and type iv pili. Mol. Microbiol.,30 (2): 285–94.

110. Prtingent–Combaret C, Vidal. O, Dorel. C, Lejeune. P (1999) Abioitic surface sensing and biofilm development regulation of gene expression in *E. coli*. J. Bacteriol. 181(19): 5993–6002.

111. Purcell A. H, Hopkins D. L (1996) Fastidious xylem–limited bacterial pathogens. Annu. Rev. Phytophatol. 34: 131–151.

112. Ramey B. E, Matthysse A. G, Fuqua C (2004) The FNR type transcriptional regulator SinR controls maturation of *Agrobacterium tumefaciens* biofilms. Mol. Microbiol. 52: 1495–1511.

113. Ramos C, Molbak L and Molin S (2000) Bacterial activity in the rhizosphere analyzed at the single cell level by monitering ribosome contents and synthesis rates. Appl. Environ. Microbiol. 66: 801–809.

114. Rasmussen. B (2000) Filamentous microfossils in a 3.235 million year old volcanogenic massive sulphide deposit. Nature. 405: 676–679.

115. Reysenbach A. L and Cady S. L (2001) Microbiology of ancient and modern hydrothermal systems. Trends. Microbiol 9: 79–86.

116. Roberts A. P, J. Pratten, M. Wilson, P. Mullany (1999) Transfer of a conjugative transposon Tn5397 in a model oral biofilm, FEMS Microbiol, Lett,177:63–66.

117. Rojas C. M, Ham J. h, deng W. L, Doyle J. J, Collmer A (2002) HecA, a member of a class of adhesins produced by diverse pathogenic bacteria, contributes to the attachment, aggregation, epidermal cell killing and virulence phenotypes of *Erwinia chrysathemi* EC16 on *Nicotiana clevelandii* seedlings. Proc Natl Acad USA.99: 13142–13147.

118. Romling U (2002) Molecular biology of cellulose production in bacteria. Res. Microbiol. 153: 205–12.

119. Russo D. M, Wiliams A, Edwards A, Posadas D. M, Finnie C, *et al*. (2006). Proteins exported via the PrsD–PrsE type I secretion system and the acidic exopolysacchride are involved in biofilm formation by *Rhizobium leguminosarum*. J. Bacteriol. 188: 4474–86.

120. Sauer K, Camper A. K, Erhlich G. D, Costerton J. W, Davies D. G (2002) *Pseudomonas aeruginosa* displays multiple phenotypes during development as a biofilm. J. Bacteriol, 184: 1140–1154.

121. Schink B (1997) Energetics of syntropic cooperation in methanogenic degradation, Micrbiol. Mol. Biol. Rev. 61: 262–280.

122. Shirtiff M. E, Mader J. J and Camper A. K (2002) Molecular interactions in biofilms. Chem. Biol. 9: 859–871.

123. Smit G, Logman T. J. J, Boerrigter METI, Kijne J. W, Lugtenberg B. J.J, 1989, Purification and partial characterization of the $Ca^{+2}$ dependent adhesin from *Rhizobium leguminosarum* biovar *viciae*, which mediates the first step in attachment of Rhizobiaceae cells to plant root hair tips, J. Bacteriol. 171: 4054–4062.

124. Smit G, Swart S, Lugtenberg B. J. j, Kijne J. W (1992) Molecular mechanisms of attachment of *Rhizobium* bacteria to plant roots. Mol. Microbiol. 6: 2897–2903.

125. Spiers A. J, Rainey P. B (2005) The *Pseudomonas fluorescens* SBW25 wrinkly spreader biofilm requires attachment factor, cellulose fiber and LPS interactions to maintain strength and integrity. Microbiology.151: 2829–2839.

126. Stintzi. A, Browsw. J, (2000), The Arabidopsis male–aterile mutant,op3, lacks the 12–oxophytodienoic acid reductase required for jasmonate synthesis. Proc. Natl. Acad. Sci. USA 97: 10625–10630.

127. Stoodley P, Sauer K, Davies D, G, Costerton J. W (2002) Biofilms as complex differentiated communities. Annu. Rev. Microbiol., 56: 187–209.

128. Stotz H. U, Pittendrigh B. R, Kroymannn J, Weniger K, Fritsche J, Bauke A, Mitchell–Olds T (2000) Induced plant defense responses against chewing insects. Ethylene signaling reduces resistance of *Arabidopsis* against Egyptian cotton worm but not diamondback moth. Plant. Physiol. 124: 1007–1018.

129. Sutherland I. W (1999) Biofilm Exopolysaccharide In Microbial Extracellular Polymeric Substances. (Wingender, J *et al.*, eds.) 73–92, Springer.

130. Swart S, Smit G, Lugtenberg B.J.J, Kijne J. W (1993) Restoration of attachment, virulence and nodulation of *Agrobacterium tumefaciens chvB* mutanat by richdhesin. Mol. Microbiol, 10: 596–605.

131. Taylor C. D, Wirsen C. D and Gaill F (2001) Rapid microbial production of filamentous sulfur mats on hydrothermal vents. Appl. Environ. Microbiol. 67: 5179–5189.

132. Timmusk S, Grancharova N, Gerhart E, Wagner H. (2005) *Paenibacillus polymyxa* invades plant roots and forms biofilms. Appl. Environ. Microbiol. 71: 8008–15.

133. Travis S. W, Harsh P. B, Erich. G, Jorge. M. V (2003) Root Exudation and Root Biology Plant Physiology, 132: 44–51.

134. Turnbull G. A, Morgan JAW, Whipps J. M and Saunders J. R (2001a). The role of motility in the in vitro attachment of *Peudomonas putida* PaW8 to wheat roots. FEMS Microbiol. Ecol. 35: 57–65.

135. Turnbull G. A, Morgan JAW, Whipps J. M and Saunders J. R (2001b). The role of motility in the survival and spread of *Pseudomonas fluorescence* in soil and in the attachment and colonization of wheat roots. FEMS Microbiol. Ecol. 36: 21–31.

136. Ude S, Arnold D. L, Moon C. D, Timmus–Wilson T, Spiers A. J (2006) Biofilm formation and cellulose expression among diverse environmental *Pseudomonas* isolates. Environ. Microbiol 8: 1997–2011.

137. Udu W.M, M. K. Jain and J. G. Zeikus, (1996) Formation of fatty acid degrading, anaerobic granules by defined species. Appl. Environ. Microbiol, 62: 2037–2044.

138. Van Bastelaere E, Lambrecht M, Vermeiren H, Van Dommelen A, Keijers V, Proost P and Vanderleyden J (1999) Characterization of sugar–binding protein from *Azospirillum brasilence* mediating chemotaxis to and uptake of sugars. Mol. Microbiol. 32: 703–714.

139. Vaneura V (1980) Fluorescent pseudomonas in the rhizosphere of plants and their relation to root exudates. Folia Microbiol. 25: 168–173.

140. Vande Broek A, Veenderleyden J (1995) The role of bacterial motility, chemotaxis, and attachment in bacterial–plant interactions, Mol. Plant. Microbe. Interact.8: 800–810.

141. Veiga M. C, M. K. Jain, W. M Wu, R. L. Hollingsworth and J. G. Zeikus (1997) Composition and role of extracellular polymers in methanogenic granules. Appl. Environ. Microbiol, 63: 403–407.

142. von Bodman S. B, Bauer W. D, Coplin D. L (2003) Quorum sensing in plant pathogenic bacteria. Annu. Rev. Phytopathol. 41: 455–482.

143. Walker T. S, Bais H. P, Grotewold E, Vivanco J. M (2003) Root exudation and Rhizopshere biology, Plant Physiol, 132: 44–51.

144. Walker T. S, Bais H. P, Dezial E, Schweizer H. P, Ralmer L. G *et al*. (2004) *Pseudomonas aeruginosa*–plant root interactions. Pathogenicity biofilm formation and root exudation, Plant physiol. 134: 320–31.

145. Wang I. W, J. M. Anderson and R. E Marchant (1993) *Staphylococcus epidermidis* adhesion to hydrophobic biomedical polymer is mediated by platelates. J. Infect. Dis. 167: 329–336.

146. Watnick P and Kolter R (2000) Minireview: Biofilms, city of microbes. J. Bacteriol, 182: 2675–2679.

147. Watnick P. I, K. J Fullner and R. Kolter (1999) A role of the mannose sensitive haemagglutinin in biofilm formation by *Vibrio cholerae* E1 Tor. J. Bacteriol. 181: 3606–3609.

148. Watnick P. I, Kolter R (1999) Steps in the development of a *Vibrio cholerae* E1 Tor biofilm. Mol. Microbiol. 34 (3): 586–595.

149. Watt M, Hugenholtz P, White R, Vinall K (2006) Numbers and locations of native bacteria on field–grown wheat roots quantified by fluorescence in situ hybridization (FISH) Environ. Microbiol. 8: 871–884.

150. Webb J. S, Givskov M, Kjelleberg S (2003) Bacterial biofilms prokaryotic adventures in multicellularity. Curr. Opin. Microbiol, 6: 578–585.

151. Westall, P *et al.*, Early Archean fossil bacteria and biofilms in hydrothermally influenced sediments from Barberton green house belt. South Africa. Precambrian. Res, 2001, 106: 93–116.

152. Williams S. T (1985) Oligotroph in soil: fact or fiction. P 81–110 In M. Feltcher and G. D Floodgate (ed). Bacteria in their natural environment. Academic Press Inc. Orlando. Fla.

153. Whipps J. M, 1990, Carbon economy, p, 59–97, In J. M Lynch (ed). The rhizosphere. John Wiley and Sons. West. Sussex. U.K

154. Whitman W. D, D. C Coleman and W. J. Wiese (1998) Prokaryotes the unseen majority. Proc.Natt. Acad Sci USA 95: 6578–6583.

155. Wolfaardt G. M, J. R Lawerence, R. D. Robarts and D. E.Caldwell (1998) In situ characterization of biofilm exopolymers involved in the accumulation of chlorinated organisms. Microbiol. Ecol. 35: 213–223.

156. Wozniak D. J, Wyckoff T. J, Starkey M, Keyser R, Azadi P *et al.* (2003) Alginate is not a significant component of the extracellular polysacharide matrix of PA14 and PAO1 *Pseudomonas aeruginosa* biofilm. Proc. Natl. Acad. Sci. USA, 100: 7907–12.

157. Wu T, Wittkamper J, Flores h.E (1999) Root herbivory *in vitro*: interaction between root and aphids grown in aseptic cocluture. In vitro Cell Dev Biol Plant 35: 259–264.

158. Yao J, Allen C (2006), Chemotaxis is required for the virulence and competitive fitness of the bacterial wilt pathogen *Ralstonia solanacearum* J. Bacteriol. 188: 3697–708.

159. Yildiz F. H, Schoolnik G. K, (1999), *V. cholerae* O1E1 Tor: identification of a gene cluster required for the rugose colony type, exopolysacchride production, chlorine resistance and biofilm formation. Proc. Natt. Acad. Sci. USA.96: 4028–33.

160. Yoder J. I (2001) Host plant recognition by parasitic Scrophulariaceae. Curr. Opin. Plant. Biol 4: 359–365.

161. ZoBell C. E (1937) The influence of solid surface as the physiological activities of bacteria in sea water. J. Bacteriol. 33: 86

# Chapter 15

# Reactive Extraction of Glycolic Using Tri-N-Octyle Amine: Cleaner Technology

**Manu Agarwal[1], Amit Keshav[2] and Kailas L. Wasewar[3]***
**[1]M.Tech. Student; [2]Research Scholar; *Assistant Professor**
**Department of Chemical Engineering, Indian Institute of Technology (IIT)**
**Roorkee – 247 667, Uttarakhand**

## ABSTRACT

Glycolic acid is used as cleaning agent in food and dairy industries, as an anti aging formulation in cosmetic industry, as a solvent for dissolving hard water scales and in production of resorbable polymer sutures, etc. It is the smallest α-hydroxy carboxylic acid present in fruits, vegetables, beverages etc. The commercial production of the acid is from oxalic acid, ethylene glycol, and glyoxal. The concentration of acid produced is relatively low. Involving any kind of conventional process will lead to high consumption of energy and time. Reactive extraction is method of intensifying process to recover these acids once produced. Reactive extraction with specific extractant is the promising technique for glycolic acid recovery. Of different extractants available for extraction, aliphatic amines were found to be more effective and less expensive. In present work, the extraction of the acid with tri-n-octyl amine dissolved in MIBK and 2-octanol was investigated for various concentrations of the acid. The overall distribution coefficient ($K_D$), loading factor ($z$) and equilibrium complexation constants ($K_{11}$, $K_{12}$) were evaluated. The

* Corresponding Author: E-mail: k_wasewar@rediffmail.com, klw73fch@iitr.ernet.in

comparison between the diluents was made and 2-octanol was found to be a better extractant. The data obtained is useful in the designing of reactive extraction process for the recovery of glycolic acid.

***Keywords:*** *Glycolic acid, Reactive extraction, Equilibrium, Tertiary amine, Diluents, Clean technology.*

## Introduction

Glycolic acid (pKa = 1.83, $CH_2(OH)COOH$, Mol. Wt. = 76.05) is the smallest α-hydroxy carboxylic acid present in very small concentration (56 ppm) in fruits, vegetables, beverages, etc. It is an important chemical used as food and dairy equipment cleaning agent, as an anti-aging formulation, as solvent for dissolving hard water scales, manufacture of resorbable polymer sutures, production of intermediates in pharmaceutical industries, etc.

The production of glycolic acid is by electrochemical reduction of oxalic acid, Cannizzaro reaction of glyoxal and Catalytic oxidation of Ethylene Glycol. Acid produced from these methods is in relative low concentration. The conventional methods used in the industries, to recover the acid are time consuming and energy inefficient. To recover acid from such dilute streams is still a challenging problem for the researchers. Reactive extraction with specific extractant is the promising technique for recovery of organic acids from the aqueous solution and has also received increasing attention.[1]

Extractants such as phosphorus bonded oxygen containing extractants, tri-butyl phosphate, tri-alkyl phosphine oxide and long chain aliphatic amines, tri-octyl amine and tri-alkyl methyl ammonium chloride, are more effective for extraction of carboxylic acids and have been used in the separation of acetic acid,[2–6] lactic acid,[7–10] propionic acid,[8–11] butyric acid,[8–12] succinic acid,[13] and so forth. A strong interaction between the acid and extractant allows for the formation of acid–extractant complexes and thus provides for high equilibrium distribution ratio.

## Materials and Methods

### Chemicals

Tri-Octyl Amine ($C_{24}H_{51}N$) (ACROS, India), a tertiary amine is a light colourless liquid with a molecular weight of 353.66 and density of 0.809 g/cc. Glycolic acid (70 per cent) (HIMEDIA, India), a hydroxy-acetic acid is a colourless liquid with a molecular weight of 76.05 and density of 1.254 gm/cc and diluents 2-Octanol (S. d. fine chem. Ltd, Mumbai, India) and MIBK (S. d. fine chem. Ltd, Mumbai, India) are of technical grade and were used without pretreatment. Distilled water was used to prepare the various concentrations of glycolic acid solutions.

### Extraction Experiments

All extraction experiments were conducted with 100 ml flasks at room temperature. Solvent 20 ml with 0.5 mol $l^{-1}$ extractant and 20 ml carboxylic acid aqueous solution were added to each flask as organic and aqueous phases. The flasks containing the mixture were shaken for 6 hour in a water bath shaker with vibrating rate of 200 rpm, and then left to settle for (2 to 3) hour. An aqueous phase sample from the bottom layer was taken for solute concentration analysis.

## Sample Analysis

The aqueous phase sample was analyzed for acid concentration by titrating them against 0.02 N sodium hydroxide solution in duplicate. Always a fresh sodium hydroxide solution was prepared for analysis. The concentration of organic acid in the organic phase was calculated by material balance.

## Theory

The reactive extraction equilibrium for polar diluent solution of organic solutes could be described by mass action law in which the equilibrium behaviour can be modeled by postulating the formation of various stoichiometric complexes of acid and amine.[14] A reactive equilibrium description of the system can be written as a set of equilibria involving the dissociation equilibrium of the acid in water and formation of complex with acid and extractant by ion pair association and hydrogen bonding.[15] It is assumed that the reactive extraction between carboxylic acid and extractant takes place at the organic–aqueous interface and two types of complexes, (1, 1), (1, 2), *i.e.*, HAS and $HAS_2$ are formed in a step wise manner.

The distribution coefficient, $K_D$, is defined as the ratio between total molar concentration of carboxylic acid (in all forms) in the organic phase and that in the aqueous phase. By appropriate mass balance distribution coefficient can be derived as

$$K_D = \frac{C_{org}}{C_{aq}} = \frac{[HA]_{org} + [HAS]_{org} + [HAS_2]_{org}}{[HA]_{aq} + [A^-]_{aq}} = \frac{\phi m + K_{11}[S]_{org} + K_{11}K_{12}[S]_{org}^2}{1 + 10^{pH-pKa}} \quad (1)$$

$$z = \frac{C_{org}}{[S]_{org}} \quad (2)$$

The complexation constants can be obtained from following equations

$$\frac{z}{(1-z)} = K_{11}C_{aq} \quad (3)$$

$$\frac{z}{(2-z)} = K_{12}C_{aq}^2 \quad (4)$$

*where,* HA and S represent the carboxylic acid and extractant respectively. The concentration of species is denoted by square brackets and is expressed in molar terms. $A^-$ stands for the anion of carboxylic acid. $K_D$ is the distribution coefficient of acid between the phases. $C_{org}$ and $C_{aq}$ are the total acid concentrations in the organic and aqueous phases at equilibrium conditions respectively. *pKa* is the dissociation constant of the acid in the aqueous phase. *m* is the physical extraction constant of acid in between pure diluent and water. $\phi$ is the volume fraction of diluent in the organic phase. $S_0$ is the initial concentration of extractant in the organic phase. $K_{11}$ and $K_{12}$ are apparent equilibrium constants of (1, 1) and (1, 2) complexes respectively, the subscripts aq and org stand for aqueous and organic phase respectively.

## Results and Discussion

The various experiments for reactive extraction equilibrium of glycolic acid using Tri-Octyl Amine in MIBK and 2-Octanol were studied at room temperature (305 K) and various acid concentrations. The equilibrium data is given in Tables 15.1 and 15.2 for glycolic acid extraction using TOA in MIBK and 2-Octanol. The values of z/(1-z) were plotted against the values of $C_{aq}$ to obtain the value of $K_{11}$ (Figure 15.2). The slope of the line obtained from the line equation gave the value of equilibrium constant ($K_{11}$) for (1, 1) acid-amine complex. The value of $K_{11}$ for MIBK and 2-octanol were obtained to be 25.906 l $mol^{-1}$ and 49.305 l $mol^{-1}$ respectively.

**Table 15.1: Equilibrium Data of Glycolic Acid between Water and Tri-n-Octyl Amine in MIBK**

| $C_{aq}$ i mol $l^{-1}$ | $C_{aq}$ mol $l^{-1}$ | $C_{org}$ mol $l^{-1}$ | $K_D$ | z | z/(1-z) | z/(2-z) |
|---|---|---|---|---|---|---|
| 0.05 | 0.0013 | 0.0486 | 36.0782 | 0.0973 | 0.1077 | |
| 0.10 | 0.0035 | 0.0965 | 27.5714 | 0.1930 | 0.2391 | |
| 0.15 | 0.0118 | 0.1382 | 11.7454 | 0.2765 | 0.3821 | |
| 0.20 | 0.0207 | 0.1793 | 8.6731 | 0.3586 | 0.5592 | |
| 0.30 | 0.0470 | 0.2530 | 5.3830 | 0.5060 | 1.0242 | 0.3386 |
| 0.40 | 0.0790 | 0.3210 | 4.0617 | 0.6419 | 1.7929 | 0.4727 |
| 0.45 | 0.0950 | 0.3550 | 3.7368 | 0.7100 | 2.4482 | 0.5504 |
| 0.50 | 0.1170 | 0.3830 | 3.2735 | 0.7660 | 3.2735 | 0.6207 |

**Table 15.2: Equilibrium Data of Glycolic Acid between Water and Tri-n-Octyl Amine in 2-Octanol**

| $C_{aq}$ i mol $l^{-1}$ | $C_{aq}$ mol $l^{-1}$ | $C_{org}$ mol $l^{-1}$ | $K_D$ | z | z/(1-z) | z/(2-z) |
|---|---|---|---|---|---|---|
| 0.05 | 0.0006 | 0.0494 | 83.7458 | 0.0988 | 0.1906 | |
| 0.10 | 0.00135 | 0.09865 | 73.0741 | 0.1972 | 0.2457 | |
| 0.15 | 0.0024 | 0.1476 | 61.5000 | 0.2952 | 0.4188 | |
| 0.20 | 0.0047 | 0.1953 | 41.5532 | 0.3906 | 0.6409 | |
| 0.30 | 0.0165 | 0.2835 | 17.1818 | 0.5670 | 1.3094 | 0.3956 |
| 0.40 | 0.0450 | 0.3550 | 7.8889 | 0.7100 | 2.4482 | 0.5504 |
| 0.45 | 0.0670 | 0.3830 | 5.7164 | 0.7660 | 3.2735 | 0.6207 |
| 0.50 | 0.0930 | 0.4070 | 4.3763 | 0.8140 | 4.3763 | 0.6863 |

**Table 15.3: Equilibrium Constants ($K_{11}$, $K_{12}$) of (1, 1) and (1, 2) Acid-amine Complexes Formed**

| Diluents | $K_{11}$ (l $mol^{-1}$) | $K_{12}$ ($l^2$ $mol^{-2}$) |
|---|---|---|
| MIBK | 25.906 | 4.4289 |
| 2-Octanol | 49.305 | 4.2905 |

The high value of $K_{11}$ for 2-octanol was observed compared to that of MIBK by around two folds in magnitude. This indicates that 2-octanol acts as a better diluent with TOA for extraction of glycolic acid. The difference in the values $K_{11}$ for same acid in two diluents indicate that the salvation of acid-amine complex by the diluent is a critical factor in extraction of acid. The interaction between the complex and the solvent can be divided into 'general salvation' and 'specific interaction' of the diluent with the complex. Polar diluents like MIBK promote extraction by providing good salvation medium for the ion pair. However, polarity alone does not completely account for salvation ability. Alcohols usually give high equilibrium constants, higher than expected. Specific hydrogen bonding

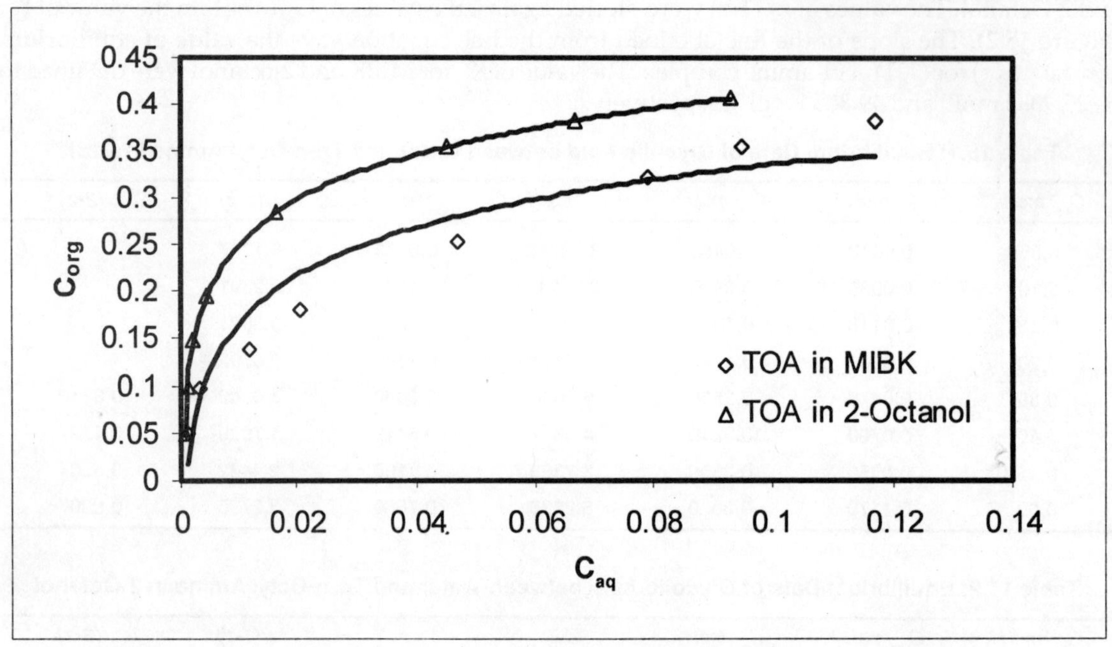


**Figure 15.1: Equilibrium Concentration Curve of Glycolic Acid between Water and Tri-n-octyl Amine in MIBK and 2-Octanol**

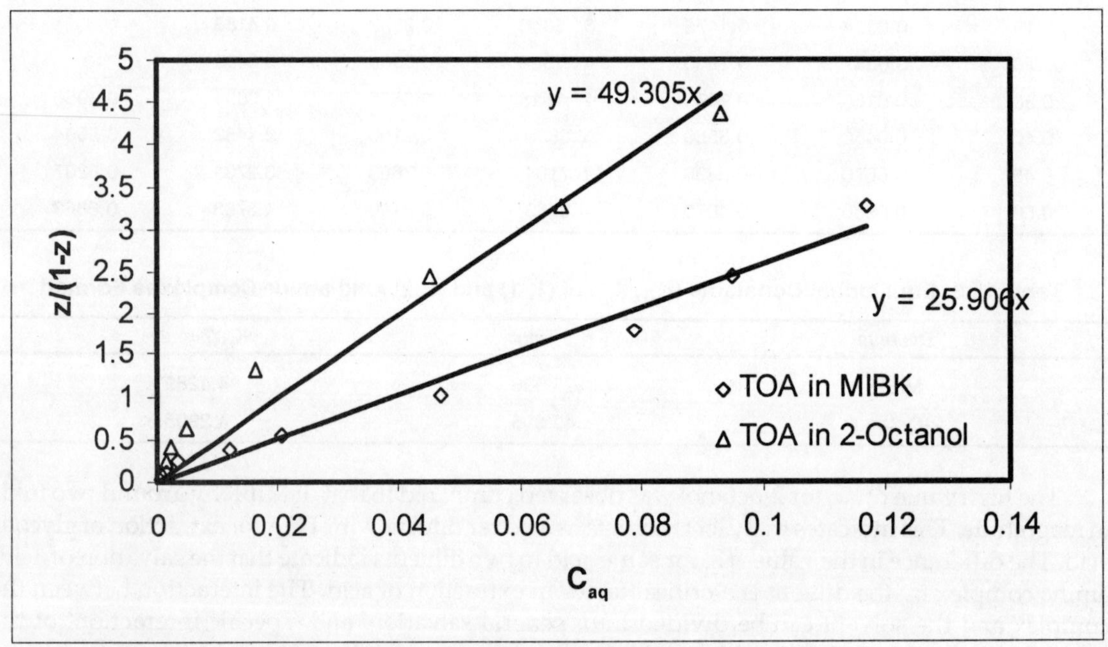


**Figure 15.2: Equilibrium Data of Glycolic Acid Between Water and Tri-n-octyl Amine in MIBK and 2-Octanol Fit in the eqn. 3**

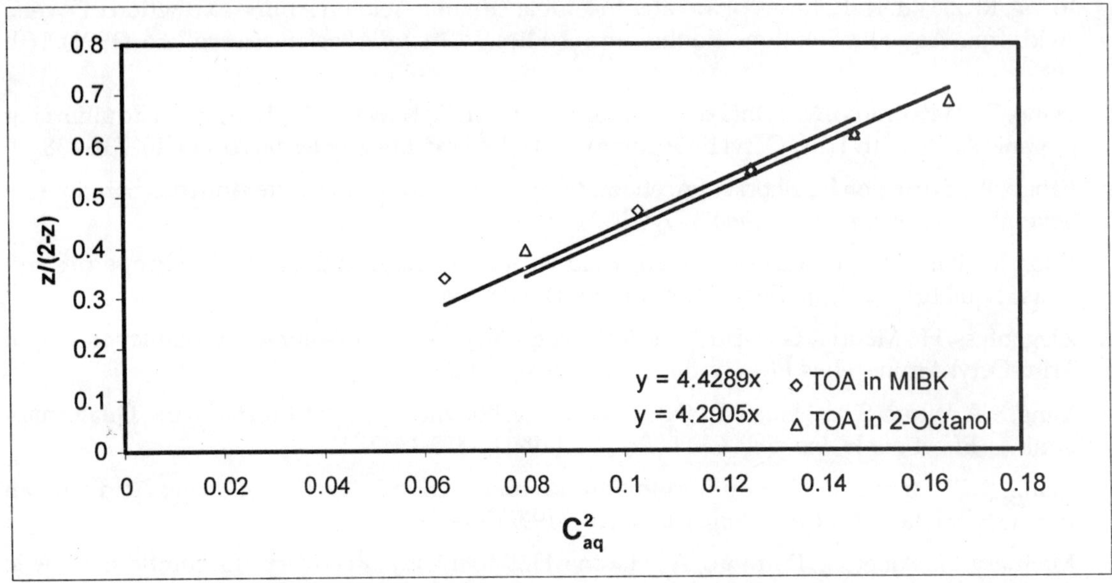


**Figure 15.3: Equilibrium Data of Glycolic Acid Between Water and Tri-n-octyl Amine in MIBK and 2-Octanol Fit in the eqn. 4**

between the proton of diluent and the acid-amine complex is thought to explain the extra salvation of these diluents.

The values of z/(2-z) were plotted against the values of $C_{aq}^2$ (Figure 15.3), The slope obtained gave the value of equilibrium constant ($K_{12}$) for (1, 2) complex. The values of $K_{12}$ (MIBK: 4.4289 $l^2$ $mol^{-2}$ and 2-Octanol: 4.2905 $l^2$ $mol^{-2}$) for both the diluents were almost equal within the error range. This indicates that the formation of (1, 2) complex is independent of diluent.

## Conclusion

The extraction of the acid with tri-n-octyl amine dissolved in MIBK and 2-octanol was investigated for various concentrations of the acid. The overall distribution coefficient ($K_D$), loading factor (z) and equilibrium complexation constants ($K_{11}$, $K_{12}$) were evaluated. The comparison between the diluents was made and 2-octanol was found to be a better extractant. The value of $K_{11}$ for MIBK and 2-octanol were obtained to be 25.906 l $mol^{-1}$ and 49.305 l $mol^{-1}$ respectively. The values of $K_{12}$ (MIBK: 4.4289 $l^2$ $mol^{-2}$ and 2-Octanol: 4.2905 $l^2$ $mol^{-2}$) for both the diluents were almost equal within the error range. The data obtained is useful in the designing of reactive extraction process for the recovery of glycolic acid.

## Acknowledgement

Department of Science and Technology (DST), INDIA, Young Scientist Project, SR/FTP/ETA-43/2005, Reactive Extraction of Propionic acid (Dr. Kailas L. Wasewar, IIT Roorkee, India).

## References

1. Kertes, A. S., King, C. J., Extraction Chemistry of Fermentation Product Carboxylic acids, *Biotechnol. Bioengg.*, 28, (1986), 269-282.

2. Juang, R. S., Wu, R. T., Effect of a Water Insoluble Organic Acid on Amine Extraction of Acetic Acids from Aqueous Solutions: Equilibrium Studies, *J. Chem. Technol. Biotechnol.*, 66, (1996), 160-168.

3. Hano, T., Matsumoto, M., Ohtake, T., Sasaki, K., Hori, F., Kawano, Y., Extraction Equilibria of Organic Acids with Tri-n-Octyl Phosphine Oxide, J. *Chem. Engg. of Japan*, 23, (1990), 734-738.

4. Fahim, A., Extraction Equilibria of Acetic and Propionic Acids from Dilute Aqueous Solutions by Several Solvents, *Sep. Sci. Technol.*, 27, (1992), 1809-1821.

5. Roos, M., Bart, H. J., Extraction of acetic Acid with Tri-n-Octyl Amine: Physical Properties and Phase Equilibria, *J. Chem. Engg. Data*, 46, (2001), 1198-1202.

6. Ziegenfuss, H., Maurer, G., Distribution of Acetic Acid between Water and Organic solution of Tri-n-Octyl Amine, *Fluid Phase equilib.*, 102, (1994), 211-255.

7. Yang, S.T., White, S.A., Hsu, S. T., Extraction of Carboxylic Acids with Tertiary and Quaternary Amines: Effect of pH, *Ind. Eng. Chem. Res.*, 30, (1991), 1335-1342.

8. Juang, R. S., Huang, R. H., Equilibrium Studies on Reactive Extraction of Lactic Acid with an Amine Extractant, *The Chem. Eng. Journal*, 65, (1997), 47-53.

9. Malmary, G., Albert, J., Putranto, A., Hanine, H., Molinier, J., Recovery of Aconitic and Lactic Acids from Simulated Aqueous Effluents of Sugar-Cane Industry through Liquid-Liquid Extraction, *J. Chem. Technol. Biotechnol.*, 75(12), (2000), 1169-1173.

10. Jarvinen, M., Myllykoski, L., Keiski, R., Sohlo, J., Separation of Lactic Acid from Fermentation Broth by Reactive extraction, *Bioseparation*, 9(3), (2000), 163-166.

11. Yoshizawa, H., Equilibrium of Aqueous Propionic Acid with Tri-n-Octyl Amine in Dodecane, *J. Chem. Engg. Data*, 39, (1994), 777-780.

12. Sabolova, E., Schlesser, S., Martak, J., Liquid-Liquid equilibrium of Butyric Acid in Water + Solvent Systems with Tri-n-Octyl Amine as Extractant, *J. Chem. Engg. Data*, 46(3), (2001), 735-745.

13. Juang, R. S., Huang, R. H., Comparison of Extraction Equilibria of Succinic and Tartaric Acids from Aqueous Solution with Tri-n-Octyl Amine, *Ind. Engg. Chem. Res.* 35, (1996), 1944-1950.

14. Tamada, J. A., Kertes, A. S., Extraction of Carboxylic Acids with Amine Extractants. (1)Equilibria And Law of Mass Action Modeling., *Ind. Eng. Chem. Res.*, 29, (1990), 1319-1326.

15. Li, Z. Y., Extraction Mechanism of Carboxylic Acid by Trioctyl-Amine, Ph.D. Dissertation, Tsinghua University, Beijing, 2001.

16. Tamada, J. A., Kertes, A. S., King, C.J., Extraction of Carboxylic Acids with Amine Extractants. 2. Chemical Interactions and Interpretation of Data, *Ind. Eng. Chem. Res.*, 29, (1990), 1327-1333.

17. Qin, W., Zhang, Y., Li, Z., Dai, Y., Extraction Equilibria of Glycolic and Glyoxylic Acids with Trialkylphosphine Oxide and Trioctylamine as Extractant, *J. Chem. Engg. Data*, 48, (2003), 430-434.

# Chapter 16

# Microbial Detoxifying Enzymes in Bioremediation of Organic Pollutants

**Pradnya Pralhad Kanekar* and Seema Shreepad Sarnaik**
**Microbial Sciences Division, Agharkar Research Institute, G.G. Agarkar Road, Pune – 411 004, M.S.**

## ABSTRACT

New chemical compounds are being developed to fulfill the requirements for betterment of human life. This rapid industrial development has resulted in contamination of air, water and soil to a great extent. Some of the industries like, petrochemical industries, oil refineries, distilleries and the units manufacturing and handling toxic chemicals like dyes, pesticides, nitroexplosives, synthetic polymers etc. are the major cause of environmental pollution. These industries cause pollution because of large volume of wastewater generated during production and presence of unreacted chemicals used as raw materials alongwith the unrecovered final product in the wastewater. Some of the chemicals appearing in the wastewater *e.g.* phenols, nitrocompounds, dyes, pesticides, plasticizers, sulfates, chlorides etc. are known to be toxic to aquatic as well as terrestrial life. Different methods therefore have been developed to remove or detoxify these contaminants. Biological methods using natural resources like plants, microorganisms seem to be economical and ecofriendly.

Microorganisms are able to secrete different enzyme systems useful for breaking down toxic chemicals. Due to ease of cultivation of microorganisms and their limited nutritional requirements, use of microbial enzymes appears to be a better solution for this purpose. In literature, degradation of pesticides by *Rhodococcus* sp. *Arthrobacter* sp. etc. and the enzyme systems like dehalogenases, dehydrogenases and cytochrome p-450 are reported. Active participation of organophosphorus hydrolase was demonstrated in biodegradation of

* Corresponding Author: E-mail: ppkanekar@aripune.org; kanekarp@rediffmail.com

organophosphorus pesticides. Laccases and peroxidases from fungi have been reported to play an active role in degradation of azodyes, triphenylmethane dyes, heterocyclic dyes, polymeric dyes etc. Monooxygenases, dioxygenases and nitroreductases from microorganisms degrading nitro aromatic compounds have been described in the literature. Role of microbial cytochrome p-450 has been demonstrated to have role in degradation of other xenobiotic compounds *e.g.* camphor, cyclohexane, alkanes, etc. Thus, ecofriendly technologies could be developed for bioremediation purposes using these microbial enzyme systems.

***Keywords:*** *Biodegradation, Bioremediation, Oxygenase, Cytochrome P-450, Microbial enzymes.*

## Introduction

Rapid industrialization has resulted in rising up of chemical industries for manufacturing a variety of chemicals to fulfill the demands of public. The high growth of chemical industries, both manufacturer and user industries led to pollution of the environment *i.e.* air, water, soil etc. to a greater extent. The industries responsible for environmental pollution include petrochemical industries, oil refineries, distilleries and the units manufacturing chemicals like pesticides, explosives, detonators, dyes, paints, detergents, soaps etc. These industries cause environmental pollution due to high volume of their effluents and the residual chemicals appearing in these effluents. The chemicals like nitro compounds, organochlorine compounds including pesticides are known to be harmful to both aquatic and terrestrial life including human beings. All these toxic compounds tend to remain in the environment for a long time hence their detoxification is necessary.

Microorganisms play an important role in the degradation of man-made chemicals that are released in the environment. These microorganisms are endowed with the property of secreting enzymes required for degradation or detoxification of different chemicals like nitro compounds, pesticides especially the organochlorine pesticides etc. The ability of these microbes to utilize or metabolize different chemical compounds including toxic chemopollutants like pesticides, dyes etc. lead to their application for bioremediation purposes. Different enzymes are involved in break down of these chemopollutants. However, involvement of oxidoreductases or also called as oxygenases is found to be major in biodegradation of chemopollutants.

Oxidoreductases are widely found in plants and animals with higher occurrence in microorganisms. These enzymes catalyze the reactions involving electron transfer, hydrogen extraction, oxygen insertion, etc. (Xu, 2005). These enzymes can be classified according to their sequence or three-dimensional structure, however for application purpose, these oxidoreductases are classified according to their signature catalysis and coenzyme-dependence *e.g.* thiol oxidases have Iron (Fe), Copper (Cu) or Cystiene (Cys) + Flavin Adenine Dinucleotide (FAD) as an active center, heme peroxidases have heme as active center, other peroxidases have Cys or Serine (Se)-Cys as active center, Flavin monooxygenases/Flavin hydroxylases have Flavin Mononucleotide (FMN)/FAD as active centers and Cu containing oxygenases and hydroxylases have Cu as active center (Xu, 2005). These enzymes carry out hydroxylation of substances in addition to the reactions like N-oxidation, sulfoxidation, epoxidation, N-, S-, and O-dealkylation, peroxidation, deamination, desulfurization, dehalogenation etc. (White and Coon, 1980).

These oxidoreductases have wide range of applications–a) environmental applications for environment protection by degrading numerous man made hazardous chemical compounds; b) in textile and detergent industries; c) in food and dairy industries; d) as biosensors and bioreporters involving reactions with enzyme co-substrates forming chromogenic, fluorogenic, electroactive products and e) in synthesis of various products, organic chemicals having asymmetric/chiral structure, nanomaterials, polymers, and other substances having medical applications.

## Enzyme Systems Used as Detoxifying Agents

Microorganisms are endowed with the property of secreting several enzyme systems for metabolizing different man made chemical compounds *i.e.* xenobiotic compounds. Hence, microbially assisted degradation/detoxification/mineralization of toxic chemical compounds seems to be a good solution for bioremediation purpose. Microorganisms degrading xenobiotic chemicals elaborate enzyme systems such as phosphatases, esterases, hydrolases/hydroxylases, dechlorinases/ dehalogenases, deaminases and oxygenases etc. These enzymes play an important role in metabolism of variety of chemicals of agricultural and environmental concern.

## Enzymatic Degradation of Pesticides

Role of different types of enzyme systems elaborated by microorganisms for degradation of some pesticides are described in literature to some extent.

Hydrolyses and dioxygenases from *Rhodoferax* sp. and *Delftia* (*Comamonas*) *acidovorans* degrading phenoxypropionate and phenoxyacetate herbicides (Muller *et al.*, 2001). Phenylurea hydrolase (puhA) from *Arthrobacter globiformis* capable of degrading phenyl urea herbicides was reported by Turnbull *et al.* (2001). Strauber *et al.* (2003) showed cleavage of ether bond of phenoxybutyrate herbicides and other phenoxybutyrate substrates catalyzed by the cytochrome p-450 present in *Rhodococcus erythropolis*.

Involvement of enzymes like dehydrogenases (Nagy *et al.*, 1995 a) and cytochrome p-450 system in degradation of herbicides EPTC (S-ethyldipropyl thio carbamate) and atrazine by *Rhodococcus* sp. are reported by some researchers (Nagy *et al.*, 1995 b; Shao and Bekhi, 1996). Puranik (2000) carried out some work on complete mineralization of ethylenethiourea (ETU), a metabolite of carbamate group of pesticide Mancozeb by the soil isolates *Arthrobacter sulfureus, Microbacterium lacticum* and *Pseudomonas aeruginosa* to carbon dioxide and ammonia through formation of intermediate metabolites *viz.* ethylenediamine and ethyleneurea. The author also reported the involvement of enzymes like decarboxylase, deaminase etc. Carbaryl hydrolase degrading carbaryl was purified by Hoshimoto *et al.* (2002).

Organophosphorus hydrolase (OPH) is a bacterial enzyme that has been shown to degrade a wide range of organophosphates. Kanekar *et al.* (2004) have reviewed thoroughly the biodegradation of some commonly used organophosphorus pesticides like parathion, malathion, chlorpyriphos, quinalphos etc. and the enzyme systems involved in degradation of these pesticides. Biodegradation of organophosphorus pesticides by organophosphorus hydrolase was reported in the literature (Munneck 1976; Richins *et al.*, 1997; Mulbry 2000; Cho *et al.*, 2002; Zhang *et al.*, 2005; Singh and Walker 2006 and Mee-Hie *et al.*, 2006). Other enzymes involved in degradation of organophosphorus pesticides are phosphatase or phosphotriesterases which are reported to be involved in degradation of pesticides like paraoxon, parathion, diazinon, methyl parathion, cyanophos, fensulfothion, monocrotophos, dimethoate, and coumaphos (Dumas *et al.*, 1989; Rabie 1995; Liu *et al.*, 2001; Cai and Xun 2002; Das and Singh 2006). Bhadbhade *et al.* (2002) have described the biomineralisation of insecticide Monocrotophos (MCP) to phosphates, ammonia and carbon dioxide with the involvement of two enzymes phosphatase and esterase present in *Arthrobacter atrocyneus, Bacillus megaterium* and *Pseudomonas mendocina*. Deshpande (2002) demonstrated complete mineralization of insecticide dimethoate commonly known as Rogor by the bacterial cultures *Brevundimonas* sp., *Bacillus* sp. and *Klebsiella oxytoca* with the involvement of extracellular enzymes phosphatase and esterase.

Some data are also available on mineralization of more recalcitrant and environmentally persistent organochlorine pesticides. Bacterial dehalogenases were well studied by Villemur *et al.* (2005). Weir *et al.* (2006) described flavin-dependent mono-oxygenase from *Arthrobacter* sp. degrading both isomers of endosulfan and endosulfate. Vaishampayan (2004, 2007) studied the complete degradation of atrazine, an organochlorine herbicide by *Arthrobacter* sp. isolated from atrazine exposed soil and presence of trzN gene, as novel chlorohydrolase enzyme. Enantioselective transformation of hexachlorocyclohexane (HCH) by dehydrochlorinases LinA1 and LinA2 from soil bacterium *Sphingomonas paucimobilis* were reported by Suar *et al.* (2005). The dehalogenase enzyme system from *Sphiongobium indicum* causing transformation of β- and δ-hexachlorocyclohexane was studied by Sharma *et al.* (2006). Mulbry (1994) had isolated, purified and characterized an inducible s-triazine hydrolase from *Rhodococcus corallinus* degrading several triazine compounds.

## Enzymatic Degradation of Dyes Specifically Triphenylmethane (TPM) Dyes

Laccases and peroxidases have applications in textile industries mainly for cotton fiber whitening, dye finishing and waste treatment. Laccase-catalysed textile dye-bleaching is useful in dyed cotton fabric and being commercialized in recent years. Some researchers have worked on microbial especially fungal peroxidases and laccases used in degradation of dyes and in treatment of dye industry wastewaters. The enzyme systems elaborated by organisms like *Pleurotus pulmonarius, Trametes modesta, Trametes hirsute, Trametes versicolor, Sclerotium rolfsii, Phanerochaete chrysosporium and Geotrichum candidum* are reported by Ollikka *et al.* (1998); Kim and Shoda (1999); Abdulla *et al.* (2000); Zilly *et al.* (2002); Nyanhongo *et al.* (2002) for degradation and/or decolorization of dyes like azo dyes, TPM dyes, heterocyclic dyes, polymeric dyes etc.

## Enzymatic Degradation of Nitro-Explosives

Microorganisms are known to degrade nitroaromatic compounds by aerobic as well as anaerobic mechanisms. Aerobic mechanism involves addition of oxygen atoms by monooxygenase or dioxygenase to eliminate nitro group. However, anaerobic mechanism involves reduction of nitro group via nitroso and hydroxylamino intermediates to the corresponding amines by anaerobic microorganisms (Spain 1995).

Meager data are available about enzymic basis of biodegradation of nitro explosives. Under aerobic conditions, cytochrome p-450 enzymes are known to accept polynitro compounds as electron acceptors. Resorsinol 2,3-dioxygease enzyme from *Candida pulcherima* and nitro aryl reductase from *Streptomyces aminophilus* degrading dinitrobenzene (DNB) were purified and characterized by Dey (1991, 1999). Degradation of nitro explosives by the reductive pathway could be due to non-specific nitroreductase enzymes present in both aerobic and anaerobic organisms (Rodgers and Bunce 2001). Some of the enzyme systems like nitroreductases, manganese peroxidase, laccases etc are reported in the literature using some of the Enterobacteriaceae strains, white rot fungus *Phelbia radiata* etc. (Rodgers and Bunce 2001; Coleman *et al.*, 2002). Kanekar *et al.* (2003) have reviewed the degradation of nitro explosives and the oxidoreductase like enzyme systems involved therein.

## Enzymatic Degradation of Other Xenobiotics Chemicals

The oxidative degradation of some organic substrates or their intermediates by *Pseudomonas* occasionally involves the participation of oxygenases. Sariaslani (1991) reviewed thoroughly the microbial cytochrome p-450 and their active role in degradation of variety of xenobiotics compounds. The author has given detailed account of enzyme systems from different microorganisms *e.g.* enzymes involved in degradation of camphor by *Pseudomonas putida* with structure of enzyme protein; *Bacillus megaterium* hydroxylating long chain fatty acids and their aldehydes and alcohols; *Xanthomonas* sp. performing NADPH-dependent hydroxylation of cyclohexane to cyclohexanol; *Acinetobacter calcoaceticus* carrying out oxidation of alkanes; several cytochrome p-450 of unspecified function from *Rhizobium japonicum* etc. along with some enzyme systems from actinomycetes, yeasts and fungi. Studies on degradation of caprolactam, a genotoxic compound which is being used in the manufacture of synthetic fibre nylon-6 and appearing as major pollutant in the effluent were carried out by Kulkarni (1998). The author reported the complete degradation of caprolactam by *Pseudomonas putida* to carbon dioxide through intermediate metabolites namely aminocaproic acid, amino acid, adipic acid, caproic acid, propionic acid and acetic acid and also the enzyme systems involved therein. Hawkes *et al.* (2002) carried out isolation, expression and characterization of cytochrome p-450 $_{cin}$ from *Citrobacter braakii* using monoterpene cineole 1 as sole source of carbon and energy. Oxygenases acting on aliphatic compounds such as alkanes may be part of complex oxidative systems.

## Future Prospects

In comparison to mammalian enzymes, microbial cytochrome enzymes are studied to a limited extent. Presently ~500 enzymes are investigated for about 50 different applications among which oxidoreductases contribute to a very small extent. Oxidoreductases are the more preferred enzymes as compared to other enzymes like hydrolases, lyases or transferases, may be due to their higher stereoselectivity and ability to act on substances like hydrocarbons which are considered as inert to other enzymes. Presently these enzymes are used effectively for bioremediation purpose for detoxifying the chemopollutants and also in food industries, textile industries etc. The oxidoreductases have very low specific activity and need cofactors. These limitations could be overcome by applying modern techniques like gene cloning, protein engineering etc. Although some information is available globally on microbial degradation of xenobiotic compounds like pesticides, dyes, nitroaromatics, etc. little work has been carried out on the enzyme systems involved therein. In India, very little attention has been paid ton degradation of these xenobiotic compounds by bacterial isolates, their metabolites and the enzymes involved in their degradation. It seems to be necessary to study the intermediate metabolites formed so as to ensure detoxification of these xenobiotics and the enzyme systems involved in degradation of these compounds with special reference to cytochrome p-450. Some of the enzymes possess very high substrate specificity while others are capable of oxidizing a wide range of structurally diverse substrates. The studies on these enzyme systems involved in degradation of xenobiotic compounds will help in establishing the metabolic pathways of degradation of structurally similar and dissimilar compounds. The identification of the intermediate metabolites will also help to establish these pathways.

Likewise, microorganisms producing enzyme dehalogenase, may be useful in removal of chlorine from the organochlorine compounds and degradation of these recalcitrant and persistent compounds. This will help in reducing the environmental pollution especially the pollution of soil and in turn that of ground water.

Development of microbial amperometric biosensor based on a carbon paste electrode containing genetically engineered cells expressing organophosphorus hydrolase (OPH) on the cell surface for the direct measurement of organophosphate nerve agents as described by Mulchandani *et al.* (2001) seems to be interesting for exploration. The success of directed evolution of OPH for improved hydrolysis of methyl parathion, can be easily extended in creating other OPH variants with improved activity against poorly degraded pesticides such as diazinon and chlorpyriphos. The researchers can attempt to obtain such OPH variants.

The oxidoreductases or oxygenases are widely distributed in nature *i.e.* in plants, animals, microorganisms but commercialized to a very limited extent. Thus there is a wide scope for exploring oxygenases for commercialization purposes.

## Conclusion(s)

Microorganisms play an important role in the degradation of these man-made chemicals which are released in the environment. The property of microorganisms to secret enzymes required for degradation or detoxification of toxic chemicals like nitro compounds, pesticides especially the organochlorine pesticides etc. can be utilized effectively for detoxification of chemopollutants. In general, the enzymes involved in degradation of chemopollutants include oxidases, hydrolyses, dehalogenases, dealkylases, deaminases etc. play an important role in oxidative metabolism of a variety of chemicals of agricultural and environmental concern *e.g.* pesticides. In comparison to mammalian enzymes, microbial cytochrome enzymes are studied to a limited extent. There is a wide scope for researchers to explore novel applications of these oxidoreductases in medical field, as biosensors, and also in detoxifying the newer toxic chemicals manufactured as on demands. The studies on these enzyme systems involved in degradation of xenobiotic compounds will help in establishing the metabolic pathways of degradation of structurally similar and dissimilar compounds. Thus, ecofriendly technologies could be developed for bioremediation purposes using these enzyme systems.

## References

1. Abdulla E, Tzanov T, Costa S, Robra KH, Cavaco-Paulo A, Gubitz GM, 2000, Decolourization and detoxification of textile dyes with a laccase from *Trametus hirsute, Applied and Environmental Microbiology*, 66, 3357–3362.

2. Bhadbhade BJ, Sarnaik SS, Kanekar PP, 2002, Biomineralization of an organophosphorus pesticide Monocrotophos by soil bacteria. *Journal of Applied Microbiology*, 93 224-234.

3. Cai M, Xun L, 2002, Organization and regulation of pentachlorophenol degrading genes in *Spingomonas chllorophenolicum* ATCC 39723, *Journal of Bacteriology*, 184, 4672–4680.

4. Cho CM, Mulchandani A, Chen W, 2002, Bacterial cell surface display of organophosphorus hydrolase for selective screening of improved hydrolysis of organophosphate nerve agents, *Applied and Environmental Microbiology*, 68, 2026-2030.

5. Coleman NV, Spain JC, Duxbury T, 2002, Evidence that RDX biodegradation by *Rhodococcus* strain DN22 is plasmid-borne and involves a cytochrome P-450, *Journal of Applied Microbiology*, 93, 463–472.

6. Das S, Singh DK, 2006, Purification and characterization of phosphotriesterases from *Pseudomonas aeruginosa* F10B and *Clavibacter michiganense* subsp. *insidiosum* SBL11, *Canadian Journal of Microbiology*, 52, 157–168.

7. Deshpande N M, 2002, *Biodegradation of Dimethoate–a carbamate group of organophosphorus insecticides*. Ph. D. Thesis, University of Pune, M.S. India.

8. Dey S, 1991, Purification and properties of resorcinol 2,3-dioxygenase from *Candida pulcherima. Indian Journal of Applied and Pure Biology*, 6, 107–111.

9. Dey S, 1999, Cloning and expression of nitroaryl reductase gene from *Streptomyces aminophilus* to *E. coli* and *S. lividans* TK64. *Indian Journal of Experimental Biology*, 37, 787–792.

10. Dumas DP, Caldwelol SSR, Wild JR, Raushel FM, 1989, Purification and properties of the phosphotriesterase from *Pseudomonas diminuta*, *Journal of Biologial Chemistry*, 264, 19659–19665.

11. Hashimoto M, Fukui M, Hayano K, Hayatsu M, 2002, Nucleotide sequence and genetic structure of a novel carbaryl hydrolase gene (cehA) from *Rhizobium* sp. strain AC100, *Applied and Environmental Microbiology*, 68, 1220–1227.

12. Hawkes DB, Adams GW, Burlingame AL, Ortiz de Montellano PR, De Voss JJ, 2002, Cytochrome p-450cin (CYP176A), isolation, expression and characterization, *Journal of Biological Chemistry*, 277, 27725–27732.

13. Kanekar P, Dautpure P, Sarnaik S, 2003. Biodegradation of nitro-explosives. *Indian Journal of Experimental Biology*, 41, 991-1001.

14. Kanekar PP, Bhadbhade BJ, Deshpande NM, Sarnaik SS, 2004, Biodegradation of organo-phosphorus pesticides, *Proceedings of Indian National Science Academy*, (Part B), 70, 57-70.

15. Kim SJ, Shoda M, 1999, Purification and characterization of a novel peroxidase from *Geotrichum candidum* Dec 1 involved in decolorization of dyes. *Applied and Environmental Microbiology*, 65, 1029–1035.

16. Kitagawa W, Kimura N, Kamagata Y, 2004, A novel p-nitrophenol degradation gene cluster from a gram positive bacterium, *Rhodococcus opacus* SAO101, *Journal of Bacteriology*, 186, 4894–4902.

17. Kulkarni RS, 1998, *Microbial degradation of caprolactam from Nylon-6 wastewater*. Ph. D. Thesis, University of Pune, Pune, M.S. India.

18. Liu Y, Chung Y, Xiong Y. 2001, Purification and characterization of a dimethoate-degrading enzyme of *Aspergillus niger* ZHY 256, isolated from sewage, *Applied and Environmental Microbiology*, 67, 3746–3749.

19. Mee-Hie CC, Mulchandani A, Chen W, 2006, Functional analysis of organophosphorus hydrolase variants with high degradation activity towards organophosphate pesticides. *Protein Engineering, Design and Selection*, 19, 99–105.

20. Mulbry WW, 1994, Purification and characterization of an inducible s-triazine hydrolase from *Rhodococcus corallinus* NRRL B-15444R, *Applied and Environmental Microbiology*, 60, 613–618.

21. Mulbry W, 2000, Characterization of a novel organophosphorus hydrolase from *Nocardiodes simplex* NRRL B-24074, *Microbiological Research*, 154, 285–308.

22. Mulchandani P, Chen W, Mulchandani A, Wang J, Chen L, 2001, Amperometric microbial biosensor for direct determination of organophosphate pesticides using recombinant microorganism with surface expressed organophosphorus hydrolase, *Biosensors and Bioelectronics*, 1666, 433–437.

23. Muller RH, Kleinsteuber S, Babel W, 2001, Physiological and genetic characteristics of two bacterial strains utilizing phenoxypropionate and phenoxyacetate herbicides, *Microbiological Research*, 156, 121–131.

24. Munnecke DM, 1976, Enzymatic hydrolysis of organophosphate insecticides, a possible pesticide disposal method, *Applied and Environmental Microbiology*, 32, 7–13.

25. Nagy I, Compernolle F, Ghys K, Vanderleyden J, de Mot R, 1995 (a), A single cytochrome p-450 system is involved in degradation of the herbicides EPTC (S-ethyldipropyl thio carbamate) and atrazine by *Rhodococcus* sp. strain N186/21, *Applied and Environmental Microbiology*, 61, 2056–2060.

26. Nagy I, Schoofs G, Compernolle F, Proost P, Vanderleyden J, de Mot R, 1995 (b), Degradation of the thiocarbamate herbicide EPTC (S-ethyldipropylcarbamothioate) and biosafening by *Rhodococcus* sp. strain N186/21 involve an inducible cytochrome p-450 system and aldehyde dehydrogenase, *Journal of Bacteriology*, 177, 676–687.

27. Nyanhong GS, Goves J, Gubitz G, Zvauya R, 2002, Production of laccase by a newly isolated strain of *Trametes modesta*, *Bioresource Technology*, 84, 259–263.

28. Ollikka P, Harjunpaa T, Palmu K, Mantsala P, Suominon I, 1998, Oxidation of Crocein Orange G by lignin peroxidase isoenzymes kinetics and effect of $H_2O_2$, *Applied Biochemistry and Biotechnology*, 75, 307–321.

29. Puranik KP, 2000, *Microbial bioremediation of ethylenethiourea from mancozeb pesticide industrial effluent.* Ph.D. Thesis, Pune University, Pune, M.S., India.

30. Rabie GH, 1995, Biodegradation of the organophosphorus insecticide monocrotophos by *Penicillium corylophilum*, Zagazig *Journal of Pharmaceutical Sciences*, 4, 14–19.

31. Richins RD, Kaneva I, Mulchandani A, Chen W, 1997, Biodegradation of organophosphorus pesticides by surface-expressed organophosphorus hydrolase, *Natural Biotechnology*, 15, 984–987.

32. Rodgers JD and Bunce NJ (2001) Treatment methods for the remediation of nitroaromatic explosives, *Water Research*, 35, 2101–2111.

33. Sariaslani FS, 1991, Microbial cytochromes p-450 and xenobiotics metabolism, *Advances in Applied Microbiology*, 35, 133–178.

34. Shao ZQ, Bekhi R, 1996, Characterization of the expression of the thcB gene, coding for a pesticide degrading cytochrome p-450 in *Rhodococcus* strains, *Applied and Environmental Microbiology*, 62, 403–407.

35. Sharma P, Raina V, Kumari R, Malhotra S, Dogra C, Kumari H, Kohler H-PE, Buser HR, Holliger C, Lal R, 2006, Haloalkane dehalogenase LinB is responsible for b- and d-hexachlorocyclohexane transformation in *Sphingobium indicum* B90A, 2006, *Applied and Environmental Microbiology*, 72, 5720–5727.

36. Singh BK, Walker A, 2006, Microbial degradation of organophosphorus compounds. *FEMS Microbiological Review*, 30, 428–471.

37. Spain JC, 1995, Biodegradation of nitroaromatic compounds, *Annual Review of Microbiology*, 49, 523–555.

38. Strauber H, Muller RH, Babel W, 2003, Evidence of cytochrome p-450 catalyzed cleavage of the ether bond of phenoxybutyrate herbicides in *Rhodococcus erythropolis* K2-3, *Biodegradation*, 14, 41–50.

39. Suar M, Hauser A, Poiger T, Buser HR, Muller MD, Dogra C, Raina V, Holliger C, van der Meer JR, Lal R, Kohler HP, 2005, Enantioselective transformation of HCH by dehydrochlorinases LinA1

and LinA2 from soil bacterium *Sphingomonas paucimobilis* B90A. *Applied and Environmental Microbiology*, 71, 8514–8518.

40. Turnbull GA, Ousley M, Walker A, Shaw E, Morgan JA, 2001, Degradation of substituted phenyl urea herbicides by *Arthrobacter globiformis* strain D47 and characterization of a plasmid associated hydrolase gene puhA, *Applied and Environmental Microbiology*, 67, 2270–2275.

41. Vaishampayan PA, 2004, *Bioremediation of a herbicide atrazine from soil.* Ph.D. Thesis, Pune University, Pune, M.S., India.

42. Vaishampayan PA, Kanekar PP, Dhakephalkar PK (2007) Isolation and characterization of *Arthrobacter* sp. strain MCM B-436, an atrazine degrading bacterium from rhizospheric soil. *International Biodeterioration andBiodegradation*, 60, 273–278.

43. Villemur R, Beaudet R, Lanthier M, Gauthier A, Boyer A, Thiboodeau J, Lepine F, Duguay M, Page-Belanger R, 2005, Molecular analysis of *Desulfitobacterium frappieri* pcp-1 involved in reductive dehalogenation of pentachlorophenol, *Water Science and Technology*, 52, 101–106.

44. Weir KM, Sutherland TD, Horne I, Russell RJ, Oakeshott JG, 2006, A single monooxygenase, Ese is involved in metabolism of the organochlorides Endoşulfan and endosulfate in an *Arthrobacter* sp., *Applied and Environmental Microbiology*, 72, 3524–3530.

45. White RE, Coon MJ, 1980, Oxygen activation by cytochrome p-450. *Annual Review of Biochemistry*, 49, 315–356.

46. Xu F, 2005, Applications of oxidoreductases: Recent progress. *Industrial Biotechnology* pp 38-50.

47. Zhang R, Cui Z, Jiang J, He J, Gu X, Li S, 2005, Diversity of organophosphorus pesticide-degrading bacteria in a polluted soil and conservation of their organophosphorus hydrolase genes, *Canadian Journal of Microbiology*, 51, 337–3343.

48. Zilly A, Souza CG, Barbosa-Tessmann IP, Peralta RM, 2002, Decolorization of industrial dyes by a Brazilian strain of *Pleurotus pulmonarius* producing laccase as the sole phenol oxidizing enzyme. *Folia Microbiology* (Praha), 47, 273–277.

# Chapter 17

# Challenges and Critical Issues in Successful Implementation of Water and Wastewater Treatment Schemes

**Mohammad Jawed***
**Associate Professor (Environmental Engineering),**
**Department of Civil Engineering, Indian Institute of Technology,**
**Guwahati – 781 039**

## ABSTRACT

The mankind is likely to face the "biggest challenge" of ensuring sustainability of potable water in the coming future since all the available water is ending up as polluted water. This has encouraged installation of water and wastewater treatment schemes at many a places–some of them are giving very good performance while others are failing miserably. This chapter makes an effort in identifying the challenges and relevant critical issues involved in successful implementation of water and wastewater treatment schemes.

*Keywords: Water, Wastewater, Treatment scheme.*

## Introduction

Water is one of the precious natural resources available to the mankind. However, rapid increase in population has shown drastic decline in availability of water. The per capita water availability for

* Corresponding Author: E-mail: jawed@iitg.ernet.in

India was over 5000 $m^3$ per annum in 1950, which now stands around 2000 $m^3$ and is projected to be only 1500 $m^3$ in 2025 (Swaminathan, 2001). Such quantitative data does not take into consideration the continued pollution of rivers and groundwater–major sources of available water. The water consumption trends indicate that water demand could double within next 25 years (Swaminathan, 2001). The need for water will have to be assessed both locally and globally. While projections of global demand are useful, what matters to individuals is the local availability of safe and hygienic water for their daily needs. The health costs of unsafe drinking water for India alone have been conservatively estimated at $4650 million for 1993 (Biswas, 2001). A UN survey report has ranked India a poor 133rd out of 185 countries in respect of safe drinking water availability. The parliamentary standing committee attached with the Department of Drinking Water Supply, Ministry of Rural Development, Government of India in its 46th report expressed a deep sense of outrage that even 55 years after independence, safe drinking water is not available to all people (Chandra, 2003). The committee has warned that ensuring sustainability and quality of drinking water sources would be the biggest challenge before the country in the coming future as the water quality conditions will continue to deteriorate since all water supplied ends up as wastewater. In addition, water quality monitoring is poorly organized and managed in nearly all developing countries (Biswas, 2001). Even expertise available for proper water quality management is not adequate (Biswas, 2001). Thus, effective water quality management becomes much more challenging even with the availability of funds to construct newer treatment schemes to improve the quality of water–for various beneficial uses, as well as collected wastewater–for treatment and its safe disposal. In the recent past, many a treatment schemes have been designed, constructed and commissioned. However, some of them are giving very good performance, while the others are failing miserably. Therefore, it is imperative to analyze and understand the reasons for the success or failure of such water and wastewater treatment schemes and identify challenges and critical issues involved thereof. This article makes an attempt in this direction based on personal experience of the author.

## Treatment Scheme Selection vis-à-vis Site Specific Conditions

The site-specific conditions (such as weather, topography, groundwater table level etc.) need careful consideration in selection of an appropriate treatment scheme–be it for water or wastewater treatment schemes. The usual practice is to engage a (competent) consultant/consultancy firm for selection and design of an appropriate treatment scheme for implementation. Depending upon the available expertise with the consultant/consultancy firm and the preliminary information available to them, a number of treatment schemes are drawn up and ranked in order of preference for consideration of the owner. At this stage, it becomes extremely important to review the proposed schemes independently keeping in mind the site-specific conditions. In some of the instances, the site-specific conditions may require thorough modification/re-ranking of the proposed schemes and/or re-drawing altogether a new treatment scheme. Let us take an example of sewerage scheme for a new residential campus of an elite technical institute located in the northeastern region of the country. The consultancy firm–which had drawn up the master plan of the entire campus as well as has provided the design and detailed working drawings for academic complex, administrative building, hostels, guest house and residential buildings for staff–submitted a proposal of sewage treatment using activated sludge process which included grit chamber, primary sedimentation, activated sludge process (comprising of aeration tank and secondary sedimentation), chlorination and sludge drying beds. The firm had also provided a sewer network design for the campus based on gravity flow (without provision of lifting stations in-between) up to the proposed treatment site–which entailed laying of sewer lines at a depth of 6 m or more below the ground level for a considerable stretch of sewer line. These proposals were reviewed at the concerned technical institute keeping in mind the site-specific conditions such as: (a) likely quality of collected wastewater, (b) duration of wet weather conditions, (c) present and future manpower constraints, (d) non-availability of on-site laboratory support and (e) non-availability of un-interrupted power supply. Since the water consumption is expected to be higher in the campus, the wastewater so produced will definitely be dilute in nature and, therefore, applicability of activated sludge process is open for question and detailed discussion. The rainy season in this region usually sets in the month of March and continues up to November–extending the wet weather conditions for 7 to 9 months in a year. It has a serious impact on the design of sludge drying beds–needing a much larger area due to requirements of relatively longer drying periods. How long drying periods one has to consider in the design of sludge drying bed is difficult to adopt due to 7-9 months of wet weather conditions. Even it will be necessary to remove and store the dried sludge cake in a dry space before its eventual disposal. In view of the changed policy decisions to reduce the number of support staffs on the payroll of the institute, the entire operation and maintenance work of sewage treatment works will have to be outsourced in the coming future. Additionally, uninterrupted power supply for the treatment plant could not be assured at the time of finalization of design. Under these site-specific conditions, the institute decided to opt for a simple but rugged treatment scheme. Therefore, a treatment scheme comprising of pond system which included a grit chamber and two facultative ponds in series followed by a polishing pond was selected for implementation. The pond system of treatment do not required skilled operation and maintenance staff and cannot go wrong even if there is frequent interruptions in power supply as it is a natural form of treatment (Arceivala, 1999). In respect of sewer line, the wet weather conditions make the soil saturated with water and digging any trench of 1 m depth or more yields considerable amount of groundwater. Hence, sewer lines at deeper depths are likely to contribute more infiltration/inflow in the collected sewage, which will further dilute the wastewater. Even laying out the sewer lines at deeper depths will pose a serious challenge. Therefore, sewer network design was also reviewed and modified so as to terminate the sewer line with provision of a sump-cum-lifting-pumping station if the sewer invert level was more than 3.5 m below the ground level. The entire sewer network contained 7 sump-cum-lifting-pumping stations with automatic water-level sensor to activate pumps to pump the collected sewage directly to a sump just before the grit chamber. The sewerage scheme has already been constructed, commissioned and in operation for more than two years and giving a satisfactory performance.

## Advanced vis-à-vis Conventional Treatment Schemes

Some of the treatment schemes proposed by consultants/consultancy firms in the recent past have included advanced treatment units in the treatment chain. Let us see an example of proposed treatment schemes for water and wastewater for a small residential campus of armed forces located in northeastern region. The water treatment scheme involves withdrawal of raw water from a small perennial stream and proposes the following scheme: raw water → chlorine dosing system → surface aeration tank → alum dosing system → reaction tank → clarifier → clarified water tank → filter feed pump → pressure sand filter → activated carbon filter → micron filter → RO plant → UV system → treated water tank. This scheme is to be operated to produce treated water at a rate of 80 KL/h for 16.5 h operation everyday. The wastewater treatment scheme proposes the following scheme: grit screen chamber → collection sump → fluidized aerobic bioreactor (attached growth process requiring air blowers and coarse bubble membrane aerator) → clarifier (with a provision of sludge drying beds) → clarified water tank → disinfection → pressure sand filter → activated carbon filter → treated wastewater disposed to the same stream at a rate of 520 KL/d. These two treatment schemes include advanced units such as activated carbon filter, micron filter, RO plant, UV system of disinfection, coarse bubble membrane reactor etc. Let us take the case of another residential campus of a public sector authority–located in northeast region–having approximately 350 residential units and presently using groundwater as the source of water for domestic supply, which contains iron more than the permissible limit. Now they intend to install treatment scheme to reduce iron content from the groundwater. The well established conventional iron removal scheme is: groundwater → natural air aeration → sedimentation → disinfection → storage and supply, whereas an advanced scheme for the same may comprise: groundwater → pH correction → resin based catalysis (to convert $Fe^{2+}$ to $Fe^{3+}$ and then to serve as a filter for $Fe^{3+}$ separation with provision for backwashing to take out deposited $Fe^{3+}$ from the resin-bed) → disinfection → storage and supply. This particular resin is not available locally and the catalytic reaction has to be carried out in an enclosed steel reactor with many valves making the system more complex to be handled by a semi-skilled operator. Therefore, one has to look into the sustainability of such schemes units vis-à-vis adoption of the conventional scheme at the particular location with a view to carry out routine operation and maintenance work with the help of local semi-skilled manpower using local material resources. If there is an iota of doubt regarding sustaining such schemes in the absence of material and skilled manpower support, it may be better to consider adoption of conventional treatment schemes.

## Unbiased Expert Interface between Treatment Scheme Owner and the Consultant

The (competent) consultant/consultancy firm provides the treatment scheme with complete design and detailed working drawings along with the budgetary estimates upon request of the owner of the treatment scheme. The owner may not be familiar with the various aspects of treatment units and therefore, may not be in a position to visualize all the essential requirements of sustaining the proposed treatment scheme upon its implementation. On the other hand, the consultant/consultancy firm may have specific expertise with them and, therefore, may not be able to propose the most appropriate treatment scheme considering the constraints and requirements of the owner. Under these circumstances, if a treatment scheme is adopted for implementation without giving considerable thought to its sustainability, the scheme may not serve the required purpose or may become a financial burden in the long run leading to its neglect. Therefore, it is important to have an interface of unbiased but right expert between the consultant/consultancy firm and the owner to avoid such situations. This unbiased and right expert should be competent enough to scrutinize the design of all the major units minutely so that the treatment scheme meets the quality requirements at the end of the day. Let us see the case of water treatment scheme of the residential campus of an elite technical institute located in northeastern region. The water treatment scheme at this particular institute comprises of floating barge (in a large perennial river) → holding tank → pre-settling tank → cascade type aeration unit → coagulant dosing unit with flash mixer → circular clariflocculator → pressure sand filter (backwashing provision with overhead stored clear water) → clear water reservoir → chlorine dosing unit (before pumping to overhead tanks for supply). This treatment scheme was adopted after deliberations with the consultancy firm, engineers of the institute and two of the faculty members having expertise in water resources and hydraulics engineering. No faculty member having expertise in environmental engineering was involved in the selection of the treatment scheme. Let us evaluate the adopted scheme with an environmental engineering perspective. The purpose of holding tank is to provide 2 h detention time to raw water pumped from the river so as to reduce the concentration of settleable solids. The subsequent unit–the pre-settling tank–has a detention time of only 11 minutes, and one wonders the justification of this unit as far as the pre-settling of settleable mass is concerned, which has already gone for pre-settling for 2 h in the holding tank. The raw river water does not contain any trace of dissolved iron and manganese and, therefore, the utility of passing the water through cascade type aeration unit need justification. The circular clariflocculator has been provided with sharp-crested weir (not adjustable V-notches, as the usual practice followed in water treatment plants) along its periphery for withdrawal of clarified water. The level of sharp-crested weir is not same through out its periphery (despite the best possible efforts of the contractor to level it during its construction) and one can clearly observe passage of clarified water through parts of the weir. Similarly, the backwashing has to be carried out with the stored overhead clear water–needing larger amount of water for backwashing process along with manual stirring of the top layers of the filter beds to loosen the bed to aid in cleaning during backwashing operation. Instead, if air compressors were provide to loosen the beds before using the overhead stored water for backwashing, which might have reduced the amount of backwash water compared to the designed system. The clear water reservoirs have been designed to store the treated water for a period of 7 days. However, it was never felt to disinfect the filtered water before storage to ensure protection against growth of microorganisms during storage and chlorination was not adopted just after filtration. These design related major issues might have been taken care of if right experts have served as an interface between the consultancy firm and the owner of the scheme.

## Construction Quality vis-à-vis Commissioning Issues

In order to avoid the problems related with construction quality and commissioning issues of water and wastewater treatment schemes, the present trend is to award the contract of entire scheme to a competent and qualified contractor on built (as per the adopted treatment scheme), commission, operate, train and transfer (BCOTT) basis. The contractor is provided with the detailed working drawings of the adopted treatment scheme. Since the contractor has to commission the plant and operate it for a period of time (varying 2 to 5 years after commissioning) and then hand it over to the owners' representative after imparting on-site training in operation and maintenance. Due to these requirements, the contractor scrutinizes the design carefully and brings out any flaws to the notice of the owner. The mechanism has helped in reducing much of the post-commissioning headache related with quality of the construction and problems in commissioning. If there is any problem with the quality of the construction and commissioning, the contractor has to sort it out at its own cost. This will ensure success of the treatment scheme and will provide a big relief to the owner of the treatment scheme. This approach has been adopted successfully for the water and wastewater treatment schemes for domestic as well as industrial sectors at many a places. However, in the absence of unbiased expert, even one can think of design, built, commission, operate, train and transfer (DBCOTT) model. In this model, the qualified contractor will be required to select a particular treatment scheme depending upon the influent and effluent quality parameters and other constraints as specified by the owner; get the design and detailed working drawings from their in-house or hired designer; construct, commission and operate the scheme; train staff of the owner and then transfer the scheme in running conditions.

## Outsourcing Operation and Maintenance of Treatment Schemes

Outsourcing of services has become one of the key concepts in the present age. This concept is also adopted for the operation and maintenance works of water and wastewater treatment schemes at many places–including industrial wastewater treatment schemes. Theoretically, outsourcing of operation and maintenance works of treatment scheme to a competent private party will ensure proper working of various treatment units with full attention on overall performance of the scheme. However, the ground realities are not in tune with the theoretical concept. This is mainly due to engagement of unskilled personnel for operation and maintenance related works. These personnel are only instructed when to and how long to switch on a particular switch and they are not made aware of any other aspects of the treatment scheme. Unfortunately, even the technical support staffs and engineers of the treatment scheme owner are not much aware of the operation and maintenance work of the treatment scheme and therefore, are unable to keep a constant watch to see whether the operation and maintenance works are going on satisfactorily. Let us see the case of one of the refineries situated in the northeast region, which has installed a domestic wastewater treatment plant for its residential campus. The treatment scheme comprises of grit channel → activated sludge process (aeration tank + secondary sedimentation, excess sludge to sludge thickener and thickened sludge to sludge drying beds) → treated effluent disposed of to a nearby drain. The operation and maintenance works of this treatment plant has been outsourced to a private party. Upon a visit to this plant, it was observed that sludge-drying beds were not in use for long (as thick vegetal growth of 30 cm height or more was observed in all the beds). When 1 L sample was drawn from the aeration tank to make a visual inspection of mixed liquor suspended solids (MLSS), no settleable solids were observed in the sample indicating that the activated sludge process was not working at all. Upon further enquiry, it was informed that the operators have been just given a schedule when to switch on the electrical switch for aerator(s), sludge re-circulation pump(s), thickened sludge to drying beds etc. The operators were not aware of the functioning of various units and they did not have any laboratory support to find out whether the treatment scheme is working satisfactorily or not. During a surprise visit to the treatment plant in late hours on a previous day, the aerators were found in switched-off mode. It appeared that aerators were stopped during 8.30 pm to 6.00 am every day and the reasons for the same were not explained. Similar observations were also made during a visit to domestic wastewater plant (based on activated sludge principles) installed to take care of township wastewater in the premises of one of the atomic power plant located in northern India way back in 1986. The reason for stoppage of aeration for more than 4-6 hours every day was meant to "give rest to aerators". Once the aerators are stopped for longer duration in an activated sludge process, the characteristics of entire active biomass will change completely and it will lead to instability and erratic performance of the treatment scheme.

## Owner Awareness for Beneficial Aspects of Treatment Scheme

Some of the industrial wastewater treatment plants (treating organic waste to produce energy) have been observed to be successful. The main reason for its success is awareness of owner to the beneficial aspect of the treatment scheme. In early '90, a distillery unit located near Kanpur was approached for permission to collect around 400 L of distillery wastewater everyday over a period of two years for doctoral research work of the author. However, the industrial unit refused to part with their waste, which was a useful resource for energy production. This particular industrial unit had put in a conventional suspended type anaerobic reactor for production of biogas (mixture of methane and carbon dioxide). The treatment scheme comprised of chemical pre-treatment (for initial pH adjustment) → acidogenesis → pH adjustment → methanogenesis. The treatment scheme was being operated in a professional manner to generate sufficient biogas to meet daily requirement of energy to operate the distillery unit. The biogas so produced was used in an imported boiler to produce steam. Therefore, the objective of the owner was to use all the organic waste to produce biogas and at the same time to reduce their dependency on coal procured from a distant place. It not only helped in saving cost of procurement of coal, but they were also able to give treatment to their organic wastewater. In fact, the wastewater treatment scheme was further extended to include aerobic pond system to impart additional treatment after the anaerobic treatment. It appeared that the owner of the distillery unit has undertaken thorough economic analysis of their energy requirements and energy likely to be produced through their wastewater, and hence the treatment scheme was given due weightage in the overall scheme of distillery operation.

## Water Quality Parameter Testing Facilities in Private Sector

The success or failure of a treatment scheme is very much dependent on regular performance monitoring. The major hurdle in performance monitoring work is non-availability of quality parameter testing facilities in private sector. Presently, the owner of the plant or the party outsourced to look after operation and maintenance work of the treatment scheme have to depend upon some of the academic institutes to get the water quality tested occasionally. Even the existing testing rules at some of the academic institutes are so framed that the persons actually carrying out the testing work do not get remuneration commensurate to the efforts put in by them. This serves as a serious impediment to utilize the testing facilities available at the academic institutes. Therefore, there is an unquestionable need to encourage development of water quality parameter testing facilities in the private sector. Unfortunately, water quality testing facilities are non-existent in almost all the cities. On the other hand, when a person suffers from diseases caused by drinking water of poor quality–the root cause of most of the common illness we suffer from, there are many pathological laboratories available in the private sector to test urine/blood/sputum/stool and to find out the exact cause of illness and to suggest a line of treatment. Therefore, water quality parameter testing facilities in private sector needs top most priority and must be made part and parcel of infrastructure development related with water and wastewater treatment schemes.

## Shortage of Qualified Technical Manpower

Water and wastewater treatment schemes need involvement of qualified manpower for its successful implementation–be at design, construction, commissioning; or operation and maintenance works levels–either in-house or outsourced to a private party. Probably, a qualified environmental engineer with a basic degree in Civil Engineering may be required to design a treatment scheme–simple and rugged design taking into considerations local constraints, to supervise civil construction works and commissioning of the plant. Many an academic institutions are offering masters' programme in environmental engineering. However, there appears to be a general lack of interest in graduating students to pursue masters' programme in environmental engineering–as their top three choices will always be amongst structural engineering, transportation engineering and geotechnical engineering; a reluctant fourth or fifth will be environmental engineering. The two main factors responsible for this situation are: (a) non-availability of specialist teaching faculty in the area of environmental engineering (or public health engineering as it was called in earlier days) and (b) re-learning chemistry again and again; and learning biology afresh (two very important subjects in this area–usually not preferred by most of the civil engineering student) as well as they have to dirty their hands with water and wastewater samples in the laboratory classes. Hence, sustained efforts must be made to encourage graduating students to pursue masters' programme in environmental engineering along with a change in methodology to teach chemistry and biology in an engineering way–with more emphasis on the applied aspects required for designing and managing treatment schemes and with lesser emphasis on the conventional aspects of these two subjects. In addition to production of qualified Environmental Engineers, there is a tremendous need of Environmental Technicians–mostly required to take care of operation and maintenance works of the treatment schemes as well as to carry out the monitoring of water quality parameters on regular basis. Now the question is how to go about producing qualified Environmental Technicians. One way is to have a diploma level of courses designed to produce Environmental Technicians. However, if prospective candidates–after passing their 12$^{th}$ standard–are not aware of job potential, they will hardly show any interest to enroll in this diploma course. The other way is to have institutional support to impart training as an Environmental Technician to a fresh candidate hired to look after operation and maintenance work of a particular type of a treatment scheme. In this mode, the basic qualification may be higher such as B Sc (with chemistry and/or biology) and the candidate is explained the basic principles of the particular treatment scheme with hands on training in routine monitoring works. A trained Environmental Technician may be an invaluable asset in successful and efficient operation of treatment schemes.

## Regulators as Friend and Advisor to Treatment Scheme Owners

Most of the pollution regulators (pollution control boards) are having qualified technical manpower duly supported by well-equipped laboratories to monitor water quality parameters. However, the regulators are generally perceived as prosecutor of offending units and, therefore, a safe distance is maintained by treatment scheme owners (if there is no other motive involved to come together). In the past, number of litigation related with pollution offences has increased and many a legal battles between the regulators and offending units are being fought in the court of law. This does not reduce and prevent the pollution whereas the litigation continues for years starting from a lower court to the highest court of the land through appeal at various stages. Therefore, a change in approach of regulators will go a long way in ensuring proper operation and maintenance of treatment schemes. The regulators' helping and friendly approach will help the treatment scheme owners to open up with their problems related with treatment scheme and they will be much more serious in considering and implementing the advice of regulators for making an overall improvement in the treatment scheme.

## Infrastructure Requirements for Water and Wastewater–Urban vis-à-vis Rural

The distribution of limited financial resources for the development of infrastructure in the area of water and wastewater treatment schemes are highly uneven–much more resources are allocated to urban centres while rural areas are being apparently overlooked. Ideally, the resources must be allocated in such a manner to take care of water and wastewater infrastructure development for rural as well as urban areas. However, one must carefully evaluate the type of infrastructures required in the urban and rural sectors–the same type of infrastructure may not be appropriate and cost effective in both the sectors. Let us take one example each related with water and wastewater treatment schemes for rural sector.

The groundwater is the major source for domestic water needs of people living in rural area. The groundwater of Assam contains very high concentration of dissolved iron (1 to 10 mg/L or above)–far beyond the permissible limit of 0.3 mg/L. Iron removal has been a subject of intensive research for the last couple of decades in developing countries. This has yielded many different designs and configurations of iron removal units to be used/applied at community levels–even our own country, India, is no exception to this. But surprisingly, none of these community level units has found application in Assam. Instead, the rural population of Assam uses indigenous household iron filter units, which is fabricated solely based on experiences gained over generations using local materials. One of the most popular household iron filter units is made of reinforced cement concrete (RCC) frame as shown in Figure 17.1. It is fabricated using two circular RCC pipes (60 cm internal diameter and a height of 60 cm), placed one over the other giving a total height of 120 cm. The top RCC pipe holds the filter media (up to a depth of 50 cm) while the bottom one stores the filtered water. The filtering medium usually comprises of three different materials available locally and is placed in layers in the top RCC pipe: river sand (on top), wooden charcoal (in middle) and gravel (at bottom). At the interface of sand and charcoal layers, nylon net with fine openings is provided to protect sand from getting washed into charcoal medium. Upon continuous use, charcoal gets exhausted needing replacement with new stocks while sand and gravel are re-used after washing and drying. Wooden charcoal is believed to be the main component responsible for iron removal. Even though, indigenous iron filter units are in use

Filter Unit

Filter media

**Figure 17.1: RCC Frame Filter Unit for Iron Removal**

for generations, very limited scientific studies have been carried out to understand the role of wooden charcoal in removal of dissolved iron from the groundwater and its potential to remove/reduce concentration of fluoride and arsenic–also present in the groundwater.

In the rural sector, sanitation is mainly required to take care of human wastes, animal dropping and poultry excrement on an individual household basis. In addition, farming may produce significant quantity of stalks of agricultural plants, chaff and husks, green grass and tree leaves–needs appropriate disposal. Farmers in the rural areas of China employ "Hydraulic Biogas Digester" for biogas fermentation (Figure 17.2) in their backyard using agricultural residues and household wastes including human and animal excrement. The biogas produced is used for cooking food, lighting gas lamp while digested slurry is used for manure in the field, which enriches the soil with organic carbon. In Indian context, the "Gobar Gas" plants are widely used. However, efforts are required to utilize human excrement and other agricultural residues for biogas production as well as to take care of rural sanitation on an individual household basis.

## Summary

The management of freshwater and wastewater can make the difference between prosperity and decline of an area. In order to sustain the management of freshwater and wastewater in future, it is essential to analyze the challenges and critical issues involved in success or failure of existing water and wastewater treatment scheme so as to ensure absolute zero failures for newer treatment schemes. This will not only ensure protection of freshwater but will ensure better utilization of treatment schemes developed at a considerable cost.

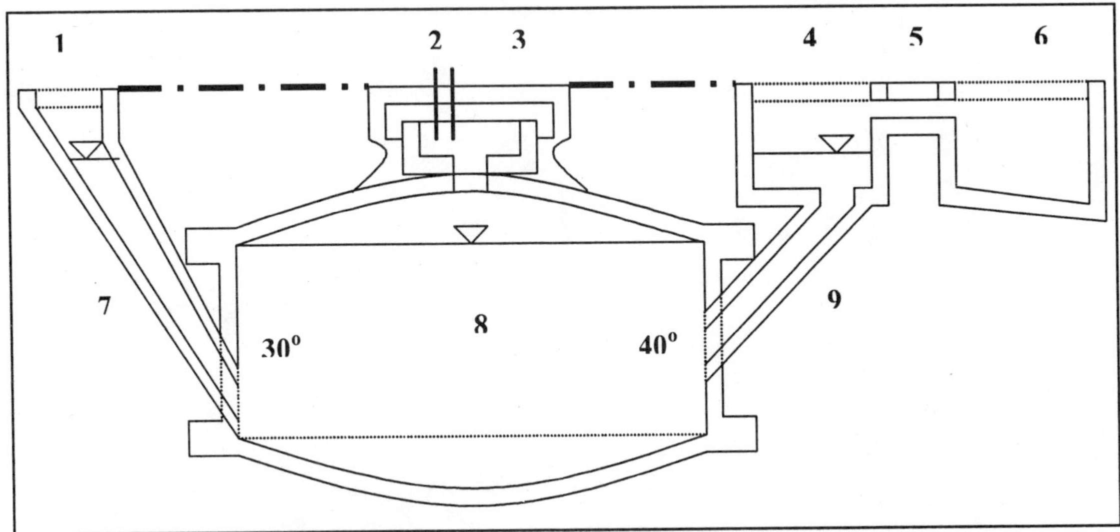


**Figure 17.2: Schematics of Chinese Hydraulic Biogas Digester (*Source*: Yongfu, 1989)**

**1: Inlet; 2: Gas Pipe; 3: Movable Cover; 4: Hydraulic Chamber; 5: Overflow Pipe; 6: Manure Storage Tank; 7: Inlet Pipe; 8: Fermentation Chamber; 9: Outlet Pipe**

## References


1. Arceivala, S J (1999) *Wastewater treatment for pollution control*, 2nd Ed., Tata McGraw-Hill, New Delhi.
2. Biswas, A K (2001) "Missing and neglected links in water management", *Water Science and Technology*, Vol. 43, No. 4, pp 45-50.
3. Chandra, P (2003) "Don't fudge water data, MPs tell govt.", *The Times of India*, New Delhi Ed., April 25, 2003.
4. Swaminathan, M S (2001) "Ecology and equity: Key determinants of sustainable water security", *Water Science and Technology*, Vol. 43, No. 4, pp 35-44.
5. Yongfu, Y (1989) *The Biogas Technology in China*, Agricultural Publishing House, Beijing, China, ISBN 7-109-01777-X/S.1174.

# Chapter 18

# Membrane Bio-Reactors: The Future of Wastewater Technology

**V.S. Sapkal, M.B. Dukare and S.S. Balkar**
**Chemical Engineering Department,**
**Anuradha Engineering College, Chikhli Distt. Buldana – 443 201, M.S.**

## ABSTRACT

With the current focus on water reuse projects and the role they play in the water cycle, the search for cost competitive advanced wastewater treatment technologies has never before been so important. Membrane technology has led to a revolutionary new concept in wastewater treatment. Membranes act as a barrier to bacteria and suspended solids to produce a low turbidity treatment plant effluent with very low bacteria counts. Membrane bioreactor systems offer several operational and economic advantages over conventional Wastewater Treatment Plants. This chapter discuss the Membrane Bioreactor (MBR) process and its suitability for water reuse applications.

The MBR process involves a suspended growth activated sludge system that utilizes micro porous membranes for solid/liquid separation in lieu of secondary clarifiers. This very compact arrangement produces a MF/UF quality effluent suitable for reuse applications or as a high quality feed water source for Reverse Osmosis treatment.

The MBR process is an emerging advanced wastewater treatment technology that has been successfully applied at an ever increasing number of locations around the world. In addition to their steady increase in number, MBR installations are also increasing in terms of scale. A number of plants with a treatment capacity of around 5 to 10 ML/d have been in operation for several years now whilst the next generation (presently undergoing commissioning or under contract) have design capacities up to 45 ML/d.

This chapter aims to discuss MBR design considerations. It includes discussion on how applicable this technology may be and it lists some of the local opportunities and local barriers that this technology may experience. This chapter also highlights some of the difficulties that may be experienced in terms of MBR scale-up.

***Keywords:*** *Wastewater, Membrane technology, Membrane bioreactor, Wastewater treatment, Bacteria.*

## Introduction

As community awareness of the impacts of discharging treated effluent to the environment increases, so is the need to improve the performance of existing treatment plants. Not only does a high effluent quality reduce negative impacts of discharging a treated effluent to inland waterways, but the range of possible reuse options is wider when a high quality effluent is available[(1)]. An option that may be considered in order to achieve additional treatment of sewage effluent is membrane technology. The use of membrane technology for effluent treatment may be appropriate and superior to that of existing process options. However, there are situations where membranes can only be used as assisting technology to existing treatment processes.

There are broadly four categories of membrane types, with classification being dependent on the pore size of the membrane. These categories, from smallest to largest pore size, are reverse osmosis (RO), nanofiltration (NF), ultrafiltration (UF) and microfiltration (MF)[(2)].

Historically, membranes have not been commonly used for the treatment of sewage effluents. Today, however, there are several large-scale membrane treatment plants being used for sewage treatment. One of the most promising newer technologies is the membrane bioreactor (MBR)[(3)]. a process that couples membrane filtration with biological treatment to achieve excellent effluent quality with a small design footprint. Several years ago, there were virtually no MBR systems in operation and the technology was generally unknown in the marketplace. The main reasons why the technology was not being utilized were generally cited as:

- ☆ Untested, complex and small scale
- ☆ High costs
- ☆ High operator skill required
- ☆ Unknown maintenance and labour requirement
- ☆ Membrane failure rate unknown
- ☆ No requirement for high effluent quality

## Overview of the Technology

The Membrane Bioreactor (MBR) process is an emerging advanced wastewater treatment technology that has been successfully applied at an ever increasing number of locations around the world. In addition to their steady increase in number, MBR installations are also increasing in terms of scale. A number of plants with a treatment capacity of around 5 to 10 ML/d have been in operation for several years now whilst the next generation (presently undergoing commissioning or under contract) have design capacities up to 45 ML/d.

Recent technical innovation and significant membrane cost reduction have Pushed membrane bioreactors (MBRs) to become an established process option to treat wastewaters. The combination of membrane separation with a suspended growth bioreactor is now widely used for municipal and industrial waste treatment[4].

When used with domestic wastewater, MBR processes could produce effluent of high quality enough to be discharged to coastal, surface or brackish waterways or to be reclaimed for urban irrigation.

Other advantages of MBRs over conventional processes include small footprint, easy retrofit and upgrade of old wastewater treatment plants into MBRs.

However, membrane fouling remains a major drawback, limiting the wider application of this process (Figure 18.1)[5].

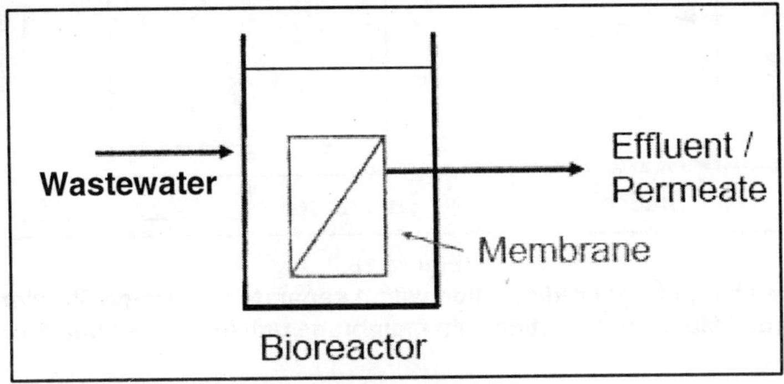


**Figure 18.1**

In recent years, membrane bioreactor (MBR) technology has gained popularity in wastewater treatment and especially for decentralized and reuse applications[6–8].

The technology consists of a compact unit which combines activated sludge treatment for the removal of biodegradable pollutants and a membrane for solid/liquid separation. MBRs have a small footprint making them attractive where space is limited and water treatment for internal recycling is desirable. Further advantages of the system are the high quality of its effluent. However, investment and operational costs are high.

## Configuration of MBR Systems

In an MBR, membrane filtration occurs either within the bioreactor [submerged configuration, Figure 18.2(a)] or externally through recirculation [Figure 18.2(b)], subject to a pressure drop across the membrane driven by either the hydraulic head or a pump. The UF or MF membranes utilized by MBRs have pore sizes such that water and most solute species pass through the membrane whilst other larger species, such as solids and microorganisms, are retained. The choice between operating options is dependent upon the application, as both systems have advantages and disadvantages (Table 18.1).

Aeration within the bioreactor provides the required oxygen transfer for growth of the biomass and mixing of the reactor. In the submerged configuration a coarse bubble diffuser is generally used. This system does not offer very efficient oxygen transfer but the rising bubbles provide a turbulent cross flow velocity (approximately 1 m/s) over the surface of the membrane. This helps to maintain the flux through the membrane, by reducing the build up of material at the membrane surface, and thereby increases the operational cycle of the system.

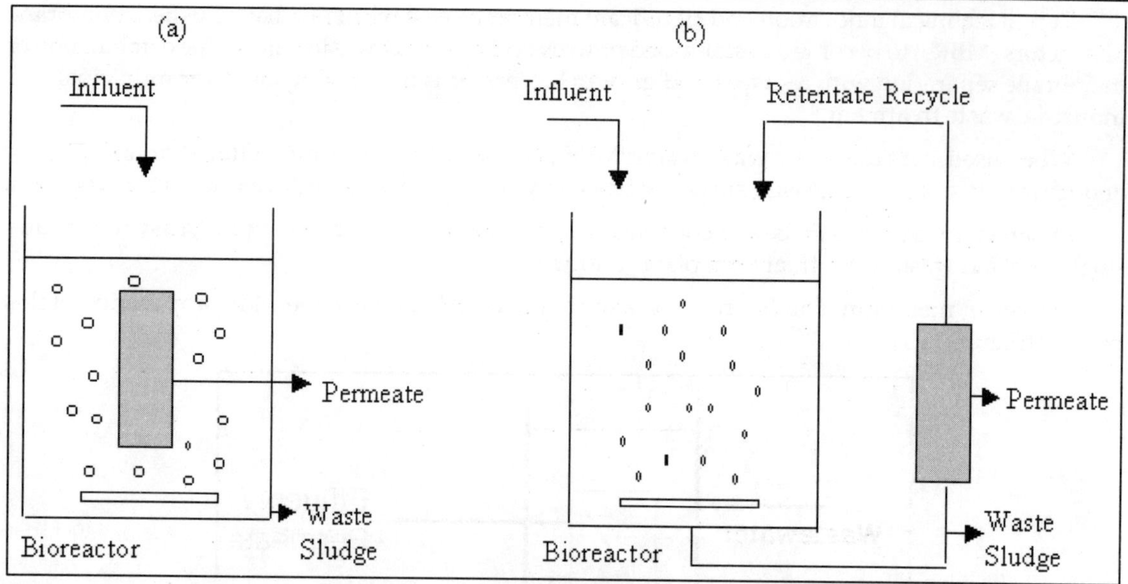


**Figure 18.2**

**(a) Side stream MBR configuration with a separate membrane filtration unit; (b) Submerged MBR configuration with membrane unit integrated into the bioreactor**

**Table 18.1: Advantages and Disadvantages of MBR Configurations**

| *Submerged MBR* | *Side-stream MBR* |
|---|---|
| Aeration cost high (~90 per cent) | Aeration cost high (~90 per cent) |
| Very low liquid pumping costs (higher if suction pump is used ~28 per cent) | Higher pumping costs (60-80 per cent) |
| Lower flux (larger footprint) | Higher flux (smaller footprint) |
| Less frequent cleaning required | More frequent cleaning required |
| Lower operating costs | Higher operating costs |
| Higher capital costs | Lower capital costs |

## Major Drawback

In recent reviews covering membrane applications to bioreactors, it has been shown that, as with other membrane separation processes, membrane fouling is the most serious problem affecting system performance[(9)]. Fouling leads to a significant increase in hydraulic resistance, manifested as permeate flux decline or transmembrane pressure (TMP) increase when the process is operated under constant-TMP or constant-flux conditions respectively. Frequent membrane cleaning and replacement is therefore required, increasing significantly the operating costs.

## Submerged MBR

With the membrane directed immersed into the bioreactor, submerged MBR systems are usually preferred to side stream configuration (with the membrane) situated into a pressurized external loop), especially for domestic wastewater treatment. The submerged configuration relies on coarse bubble aeration to produce in-tank recirculation and limit fouling. The energy demand of the submerged system can be up to 2 orders of magnitude lower than of the side stream systems[3]. And submerged systems operate at a lower flux, demanding more membrane area.

## MBR Process Description

The MBR process is a suspended growth activated sludge system that utilises microporous membranes for solid/liquid separation in lieu of secondary clarifiers. The typical arrangement shown in Figure 18.3 includes submerged membranes in the aerated portion of the bioreactor, an anoxic zone and internal mixed liquor recycles. Incorporation of anaerobic zones for biological phosphorus removal has been the focus of recent research, and there is at least one full scale facility of this type being designed presently in North America. As a further alternative to Figure 18.3, some plants have used pressure membranes (rather than submerged membranes) external to the bioreactor.

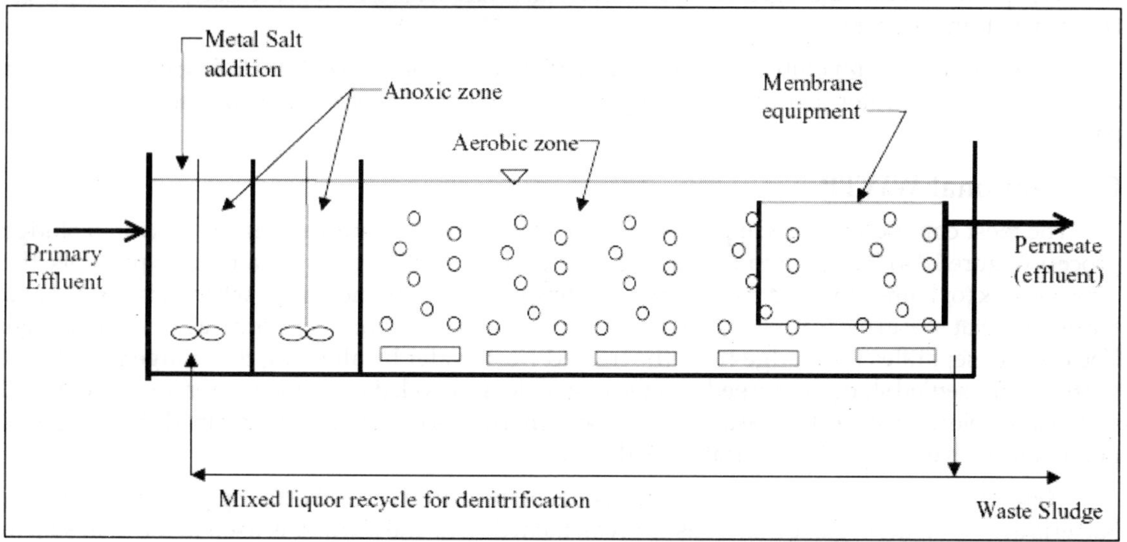


**Figure 18.3**

The ZeeWeed membrane bioreactor (MBR) process is a technology that consists of a suspended growth biological reactor integrated with an ultrafiltration membrane system, using the ZeeWeed hollow fiber membrane. Essentially, the ultrafiltration system replaces the solids separation function of secondary clarifiers and sand filters in a conventional activated sludge system.

ZeeWeed ultrafiltration membranes are immersed in an aeration tank, in direct contact with mixed liquor. Through the use of a permeate pump, a vacuum is applied to a header connected to the membranes. The vacuum draws the treated water through the hollow fiber ultrafiltration membranes. Permeate is then directed to disinfection or discharge facilities. Intermittent airflow is introduced to the bottom of the membrane module, producing turbulence that scours the external surface of the hollow fibers. This scouring action transfers rejected solids away from the membrane surface.

ZeeWeed membrane bioreactor technology effectively overcomes the problems associated with poor settling of sludge in conventional activated sludge processes. ZeeWeed MBR technology permits bioreactor operation with considerably higher mixed liquor solids concentrations than conventional

such as digesters or UV disinfection can also be eliminated/minimised (dependent upon governing regulations).

- Unlike secondary clarifiers, the quality of solids separation is not dependent on the mixed liquor suspended solids concentration or characteristics. Since elevated mixed liquor concentrations are possible, the aeration basin volume can be reduced, further reducing the plant footprint.
- No reliance upon achieving good sludge settleability, hence quite amenable to remote operation.
- Can be designed with long sludge age, hence low sludge production.
- Produces a MF/UF quality effluent suitable for reuse applications or as a high quality feed water source for Reverse Osmosis treatment. In addition, MF/UF provides a barrier to certain chlorine resistant pathogens such as Cryptosporidium and Giardia.
- The resultant small footprint can be a feature used to address issues of visual amenity, noise and odour. Example MBR plants exist where the entire process is housed in a building designed to blend in with its surrounding land use. This can reduce the buffer distance required between the plant and the nearest neighbour and can increase the surrounding land values.[10, 11]
- Lowers Capital Costs than conventional treatment.
- Non-fouling membranes can be designed for high TSS and collides.

**Membrane Bioreactor Removes**

Silicates, colloids, particulate matter, algae, bacteria, viruses, giardia, proteins.

**Membrane Bioreactor Reduces**

Total organic carbon, colour components.

**Membrane Bioreactor can not Remove**

Salts, gasoline, sugars, alcohols.

The two main suppliers of MBR systems for wastewater treatment are Kubota (Japan) and Zenon (USA). Other suppliers are Degremont (France), Membratek (S. Africa), Orelis/Mitsui (Japan), US Filter (USA) and Wehrle Werk (Germany)[3].

## Local Opportunities

Particular local considerations such as water reuse potential, effluent license targets, wastewater characteristics, wet weather hydraulic peaking factors, climatic considerations, land availability, and the characterization of existing infrastructure are all worthy of attention.

### Water Reuse

Largely we are dependent on rainfall water and generally situation of draught and short of water arises from time to time. Water restrictions and discontinuity of supply in many parts of the country are commonplace during periods of drought.

High quality, effectively disinfected effluent from advanced wastewater treatment systems (such as MBR) is suitable for agriculture, river flow replenishment and many other reuse markets.

## Climatic Considerations

A MBR plant located in a warm climate will be less costly to construct than one with an identical capacity located in a cold climate. This is due to the effect that liquid viscosity has on the flow rate of a liquid through the membrane pores.

The minimum wastewater temperature is therefore a major factor in determining the number of membrane modules required to meet a given MBR treatment capacity. Fewer membranes translate to lower costs. Warmer climate is therefore an advantage which will help the technology be cost competitive in this country.

## Conclusions

The global trend is for an increase in the number of MBR installations, largely due to the declining membrane costs and the increasing demand for water.

It appears that the projects most likely to favour MBR have an alignment of factors such as:

- Effluent quality is consistently high and generally independent of the influent quality.
- Good disinfection capability, with significant bacterial and viral reductions achievable using UF and MF membranes.
- Longer retention of nitrifying bacteria within the bioreactor results in greater nitrification than in a conventional ASP. Denitrification can be achieved by utilizing a second anoxic vessel.
- Long sludge ages result in lower sludge production.
- Small footprint technology.
- Proven reliability and easy operation.

## References

1. Chiemchaisri, C., Womg, Y. K., Urase, T. and Yamamoto, K. (1992). *Organic stabilisation and nitrogen removal in membrane separation bioreactor for domestic wastwater treatment. Water Science and Technology,* 25 (10), 231-240.

2. Kolega, M., Grohman, G. S., Chiew, R. F. and Day A. W. (1991). *Disinfection and clarification of treated sewage by advanced microfiltration. Water Science and Technology,* 23 (7-9), 1609-1618.

3. Gander, M., Jefferson, B. and Judd, S. (2000). *Aerobic MBRs for domestic wastewater treatment: a review with cost considerations. Separation and Purification,* 18, 119-130.

4. S. Judd, The MBR book (2006) Principles and applications of membrane bioreactors in water and wastewater treatment, Elsevier, Oxford.

5. Z.F. Cui, S. Chang, A.G. Fane, The use of gas bubbling to enhance membrane process, J. Memb. Sci. 2211 (2003) 1-35.

6. B. Jefferson, S. Judd, C. Diaper. Treatment methods of grey water. In: P. Lens, (ed). *Decentralised Sanitation and Reuse–Concepts, Systems and Implementation,* IWA publishing, (2000), Chapt. 17.

7. M. Kraume, R. Gildemeister. Grauwasserrecycling unter Anwendung von Membranverfahren. Germany (2004).

8. C. Abegglen, H. Siegrist. Domestic wastewater treatment with a small-scale membrane bioreactor, *Wat.Sci.Tech.* 53(3) (2006) 69-78.

9. P. Le-Clech, V. Chen, A.G. Fane, Fouling in membrane bioreactors used for wastewater treatment–A review. J. Memb. Sci. 284 (2006) 17-53.

10. George Crawford *et al.* (2001) *The Evolution of Membrane Bioreactor System Designs for Wastewater Treatment.* IWA Sept 2001.

11. Ed Fleischer *et al.* (2001) *Evaluating the Next Generation of Water Reclamation Processes.* WEFTEC 2001.

# Chapter 19

# Cleaner Production Practices for Sustainable Development of Pulp and Paper Industries

**Indra Deo Mall*, Vimal Chandra Srivastava, Saurabh Singh and Anand Pachouri**

**Department of Chemical Engineering, Indian Institute of Technology Roorkee, Roorkee – 247 667, Uttarakhand**

## ABSTRACT

Pulp and paper industry is one of the oldest industries and has made continuous strides during recent years. However, pulp and paper industries are amongst the top highly polluting industries. Present chapter deals with the profile and environmental pollution problems of the pulp and paper industry. This chapter deals with environmental impact of pulp and paper production, the technological developments during the last two decades in the area of development of cleaner technologies for the environmental protection and sustainable development of the pulp and paper industries.

*Keywords:* *Sustainable development, Cleaner technologies, Pulp and paper mill, Paper manufacturing, Fibre recycling.*

* Corresponding Author: E-mail: id_mall2000@yahoo.co.in

## Introduction

Cleaner technologies have received considerable interest during last two decades in order to minimise the damage caused to the environment due to emission/discharge of hazardous waste, and various other types of gaseous, liquid and solid wastes. The problem of environmental pollution is increasing consistently in spite of continuous development in the process technology. Fast development and large scale industrialisation have caused depletion of natural life sustaining resources. There is urgent need to minimise further damage to the environment through cleaner technologies using principle of recover, reduction, reuse, recycle and retrofit; use of renewable resources; preventive environmental technologies; and development of environmental friendly microbial system. For ecological sustainable development, cleaner technology will play significant role in future.

History of paper making goes back to over 2000 years, and is one of the oldest core sector industries all over the world. Per capita consumption of paper is sometimes yardstick for measurement of industrialization and economic growth of a country. Paper making like many other industries, has undergone a sea change in the last few decades from being a small scale and totally manual operations to a large scale and substantially mechanized operation. Today, the paper industry accounts for about 2.5 per cent of world's production and 2 per cent of world trade. Global demand for paper and board is forecast to grow from current level of 353 million tonnes to over 410 million tonnes by the year 2010 at an average growth rate of 2.2 per cent per annum [Mittal, 2005]. There are about 500 kraft mills and many thousands of other types of pulp and paper mills all over the world.

Pulp and paper industry is one of the oldest industries in India. With a 20 mills and installed capacity of one lakh tonne per annum at the time of independence, the paper industry has made continuous strides and now there are about 700 mills with a production of 6.8 million tonne of paper, paper board and newsprint. Indian paper industry requires more raw material and energy, generates more waste and has low resource recovery per unit of product as compared to the global paper industry. In this era of globalization, Indian paper industry needs to match global capacities, efficiencies, and quality standards with minimum degradation of environment.

In spite of its significant contribution towards social, cultural as well as economic development of a nation paper industry is amongst the highly polluting industries and is listed in red category. Pulp and paper making consumes huge amount of water and generated large quantity of wastewater containing high BOD, COD, pH, solid content, colour, foam, AOX and furan; air pollutant containing mercaptans, $H_2S$, SOx, NOx, chlorine; solid waste containing bark, screen rejects, ETP sludge, flyash. A relatively large number of toxic compounds have been identified in the pulp mill effluent, which has been cause of major concern. With discovery of dioxins and furans in pulp mill effluent, there has been wide spread environmental concern over the adverse impact of AOX and environmental pressure has generated for removal of the AOX and other toxic compounds. It is also responsible for emission of gases causing green house effect, acidification, and photochemical smog. The problem of environmental pollution in agro-based paper mills is more serious due to obsolete process technology and lack of chemical recovery and many mills have shifted towards more and more waste paper. Figure 19.1 shows the environmental impact of pulp and paper manufacture on global as well as local environment.

Sustainable development demands attaining the three dimensional balance between economic growth and social and environmental responsibility, therefore sustainability entails the continuous fulfillment of the varying needs and some-times conflicting objectives of the multiple stakeholders [Alyousfi, 2004].

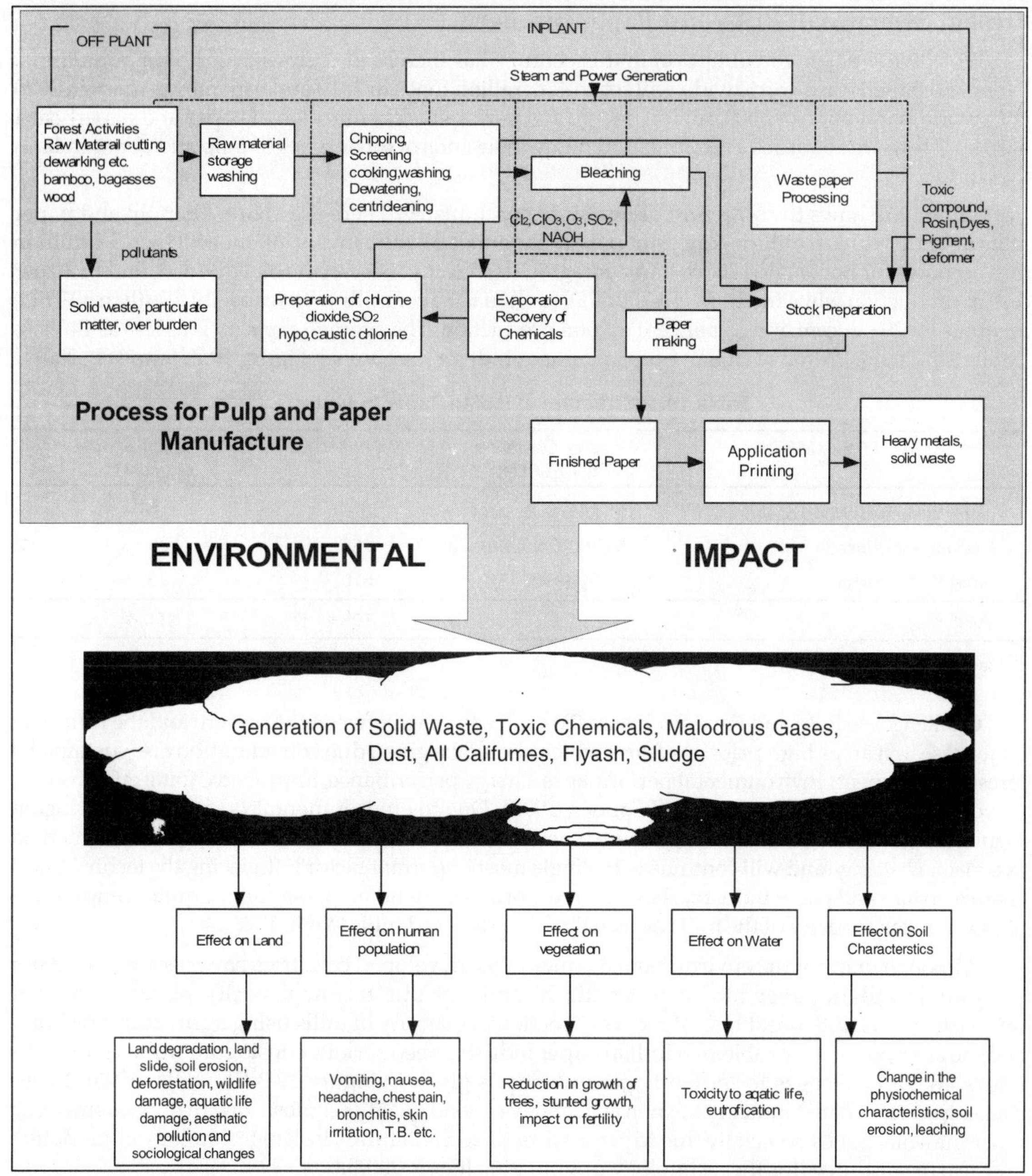


**Figure 19.1: Environmental Impact of Pulp and Paper Manufacture**

The present chapter deals with the various aspects of cleaner technologies in pulp and paper mills to meet the gap between present and future for sustainable development.

## Present Secnario of Pulp and Paper Industry

Worldwide paper consumption in this century has increased 4 times faster than population. Paper and paperboard worldwide will reach 476 million tones in 2010 and 640 million tones in 2020 with world population of 8000 million people and per capita consumption of paper and paperboard of 80 kg. There are about 500 kraft mills and many thousands of other types of pulp and paper mills in the world.

Table 19.1 shows the structure of Indian paper industry. There are about 700 pulp and paper mills in India with 6.8 million tonnes of paper, paper board and newsprint out of which 5.8 million tons accounts for paper and board. Large integrated (capacity more than 100 TPD) account for 30 per cent of production while medium (50-100TPD) and small paper mills with (capacity less than 50 TPD) accounts for 29 per cent and 35 per cent of total production. The average capacity is as less as 14 TPA. [Mall *et al.*, 2005]. Profile of Indian pulp and paper industry is given in Figure 19.2 [Jain *et al.*, 2007].

**Table 19.1: Structure of Indian Paper Industry**

| *Type of Paper Mill* | *Scale of Operation, Tons Per Day* | *No. of Mills* | *Production Share per cent* |
|---|---|---|---|
| Large Integrated Wood based | 101-800 (Avg. 300) | 33 | 36 |
| Medium Agro Based | 50-100 (Avg. 60) | 165 | 29 |
| Small Waste Paper | 5-50 (Avg. 15) | 510 | 35 |
| Total | | 708 | 100 |

*Source*: Jain *et al.*, 2007.

Challenges to Pulp and Paper Industry: To bridge the gap between the present and the future six major research areas high priority efforts which should be given due consideration are: sustainable forest management, environmental performance, energy performance, improved capital effectiveness recycling and sensors and controls [Copoer, 1997]. Due to environmental pressure and stringent regulatory parameters, and market pressure for eco-friendly product environmental issues has emerged as a major challenge and will continue to be single most important factor influencing the technological change in pulp and paper industry. However, the problem in Indian paper mills is quite complex due to low average capacity of the mills especially the agro-based mills [Mall, 1998].

Although many of the environmental issues in the developed countries have been resolved but situation in Indian paper industry is entirely different due to low capacity plants with poor infrastructure facilities and lack of recovery section in majority of mills using agricultural residues and the environmental problem in Indian paper industry need serious considerations to resolve the issues. Major challenges to Indian Paper industry is given in Figure 19.3 [Rakhilla, 2005]. Major challenges before paper industry are non availability of good quality cellulosic raw materials, emerging environmental issues especially due to presence of dioxin and furan, lack of technology upgradation and also the resources for the technology upgradation [Garg, 2007].

## Cleaner Production

Cleaner production is the continuous application of an integrated preventive strategy to processes, products and services, to increase eco-efficiency and to reduce risks to humans and the environment [UNEP-IE, April 1998]. Cleaner production is now need of time as it offers lower waste treatment costs

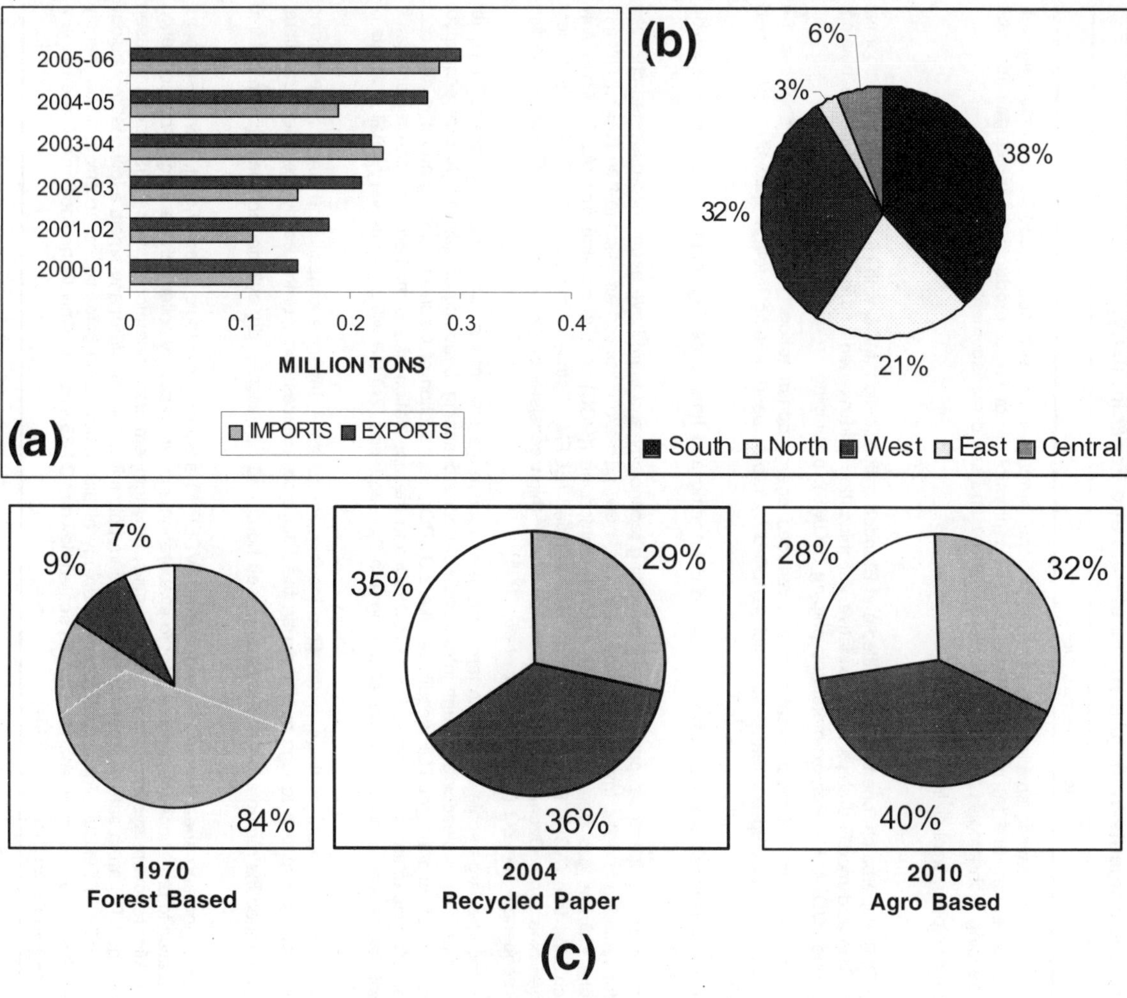


**Figure 19.2: Profile of Indian Pulp and Paper Industry**

**a: Imports and exports of various items in India; b: Geographic Spread of Newsprint Capacities in India; c: Trends in Use of Raw Material for Pulp and Paper production in India.**

and increased production efficiency, lower costs, conservation of raw materials and energy, better access to finance, marketing requirements of environment friendless products, improved environment and better compliance with environmental regulations.

Due to environmental pressure and stringent regulatory parameters, and market pressure for eco-friendly product environmental issues has emerged as a major challenge and will continue to be single most important factor influencing the technological change in pulp and paper industry. However, the problem in Indian paper mills is quite complex due to low average capacity of the mills especially the agro-based mills. Some of the major developments in various areas of pulp and paper industry are given in Table 19.2.

**Table 19.2: Major Developments in Various Areas of Pulp and Paper Industry.**

| *Area* | *Major Development* |
|---|---|
| Forest management | Sustainable forest productivity with fast growing species with high yield, clonal propagation of soft wood, good harvest management. Main focus areas in forest management are (a) biotechnology, (b) basic physiology of forest productivity, (c) sustainable soil productivity, and (d) remote sensing technologies to improve forest inventory and stand management [Cooper, 1997] |
| Chipping | Washing system, High capacity chipper, minimizing dust |
| Pulping | |
| Chemical Pulping | Rapid displacement heating (RDH), superbatch, enerbatch, Modified and Etended delignification; MCC with Hi heat washing, isothermal cooking, black liquor impregnation and modified chips bins, oxidative delignification with oxygen and ozone, anthraquinone pulping, Octahydrodimethyl-anthraquinone (ODIMAQ), Polyoxometalates pulping (POMS), biopulping |
| Mechanical pulping | Biopulping: biopulping improves the quality of mechanical pulp and reduces energy consumption. TMP, RTS. RTS shortens the retention time, lowers the temperature of chip pretreatment, which is folled by rapid speed refining through specially configurated refiners plate [Atalla *et al.*, 2004]. |
| Washing and screening | Drum displacer, Pressure diffuser, Displacement presses, combined deknotting and fine screening, high temperature screening before washing, reverse cleaners. |
| Bleaching | Pulp and paper market forces are driving the development of pulp bleaching technology and there have been unprecedented changes in the bleaching technology for reducing toxicity. Instead of dioxin free pulp, now, pulp are classified as chlorine free (CGF), chlorine chemical free (CCF), Molecular chlorine gas free (MCGF), Non chlorine compound ((NCC), Active chlorine free (ACF), Absolutely chlorine free (ACF), Almost chlorine free (ACF), Elemental chlorine free modified (ECFM), Totally chlorine free (TCF), Totally effluent free (TEF), ECF, TCF, Oxygen bleaching, Ozone bleaching, Mesto Ze Trac$^{TM}$ (high consistency ozone system): ECF-sequence like (Ze)DP or (Ze)DD or a TCF sequence- like ZQ (PO) [Norden *et al.*, 2005].<br><br>The changes that have experienced a large scale benefits in terms of these reduce the effluent load from bleaching, are to eliminate/ reduce chlorine and chlorine compounds in bleaching through introduction of ozone beaching, beaching of pulp in light ECF sequence, like (Ze) DP or (Ze) DD, beaching of pulp in light TCF sequence, like ZQ (PO), introduction of Ozone bleaching and it is also cost effective compared with other ECF bleaching sequences, like D (EOP)D and the application of this methodology can replace large amounts of chlorine dioxide with moderate amount of ozone, bleaching with the sequence (Ze) DP will generate very small amounts of AOX. |
| Fibre modification | Enhancing the opacity of fibre through pigment particles, better fibre loading, Pressurised sensitive adhesives (PSAs). Fibre loading contributes opacity, smoothness and brightness to high quality paper; this filler also replaces some of the costly bleached kraft fibre. |
| Paper making | High speed machine with sophisticated instrumentation and control; closed hood and insulation of dry cylinder end; installation of trinip press. |
| Evaporation and recovery | Free flow falling film evaporator, vibrating shear enhanced processing (VSEP$^{TM}$) system [New logic International, www.vsep.com] using ultra fitration and nano-filtration to pre-concentrate solids prior to evaporation; modern recovery boiler with black liquor concentration 85 per cent; installation of lime kiln, recovery system with agro based mills. Low temperature incinerator system (LTIS) for agro based paper mills. Some of the advantage LTIS are a safe process for small medium pulp mills, does not form any smelt, a low cost technology of small/medium pulp and paper mills, a flexible widely applicable technology alkali/neutral sulfite black liquor, an environment friendly clean technology with least problem, minimum maintenance, Can be scaled down to 30-35 TPD pulp mill technology and economics are viable, very low generation time [Gupta and Ramani, 2004]. |

*Contd...*

**Table 19.2–Contd...**

| *Area* | *Major Development* |
|---|---|
| Recycling of fibre | Improved recycling reduces pressure on wood and often results in lower energy consumption, less chemical usage and lower pollutant production. |
| Deinking | Enzymatic deinking using celluloses. Celluloses improve de-inking by removing fibrills from fibre substances and facilitate the removal of tonners particles, increase the freeness and flexibility and reduce the coarseness of dried fibres. Cellulose treatment of recycled fibres reduce disintegration time and increase bleachability of low quality, recyled paper [Atalla *et al.*, 2004]. |
| Effluent treatment system | Membrane separation technology (MST) for treatment of effluent and waste management. Membrane separation technology has been applied in pulp and paper industry in following areas: Kraft lignin and lignosulfonate separation, recovery of inorganic chemicals from kraft black liquor, Recovery of inorganic chemicals and lignin from black liquor from agricultural residue, Concentration of spent kraft and sulfite liquor in pre concentration stage, Wastewater treatment for decolourization and detoxification. Vibrating shear enhanced processing (VSEP™) using ultra filtration and nano-filtration avoiding fouling problem exhibited by conventional membrane system, pressurized ozone/ultrafiltration membrane systems for removing total solids from paper mill process water by converting it to total suspended solids; Hybractor™ is a hybrid reactor which combine best features of both suspended bed and fixed bed film technologies. |

*Source*: Atalla *et al.*, 2004.

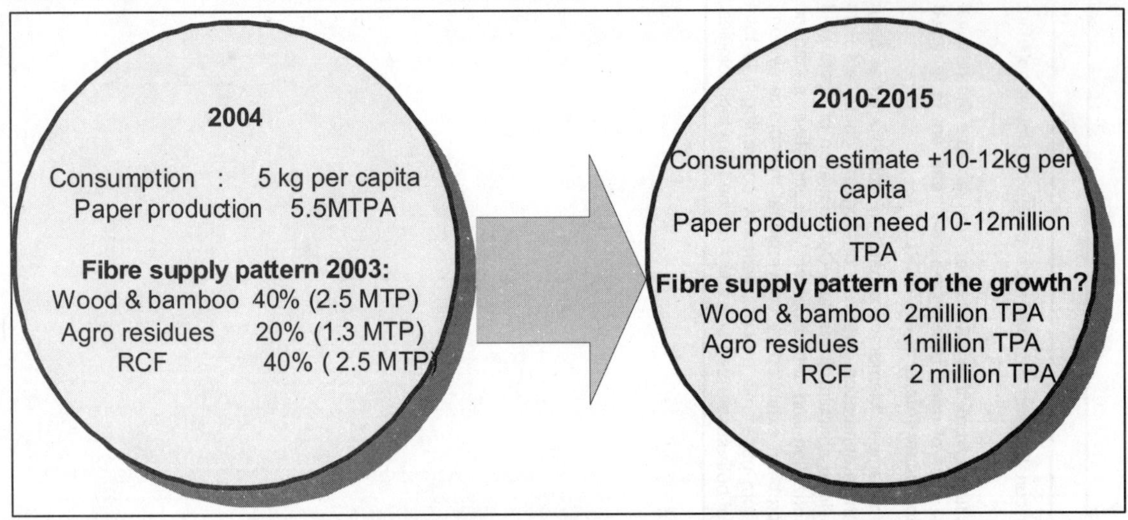


**Figure 19.3: Challenges to Indian Pulp and Paper Industry**

## Innovation in Technology for Waste Management for Cleaner Production

There has been significant innovation and developments in technology in pulp and paper manufacture to improve quality, minimize the energy consumption, reduce the toxicity of bleach plant effluent, and wastewater generation, reduce the colour in the effluent, to improve better recycle. Figure 19.4 shows the five R initiatives for cleaner technologies implementation. Figure 19.5 shows cleaner

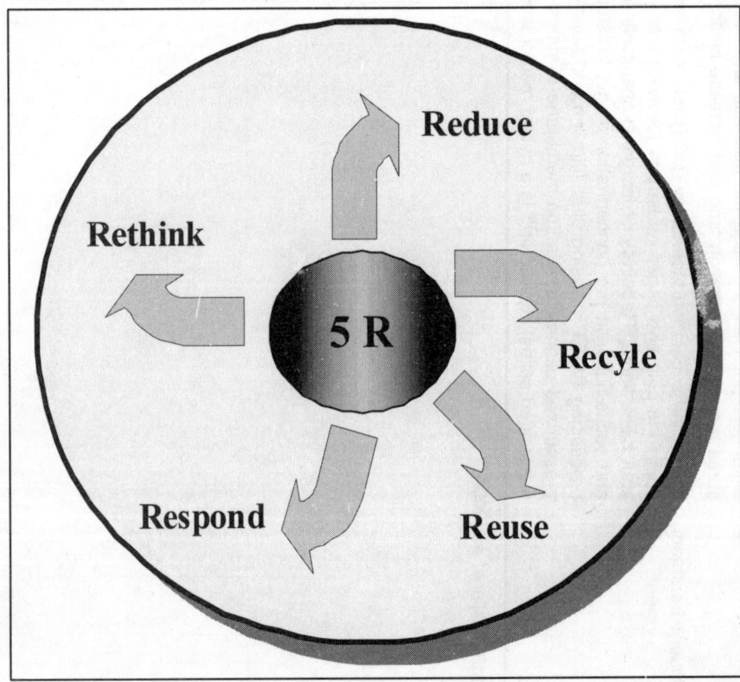


**Figure 19.4: Five R Initiatives for Cleaner Technologies Implementation**

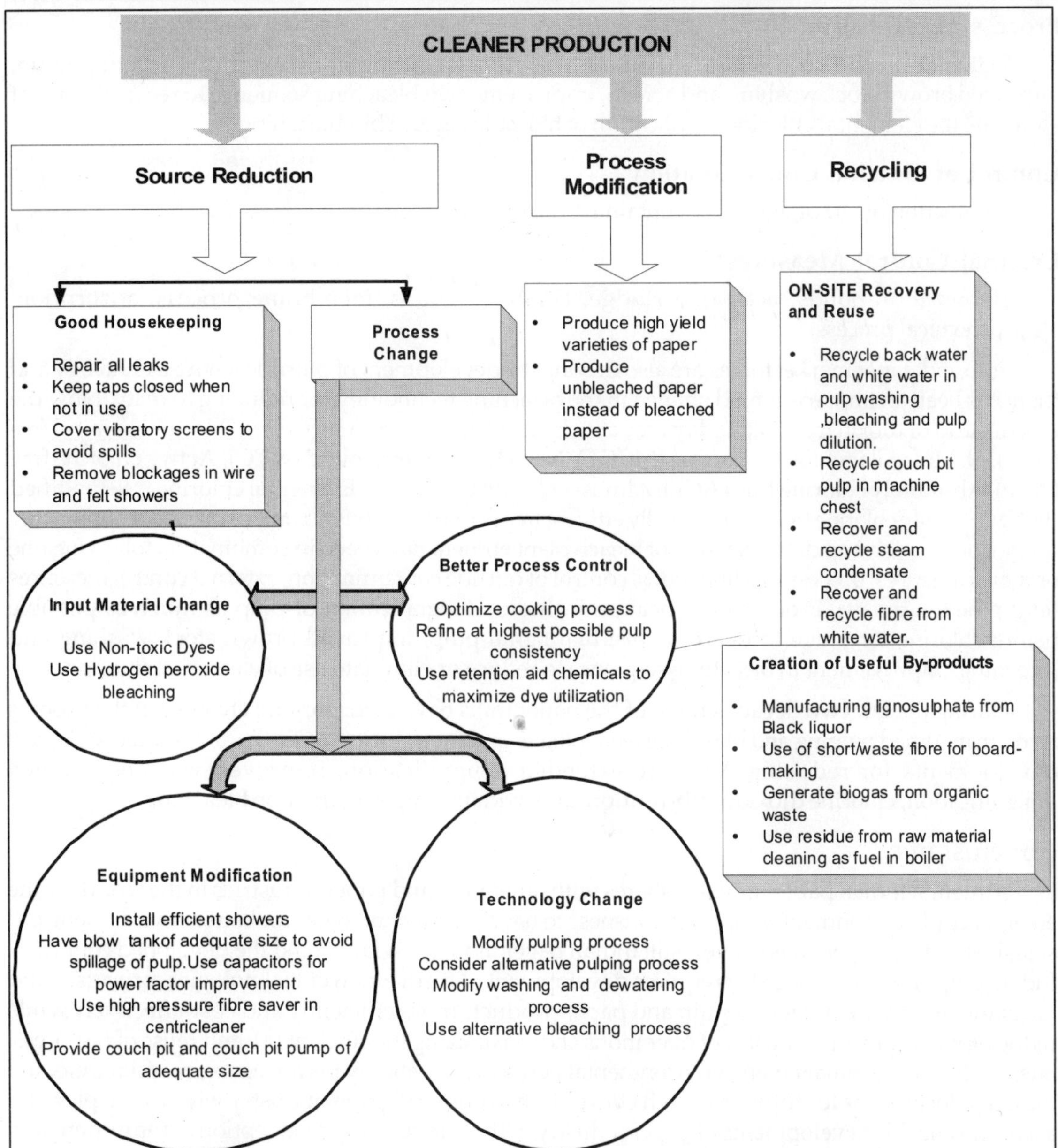


**Figure 19.5: Cleaner Production Techniques in Pulp and Paper Mills [UNEP Manual, 1999]**

production techniques that may be applied in pulp and paper mills [UNEP Manual, 1999]. Pulping and bleaching has been one of the major areas where major technological development has taken place in the pulp and paper industry. Some of the technological development in pulping and bleaching for reducing toxicity and waste generation are:

### Process Modification

Delignification of pulp as much as possible before chlorination stage to decrease residual lignin, improved brown stock washing and screen, improvement in bleaching sequence to reduce or use of eliminate the elemental chlorine, application of bio-pulping and bio-bleaching.

### Control of Outside Contamination

Elimination of use of defoamer containing dioxin precursors.

### External Control Measures

Stabilization ponds, activated sludge, aerated lagoons, membrane process, adsorption, electrochemical process.

Pulp and paper market forces are also driving the development of pulp bleaching technology and there has been an unprecedented change in the bleaching technology for reducing toxicity. Now the pulp instead of dioxin free pulp today they are classified as chlorine free (CGF), Chlorine chemical free (CCF), N, Molecular chlorine gas free (MCGF), Non chlorine compound (NCC), Active chlorine free (ACF), Absolutely chlorine free (ACF), Almost chlorine free (ACF), Elemental chlorine free modified (ECFM), Totally chlorine free (TCF), Totally effluent free (TEF) [Pyre, 1991]. Some of the major approaches for de-colorization and detoxification of bleach plant effluent have been in combination following one or more measures: process modifications, control of outside contamination, external control measures (end pipe treatment). Process modification includes: delignification of pulp as much as possible before chlorination stage to decrease residual lignin pulp, improved brown stock washing and screening, improvement in bleaching sequence to reduce or eliminate use of elemental chlorine.

During the last two decades most of the paper mills have incorporated changes in the process through modified pulping and bleaching technologies and for reducing AOX. Some of the technological developments for reducing AOX are extended delignification, improved washing, oxygen delignification, chlorine dioxide substitution, and oxidative alkali extraction bleaching.

## Conclusions

Although Indian paper industry shares with other pulp and paper industries in the world for the emergence of environmental conscientiousness to have clear technologies for better environment and sustainable development, however, still Indian paper industry is far behind the developed countries. Indian pulp and paper industry requires development of further newer technologies, processes and measurements to manufacture pulp and paper products more efficiently and cost-effectively while reducing environmental impacts of operations and maximizing the efficient use and reuse of resources. Sustainable forest management, environmental performance, energy performance, recycling, etc., are the major focus areas for future research efforts. Indian pulp and paper industry will have to plan and act for sustainable developments of paper industry with clear vision and perception of future demand; vision without action and action without vision will be disaster.

## References

1. Alyousfi AB. Practice periodical of hazardous, toxic, and radioactive waste managements. October 2004, p. 265.
2. Atalla RH, Reiner RS, Houtman CJ. New technology in pulping and bleaching. Springer, E.L. Elsevier Ltd. 2004.

3. Cooper JS. The basics of Agenda 2020. TAPPI Journal, February 1997, p. 32-35.
4. Garg NK. IPPTA Journal, 19 (4), October-Dec. 2007, p. 2.
5. Gupta D, Ramani S. Low temperature incinerator System for agrobased paper mills" IPPTA J. Vol 16 No.4 Oct2004, p. 47.
6. Jain RK, Singh K, Roy TK. Indian Paper Industry Raw material Scenario, Growth Prospects and Path ways. IPPTA Journal, 3, July-Sep 2007, p. 129.
7. Mall ID, Srivastava VC, Singh B. Environmental protection in pulp and paper manufacture through cleaner technologies. XXI National Convention of Chemical Engineers, Indian Chemical Industry Vision 2020, Udaipur, INDIA, September 2-4, 2005.
8. Mall ID. Emerging Scenario in Indian paper Industry: Responding to the Challenges of Globalisation. IPPTA Convention issue, Dec. 1998, p. 91.
9. Mittal SK. Strategic moves of pulp and paper industry. Towards achieving sustainable development. IPPTA Convention issue Vol. 17, No 2, ii, Jan 2005.
10. New logic International www.vsep.com
11. Norden, S., Carre, G., Lundahl, Y., Jain, N.K, "New bleaching processes and fibre line with state-of-the-art technology" IPPTA Convention Issue, Vol. 17, No 2, Jan, 2005, p. 29-34.
12. Pryke, D.C., in Chapter 2.1 of "Environmental solution for the pulp and paper industry" Edited by Kelly, Ferguson, Miller Freeman Inc. US, 1991, p. 67.
13. Rakkhila, P. A global view on chemical pulping industry and it's recent developments. IPPTA Convention Vol. 17, No 2, Jan, 2005, p. viii-xii.
14. UNEP Manual. Cleaner Production in the Pulp and Paper Industry: Technology Fact Sheets. UNEP-IE, 1999.
15. UNEP-IE manual. Cleaner Production in the Pulp and Paper Mills. A training resource package. UNEP-IE, April 1998.

# Chapter 20

# Reactive Extraction of Butyric Acid: A Clean Technology

**Amit Keshav[1], Abhimanyu Kumar[2], Shri Chand[3] and Kailas L. Wasewar[4]***

**[1]Research Scholar, [2]UG Student, [3]Professor and Head, [4]Assistant Professor**
**Department of Chemical Engineering, Indian Institute of Technology (IIT)**
**Roorkee – 247 667, Uttrakhand**

## ABSTRACT

Butyric acid is an important industrial chemical used in production of cellulose acetate butyrate plastics, as flavors in soft drinks and chewing gum industries, in plasticizers, leather tanning and pharmaceutical industries. Commercial production by synthetic route has many limitations. Fermentation is clean alternative technology for butyric acid production. However energy budget studies and economic analysis have indicated that product acid separation has proven to be energy intensive and expensive stage in fermentation. Reactive extraction emerges to be the successful technology for separation of acid from waste streams and fermentation broths. Reactive extraction involves an extractant in diluent system, which chemically complexes with acid in the system. The reaction was proved to be exothermic in nature. Thus it becomes important to study temperature characteristic of extraction. In present work effect of temperature on reactive extraction of butyric acid by 20 and 30 per cent Aliquat 336 in oleyl alcohol was studied. Distribution coefficients, equilibrium complexation constants, loading and degree of extraction were presented. The study will be useful for optimizing the temperature of the

* Corresponding Author: E-mail: k_wasewar@rediffmail.com, klw73fch@iitr.ernet.in

reaction mass for maximum recovery of acid for the source streams. Also an extractive fermentation process (fermentation–extraction–back extraction–regeneration), as a clean process for the production of butyric acid has been proposed.

***Keywords:*** *Butyric acid, Extractive fermentation, Reactive extraction, Temperature, Distribution coefficient, Clean technology.*

## Introduction

Butyric acid ($CH_3CH_2CH_2COOH$) is a light colourless liquid with boiling point 163.5 °C and density 0.959 g/cm³. It is used in the production of cellulose acetate butyrate plastics, which have myriad of uses, especially where resistance to high temperature and sunlight is required; as flavours in soft drinks and chewing gum industries; as plasticizers (glycerol tributyrate); in leather tanning (calcium butyrate) and pharmaceutical industries (antithyroid and vasoconstrictor drug)(1).

Most butyric acid is manufactured either by Oxo process or by liquid phase oxidation of propane or butane. It is also formed during the air oxidation of cyclohexane in the manufacture of cyclohexanone/nol in a caprolactum plant, as byproduct(2). Currently, there is an active interest in developing a continuous fermentation process for producing butyric acid. In both industial production and fermentation broth, recovery of acid is important process limiting step. In caprolactum plant, butyric acid is separated from reaction mixture by extraction with an aqueous caustic to form water soluble, sodium salt of the acid. The resulting aqueous stream is then neutralized with $H_2SO_4$ to liberate the organic acid. Recovery from fermentation broth is also carried out in similar way. Though the above separation method is simple, yet it consumes lot of lime and $H_2SO_4$ and produced waste sludge. Thus an alternative recovery process is needed to be looked for.

Numbers of methods are available eg. precipitation, adsorption, membranes, dialysis, distillation, ion exchange, reactive extraction to recover carboxylic acids from fermentation broths. All these above methods have their advantages and disadvantages(3). Electrodialysis and dialysis have the problem of membrane fouling, which requires frequent cleaning of the dialyzer. Moreover, large-volume dialysis units, even greater than the volume of the fermentor vessel, would be required in a commercial-scale unit. Electrodialysis gives a higher extent of acid separation but with increased power and energy consumption. Also, large amounts of byproduct salts from the ion-exchange regeneration are formed. Adsorption or the ion-exchange process requires regeneration of an ion-exchange resin and adjustment of the feed pH to increase the sorption efficiency, requiring large amounts of chemicals. During direct distillation, high-boiling internal esters as dimers and polymers can be formed. In the extraction of carboxylic acids from fermentation broth with microporous hollow fiber membranes, there is a tendency to form an emulsion. Liquid surfactant membrane extraction exhibits a high complexity of operation due to swelling of the membranes in liquid surfactants. Supported liquid membranes often suffer from membrane instability. Reactive extraction with a specified extractant giving a higher distribution coefficient has been proposed as a promising technique for the recovery of carboxylic and hydroxycarboxylic acids. With proper selection of extractant and diluent combination, reactive extraction can profoundly result in higher recovery and selectivity of butyric acid.

## Reactive Extraction

Reactive extraction is relatively new technology that has been shown to be advantageous for alcohol and organic acid fermentations(4–5). Reactive extraction removes the inhibitory, acidic product from the bioreactor reactor and therefore increased productivity, provides ease in reactor pH control without requiring base addition, and allows use of a high-concentration substrate as the process feed to reduce process wastes and production costs. Figure 20.1 proposes a reactive extraction unit for separation of butyric acid from fermentation broth.

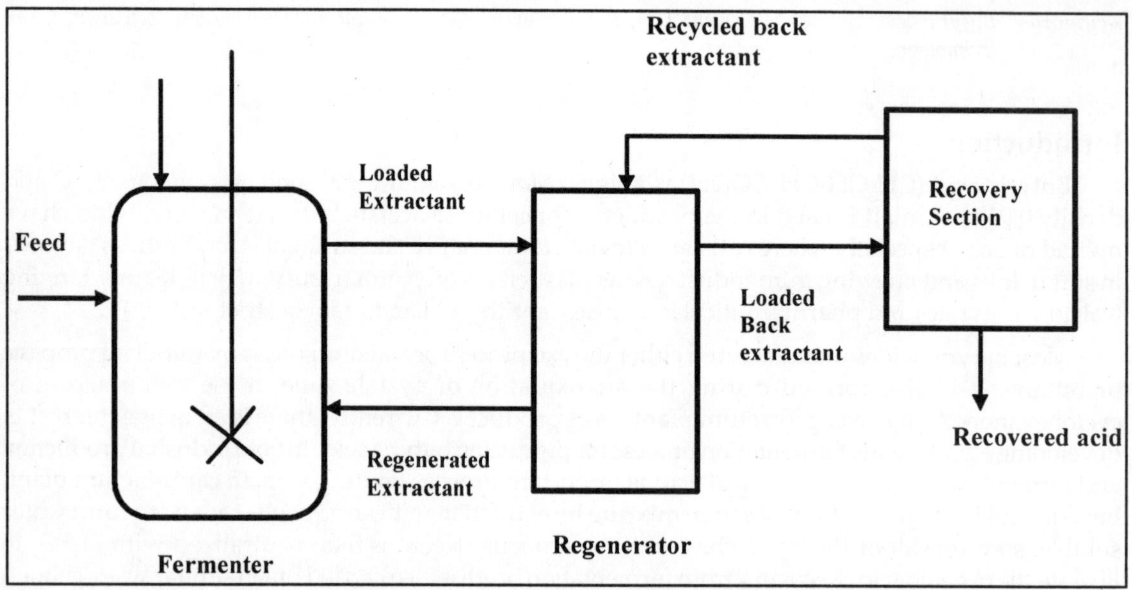


**Figure 20.1: Conceptual Flow Sheet for Production of Butyric Acid (Fermentation-Extraction-back extraction)**

There is a challenging problem in developing an extractive fermentation for organic acid production: It is difficult to find a biocompatible solvent that also has high extraction coefficient or distribution coefficient value for the product. Since most of solvents with high distribution coefficient values are usually also toxic to bacteria, they either inhibit or stop bacterial growth when recovery is to be carries out *in situ* from fermentation broth. Therefore a search of a nontoxic extractant diluent combination is the essentiality for establishing reactive extraction for butyric acid separation from fermentation broth.

Further a design of reactive extraction unit for butyric acid separation, need the data on distribution coefficient, equilibrium complexation constant, loading ration, effect of pH, temperatures, water co-extraction etc., with this aim, effect of temperature was studied in present text.

## Material and Methods

### Reagent and Solutions

The commercial solvent Aliquat 336 (Methyltricaprylammonium chloride), a mixture of $C_8$-$C_{10}$ with a minimum assay of 80 per cent and with molecular weight of 404.17 and density of 0.888 g/cm$^3$ was used for the extraction of acid. Butyric acid (99 per cent) (Himedia, India) and the diluent oleyl alcohol are of technical grade and were used without pretreatment. Distilled water was used to prepare the solutions of various concentrations of butyric acid solutions. NaOH used for titration is of analytical grade and was supplied by Ranbaxy, India. For the standardization of the NaOH, oxalic acid (99.8 per cent) was obtained from S. D. Fine-Chem Ltd, India. Phenolphthalein solution (pH range 8.2–10.0) was used as indicator for titration and was obtained from Ranbaxy, India. The initial aqueous acid concentrations range ($[HA]^o$) of 0.05-0.4 $kmol.m^{-3}$ were used. Low concentration was used because butyric acid concentration in the fermentation broth or waste streams are not greater than 0.5 $kmol.m^{-3}$

## Experimental

Desired volumes of aqueous solution containing acid and extractant were equilibrated in a temperature controlled water bath shaker at 305K for 12 hours. After the phase disengagement, the aqueous phase was separated and the acid concentration was determined by titration with NaOH. The organic phase concentration was calculated from the difference between concentration of acid in the aqueous phase before and after extraction.

## Results and Discussion

Quaternary ammonium chloride: Aliquat 336 is a selective and powerful extractant for separating carboxylic acids from their dilute solution. Aliquat 336 extracts both the dissociated and undissociated forms of acids[(6)]. The distribution coefficient using Aliquat is higher by 1 or 2 magnitude compared to those for extraction with conventional solvents like alcohols, ethers, esters etc. High values of distribution coefficient provide advantage of high yield, selectivity, low phase ratio between solvent and feed solution and lesser number of stages for extraction unit. Since a overall reactive extraction involves both extraction and back extraction steps, it poses a problem of recovering acid from the loaded acid-amine complexed phase. The solution to this problem was suggested by Baniel *et al.*[(7)], who found that distribution coefficient is affected by temperature. They suggested temperature swing process for recovery of carboxylic acid form the loaded organic phase. In temperature swing extraction/regeneration scheme, the extraction is carried out at a relatively low temperature, producing an acid loaded organic extract and an aqueous raffinate waste stream containing the unwanted feed components. During regeneration, the extract is contacted with a fresh aqueous stream at a higher temperature to produce an acid laden aqueous product and an acid free organic phase.

The mass law equilibria describing the extraction of butyric acid ($[HA]_{aq}$) by Aliquat 336 (T) in oleyl alcohol can be represented by the following set of equations[(8)]:

$$[HA]_{aq} + [T]_{org} \xleftrightarrow{K_{11}} [T:HA]_{org} \tag{1}$$

$$[HA]_{aq} + [T:HA]_{org} \xleftrightarrow{K_{21}} [T:(HA)_2]_{org} \tag{2}$$

$$\vdots$$

$$[HA]_{aq} + [T:(HA)_{n-1}]_{org} \xleftrightarrow{K_{n1}} [T:(HA)_n]_{org} \tag{3}$$

Equilibrium extraction constant and the number of reacting molecules of extractant were computed by applying law of mass action that was the ratio between concentrations of reactant molecules and concentration of the product species, according to the general equation of interaction between the extractant and the extracted species

$$K_{En} = \frac{[T:(HA)_n]_{org}}{[T]_{org}[HA]_{aq}^{\,n}} \tag{4}$$

$K_{En}$ was expected to depend on properties of the acid and the solvation efficiency of diluent used. The experimentally accessible distribution ratio $K_D$ can be given as:

$$K_D = \frac{[HA]_{org}}{[HA]_{aq}} = \frac{[HA.T]_{org}}{[HA]_{aq} + [A^-]_{aq}} \quad (5)$$

The extent to which the organic phase (amine + diluent) can be loaded with carboxylic acid was expressed as the loading ratio, z;

$$z = \frac{[HA]_{org}}{[T]_{o,org}} \quad (6)$$

The value of *z* in equation (6) depends on extractability of acid (strength of the acid-base interaction) and its aqueous concentration. The stoichiometry of the overall extraction reaction depends on the loading ratio in the organic phase, *z*, as:

$$\frac{z}{1-z} = K_E[HA]_{aq} \quad (7)$$

From the above equation plot of *z*/1-*z* versus $[HA]_{aq}$ provides the equilibrium complexation constant value.

Figure 20.2 presents distribution curves for the extraction of butyric acid by 0.44 kmol/m³ Aliquat 336 in oleyl alcohol. The curves show that extraction decreases as temperature is increased. Through

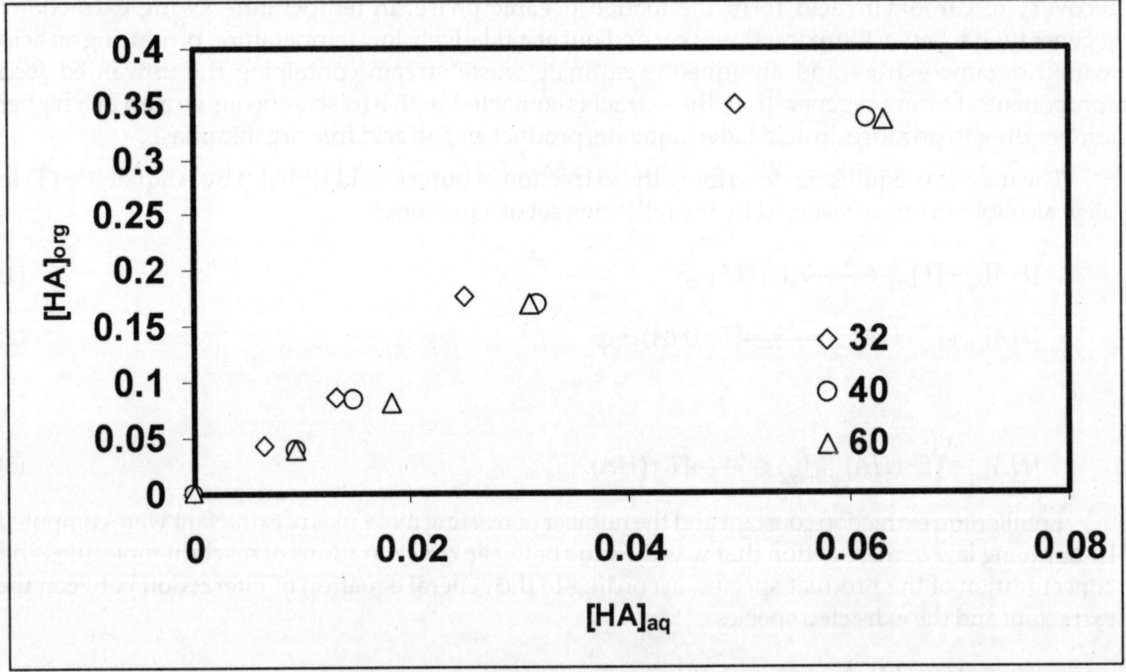


**Figure 20.2: Chemical Equilibria Isotherms for Extraction of Butyric Acid by 0.66 kmol/m³ Aliquat 336 in Oleyl Alcohol at Various Temperatures**

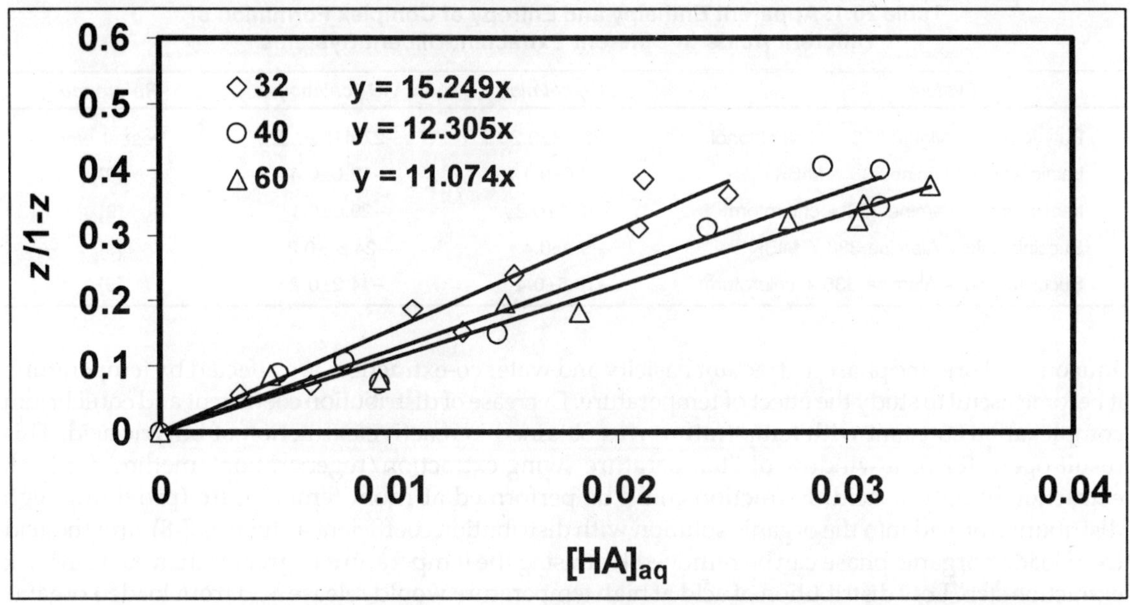


**Figure 20.3: Plot of $z/1$-$z$ vs. $[HA]_{aq}$ for the Estimation of Acid–Aliquat 336 Equilibrium Complexation Constant at Various Temperatures**

the entire tested aqueous phase concentration ranges. Equilibrium complexation constants for the system were calculated at different temperatures by the plot of equation[7] (Figure 20.3). A similar trend of decrease of equilibrium complexation constant with temperature was observed. The decrease of the above two parameters with increasing temperature, signifies that temperature elevation could be considered as the back extraction step into the fresh aqueous phase without overall dilution of the acid.

The complexation reaction in the organic phase involves proton transfer or hydrogen bond formation and is expected to be exothermic[9]. Formation of a complex makes the system more ordered and thus decreases the entropy. Therefore as the temperature is increased, the amount of acid extracted decreases. If enthalpy and entropy of reaction are assumed to be constant over the temperature range,

$$\ln K = \frac{-\Delta H}{RT} + \frac{\Delta S}{R} \tag{8}$$

predicts that plot of ln K versus 1/T would provide enthalpy as slope and entropy as intercept. The value of enthalpy and entropy of extraction of butyric acid by Aliquat 336 in oleyl alcohol and that found for other acids in literatures are indicated in Table 20.1. As can be seen from table, the apparent enthalpies of association are more exothermic for butyric acid.

Oleyl alcohol has desirable property for use as a diluent in temperature swing regeneration of butyric acid. KD for extraction are high enough for facile extraction, and at higher temperature, the acid readily distributes back into the aqueous phase.

## Conclusion

Reactive extraction of butyric acid using Aliquat 336 in oleyl alcohol was studied to see the effect of temperature. Since parameters like acid $pK_a$, extractant-acid interaction, solubility of the acid in

**Table 20.1: Apparent Enthalpy and Entropy of Complex Formation of Different Acids in Different Extractant-Diluent Systems**

| *System* | *ΔH (kcal/mol)* | *ΔS (cal/mol)* | *Reference* |
|---|---|---|---|
| Butyric acid + Aliquat 336 + oleyl alcohol | –36.83±0.2 | –27.412±0.3 | Present work |
| Lactic acid + Alamine 336 + MIBK | –5.6±0.4 | –12.6±0.4 | [9] |
| Lactic acid + Alamine 336 + chloroform | –12.1±0.2 | –29.0±0.1 | [9] |
| Succinic acid + Alamine 336 + MIBK | –9.2±0.4 | –24.6 ±0.2 | [9] |
| Succinic acid + Alamine 336 + chloroform | –16.5±0.4 | –44.2±0.2 | [9] |

aqueous and organic phase, extractant basicity and water co-extraction, are affected by temperature, it become useful to study the effect of temperature. Decrease of distribution coefficient and equilibrium complexation constant with temperature was obtained in reactive extraction of butyric acid. This result opens for us a window of "temperature swing extraction/regeneration" method for back extraction of butyric acid. Extraction could be performed at room temperature (providing high distribution of acid into the organic solution with distribution coefficient as high as 7-8), and the acid from loaded organic phase can be removed by raising the temperature to greater than 60 °C in back extraction step. Low distribution of acid at high temperature would release acid from loaded organic phase. Thus a complete reactive extraction process for butyric acid separation from fermentation broth can be designed.

## References


1. Playne, M. J. Propionic and butyric acids, in: Moo-Young M. (Eds) Comprehensive Biotechnology, Vol. 3, Pergamon, New York, 1985, 731-759.
2. Ingale, M.N.. Mahajani, V.V. Recovery of butyric acid, valeric acid, and caproic acid (BVC acids) from an aqueous waste stream using tributylphosphate (TBP) as an extractant. *Sep. Technol.* 1994, 4, 252 -257.
3. Wasewar, K. L. Separation of lactic acid: Recent advances. *Chem. Biochem. Engg.* 2005, *19(2)*, 159-172.
4. Wasewar, K. L., Heesink, A. B. M., Versteeg, G. F., Pangarkar, V. G. Intensification of enzymatic convertion of glucose to lactic acid by reactive extraction, *Chem. Engg. Sci.* 2003, *58*, 3385-3393.
5. Weilnhammer, C., Blass, E. Continuous fermentation with product recovery by in situ extraction. *Chem. Eng. Technol.*1994, *17*, 365-373.
6. Kyuchoukov, G., Yankov, D., Marinova, M., Molinier, J., Albet, J. Mechanism of lactic acid extraction with quaternary ammonium chloride (Aliquat 336). *Ind. Eng. Chem. Res.* 2005, *44*, 5733-5739.
7. Baniel, A. M., Blumberg, R., Hajdu, K. Recovery of acids from aqueous solutions. U.S. Patent 4,275,234, 1981; *Chem. Abstr.* 1982, *97*, 109557.
8. Wasewar, K. L., Pangarkar, V. G. Intensification of propionic acid production by reactive extraction: Effect of diluents on equilibrium. *Chem. Biochem. Engg. Q.* 2006, *20*, 1-7.
9. King, C.J., Tamada, J. A. Extraction of carboxylic acids with amine extractants. (3)Effect of temperature, water co-extraction, and process considerations. *Ind. Eng. Chem. Res.* 1990, *29*, 1333-1338.

# Chapter 21

# Energy Consumption and its Environmental Consequences

**P. Nema[1]* and M.F. Ansari[2]**
**[1]Deputy Director and Head; [2]Project Fellow**
**Air Pollution Control Division,**
**National Environmental Engineering Research Institute,**
**Nehru Marg, Nagpur – 440 020, M.S.**

## ABSTRACT

The world energy supply is projected to grow by 52 per cent between 2004-2030. With fossil fuel remaining at 81 per cent of total primary energy supply. Carbon dioxide emission are consequently expected to continue their growth unabated reaching 40.4 Gt $CO_2$ by 2030 from 26.6 Gt in 2004. Electricity demands is growing at an even faster rate in developing countries coupled with rising prices in oil and other energy supplies and continuing market instability present an uncertain future.

India is far behind in per capita energy consumption as compared to developed countries. It is just 4 per cent of USA and 20 per cent of the world average. Rapid expansion of energy production and consumption has brought with it a wide range of environmental issues at the local, regional and global levels. With respect to global environmental issues, India's $CO_2$ emissions have grown along with its energy consumption. $CO_2$ and carbon monoxide (CO) are the main GHGs associated with global warming. At the present time, coal is responsible for 30 -40 per cent of the world $CO_2$ emissions from fossil fuels. Sulfur dioxide ($SO_2$) and NOx contribute to acid rain.

Fossil fuel combustion is the single largest human activity, having an impact on climate change, accounting for approximately 80 per cent of anthropogenic greenhouse gas emissions.

* Corresponding Author: E-mail: p_nema@neeri.res.in

The increased concentrations of key GHGs *e.g.* $CO_2$, methane ($CH_4$) and nitrous oxide ($N_2O$) are a direct consequence of human activities. Carbon dioxide has been significantly increasing compared to the rather steady level of about 280 parts per million in volumes (ppmv) during the pre-industrial era and was about 377 ppmv in 2004, with the fastest growth occurring in the most recent years. Comparable growth has also occurred in levels of $CH_4$, $N_2O$ and $CO_2$ emissions from fuel combustion during 1971-2004. While industrialized countries have so far dominated historical emissions, rapid growth in energy demand of developing economies, led by China and India may soon surpass their emissions beyond those of industrialized countries.

Ozone acts as a shield to protect the earth's surface by absorbing harmful ultraviolet radiation. Ozone depletion may cause eye and skin cancers and cataracts, the major causes of blindness in the world. A sustained 10 per cent thinning of the ozone layer is expected to result in almost two million new cases of cataracts per year, globally. Food production could be reduced by 10 per cent with every 1 per cent increase in UV-B radiation.

This chapter highlights energy consumption and its consequences on global environment with respect to greenhouse gas emission and ozone depletion.

## Introduction

The combustion of gasoline and other fuels in automobiles and jet airplanes produce several primary pollutants: nitrogen oxides, gaseous hydrocarbons, and carbon monoxide, as well as large quantities of particulates. In the presence of sunlight, nitrogen oxides combine with hydrocarbons to form secondary pollutants *viz.* ozone and the eye-stinging peroxyacetyl nitrate (PAN). Nitrogen oxide also reacts with oxygen in the air to form nitrogen dioxide, a foul-smelling brown gas. In urban areas, transportation is the main cause of air pollution where nitrogen dioxident the air, blending with other contaminants and the atmospheric water vapor to produce brown smog. Although the use of catalytic converters has reduced smog-producing compounds in motor vehicle exhaust emissions, recent studies have shown that in doing so the converters produce nitrous oxide, which contributes substantially to global warming.

In cities, air may be severely polluted not only by transportation but also by the burning of fossil fuels in power generating stations, factories, office buildings, and homes and by the incineration of garbage. The massive combustion produces tons of ash, soot, and other particulates responsible for the gray smog of cities, along with enormous quantities of sulfur oxides (which also may be result from burning coal and oil). These oxides rust iron, damage building stone, decompose nylon, tarnish silver, and kill plants. Air pollution from cities also affects rural areas for many miles downwind as the air pollutants disperse in the dynamic atmosphere with time.

Every industrial process has its characteristic emissions. Petroleum refineries emit hydrocarbon and particulate pollution extensively. Iron and steel mills, metal smelters, pulp and paper mills, chemical plants, cement and asphalt plants all discharge vast amounts of various particulates. Uninsulated high-voltage power lines ionize the adjacent air, forming ozone and other hazardous pollutants. Airborne pollutants from other sources include insecticides, herbicides, radioactive fallout, and dust from fertilizers, mining operations, and livestock feedlots.

The energy requirement will grow greater in the years ahead as the rising economies; specifically China and India expand at remarkable rates. According to the United States, Energy Information Administration (EIA), China's oil consumption increased by almost half a million barrels per day in 2006, or 38 per cent of total growth in world oil demand. India's electricity consumption is estimated to grow from 519 billion kilowatt-hours in 2003 to 845 billion kilowatt- hours in 2010. EIA forecasts that worldwide oil consumption will rise from 80 million barrels per day in 2003 to 98 and 118 million in 2015 and 2030 respectfully.

The consumption of resources and energy have also increased rapidly as a consequence of explosive increase in the population, The IEA (1998) projects world energy demand to grow by 65 per cent and $CO_2$ emissions by 70 per cent between 1995 and 2020, unless new policies are put in place. Two-thirds of the increase in energy demand over the period 1995–2020 is expected to come from China and other developing countries. Fossil fuels are expected to meet 95 per cent of additional global energy demand from 1995 to 2020. Oil is being used increasingly to fuel rapidly growing demands for road and air transport. Coal remains important in power generation because of its low cost, in both developed and developing countries.

## Energy Use and Greenhouse Gases

Energy use accounts for over 80 per cent of the global anthropogenic GHG emissions with emission resulting from the production, transportation, handling and consumption of all kinds of energy commodities. Responsible for about 95 per cent of the energy–related emission, $CO_2$ from energy represent about 80 per cent of the global anthropogenic GHG emission. Smaller shares correspond to agriculture, producing mainly $CH_4$ and $N_2O$ from domestic livestock and rice cultivation and to industrial process not related to energy, producing mainly fluorinated gases and $N_2O$. Despite the growth of non-fossil fuel energy (such as nuclear and hydropower), considered as non-emitting fossil fuels have maintained their shares of the world energy supply relatively unchanged over the past 35 years. The link between climate change and energy is a part of the larger challenge of sustainable development. The socioeconomic and technological characteristics of development paths will strongly affect emissions, the rate and magnitude of climate change, climate change impacts, the capability to adapt, and the capacity to mitigate the emission themselves

## Sources of $CO_2$ Emission

The global increase between 2003 and 2004 was more than 1.2 Gt $CO_2$. This dramatic rise was largely (86 per cent) driven by energy demand of developing countries. Coal alone was responsible for almost 60 per cent of the global increase in emission.

In the future, coal is expected to satisfy much of the growing energy demand of those developing countries such as China and India, where energy intensive industrial production is growing rapidly and large coal reserve exist with limited reserve of other energy sources. In addition coal demand may rise as price of oil and natural gas increase consequently to growing demand and pressure on the reserve of these two fuels (IEA, 2006c). Inevitably, intensified use of coal would increase the emission of $CO_2$ even with the development of carbon sequestration technologies. Industrialized countries emit more amount of $CO_2$ per capita than the world average. However, the rapidly expanding economies of China and India increased their emission per capita by 82 per cent and 48 per cent, respectively, over the period 1990-2004.These two countries contribute much to the 6 per cent increase of global per capita emission. Power generation and transport challenge the sustainability of both global economy and the environment. This is particularly pronounced for developing countries since increase in their emissions from these two sectors is faster.

## Energy Consumption in India

Coal is the predominant energy source for power production in India, generating approximately 70 per cent of total domestic electricity. Energy demand in India is expected to increase over the next

10-15 years; (Table 21.1). Although new oil and gas plants are planned, coal is expected to remain the dominant fuel for power generation. Despite significant increases in total installed capacity during the last decade, the gap between electricity supply and demand continues to increase. The resulting shortfall has had a negative impact on industrial output and economic growth. However, to meet expected future demand, indigenous coal production will have to be greatly expanded. Production currently stands at around 290 million tonnes per year, but coal demand is expected to more than double by 2010. Indian coal is typically of poor quality and as such requires beneficiation to improve the quality. Coal imports will also increase dramatically to satisfy industrial and power generation requirements.

**Table 21.1: Demand for Commercial Energy for Final Consumption**

| *Source* | *Units* | *1994-95* | *2001-02* | *2006-07* | *2011-12* |
|---|---|---|---|---|---|
| Electricity | Billion Units | 289.36 | 480.08 | 712.67 | 1067.88 |
| Coal | Million Tons | 76.67 | 109.01 | 134.99 | 173.47 |
| Lignite | Million Tons | 4.85 | 11.69 | 16.02 | 19.7 |
| Natural Gas | Million Cubic Meters | 9880 | 15730 | 18291 | 20853 |
| Oil Products | Million Tons | 6.355 | 99.89 | 139.95 | 196.47 |

*Source*: Planning Commission; BAU: Business as Usual.

## Greenhouse Gases and Global Warming

The increased concentrations of key GHGs are a direct consequence of human activities. Since anthropogenic GHGs accumulate in the atmosphere, they produce net warming by enhancing the natural "greenhouse effect". The amount of heat trapped in the troposphere depends mainly on the concentrations of heat-trapping greenhouse gases and the length of time they stay in the atmosphere.

Carbon dioxide ($CO_2$) has been significantly increasing compared to rather steady level of the pre-industrial era (about 280 ppmv) and touched a level of 377 ppmv 2004, with fastest growth occurring in the most recent years. Comparable growth has also occurred in levels of methane ($CH_4$) and nitrous oxide ($N_2O$).

Some impacts of the increased GHG concentrations may be slow to become apparent since inertia is an inherent characteristic of the interacting climate, ecological, and socio-economic systems. The time scales of some of the effects of climate change are illustrated in Figure 21.1. Even after stabilization of the atmospheric concentration of $CO_2$, surface air temperature is projected to increase for a century or more, while sea level is projected to rise for many centuries. Some changes in the climate system would be effectively irreversible.

## Impacts of Global Warming

- ☆ By the 2050s, freshwater availability in Central, South, East and South-East Asia, particularly in large river basins, is projected to decrease.
- ☆ Coastal areas, especially heavily populated mega delta regions in South, East and South-East Asia, will be at greatest risk due to increased flooding from the sea and, in some mega deltas, flooding from the rivers.
- ☆ Climate change is projected to compound the pressure on natural resources and environment, associated with rapid urbanization, industrialization and economic development.

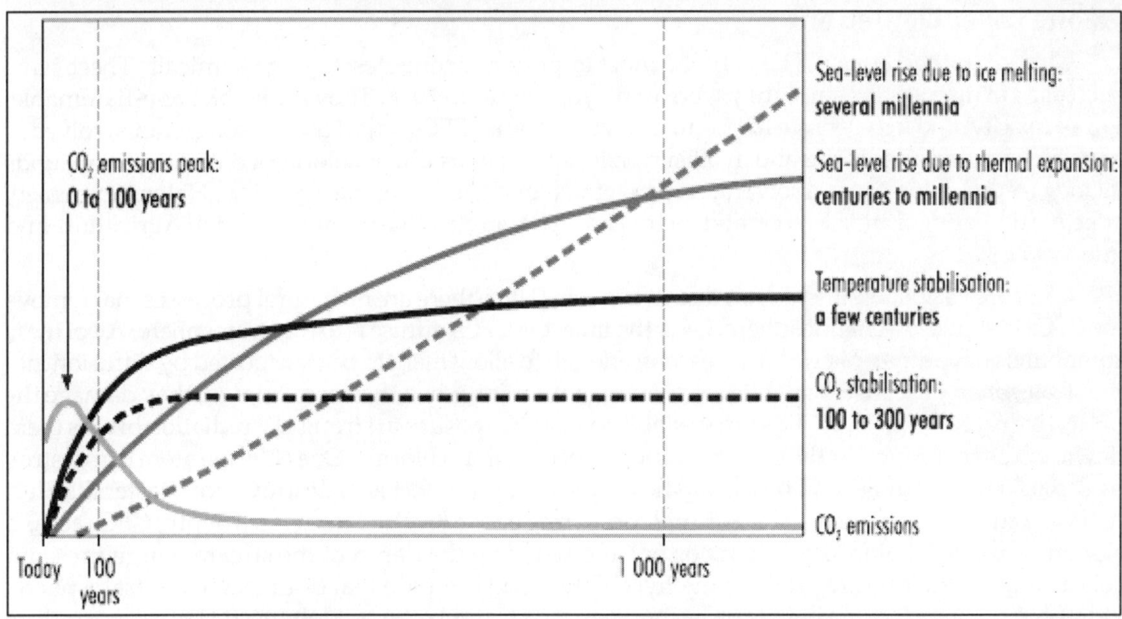


**Figure 21.1: Time Scales of Climate Change Effects**
**(Generic illustration for stabilization of $CO_2$ concentration at any level between 450 and 1000 ppmv)**

☆ Endemic morbidity and mortality due to diarrheal diseases primarily associated with floods and droughts are expected to rise in East, South and South-East Asia due to projected changes in the hydrological cycle.

## Climate Change and its Consequences

Climate change and associated threats have also become one of the main global challenges to humanity. The problems climate change raises involve science and economics, decision-making under uncertainty, and the balancing of the interests of the current and future generations. The third assessment report (TAR) of the IPCC released in 2001 confirmed the inevitability of climate change. The IPCC latest report (2007) predicts temperature rise of 1.1–6.4 °C by 2100 and attributes most of the warming observed in the last 50 years to human activity. The report also mentions the UN Secretary-General's warning in his "In Larger Freedom" report of 2005 that "One of the greatest environmental and development challenges in the twenty-first century will be that of controlling and coping with climate change." Climate change could cause serious and far-reaching problems for the Earth's inhabitants. At present some 46 million people live in areas at risk of flooding due to storm surges. Scientists estimate that a 50 cm rise in sea level would increase this number to 92 million and a one-metre rise would put 118 million in dangerous situation. If the global ocean level went up by one meter, Egypt would lose 1 per cent of its land area, the Netherlands would lose 6 per cent, Bangladesh would lose 17.5 per cent, and on the Majuro Atoll in the Pacific Marshall Islands some 80 per cent of the land would disappear under water. Populations particularly under threat live in Sub-Saharan Africa, South East Asia and tropical areas of Latin America.

## Ozone Layer Depletion

Chlorofluorocarbons (CFCs) are the most important ozone-destroying chemicals. These have been used in many ways since they were first synthesized in 1928. They are stable, non-flammable, low in toxicity, and inexpensive to produce. Over the time, CFCs found uses as refrigerants, solvents, foam blowing agents, aerosols and in other smaller applications. Other chlorine–containing compounds include methyl chloroform, a solvent, carbon tetrachloride, an industrial chemical, Halons, extremely effective fire extinguishing agents, and methyl bromide, an effective fumigant used in Agriculture and grain storage.

CFCs are very stable and do not dissolve in rain. Thus, there are no natural processes that remove the CFCs from the lower atmosphere. Over the time, the CFCs diffuse into the stratosphere. All of these compounds have atmospheric lifetimes long enough to allow them to be transported by diffusion into the stratosphere. Because they release chlorine or bromine when they break down, they damage the protective ozone layer. The CFCs are so stable that only exposure to strong UV radiation breaks them down. When that happens, the CFC molecule releases atomic chlorine. One chlorine atom can destroy over 100,000 ozone molecules by a catalytic process. The net effect is to destroy ozone faster than it is naturally created. To return to the analogy comparing ozone levels to a stream's depth, CFCs act as a siphon, removing water faster than normal and reducing the depth of the stream. Large fires and certain types of marine life produce methyl chloride and bromide that does reach the stratosphere. However, numerous experiments have shown that CFCs and other widely-used chemicals produce roughly 85 per cent of the chlorine in the stratosphere, while natural sources contribute only 15 per cent, as per United States Environmental Protection Agency.

## Environmental Effects of Ozone Depletion

Ozone acts as a shield to protect the Earth's surface by absorbing harmful ultraviolet radiation. If this ozone becomes depleted, then more UV rays will reach the earth. Exposure to higher amounts of UV radiation could have serious impacts on human beings, animals and plants, such as the following:

- ☆ More skin cancers, sunburns and premature aging of the skin.
- ☆ Cataracts (a clouding of the lens) are the major cause of blindness in the world. A sustained 10 per cent thinning of the ozone layer is expected to result in almost two million new cases of cataracts per year, globally (Environment Canada, 1993).
- ☆ Several of the world's major crop species are particularly vulnerable to increased UV, resulting in reduced growth, photosynthesis and flowering. Many agricultural crops are sensitive to the burning rays of the sun, including the world's main food crops, rice, wheat, and corn and soya bean.
- ☆ Decreases in plankton could disrupt the fresh and saltwater food chains, and lead to a species shift.
- ☆ Species of marine animals in their growing stage, including young fish, shrimp larvae and crab larvae, have been threatened in recent years by the growing UV-B radiation under the Antarctic ozone hole. Loss of biodiversity in our oceans, rivers and lakes could reduce fish yields for commercial and sport fisheries.
- ☆ In domestic animals, UV over exposure may cause eye and skin cancers.
- ☆ Wood, plastic, rubber, fabrics and many construction materials are degraded by UV radiation.
- ☆ The economic impact of replacing and/or protecting materials could be significant.

## Conclusion

Fossil fuel combustion is the single largest human influence on climate. Over the last two decades the global community has recognized the pressure need to address and reduce $CO_2$ emissions from combustion that enters the atmosphere.

Two growing sectors, both growing represent the bulk of $CO_2$ Emission from fuel: electricity and heat generation sector and the transport sector. Improving the energy efficiency and reducing the carbon intensity of both sectors could significantly diminish their contribution to global climate change.

Since the industrial revolution, most of the $CO_2$ emissions have originated from the industrialized country. However this dominance appears short–lived due to the size of some of developing economies and the growth in their energy needs. Effective emission mitigation will require all countries regardless of energy demand and infrastructure to use energy and manage natural resources in a sustainable manner to achieve economic growth without undue GHG emission. There is a growing concern that long-run sustainable development may be compromised unless measures are taken to achieve balance between economic, environmental and social outcomes. Three specific issues of sustainable development that are of particular importance for any organization are addressing climate change, reducing air pollution and ensuing sustainable use of natural resources.

## Bibliography

1. Australian Institute of Health and Welfare (AIHW) and Australasian Association of Cancer Registries (AACR). Cancer in Australia 1998. AIHW cat. no. CAN12. Canberra: AIHW, 2001. http://www.aihw.gov.au/publications/can/ca98/
2. Department of Environment and Land Management (DELM) (1993). Understanding Ozone: ozone depletion, its consequences and actions. 2nd ed., DELM, Adelaide.
3. Gies, P. Roy, C., Toomey, S. and Tomlinson, D. (1999). 'Ambient solar UVR, personal exposure and protection', Journal of Epidemiology, Vol. 9, No. 6.
4. Atmosphere, Australia State of the Environment Report 2001 (Theme Report). CSIRO Publishing on behalf of the Department of the Environment and Heritage, Canberra. http://www.deh.gov.au/soe/2001/atmosphere/index.html
5. Mathers, C. Penm, R. Sanson-Fisher, R. Carter, R. and Campbell, E. (1998). 'Health system costs of cancer in Australia 1993–94'. AIHW cat. no. HWE 4. Canberra: Australian Inst. Health and Welfare and National Cancer Control Initiative (Health and Welfare Expenditure Series No.4). http://www.aihw.gov.au/publications/hwe/hscca93-94/
6. Udelhofen, P., Gies, P., Roy, C. and Randel, W. (1999). 'Surface UV radiation over Australia, 1979-1992: Effects of ozone and cloud cover changes on variations of UV radiation', Journal of Geophysical Research, Vol. 104, No. D16, pp 19, 135-19, 159, Aug. 27, 1999.
7. World Health Organization (WHO) (2002). Global solar UV index: a practical guide. Geneva, Switzerland, WHO http://www.who.int/mediacentre/factsheets/who271/en/
8. http://www.global-greenhouse-warming.com
9. Roberta Quadrelli, Sierra Peterson (2007). The energy–climate challenge: Recent trends in $CO_2$ emissions from fuel combustion.

# Chapter 22

# Phytoremediation of High Nitrate Containing Wastewater Using Aquatic Plants

**P.P. Kanekar*, S.S. Sarnaik, S. Roy, P.N. Shede and M.R. Kedargol**

**Microbial Sciences Division, Agharkar Research Institute, G.G. Agarkar Road, Pune – 411 004, M.S.**

## ABSTRACT

Nitrates, the essential nutrients become nuisant when present in high concentration. The industrial units manufacturing and handling nitroexplosives, dyes, pesticides, plastics etc. generate wastewater containing high amounts of nitrate, crossing the limits laid down by Water Pollution Control Board. Since presence of high levels of nitrates in water cause methemoglobinemia, nausea, disturbances in functioning of liver and kidney etc., the removal of nitrates from such wastewaters becomes necessary. Phytoremediation is a new emerging technology making use of natural resources like plants for removal of pollutants. Present investigation explores possibility of aquatic plants for removal of nitrates from high nitrate containing wastewater generated during production of Nitroexplosive–High Melting Explosive (HMX).

Hydroponics experiments using aquatic plants were carried out to see the removal of nitrate with simultaneous uptake from the wastewater having initial concentration of nitrate as >40000 mg/l. The plant species like *Ceratophyllum, Salvinia, Vallisneria, Najas* and *Pistia* were used in the experiment. *Ceratophyllum, Vallisneria, Salvinia* and *Hydrilla* showed tolerance to nitrate concentration as high as 8,000 mg/l for eight days while *Najas, Pistia* showed wilting during the experimental period. *Ceratophyllum, Vallisneria* and *Salvinia* exhibited 50–55 per cent nitrate removal

* Corresponding Author: E-mail: kanekarp@rediffmail.com

from the wastewater and 40 to 66 per cent uptake of nitrate. There was ~ 65 per cent reduction in chlorophyll content of the plants due to high nitrate content of the wastewater. Constructed wetland experiments carried out using gravel/soil as supporting medium for *Hydrilla* plants resulted in 61 per cent removal of nitrate from the wastewater within 11 days and 63 per cent uptake of nitrate by leaves with 56 per cent reduction in chlorophyll content of the leaves.

Thus aquatic plants like *Ceratophyllum, Vallisneria, Salvinia* and *Hydrilla* could be the suitable candidates for developing phytoremediation technology for treatment of high nitrate containing wastewaters.

***Keywords:*** *Phytoremediation, Aquatic plants, Nitrate removal, Nitrate uptake, Chlorophyll content.*

## Introduction

Indiscriminate discharge of sewage and industrial wastewaters into natural water bodies is posing a serious problem of water pollution. With increase in urbanization and industrialization human being is introducing new and complex chemicals without any rigorous bio-assessment of their toxicity. Nitrates are present in the environment as both organic and inorganic compounds. These nitrates are known to be the essential chemicals required for well maintenance of life. However, presence of nitrates in high concentrations, results in pollution of the environment and cause problems to aquatic as well as terrestrial life.

### Sources of Nitrates Polluting the Environment

The nitrates enter in the environment under a variety of conditions both natural *i.e.* decaying of plant materials, animal and human activities etc. The major nitrate contributing industries include agrochemical *i.e.* pesticides and fertilizers manufacturing and user industries; dye industries especially those producing nitroso dyes; metal processing industries like steel mills; food processing industries especially the meat industries and the nitro explosive producer and user units etc. Different sources of nitrates responsible for contamination of water bodies are described by Pinar *et al.* (1997) which include the wastewaters arising from manufacturing units of chemicals, fertilizers, explosives etc.

### Toxicity of Nitrates

Most of these sources cause nitrates to leach out into ground water. Amongst various contaminants present in the environment, nitrates play significant role in ground and surface water contamination. Short-term exposure to drinking water with a nitrate level at or just above the health standard of 10 mg/l nitrate-N is a potential health problem primarily for infants *i.e.* methemoglobinemia. Excessive levels of nitrate in drinking water have caused serious illness and sometimes death. Nitrates and nitrites have the potential to cause dieresis, increased starchy deposits and hemorrhaging of the spleen when there is a long-term lifetime exposure at levels above the MCL. Epidemiological and toxicological aspects of nitrates and nitrites in the environment are described by Shuval and Gruener (1972). Wigington (2001) stated the ill effects of excess concentration of nitrates in drinking water which can cause kidney trouble, jaundice, methemoglobinemia and their carcinogenic nature. In general the toxicity of the explosives and other nitro aromatic compounds is well described by Kanekar *et al.* (2003). The toxicity and polluting ability of nitrates and nitro aromatic compounds especially the nitroexplosives necessitates their remediation from the nitrate containing wastewaters.

## Treatment Methods Presently Used

Many physical and physicochemical techniques like settling, filtration, UV radiation, incineration etc. are available to treat water contaminated with nitro compounds. Electro Dialysis Reversal (EDR) seems to be an attractive technology as compared to Reverse Osmosis (RO), developed recently for treating waters contaminated with high levels of nitrates. However all these methods are highly expensive and in the long run will not be economically feasible. Treatment methods developed using biological activities of natural resources like microorganisms or plants will be economically viable.

Aiming at the long term and eco-friendly approach, bioremediation is the most suitable technique. Some reports are available on biodegradation and/or bioremediation of nitro compounds, their transformed products and wastewaters containing these chemicals using microbial activities. Microbial degradation of nitroaromatic compounds like Trinitrotoluene (TNT) and wastewater generated during its production was studied by Kanekar and Godbole (1983 and 1984). Degradation of Dinitrobenzene (DNB) and wastewater generated during its production by microorganisms was demonstrated by Dey *et al.* (1986). Glass and Silverstein (1999) reported the bioremediation of groundwater having high nitrate and high salinity. Biodegradation of nitroaromatic compounds with special reference to Nitroexplosives is well described by Kanekar *et al.* (2003).

## Phytoremediation Technique

Phytoremediation is an emerging technology wherein natural plant physiological processes are exploited to remove or destroy the contaminants. It is an aesthetically pleasing mechanism that can reduce remedial costs, restore habitat, and clean up contamination in place rather than entombing it in place or transporting the problem to another site. Nitrogen is the most important element for plants. Plants can use nitrogen in various ways and in different forms. Most common way is uptake of nitrate and ammonia forms by plant roots. Nitrate ions are absorbed quickly by plant roots, but leach easily. Plants directly take nitrates from the soil to achieve nutrition. Phytoremediation of nitrate contaminated sites can be an attractive option for removal of nitrates from the contaminated sites.

The phytoremediation processes mediated by plants include: *Phytoextraction*–uptake and concentration of substances from the environment into the plant biomass; *Phytostabilization*–reducing the mobility of substances in the environment, for example by limiting the leaching of substances from the soil; *Phytotransformation*–chemical modification of environmental substances as a direct result of plant metabolism, often resulting in their inactivation, degradation (phytodegradation) or immobilization (phytostabilization); *Phytostimulation*–enhancement of soil microbial activity for the degradation of contaminants, typically by organisms that associate with roots. This process is also known as *rhizosphere degradation; Phytovolatilization*–removal of substances from soil or water with release into the air, sometimes as a result of phytotransformation to more volatile and/or less polluting substances; *Rhizofiltration*–filtering water through a mass of roots to remove toxic substances or excess nutrients. The pollutants remain absorbed in or adsorbed by the roots. Nitrate removal by plants was studied by few research workers. Best *et al.* (1997) carried out some work on removal of nitro compounds *i.e.* explosives from groundwater using aquatic plants. Work on phytoremoval of Trinitrotoluene (TNT) was carried out by Medina *et al.* (2000). Shrivastava *et al.* (2001) studied the nitrate removal by water hyacinth and Ipomea leaves. Kanekar *et al.* (2003) have reviewed use of some aquatic and terrestrial plant species for phytoremediation of Nitroexplosives TNT, HMX etc. Vanek *et al.* (2007) carried out the phytoremediation of nitro explosives.

## Constructed Wetlands

Wetland encompasses a broad range of ecosystems ranging from submerged coastal grass beds, to salt marshes, swamp forests and boggy meadows. They are the transitions between the terrestrial and the aquatic systems. Williams (2002) has discussed the progress and potential of wetland ecosystems in phytoremediation.

Few reports are available on phytoremediation of soils contaminated with toxic organic chemopollutants. Susarla *et al*. (2002) have stated that phytoremediation will be an ecological solution to any organic chemical contamination. Phytoremediation of soils contaminated with organic pollutants was carried out by Cunningham and others (1995, 1996). Singh and Jain (2003) used phytoremediation technique for decontaminating the soil contaminated with toxic aromatic pollutants.

A phytoremediation method developed for treatment of high nitrate containing wastewater generated during production of High Melting Explosive (HMX) using aquatic plants is presented.

# Materials and Methods

## Characterization of Wastewater

### Collection of Wastewater Samples

Wastewater generated during production of HMX was collected from High Energy Material Research Laboratory (HEMRL) situated nearby Pune, as and when required and preserved in the laboratory at ambient temperature (28±2°C) for further studies. The characterization of wastewater is needed for developing a suitable phytoremediation method.

### Chemical Characterization of Wastewater

The wastewater was characterized for pH, concentration of nitrates and Chemical Oxygen Demand (COD) as per Standard Methods (Greenberg *et al.*, 1992). The analysis for each parameter was performed for four samples.

### pH

pH of the wastewater was measured with the help of pH meter (Control Dynamics, Mumbai, India).

### Nitrate

Nitrate estimation was carried out using spectrophotometric method by measuring absorbance at 220 nm and 275 nm as described in Standard Methods (Greenberg *et al.*, 1992) and modified by Shivran *et al*. (2006).

### Chemical Oxygen Demand (COD)

COD of the wastewater was determined by Open Reflux Method as per Standard Methods (Greenberg *et al.*, 1992). Suitably diluted samples were refluxed with Potassium dichromate and concentrated sulfuric acid for 2 hours at 150°C and were allowed to cool. The samples were then titrated with Eerrous Ammonium Sulfate using ferroin indicator.

## Phytoremediation Studies

### Collection and Acclimatization of Aquatic Plants

Aquatic free floating or submerged type of plants namely *Hydrilla verticillata*, *Ceratophyllum demersum*, *Salvinea molesta*, *Vallisneria* sp., *Najas* sp., *Potamogeton* sp. and *Pistia* sp. were collected from different places in and around Pune city and also from Nasik and Satara districts.

These plants were then introduced in modified Hoagland's medium for establishment for a period of one week and then used for hydroponics experiment. Chemical contents of Hoagland's medium in mg/l include: Ammonium phosphate monobasic–115.03, Boric acid–2.86, Calcium nitrate–656.4, Cupric sulfate, 5 $H_2O$–0.08, Ferric citrate–5.32, Magnesium sulfate–240.76, Manganese chloride, $4H_2O$–1.81, Molybdenum trioxide–0.02, Potassium nitrate–606.6 and Zinc sulfate–0.22 (Gujrathi, 2005). pH of the medium was maintained at seven and the same medium was used for dilution of the wastewater. After 7 days of stabilization in Hoagland Medium, the plants were tested for their survival in high nitrate containing wastewater.

**Screening of Aquatic Plants for their Survival in High Nitrate Containing Wastewater**

Since all the above mentioned plant species showed tolerance to nitrate concentration of 8000 mg/l (1:5,v/v dil.) in earlier experiment, they were tested for their survival in wastewater. Three plants of about same size of each plant species were introduced in containers filled with 750 ml of diluted wastewater. Water losses due to evaporation were replenished everyday. The plants were monitored for their survival in the wastewater daily and observations were recorded. The experiment was carried out in duplicate.

**Hydroponics Experiment to Study Ability of Plants for Nitrate Removal and Uptake**

The plants tolerating nitrate concentration of 8000 mg/l for about a week and more than that were selected to study the uptake and removal of nitrate from the wastewater. Wastewater was diluted suitably with Hoagland's medium devoid of nitrate source so as to have nitrate concentration of nitrate in the wastewater as 8000 mg/l. Plants of about equal length were weighed and were added to the containers in triplicates. Wastewater samples were drawn at 24 h interval to see the removal of nitrate by the plants from the wastewater after suitable dilution and spectrophotometric analysis as described earlier. For studying the nitrate uptake by the plants, leaf samples were collected after initiation of the experiment and on last day of the experiment at the time of termination. Leaf samples were dried and crushed with Milli Q water. The tubes were centrifuged for 10 minutes at 10000 rpm to remove insoluble suspended matter and the supernatant samples were diluted suitably for estimation of nitrate as stated above.

**Effect of Nitrates from Wastewater on Chlorophyll Content of the Plants**

The chlorophyll content of leaf samples was estimated after extraction of leaves in 80 per cent acetone till the leaves become colourless, centrifuged to remove insoluble suspended matter and supernatant used for estimation of chlorophyll content by recording absorbance at 652 nm.

**Constructed Wetlands**

The plants showing satisfactory uptake and removal of nitrates from the hydroponics experiments were used for constructed wetland experiments. Three rectangular glass containers were used for the experiment and filled with the mixture of gravel and soil. The particulars of the experimental setup were as follows:

| *Particulars* | *Set up* | *Particulars* | *Set up* |
|---|---|---|---|
| Dimensions of the tank | 30 cm x 17 cm x 17.5 cm | Volume of the tank | 8925 $cm^3$ |
| Quantity of garden soil | 0.6 kg | Quantity of gravel | 2.4 kg |
| Total height of gravel + soil layer | 6.1 cm | Volume of waste (1: 5 dil.) | 5 L |
| Nitrate concentration, mg/l | 8000 | Head space | 5.2 cm |
| Weight of plants | 25 gm | | |

Removal and uptake of nitrates by the plants: Wastewater samples were collected immediately after introduction of plants and after 48 h. (alternate day) for estimation of nitrate so as to know the nitrate removal by the plants and leaf samples were collected for determination of uptake of nitrate by the plant and the effect of nitrates on chlorophyll content of the leaves. The procedure for the analysis was same as mentioned above.

## Results and Discussions

Characterization of wastewater under study: Chemical characterization of high nitrate containing wastewater with respect to pH, COD and nitrate concentration is detailed in Table 22.1.

**Table 22.1: Characterization of High Nitrate Containing Wastewater**

| *Parameters* | *Concentration* | *MPCB Limits* |
|---|---|---|
| pH | 4.52 | 6.5–7.5 |
| Nitrate, mg/l | 39006 | 45 mg/l |
| Chemical Oxygen Demand, mg/l | 126240 | 250 mg/l |

From the table it can be seen that the wastewater under study is highly acidic in nature having pH value around 4.5. Nitrate content of the wastewater is high of the magnitude of ~ 40,000 mg/l which crosses the limits laid down by MPCB. COD of the wastewater was also found to be very high *i.e.* ~ 126240 mg/l and cross the limits laid down by MPCB which are 250 mg/l. Hence before discharging of the wastewater in the environment, its remediation is necessary so as to have the concentration of nitrates and COD values near about the MPCB limits.

In the present investigation, phytoremediation technique is used. It was necessary to neutralize the wastewater so as to avoid any adverse effect of wastewater on plants due to highly acidic pH. Nitrates are present in higher concentration hence nitrates were considered as major pollutant and estimation of nitrates was given more emphasis during the studies. All the results are expressed in terms of removal of nitrates from the wastewater in which plants were exposed and as uptake of the nitrates by the leaves of exposed plants.

## Phytoremediation Studies

### Screening of the Plants for their Survival in Wastewater having High Concentration of Nitrate

Among the aquatic plants under study, *Potamogeton* sp. was found to be highly sensitive to high concentration of nitrate from the wastewater as shown by browning and wilting of the plants within 2 days of exposure to high nitrate containing wastewater. *Pistia* sp. and *Najas* sp. were also found to be considerably sensitive to the higher concentration of nitrates present in the wastewater as shown by wilting and browning of the plants after their exposure for 3-4 days. *Vallisneria* sp., and *Ceratophyllum demersum* showed appreciable resistance to nitrate concentration of the wastewater as high as ~ 8000 mg/l. *Hydrilla* plants were quite resistant to this high concentration of nitrate present in the wastewater as they were fresh till 10 days of exposure with only slight browning of the leaves. The results are detailed in Table 22.2.

The results indicated that browning was observed almost in all the plant species, however the extent of browning was different depending upon the resistance of the plants. The period required for complete browning was different for different plant species. Thus, survival of plants in wastewater was found to be maximum for 10 days in case of *Hydrilla verticillata*, and in case of *Ceratophyllum demersum, Salvinea molesta, Vallisneria* sp. it was for nearly one week *i.e.* 8,7, and 6 days respectively. *Potamogeton* sp. was found to be most sensitive one showing survival for only 2 days, *Pistia* sp. and *Najas* sp. for 3–4 days. Thus *Hydrilla verticillata, Ceratophyllum demersum, Salvinea molesta, Vallisneria* sp. could be the suitable candidates for further studies on removal and uptake of nitrates from the wastewater.

**Table 22.2: Survival of Aquatic Plants in Wastewater Containing High Nitrate Concentration**

| Sl.No. | Plant Species Studied | Survival of Plants, Days at ~ 8,000 mg/l Nitrate Concentration | Observations |
|---|---|---|---|
| 1. | *Hydrilla verticillata* | 10 | Very little browning |
| 2. | *Ceratophyllum demersum* | 8 | Slight browning |
| 3. | *Salvinea molesta* | 7 | Slight browning |
| 4. | *Vallisneria* sp. | 6 | Greenish brown |
| 5. | *Najas* sp. | 4 | Complete browning |
| 6. | *Pistia* sp. | 3 | Complete decolorization and wilting |
| 7. | *Potamogeton* sp. | 2 | Complete browning and wilting |

## Hydroponics Experiments to Study Ability of Plants for Nitrate Removal and Uptake

Based on the observations of earlier experiments, *Hydrilla, Ceratophyllum, Vallisneria* and *Salvinea* plants were selected for hydroponics experiments. The results on nitrate removal by the plant species and the uptake of nitrates by these plants are detailed in Table 22.3.

**Table 22.3: Removal and Uptake of Nitrate by the Aquatic Plant Species Under Study**

| Plant Species Under Study | Per cent Nitrate Removal from Wastewater | Per cent Nitrate Uptake by the Leaves | Per cent Reduction in Chlorophyll Content |
|---|---|---|---|
| *Hydrilla verticillata* | 53.78 | 66.29 | 66.48 |
| *Ceratophyllum demersum* | 54.47 | 53.65 | 86.07 |
| *Salvinea molesta* | 49.44 | 40.61 | 69.65 |
| *Vallisneria* sp. | 54.20 | 41.30 | 66.90 |

It was observed that all the four plant species under study *i.e. Hydrilla verticillata, Ceratophyllum demersum, Salvinea molesta* and *Vallisneria* sp showed around 50 per cent removal of nitrate from the wastewater. The uptake of nitrate by all the four plant species was more than 50 per cent, maximum of 66 per cent was exhibited by *Hydrilla verticillata*. However, the uptake of high nitrate content (~ 50 per cent *i.e.* ~ 4000 mg/l) by the leaves had significant adverse effect on chlorophyll content of *Ceratophyllum demersum* which was indicated by reduction in chlorophyll content of the leaf sample. Slightly lesser adverse effect was observed in case of *Salvinea molesta* and *Vallisneria* sp. In case of *Hydrilla verticillata* plants, there was good removal and uptake of nitrate by the leaves and considerably less adverse effect on chlorophyll content of the leaves of *Hydrilla verticillata*. Hence *Hydrilla verticillata* can be good candidate for phytoremediation studies of high nitrate containing wastewaters and these were selected for further experiments using constructed wetlands.

*Constructed Wetland studies:* The results on efficiency of *Hydrilla verticillata* plants in removal and uptake of nitrates in constructed wetland system are detailed in Table 22.4.

**Table 22.4: Removal and Uptake of Nitrates from Wastewater by *Hydrilla verticillata* Plants in Constructed Wetland System**

| *Parameter* | *Per cent Reduction* |
|---|---|
| Per cent Nitrate removal | 60.91 |
| Per cent nitrate uptake | 62.85 |
| Per cent reduction in chlorophyll content | 56.46 |

The results showed that efficiency of *Hydrilla verticillata* to remove nitrates from the wastewater was increased in constructed wetland upto 61 per cent as compared to 54 per cent in Hydroponic experiments. Chlorophyll content of the plant was found to get less affected in case of wetland experiments than hydroponics experiments. There was not much effect on uptake of nitrate due to change in experimental condition *i.e.* in hydroponics experiment, it was 66 per cent and in constructed wetland it was 65 per cent. Constructed wetland system using gravel and soil in the proportion of 4:1 (w/w) were found to be more favourable for *Hydrilla verticillata* to remove nitrate from the wastewaters generated during production of HMX having nitrate concentration as high as 8000 mg/l.

## Conclusion(s)

The wastewater under study was highly acidic in nature (pH having high COD value indicating high carbon content in terms of chemical compounds and also very high nitrate content crossing the limits prescribed by Water Pollution Control Board. Submerged type of aquatic plants like *Hydrilla, Ceratophyllum* and *Vallisneria* were found to remove considerable amount of nitrate from the wastewater. These plants can be suitable candidates for constructing wetlands for treatment of wastewaters containing high amounts of nitrates depending upon their availability. Constructed wetland system using gravel and soil in the proportion of 4:1 (w/w) seems to be favorable for *Hydrilla* plants in removing nitrates from wastewater containing nitrates up to the concentration of 8000 mg/l.

## Acknowledgements

The work was carried out under Indo-US R&D Networked Joint Center on Bioremediation and Phytoremediation of the chemopollutants sponsored by Indo-US Science and Technology Forum (IUSSTF), Govt. of India and US Govt. The authors thank authorities of HEMRL, Pune for providing HMX wastewater and related information.

## References

1. Best E.P., Zappi, M.E., Fredrickson, H.L. Sprecher, S.L., Larson, S.L., and M. Ochoa (1997). "Screening of aquatic and wetland plants for phytoremediation of explosives- contaminated groundwater from the Iowa Army Ammunition Plant" *Annals of the New York Academy of Sciences*. Vol. 829, pp. 179-194.

2. Cunningham, S. D., Berti, W. R., and Huang, J W. (1995). Phytoremediation of contaminated soils. *Trendsbiotech.* 13:9 p.393-397.

3. Cunningham, S. D., Anderson, T. A., Schwab, A. P., and Hsu, F. C. (1996) Phytoremediation of soils contaminated with organic pollutants. *Adv-agron*. San Diego, Calif.: Academic Press. p. 55-11.

4. Dey S, Kanekar P, Godbole SH (1986) Aerobic microbial degradation of m-Dinitrobenzene. Indian J. of Environ. Health,29 (2): 118-128.

5. Glass C and Silverstein J (1999) Denitrification of high nitrate, high salinity wastewaters. Water Research, 33(1): 223–229.

6. Greenberg, A.E., Clesceri, L.S. and Eaton, A.D. (1992) In Standard methods for the examination of water and wastewater. American Public Health Association, American Water Works Association, Water Environment Federation, Washington DC.

7. Gujrathi, N.P. (2005) Phytoremediation of tetracycline and oxytetracycline Ph.D. Thesis, Colorado State University, Fort Collins, US.

8. Kanekar P, Godbole SH (1983) *Pseudomonas trinitrotoluenophila* sp. nov. isolated from soil exposed to explosive waste. Biovigyanam 9: 5-8.

9. Kanekar P, Godbole SH (1984) Microbial degradation of Trinitrotoluene (TNT). Indian J. of Environ. Health, 26: 89-101.

10. Kanekar, P., Dautpure, P., and Sarnaik, S. (2003) Biodegradation of Nitroexplosives, Indian Jour. Expt. Biol. 41, 991-1001.

11. Medina, V.F., Larson, S.L., Bergstedt, A.E. and McCutcheon, S.C. (2000) Phyto removal of trinitrotoluene from water with batch kinetic studies. Water Res. 34, 2713-2727.

12. Pinar, G., Duque, E., Haidour, A., Oliva, J.M., Sanchez-Barbero, L., Calvo, V. and Ramos, J.L. (1997). Removal of high concentrations of nitrate from industrial wastewaters by bacteria. Appl. Environ. Microbiol. 63(5): 2071–2073.

13. Shivran, H.S., Kumar, D. and Singh, R.V. (2006), Improvement of water quality through biological denitrification, J. Environ. Sci. Engin., 48, 57-60.

14. Shuval, H.I., and N. Gruener. 1972. Epidemiological and toxicological aspects of nitrates and nitrites in the environment. Amer. J. of Public Health, 62:1045-51.

15. Singh, O.V. and Jain, R.K. (2003) Phytoremediation of toxic aromatic pollutants from soil. Appl. Microbiol. Biotechnol. 63, 128–135.

16. Srivastava RK, Ayachi AK, Sehgal V and Sen A (2001) Studies on the nitrate removal by water hyacinth and Ipomea leaves. *Plant Arch.* 1(182): 81-85.

17. Susarla S., Medina VF, McCutcheon SC (2002). Phytoremediation: an ecological solution to organic chemical contamination. *Ecological Engineering*, 18 (5): 647-658.

18. Vanek, T., Gerth, A., Vakrikova, Z., Podlipna, R. and Soudek, P. (2007) Phytoremediation of Explosives: In Advanced Science Technology for Biological Decontamination of Sites affected by Chemical and Radiological Nuclear Agents, Part III pp 209-225, Springer, Netherlands.

19. Wigington P.S. (2001) Monograph Series Published by Agency for Toxic Substances and Diseases Registry (ATSDR), US Dept. pf Health and Human Services.

20. Williams JB (2002) Phytoremediation in wetland ecosystems: progress, problems and potential. Critical Reviews in Plant Sciences, 21 (6): 607–635.

# Chapter 23

# Issues of Hazardous Waste Management in India with Specific Reference to Reduce, Reuse, Recycling and Recovery

**T. Chakrabarti**
**Acting Director,**
**National Environmental Engineering Research Institute,**
**Nehru Marg, Nagpur – 440 020, Mh.**

## Hazardous Wastes

Hazardous Waste is defined 'as any waste which by reason of any of its physical, chemical, reactive, toxic, flammable, explosive or corrosive characteristics causes danger or is likely to cause danger to health or environment, whether alone or in contact with wastes or substances. Hazardous wastes are those which answer to the attribute/attributes listed in the Schedules 1 and II of the HMW, Rules of 1989 as amended in 2003.

## Sources and Sinks of Hazardous Wastes in India

Andhra Pradesh, Gujarat, Karnataka, Maharashtra and Tamil Nadu, taken together, generate more than 80 per cent of the hazardous wastes of the country. In addition, sources include:

1. Hazardous wastes imported in the country under Basel Convention for recycle
2. Illegally imported hazardous wastes impounded at ports and inland container depots (ICDs).These are (in alphabetical order)
   - ☆ CFS, Ludhiana
   - ☆ Chennai Port Trust

- ☆ Cochin Port Trust
- ☆ ICD, Faridabad
- ☆ ICD, Whiteland, Bangalore
- ☆ Jawaharlal Nehru Port Trust (JNPT)
- ☆ Kandla Port Trust
- ☆ Kolkata Port Trust
- ☆ Mumbai Port Trust
- ☆ New Customs House, Ballard Estate, Mumbai
- ☆ New Mangalore Port Trust
- ☆ Tuticorin Port Trust

3. The consignments lying in these ports are mostly
   - ☆ Waste and used oil
   - ☆ Lead scrap
   - ☆ Battery scrap
   - ☆ Zinc Scrap
   - ☆ Metal ash; slag and Precipitate
   - ☆ LPPE film scrap
   - ☆ Low sulphur waxy residues

The sinks for the hazardous wastes are:

1. Stored in generator's premises
2. Treatment, storage and disposal facilities (TSDFs)
3. Illegal dump sites
4 Surface water through nonfunctional effluent treatment plants (ETPs) and common effluent treatment plants (CETPs)
5. Groundwater where untreated effluents are introduced through injection wells in order to avoid treatment and generation of hazardous waste.

## Significance of SMEs in Industrial Output and Pollution

Nearly fifty percent of the total industrial output in India is contributed by SMEs. They also account for 60 to 65 percent of the total industrial pollution. Most of these industries generate hazardous wastes, which find their way uncontrolled into the environment. According to the National Productivity Council, New Delhi (India), there are more than 3 million small and medium scale industries, which are spread throughout the country in the form of clusters/industrial estates. Capacity building to support increased waste reuse and recycling by the SMEs should focus on the following areas:

1. Making operational national policies and incentives for waste management
2. Enabling SMEs to carry out waste reuse and recycling operations and undertaking waste management planning that incorporates resource recovery practices.

SMEs in India, in general, cannot afford to adopt and maintain adequate hazardous waste treatment and disposal technologies. Under such circumstances, the SMEs must use the available common hazardous waste treatment and disposal facility in the area. The Government should look into the possibility of setting up of new industries keeping in mind the concept of environmentally balanced industrial complexing (EBIC) where the waste of one industry becomes the resource for the others. EBICs are often referred as industry ecology. Industries, especially SMEs will have a future if sustainability is achieved with recourse to waste minimisation through concepts of environmentally balanced industrial complexing (EBIC) and green productivity (GP) alternatives.

## Supreme Court Monitoring Committee (SCMC)

- ☆ SCMC was set up in November, 2003 pursuant to Directions issued by the Supreme Court in its order and judgment dated, October 14, 2003 in Writ Petition No.657 of 1995.
- ☆ SCMC was to monitor the implementation of several directions contained in the order. So far, the committee has filed three quarterly report
- ☆ SCMC was holding regular meetings with CPCB, SPCBs/PCCs to review the status of compliance with the numerous directions
- ☆ SCMC had made visits of short durations to the states through its various sub-committees. So far, the sub-committee has visited Chattisgarh, Delhi, Gujarat, Kerala, Madhya Pradesh, Maharashtra and Tamil Nadu
- ☆ Sixteen states viz, Assan, Bihar, Chattisgarh, Gujarat, Haryana, Jharkhand, Karnataka, Kerala, Meghalaya, Mizoram, Nagaland, Punjab, Rajasthan, Tamil Nadu, Tripura and Uttaranchal and five UTs, viz, Chandigarh, Dadra and Nagar Haveli, Daman, Lakshadweep and Pondicherry have completed the inventory of hazardous waste generating units. CPCB is to randomly check the authenticity or otherwise of such inventories which are in progress
- ☆ Invenory of illegal hazardous waste dumpsite is in progress in Andhra Pradesh (42), Gujarat (72), Karnataka (18), Maharashtra (10 in MIDC area) Orissa (7), Punjab (14), Rajasthan (1) Tamil Nadu (2), Uttar Pradesh (10) and West Bengal (Not disclosed)
- ☆ SCMC called for commissioning of CETPs in Gujarat, Maharashtra, Andhar Pradesh, Tamil Nadu and Delhi at the earliest
- ☆ SCMC had taken steps to arrange for hazardous waste contaminated site remediation exercise. Contaminated sites in Tamil Nadu, MP Gujarat, Maharashtra and other States are to be remediated as soon as possible
- ☆ SCMC directed SPCBs/PCCs for closure of all unauthrised hazardous waste generating units in the first week of July, 2004. So far, 957 industries in Delhi and 195 industries in Kerala were issued specific directions of closure
- ☆ SCMC had requested SC to direct Hindustan Coca Cola, Binani Zinc, Kerala Mines and Minerals (all in Kerala) and State governments Gujarat and Madhya Pradesh to supply water to the communities affected by ground water pollution
- ☆ SCMC recommended to the Supreme Court for setting up of a "National Loss of Ecology Fund" for the entire country which will be financed partly from the fine levied against units found violating Hazardous Waste Rules and Supreme Court directions with matching contributions from the respective State Governments/Union Territories
- ☆ SCMC considered recommendation of a complete ban on import of all hazardous wastes
- ☆ Out of 30 sites chosen for setting up common facilities for hazardous waste disposal, common facilities have been/is being developed and made/being made functional only at 12 sites —

01 in Andhra Pradesh, 02 in Maharashtra 09 in Gujarat.The upcoming facilities are in the states of Karnataka (Debbaspet), West Bengal (Haldia), Punjab (Nimbua), Tamil Nadu (Gummidipoondi), Madhya Pradesh (Indore) and Maharashtra (Buti bori).

- ✰ Hon'ble Supreme Court has directed that immediate steps should be taken to expedite setting up/completion of facilities at the remaining sites. Although more than two years have passed since then, common hazardous waste disposal facilities are yet to be developed and made operational in several States/UTs.
- ✰ Due to non-availability of common facilities for treatment, storage and disposal of hazardous waste in State/UT, temporary storage of hazardous waste has been permitted within industry premises and that this has the potential to cause adverse impacts on the environment.
- ✰ As per the SCMC decision taken in the 52nd Conference of Chairpersons and Member Secretaries of State Pollution Control Boards (SPCBs) and Pollution Control Committees (PCCs) of Union Territories held at Mumbai on January 5, 2006.

I. All States/UTs should ensure setting up Common Facilities for secured land filling of Hazardous Waste, latest by June 2006

II. States/UTs which propose to utilize Common landfill facilities set up/being set up in neighboring States, should finalize formal agreements by June, 2006

III. States/UTs generating incinerable wastes exceeding 5000 TPA, should ensure setting up of Common Facility for incineration by December 2006

IV. States/UTs proposing to utilize Common Incineration facilities available in neighboring States should finalize formal agreements by December 2006.

## Inventorization and Quanta of Hazardous Wastes Generated in India

Inventorization of hazardous waste has been made through compilation of information in uniform format for inventory available at SPCBs/PCCs as per applications for authorization submitted by generators as well as information obtained by SPCBs/PCCs by engaging consultants. These inventories are cross-checked by CPCB. The SPCBs/PCCs are required to updated these inventories periodically and prepare inventory report regularly. Quanta of hazardous waste currently generated in India is around 82,61,640 TPA from 29,716 industries.

## Destination of Hazardous Waste

Disposal strategy for imported hazardous wastes are:

- ✰ Those which are totally banned and cannot be recycled/reused are to be re-exported to the country of origin at importers cost
- ✰ Rest are to be disposed off in an environmentally friendly manner in the TSDFs
- ✰ Those which can be recycled should be auctioned to authorized (registered) recycling and reprocessing units having environmentally sound technologies (ESTs) within the country.

Most of hazardous waste is sent to secure landfills, or incinerators specially designed to thermally destroy hazardous wastes mostly located in centralized treatment storage disposal facilities (TSDFs). The National Environmental Policies 2006" published by MOEF requires development and implementation of viable models of public-private partnerships for setting up and operating secure landfills, incinerators and other appropriate techniques for treatment and disposal of hazardous wastes and biomedical wastes, on payment by users (local community concerns are to be taken into account).

However, TSDFs exist only in some states and are being developed in many states. Therefore there are several instances of illegal hazardous waste dumping in the country.

## Goal

The goal of hazardous waste management is to channelize all hazardous wastes generated to TSDFs and to conserve valuable material and energy resources wherever techno-economically feasible and to promote the protection of health and the environment.

## Sustainable Hazardous Waste Management Strategy

Reducing (through prevention and minimization) the amount of waste the country produces is by far the most effective way to battle the flow of hazardous waste into the facility. Recycling is just one way to reduce wastes. Sustainable management of hazardous waste requires to incorporate the 4Rs Reduce, Reuse, Recycle and Recover in the strategy.

### Advantages

- ☆ Reducing the amount of waste produced can cut disposal costs.
- ☆ Reusing material previously considered as waste saves energy and raw materials.
- ☆ Recycling materials can minimize the burden of waste collection, transportation and disposal.
- ☆ Recovery can provide a cash flow easing the investment payback.

The waste hierarchy is shown in Figure 23.1.

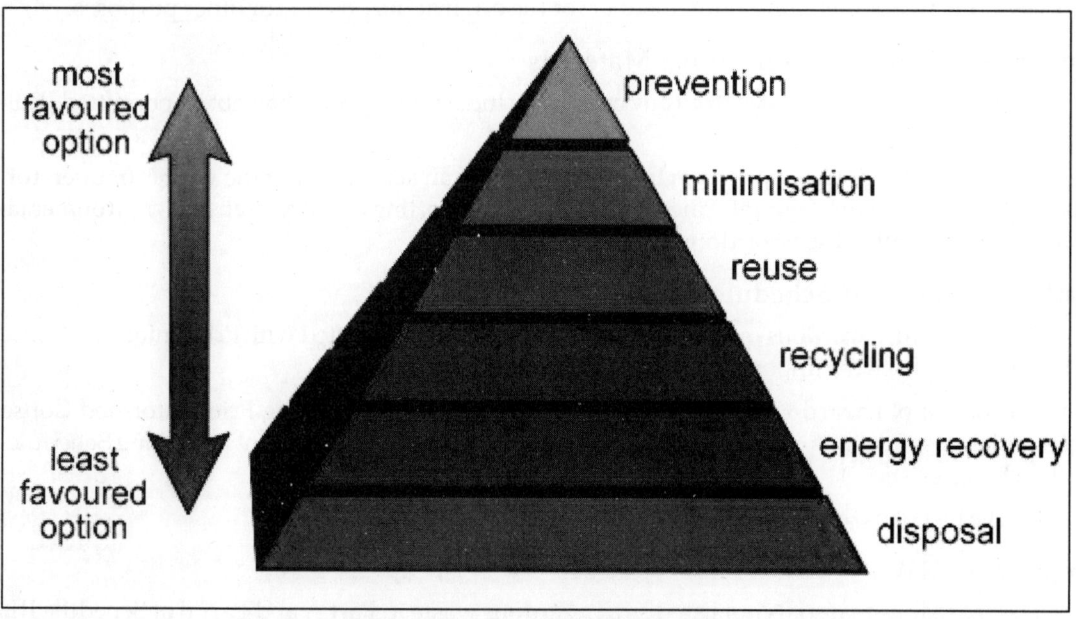


**Figure 23.1: The Waste Hierarchy**

## The Existing Disposal Strategy

Most of the hazardous waste is sent to secure landfills, or to incinerators specially designed to thermally destroy hazardous wastes. However, with more and more people producing more and more waste, landfills are filling up faster than we can find new sites for them. Hence landfilling cannot be a permanent solution for hazardous waste management.

Incinerators, retrofitted with energy recovery system(s) are sometimes used to thermally destroy hazardous wastes under controlled condition. They do reduce the stress on landfills, but they create other environmental problems. The ashes must be disposed of, either at a landfill, or, if they are toxic, at a hazardous waste facility.

Thermally destroying hazardous wastes also produces acid gases, carbon dioxide and toxic chemicals that must be treated with expensive air pollution control equipment to avoid contributing to acid rain, ozone depletion and adverse health impacts.

To minimize or eliminating formation of products of incomplete combustion, such as dioxins and furans, hazardous waste incinerators should be operated at 1200 C which makes its use quite expensive.

## Draft Rules, 2007: Recycling and Recovery

The Hazardous Materials (Management, Handling and Transboundary Movement) Rules, 2007, was circulated as draft on on September 28, 2007, with a view to review the rules published earlier, to enable the recovery and/or reuse of useful materials from hazardous materials generated from a process, thereby, reducing the hazardous wastes destined for final disposal and to ensure the environmentally sound management of all hazardous materials.

### Definition of Recycle

"Recycling" means reclamation and reprocessing of hazardous materials from a production process in an environmentally sound manner for the original purpose or for other purposes.

### Import and Export of Hazardous Materials

Import of hazardous wastes from any country to India for disposal shall not be permitted [Rule 5 (1)].

Export of hazardous wastes from India is allowed to an actual user of the wastes or operator of disposal facility with the Prior Informed Consent of the importing country to ensure environmentally sound management of the hazardous waste in question.

### Parts A, B and C of Schedule III

Part A: Lists of hazardous materials where import shall be permitted with Prior Informed Consent (PIC) in writing from the importing country.

Part B: List of hazardous materials for import/Export not requiring Prior Informed Consent unless otherwise contains or contaminated with List A materials to the extent exhibiting Schedule III Part C characteristics.

Part C: List of hazardous characteristics.

### Importable HW

The hazardous materials not listed as a hazardous waste in Part A and Part B of Schedule III, of these rules and which do not exhibit hazard characteristics specified in Part C of that Schedule shall be allowed for imports to actual users licensed by Central Pollution Control Board for its reuse, recycling or reprocessing under these rules.

## Not Importable HW

The import of hazardous materials listed in Part A and B of Schedule III which are containing or contaminated with hazardous materials and which exhibit one or more hazards characteristics specified in Part C of that Schedule and are not safe for recycling or reuse shall be considered as hazardous waste and shall not be permitted for imports in any circumstances.

The Schedule IV containing "list of hazardous materials for recycling, reprocessing and reuse" is only an indicative list of hazardous materials suitable for recycling, reprocessing and reuse and in case any other hazardous material is proposed to be imported, the Ministry of Environment and Forests shall determine whether the hazardous material is a waste, in that case the same shall not be allowed to be imported and in case the hazardous material is not a waste, it may be allowed in accordance with the provisions of these rules.

The import of hazardous materials listed in Schedule V shall not be permitted.

## No Objection Certificate

The Central Government grants 'No Objection Certificate' for the import and authenticate copy of Form 1 subject to the following:

- ☆ The hazardous material in question is covered under Basel Convention (Basel Convention" is the United Nations Environment Programme (UNEP) Convention on the Control of Transboundary Movement of Hazardous Wastes and their Disposal)
- ☆ The importer submits a valid Prior Informed Consent letter along with authenticated copy of Form 1 from the concerned authority of the exporting country
- ☆ Possesses a valid license from Central Pollution Control Board for undertaking the import in question.

In case of illegal import of hazardous wastes the importer shall re-export the waste in question within a period of thirty days from the date of its arrival into India.

The Central Government shall forward a copy of the 'No Objection Certificate' granted, to the Central Pollution Control Board, the concerned State Pollution Control Board or Pollution Control Committee of UT and the concerned Port and Customs authorities for ensuring compliance of the conditions of imports and safe handling of the hazardous materials.

## Confusing Issues

### Freely Importable Materials for Recycling

Hazardous materials listed under an 'Open General License' of the Directorate General of Foreign Trade shall only be imported by those units who have been granted license for recycling by the Central Pollution Control Board or any other authority or agency designated by the Central Government in the Ministry of Environment and Forests in accordance with the procedure laid down under rule 7. Are we not creating two authorities (MoEF and DGFT) for carrying out similar functioning?

"Safe for recycling" means any hazardous material containing hazardous constituents more than sixty percent of the recoverable materials and recycled using environmentally sound technologies. In that case what will happen to the remaining 39 or less percentage of residue which will be destined to go to the country's secured landfill site(s). Are we not importing a part of HW for disposal which is banned by Rule 5(1)?

## Hazardous Waste Handling in India

- ☆ Existing dump sites have been identified and assessment initiated
- ☆ Industries are to store hazardous waste till common facility for treatment, storage and disposal facility (TSDF) does not come up in their state. Where common facilities have come up in states, manifest system is practiced to transport hazardous waste for treatment and disposal
- ☆ TSDFs exist in some states and being developed in many states
- ☆ Industries are adopting recycle/reuse of hazardous wastes
- ☆ Attempts are made to minimize hazardous wastes and studies/trial runs are conducted for reuse or co-incineration in cement kiln
- ☆ Registration of recyclers/refiners of used oil/waste oil and non-ferrous metals containing hazardous waste for adopting EST is in practice
- ☆ Guidelines prepared for identification and characterization of hazardous wastes
- ☆ Industry-wise document prepared for some sectors.

## Ten FAQs on Hazardous Waste (HW) Management

***Q1. Do all industries generating HW require Authorization from the State Pollution Control Board?***

A1. Yes, as per the Supreme Court Order dated November 14, 2003.

***Q2. How to determine as to whether an Industry is generating HW or not?***

A2. The definition and the exclusion list provided in the Hazardous Waste (Management and Handling) Rules 1989 as amended on 2003 shall determine as to whether a waste is hazardous or otherwise.

***Q3. What does the asterisk provided in Schedule 1 of HM Rules stand for?***

A3. The waste is hazardous unless proved otherwise by the occupier based on sampling and analysis carried out by a laboratory recognized under the Act not to contain any of the constituents mentioned in Schedule 2 to the extent of concentration limits specified therein.

***Q4. Can the industry use its hazardous waste for recovery of valuable materials?***

A4. Yes, only if it uses environmentally sound technology(ies) for processing its hazardous waste(s) for such purposes.

***Q5. How to determine whether a waste is hazardous or otherwise?***

A5. All wastes mentioned in column (3) of Schedule-1 are hazardous wastes irrespective of concentration limits given in Schedule-2 except as otherwise indicated and Schedule-2 shall be applicable only for wastes or waste constituents not covered under column (3) of Schedule-1

***Q6. When immobilization and solidification/stabilization become prerequisite for HW disposal?***

A6. This is determined by analyzing leachate quality based on water leaching test. The S/S matrix should have a transversal strength (Vane testing) of > 25 KN/square metre, unconfined compression strength of >50 KN per square metre and axial deformation of < 20 per cent.

***Q7. What is the leaching test recommended for HW in India?***

A7. Water Leach Test (and not TCLP) is only recommended.

***Q8. Is TSDF is mandatory for all States and Union Territories in India?***

A8. Only those who are producing over 20,000 MTA require TSDF facility.Other states where the production is less may use the facility located in the nearby state(s) with prior permission to do so.

***Q9. What kind of "High volume low effect" wastes are exempted from HW category in India?***

A9. Fly ash, red mud, phosphogypsum and phosphochalk are exempted from HW categories.

***Q10. Are CRIT criteria included in HM Rules with quantifiable attributes?***

A10. No. Only a casual mention about corrosivity, reactivity, ignitability and toxicity are made. One may look into USEPA definitions for CRIT.

# Chapter 24

# Microwave Digestion Process for the Hydrolysis of Cellulosic Residues

**Rasika R. Harkare, Ancy E. Mathew, S.P.M. Prince William, A.N. Vaidya, J.K. Bhattacharyya and T. Chakrabarti**

**National Environmental Engineering Research Institute (NEERI), Nehru Marg, Nagpur – 440 020, Mh.**

## ABSTRACT

Utilization of cellulosic residues offers a great potential for the production of cellulosic ethanol. However, the major hindrance for this technology is the hydrolysis of cellulose into glucose. Acid hydrolysis by conventional methods is limited by long digestion period, high acid consumption, and contamination risk. The use of microwave digester is comparatively effective for the reasons of reduced acid consumption, no loss of reaction mixture due to evaporation, increased speed of digestion and higher temperatures achieved under pressure. In the present study an attempt was made to understand the effectiveness of both conventional and microwave digestion processes for the hydrolysis of different cellulosic waste materials (garden biomass, vegetable waste and collapsed carton box waste). The feedstocks were pretreated using varying (0 per cent to 10 per cent) concentrations of sulphuric acid in an autoclave for 5 to 30 minutes. The pretreated materials were then subjected to hydrolysis after necessary dilution using both conventional (under different temperatures 50°C to 175°C) and microwave digestion using different programmes (100 watts to 500 watts) and with different retention times ranging from 5 to 50 minutes. The initial and final cellulose concentrations and the glucose concentration of each of the hydrolysate were estimated using standard protocol.

The results obtained showed the increased glucose yield for all the three materials, *viz.*, Garden biomass, Collapsed carton box waste, and Vegetable waste; when treated in microwave digestion rather than conventional digestion processes. In microwave digestion, the effective hydrolysis was achieved at an acid concentration of 1 per cent with 19.20 per cent, 24.48 per cent and 25 per cent of glucose for Garden biomass, Collapsed carton box waste, and Vegetable waste respectively. From the present investigation it can be concluded that compared to conventional digestion process, the microwave digestion process is suitable for an efficient hydrolysis of cellulosic residues into glucose and to maximize the yield of cellulosic ethanol.

## Introduction

Segregation of organic material (recycled organics) from the general MSW stream and utilizing them for value added application is the recent advancement in bioprocessing of MSW. Bioprocessing has been a preferred method for effective treatment of biodegradable waste. It is a sustainable technology as it has provision for energy recovery. Compared to other processing technologies, such as physical and chemical treatments, biotreatment is very cost effective and viable also. There are options to bioprocess organic wastes into energy. Various raw materials have already been tested successfully for reaping energy to different forms. However, lack of advancement in process technology and maximization of product yield are the major stumbling blocks for the expansion of MSW bioprocessing technology (Lee, 1997).

Cellulose is one of the important organic materials that is present abundantly in solid waste. The commonly available cellulosic resources are paper, cardboard, wood, plant remnants, vegetable waste and other fibrous plant materials. Cellulose is the structural component of plant cells and therefore, any organic waste originated from plant source is expected to have maximum portion of cellulosic moiety.

In recent times, lot of emphases is given to the production of cellulosic ethanol. Given investment and technology development, it is possible to replace at least 25 per cent of the gasoline supply with cellulosic ethanol, requiring about 50 billion gallons. An additional benefit is that cellulosic ethanol made from biomass is likely to produce smaller amounts of greenhouse gases than corn ethanol, and far less greenhouse gases than the gasoline it will replace.

Cellulosic ethanol is an economically compatible energy source. However, the art of making cellulosic ethanol from variety of resource materials have not been fully explored. The existing technologies for cellulosic ethanol production are not enough to harvest the maximum from various resources. Among various factors determining the production of ethanol from cellulosic substrate, saccharification is very important. Saccharification is the process, which delivers essential monomeric sugars from fermentation. Saccharification in turn depends on the pretreatment and the nature and intensity of hydrolysis.

Many reports are available on the hydrolysis of cellulose. However, for a better hydrolysis and saccharification rate, it is essential to optimize hydrolysis conditions for various kinds of cellulosic resources. In this direction, works have already been carried out in NEERI for selected cellulosic waste materials such as garden biomass, paper waste and vegetable waste. Different kinds of pretreatment and hydrolysis conditions have also been worked out to establish the suitability of these cellulosic resources for enhanced recovery of cellulosic ethanol.

In the present investigation, an attempt has been made to increase the rate of saccharification for enhanced recovery of glucose using microwave digester.

## Materials and Methods

Three different materials *viz.*, vegetable waste, carton waste and garden biomass were collected from different sources. Vegetable waste was collected from professional caterers. The paper (empty carton boxes) and biomass wastes were collected from the stores of NEERI and botanical garden, respectively.

All the samples, except paper waste were sun dried at 60°C for 48 hours. The dried materials were then powdered and sieved through 4 mm sieve. The paper waste (empty carton boxes) was cut into pieces and was soaked in water for about 5 hours. The soaked material was removed from the water in the form of a wet pulp and was dried at 60°C for 48 hours. The initial cellulose content of each of these materials were estimated by gravimetric method.

8 gm of each sample was taken and slurried with 92 ml of varying concentration (0 per cent, 0.25 per cent, 0.50 per cent, 0.75 per cent and 1 per cent) of sulphuric acid. Each slurried sample was then kept inside the microwave digesters for 30 minutes. The filtrate was then collected and neutralized with $CaCO_3$. Representative samples were then estimated for glucose by dinitrosalicyclic acid (DNS) method.

## Results and Discussion

The results of the present investigation revealed that in all the cellulosic residues (paper waste, biomass and vegetable waste) tested, effective hydrolysis was observed at an acid concentration of 1 per cent excepting at 100 and 200 watts, where maximum hydrolysis was observed at 2 per cent. An increase in acid concentration above 1 per cent reduced the rate of hydrolysis, which was reflected as the decrease in glucose concentration, the product of hydrolysis. The effect of acid concentration on hydrolysis of cellulosic materials is very prominent in all the materials. However, in the case of biomass, the decrease in glucose concentration after optimum (1 per cent) acid concentration was not very sharp.

As regards watts of digestion power used, in case of paper waste and biomass, 300 watts of power showed better hydrolytic efficiency, followed by 400 and 500 watts. Hydrolysis of paper and biomass bellow 300watts was not productive.

In the case of vegetable waste, unlike paper and biomass waste, maximum hydrolysis was observed at 400 watts, followed by 500 and 300 watts. Here also hydrolysis at lower powers (100-200 watts) was not productive, which ultimately yielded lower quantities of glucose.

Different cellulosic residues at different reaction temperatures responded differently. At 100 watts power, hydrolysis was maximum in case of paper waste followed by vegetable waste and biomass. At 200 and 300 watts also, a more or less same trend was noticed. Whereas, at 400 and 500 watts of power, maximum hydrolysis was observed in vegetable waste followed by paper waste and biomass. These results indicate the relative resistance of various cellulosic residues for different watts of power. One more interesting observation is that, among the different watts of power tested, only at and above 300 watts a clear-cut concentration dependant change in the rate of hydrolysis was observed.

## Conclusion(s)

In the present investigation all the three cellulosic residues selected for hydrolysis showed promising results for their use as low cost feed stocks for cellulosic ethanol. However, successful conversion of the materials into fuel ethanol rest with the rate of hydrolysis, which ultimately yield fermentable sugars. Among the three material tested, both paper and biomass waste responded similarly that maximum hydrolysis was done at 300 watts followed by 400 and 500 watt of power. However, in case of vegetable waste maximum hydrolysis occurred only at 400 watts, followed by 500 and 300 watts. This indicated the relative strength of vegetable waste over paper and garden biomass. Vegetable waste is relatively rich in lignin and other such molecules that may offer relatively higher hydrolytic resistance.

# Chapter 25

# Towards Green and Clean Environment in Dairy Industry: A Pragmatic Analysis

**A.B. More[1], S.N. Kaul[2] and V.A. Bugade[3]**
**[1]Assistant Professor, Department of Civil Engineering, JSPM's Imperial College of Engineering and Research, Wagholi, Pune**
**[2]Principal, MAEER's MIT College of Engineering, Pune**
**[3]Director, JSPM's, Wagholi Campus, Wagholi, Pune**

## ABSTRACT

We have entered in the 21st century with the gifts of urbanization and industrialization from the early century. All such developmental processes have led to generation of waste material and have become environmental sensitive issues for the industry and the governments. "Zero Waste Technique" is one of the very promising technology recently implemented to fulfill the aim of the minimization of waste generated in industry ideally to have greener and cleaner environment. Integrated approach helps while considering all types of waste minimization up to zero level. Dairy Industry represents one of the important agro-based industries that produce milk and milk products for human consumption. The industry involves the activities like collection, storage, process, packaging, distribution, etc. In all these activities one may expect the generation of waste which in turn having the impact on the environment and the human health and finally on quality of life. In this chapter "Pune Zillha Sahakari Dudh Utpadak Sangh Maryadit", Pune (Katraj Dairy) was selected for the study of application of the integrated approach using "Zero Waste Technique". This represents one of the pioneer dairy industries in Pune run on co-operative basis since 1960. The capacity of milk processing is 2 lakh lit/day and the quantity of raw milk is 1 lakh lit/day. Water used for the different process will be 2 lakhs lit/day and the wastewater

* Corresponding Author: E-mail: 1: prashmorye@yahoo.co.in; 2: principal_mitcoe@mitpune.com

generated is around 1.5 lakhs lit/day. Dairy industries are the source of water pollution, air pollution, noise pollution and also generate solid waste. The pollution level can be reduced at the source itself by using good housekeeping and by applying 3 R principals (Reduce, Recycle and Reuse). The application of the integrated approach to dairy industry using "zero waste technique" results in the economical growth of the industry.

***Keywords:*** *Zero waste technique, Integrated approach, Water pollution, Air pollution, Noise pollution, Solid waste, Green and clean environment.*

## Introduction

Zero Waste Technique means "designing and managing products and processes to reduce the volume and toxicity of waste and materials, conserve and recover all resources, and not burn or bury them". The integrated approach used along with zero waste techniques and applied for different environments like air, water, noise, and solid waste to reach at cleaner environment. Cleaner Production refers to a mentality of how goods and services are produced with the minimum environmental impact under present technological and economic limits. *It is the continuous application of an integrated preventive environmental strategy to processes, products, and services to increase overall efficiency, and reduce risks to humans and the environment. Cleaner production can be applied to the processes used in any industry, to products themselves and to various services provided in society.*

Dairy industry represents one of such important agro based industry to produce milk and milk products for human consumption. It involves the activities like collection, storage, process, packaging and distribution. At all these steps one may except the generation of waste that is creating impact on the environment and the health and finally on quality of life. "Pune Zillha Sahakari Dudh Utpadak Sangh Maryadit", Pune (Katraj Dairy), located on the Pune-Satara road is selected to study the application of the integrated approach using "Zero Waste Technique". This dairy represents one of the pioneer dairy industries in Pune run on co-operative basis. The capacity of milk processing is 2 lakhs lit/day and the quantity of raw milk is 1 lakh lit/day. Water used for the different process will be 2 lakhs lit/day and the quantity of wastewater generation is about 1.5 lakhs lit/day.

## Water Environment

Water which is required for the different processes in dairy industry mainly requires for the sanitary purpose like washing of utensils and floor washing, steam generation in boiler house for heating of milk and cooling of water in chilling house. From the data collected, the quantity of raw water used is about 2.1 lakhs lit/day, about 25 per cent of it contributes for can and crate washing. All the sections where after water use it turns to the wastewater about 1.6 lakhs lit/day with the following source characteristics (Table 25.1).

Dairy purchase daily 2 lakhs liters of water from the Pune Municipal Corporation and conveyed by tanker having the capacity of 10,000 lit. The cost of each tanker is about Rs. 400. Therefore the total daily expenditure is Rs 8000. The consumption of water in dairy is more due to the poor housekeeping and less awareness among workers about the water use.The volume reduction of the wastewater should be carried out at source itself by the proper usage of water and it will reduce the generation of wastewater by using the 3Rs principal like Reuse, Recycle, Reduce. The proper usage of water quantity and strength reduction of wastewater should be carried so as to reduce the load on Effluent Treatment Plant (ETP).

**Table 25.1: Wastewater Characteristics at Source**

| Sl.No. | Sample | PH | Chlorides mg/lit | COD mg/lit | BOD mg/lit | TDS mg/lit | TSS mg/lit | Alkalinity mg/lit | O&G Mg/lit | Sulphates mg/lit |
|---|---|---|---|---|---|---|---|---|---|---|
| 01 | 02 | 03 | 04 | 05 | 06 | 07 | 08 | 09 | 10 | 11 |
| 1. | Can Washing | 5.30 | 101.21 | 2988.3 | 2100 | 1389 | 455 | 385.6 | 405 | 68.3 |
| 2. | Crate Washing | 6.20 | 74.55 | 1445 | 936.6 | 1269.3 | 405 | 292 | 372.34 | 70.3 |
| 3. | Processing Unit | 12.6 | 137.74 | 2883.3 | 1250 | 1072.7 | 211.3 | 679 | 430.34 | 71 |
| 4. | Ghee House | 12.4 | 165.2 | 2585.6 | 1183.3 | 4720.3 | 110.6 | 679.3 | 538 | 71.6 |

**Table 25.2: Generation of the Wastewater**

| Sl.No. | Dairy Process/Section | Quantity of Water Use Lit. | Quantity of Wastewater Generated Lit. |
|---|---|---|---|
| 1 | 2 | 3 | 4 |
| 1. | Can | 20,000 | 17,000 |
| 2. | Weighing Bowl/Dump tank/Mechanical filters | 2,000 | 1,500 |
| 3. | Pipes and values | 20,000 | 1,71,000 |
| 4. | Parturition | 7,000 | 6,000 |
| 5. | Homogenization | 8,000 | 7,000 |
| 6. | Silo | 5,000 | 4,000 |
| 7. | Storage tank | 5,000 | 4,000 |
| 8. | Packaging machine | 15,000 | 13,000 |
| 9. | Crate washing | 30,000 | 23,000 |
| 10. | Ghee, Butter, Shrikhand unit | 2,000 | 1,500 |
| 11. | Boilers | 20,000 | 15,000 |
| 12. | Cold storage | 5,000 | 4,000 |
| 13. | Floor washing | 15,000 | 14,000 |
| 14. | Chilling plant | 15,000 | 12,000 |
| 15. | Tanker washing | 7000 | 5,000 |
| 16. | Domestic waste & Lab use | 25,000 | 20,000 |
| | **TOTAL** | **2,10,000** | **1,60,000** |

In ETP flow of the wastewater varies by time to time. The flow of wastewater is maximum in the morning hours *i.e.* 8.00 A.M. to 1.00 P.M., where as in the afternoon flow is less and in the evening the flow of wastewater is again increases. The wastewater contains oil and grease because milk contains fats (6.50 per cent) and proteins (4.30 per cent). It should be removed in the oil and grease traps after that the water goes to the holding tank where the wastewater is hold for some time and send it into the aeration tank where air is supplied and the micro-organisms converts the organic matter into the simple form and the separation of the sludge and effluent is done in the secondary clarifier. Then sludge is sending to the drying beds after that it is used as manure. By giving the proper treatment

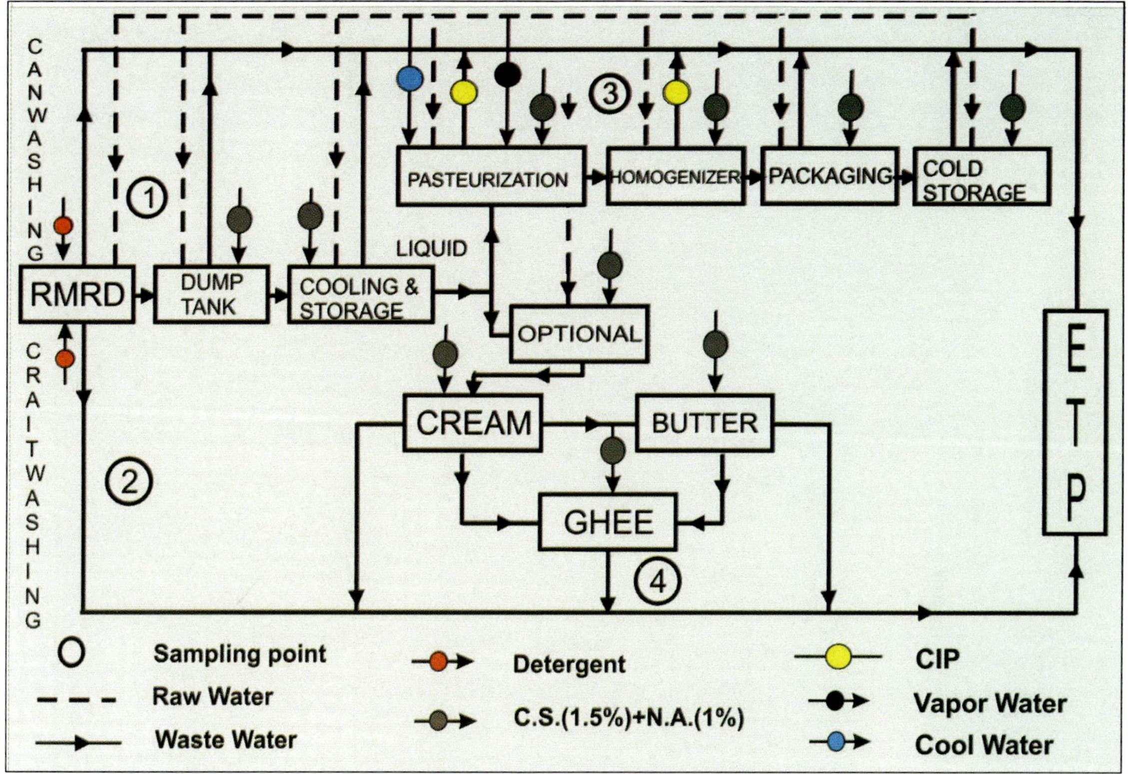


**Figure 25.1: Water Use and Wastewater Generation**

through efficient and effective utilization of existing treatment facilities, the effluent may be reused for gardening or floor washing.

For green and clean environment, the minimization of the wastewater at source should be carried out as follows.

1. The crate washing is carried out by batch process instead of continuous process. The batch process will save approximately 30 per cent of generation of the wastewater *i.e.* about 9000liters/day.
2. Wastewater generation in the CIP and Ghee House will highly alkaline having the pH 12, so it is necessary to reduce the pH and it should be carried out by providing a separate tank for the collection of wastewater from this unit and neutralized by the addition of acid as HCl.
3. Proper housekeeping should be done. It should be carried out by controlling the leakages in the process, there are couple of leakages *i.e.* Pasteurization unit and cream separator will reduce approximately 4000liter/day.
4. The water used for the cooling of the ice traps (for removing of ice) should be reused again. It will reduce approximately 3000liters/day of quantity of water use.
5. Use of pressure guns instead of the hosepipe. The hosepipe is mainly used for the tanker washing. The water used for the tanker washing is given in a table (Column No.5) from that table; the water used is about 7000liters/day. By using the pressure guns it should be

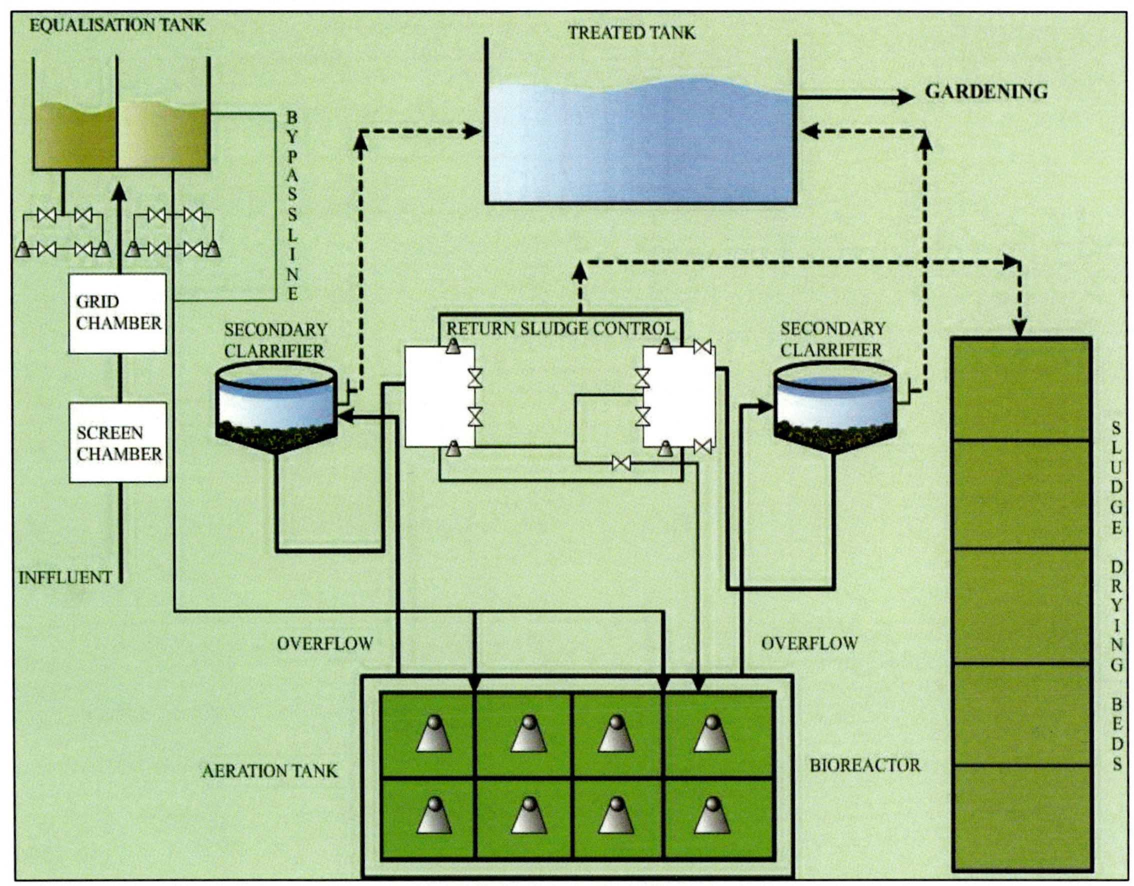


**Figure 25.2: ETP Layout**

minimize approximately 30 per cent of generation of the wastewater *i.e.* about 2100liters/ day.

6. The low concentrated water should be reused for floor washing.
7. The total quantity of water used for steam generation in the boiler is given in a Table 25.2 (Sr. No. 5) *i.e.* 15,000liters/day. The steam generation in boiler house used in pasteurization unit for heating of milk. During the transportation of the steams, some amounts of steam get condensed and removed from the steam traps automatically. After use of steams it should be condensed and reused again *i.e.* about 10,000liters/day.

Approximately minimum saving of the water is about 28,000lit/day will give the around Rs 2100/day. It will also reduce the load on the ETP up to 1.3 lakhs lit/day.

## Noise Environment

$L_{eq}$ is the concept used for the measurement of noise level. The time selected for monitoring the noise level is in between 7A.M. to 10 P.M. for three days at each source and the $L_{eq}$ noise levels are calculated at different sources like RMRD (65-97dB), packaging section (83-98dB), processing Section

3000
2500
2000
1500
1000
500
0
Values
pH
TSS
TDS
Chlor
COD
BOD
O & G
Sulph
Parameters
Influent
Eq.tank (inlet)
Aeration tank(inlet)
SST(inlet)
Treated tank

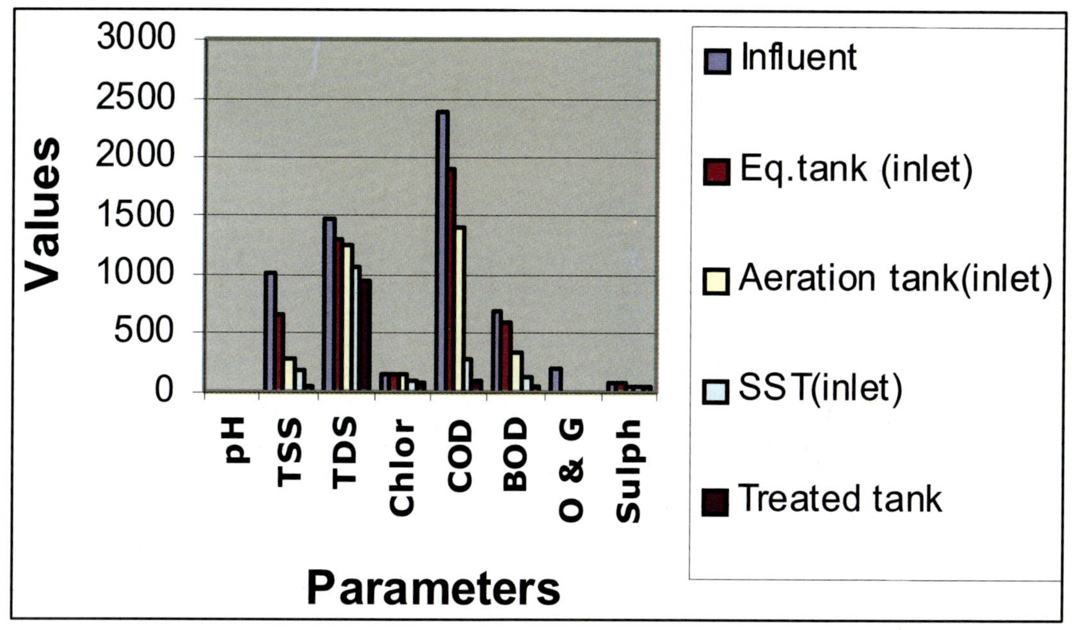

**Figure 25.3: Wastewater Characteristics in ETP**

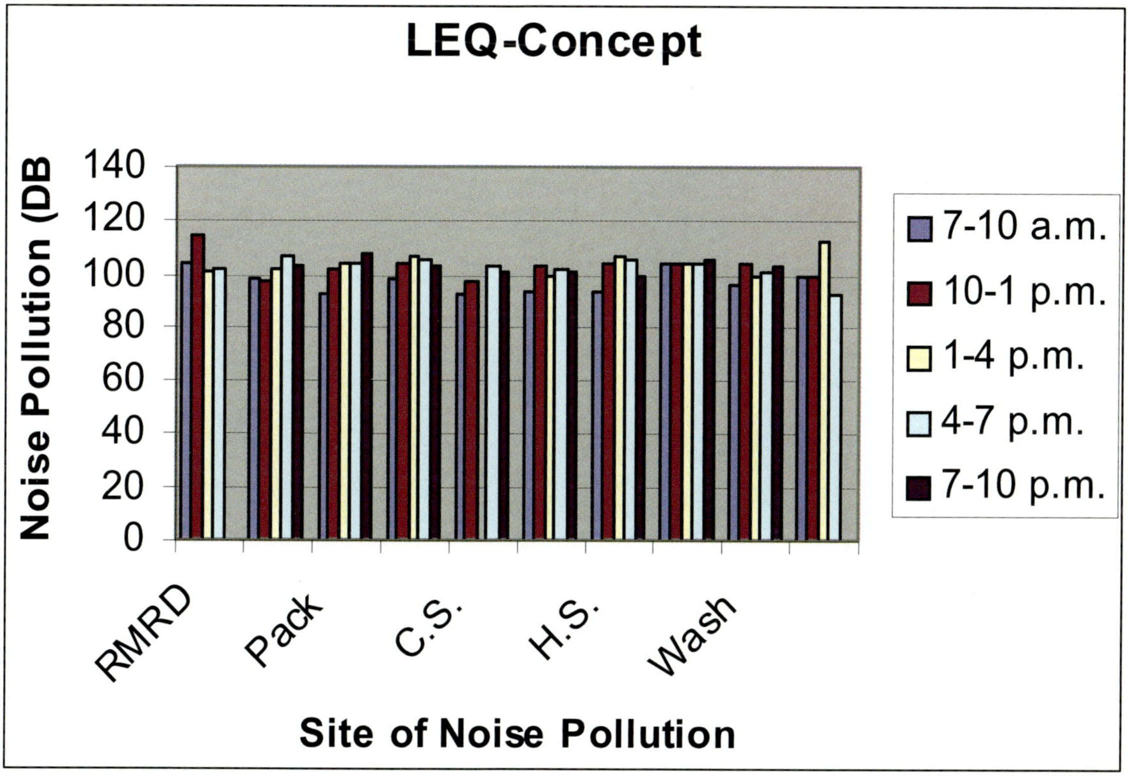

LEQ-Concept
140
120
100
80
60
40
20
0
Noise Pollution (DB
RMRD
Pack
C.S.
H.S.
Wash
Site of Noise Pollution
7-10 a.m.
10-1 p.m.
1-4 p.m.
4-7 p.m.
7-10 p.m.

**Figure 25.4: Noise Level at Different Sections**

Figure 25.5: Generation of Wastewater at Can Washing

Figure 25.6: Generation of Wastewater at Crate Washing

(67-92dB), boiler house (85-98.5dB), chilling house (82-94dB), homogenizer (87-91dB), generator (98-107dB), washing of crate (82-89dB) and ghee house (78-86dB). It is observed that the noise levels are moderately high in generator campus, the possibilities of reduction of noise level by providing the body cover generally called canape, which reduces the noise level up to 30-40 per cent. Generator should be placed separately instead of current location in maintenance section in between the boiler and the chilling house. Noise level in chilling section (88dB), is mainly due to the compressor, used for the packaging of milk. Only general ways helps to keep control on noise pollution. In RMRD section, the main cause of the noise pollution is when the can is empty. After washing, it is thrown on the floor. The peak time of the noise pollution in RMRD section is in the morning at 8.30 A.M to 1.00 P.M. and it can be minimized by putting rubber sheet on floor. Packaging machines may also cause the noise pollution because of the improper handling and poor maintenance of machine. All the noise levels are beyond the limit or standards for industry during day and night time.

Result data got by asking questions to the workers in dairy are summarized, in that about 75 per cent of workers are educated up to SSC and 85 per cent are not aware about the noise pollution. They all are having 8hrs working duty and no precaution is taken by them to protect themselves from noise pollution. Concerned authorities have not taken any preventive measures taken about the issue. Majority of respondents had indicated generator as a major source of noise.

The effects of noise pollution are observed on the health of the workers because of symptoms like feeling ringing tone in the ear, suffering from headache, irritation, loss of efficiency, etc and also reduce the efficiency of the worker. These effects should be minimized at the source in the following general ways.

1. Use of personal protective equipments (PPE) like Ear Plugs and Ear Muffs.
2. Proper maintenance of the machinery should be carried out at regular intervals.
3. Workers should be trained to apply the recent advances in noise pollution control technologies.

## Solid Waste Management

Solid waste mainly contains plastic, papers and cardboard etc. Plastic is used for packaging of milk and milk products. Generation of plastic waste is about 36.26 kg/day. All the plastic is collected in the morning is then brought into a separate room, which is near the chilling house and it is sold in the market by tendering process. The plastic waste is sold at the rate of Rs 12/- per kg. Sources of these plastic wastes are the packaging section and the sale outlet (three in number). The weight of empty pouch is about 5gms. 1kg of plastic waste may contain about 200 empty pouches. Single pouch is sold at the rate of 6 paisa. And the manufacturing cost of one new pouch is about 11 paisa. Therefore per kg new plastic bag cost is about Rs 22. The causes of the plastic waste in packaging section are due to poor housekeeping and improper maintenance of packaging machine. This produces high generation of solid waste. In packaging section 12-15 kg of plastic waste is generated. By proper maintenance and well operation of packaging machine, the generation of plastic waste from the packaging is reduced by 50 per cent. Economically it will save more then Rs 130/- per day.

Another source of the solid waste is paper and cardboard with generation of about 16.75 kg/day. The paper and the cardboard waste sold at the rate of Rs 4.50/kg. The cardboard also comes with raw packaging plastic sheets, which find its way to solid waste. Reusing the photocopy pages from printer should minimize the paper waste. Few quantity of residential and gardening waste is generated and it can be minimized as follows.

1. Applying proper disposal methods for agriculture and gardening waste.
2. Dustbin should be provided in the residential area.

## Air Environment

Ambient air monitoring was carried out in the campus of Katraj dairy with the help of HVS (High Volume Sampler). The main source of air pollution is the boiler (fire tube) house, which is used for the generation of steam for the heating of the milk in pasteurization process. There are two boilers in stand by systems, the capacity of the each boiler is 3 tons/hour. For boiler operation furnace oil is used as lubricant. To maintain 12kg vapor generation, automatic start up mechanisms is provided. The internal diameter is about 760 mm, height of stack is about 30 meters (75 feet) and stack gas temperature is about 150-200°C. The raw water required for the steam generation is about 18,000 lit/day. The three receptor points selected for the air quality monitoring in the campus and are near to: spot no. (1) RMRD, (2) ETP, (3) Chilling House.

Intermediate type of sampling method is selected with 3 intervals of 4 hours each as 6.00-10.00 A.M., 10.00-2.00 P.M., 2.00-6.00 P.M. at each location for two days. The concentration of air pollutants is given in Table 25.3. The concentration of NOx (8.04µg/m$^3$) SOx (3.21µg/m$^3$) and PM (49.24µg/m$^3$) is within the limit or standards but from the table, it is observed that suspended particulate matter is beyond the limit in residential, rural and other area and sensitive area which is having the limit of 140µg/m$^3$ and 70µg/m$^3$ respectively.

**Table 25.3: Concentration of Air Pollutants**

| *Sl.No.* | *Spot No.* | *Duration* | *SOx* µg/m$^3$ | *NOx* µg/m$^3$ | *PM$_{10}$* µg/m$^3$ | *Suspended Particles* µg/m$^3$ |
|---|---|---|---|---|---|---|
| *01* | *02* | *03* | *04* | *05* | *06* | *07* |
| 1 | 1 | 6.00am-10.00am | 3.21 | 6.56 | 30.68 | 219.84 |
| | | 10.00am-2.00pm | 2.08 | 6.02 | 46.21 | 140.52 |
| | | 2.00pm-6.00pm | 3.11 | 5.88 | 32.38 | 159.65 |
| 2 | 2 | 6.00am-10.00am | 2.85 | 2.49 | 18.37 | 107.95 |
| | | 10.00am-2.00pm | 1.7 | 3.77 | 26.13 | 104.92 |
| | | 2.00pm-6.00pm | 2.99 | 4.46 | 14.57 | 135.22 |
| 3 | 3 | 6.00am-10.00am | 1.33 | 4.63 | 24.80 | 122.15 |
| | | 10.00am-2.00pm | 1.79 | 5.58 | 34.46 | 121.97 |
| | | 2.00pm-6.00pm | 1.51 | 6.68 | 31.24 | 103.02 |

Following ways should minimize the concentration of suspended particulate and help to maintain the levels of NOx and SOx:

1. Usage of good quality of raw material during boiler activity.
2. Primary treatment units like settling chamber may be provided before emission of air pollutants in the atmosphere.

## Conclusion

The quantity of raw water required is 2 lakhs liters per day for various processes in dairy and the wastewater generation is about 1.5 lakhs liters per day. The volume of water used and wastewater should be reduced at the source itself by using good housekeeping and 3 Rs principals like reuse, recycle and reduce. Strength reduction of wastewater should be carried out so as to reduce the load on ETP. Approximately minimum saving of water if it is 28500 lit./day then the saving of cost is about Rs 2100/day and reduces the wastewater generation upto 1.3 lakhs lit/day ultimately less load on ETP and there fore the maintenance and operating cost of ETP reduces.

The main source of noisc pollution is generator section (107dB) and other sources like the packaging section, boiler house, chilling house, RMRD having range of 85-95dB. It will affect the health of the workers and efficiency of the working. Noise pollution should be reduced at source itself. The use of canopy (body cover) to generator will reduce 30-40 per cent of noise and the effect of the noise pollution should be minimized by applying the personal protective equipments (PPE) like earplugs and earmuffs. So as to minimize the effect on health of the workers and efficiency should be increased.

The solid waste generates in the form of plastic (36.23kg/day) paper and cardboard (16.75kg/day) collected waste is stored and sold by tendering processes and sold for recycling. Possibilities of controlling generation of waste plastic bags were checked and it is observed that around 40 per cent reduction in waste generation which saves cost around Rs 70–80/day of plastic waste.

The main source of air pollution is the boiler having the capacity of 3 tons per day. The stack height is about 30meters and the internal diameter is 760mm. The concentration of NOx (8.04$\mu g/m^3$), SOx (3.21$\mu g/m^3$) and PM (49.24$\mu g/m^3$) having within the limit. Whereas the SPM (255.98$\mu g/m^3$) beyond the standards of national ambient air pollution standards and should be minimized by selecting and implementing proper technologies.

The applications of the integrated approach using zero waste technique reduce the volume of waste and material to conserve and recover all resources and result in the overall economical growth of the industry. Reducing the generation of waste at the source itself can be achieved by best practices of waste minimization technique, which encourages greener and cleaner environment in and around industry. After exhausting the source reduction opportunities, attempts should be made to recycle the waste within the dairy. Finally, modification or reformulation of products so as to manufacture it with least waste generation should be considered.

## References

1. W. Wesly Eckenfelder, Jr, Industrial Water Pollution Control, McGraw Hill- Series in Water Resources and Environmental Engineering.
2. Mackenzie L. Davis, David A. Cornwell, Introduction to Environmental Engineering, McGraw Hill- Series in Water Resources and Environmental Engineering.
3. Metcalf and Eddy, INC, Wastewater Engineering, Tata McGraw- Hill.
4. Rao M. N., A. K. Dutta, Wastewater Treatment, Oxford and IBH publishing Co. Pvt. Ltd.
5. J. Glynn Henry, Gary W. Heinke, Environmental Science and Engineering, Published by Pearson Education (Singapore) pvt. Ltd., Indian Branch, Delhi, India.
6. M. N. Rao, Air Pollution, Tata McGraw-Hill Publishing Company Limited, New Delhi.

7. S. Eswaramoorthi, K. Dhanapal and J. Karpagam, Zero Discharge–Treatment Options for Textile Dye Effluent:A Case Study at Manickapurampudur Common Effluent Treatment Plant, Tirupur, Tamil Nadu, India.

8. Reports Published by Maharashtra Pollution Control Board.

9. http://www.targetzerocanada.org, http://www.zerowaste.co.nz, http://www.zeria.at.

# Chapter 26

# Electron Waveguide: A Revolutionary Device for Green Nanotechnology

**Shalini Garg**
**MAEER's MIT College of Engineering, Pune, M.S.**

## ABSTRACT

In the rapidly developing field of nanotechnology- nanodevices are becoming an increasingly key component in everything from drug delivery to improving or even replacing the microprocessors in computers or optical switches in telecommunication networks. The environment research focuses on ecosystem restoration and applied ecology. It relates to the use of nanodevices in the study of cellular and molecular interaction of nanoparticles. This chapter provides a review of a basic electron waveguide which may further be used in designing nanodevices such as nanofilters, nanoswitches, nano directional couplers etc. for their application for green and clean environment.

***Keywords:*** *Nanodevices, Electron waveguide, Nanofilter, Green nanotechnology.*

## Introduction

Nanotechnology is used to address several environmental issues like generating clean power, reducing existing power consumption, providing drinkable water, cleaning contaminated land, reducing harmful emissions, environmental monitors and enabling long-life portable power. By introducing electron waveguides, the author suggests that goods could be produced and distributed locally, limiting the amount of worldwide transport that takes place annually. This in turn will reduce pollution and create a sustainable environment.

In nanoscale structured devices, one or more device dimensions are smaller than mean free path of the carriers and so they have ballistic behaviour along these dimensions. Depending upon the number of dimensions, these devices are classified as quantum wells, quantum wires and quantum dots. Electron waveguide is a thin wire to which the electrons are confined by such a potential that the electrons can move freely only in one dimension. The most important characteristic of electron waveguides is that their width be comparable to the de Broglie wavelength in the material. By reducing the device size, the packing density and speed has increased by several orders of magnitude over the years. This development is likely to continue until hampered by physical limits. Electron waveguides are currently in the research and development stage. The motion of electrons is analyzed using quantum mechanical plane waves. These quantum mechanical waves do not experience any elastic or in-elastic scattering events and in this case electron propagation is considered to be ballistic. Ballistic electrons account for more than half of the current in small devices. The steady state motion of the electron wave is governed by the well known schrodinger equation

$$\left(-\frac{\hbar^2}{2m^*}\nabla^2 + U(\vec{r})\right)\psi(\vec{r},t) = i\hbar\frac{\partial}{\partial t}\psi(\vec{r},t) \tag{1}$$

Here, $\psi(\vec{r},t)$ is the electron wave function, $U(\vec{r})$ is the potential energy of the electron at coordinate $\vec{r}$, $\hbar$ is the Planck's constant divided by $2\pi$ and $m^*$ is the electron mass. The potential, $U(\vec{r})$ is a function of only the transverse coordinates (x and y), while it is constant along the length of the waveguide(z). A semiconductor quantum well can be designed to confine these electron waves and act as electron waveguides.

## Electron Wave Optics

Electron wave optics gives the necessary base for designing the MQW electron waveguide. Quantum mechanical electron waves in semiconductors and electromagnetic optical waves in dielectrics exhibit transmission, reflection, and propagation characteristics that are analogous.

## Reflection and Refraction of Electron Waves at a Potential Energy Boundary

There are various analogies between optical plane waves in electromagnetics and electron plane waves in quantum mechanics. The wave vector for the optical wave is given by

$$k = \frac{2\pi n}{\lambda} \tag{2}$$

*where,* $\lambda$ is the wavelength of light in the medium of refractive index $n$. The corresponding value of wave vector for electron plane waves is given by

$$k = \left[2m^*(E-V)\right]^{1/2} / \hbar \tag{3}$$

*where,* $m^*$ is the electron effective mass, $E$ is the total electron energy, $V$ is the potential energy of the electron and $\hbar$ is the Planck's constant divided by $2\pi$. The phase effects for both types of waves can be described using the corresponding wave vector. These effects occur in devices based on path differences, wave interference etc. For phase effect, the electron wave phase refractive index, $n_e$ (phase), is proportional to the square root of the product of the effective mass and the kinetic energy. That is,

$$n_e\,(\text{phase})\ \alpha \left[m^*(E-V)\right]^{1/2} \tag{4}$$

Amplitude effects such as transmittivity and reflectivity may be described in terms of the wave function amplitude for an electron wave and in terms of the electric field amplitude (for TE polarization) or magnetic field amplitude (for TM polarization) for an optical wave incident upon a boundary. Continuity of the wave function across a potential energy boundary is analogous to the continuity of the tangential component of the electric field (for TE polarization) or of the magnetic field (for TM polarization) across a boundary between dielectrics. Similarly, conservation of electron current normal to the potential energy boundary is analogous to conservation of power flow normal to a boundary between dielectrics. The electron wave and optical wave reflectivities and transmittivities are made equivalent by introducing an analogous characteristic impedance or index of refraction for electron waves. Thus, the electron wave amplitude refractive index $n_e$ (amplitude) is proportional to the square root of the ratio of the kinetic energy to the effective mass. That is,

$$n_e\,(\text{amplitude})\ \alpha \left[(E-V)/m^*\right]^{1/2} \tag{5}$$

In the transmittivity and reflectivity expressions, only dimensionless ratios of the electron wave amplitude refractive index appear. Both types of electron wave refractive indices exhibit normal dispersion. That is, they increase with decreasing wavelength.

The ballistic electrons are quantum-mechanical de Broglie waves and can be reflected, refracted, diffracted, guided and interfered like optical waves in dielectrics. An asymmetric potential energy well and it's depiction as a slab waveguide are shown in Figure 26.1.

Using standard waveguide terminology, the three regions are denoted substrate(s), film(f) and cover(c), and the direction perpendicular to the waveguide surfaces is x. The electron potential energy at the bottom of the quantum well (film) is $V_f$. The potential energy barrier heights associated with the substrate and cover are $V_s$ and $V_c$, respectively. This quantum well acts as an asymmetric slab waveguide. The thickness of the waveguide is d. The direction of guided mode propagation is z. The angle of incidence of the two plane wave components that constitute the guided wave is the zig-zag angle θ. For a symmetric waveguide, $V_c = V_s$.

The magnitude of the electron wave vector based on equation(3) in any of the three regions may be expressed as

$$k = [2m^*(E-V)]^{\frac{1}{2}}/\hbar \tag{6}$$

An electron of total energy E is incident upon the film-cover interface. The kinetic energy of the incident electron is

$$E - V_f = \frac{\hbar^2}{2m_f^*}\left(k_{fx}^2 + k_{fy}^2\right) = \frac{\hbar^2}{2m_f^*}k_f^2 \tag{7}$$

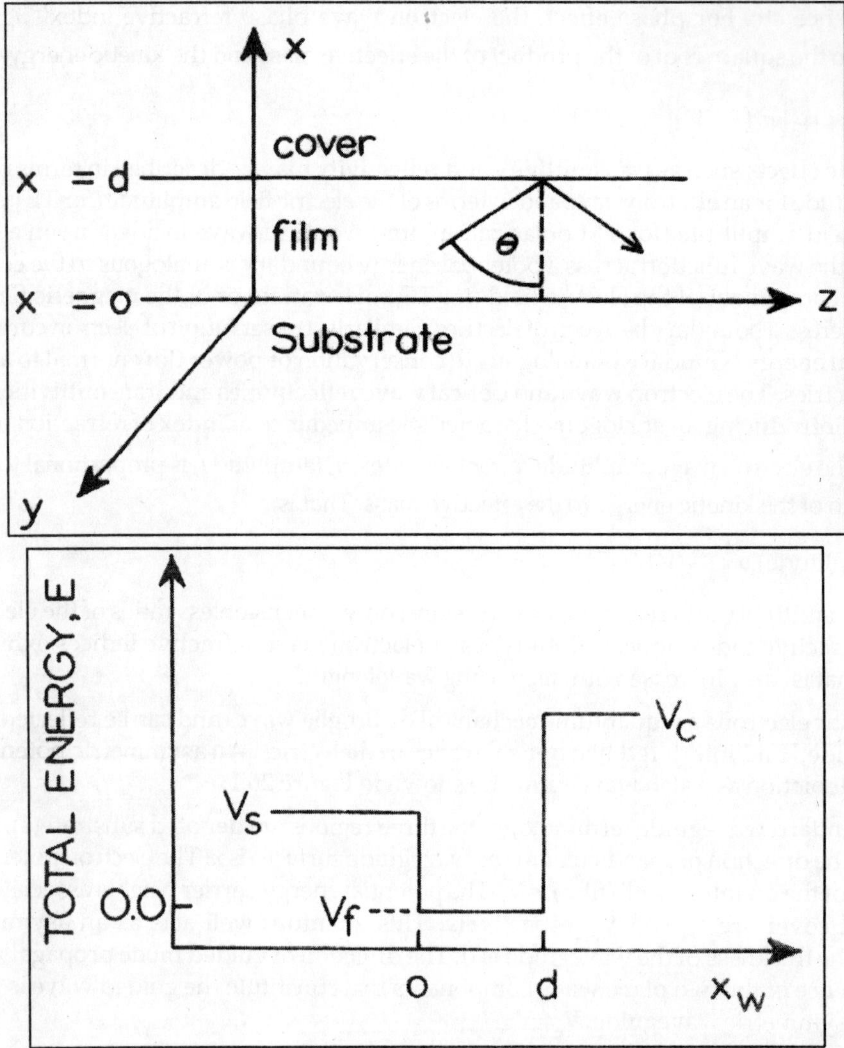


**Figure 26.1: An Asymmetric Potential Energy well and it's Depiction as a Slab Waveguide**

*where,* $m_f^*$ is the electron effective mass in the film region. $k_{fx}$ and $k_{fy}$ are the components of the incident electron wave vector normal and parallel to the boundary, respectively and $k_f$ is the magnitude of the incident electron wave vector which is given by

$$k_f = [2m_f^*(E - V_f)]^{\frac{1}{2}} / \hbar \tag{8}$$

The kinetic energy of the transmitted electron is

$$E - V_c = \frac{\hbar^2}{2m_c^*}\left(k_{cx}^2 + k_{cy}^2\right) = \frac{\hbar^2}{2m_c^*} k_c^2 \tag{9}$$

*where,* $k_{cx}$ and $k_{cy}$ are the components of the transmitted electron wave vector normal and parallel to the boundary, respectively, and $k_c$ is the magnitude of the transmitted wave vector in the cover region which is given by

$$k_c = [2m_c^*(E - V_c)]^{\frac{1}{2}} / \hbar \tag{10}$$

The kinetic energy of the reflected electron is

$$E - V_f = \frac{\hbar^2}{2m_f^*}\left(k_{fx}^2 + k_{fy}^2\right) = \frac{\hbar^2}{2m_f^*}k_f^2 \tag{11}$$

*where,* $k_f$ is the magnitude of the reflected wave vector.

The phase of the transmitted and reflected electron waves along the boundary must be identical to that of the incident electron wave. This phase-matching requirement means that the component of the wave vector parallel to the boundary must be same immediately before and after reflection and refraction.

$$k_c \sin\theta_c = k_f \sin\theta_f \tag{12}$$

This gives rise to the equivalent of Snell's law for electrons.

$$\frac{\sin\theta_c}{\sin\theta_f} = \frac{k_f}{k_c} = \frac{[m_f^*(E - V_f)]^{\frac{1}{2}}}{[m_c^*(E - V_c)]^{\frac{1}{2}}} \tag{13}$$

The onset of total internal reflection occurs when $\theta_c = 90°$. This happens when the angle of incidence is equal to the critical angle θ'. Thus from equation (13), the cover-film critical angle is

$$\theta_{cf}' = \left\{\sin^{-1}[m_c^*(E - V_c) \ / m_f^*(E - V_f)]^{\frac{1}{2}}\right\} \qquad \text{for } (E - V_c) > 0$$

$$= 0 \qquad \text{for } (E - V_c) \le 0 \tag{14}$$

*where,* $V_c < E$.

For an electron wave incident at an angle greater than $\theta_{cf}'$, the wave suffers total internal reflection for an infinitely thick barrier. At steady state, entire electron current is reflected back into the film. The electron wave function decays exponentially into the cover. If the kinetic energy $E - V_c \le 0$, then total internal reflection occurs for any angle of incidence including normal incidence. This is in contrast to the electromagnetic case where total internal reflection can never occur at normal incidence due to the nonzero value of the refractive index. Similarly the substrate-film critical angle is

$$\theta'_{sf} = \left\{ \sin^{-1}[m_s^*(E - V_s) \,/\, m_f^*(E - V_f)]^{\frac{1}{2}} \right\} \tag{15}$$

*where,* $V_s < E$

The fraction of electron current transmitted and reflected at a potential interface is evaluated by solving the Schrodinger wave equation using a steady state time dependence of the form $\exp(-j\omega t)$. The electron wave functions for incident, transmitted and reflected wave are given by

$$\psi_i(x,y) = A\exp[j(k_{ix}x + k_{iy}y)] \tag{16}$$

$$\psi_t(x,y) = B\exp[j(k_{tx}x + k_{ty}y)] \tag{17}$$

and

$$\psi_r(x,y) = C\exp[j(k_{rx}x + k_{ry}y)] \tag{18}$$

respectively, where A, B and C are the amplitude constants. $k_{ix}$ and $k_{iy}$ are the values of the incident wave vector along the x and y-axis respectively. $k_{tx}$ and $k_{ty}$ are the values of the transmitted wave vector along the x and y-axis respectively. $k_{rx}$ and $k_{ry}$ are the values of the reflected wave vector along the x and y-axis respectively. Here, $k_{rx} < 0$ because the direction of propagation of the reflected wave is in the negative x-direction. The requirement that the wave-function be continuous across the potential energy boundary at x = 0 gives

$$\psi_i(0,y) + \psi_r(0,y) = \psi_t(0,y) \tag{19}$$

Since $k_{iy} = k_{ry} = k_{ty}$, then A + C = B or $1 + r_e = t_e$

*where,* $r_e = C/A$ and $t_e = B/A$ are the values of the electron amplitude reflectivity at the potential barrier and the electron amplitude transmittivity at the potential barrier respectively. The probability current density normal to the boundary is

$$J_x = \left(\frac{\hbar}{2m^* j}\right)\left[\psi^*\left(\frac{\partial \psi}{\partial x}\right) - \psi\left(\frac{\partial \psi^*}{\partial x}\right)\right] \tag{20}$$

In quantum mechanics, the probability current density normal to a potential energy boundary is conserved, there-by implying

$$J_{ix} + J_{rx} = J_{tx} \tag{21}$$

*where,* $J_{ix} = (\hbar/m_1^*)AA^*k_{ix}$, $J_{tx} = (\hbar/m_2^*)BB^*k_{tx}$ and $J_{rx} = (\hbar/m_1^*)CC^*k_{rx}$ are the individual current densities for incident, transmitted and reflected electron waves. Substituting the above values in equation (21), the following result is obtained.

$$[(m_1^*/k_{ix})/(m_2^*/k_{tx})]\ t_e^2 + r_e^2 = 1 \tag{22}$$

comparing equation (22) with the expression, $1 + r_e = t_e$ obtained from equation (19), and using the values of the wave vectors $k_{ix} = k_i \cos\theta_i$ and $k_{tx} = k_t \cos\theta_t$, the expression for the electron amplitude transmittivity takes the form

$$t_e = \frac{2(E/m_i^*)^{1/2}\cos\theta_i}{(E/m_i^*)^{1/2}\cos\theta_i + [(E-V)/m_t^*]^{1/2}\cos\theta_t} \tag{23}$$

and the expression for electron amplitude reflectivity

$$r_e = \frac{(E/m_i^*)^{1/2}\cos\theta_i - [(E-V)/m_t^*]^{1/2}\cos\theta_t}{(E/m_i^*)^{1/2}\cos\theta_i + [(E-V)/m_t^*]^{1/2}\cos\theta_t} \tag{24}$$

comparing equation (22) with the well known expression $T + R = 1$, the expression for the fraction of the electron current transmitted is given by

$$T_e = \left[\frac{(E-V)_t}{(E-V)_i}\right]^{1/2} |t_e|^2 \tag{25}$$

and the fraction of the electron current reflected is given by

$$R_e = |r_e|^2 \tag{26}$$

## Reflection and Transmisson of Electron Wave by an Electron Waveguide Film

The system consists of three layers with the potential energy values of $V_1$, $V_2$ and $V_3$. The electron wave vectors corresponding to the media in the three regions are given by $k_1$, $k_2$ and $k_3$ respectively. The region 2 is called the film layer with a thickness d. The angle of incidence is $\theta_1$. $r_1$ and $r_2$ are the amplitude reflection coefficients at the first and second interface while $t_1$ and $t_2$ are the amplitude transmission coefficients at the first and second interface. Corresponding to the potential energies of each of the three regions, the wave vectors are calculated by using the relation given in equation (6). The objective is to analytically calculate the amplitude reflection and transmission coefficients of the film region. The incident electron wave functions in region 1 and region 3 are $\psi_{i,1}$ and $\psi_{i,3}$ respectively while the reflected electron wave functions in region 1 and region 3 are $\psi_{r,1}$ and $\psi_{r,3}$ respectively. The relation between these wavefunctions is given by

$$\begin{pmatrix}\psi_{i,1}\\ \psi_{r,1}\end{pmatrix} = S\begin{pmatrix}\psi_{i,3}\\ \psi_{r,3}\end{pmatrix} = \begin{pmatrix}a & b\\ c & d\end{pmatrix}\begin{pmatrix}\psi_{i,3}\\ \psi_{r,3}\end{pmatrix} \tag{27}$$

*where,*

$$a = \frac{1}{t_1 t_2}\left(e^{i\delta_2} + r_1 r_2 e^{-i\delta_2}\right) \tag{28}$$

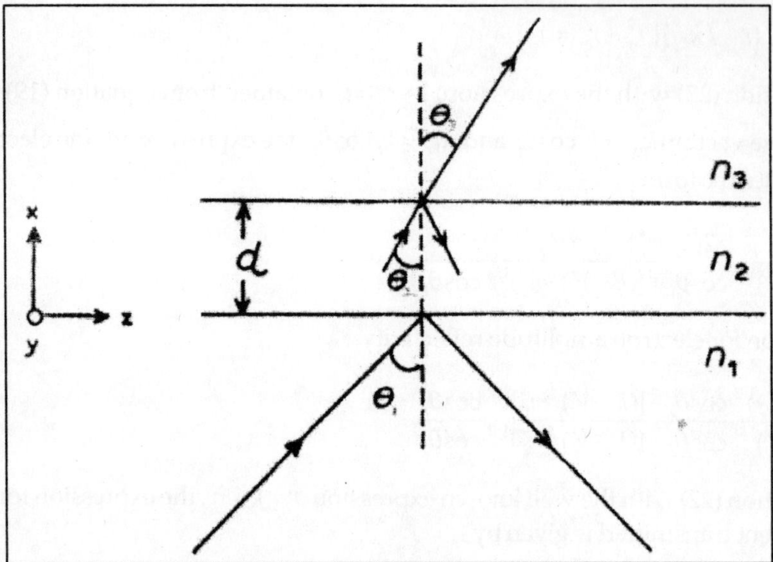


**Figure 26.2: Oblique Incidence of a Beam on an Electron Waveguide Film**

$$b = \frac{1}{t_1 t_2}\left(r_2 e^{i\delta_2} + r_1 e^{-i\delta_2}\right) \tag{29}$$

$$c = \frac{1}{t_1 t_2}\left(r_1 e^{i\delta_2} + r_2 e^{-i\delta_2}\right) \tag{30}$$

$$d = \frac{1}{t_1 t_2}\left(r_1 r_2 e^{i\delta_2} + e^{-i\delta_2}\right) \tag{31}$$

and $\delta_2 = k_2 d \cos\theta_1$ (32)

The amplitude reflection coefficient and the corresponding reflectivity, R of the film is given by

$$r = \frac{c}{a} \tag{33}$$

and

$$R = |r|^2 = \frac{r_1^2 + r_2^2 + 2r_1 r_2 \cos 2\delta_2}{1 + r_1^2 r_2^2 + 2r_1 r_2 \cos 2\delta_2} \tag{34}$$

respectively.

The amplitude transmission coefficient and the corresponding transmittivity, T of the film is given by

$$t = \frac{1}{a} \tag{35}$$

and

$$T = \left[\frac{(E-V)_3}{(E-V)_1}\right]^{1/2} |t|^2 = \left[\frac{(E-V)_3}{(E-V)_1}\right]^{1/2} \frac{t_1^2 t_2^2}{1 + r_1^2 r_2^2 + 2r_1 r_2 \cos 2\delta_2} \tag{36}$$

respectively.

## Reflection and Transmission of Electron Waves by a Superlattice Structure

In a semiconductor superlattice, an electron wave with energy above the potential energy barriers may be totally transmitted (analogous to an anti reflection coating in optical thin film devices) or totally reflected (analogous to a dielectric-coated mirror). The arrangement as shown in figure 3 is considered.

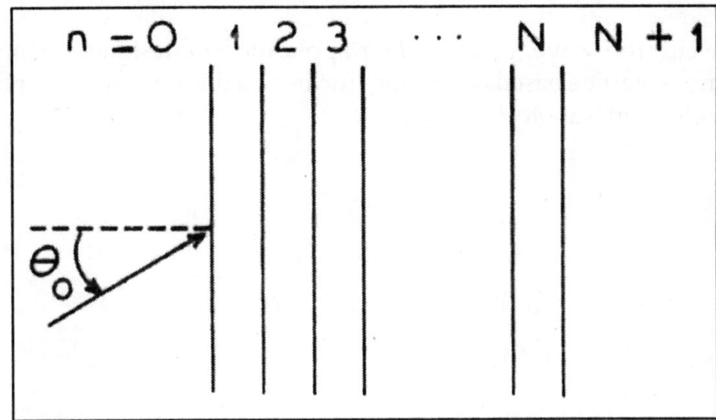


**Figure 26.3: An Electron Wave in the Region n = 0, Incident at an Angle, $\theta_0$ upon a Superlattice Structure Consisting of N, Number of Quantum Wells**

It shows an electron wave in the region $n = 0$, incident at an angle $\theta_0$, upon a stack of N layers of differing electron potential energy. The amplitudes of electron wave incident $(\psi_{i,n})$ upon and reflected $(\psi_{r,n})$ from the n[th] layer may be expressed in terms of the amplitudes incident $(\psi_{i,n+1})$ upon and reflected $(\psi_{r,n+1})$ from the (n+1)[th] layer, as

$$\begin{bmatrix}\psi_{i,n}\\ \psi_{r,n}\end{bmatrix} = \frac{1}{t_n}\begin{bmatrix}1 & r_n\\ r_n & 1\end{bmatrix} \times \begin{bmatrix}\exp[jk_n d_n(\cos\theta_n)]\ 0 & \\ 0 & \exp[-jk_n d_n(\cos\theta_n)]\end{bmatrix} \times \begin{bmatrix}\psi_{i,n+1}\\ \psi_{r,n+1}\end{bmatrix} \tag{37}$$

*where,* $t_n$ is the amplitude transmittivity of the interface between layer n-1 and layer n, $r_n$ is the amplitude reflectivity at the same interface. $k_n$ is the magnitude of wave-vector and $d_n$ is the thickness of n[th] layer. $\theta_n$ is the incident angle of the wave vector in the n[th] layer. Generalizing the above analysis for N films and for oblique incidence, the analysis of propagation of electron waves in multilayer superlattice structures can be performed. By chain multiplying a total of N+1 versions of equation (37), we obtain a relation between the total normalized transmitted electron wave amplitude in region N+1, $\psi_{t,N+1}$ and the total normalized reflected electron wave amplitude in region 0, $\psi_{r,0}$ as

$$\begin{bmatrix} 1 \\ \psi_{r,0} \end{bmatrix} = \prod_{n=1}^{N} \frac{1}{t_n} \begin{bmatrix} 1 & r_n \\ r_n & 1 \end{bmatrix} \times \begin{bmatrix} \exp[jk_n d_n(\cos\theta_n)] & 0 \\ 0 & \exp[-jk_n d_n(\cos\theta_n)] \end{bmatrix}$$

$$\times \frac{1}{t_{N+1}} \begin{bmatrix} 1 & r_{N+1} \\ r_{N+1} & 1 \end{bmatrix} \begin{bmatrix} \psi_{t,N+1} \\ 0 \end{bmatrix} \tag{38}$$

These equations serve as the basis for the design and simulation of the electron waveguide based devices.

## Conclusion

Analysis of the electron waveguide for their application in designing nanodevices has been done. These electron waveguide based devices are further used in environment related applications and is the backbone of Green Nanotechnology.

# Chapter 27

# Risk-Based Management of Contaminated Sites: Methodological Background and Advanced Decision Support Systems

**Andrea Critto[1] and Antonio Marcomini[2]**
**[1]Department of Environmental Sciences, University Ca' Foscari of Venice, Calle Larga S. Marta 2137, I-30123 Venice, Italy**
**[2]CVR (Venice Research Consortium), Viale della Liberta' 5-12, Marghera-Venice, Italy**

## ABSTRACT

Several research and application activities are undergoing worldwide for the development of decision support systems that allow management of multiple and different data in order to aid decision-making and communication processes. These activities are particularly focused on Decision Support Systems (or DSSs) that address problems of contaminated sites or brown fields, where economic, social and environmental information need to be integrated and evaluated together, also in the light of stakeholders' preferences and in a sustainable perspective. In fact, contaminated sites include a large variety of management issues, from fresh and marine waters contamination to soil pollution, from brown fields within urban areas to large industrial sites (*i.e.* megasites). Many methodologies may be applied to manage these problems, such as Multi-criteria Decision Analysis or economic and environmental analyses. A combined evaluation of these methodologies and discussion on the benefits and limits within the development of Decision Support Systems may be beneficial in terms of scientific progress as well as policy making. In

* Corresponding Author: E-mail: 1: critto@unive.it, marcom@unive.it; 2: amb.cvr@vegapark.ve.it

order to manage the complex decision-making process for the risk-based assessment and management of contaminated land and water, four DSSs called SYRIADE, DESYRE, ERA-MANIA and MODELKEY are described. These DSSs were developed by the University Ca' Foscari of Venice and Venice Research Consortium, within national and European funded project.

## Introduction

Within the more general theme of sustainable development, the remediation and the subsequent reuse of contaminated sites may play an important role. In fact, the worldwide relevance of the contaminated sites theme is proved by the extension of the problem. Only in Europe the number of estimated potentially contaminated sites was around 1 400 000 in 2000, which determined an economic impact for remediation costs of approximately 59.000 to 109.000 million Euros (Schamann *et al.*, 2003).

Sustainability is commonly referred to as the balance among the social, the environmental and the economic aspects, in order to promote a development that may be beneficial for present as well as future generations (World Commission on Environment and Development, 1987). But what are the positive paybacks that the contaminated site management may provide to sustainability?

First of all, it is widely recognized both in Europe and in USA that contaminated or abandoned sites redevelopment is *per se* a sustainable process, since it is directed toward the reuse more than the use of new land and water (or "green" sites). This is an important aspect of preservation of existing natural ecosystems and maximization of resources already in use. In addition, many other social and economical benefits may derive from contaminated sites rehabilitation, such as future economic returns (especially for private developers), new jobs creation or social well being. All these benefits equally cover the three main pillars of sustainability (environment, society and economy), thus making contaminated sites management a critical component of the world progress towards sustainability.

If these can be considered the benefits of contaminated sites management in the light of the achievement of sustainable objectives, the next question that can be posed is how the management itself of these sites can be considered really a sustainable process, particularly when decisions on remediation and future use of the site have to be taken.

In general, the management process for site redevelopment consists of three separated phases: the assessment, the remediation and the reuse.

In the first phase, specific analysis have to be undertaken in order to define the present environmental or socio-economic conditions and the future use of the site. The environmental investigations, mainly risk assessment procedures, are conducted jointly with the economic assessment that define the benefits and costs of the redevelopment options and also with the involvement of interested stakeholders.

Then, the remediation phase includes all the technical activities to support the containment and reduction of the contamination. In this phase, the technical and technological aspects have to be explored in order to select the most appropriate remediation technologies and perform the clean-up activities. At this point, the site is ready for reuse.

If this three-phase management process is merged with sustainability principles, the main factors that characterize the sustainable management of contaminated sites emerge: economic drivers and funding opportunities; stakeholders involvement and responsibility; legal and regulatory framework consideration; estimation of human health and ecological risk and risk management; choice of reliable remediation techniques; costs and benefits evaluation (Bardos *et al.*, 2001).

Given these general considerations on contaminated sites management, what becomes urgent in order to put sustainability principles into a realistic decision-making process is to develop efficient and suitable Decision Support Systems, in order to promote dialogue and consultation between experts and public and to encourage largely shared decisions.

A Decision Support System or DSS is a computerized system that helps decision makers in structuring and evaluating decisions, by providing easy-to-use and integrated tools for information elaboration and displaying (Shim *et al.*, 2002; Jensen *et al.*, 2002; Lahmer, 2004).

DSSs not only provide information but also suggest or make recommendations (Workmann, 2005); they have the common objective of "provide timely information that supports human decision makers–at whatever level of decision making" (Loucks, 1995). They create an intensive human-computer interaction that evolves through a series of stages to perform complex analyses and derive computer-generated solutions to problems (Workmann, 2005).

DSSs are instruments that, applied in any type of decision-making for environmental problems solving, may guarantee transparency, efficiency, communication, all-inclusiveness and active participation within the whole decision process.

In order to manage the complex decision-making process for the risk-based assessment and management of contaminated land and water, four DSSs called SYRIADE, DESYRE, ERA-MANIA and MODELKEY will be described within the following paragraphs.

## Spatial Decision Support System for Regional Risk Assessment of Degradated Land: SYRIADE

The recent directive on EU Mining Directive (2006/21/EC) as well as the European Soil Thematic Strategy poin out the necessity of the development of spatial environmental risk-based methodologies for sustainable management of contaminated sites and waste mining sites at regional scale. Whith this porpouse, a Spatial decision support sYstem for Regional rIsk Assessment of DEgradated land called SYRIADE, was developed by the European Commission JRC in collaboration with Consorzio Venezia Ricerche and University Ca' Foscari of Venice. SYRIADE is intended to be an aid for national and regional competent authorities in the inventory and assessment of contaminated sites and mining waste sites at regional scale. The system implements a regional risk assessment methodology for the evaluation of the impacts occuring at regional scale on species, populations or ecosystem that are widespread. The identification and characterization of the spatial relations between the sources and the receptors are performed by the application of GIS spatial functionalities, while the ranking process is based on specific risk indices which are aggregated by the use of Multicriteria Decision Analysis.

The overall objective of the system is the ranking of potentially contaminated sites for priority of investigation, when no information on characterization and risk by site specific methodologies is available. To reach the objective, the system performs a prioritization of contaminated sites according to a regional risk assessment, and on a socio-economic analysis. Three main components of the system were identified: the regional risk assessment and the socio-economic assessment components both leading to the ranking of the potentially contaminated sites according to environmental and socio-economical aspects respectively and the integrate management component which allows the identification of homogeneous areas for integrated management.

Within the regional risk assessment methodology, the analyzed potential exposure pathways were the leaking of the contaminants of concern through the vadose zone to the groundwater, the volatilization and the wind transport and finally the migration of the contamination to the surface water. The main recognized receptors were humans, surface water, groundwater and protected areas. Surface water and groundwater are both a receptor and a contamination transport medium towards other receptors such as protected areas. The GIS functionalities, integrated with mathematical approaches, allowed to take into consideration all in once the multiplicity of sources and impacted receptors within the regional territory and therefore to assess risks of all the contaminated sites in the region, thus supporting the final ranking objective. Multicriteria Decision Analysis methodologies were incorporated in the system in order to address the experts and stakeholders points of view in the definition of the scores of each parameter used for the characterisation of the risk assessment components (hazard analysis, pathway analysis and sensitivity analysis), in the estimation of the risk assessment components scores and in the estimation of the final ranking of the potentially contaminated sites according to environmental aspects. Finally, in the integrated management component, GIS spatial functionalities and Multicriteria Decision Analysis procedures are combined in order to support the integration of the risk assessment results with the socio-economic results and the identification of the homogeneous areas for integrated management.

SYRIADE has been applied to the case study of the Upper Silesia region (Poland).

## Decision Support System for Rehabilitation of Contaminated Sites: DESYRE

In order to provide a decision support tool for the remediation of contaminated mega-site, encompassing the planning and the intervention project phase, the software DESYRE (DEcision Support sYstem for the REqualification of contaminated sites) has been developed.

The DESYRE specific main objective is the creation and comparison of different remediation scenarios in terms of residual risk, technological choices and socio-economical benefits.

The software users are supposed to be a multi-disciplinary team of experts, involving risk assessors, socio-economists and technology engineers. DESYRE provides support to the experts along several steps for contaminated sites rehabilitation, encompassing characterisation, socio-economic evaluation, risk assessment and technologies ranking. It allows the integration, spatial evaluation and analysis of available information by using GIS functions and a number of decision support tools implemented in the system. Finally, the DESYRE final output is addressed to decision makers for comparing benefits, costs and impacts of different strategies and technical solutions.

DESYRE is structured into five modules integrated into a GIS framework: four analytical modules, Socio-economical, Characterisation, Risk assessment and Technological, and a Decision module. The four analytical modules are supposed to be used by the multi-disciplinary expert team, while the decision module is addressed to stakeholders such as public decision makers, planners, site owners, investors and associations.

The *socio-economic module* addresses the socio-economical constraints generated by different land uses. The resulting scenarios are defined and ranked with Fuzzy Logic Analysis. The module allow the experts to select the most attractive, from a socio-economical point of view, land uses to be considered in risk assessment.

The *site characterization module* supports the analysis of contaminant spatial distributions and provides as much of the data necessary to run fate and transport and risk assessment models. The spatial distribution of contaminants are analysed by using geostatistical methods, including variography and Kriging.

The *risk assessment module* provides tools for human and environmental risk analysis of contaminants in soil and groundwater. Risk calculation is based on standard algorithms, while a novel procedure has been developed to investigate the spatial distribution of risk. Moreover, the final risk estimation is intended to be used for the selection of remediation technologies.

The *technological module* provides a ranking system for the remediation technologies based on multi-criteria decision analysis (MCDA) methods. Based on the technologies ranking and actual risk, the expert defines alternative technological sets, *i.e.* combinations of technologies extended spatially and temporally. For each set, a simulation of the interventions performances produces residual risk maps.

The *decision module* provides the description of alternative remediation scenarios. A scenario is an inclusive and suitable solution for the rehabilitation of a contaminated site, including the technology set, the associated human health and environmental feedback, in terms of residual risk, and the socio-economic benefits of the site rehabilitation for the specific use. Technological, environmental and socio-economical aspects are described by macro-indexes that are derived directly from the technological, risk assessment and socio-economical modules, respectively. Multi Criteria Analysis (MCDA) methodologies are applied to define indexes based on criteria and rules. A scoring system is provided for the stakeholders to rank alternative scenarios according to their major sensitivities, *e.g.* to environmental benefits, or costs, or time etceteras.

A DESYRE prototype software has already implemented. It is based on a GIS software (ArcView for Windows) and supported by Visual Basic to develop the different applications. A friendly interface makes DESYRE an easy tool also for not GIS-trained users. A clear guideline makes the decision flow transparent and reproducible, by ensuring that the process itself is documented (*i.e.* all parameters, assumption, and data used to reach the decision should are clearly schematized).

DESYRE has been tested by using the mega-site of Porto Marghera, Venice, as case study. Porto Marghera is c.a. 3600 ha industrial zone, with c.a. 500 ha accounted for surface water.

## Decision Support System for Site-Specific Ecological Risk Assessment for Contaminated Site: ERA-MANIA DSS

The diffuse soil contamination caused by both organic and inorganic compounds is a problem of actual interest. As result, the increasing demand for the assessment and management of contaminated soils requires the development of new and effective tools to be used in the site-specific Ecological Risk Assessment (ERA) procedure. For this purpose the ERA-MANIA DSS is proposed as a Decision Support System that aims at improving the ERA procedure and supporting the expert/decision maker in its assessment. It was developed according to the Weight of Evidence (Burton *et al.*, 2002) and the TRIAD (Rutgers *et al.*, 2005) approaches, where the results provided by a set of Lines of Evidence (LoEs) are gathered and compared to support the assessment and evaluation of the ecosystem impairment caused by the stressor(s) of concern. In particular, in the TRIAD approach the LoEs refer to the following three areas of investigation (*i.e.*, leg): environmental chemistry (*e.g.* estimate of the total and the biovailable contaminant concentrations in soil), ecotoxicology (*e.g.*, application of bioassay and biomarkers) and ecology (*e.g.*, biodiversity evaluation). The rationale behind is that integrating the evidences highlighted by the three different investigated areas the uncertainty in the risk estimation and evaluation will be pragmatically reduced.

The ERA-MANIA DSS, developed on the basis of the TRIAD approach, consists of two modules: "Comparative Test Tables" and "Integrated Ecological Risk Indexes". The "Comparative Test Tables" module aims at comparing the different tests or Lines of Evidence (LoEs) belonging to each of the investigated area or leg (chemistry/bioavailability, ecology and ecotoxicology) to guide the expert/

decision maker in the choice of the most suitable set of tests to be applied to the case study. These tables can first of all be regarded as database for the collection of information related to the specific LoEs.

In addition the tables support experts in: excluding those LoEs which cannot be applied at the case study; comparing the tests belonging to each TRIAD leg in order to define their suitability to be applied at the first, the second or the third TRIAD tier; selecting the best set of LoEs for the case-study. Using Multi-Criteria Analysis, including expert judgment and specific comparative criteria, the selected LoEs will be ordered to form a ranking from which the expert/decision maker will choose the most suitable set of tests to be applied to the case study. In addition, the system can evaluate the concordance among the values assigned by the multiple involved experts.

The "*Integrated Ecological Risk Indexes*" module provides qualitative and quantitative tools that allow to assess the terrestrial ecosystem impairment (*i.e.* the impairment occurring on biodiversity and functional diversity of terrestrial ecosystem) by integrating the heterogeneous information obtained by the LoEs application. It includes the following tools: the Integrated Effect Index (IEI) and the Ecosystem Impairment Matrix (EcoIM). The IEI is a multicriteria analysis-based quantitative estimation of terrestrial ecosystem impairment that integrates the "impairment level" highlighted by each applied Line of Evidence. According to the expert judgment the LoE results belonging to the three TRIAD legs are evaluated in terms of "evidence of impairment" and integrated separately using multicriteria analysis procedures, taking into account the relevance of the information provided at ecosystem level. The comprehensive and immediate understanding of the quantitative estimation provided by the calculated IEIs is supported by the EcoIM. In fact, this tool provides a qualitative assessment of the impairment at the ecosystem level evaluating both biodiversity and functional diversity of terrestrial ecosystem and highlighting the ecosystem factors stressed by soil contamination. Its development is based on soil ecological knowledge and multicriteria analysis methods and the obtained results can drive the risk management process.

Finally, the ERA-MANIA DSS was implemented in a software and applied, within the ERA-MANIA research project, to the ACNA di Cengio (Italy) contaminated megasite, that has been identified by the Italian Ministry of the Environment as a site of national interest.

## Risk Based Decision Support System to Assess Ecological and Chemical Status of Water Bodies: MODELKEY DSS

The European Water Framework Directive (WFD; EC, 2000) prescribes a series of tasks for properly assessing and managing river basins with the ultimate aim of protecting and restoring the overall quality status of European surface waters. It specifically recommends the achievement of good ecological and chemical status of surface waters by the end of 2015. The ecological status of each water body has to be evaluated and classified by using biological quality elements as key parameters and physico-chemical as well as hydromorphological quality elements as supportive parameters. Moreover, the chemical status has to be evaluated by comparison of measured concentrations of priority substances in water, sediment and biota compartments with Environmental Quality Standards (EQS).

Within the European integrated project called MODELKEY a risk-based Decision Support System (MODELKEY DSS) is under development. The system aims at integrating different analytical tools and exposure/effect models in order to assist water managers in assessing and managing river basins according to WFD obligations. The main functionalities of the DSS will concern:

- Information gathering, processing and integration;
- Communication;
- Preliminary support to the setting up of management strategies.

In fact, the system will provide an integrated assessment of the river quality conditions. This assessment, through the definition of areas of concern (hot spots) and their prioritization, can be used by the river authorities and decision-makers to establish where to intervene promptly and actively. Besides, more specific investigations can be supported at the site-specific level.

As far as management functionalities, the DSS can support the definition of possible options by analysing the effects of different changes in input parameters (due to management choices) in the outputs variables, *i.e.* the calculated indices.

Due to the specific questions that the MODELKEY DSS can help to address, by including the models and tools that perform the abovementioned tasks, the MODELKEY DSS, differently from other existing tools, analyses biodiversity-related issues at basin scale in an integrated way and links them with the toxicants present in the river.

As far as the structure is concerned, MODELKEY DSS will include databases (DB) for river basin data, analytical models (AM) aimed at integrated environmental and socio-economic assessment, spatial analysis (SA) performed through GIS platform included in the system and a user-friendly and easy interface (GUI) combining Web performances with downloadable options. The simulation functionalities (SOM) will be included in the models themselves.

In addition, it is foreseen that the MODELKEY DSS will be constructed in such a way to allow the possibility of integration with other existing models and systems, in order to provide the end-users, *i.e.* decision-makers and stakeholders, with a collection of connected systems that can support the overall process of decision-making for river basin management.

Finally, the whole DSS will be applied to three case-studies within the European integrated project MODELKEY: Elbe, Scheldt and Llobregat river basins.

## References


1. Agostini P, Semenzin E, Critto A, Micheletti C, Nunes P, Ghermandi A, Gottardo S, Giove S, de Zwart D, Brack W and Marcomini A (2006). Decision Support System for the assessment and evaluation of ecological impacts on aquatic ecosystems: objectives and framework. In: Voinov A, Jakeman A, Rizzoli A editors, *Proceedings of the iEMSs Third Biennial Meeting: "Summit on Environmental Modelling and Software"*. International Environmental Modelling and Software Society, Burlington, USA, July 2006. CD ROM. Internet: http://www.iemss.org/iemss2006/sessions/all.html, pp 1-6.

2. Bardos, P.R., C. Mariotti, F. Marot and Sullivan T. 2001. Framework for decision support used in contaminated land management in Europe and North America. Land contamination and reclamation, 9, 149-163.

3. Burton, Jr., G.A., P.M. Chapman, E.P. Smith, 2002. Weight-of-Evidence approaches for assessing ecosystem impairment. Human and Ecological Risk Assessment 8: 1657-1673.

4. Carlon, C., Critto, A.; Ramieri, E.; Marcomini, A. DESYRE: Decision Support System for the Rehabilitation of Contaminated Megasites. *Integr Environ Assess Manag*. 3, 211-222; 2007.

5. Critto A, Torresan S, Semenzin E, Giove S, Mesman M, Schouten AJ, Rutgers M, Marcomini A. Development of a site-specific Ecological Risk Assessment: Part I: A multi-criteria based system for the selection of ecotoxicological tests and ecological observations. Sci Tot Env 2007;379:16-33.

6. Critto, A.; Cantarella, L.; Carlon, C.; Giove, S.; Petruzzelli, G.; Marcomini, A. Decision Support-Oriented Selection of Remediation Technologies to Rehabilitate Contaminated Sites. *Integr Environ Assess Manag*. 3:273-285; 2006.

7. Jansen, R A, Krejcik J, Malmgren-Hansen A, Vanecek, S, Havnoe K and Knudsen J (2002) River Basin Modelling in the Czech Republic to optmise interventions necessary to meet the EU environmental standards. Proceedings of the Internation Confernce of Baisn Organizations, Madrid 4-6 November 2002, DH ref 39/02.

8. Lahmer, W. (2004). Multi-disciplinary approaches in River Basin Management–an example. Presentation at the International Conference on Water Observation and Information System for Decision Support, Ohrid, Macedonia, 25-29 May 2004.

9. Loucks, D P (1995) Developing and implementing decision support systems: a critique and a challenge. Water Resources Bulletin 31 (4): 571-582.

10. Rutgers M., Den Besten P., 2005, Approach to legislation in a global context, B. The Netherlands perspective–soils and sediments. In: Environmental toxicity testing, Thompson K.C., Wadhia K., and Loibner A.P., eds., Blackwell Publishing CRC Press, Oxford. pp. 269-289.

11. Schamann, M., A. R. Gentile, G. Prokop and Wepner M. 2003. The European Environment Agency's framework for the development of European-wide policy-relevant indicators on local soil contamination. Annokkée, G.J., Arendt, F., Uhlmann, O. (Ed.) Consoil 2003 Conference Proceedings, pp. 28-34. Forschungszentrum Karlsruhe publishing, Karlsruhe.

12. Semenzin E, Critto A, Carlon C, Rutgers M, Marcomini A. Development of a site-specific Ecological Risk Assessment: Part II: A multi-criteria based system for the selection of bioavailability assessment tools. Sci Tot Env 2007; 379: 34-45.

13. Semenzin E, Critto A, Rutgers M, Marcomini A. Integration of bioavailability, ecology and ecotoxicology by three lines of evidence into ecological risk indexes for contaminated soil assessment. Sci Tot Env 2008; in press.

14. Shim, J P, Warkentin, M., Courtney, J F, Power D J, Shards, R, Carlsson C (2002) Past, present and future of decision support technology, Decision Support Systems, n. 33, pp. 111-126.

15. Workmann, Michael (2005) Expert decision support use, disuse and misuse: a study using theory of planned behaviour. Computers in Human Behaviour, 21: 211-231.

# Chapter 28

# Recycling of Whey and Agricultural Residues for Preparation of Carrier Based Plant Growth Promoting Bioinocula

**V.G. Gadekar, N.N. Nawani and B.P. Kapadnis**
**Department of Microbiology, University of Pune, Pune – 411 007, M.S.**

## ABSTRACT

Whey, a byproduct of dairy industry is an important effluent produced during cheese and casein manufacture. The lactose and proteins are both potential biochemical sources, which cause pollution problems. Whey produced in huge amount in dairy industry and its disposal requires large amount of capital investment. Most of the milk plants do not have proper treatment system for disposal of whey. Without prior treatment, dumping of whey constituents is a significant loss of bioresource. Whey is also a serious pollutant because it imposes high biochemical oxygen demand $(BOD)_5$ of 30,000-50,000 mg/l and chemical oxygen demand (COD) of 60,000-80,000 mg/l. Disposal of whey leads to serious environmental problems by affecting physical and chemical structure of soil. At the same time India is agriculture based country and generating huge quantity of agricultural residues like paddy husk, wheat bran, bagasse etc. Maximum 2.1 per cent (dry weight basis) of actinomycetes biomass was produced. With various strains of actinomycetes significant increase in wheat plant was observed *in vitro* experiments. Wheat bran based bioinocula was the best among tested carriers. Use of whey for growth of plant growth promoting actinomycetes and use of agricultural residues as a carrier in bioinocula preparation, will be one of the alternative method for disposal of these agricultural based wastes.

* Corresponding Author: E-mail: bpkap@unipune.ernet.in

## Introduction

Various aspects have been studied with reference to utilization of whey (Ahn *et al.*, 2000; Chartrain and Zeikus 1986; Tahoun *et al.*, 1987; Sorba *et al.*, 1993; Maullu *et al.*, 1999; Holsinger *et al.*, 1974; Mishra *et al.*, 2000). Lactose, casein and lactic acid are major ingredients of whey, which are useful for growth of microorganisms. Various microorganisms have been studied for their ability to utilize whey ingredients (Meyrath and Bayer 1979). However, there was lack of reports on utilization of whey as a substrate for propagation of acidophilic actinomycetes. The objective of the present study was to investigate the possibility of using whey as a substrate for propagation of acidophilic actinomycetes having plant growth promoting potential. The whey, a waste of dairy industry can be a simple and economically viable medium for propagation of acidophilic actinomycetes. This biorecycling approach will prove to be an alternative method of disposal of whey. In the present study, various agricultural residues like paddy husk, wheat bran and bagasse were screened as carriers of prepared bioinocula. India has a vast potential of agricultural and industrial waste. With introduction of high yielding varieties and adaptation of intensive systems of cropping, large quantities of wastes generate every year. There are references available on use of agricultural residues *viz.*, food, formation of single cell etc. Use of these agro-based wastes for bioconversion into useful products is one of the environmental and eco-friendly novel approaches.

## Materials and Methods

### Collection of the Whey Samples

The cheese whey collected from Parag Milk, Gowardhan dairy, Awasari, Tal. Ambegaon, Dist. Pune, India.

Whey samples were collected in five-litre polypropylene cans at low temperature. Samples were transported to the laboratory immediately and stored at 4°C until use. The sampling was continued throughout the research work.

#### Cheese Whey Analysis

The proteins were determined in terms of bovine serum albumin (BSA) by the Lowry's method (1951), Reducing sugar was determined by Miiler's DNS method (1959). The fatty acids in terms of lactic acid, by Titrable acidity method (APHA, 1995) and COD was determined by Closed reflux method with an automatic COD analyzer (Machery- Nagel, Germany). The amount of protein, lactic acid and reducing sugar were extrapolated from the respective standard graph.

### Media

*Starch Casein medium (SCM).* The SCM of pH 4.5 was made as per standard protocol (Mackey 1977).

*Soil extract agar (SEA).* 1000 g of air dried soil was suspended in 1000 ml of tap water, sterilized at 121°C for 20 min. and kept at room temperature for sedimentation of particulate matter. To the supernatant $CaCO_3$, 0.05g; glucose, 1g and $KH_2PO_4$, 0.5g were added. The final pH was adjusted to 4.5 (Subba Rao, 1986) with 1N HCl. Solidified with 2 per cent agar.

*Tap water agar (TWA).* Agar 20g per liter of tap water. The final pH was adjusted to 4.5 with 1N HCl.

*Whey medium (WM).* Various concentrations of whey *viz.*, 20 per cent, 40 per cent in tap water (v/v) were selected. The final pH was adjusted to 4.5 with 2N NaOH or 2N HCl and sterilized at 110°C for 15 minutes. The diluted whey media were solidified with 2.5 per cent of agar (Himedia Laboratories, Mumbai, India) to prepare whey agar (WA).

## Micro-organisms

*Acidophilic Actinomycetes (AA).* 100, AA were isolated from the acid soils of Mahableshwar (India) on starch casein agar.

## Screening of Acidophilic Actinomycetes on the Whey Agar

All AA were inoculated on whey agar. Besides, they were inoculated on to conventional media *viz.* SCA, SEA and TWA for comparison. All cultures were incubated at 37°C for 72 h.

## Growth of Acidophilic Actinomycetes in the Whey Medium

100 ml each of whey medium WM, was dispensed in two sets, each of eight 500ml conical flasks. One of the flasks from each set was control *i.e.* without inoculation. Rests of the flasks were inoculated with respective actinomycete culture (one 8 mm agar disc with actinomycete growth per flask). These were incubated on shaker (150rpm) at room temperature for 96h.

## Evaluation of Bioinocula

### Bioinocula without Carrier

#### *Sterilization of Soil*

The poita soil was air dried and passed through sieve of mesh size <1mm, dispensed in polypropylene bags and autoclaved (121°C) for 1 h, for three consecutive days. Sterility was tested by plating the soil on nutrient agar.

#### *Evaluation of Actinomycetes Inocula on Wheat (Triticum aestivum L.)*

#### *Pot Experiments*

Experiments (sterile and nonsterile soil) were carried out in 16 pots (10 cm diameter) in duplicate, out of which eight pots were filled with 135 gm of sterile poita soil and eight with natural poita soil. The prepared biomass [4gm/pot (wet wt., resuspended in 5 ml saline)] was mixed with upper 5 cm layer in both the soils aseptically. In each pot, five wheat seeds (Variety, Lokwan) were sown. The plants were harvested after 27 days. The harvested plants were statistically analyzed for the wet weight, average root- shoot length and weight of the plant. Results were based on average value of all plants.

#### *Statistical Analysis*

The data on effect of plant growth promoting bacteria on plant growth was subjected to analysis of variance, T-test to know wether there was significant increase at 0.05 significant level (Bailey 1965).

## Preparation of Bioinocula

### Actinomycetes Liquid Inocula

Fifteen 500 ml conical flasks, each with 200 ml of neat sweet whey, were autoclaved at 110°C for 15 min. Each culture was inoculated (two 8 mm agar discs with actinomycetes growth per flask) in duplicate. These flasks were incubated on shaker (150 rpm) at 37°C for 5 days. After incubation the biomass was harvested by centrifugation (7000g 20 min) as well as filtration through 0. 45 mm pore size membrane filter and the biomass resuspended in 5ml sterile saline. This constituted liquid bioinoculum (without carrier).

## Carrier Based Bioinocula

### Screening of Carriers

Various agricultural residues *e.g.*, wheat bran, paddy husk and bagasse, were tested as carriers in the WM. Thirty two 500 ml conical flasks, were with WM. Similar sets, each of 32 flasks were prepared for these different carriers. In each set, seven flasks had respective culture, carrier and whey, one flask was with respective whey and carrier as a control. The WM and carriers were autoclaved separately twice at 110°C for 15 min. 1 per cent of each carrier was incorporated into eight out of sixteen flask of each medium. Each flasks was inoculated with seven actinomycete culture (two, 8mm disc with actinomycetes culture) in to two flask of each medium, one of which was with carrier. The same procedure repeated for remaining carriers. All these flasks were incubated at 37°C temperature under static conditions for 96 h. The biomass from each flask was harvested by centrifugation (10,000g) for 10 min followed by drying at 60°C until constant weight. The actinomycete biomass in presence of carrier was determined in terms of relative cell mass with reference to control (without carrier). The biomass was determined in terms of dry weight after spinning the culture at 10,000g, followed by washing and drying at 60°C until constant weight. For determination of achieved biomass, the weight of the carrier was subtracted from total weight of the biomass.

# Results and Discussion

## Collection of the Whey Samples

Whey samples were transported to the laboratory immediately and stored at 4°C until use. The sampling was continued throughout the research work.

## Microorganisms

On the basis of their ability to grow on WM, seven strains were selected (Kanavade and Kapadnis, 2000). Further they were also tested for their plant growth promoting abilities (Kanavade and Kapadnis, 2002). Selected seven strains were denoted as KKA, KKB, KKC, KKD, KKE, KKF and KKG.

**Table 28.1: Composition of Cheese Whey**

| *Parameters** | *Whey†* |
|---|---|
| PH | 5.6 |
| Protein | 0.072 |
| Lactose | 4.5 |
| Fatty acids | 0.42 |

* All parameters in per cent except pH; † Whey from Gowrdhan Dairy, Pune.

## Cheese Whey Composition

The cheese whey from two different sources was analyzed for their ingredients. The results are presented in the Table 28.1. The results indicate that the whey is potent source of carbohydrate, protein etc. On the basis of results the collected whey was categorized as sweet whey.

## Growth of Acidophilic Actinomycetes on Whey Media and Conventional Media

All seven strains of AA were inoculated at varied concentrations of whey*viz.* 10, 20, 40, 60, 80 and 100. All of the isolates could grow luxuriently at 37°C (Table 28.2).

**Table 28.2: Growth* of Acidophilic Actinomycetes on Different Media**

| Acidophilic Actinomycetes Strains | WA with Whey (per cent) | | | | | | TWA | SEA | SCA |
|---|---|---|---|---|---|---|---|---|---|
| | 10 | 20 | 40 | 60 | 80 | 100 | | | |
| KKA | + | ++ | ++ | +++ | +++ | +++ | + | + | +++ |
| KKB | + | ++ | ++ | +++ | +++ | +++ | + | + | +++ |
| KKC | + | ++ | ++ | +++ | +++ | +++ | + | + | +++ |
| KKD | + | ++ | ++ | +++ | +++ | +++ | + | + | +++ |
| KKE | + | ++ | ++ | +++ | +++ | +++ | + | + | +++ |
| KKF | + | ++ | ++ | +++ | +++ | +++ | + | + | +++ |
| KKG | + | ++ | ++ | +++ | +++ | +++ | + | + | +++ |

* The strains were grown at 37°C for 72 h.

–: No growth; +: Less growth, ++: Moderate growth; +++: Good growth; WA: Whey agar; TWA: Tap water agar; SEA: Soil extract agar.

Subsequently these promising cultures were inoculated on whey agar with 10, 20, 40, 60, 80 and 100 per cent whey and incubated for 72 h. All actinomycete strains could grow on whey agar. However, the growth was significant on whey was comparable on SCA, a conventional medium used for isolation of actinomycetes. While growth on TWA and SEA was scanty.

## Growth of Acidophilic Actinomycetes in Liquid Whey Media

WM was found to be a better substrate for production of AA biomass (Table 28.3). With various strains of AA various quantity of biomass produced, indicated variation in utilization of whey ingredients (Sorba *et al.*, 1993; Maullu *et al.*, 1999). Biomass production in whey differs from various dairy plant to plant, because it depends on the type of whey generate ie. sweet whey, sour whey etc. as well as the selected organisms. variation in biomass production may vary from strain to strain.

## Effect of AA Liquid Bioinocula on Wheat

Wheat inoculation experiment carried out for 27 days showed significant increase in wheat plant (Tables 28.4 and 28.5) with some selected isolates. By AA strain KKG maximum root, shoot length and wt of the plant increased significantly, at the end of the experiment *i.e.* 27[th] day.

Bioinocula prepared with AA strain KKD and KKE not showed significant increase in some of the parameters of plants grown on sterile soil. But in nonsterile soil weight of the plant significantly increased in natural soil. It may be due to nonavailability of essential nutrients in natural soil due to autoclaving. There are very few reports available on actinomycete bioinocula. The available reports seen positive effect of these

**Table 28.3: Cell Mass* of Acidophilic Actinomycetes in Whey†**

| Acidophilic Actinomycetes Strains | Cell mass (%) in WM |
|---|---|
| KKA | 1.35 |
| KKB | 2.10 |
| KKC | 1.59 |
| KKD | 1.65 |
| KKE | 1.10 |
| KKF | 1.79 |
| KKG | 1.80 |

* Biomass in terms of dry weight; † Grown in 500 ml flask with 100 ml respective medium, on shaker at 37°C for 96 h; WM: Whey medium.

organisms on various plant recently (You *et al.*, 1996; El- Tarabily *et al.*, 2000). Present investigations also report plant growth promoting potential of actinomycetes inocula.

On the basis of the results obtained at 27[th] day, it can be seen that these organisms are competent with other natural flora of soil. There is need to carry out such experiments in the field.

**Table 28.4: Effect of Acidophilic Actinomycete Strains Liquid Inocula on Wheat* in Sterile Soil**

| *Response Variables →* | | | *Avg Root Length (cm)* | *Avg Shoot Length (cm)* | *Avg Plant Wet Wt (g)* |
|---|---|---|---|---|---|
| Value for | Strain | Control | 7.9±0.1 | 21.5±3.6 | 0.09±0.0 |
| variable with Mean±S.D. | | KKA | 11.11±2.0 (0.0548)[a] | 23.66±1.8 (0.2121)[c] | 0.17±0.0 (0.0284)[a] |
| | | KKB | 10.5±2.09 (0.0182)[a] | 22.16±3.0 (0.4091)[c] | 0.33±0.0 (0.0023)[a] |
| | | KKC | 8.35±2.6 (0.3972)[c] | 27.16±1 (0.0514)[a] | 0.34±0.0 (0.0022)[a] |
| | | KKD | 8.08±1.4 (0.4248)[c] | 27±1.8 (0.0504)[a] | 0.27±0.0 (0.0025)[a] |
| | | KKE | 8.74±1.5 (0.2253)[c] | 26.16±1.4 (0.0708)[ab] | 0.21±0.0 (0.0095)[a] |
| | | KKF | 12.3±0.5 (0.0022)[a] | 24.5±0.8 (0.1420)[ab] | 0.34±0.0 (0.0000)[a] |
| | | KKG | 13.66±1.5 (0.0111)[a] | 26.83±3.0 (0.0614)[ab] | 0.45±0.0 (0.0000)[a] |
| | F | | 0.006[a] | 0.057[a] | 0.0001[a] |

* Harvesting after 27 days; a: Significant increase; ab: Weakly significant; c: Not significant increase.

Figures in parentheses, *P* value of 't' test; F, observed value of ANOVA.

## Carrier Selection

The results of screening of carriers in whey medium are represented in Figure 28.1. In whey, the growth of actinomycetes was comparatively high when supplemented with wheat bran. Rice husk, bagasse are mainly rich with cellulose. Wheat bran contained protein 15 per cent, fat 3 per cent and fiber 11 per cent.

There are also number of references available on various carriers like lignite, coal, cellulose, bagasse, wheat straw, coconut shell etc. individually and also in combination (Chao and Alexander 1984; Kremer and Peterson,1983; Peterson and Loynachan 1981; Kandaswamy and Prasad 1971).

In present study the biomass production was seen more in whey supplemented with wheat bran than paddy husk and bagasse. The non-availability of adequate nutrients in bagasse may be one of the reasons of less biomass production. The growth of *Streptomyces* sp. KKB in whey medium supplemented with 1 per cent bagasse was least amongst all. The pH of the medium supplemented with bagasse was significantly reduced after fermentation, which may have adversely affected biomass production. Lignite and peat soils are the conventional carries used for bioinocula preparation. However, their resources are decreasing day by day. Besides, they are expensive. Therefore, wheat bran will be an alternative to conventional carriers. Wheat bran was preferred carrier in subsequent experiments

**Table 28.5: Effect of Actinomycete Strains Liquid Inocula on Wheat* in Nonsterile Soil**

| | *Response Variables →* | | *Avg Root Length (cm)* | *Avg Shoot Length (cm)* | *Avg Plant Wet Wt (g)* |
|---|---|---|---|---|---|
| Value for variable with Mean ± S. D. | Strain | Control | 6.23±1.41 | 19.8± 4.1 | 0.089±0.0 |
| | | KKA | 10.87±1.40 (0.0006)[a] | 24.83±1.2 (0.0834)[ab] | 0.383±0.05 (0.040)[a] |
| | | KKB | 10.44±1.65 (0.0128)[a] | 23.33±3.5 (0.0869)[ab] | 0.362±0.04 (0.0005)[a] |
| | | KKC | 7.69±0.20 (0.0417)[a] | 25.66±0.5 (0.0563)[a] | 0.327±0.0 (0.0000)[a] |
| | | KKD | 6.33±1.15 (0.457)[c] | 19.8±4.2 (0.5)[c] | 0.230±0.0 (0.0000)[a] |
| | | KKE | 6.70±0.43 (0.248)[c] | 20.16±4.1 (0.0584)[a] | 0.236±0.0 (0.0000)[a] |
| | | KKF | 9.13±2.85 (0.0418)[a] | 24.8±1.8 (0.0826)[ab] | 0.322±0.1 (0.0000)[a] |
| | | KKG | 14±2.64 (0.0123)[a] | 26.5±0.1 (0.0670)[ab] | 0.464±0.0 (0.0000)[a] |
| | F | | 0.001[a] | 0.110 [c] | 0.001 [a] |

* Harvesting after 27 days; a: Significant increase; ab: Weakly significant; c: Not significant increase.

Figures in parentheses, *P* value of 't' test; F, observed value of ANOVA.

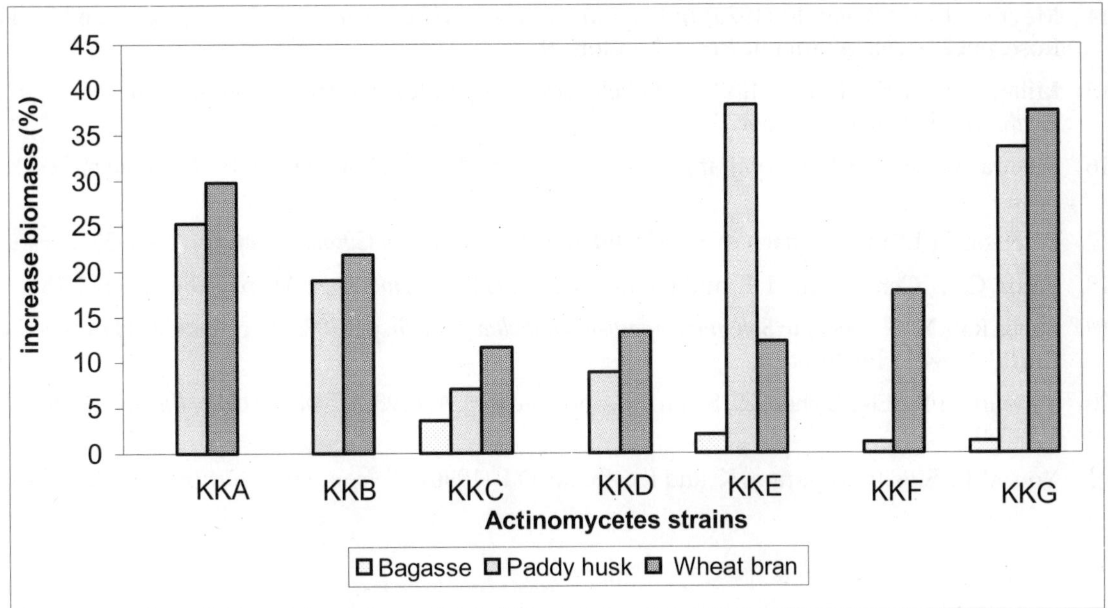


**Figure 28.1: Screening of Carriers Whey Media for Carrier Based Bioinocula of Acidophilic Actinomycete (Biomass per cent Relative to that in absence of carrier)**

## References


1. Ahn, W.S., Park, J. and Sang, Y.L. (2000). *Applied Environmental Microbiology* 66, 3624-3627.
2. APHA-AWWA. WPCF (1995). In Standard Methods for Examination of Water and Wastewater. 19th ed. American Public Health Association, Washigton, D. C.
3. Bailey N. T. J. (1965). In *Statistical methods in Biology*. E.L.B.S. edn. 1-200, The English Universities press Ltd., Great Briton.
4. Chao, W.L. and Alexander, M. (1984). *Applied and Environmental Microbiology* 47, 94-97.
5. Chartrain, M. and Zeikus, J.G. (1986). *Applied and Environmental Microbiology* 51, 188-196.
6. El-Tarabily, K.A., Soliman, M.H., Nasar, A.H., Al-Hassani, H.A., Sivsithumparam, K., McKenna, F. and Hardy, G.E. St. J. (2000). *Plant Pathology* 49, 573-583.
7. Holsinger, V.H., Posati, L. P. and Devilbiss, E.D. (1974). *Journal of Dairy Science* 57, 849.
8. Kanavade V. L. and Kapadnis B. P. (2000). In *International Conference on Microbial Technology, Trade and Public Policy* 15–17th July, 2000, Hyderabad, India.
9. Kanavade V. L. and Kapadnis B. P. (2002). *Journal of Agricultural universities*. 27, 168-172.
10. Kandaswamy, R. and Prasad, N.N. (1971). *Current Science* 40, 496.
11. Kremer, R.J. and Peterson, H.L. (1983). *Applied and Environmental Microbiology* 45, 1790-1794.
12. Lowry, O. H., Rosebrough, N.J., Farr, A.L. and Randall, R.J. (1951). *The Journal of Biological Chemistry* 193, 265-275.
13. Mackay, S. J. (1977) Improved enumeration of *Streptomyces* sp. On a salt casein starch medium. *Applied Environmental Microbiology* 33, 227-230.
14. Meyrath, J. and Bayer, K. (1979) Biomass from whey. In *Economic Microbiology* Vol. 4., ed. A. H. Rose. pp. 207-268. Academic Press. London. N. Y.
15. Miller, G.L. (1959) Use of dinitrosalicyclic acid reagent for determination of reducing sugars. *Analytical chemistry* 31, 426-428.
16. Mishra, S., Barik, S. K., Ayyapan, S. and Mohapatra, B.C. (2000). *Bioresourse Technology* 72, 213-218.
17. Peterson, H.L. and Loynachan, (1981). *International Review of Cytology Suppl.* 13, 311-331.
18. Sorba, C.N., Cerevellini M. I., and Balatti, A. P. (1993) *Review Argent Microbiology* 25, 171-184.
19. Subba Rao, N. S. (1986). In *Soil microorganisms and plant growth*. pg. 388. Oxford and IBH publishing Co. ltd. New Dehli, India.
20. Tahoun, M.K., El–Merheb, Z., Salam, A. and Youssef, A. (1987). *Biotechnology Bioengineering* 29, 358-360.
21. You, M.P., Sivsithumparam, K. and Kurtboke, D.I. (1996). *Biology and Fertility of Soil* 22, 237-242.

# Chapter 29

# Green Planning for Emerging Urban Growth Centres: Important Issues

**D. Mohan[1]* and S.P. Sharma[2]**

**[1]Environmental Engineering Division, Department of Civil Engineering, Institute of Technology, B.H.U., Varanasi – 221 005**

**[2]Graduate Student, Department of Civil Engineering, Indian Institute of Technology, Delhi**

## ABSTRACT

Towns and cities emerging as new growth centres in the developing world are facing the problem of suddenly enhanced alarming levels of atmospheric pollution due to rapid urbanization and industrialization and increased emissions from automobiles employed for surface-transportation. Recently, the automobile sector has emerged as one of the biggest sources of air pollution in such places. In spite of various efforts to curb the menace of automobile-traffic generated air pollution, the improvements achieved in the quality of urban atmospheric environment have continued to remain far away from the satisfactory level. In fact, various physiological threats, which may arise, to human health, due to this source of pollution are not given appropriate weightage at the time of planning of surface transportation infrastructure in urban areas.

This chapter emphasizes that there is an urgent need for inclusion of vehicular traffic generated atmospheric pollution as one of the parameters in planning of surface transportation infrastructure in large cities as well as in emerging growth centres in urban areas in India so that an improved level of safety (from the probable risks caused due to the air pollution) to health of the inhabitants may be ensured. It also discusses certain issues, which need to be addressed

* Corresponding Author: E-mail: devmohan9@gmail.com

while planning a new growth centre, for maintaining the automobile-traffic generated pollution at the lowest-feasible level. At the end, it illustrates in brief the case of Varanasi city, which is a very fast developing new growth centre so far as commercial, industrial, educational and various other types of activities related to development of society are concerned.

***Keywords:*** *Automotive exhaust, Growth centres, Atmospheric pollution, Environmental and physiological impacts, Designed plantation, Infrastructure planning, Water transport.*

## Introduction

Rapid urbanization coupled with greater demand for transportation for day-to-day activities have created near chaotic traffic conditions in a number of cities and towns in many developing countries of the world. Over the years, there has been a substantial increase in air pollution caused by vehicular exhaust emissions due to addition of more and more vehicles on roads to meet the rising transportation demand (Mayer, 1999; Sharma, 1997; Sharma and Khare, 2001). According to the findings of Global Workshop on Air Pollution, large cities in South Asia have reached alarming levels of air pollution due to rapid urbanization, industrial activities and ever-increasing emissions from automobiles. The expenditure incurred on health-related problems has been estimated to be about U.S. $ 200 millions per annum for Dhaka, the capital city of Bangladesh, which has been found to be highly polluted due to vehicular exhaust in a study carried out continuously for four weeks by Paul and Quader (2002). It was found there that an average concentration of particulates with size less than or equal to 10 mm was 422.7 mg/m$^3$ in ambient air, which is much in excess of the permissible limits set by Department of Environment, Bangladesh or United States Environmental Protection Agency.

Delhi has been found to be the fourth most polluted city in terms of concentration of the suspended particulate matter (SPM) in ambient air (WHO, 1992). Mumbai had also been observed to be the fourth highest $NO_x$ polluted city in the world (Srinivas, 1999). Table 29.1 presents the trend of rise in numbers of different types of automobiles in various important cities of India and Table 29.2 shows estimated quantities of different automotive exhaust pollutants, emitted in the atmosphere of Delhi, a centre, which has experienced a continuous growth in various arenas of development.

Due to mainly two- and three-wheeler automobiles, even a small city like Silchar in the state of Assam in India has been found to have significant level of air pollution due to automotive traffic and it has been reported that a sizeable portion of the urban population suffered from its adverse effects on human health (Pandey and Pandey, 2003). In a number of studies conducted in important Indian cities, it has been found that the quality of air has deteriorated to a perceptible degree due to pollution originating from mobile sources (Mariappan *et al.*, 2000; Pandey *et al.*, 1998; Pandit, 1974; Ramamurthy and Thirumaran, 2002; Sivacoumar, 2000; Srinivas, 1999).

Carbon monoxide (CO) is one of the most abundant of the gaseous pollutants found in automobile exhaust and more than $4 \times 10^{12}$ kg of this gas has been estimated to be emitted globally every year (Li *et al.*, 2000). In large cities, due to largely obstructed airflow, prolonged periods of stagnation lead to inadequate dispersion and consequently, result in enhancement of ambient concentration of CO in air. Community CO levels vary with proximity to traffic, traffic volume, type of traffic and many meteorological variables. As CO emission increases with decreasing engine speed, traffic congestion leads to enhancement in ambient CO concentration, both by increasing the number of vehicles on given stretch of road and also, by enhancing the value of emission rate per unit road length travelled

**Table 29.1: Automobiles Registered in some Fast Growing Cities in India**

| *Time* | *Two-wheelers* | *Autos/ Tempos* | *Cars/Cabs* | *Buses* | *Goods Carriers* | *Tractors/ Others* | *Total* |
|---|---|---|---|---|---|---|---|
| | | | **Chennai** | | | | |
| 01.04.1985 | 122123 | 6115 | 55529 | 2945 | 12337 | 798 | 199847 |
| 01.04.1990 | 338486 | 5580 | 113783 | 3657 | 28917 | 1871 | 492294 |
| 01.04.1995 | 565451 | 20849 | 148896 | 4317 | 23475 | 1460 | 764448 |
| 01.04.2002 | 988630 | 44771 | 250080 | 4541 | 61459 | 6202 | 1355683 |
| | | | **Delhi State** | | | | |
| 01.04.1985 | 579064 | 30017 | 166263 | 13522 | 52370 | * | 841236 |
| 01.04.1990 | 1113236 | 58934 | 354810 | 17844 | 92778 | * | 1637602 |
| 01.04.1995 | 1617732 | 74981 | 588309 | 26202 | 26202 | * | 2432295 |
| 01.04.2002 | 2265955 | 86985 | 989522 | 47578 | 161650 | * | 3551690 |
| | | | **Greater Mumbai** | | | | |
| 01.04.1985 | 167338 | 32351 | 236186 | 22506 | 46840 | 3427 | 508648 |
| 01.04.1990 | 305099 | 41814 | 302122 | 10878 | 75405 | 3623 | 738941 |
| 01.04.1995 | 434802 | 101597 | 313353 | 16291 | 83517 | 6070 | 955630 |
| 01.04.2002 | 787527 | 212862 | 547224 | 20718 | 124718 | 8215 | 1701264 |
| | | | **Kolkata** | | | | |
| 01.04.1985 | 160556 | 4968 | 177736 | 15736 | 62514 | 10375 | 431885 |
| 01.04.1990 | 217304 | 7100 | 219079 | 18330 | 75083 | 11480 | 548376 |
| 01.04.1995 | 294110 | 10146 | 270039 | 21353 | 90179 | 12703 | 698530 |
| 01.04.2002 | 467756 | 27003 | 380079 | 28923 | 105687 | 28003 | 1037451 |
| | | | **Varanasi (District)** | | | | |
| 01.04.1985 | 47220 | 3000 | 2940 | 304 | 1274 | 2945 | 57683 |
| 01.04.1990 | 78700 | 4526 | 5876 | 605 | 2672 | 4703 | 97082 |
| 01.04.1995 | 133280 | 6083 | 11049 | 1701 | 4924 | 9520 | 166557 |
| 01.04.2002 | 285068 | 7391 | 22322 | 1989 | 5837 | 22648 | 345255 |

* Already included in the number of goods carriers.

**Table 29.2: Pollution Load Generated by Automobile Exhaust in Delhi (PNL, 1997)**

| *Pollutant* | *Load (thousand tonnes/year)* | | | | | | | |
|---|---|---|---|---|---|---|---|---|
| | *1991* | | *1996* | | *2000* | | *2005* | |
| | *A* | *B* | *A* | *B* | *A* | *B* | *A* | *B* |
| Carbon monoxide | 243 | - | 373 | 351 | 477 | 406 | 606 | 441 |
| Hydrocarbons | 82 | - | 123 | 113 | 156 | 135 | 198 | 149 |
| Nitrogen oxides | 139 | - | 208 | 207 | 261 | 250 | 330 | 279 |
| Sulphur dioxide | 10 | - | 15 | 15 | 18 | 16 | 23 | 18 |
| Lead | 0.19 | - | 0.297 | 0.259 | 0.384 | 0.269 | 0.489 | 0.276 |
| Particulate matter | 19 | - | 28 | 28 | 35 | 34 | 44 | 41 |

A: No control measures adopted; B: With enforcement of emission norms and fuel quality.

by each of the vehicles. CO concentrations in ambient air encountered in heavy traffic zones is about three times higher in comparison to that in commercial areas and five times higher than the value observed in residential areas. In developed countries, estimations of CO concentration in and around roadways and urban arterials are considered essential (at the stage of environmental impact assessment) to design the layout of different transportation routes (Cooper, 1987; Koushki, 1991; Ward, 1975).

## Automobile Exhaust-Pollution: Some Issues

### Parameters Affecting the Emission Characteristics

In India, surface transport poses specific problems because of mixed traffic flow, having about thirteen different incompatible types of vehicles occupying the same right of way and moving in an aggregated flow. This condition results in higher consumption of fuel for relatively less distance travelled and consequently, enhanced quantities of air pollutants are emitted into the exhaust. To make the situation further worse, operational cost of a vehicle has been estimated to go up by 20-25 per cent on bad roads, in addition to aggravation of the air pollution problem. Average mileage is estimated to be 12 to 30 kmpl in Delhi and average speed on some roads, to be as low as 10 kmph.

The exhaust composition depends on many variable parameters like engine type and operating conditions, driving mode, type of fuel used, vehicle age, make and maintenance conditions, type of emission control system, weight of vehicle and its engine capacity and many other such factors. Many studies have revealed the effects of various operating parameters on exhaust emission from vehicles (Belardini *et al.*, 1993; Hansen *et al.*, 1995; Jensen, 1995). Certain parameters related to operation, maintenance and driving behaviours can be effectively modified to a large extent by planning appropriately the surface-transportation infrastructure in rapidly developing new growth centres in India.

### Adverse Impacts

The health effects of the pollution generated by automotive traffic are yet to be fully understood, but increasing incidences of asthma, other respiratory diseases and certain types of cancer have been linked to various exhaust pollutants such as carbon monoxide, oxides of nitrogen and volatile organic compounds (Clifford *et al.*, 1997). It is shocking to note that about 40 percent of children in Kolkata have been found to be suffering from respiratory distress syndrome due to air pollution (Bhattacharya, 2001). Average figure of human morbidity and mortality has risen to about 57,000 everyday in 36 Indian cities, with about 26 million people being hospitalized every year, due to different diseases caused by atmospheric pollution (Pattnaik and Pattnaik, 2000).

The estimated cost (due to biospheric pollution) to India in 1995 was Rs. 3, 68,000 millions, out of which an amount of about Rs. 52,400 millions was attributed to adverse health effects of urban air pollution alone (IJEP, 1999). The statistics of prevalent diseases in Delhi in recent past shows that there has been a steep rise in the number of patients facing various physiological disorders related to air pollution caused by automobiles (Srinivas, 1999).

Automotive emissions not only lead to severe degradation in urban air quality on micro-scale of meteorology, but are also reported to contribute to atmospheric burden of radiatively important trace gases, which have potential to influence the global climate significantly (Deluchi *et al.*, 1987), such as $CO_2$, methane and nitrous oxide. Light-absorbing property of diesel particulate matter (DPM) found in exhaust of diesel engines also affects Earth's radiation balance. $CO_2$ levels have been increasing since the Industrial Revolution and are expected to continue like this in near future also. Mankind has been extracting carbon-rich fuel from beneath the Earth's surface and putting its combustion-products into the atmosphere, where it becomes available for conversion by plants through the process of photosynthesis (Robinson *et al.*, 1998). In India, $CO_2$ is emitted by various sectors as shown in Figure 29.1 (http://earthtrends.wri.org/pdf_library/country_profiles/cli_cou_356.pdf.).

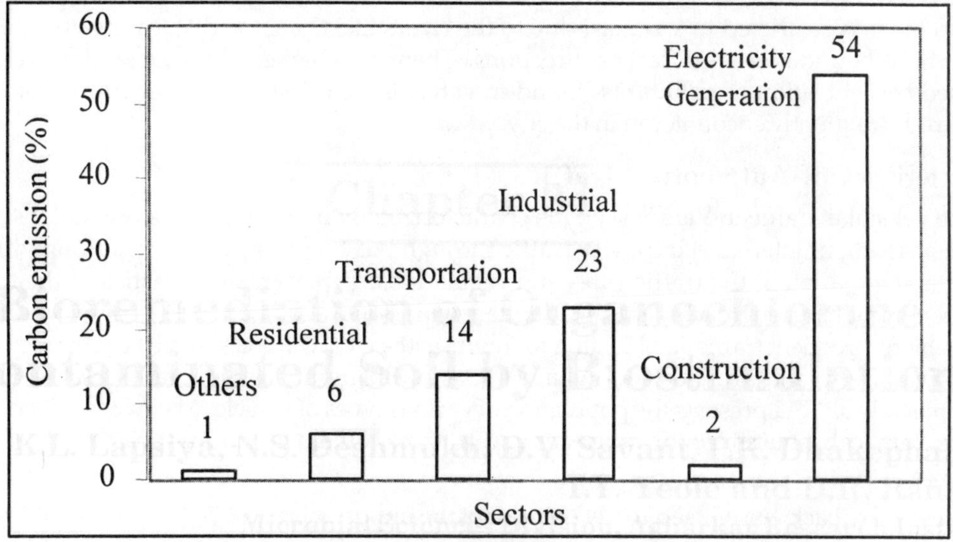


**Figure 29.1: Sectoral Share of $CO_2$ Emissions in India**

Concentration of $CO_2$ has increased by more than 25 percent since pre-industrial times. About one-half of potential global warming is being caused by $CO_2$ in the atmosphere and the remaining half by the other greenhouse gases. The Intergovernmental Panel on Climate Change (IPCC) has made a forecast that global average surface temperature will raise 1.4–5.8°C (above that of 1990 level) by 2100. If the world's carbon sinks are indeed reaching a saturation point, it is feared that actual rise in temperature could be even greater (Earth Policy Institute, 2004).

Excessive release of carbon into atmosphere may lead to under-nutrition and also, various adverse health consequences due to population-displacement and economic disruption. It is feared that average rise in global temperature may increase the likelihood of many vector-borne diseases in new areas by 2100. Effect of the climate changes on the diseases is likely to be observed at the extremes of the temperature-ranges at which the transmission occurs. Encephalitis may become a grave public health concern with more and more rise in global temperature (Githeko *et al.*, 2000).

## A Case of Varanasi City

### About the Location

District Varanasi is located in south-western portion of the state of Uttar Pradesh in India. It is geographically located between 82°15′ to 82°30′ East Longitude and 24°34′ to 25° 30′ North Latitude. It is 76.19 m above mean sea level. The district is having 2696 villages and 17 towns. Population of Varanasi district is 3,147,927 and its population density is about 1995 persons/sq. km according to 2001 census (http://www.censusindia.net/results). Varanasi is a fast growing commercial, industrial and trading centre of Eastern Uttar Pradesh and Central India. The city is well connected with most of the important cities of state and the country through highways. The major roads passing through the city are Grand Trunk road, NH–56, NH–29, a Major District Road to neighboring Azamgarh district and an Other District Road to Sindhora town.

Besides the important roads mentioned above, there are other minor categories of roads also, which connect the city with nearby places, *e.g.*, Chunar, Mirzapur and far and near villages. Expansion of urban limits is restricted in Varanasi due to the rivers like Ganga and Varuna, which make the geographical boundaries in different directions. Therefore, higher automobile-density leads to enhanced concentration of pollutants, found in vehicular exhaust because of traffic congestions, which are so frequently encountered in the city.

### Characteristics of Automotive Traffic

The vehicular traffic in Varanasi city is characterized by intermixing of slow and fast moving vehicles, passing of inter- and intra-city traffic through busy portions of the city alongwith erratic driving and violation of the traffic rules and regulations. Typical characteristics of the roads are improper geometries, encroachments, inadequate widths and road networks. These conditions are responsible for frequent traffic jams, lasting for hours in the city. The areas, which experience frequent traffic jams, are in and around Godowlia, Maidagin, Bhelupur, Sonarpura, Nadesar, Sigra and Cantonment. Table 29.1 presents the population of various types of vehicles like two wheelers, Autos/Tempos, cars/cabs, buses, goods carriages and tractors/others in Varanasi city at different time (PPAC, 2003).

A few years back, buses used to ply on most of the important roads, but at present, these are not there to provide the passenger-transportation services at any of the routes of the city. In addition, there have been tremendous rise in number of two-wheelers and also, in those of three- and four-wheelers to provide the much-required passenger-mobility. On account of the enhanced number of automobiles on roads, traffic-jams have become quite frequent. Moreover, it has become next to impossible to drive the vehicles near the least-polluting speeds and consequently, the automotive-traffic generated air pollution has increased enormously.

In a study conducted in Varanasi city, it has been observed that in the vicinity of certain traffic intersections, average speed of two-wheelers was about 12 kmph; that of autorickshaws, around 10 kmph and in case of four wheelers, about 8 kmph (Padhiari, 2003). Thus, it is clear that at these speeds, all the automobiles will spew excessively high quantities of exhaust pollutants. Efforts to abate automobile-exhaust generated air pollution without improving the traffic management may not meet significant success. Introducing bus lanes and giving priority to buses at traffic intersections have been estimated to bring down the travel-time by as much as about 25 per cent in certain megacities. Dedicated bus lanes have also been estimated to contribute towards the enhancement in the number of passenger transported (by buses) by more than 50 per cent.

## Control of Carbon Emission

The idea of providing carbon sinks by growing more number of trees is based on the carbon cycle. One of the studies of a forest in an atmosphere enriched in $CO_2$ suggests that the trees absorb carbon more efficiently and the carbon absorbed gets stored in the soil around tree roots (Ding *et al.*, 2006). Younger and therefore, fast-growing trees absorb relatively larger amounts of carbon dioxide than mature trees. $CO_2$ absorption continues till the trees grow in size and this time-period is limited. Therefore, to enable the trees continue to work as effective sinks, properly planned plantation and harvesting of rapidly growing vegetation and trees need to be given serious consideration (http://

en.wikipedia.org/wiki/Carbon_sequestration). According to a study related to the environmental implications of material consumption patterns across various income groups in India, carbon emissions for urban population in India are estimated as given in Table 29.3 (Warran and Patwardhan, 2002).

**Table 29.3: Estimated Per Capita Carbon Emission (kg/year)**

| *Income Group* | *1989-90* | *2005* | *2010 (Projected)* |
|---|---|---|---|
| Lower class | 103.4 | 114.86 | 122.5 |
| Middle class | 245.7 | 265.92 | 279.4 |
| High class | 873.6 | 898.26 | 914.7 |

In Varanasi city, $CO_2$ is mainly emitted by transportation, residential and other sectors only, which is related to 21 per cent of the total of emissions by all sectors in case of India at national level, in which (the quantity for India) transportation accounts for 14 per cent and therefore, in case of Varanasi, it can be taken as 2/3$^{rd}$ of the total emissions from all the important sources as mentioned above.

In Varanasi, plantation is very much required for maintaining the $CO_2$ concentration within sustainable limits. This option can be very effective in recycling the carbon, which is being emitted into the noosphere of Varanasi city. The trees carry out the task of $CO_2$ recycling by the process of photosynthesis, using their canopy. Therefore, the canopy area is very important so far as $CO_2$ utilization is concerned.

Almost all type of plant life on Earth can be divided into two categories, based on the manner they assimilate carbon dioxide into their metabolic systems (http://www.serc.si.edu/labs/co2/c3_c4_plants.jsp). A $C_3$ plant is one, which produces phosphoglyceric acid, as a stable intermediary in the first step in photosynthesis. It is a molecule that has three carbon atoms in it. Over 95 percent plants including all trees on Earth are $C_3$ plants. A $C_4$ plant is one that produces oxaloacetic acid, as a stable intermediary in the first step, in the photosynthesis. It is a molecule that has four carbon atoms in its molecule. Photorespiration in $C_4$ plants is more efficient in intense sunlight conditions. In India, $CO_2$ consumption rates of $C_3$ plants lie in the range of 15-35 mg of $CO_2$/hour/dm$^2$ of canopy area with an average value of about 20 mg/dm$^2$/hour (Salisbury and Ross, 1977).

A plant's photosynthetic and physiological response to $CO_2$ is related to concentration of the gas in air, immediately surrounding the leaf surface and this concentration may differ from atmospheric concentrations, measured outside the plant canopy. Amongst wide range of $C_3$ species, which have been examined, the photosynthesis capability of about 95 per cent has not reached the saturation limits by the present $CO_2$ concentration in atmosphere (Champigny and Mousseau, 2003).

Generally, $C_3$ plants show little or no benefit from $CO_2$ enrichment at low temperature conditions. As temperature rises within the range of 20-35°C, proportionate nett photosynthetic rate due to high $CO_2$ levels increases rapidly and benefit from a high level of $CO_2$ is usually observed at warm temperatures. At temperature above about 40°C, the photosynthetic response to $CO_2$ starts falling down as the activity goes down slowly by adverse effect of high temperature (Wolfe, 2003).

Required canopy area for effective recycling process can be estimated in limiting conditions of sunlight availability, $CO_2$ concentration and stomatal response. The critical design period in case of Varanasi city will be the winter season, in which the prevailing temperature will be low and there will not be significant benefit, derived from $CO_2$ enrichment due to the minimum availability of sunlight.

Calculated values of canopy areas obtained for Varanasi in winter season for different conditions are as shown in Table 29.4 below. An average value of effective sunlight-availability has been taken as not more than six hours a day for carrying out these calculations considering the case of $C_3$ type of vegetation.

**Table 29.4: Canopy Areas Required for Varanasi City (Mohan, Ankur and Sharma, 2006)**

| *Sl.No.* | *Conditions* | *Canopy Area, $m^2$* |
|---|---|---|
| 1. | Theoretical requirement (Minimum) | $16.8 \times 10^7$ |
| 2. | Clean environment | $27.5 \times 10^7$ |
| 3. | Polluted environment | $41.3 \times 10^7$ |

Lower $CO_2$ concentration enhances stomatal openings and the increase leads to their partial closure. Sensitivity of the stomata to $CO_2$ varies with species and also, with environmental conditions. The stomatal activity will decrease by 30-40 per cent, when plants are exposed to a short-term doubling of $CO_2$ concentration. Experiments have shown that $CO_2$ concentration not only affects stomatal behaviour, but can also change the stomatal densities. Stomatal densities of certain species have shown a reduction by about 40 per cent over the past two centuries as $CO_2$ in the atmosphere has been increasing in this time span (Wolfe, 2003). Therefore, the canopy area requirement should be increased by 64 per cent.

### Impacts of Other Gaseous Pollutants

Effects of exposure to atmospheric concentrations of three major pollutants like sulphur dioxide ($SO_2$), nitrogen oxides ($NO_x$) and tropospheric ozone ($O_3$), on crops and natural vegetation, have been studied extensively and it is established that high concentrations of sulphur dioxide, nitrogen oxides and ozone in the atmosphere adversely affect the plant growth and their reproduction rates. These pollutants are capable of inhibiting photosynthesis and altering carbon allocation to the roots and stems (Garner, 2002). Thus, the canopy area need to be increased further by 50 per cent to take into account the reduction in photosynthetic activity due to the pollution caused by these gases.

This canopy area should be provided preferably by using a variety of trees, having the canopies of different sizes (from very large to small), depending on availability of the space at various locations and also, the possibility of survival of the trees or plants under the prevailing conditions. In addition, shrubs, creepers, flowering plants and different types of grasses may occupy the spaces not suitable for tree plantation. The number of trees as well as vegetation can be planned on the basis of the canopy areas provided by them and certain other considerations like aesthetics, etc..

## Planning of the Plantation: Main Considerations (Street Tree Master Plan, 2002)

The following points should be kept in mind, while carrying out the planning of plantation to provide the required canopy area (Sl.No. 3 in Table 29.4) in case of the city:

- ☆ Native trees should be preferred due to their capability to withstand the prevailing environmental conditions.
- ☆ A variety of trees to avoid the problems on account of outbreak of specific diseases should be employed. A mix of evergreen and deciduous trees should be used.
- ☆ New trees should be planted in neighbourhoods of the older ones, especially those situated on or near the planting strips and road-medians.

- ☆ Interchange and crossover areas located between thoroughfares and freeways should be more fully landscaped with a variety of species.
- ☆ Some tree species should be avoided, particularly fruit-bearing and heat-sensitive varieties in parking lots and other similar landscapes as well as on roadsides.
- ☆ Size of the tree at maturity in relation to the space available should be given a serious consideration at the plantation-stage itself. Smaller trees should be used to avoid any possibility of conflicts with electric transmission or telephone lines.
- ☆ Evergreen trees and other species, which cannot be easily trimmed, must be avoided at intersections, parking lot islands and other such locations, where it is required to have a clear vision of the approaching traffic.
- ☆ The open space adjoining many streets should be used for plantation as far as possible.

Trees with large canopy areas should be selected most often in the plantation programmes to be implemented along major thoroughfares. In India, examples of such trees may be Peepal (*Ficus religiosa*), Neem (*Azaidirachta indica*), Shisham (*Dalbergia sissoo*), mango (*Mangifera indica*), Jamun (*Syzygium cumini*), tamarind (Tamarindus indica), Banyan (Ficus indicus), teak (Tectona grandis), Ber (*Zizyphus mauritiana*), Khejri (*Prosopis cineraria*), Lasura (*Cordia dichotoma*), etc..

Many native species of trees can also be used to provide medium canopy areas. Such trees can be used at many locations such as road-medians, small plantation strips as well as at sidewalk plantation and may also be employed inside the residential colonies, market places, etc.. Examples of such trees are Khair (*Acacia catechu*), Kikar (*Acacia nilotica* L.), Dhak (*Butea frondosa*), Shahtoot or mulberry (*Morus alba atropurpurea*), Amrood or guava (*Psidium guajava*) and poplar. Pine and eucalyptus may also be considered under specific conditions. Other trees in this group, which can be used, are Tun (Red Cedar), Baheda and Bel.

Trees with small canopy areas seasonally brighten the regional landscapes. These include large size flowering and also, a variety of ornamental plants. Bogen belia, Amaltash, Alistonia, Gulmohar and hedge plants like Kaner may be used. Leafless shrub species like Kair or Teat (*Capparis decidua*) and a beautiful flowering tree, Kachnar, can also be used. Land areas covered with carpet of grasses and parks would also be preferred keeping the availability of space in view. Other important plants with high medicinal values like Tulsi (*Oscimum sanctum* and *Oscimum basilicum*) should be used at places, wherever, possible (http://www.haryana-online.com/flora.htm).

## Green Planning of Transportation Infrastructure

In the present era of liberalization and globalization in India, a large number of towns and cities are developing very fast as new growth-centres due to rapidly advancing industrial activities of various types. There has been tremendous rise in number of automobiles to meet the enhanced transportation demand. Due to compulsion of accommodating the increased number of automobiles within the existing capacity of roads, average speed of traffic movement has gone down drastically, leading to an alarming rise in the air pollution caused by their exhaust. Therefore, surface transportation infrastructure need to be planned, keeping in view the severity of the threat to public-health due to the air pollution being generated there. Other pollution-control measures like green-belt development at such places should also be considered.

Properly planned and operated surface transportation infrastructure will help in achieving congestion-free traffic-flow in the range of the least polluting speeds and also, it will bring down the probability of formation of high pollution-zones, which tend to retain various air pollutants by restricting the wind movement. Consequently, a huge expenditure on account of 'the pollution induced health hazards' can be brought down effectively.

It has been observed that various primary and secondary measures for control of pollution caused by the automobile exhaust have only been partly successful in light of the ever-evolving regulations related to concentrations of various pollutants in the exhaust. Therefore, multi-pronged strategy need be adopted for effective abatement of pollution due to automobile exhaust to realize the goal of healthy atmospheric environment in large cities and various new growth centres.

In case of gasoline powered vehicles, driving at an optimum speed of 30-40 kmph in case of two/ three wheeler vehicles and 45-55 kmph in case of four wheeler vehicles can be possible only after ensuring an appropriate planning and operation of surface transportation routes (Mohan, 2004). This may also help to large extents in avoiding harsh braking, and frequent acceleration and deceleration while driving, which are highly polluting phases. For diesel-powered automobiles, optimum-driving speed should be around 40 kmph.

## Development of Water Transport

The rivers flowing on the different sides of this city, *i.e.*, Ganga and Varuna, have the potential to be put to an effective application as water-transportation channels. This will bring down the traffic-congestion on the roads, enhance the travel-speeds and as a result, the mileages achieved by various types of automobiles, and consequently lead to a significant reduction in the automobile-traffic generated air pollution in the city. A large floating population of tourists and pilgrims may also like to travel by these water-routes, leaving ample scope at roads to be utilized by the commuters for making trips in connection with the educational, commercial and for various other purposes also.

## Application of Emerging Technologies

To achieve lower emissions of automobile-exhaust pollutants further by ensuring smoother movement of the traffic-flow around the least polluting speeds, application of emerging concepts and technologies like intelligent vehicle highway system (IVHS) and Intelligent transportation system (ITS) may be considered for application tom the best possible extent in various emerging growth centres due to rapidly growing industrial activities. IVHS uses the latest technologies to squeeze more transportation from existing capacities of roads, allowing the vehicles to move in a more fuel-efficient manner. It helps to improve the road safety, reduce fuel consumption, bring down the pressure for construction of new surface-transportation-routes, make public transportation system more attractive and faster and consequently, can help in realizing effectively the reduction in quantities of different air pollutants generated.

Such technologies, if integrated into surface transportation infrastructure and vehicles, traffic-flow can be regularly monitored and efficiently managed, congestions can be minimized and alternative routes to travellers can be indicated. Advanced traffic management along with the highway advisory radio-system may be considered to monitor the traffic, optimize signal timings on major routes and control the traffic-flow in such places. Such measures will help largely in mitigation of recurring and non-recurring congestions, leading to reduction in the vehicular traffic-generated air pollution.

## Conclusion(s)

To get an appropriate and effective solution of this serious problem, a multi-pronged strategy needs to be adopted in this fast-growing city at the earliest. This may involve widening of the roads along with improvement in their surface quality; removal of bottlenecks existing at various places at present; adoption of a properly designed traffic-signalling system; provision of separate lane for pedestrians, cycles and rickshaws; and buses on both sides of the main routes, etc.. In addition, restrictions on registration of automobiles after certain age need to be strictly enforced to bring down the rate of pollutant-generation by the automotive traffic.

In Varanasi city, bus corridors need also to be identified and a shuttle service along these routes should be provided. If reliable, comfortable, fast, low-cost and easily accessible public transport is provided and proper infrastructural development ensured, the number of private vehicles on-road may go down significantly. In addition to various other measures discussed above, an intensive campaign needs to be launched to educate the road-users to modify their erratic movement-behaviour observed on roads at present. Also, the presence of cattle on roads in Varanasi city needs to be effectively controlled by the local authorities to bring down the hindrance in traffic-flow and also in number of accidents, many times caused due to such stray animals.

Another important aspect of the emerging growth centres in urban areas is that an intensive consumption of fossil fuels is localized in relatively small geographical regions. Natural recycling mechanism of $CO_2$, which is the major product of fossil-fuel-burning, is based mainly on the process of photosynthesis. As a matter of fact, there is an urgent necessity to plan the cities in such a way that the recycling capacity of appropriately designed plantation for $CO_2$ uptake at such places becomes adequate to recycle the excess gas generated at such places by the automotive-traffic. Management of this type of pollution is feasible by application of engineered plantation at a large scale to achieve a sustainable growth and also, to realize in practice the concept of a Green City for the newly developing growth-centres like Varanasi.

In addition to abatement of $CO_2$, presence of a green environment due to designed plantation in rapidly emerging growth-centres like Varanasi will also lead to a desirable psychological impact for better productivity. Thus, engineered plantation for pollution control will help in achieving a clean environment along with other benefits like aesthetic and medicinal values of the greenery for the nation in general and inhabitants of such places in particular. As a matter of fact, planning of engineered plantation should be done keeping in view locally prevailing meteorological, geographical, topographical and ecological conditions.

## References


1. Belardini, P., Bertoli, C., Giacomo, N. D., Iorio, B., 1993. Combustion and Pollutant Emissions from Light Duty Diesel Engines: The Influence of Mixing Process and Transient Operating Conditions. *The Science of the total Environment*, 134(1-3), pp. 285-293.
2. Bhattacharya, S. K., 2001 (Jan. 13-19). Protection of Environment and the Role of Law. *Employment News Weekly*, New Delhi, pp. 1-2.
3. Champigny M. and Mousseau M., 2003. Plant and Crop Response to Trends in Climate Changes, in: Pessarakli M. (Ed.), Handbook of Plant and Crop Stress, Marcel Dekker Inc., New York, pp. 1091-1097.
4. Clifford, M. J., Clarke, R. and Riffat, S. B., 1997. Local Aspects of Vehicular Pollution. *Atmospheric Environment*, 31(2), pp. 271-276.
5. Cooper, D.C., 1987. Indirect Source Impact Analysis: Carbon Monoxide Modelling. *Journal of Air Pollution Control Association*, 37 (2), pp. 1308-1313.

6. Deluchi, M. A., Johnston, R. A. and Sperling, D., 1987. *Transportation Fuels and the Green House Effect*. University Wide Energy Research Group, UERG California Energy Studies Report No. UER-182, University of California, Davis.

7. Ding, J., Wu, Y., Zheng, H., Fu, W., Reardon, R. and Liu, M., 2006. Assessing Potential Biological Control of the Invasive Plant. *Biocontrol Science and Technology*, pp. 1-20.

8. Earth Policy Institute, 2004. Detailed report available at http://www.earth-policy.org/Indicators/CO2/2004.htm.

9. Garner, J. H. B., 2002. Air Pollutants, Plants Response, Soil Microbes and Ecosystem Biodiversity. Detailed report available at http://www.geocities.com/isebindia/01_04/02-07-1.html.

10. Githeko, A. K., Lindsay, S. W., Confalonieri, U. E. and Patz, J. A., 2000. Climate change and vector-borne diseases: a regional analysis. Detailed report available at http://whqlibdoc.who. int/bulletin/2000/Number per cent 209/78(9)1136-1147.pdf.

11. Hansen, J. Q., Winther, M. and Sorenson, S. C., 1995. The Influence of Driving Patterns on Petrol Passenger Car Emissions. *The Science of the Total Environment*, 169(1-3), pp. 129-139.

12. http://earthtrends.wri.org/pdf_library/country_profiles/cli_cou_356.pdf.

13. http://en.wikipedia.org/wiki/Carbon_sequestration.

14. http://www.haryana-online.com/flora.htm.

15. http://www.serc.si.edu/labs/co2/c3_c4_plants.jsp.

16. IJEP, 1999 (June). *Indian Journal of Environmental Protection*, 19(6), pp. 462-470.

17. Jensen, S. S., 1995. Driving Patterns and Emissions from Different Types of Vehicles on Roads. *The Science of the total Environment*, 169(1-3), pp. 123-128.

18. Koushki, P. A., 1991. Evaluation of Street-Canyon Carbon Monoxide: Dispersion Simulation Model. *Journal of Transportation Engineering, ASCE*, 117(4), pp. 444-456.

19. Li, X., Yang, L., Lei, Y., Wang, J. and Lu, Y., 2000 (Oct.). A Method for Removal of CO from Exhaust Gas Using Pulsed Corona Discharge. *Journal of Air and Waste Management Association*, 50, pp. 1734-1738.

20. Mariappan, G., Rajashekhar, R. V. J. and Muthusubramanian, P., 2000. Estimation of Automobile Source Strength for Some Air Pollutants in Madurai City. *Indian Journal of Environmental Protection*, 20(8), pp. 561-565.

21. Mayer, H., 1999. Air Pollution in Cities. *Atmospheric Environment*, 33, pp. 4029-4037.

22. Mohan, D., 2004. Automobile Traffic Generated Atmospheric Pollution: Need for Consideration in Urban Infrastructure Planning. *Proceedings of the Third International Symposium on New Technologies for Urban Safety of Mega Cities in Asia*, Agra, India, pp. 183-192.

23. Mohan, D., Ankur, K. and Sharma, S. P., 2006. Carbon Dioxide Generated by Automobiles: An Abatement Strategy, *Proceedings and Souvenir of All India Seminar on Air Pollution and Health Impacts (AISAPHI)*, Gurjar, B.R. (editor), The Institution of Engineers (India), Roorkee Local Centre, Roorkee, pp. 200–211.

24. Padhiari H. K., 2003. Carbon Monoxide in Automobile Exhaust: A Case Study of On-road Vehicles in Varanasi City. M. Tech. Dissertation submitted to the Department of Civil Engineering, Institute of technology, Banaras Hindu University, Varanasi, pp. 98-121.

25. Pandey, K. M. and Pandey, A. K., 2003. Impact on Health of Usage of Non-Conventional Energy Systems for Controlling Environmental Pollution: A Case Study of Silchar City. *Proceedings of Fourth National Conference on Thermal Systems (FNCTS-2003)*, Gupta, S. N. and Prakash, O. (editors), Varanasi, pp. 47-59.

26. Pandey, V., Kumar, A., Pal, A., Singh, N. and Younus, M., 1998. Status of Ambient Air Quality in Lucknow City. *Indian Journal of Environmental Protection*, 19(3), pp. 181-184.

27. Pandit, V. I., 1974. Automobile Exhaust: Air Pollution and Its Control. *Indian Journal of Environmental Health*, 16(4), pp. 344-359.

28. Pattnaik, S. and Pattnaik, M., 2000 (Oct. 7-13). Combating Air Pollution in India. *Employment News Weekly*, New Delhi, pp. 1-2.

29. Paul, M. K. and Quader, A. K. M. A., 2002. A Study of the Air Pollution of Dhaka City. *Abstract Proceedings of Indian Chemical Engineering Congress-2002*, p. 54.

30. PNL, 1997. Pollution Control in Problem Areas. *Parivesh News Letter*, 3 (IV), pp. 1-11.

31. PPAC: Petroleum Planning and Analysis Cell (PPAC), 2003. Ministry of Petroleum and Natural Gas, *Government of India*.

32. Prakash R., Henham A. and Bhat I. K., 2005. Gross carbon emissions from alternative transport fuels in India. *Energy for Sustainable Development*, 9 (2), pp. 10-16.

33. Ramamurthy, N. and Thirumaran, M., 2002. Automobile Exhaust Level of Carbon Monoxide: A Study in Chidambaram Town. *Indian Journal of Environmental Health*, 44(1), pp. 37-40.

34. Robinson A. B., Baliunas S. L., Soon W. and Robinson Z. W., 1998. Environmental Effects of Increased Atmospheric Carbon Dioxide. Detailed report available at http://www.oism.org/pproject/s33p36.htm#Message44.

35. Salisbury F. B. and Ross C., 1977. Plant Physiology, Prentice-Hall of India Pvt. Ltd., New Delhi, pp. 627-628.

36. Sharma, P. and Khare, M., 2001. Modeling of Vehicular Exhaust-A Review. *Transportation Research*, D6, pp. 179-198.

37. Sivacoumar, R., 2000. Vehicular Pollution near Roadways in Chennai City. *Indian Journal of Environmental Protection*, 20(8), pp. 592-596.

38. Srinivas, D. S. R. K., 1999. Spatial Patterns of Air Pollution in Delhi. *Indian Journal of Environmental Protection*, 19(3), pp. 172-180.

39. Street Tree Master Plan, 2002. Knoxville Knox County Metropolitan Planning Commission. Detailed report available at http://www.knoxmpc.org/plans/treeplan/index.htm.

40. Ward, C. E., 1975. *Air Quality Manual Modifications*. Transportation Laboratory, California Department of Transportation, Sacramento, California.

41. Warran, A. and Patwardhan, A., 2002. Carbon Sequestration Potential of Trees in and around Pune City. Detailed report available at http://www.ranwa.org/carbonpune.pdf.

42. WHO, 1992. *Air Quality in Mega Cities of World*. World Health Organization Publication.

43. Wolfe D. W., 2003. Physiological and Growth Responses to Atmospheric Carbon Dioxide Concentration, in: Pessarakli M. (Ed.), Handbook of Plant and Crop Physiology, Marcel Dekker Inc., New York, pp. 225-234.

# Chapter 30

# Integrated Biotechnological Approach for Reclamation of Mine Spoil Overburden Dumps in India

**Asha A. Juwarkar*, Hemlata P. Jambhulkar, Sanjeev K. Singh and Tapan Chakrabarti**

**Environmental Biotechnology Division, National Environmental Engineering Research Institute (NEERI), Nehru Marg, Nagpur – 440 020**

## ABSTRACT

During the process of mining and mineral extraction huge amount of overburden is produced. These overburdens when dumped in unmined areas in the vicinity of mines create mine spoil dumps. These mine spoil dumps pose physico-chemical and biological constraints for sustainable vegetation. To overcome such constraints, some remedial measures are required to stabilize these mine spoil dumps for sustainable vegetation. In this endeavour, NEERI has developed a multidisciplinary approach to restore the productivity and fertility of the mine spoil dumps. The approach involves, use of industrial waste, as an organic amendment to support the nutritive capacity for vegetation establishment, biofertilizers and mycorrihzal fungi along with suitable ecologically and economically important native plant species. The results of the study indicated that integrated biotechnological approach is an excellent tool for fast restoration of the degraded ecosystem, which provides $CO_2$ sink potential, generates fuel, fiber, food, fodder, medicine etc.

***Keywords:*** *Industrialization, Overburden, Organic amendment, Biofertilizers, Integrated biotechnological approach.*

* Corresponding Author: E-mail: aa_juwarkar@neeri.res.in

## Introduction

Rapid increase in industrialization has resulted in degradation of land and denudation of forest cover. Mining is also one of the industrial activities where land gets contaminated and degraded resulting into pollution of water bodies, streams and groundwater. In India 5,510 mining leases are spread all over the country covering an area of 8, 00, 000 hectares. In 2001-2002, about 1870 million tones of overburden are being removed to obtain 623 million tones of minerals in India. (Juwarkar *et al.*, 2006). This overburden often poses major management problem due to its volume and presence of pollutants. These overburdens when dumped in unmined areas in the vicinity of mines create mine spoil dumps. The dumping of overburden destroys the surrounding vegetation and leads to severe soil and water pollution; resulting into massive damages to landscapes and biological communities of the earth. In this mine spoil dumps the geomorphic system is in disequilibria because of the destruction of the balance between landforms and the processes, which accelerates erosion rates. As soon as mining disturbs this geology- soil–plant stability circuit, ecological chaos takes place. Thus these mine spoil dumps occurs as pockets of disturbed ecosystem. (Juwarkar and Jambhulkar, 2007b).

These mine spoil dumps are extremely difficult to revegetate, due to physical or chemical limitations to plant growth. Among the possible chemical limitations is the absence of N, less available P, and presence of potentially toxic concentration of heavy metals in the spoil. The other reasons for the toxicity and difficulties of plant growth is because of its poor physical properties such as unfavorable porosity, aeration and water infiltration along with bulk density while biological constraints accounts for low or negligible microbial activity. Such constraints in mine spoil could be resolved by applying suitable waste amendments for establishment of productive rhizosphere and to achieve a good and healthy vegetation cover, (Juwarkar and Jambhulkar, 2007b).

In reclaiming such areas, it is therefore necessary to create supportive and nutritive rhizosphere. Augmenting the nutritive potential of organic amendment, the use of wastes from domestic and industrial resources viz: sewage sludge, farm yard manure (Juwarkar and Jambhulkar, 2007a), press mud from sugar industry, ETP sludge from paper mill industry (Juwarkar *et al.*, 2004) and biosludge from oil industry (Kumar *et al.*, 2007) are good ameliorative material to enhance the nutritive potential of wastelands for better growth of the plants. These wastes as an amendment material along with biofertilizer inoculants could provide cost effective solution for waste dump reclamation. Biofertilizers helps for restoration of dump fertility and improvement in vegetation cover denuded by mining and mineral extraction activities while the organically rich waste material will help in restoration of dump fertility and productivity which otherwise creates the problem of disposal.

NEERI developed Integrated Biotechnological Approach to restore the productivity and fertility of the mine spoil overburden dumps. The approach involves selection of suitable plant species, isolation and inoculation of site specific specialized nitrogen fixing strains of *Bradyrhizobium* and *Azotobacter* species and nutrient mobilizing vesicular arbuscular mycorrhizal spores of *Glomus* and *Gigaspora* species in combination with organic amendments. The different ecologically and economically important plant species used for stabilization of mine spoil dumps were, biodiesel producing plant (*Jatropha curcus*), medicinal plant species (*Emblica officinalis, Occimum sanctum, Ficus bengalensis etc.*), biopesticide producing plant (*Azadirecta indica*), timber and industrial wood producing plant species (*Tectona grandis, Delbergia sisso, Gmelina arborea* etc.) and the plants that provide raw material for cottage industries (*Dendrocalamus strictus*). This chapter describes the Integrated Biotechnological approach adopted for mine spoil dumps in different parts of the country.

## Materials and Methods

### Study Area

The sites selected for the study includes different mining sites, such as coal, manganese, zinc, iron, copper tailing and fly ash dumps in different parts of the country. Figure 30.1 shows one of the barren coal mine spoil dump used for reclamation studies at Durgapur near Chandrapur in Maharashtra state.

### Sampling Technique Adopted for Collection of Mine Spoil

Based on the slope and depth of mine spoil dump, the dump was divided into four sectors. Mine spoil samples were collected from each of these sectors in a composite manner (before reclamation); to characterize the physical, chemical, nutritional and microbiological properties of the mine spoil. Auger was used as a sampling tool to draw the sample. Before taking the sample surface litter or stones etc. were scraped away and samples were collected from each of these sectors at a depth of 0–30cm. The samples were collected by hand and intimately mixed on a clean piece of paper or cloth, brought to the laboratory for further processing. The spoil samples were dried at room temperature or in the shade and stored in plastic bags, a smaller portion of these homogenized samples (nearly 500g) were drawn and ground to pass through 2mm. Sieve, before laboratory analysis. The samples were analyzed for pH, EC, organic carbon, total NPK, soluble cations and anions.

### Physico-Chemical Characteristics of Mine Spoil

To characterize the physical, chemical and nutritional properties of the mine spoil samples were analyzed for different parameters. Physical properties of the mine spoil such as bulk density, maximum water holding capacity (using Keen's. method) and porosity were analyzed as per the methods of soil analysis, (Jackson 1973, Black, 1965), (Table 30.1). The chemical parameters of coal mine spoil were determined as per the standard procedures of, (Piper 1994 and Black *et al.*, 1965). pH was determined using pH meter in a spoil to water suspension of 1:2.5, while EC was determined using conductivity meter. Nutrient parameters such as total nitrogen, phosphorus and potash were analyzed as per the standard procedures of (Jackson 1973, Black *et al.*, 1965). Total nitrogen by Micro Kjeldahl method, total P by Van ado molybdo phosphoric yellow colour method, total K by flame photometer method. Organic carbon was estimated as per the method of Walkley and Black, (1934).

### Microbiological Properties of Mine Spoil

Different types of microbial population such as bacteria, fungi, actinomycetes and nitrogen fixing strains of *Rhizobium* and *Azotobacter* were analyzed following standard methods for soil microbial populations and were expressed in terms of colony forming units. (Black *et al.*, 1965 and Page *et al.*, 1982), (Table 30.1).

Microbiological characterization of mine spoil was conducted following serial dilution method (Clark *et al.*, 1965). For the isolation and enumeration of bacterial isolates, ten gram of mine spoil was serially diluted and 1 ml of respective dilution was used for pour plating using the following specific media. The media used were Nutrient agar medium for total bacterial count, Rose Bengal agar medium for fungi, Ken Knights Munaier agar medium for actinomycetes, Jensen's medium for *Azotobacter* and Yeast extract manitol agar medium for *Rhizobium*. For bacterial and fungi count the plates were incubated at 30° C for 24 hours and for *Azotobacter* and *Rhizobium* the plates were incubated at 30°C for 4 days. The no. of microorganisms was expressed as colony forming unit of mine spoil.

**Table 30.1: Physico-Chemical and Microbiological Characteristics of the Mine Spoil Dump**

| *Parameter* | *Mine Spoil* |
|---|---|
| **Physical** | |
| Bulk Density, g/cc | 1.29-1.42 |
| Maximum water holding capacity, per cent | 23.42-25.30 |
| Porosity, per cent | 31.25-34.56 |
| Sand, per cent | 42.80–48.30 |
| Silt, per cent | 31.45 -35.23 |
| Clay, per cent | 15.45-21.15 |
| **Chemical** | |
| pH | 6.50-7.10 |
| Ece, mS/cm | 1.48-1.90 |
| Na, meq/L | 1.82-3.25 |
| K, meq/L | 0.62-0.84 |
| Ca,meq/L | 2.14-3.12 |
| Mg, meq/L | 2.14-2.84 |
| $HCO_3$, meq/L | 1.20-2.10 |
| Cl, meq/L | 1.25-2.10 |
| **Total Nutrients** | |
| Organic carbon, per cent | 0.21-0.34 |
| Nitrogen, per cent | 0.003-0.008 |
| Phosphorus, per cent | 0.002-0.007 |
| Potash, per cent | 0.005-0.012 |
| **Microbial Groups** | |
| Bacteria, CFU/g | $14\text{-}18 \times 10^3$ |
| Fungi, CFU/g | $8\text{-}11 \times 10^1$ |
| *Actinomycetes*,CFU/g | $7\text{-}9 \times 10^1$ |
| *Rhizobium*,CFU/g | Nil |
| *Azotobacter*,CFU/g | Nil |
| VAM spores/10g | Nil |

## Isolation, Identification and Mass Prorogation of Site-Specific Biofertilizer Strains

For isolation of site specific biofertilizer strains, rhizospheric soil samples were collected from plants growing nearby mine spoil dumpsites. These soil samples were screened for vesicular arbuscular mycorrhizal (VAM) fungi by wet sieving and decanting method (Gerdman, 1963). Two genus of the VAM fungi were isolated and identified as *Glomus* and *Gigaspora* sp. Site-specific non-symbiotic nitrogen fixing strain was isolated from the rhizospheric soil samples and were collected from the vicinity of the roots of *Tectona grandis*, a non-leguminous plant growing nearby the site using nitrogen free Jensen's medium. The isolated strain was identified as *Azotobacter* species as per the Bergey's Manual of determinative Bacteriology (1984).

Symbiotic nitrogen fixing strain was isolated from the healthy root nodule of *Delbergia sisso* a leguminous plant as per the Beringer, (1974) method and was identified as *Rhizobium* species as per Bergy's Manual of determinative Bacteriology (1984). Site-specific *Azotobactor* and *Rhizobium* strains were separately propagated in 2 litre capacity fermentor (BiofloII, New Brunswick, U.S.A) using yeast extract manitol broth and Jensen's medium respectively. The fermentor was operated at pH–7.0, temperature at 30° C, agitation at 400 rpm and aeration at 2 SLPM and one and half VVP (volume of air/volume of liquid). The titer value of biofertilizer inocula developed in the fermentor was estimated by dilution and plate count technique using above specified media. *Glomus* and *Gigaspora* were separately propagated in green house in polypropylene packets filled with sterilized soil and sand (1:1) using *Bryophyllum* species as the host plant.

## Selection of Organic Amendment as an Ameliorative Material for Mine Spoil Reclamation

Mine spoil dumps are characterized by low nutritive and supportive capacity necessary for better establishment of plant growth. Therefore it is necessary to give proper source of nutrients for plants establishment on mine spoil dump. Various organic amendments of domestic and industrial origin such as pressmud, a waste product of sugar industry, ETP sludge from pulp and paper mill industry and farm yard manure were selected based on their availability nearby to mine site. These waste materials were used as potential amendments to reclaim different types of mine spoil dumps.

## Selection of Plant Species for Reclamation

The plant species selected for reclamation were based on their adaptation to climatological condition, tolerance to pH and salinity, tolerance to poorly drained conditions of mine spoil, tolerance to metal resistance. The native plant species were selected based on the basis of their commercial availability. Divergent native plant species selected are listed in Table 30.2.

**Table 30.2: List of the Different Plant Species Planted at Different Mine Spoil Dumps/Tailings**

| Sl.No. | Botanical Name | Common Name | Sl.No. | Botanical Name | Common Name |
|---|---|---|---|---|---|
| 1. | *Albezia lebbek* | Siras | 2. | *Azadirechta indica* | Neem |
| 3. | *Cassia fistula* | Amaltash | 4. | *Cassia* sp. | Cassia |
| 5. | *Casuarina* sp. | Casuarina | 6. | *Dalbergia sissoo* | Shisham |
| 7. | *Dendrocalmus strictus* | Bamboo | 8. | *Euginea jambulina* | Jamun |
| 9. | *Ficus bengalensis* | Pipal | 10. | *Mangifera indica* | Mango |
| 11. | *Pithocellobium dulcii* | Chinchbilai | 12. | *Pongamia pinnata* | Karanj |
| 13. | *Embelica officinalis* | Awla | 14. | *Tamarindus indica* | Emli |
| 15. | *Peltoforum* sp. | Peltaforum | 16. | *Delonix regia* | Gulmohar |
| 17. | *Samanea saman* | Raintree | 18. | *Melia azadirach* | Bakain |
| 19. | *Tectona grandis* | Teak | 20. | *Adathoda vesica* | Adulsa |
| 21. | *Elaeocarpus ganitrus* | Rudraksha | 22. | *Parkinsonia accleata* | Parking sonia |
| 23. | *Gmelina arborea* | Shiwan | 24. | *Ficus religiosa* | Wad |
| 25. | *Bauhinia variegate* | Kachnar | 26. | *Terminilia bellrica* | Behda |
| 27. | *Nerium* sp. | Kaneer | 28. | *Bouganvillea* sp. | Bouganvelia |
| 29. | *Morus alba* | Shahtut/Mulberry | 30. | *Psedium guajava* | Amrud |

### Plantation Technique Used for Reclamation of Mine Spoil Dump

Pitting technique was adopted for plantation at mine spoil dump. Pits of dimension of 0.6 m×0.6m ×0.6 m were excavated and were filled with four parts of mine spoil and 1 part of soil. The VAM fungal spores were applied to each pit by mixing with the organic amendment as per its availability to enhance nutrient availability, development of profuse root system of the plant and solubilisation and mobilization of nutrients. The saplings of the selected plant species like *Cassia siamea* were treated with respective site-specific cultures of *Rhizobium* while the non-leguminous were treated with *Azotobacter* species by root inoculation method, before planting in pits. Species of high economical, ecological and medicinal importance were planted on mine spoil dump. Spacing of 1.5 m between the plants and 2 m between two rows for timber species such as *Tectona grandis* and spacing of 2 m×2 m for fruit trees like *Custard apple, Emblica officinalis* etc. and 1.5 m×1.5 m for *Dendrocalamus strictus* was maintained. The most responsive treatment selected for plantation was organic amendment (50t/ha) + spoil (4) + soil (1) + Biofertilizer inoculation.

## Results and Discussion

### Changes in Physico Chemical Characteristics of Mine Spoil Due to Application of Organic Amendment and Biofertilizer

Physico-chemical and nutrient characteristics of the mine spoil were evaluated at periodic intervals. After three years of plantation changes in the chemical characteristics of mine spoil were evaluated, which indicated significant improvements in properties of mine spoil due to organic amendment and biofertilizer inoculation as depicted in Table 30.3. The mine spoil was moderately acidic having pH in the range of 6.50-7.10. To improve its pH mine spoil should be supported with some suitable amendment material, hence blending of mine spoil with organic amendment and biofertilizer improved the pH from 7.10 to 8.00. Similarly electrical conductivity improved from 1.90 to 3.10 mS/c.m. Amendment of mine spoil with organic amendment resulted in the improvement of calcium, magnesium and potassium status, which are required for the plant growth. This increase in the status of cations and anions is attributed to the organic amendment, which is enriched with nutrients. The organic carbon content also increased to 2.58 per cent as against 0.34 per cent in control, due to high organic matter content in the amendment material.

Changes in physical characteristics of mine spoil due to organic amendment decreased the bulk density to 1.26 g/cc, which was high due to massive compaction. This is well supported by the findings of Juwarkar and Jambhulkar, 2007a where they found that the addition of organic rich amendments (FYM), results in corresponding reduction in bulk density, which is related to the magnitude and frequency of waste applications. The decrease in bulk density will reduce the compaction and facilitate the aeration and better penetration and spreading of roots thereby making the rhizosphere favourable for massive root development. Similarly the maximum water holding capacity was very low 25.30 per cent, because of sandy loam texture of the mine spoil. Amendment of mine spoil significantly increased the maximum water holding capacity to 50.30 per cent, while porosity increased from 34.56 per cent to 48.56 per cent. This is because of good percentage of organic amendment's water holding capacity and porosity, which brought about significant improvement in maximum water holding capacity and porosity.

**Table 30.3: Effect of Organic Amendment and Biofertilizer on Physico-Chemical and Microbiological Characteristics of the Mine Spoil Dump**

| *Parameter* | *Mine Spoil* |
|---|---|
| **Physical** | |
| Bulk Density, g/cc | 1.20 -1.26 |
| Maximum water holding capacity, per cent | 45.40–50.30 |
| Porosity, per cent | 45.12–48.56 |
| Sand, per cent | 35.40–36.30 |
| Silt, per cent | 22.54–24.12 |
| Clay, per cent | 36.45–38.23 |
| **Chemical** | |
| pH | 7.50–8.00 |
| Ece, mS/cm | 2.54 -3.10 |
| Na, meq/L | 2.84–3.55 |
| K, meq/L | 1.70–1.96 |
| Ca, meq/L | 3. 44–3.87 |
| Mg, meq/L | 2.84–3.10 |
| $HCO_3$, meq/L | 1.40–1.80 |
| Cl, meq/L | 1.80–2.42 |
| **Total Nutrients** | |
| Organic carbon, per cent | 1.10–2.58 |
| Nitrogen, per cent | 0.3–0.4 |
| Phosphorus, per cent | 0.2–0.3 |
| Potash, per cent | 0.2–0.4 |
| **Microbial Groups** | |
| Bacteria, CFU/g | $14 \times 10^5$–$18 \times 10^6$ |
| Fungi, CFU/g | 19–$2 \times 10^4$ |
| *Actinomycetes*,CFU/g | 30–$50 \times 10^4$ |
| *Rhizobium*,CFU/g | 5–$6 \times 10^7$ |
| *Azotobacter*,CFU/g | 6–$8 \times 10^7$ |
| VAM spores/10g | 15–30 |

## Changes in Microbial Characteristics of Mine Spoil Due to Blending with Organic Amendment and Biofertilizer

The microbiological analysis of rhizospheric samples collected from different plant species established on mine spoil dump revealed improvements in the counts of microbial groups, due to blending of mine spoil with organic amendment and biofertilizer inoculation. After three years the bacterial population linearly increased from $18 \times 10^3$ to $18 \times 10^6$ CFU/gm. Similarly fungi and actinomycetes population increased from $8.0 \times 10^1$ to $28 \times 10^4$ and $9.0 \times 10^1$ to$50.0 \times 10^4$ CFU/gm. respectively, which were almost similar to that found in the productive soil. The population of bacteria, fungi, and actinomycetes in good productive soil, is found to be in the range of 1-34 x $10^6$, 1-9 x $10^5$ and 1-36 x$10^4$ CFU/g respectively. The population of inoculated strains of nitrogen fixers viz; *Rhizobium and Azotobacter* showed increase in the counts in the range of 6.0 x $10^7$ and 8.0 x$10^7$ CFU/gm respectively. Also the VAM spore count increased to 33/10g of spoil.

## Changes in Nutrient Status of Mine Spoil Due to Blending with Organic Amendment and Biofertilizer

The mine spoil is characterized by a deficit in macronutrients viz; nitrogen, phosphorus and potassium. As the spoil was low in nitrogen content, plants were inoculated with nitrogen fixing strains of microorganisms to fulfill the nitrogen requirement of the plant. These nitrogen fixers also secrete growth-promoting substances that support plant growth. The nutrient analysis of rhizospheric samples collected from different plant species established on mine spoil dump revealed significant improvements in the nutrient status with respect to nitrogen, phosphorus and potassium. After three years the nutrient status with respect to total nitrogen content significantly improved to 0.4 per cent as against 0.008 per cent in control. While phosphorus and potassium content improved to 0.3 per cent and 0.4 per cent respectively. The increase in nitrogen content is due to amendment with biofertilisers, which helps in biological nitrogen fixation, while the increase in phosphate content was due to addition of VAmycorrhizae, which helps in phosphate immobilization and solubilisation, (NEERI report, 2000). Thus the nutrient enriched organic matter act as a source and sink of plant available nutrients. It also provides a reservoir of essential macronutrients and a resource of microorganisms in energy and nutrient cycling. In general plants serve as a source for the microbial community and in turn microbes provide nutrients for plant growth through decomposition and mineralization (Juwarkar and Jambhulkar, 2007b).

## Growth Response of Different Plant Species

Plant growth of any plant species is influenced by its associated micro flora and microbial biodiversity in the rhizosphere of the plant. (Juwarkar *et al.*, 2006).Therefore, main stress was focused on isolation of site specific and species specific micro flora from the rhizospheric soil samples collected from nearby area of mine soil dumps. These strains were introduced in the rhizosphere of the sapling of different plant species at the time of plantation on the dump. The biofertilizers were propagated and developed to enable their survival under toxic conditions prevailing in the specific mine spoil dump to facilitate rapid establishment of natural biogeochemical cycles and regeneration of top soil, (Juwarkar *et al.*, 2006). The site specific, stress tolerant biofertilizers and endomycorrhizae, which were isolated from manganese, coal, iron, copper, zinc and fly ash dumps increased the biomass of the plant to a great extent. Thus blending of mine spoil with organic amendment and inoculation with biofertilizer resulted in better rhizosphere establishment and increased biomass of different plant species after three years of plantation (Figure 30.2). This increased biomass will contribute to the fertility of mine spoil and productivity of the reclaimed dump. Therefore use of such types of ameliorants will be beneficial for proper establishment of plant species under environmental stress conditions prevailing at mine spoil dumpsite (NEERI report, 2000).

## Socio-economic and Environmental Benefits of IBA

The technology of Integrated Biotechnological Approach can be commercialized and implemented in different industrial sectors. Table 30.4 shows reclamation of different mining wastelands in different industrial sectors. IBA is an excellent tool for fast recovery of degraded ecosystems. Fast recovery of the system through IBA provides oxygen production, control of soil erosion, top soil built up, and

Figure 30.1: Barren Coal Mine Spoil Dump at Durgapur

Figure 30.2: Revegetated Coal Mine Spoil Dump at Durgapur

development of $CO_2$ sink potential and control of air pollution in terms of ecological benefits. In terms of economic benefits it provides timber and industrial wood, raw material for cottage industry, fuel wood, fiber, food and fodder. Besides these IBA also helps in improving the environmental quality and generating employment. Thus IBA coupled with co-recycling of waste, will help in different mining wastelands to meet their social obligations, enabling pragmatic resource management.

**Table 30.4: Integrated Biotechnological Approach Adopted for Reclamation of Mine Spoil Dumps at Different Industrial Sectors**

| *Names of Mine Site* | *Area in Hectares* | *No. of Plant Species* | *Total No. of Plants Planted* |
|---|---|---|---|
| **Manganese Mines** | | | |
| Gumgaon, Chikla and Dongribuzurg | 62 | 12 | 1,60,000 |
| Tirodi | 10 | 16 | 16,000 |
| **Coal Mines** | | | |
| Durgapur, WCL | 61 | 12 | 1,52,500 |
| Umrer, WCL | 20 | 19 | 50,000 |
| Bishrampur, SECL | 10 | 12 | 25,000 |
| Chirimiri, SECL | 5 | 12 | 12,500 |
| Sasti, WCL | 10 | 8 | 16,000 |
| Lajkura, MCL | 10 | 15 | 30,000 |
| Padmapur, WCL | 10 | 95 | |
| **Iron Mines** | | | |
| Kudremukh, KIOCL | 4 | 10 | 10,000 |
| Kodli Mines, Goa | 5 | 30 | 8,000 |
| Copper Mine | | | |
| Malanjkhand, HCL | 5 | 28 | 10,000 |
| **Zinc Mine** | | | |
| Udaipur, HZL | 5 | 30 | 8,000 |
| Fly Ash Dump, Khaperkheda, MSEB | 10 | 16 | 16,000 |
| **Total** | **227** | **305** | **51,4000** |

WCL: Western Coalfields Limited, Nagpur; SECL: South Eastern Coalfields Limited, Bishrampur; MCL: Mahanadi Coalfields Limited, Sambalpur; MOIL: Manganese Ore India Limited, Nagpur; KIOCL: Kudremukh Iron Ore Company Limited, Mangalore; MSEB: Maharashtra State Electricity Board, Khaperkheda; HCL: Hindustan Copper Limited, Malajkhand; HZL: Hindustan Zinc Limited, Udaipur.

## Conclusion(s)

Integrated biotechnological approach is an ecosystem approach for reclamation of mine spoil dumps, which led to the development of supportive and nutritive rhizosphere and accelerates the biogeochemical cycles in a short span. The results indicated that IBA restored the fertility, productivity and stability of mine spoil dumps within three years leading to the development of self- sustainable ecosystem and turned the barren mine spoil dump into luxuriant green areas thereby mitigating the environmental hazards and improving the environmental health of the surrounding areas. Thus the technology can be implemented on different wastelands with due modifications.

## References


1. Bergey, D.H. 1984. Bergey's manual of Systematic bacteriology, vol.1. Pp. 200-204.
2. Beringer, J.E. 1974. R-factor in *Rhizobium leguminosarum.J.Gen.Microbiol.*, 84,188-198.
3. Black C.A., Evans D.D., White J.L., Ensminger L.E. and Clark F.E. 1965. *In: Methods of soil Analysis: Chemical and Microbiological properties, Agronomy* 9, Part II Medison, Wisconsin, USA.
4. Clark, F.E. 1965.Agar-plate method for total microbial counts. In C.A. Black *et al.* (Eds.), *Methods of soil analysis.* Part 2. *Chemical and microbiological properties* (pp1460-1466). Madison, WI: American Society of Agronomy.
5. Gerdmann, J.W and Nicholson, T.H, 1963. Spores of mycorrhizal endogen species extracted from soil by wet sieving and decanting. *Transactions of the British Mycological society; vol.46*, 235-239.
6. Jackson, M.L.1973. Soil chemical analysis. Printce Hall of India Private Limited, New Delhi.
7. Juwarkar A.A., Jambhulkar, H.P. and Singh. S.K., 2004. Appropriate strategies for reclamation and revegetation of coalmine spoil dumps. *In proceedings of the National Seminar on Environmental Engineering with special emphasis on Mining Environment, NSEEME-19-20 March 2004,pp.no.1-9*
8. Juwarkar A. A, Singh. S.K.and Devotta, S. 2006. Revegetation of mining wastelands with economically important plant species through biotechnological interventions. *In proceedings of the international symposium on"Environmental issues of mineral industry, Mintech, India. pp, 207-216.*
9. Juwarkar A.A and Jambhulkar, H.P, 2007. (a) Restoration of fly ash dump through biological interventions. *Environ. Monit. Assess.* (Available online).
10. Juwarkar A.A and Jambhulkar, H.P, 2007. (b) Phytoremediation of coal mine spoil dump through integrated biotechnological approach. *Bioresource. Tech.* (available online at www.sciencedirect.com)
11. Kumar, G.P., Yadav, S.K, Thawle, P.R, Singh S.K, Juwarkar, A.A. 2007. Growth of *Jatropha curcas* on heavy metal contaminated soil amended with industrial wastes and *Azotobacter*- A greenhouse study. *Bioresource Technology* (In press)
12. NEERI report, 2000. Demonstration project on restoration of manganese and coalmine spoil and fly ash dumps through an Integrated Biotechnological Approach, sponsored by department of biotechnology, New Delhi.
13. Page, A.L., Miller, R.H and Keeney, D.R.1982. *Method of soil Analysis: Chemical and Microbiological Properties, Agronomy 9. Part IIASA.SSSA,* Madison, Wisconsin, USA.
14. Piper, C.S. 1994, *Soil and Plant Analysis*: New York: Inter science.
15. Walkley, Y.A., and Black, I.A.1934. 'An examination of the method for determining soil organic matter and a proposed modification of the chromic acid titration method.' *Soil Science* 37.29

## Chapter 31

# Healthcare Waste Management: Critical Issues and Concerns in Indian Context with Special Reference to a Case Study for Kolkata

**Goutam Banerjee**
**Professor of Environmental Engineering,**
**Department of Civil Engineering, Institute of Technology,**
**Banaras Hindu University, Varanasi, U.P.**

## Need for Rational Healthcare Waste Management System

It is generally believed that the magic of management can turn waste into wealth. Recycling, recovery, and reuse can obviously minimize waste generation and greatly facilitates its treatment and disposal. But, that does not mean recycling of certain items like disposable syringe, needles, etc shall be encouraged on socio-economic ground. Good housekeeping followed by suitable treatment can save us from deadly string of medical waste. Lack of efficient and cost-effective hospital waste management system, inadequate training, lack of awareness among the employees, and ignorance and lackadaisical attitude of the hospital authorities for proper handling and disposal of hospital waste are found to be most common in our country.

In order to safeguard the environment and community health by and large, it is of paramount importance to establish a rational hospital waste management system. All the facets of (solid) waste management system, eg handling, storage, collection, transportation, and final disposal accordingly warrant a thorough improvement.

## Case Study in the City of Kolkata

All India Institute of Hygiene and Public Health (Ministry of Health and FW, Govt. of India), Kolkata conducted an R&D study on Hospital Waste Management (HWM) in collaboration with West Bengal (WB) Pollution Control Board with the general objective to find the status of HWM and to formulate a rational strategy for the same. The study had the special objectives of quantitative and qualitative assessment of solid wastes generating in hospitals in WB; review of the existing system of collection, transportation, and disposal of hospital waste; and examination of the deficiencies in the system.

### Field Survey

In the first phase of the study, the healthcare authorities were interviewed to fill in a well-designed questionnaire. The primary purpose of such survey was to assess the waste management systems prevailing at various healthcare establishments. In order to gather a clear idea of the actual state of hospital waste handling and disposal, the hospital staffs were also interviewed whenever it was needed.

In the second phase of the investigation, a quantitative assessment of the generated wastes was carried out. The objective of the quantitative analysis was to determine the total amount of wastes generated in different types of healthcare establishments such as government hospitals, private nursing homes, health centres, *i.e.* to quantify the per capita waste generation as well as the percentage of various categories of wastes generated which would facilitate the preparation of a rational guideline for an efficient waste management system.

In order to do so, large polythene sacks with tags were distributed to various wards and departments of hospitals/nursing homes' solid wastes. The solid wastes generated in 24 hours were collected in the sacks and subsequently analysed to quantify different categories of wastes.

Categorization of the collected wastes, primarily the solid waste from various medical units was done as per the Guidelines outlined in the Gazette of Government of India (1998).

### General Information of the Medical Establishments

Hospitals/Nursing Homes/Primary Health Centres selected for the study represent various categories of medical establishments in West Bengal. All together twenty five government and private hospitals, nursing homes, and specialized hospitals were surveyed. These included large government hospitals in Kolkata and in districts; private hospitals and nursing homes. Among the large hospitals surveyed in Kolkata, 50 per cent had bed strength more than 1000 and one-third of the rest (16.67 per cent) each had a bed strength ranging from 101–200; 201–500; and 500–1000. Whereas, among the hospitals selected in the districts, 50 per cent had bed strength between 210–500, 33.33 per cent had more than 501, and the rest between 101–200. Most of the private hospitals (66.67 per cent) selected for the survey found to have bed strength between 201–500, while the nursing homes had the bed strength up to 500.

The salient features of solid waste management practices at various medical establishments in Kolkata and in the districts of West Bengal are provided in Table 31.1–Table 31.5.

### Present Practice of Solid Waste Management

Though solid wastes in the hospitals and nursing homes are segregated at source in different containers conforming to the colour coding guidelines of Govt. of India in some of the major government and private/semi-private hospitals in the city of Kolkata, in most of the healthcare establishments in the districts as well as in some smaller health units in and around Kolkata, these are mixed and stored in vats or storage bins in the hospital/nursing home premises. The mixed solid waste is then generally transported by the municipal conservancy staffs to the common disposal ground, usually by means of open dumping along with the municipal solid wastes.

**Table 31.1: Frequency of Waste Collection per day in Hospitals/Nursing Homes**

| *Categories of Hospitals/Nursing Homes* | *Frequency of Collection per day* | | | |
|---|---|---|---|---|
| | *Once* | *Twice* | *thrice* | *Four Times* |
| Large Government Hospitals (in Kolkata), per cent | 0.0 | 83.30 | 16.70 | 0.0 |
| Large Government Hospitals (in Districts), per cent | 25 | 25 | 50 | 0.0 |
| Private Hospitals, (in Kolkata), per cent | 0.0 | 50 | 0.0 | 50 |
| Nursing Homes, (in Kolkata), per cent | 20 | 20 | 40 | 20 |

**Table 31.2: On-site Treatment of Hospital Waste**

| *Categories of Hospitals/Nursing Homes* | *Facilities for Treatment of Bio-Medical Wastes* | |
|---|---|---|
| | *Available (using B.P.)* | *Not Available* |
| Large Government Hospitals (in Kolkata), per cent | 17 | 83 |
| Large Government Hospitals (in Districts), per cent | 0.0 | 100 |
| Private Hospitals, (in Kolkata), per cent | 50 | 50 |
| Nursing Homes, (in Kolkata), per cent | 0.0 | 100 |

**Table 31.3: Transportation of Hospital Wastes***

| *Categories of Hospitals/Nursing Homes* | *Means of Transportation* | | | |
|---|---|---|---|---|
| | *Hydraulic Container* | *Covered Trucks* | *Open Trucks* | *Hand Carts*** |
| Large Government Hospitals (in Kolkata), per cent | 66 | 17 | 0.0 | 17 |
| Large Government Hospitals*** (in Districts), per cent | 0.0 | 0.0 | 75 | 0.0 |
| Private Hospitals, (in Kolkata), per cent | 0.0 | 100 | 0.0 | 0.0 |
| Nursing Homes, (in Kolkata), per cent | 0.0 | 40 | 60 | 0.0 |

* By Municipal Services.

** For transferring in hydraulic container at vat or transfer station.

*** Balance 25 per cent accumulates in the hospital premises and disposed off periodically.

**Table 31.4: Generation of Bio-Medical Solid Waste per Bed**

| *Categories of Medical Establishment* | *Bed Occupancy* | *Total Qty. of Waste per day (Kg)* | *Qty. of Solid Waste per Bed/day (gm)* |
|---|---|---|---|
| Large Government Hospitals (in Kolkata) | 417 | 143.97 | 345.25 |
| Large Government Hospitals (in Districts) | 210 | 87.95 | 418.00 |
| Specialized Hospitals | 200 | 101.56 | 507.80 |
| Nursing Homes | 27 | 16.20 | 600 |

Medical units surveyed: 1. ESI Hospital, Maniktala; Chittaranjan natl. Cancer Res. Inst., Kolkata; Cancer Res. Inst., Thakurpukur (Kolkata); Chandannagar Sub-Divisional Hospital, Hoogly Dist (WB); Santi Swasthalaya, Siliguri, Darjeeling Dist. (WB)

## Recommendations

1. Introduction of segregation of solid waste in hospitals/nursing homes according to different categories is considered to be essential for a rational hospital solid waste management. As per present practice, since no segregation of solid waste is carried out at source, it would be not practical to introduce segregation in 10-12 category-containers at the source-rather in our country, it would be more appropriate to introduce immediately 3-container system as detailed below in each point of generation of solid waste at various wards, laboratories, etc in hospitals.

| *Sl.No.* | *Container Colour Coding* | *Type* | *Waste Category* |
|---|---|---|---|
| A. | Green | Covered | General putrescible wastes, *viz.* Food, kitchen wastes |
| B. | Red | Covered | Microbiological, surgical, human anatomical organs, tissues, blood and bloody fluids, etc; pathological, infectious wastes; isolated discarded waste contaminated with blood, secretion, etc.; animal wastes; research laboratory wastes, etc. |
| C. | Yellow | Covered | Disposables, plastics, PVC, sharps (needles, blades, scalpels, etc.) Cardboard, thermocol, discarded glasswares, soiled cotton, dressing, beddings |

(*i*) At least twice a day collection of solid waste from different wards is recommended.

(*ii*) Salvaging of waste articles from hospitals/nursing homes need to be discouraged.

(*iii*) Open burning of solid waste, wherever being practiced shall be discouraged.

(*iv*) Transportation of solid wastes must be carried out in closed containers fitted with the prime mover. Compactor type of vehicles may be introduced for carrying hospital wastes.

(*v*) As solid wastes stored in green container conform to municipal solid waste characteristics, the same may be disposed through existing municipal services. Such solid wastes should be taken to the disposal sites through containerised collection system.

(*vi*) Hospital wastes of category A (putrescible part) must be disposed by modified sanitary landfilling method. Disposed solid wastes need to be compacted and after filling up the low lying land it should be covered by at least 30 cm earth. Necessary safeguards against groundwater as well as surface water contamination must be exercised at the landfill sites.

(*vii*) Municipal and hospital solid waste management are complimentary to each other. As a results, in order to improve the hospital waste management system, concomitant improvement in the management of both types of solid wastes is necessary.

**Table 31.5: Quantitative Analysis (in per cent) of Biomedical Waste**

| *Sl.No.* | *Categories* | *Percentage by Weight* | | | | |
|---|---|---|---|---|---|---|
| | | *Large Govt. Hospitals (Kolkata)* | *Large Govt. Hospitals (Districts)* | *Private Hospitals* | *Nursing Homes* | *Avg. Values* |
| 1. | Human anatomical wastes (human tissues, organs, body fluids, etc.) | 1.9 | 13.7 | 4.4 | 0.9 | 3.9 |
| 2. | Microbiological wastes (laboratory cultures of animal and human origin) | 2.7 | 0 | 0 | 0 | 0.67 |
| 3. | Waste sharps (needles, syringes, blades, broken glass pieces) | 0.7 | 0.4 | 1.6 | 7.9 | 2.6 |
| 4. | Discarded medicines (outdated, contaminated medicines, vials) | 0.4 | 4 | 2.3 | 0.7 | 1.9 |
| 5. | Soiled wastes (cotton, dressing, linen, bedding) | 12 | 10.5 | 14.8 | 37.1 | 18.9 |
| 6. | Discarded glasswares (bottles, equipments) | 4.3 | 0.7 | 2.3 | 2.2 | 2.4 |
| 7. | Disposable items (plastic containers and conduits) | 1.8 | 2 | 4.6 | 3.6 | 3 |
| 8. | Animal wastes (tissues, organs, carcasses) from autopsy, and research animals | 0.7 | 0 | 0 | 0 | 0.2 |
| 9. | Isolated wastes (biological wastes from human and animals isolated due to communicable diseases) | 0.6 | 0.4 | 0 | 0 | 0.3 |
| 10. | Highly infectious wastes (wastes containing highly infectious living and non-living pathogens) | 0.9 | 1.1 | 0 | 0 | 0.5 |
| 11. | General wastes (food wastes, paper, plastic, textile, earthen pots, and floor sweepings) | 74 | 67.2 | 70 | 47.8 | 65.9 |

(*viii*) Waste category B (microbiological, surgical, human anatomical organs, tissues, blood and body fluids, etc as specified in the table), that will be stored in Red Container may be anaerobically digested, either isolatedly or combinedly with the general putrescible wastes (category A).

(*ix*) Waste category C should be properly sterilized by autoclaving or adequately disinfected by microwave radiation or by some other suitable means, and then should be landfilled or recycled with utmost care. However, the ignitable part of this category such as cotton, bedding, cardboards, etc may be burnt also in properly designed and operated chamber (incinerator) entailed with appropriate air quality control system, preferably in a rather remote place near landfill sites.

(*x*) Hospitals/nursing homes/clinics, etc having bed strength less than 400 should avail the facilities of share-incinerators. It is recommended to install centralized incinerators on 'Pay and Use' at suitably located places

(*xi*) The residue or ash after incineration should be disposed by modified landfill method.

(*xii*) Wastes comprising expired, contaminated, and discarded medicines should be stored safely and to be sent back to the manufacturers concerned for ultimate safe disposal.

(*xiii*) It is strongly recommended to provide adequate protective gears against occupational health hazards (gloves, masks, gumboots, apron, soaps, disinfectants, etc) to the persons engaged in handling and transferring biomedical solid wastes.

(*xiv*) Awareness camps and training programme on hospital waste management need to be organised on regular basis before introducing the new system of hospital solid waste management.

(*xv*) It is emphasized to carry out performance monitoring and evaluation of the new hospital waste management by the competent authority.

# Chapter 32

# Utilization of Citrus Peels for Value Addition: Environmental Management Issues

**Kshama Murarkar[1], D.B. Satpute[2], Pratima Shastri[1] and D.M. Dharmadhikari[2]**

**[1]Department food technology LIT, RTM, Nagpur University, Nagpur, M.S.**
**[2]National Environmental Engineering Research Institute (NEERI), Nehru Marg, Nagpur – 440 020, M.S.**

## ABSTRACT

Processing of citrus fruits for production of juices and related products such as jam, jelly and marmalade gives rise to 25-30 per cent peel, about 15 per cent pomace and 5-7 per cent seeds. Citrus processing waste causes environmental problems that remains unsolved. Due consideration to proper utilization and disposal of solid waste generated in citrus processing is the need of the hour for sustainable development of the industry. Citrus peel constitutes a valuable source of essential oils, pectin and flavonoids. This review paper envisages the chemistry of essential components, industrial and pharmaceutical applications and utilization of citrus peel for recovery of valued products viz pectin, essential oil, bioflavonoids.

## Introduction

Citrus fruits represent the third major fruit in India, next to mango and banana. Citrus fruits comprise about 10 per cent of total fruits produced in India (4.57 million tonnes out of 44.04 million tonnes) (Indian Horticulture Database 2000). It occupies about thirteen percent of total area under fruit cultivation (0.48 million hectare out of 3.72 million hectare). The production of citrus fruit in India is about seventy four percent of the world citrus fruit production (Mishra D. *et al.*, 2005).

India produces a number of varieties of citrus fruit including sweet oranges (Mousambi), mandarin oranges, lime/lemon, galgal and Kinoo. After proper maturity, a sizable fraction of citrus fruit is consumed as table fruit and remaining portion is processed in factories. The edible portion like juicy placental segments are processed into various products *e.g.* segments in sugar syrup, RTS beverages, squashes, jam/jelly/marmalades, juices and juice concentrates. Flavado or fruit peel is mostly discarded as waste (Himanish Das *et al.*, 2001).

Solid waste generated in fruit processing industries pose increasing disposal and potentially severe pollution problems and represent a loss of valuable components. (Laufenberg *et al.*, 2003). Due consideration to proper utilization and disposal of solid waste is the need of the hour for sustainable development of the industry. Therefore, citrus processing sector needs to focus on the optimization of the processes and products with due consideration to integrated environmental protection.

Apart from the spoilage (25-30 per cent) major waste generated from citrus fruit is peel, which constitutes a valuable source of essential oil, pectin and flavonoids. The inner white spongy portion of peel, the albedo is rich in pectin and the outer colored portion, the flavedo has glands containing essential oil and bioflavonoids. This review paper envisages the chemistry of essential components, industrial and pharmaceutical applications, and utilization of citrus peel for recovery of valued products *viz.*, pectin, essential oil, bioflavonoids.

## Holistic Approach

Holistic approach comprising of reduce, reuse, recover and recycle has the ultimate goal to optimize utilization of material while minimizing pollution and environmental problems. Laufenberg *et al.* (2003) has brought holistic concept of food production, which includes product quality, production efficiency and environmental protection.

This approach tries to connect different goals, such as highest product quality and safety, highest production efficiency and the integration of environmental aspects into product development and food production. Within this concept, a special attention is given for recovery of valuable substances or products and internal process of recycling and recovery of valued products. Figure 32.1 illustrates holistic approach of food production.

## Recovery of Value Added By-products

### Pectin

Pectin is an important constituent of cell wall of fruits and vegetables. Pectin is an anionic polysaccharide mostly derived from citrus fruits. Its backbone is composed of alpha-1, 4-D galactopyranosyluronic acid units with some (1, 2)-linked L-rhamnopyranosyl units present (Voragen, Pilnik, Thibault, Axelon, and Renard 1995; Willats, Knox, and Mikkelsen, 2006). In purified form, pectin consists of partly methoxylated polygalacturonic acid and is capable of forming jelly with sugar and acid under suitable conditions (Chaliha *et al.*, 1963). In plant tissues, pectin exists in three different forms; water soluble, oxalate soluble and acid or enzyme soluble.

Pectin plays a prime role for functional properties in various foods, cosmetics, chemical and pharmaceutical industries. Pectin is an important jelling agent in making jam, jelly, marmalade and stabilizing agent in RTS beverages made from fruit juices or pulps to improve their consistency, uniform product quality, appearance and viscosity. In pharmaceutical preparations pectin is used as

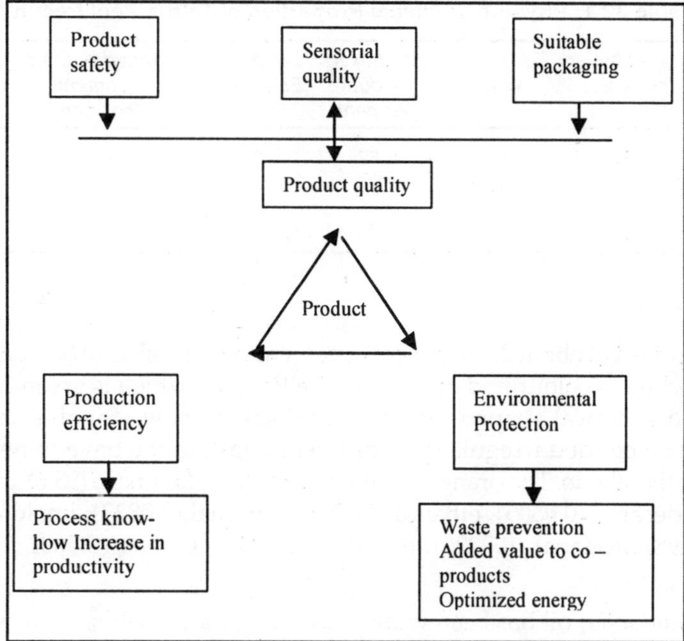


**Figure 32.1: Holistic Approach of Food Production**
***Source*: Laufenberg *et al.* (2003)**

an anti-diarrheal agent. Therefore, pectin based preparations are preferred in dietary menus as cleanser of intestine in treatment of wounds. High methoxy pectin is added in milk to avoid curdling of milk into curd at lower pH. Pectin has good emulsion property and it is used in production of mayonnaise by oil industry. Low methoxy pectin is used in manufacturing of canned fruits such as grapefruit segments, cherries and peaches and also for diabetic jams.

## Process for Pectin Recovery

Citrus peel and residues contains 2.5-5.5 per cent pectin (Ladaniya M.S.). After the extraction of essential oil from the peel and juice residue is washed with cold water. The washed material is slowly boiled with 0.015–0.02 N hydrochloric acid or sulphuric acid or 0.025 molar citric acid (pH-2.5-3.0) for 40-45 minutes. The liquid is pressed and filtered. The pectin extract is concentrated to a thick paste of 25-30 per cent solids and then treated with ethanol to precipitate pectin. The precipitate pectin is washed with ethanol, dried and powdered. The process is used in commercial way.

Takuo Sakai and Minoru Okushima (1980) developed a process for production of pectin from citrus peel by using biotechnological approach. Microorganism *Trichosporon penicillatum* which produces a protopectin solubilizing enzyme was isolated and citrus peel was suspended in water (1:2weight/volume). The organism was added and fermentation proceeded over 15 to 20 hours at 30°C. During fermentation, the pectin in the peel was extracted almost completely without macerating the peel. By this method, 20 to 25 grams of pectin was obtained per kg of peel.

**Table 32.1: Physico-chemical Properties of Citrus Peel Pectin**

| Sl.No. | Pectin Type | Ash (per cent) | Equivalent Weight | Methoxyl Group (per cent) | Degree of Esterification (per cent) | Anhydrouronic Content (per cent) | Intrinsic Viscosity (cp) | Molecular Weight |
|---|---|---|---|---|---|---|---|---|
| 1. | Lemon | 2.82 | 1452 | 8.62 | 63.2 | 77.4 | 4.4 | 92600 |
| 2. | Orange | 2.97 | 969 | 7.60 | 56.1 | 73.9 | 3.7 | 78000 |
| 3. | Sweet orange | 2.85 | 859 | 7.73 | 57.0 | 76.9 | 3.2 | 72700 |

## Essential Oils

Citrus fruit comprised of about 25-30 per cent peel. The peel consists of two parts, inner part such as albedo, which is white in colour and outer part like flavedo, which is green in colour. Peel oil is confined in oblate to spherical shaped oil glands, which is known as oil sacs. The oil sacs are intercellular cavities embedded irregularly at different depth in the flavedo beneath epicarp and hypoderm and above the albedo. The orange essential oil contains limonene (95.6) and other compound decanal (0.2765), dodecanal (0.0305), ethanol (0.0386), geranial (0.0823), linalool (0.2226), myrene (1.1030), neral (0.0498), nonanal (0.2226), ocatanal (1.1030), pinene (0.5602), sabinene (0.2277), valencene (0.4717).

Lime (*Citrus aurantifolia*) oil has been widely used in various fields, such as pharmaceuticals, cosmetics, food and biochemicals. In pharmaceuticals, lime oil has been accepted as a standard ingredient for human remedies. It also contains antiseptic properties that help fight infections and protect the body from bacteria. Lime oil is a natural deodorant and is capable of tightening skin and connective tissue. In food preparation, the essential oil of lime is used in limeade, mixed drinks, and deserts. Mandarin essential oil is extracted in small-scale units based on orange peels collected from fresh juice. Citrus oil inhibits the growth of methanogenic and nonmethanogenic microorganisms.

The oil is recovered mainly by liquid–liquid extraction, distillation, super heated steam distillation, sonication and cold extraction. Physicochemical characteristics of large number of Indian mandarins and lime oil, both cold pressed and steam distilled, varies slightly. D-limonin is a major byproduct from citrus industry (Anand, 1997). Citrus oil extract comprises two main groups. The first group is the less polar but the more volatile monoterpene, which give only little fragrance; it is then considered to be less favourable constituent. The other group gives more polar oxygenated derivative of the monoterpene compound, which contributes to highly pleasant, distinguishable flavour, even in small amount, thus providing more valuable oil in the extract.

## Bioflavonoids

The single most important group of phenolic is the bioflavonoids. These compounds are highly diverse, in both their chemical structure and biological functions (Macheix, 1990). A flavonoid skeleton is composed of two aromatic rings, which are connected through a pyrone ring in the case of flavones or a dihydropyrone ring in the case of flavanones.

Flavonoids show a strong antioxidant and radical scavenging activity and appear to be associated with reduced risk for certain chronic diseases, the prevention of some cardiovascular disorder and certain kinds of cancerous process. Flavonoids exhibit also antiviral, antimicrobial, anti-inflammatory activities and beneficial effects on capillary fragility and an ability to inhibit human platlet aggregation, antiulcer and antiallergic properties (Giuseppe G. *et al.*, 2007). Bioflavonoids viz hesperidin, naringin or rutin are able to normalize capillary permeability and vascular brittleness. Hesperidin is applied in vein dedication, acta antiviral in the flue therapy and owns artificial sweeter properties. Hydrated naringin from citrus is about 300 times sweeter than saccharose, neohesperidin almost 2000 times.

Citrus waste streams are used in horticulture (Widmer and Montanari, 1995). Mandarin peel flavonoids are interesting due to their fungistatic activity (Chkhikvshvii and Gogiya, 1995), which may be applied to naturally protect vegetables and fruits from moulding. The limonoid compound in citrus peel and seed have recently been found to have important pharmacological properties as well as potential in the use as an insect antifeedant for agriculture crops (Manthey and Grohman, 1996).

The recovery of bioflavonoids encompasses a number of steps involving extraction of flavonoids using solvent extraction and further down stream process for purification. The ultimate goal of this procedure is the preparation and extraction of uniformly enriched components of interest and free from interfering matrix. Various procedures are used at each stage reflecting the physico-chemical properties of various flavonoids. The recovery of bioflavonoids by liquid extraction has been common. The flavonoids are generally stable compounds and may be extracted from dried and ground plant material with cold or hot solvents. Suitable solvents for this purpose are aqueous mixture with ethanol, methanol, acetone and dimethylformamide. The extractions are freeze-dried to get concentrate mixture of bioflavonoids. The bioflavonoids are further purified using preparative chromatographic columns.

## Conclusion

The recovery of valuable components viz pectin, essential oil and bioflavonoids from citrus peel is economically desirable because of low cost of row material and high demand of natural valuable products. Apart from the recovery of valued product it also contributes in environmental management and pollution control.

## References


1. *Indian Horticulture Database*, National Horticulture Board, Govt. of India (2000).
2. Mishra D. and Shukla A.K., Dixit A.K. and Singh K., "Aqueous Enzymatic Extraction of Oil from mandarin Peels", *Journal of oleo science*, 2005,54 no.6 355-359.
3. Himanish Das and Jayaraman, " prospects for production of foods and feeds from cellulosic wastes of agriculture and food industry–A review", *Beverage and food world*, May 2001.
4. Laufenberg, Kunz B., Nystroem, M. (2003) Transformation of vegetable waste into value added products (A) The upgrading concept; (B) Practical implementations. *Bioresource Technology* 87 (2003) 167-198.
5. Ladaniya M.S., "Citrus processing"*Citriculture 325-338,winter school 1999.*
6. Vorage A.G.J., Polnik, W., Thbault, J.F., Axelon, M.A.V., and renard, C.M.G.C. (1995). Pectins. In A.M. Stephan (Ed.), *Food polysaccharides and their functional applications* (pp.287-339). New York; Marcel Dekker.
7. Chaliha, B. P., Barua, A. D., Gohain, A., Siddapa, G. S. (1963) Assam lemon as a source of pectin Part I Effect of method of extraction, drying and storage of peels and pomace on the recovery and quality of pectin. *Ind. Food Packer* 17(3): 8.
8. Takuo Sakai and Minoru Okushima. Microbial Production of Pectin from Citrus Peel, *Applied and Environmental Microbiology*, April 1980,Vol. 39 (4), pp 908–912.

9. Anand J.C. and Maini S.B (1997) Utilization fruit and vegetable wastes, *Indian food packer*; March April 1997: 45-64.

10. Macheix, J. J., Fleuriet, A. and Billot, J. (1990) Fruit phenols. *CRC press Boca Raton FL.*

11. Giuseppe Gattuso, Davide Barreca, Claudia Gargiulli, Ugo Leuzzi and Corrado Caristi, "flavoids composition of *citrus* juices" *Molecules* 2007, 12, 1641-1673.

12. Widmmer, W., Montanari, A. M. (1995) Citrus waste streams as a source of phytochemicals in: 107 the Annual meeting of the Florida State Horticultural society, Orlondo/Florida USA, vol. 107 pp. 284–288.

13. Chkhikvishvili,I..D., Gogiya N.N. (1995). Flavonoids of mandarin fruit wastes and their fungi static effect on the fungus *Phoma tracheiphila*. *Applied Biochemistry and Microbiology*, 31 (3), 292-296.

14. Manthey. J.A. Grohman. K., 1996. Concentration of Hesperidin and other orange peel Flavonoids in citric processing byproducts. *Journal of Agriculture and Food Chemistry* 44, 811-814.

# Chapter 33

# Biomass Energy: New Opportunity for Agriculture and Marginal Areas

**M. Bruna Zolin***

**Professor of Rural Development Economics, Department of Economics, University of Venice**

## ABSTRACT

As it is well known, energy is essential for human activities. Despite the high level of technological progress, the increase of energy consumption, satisfied mainly by fossil fuels, raises a series of issues linked to the scarcity of oil reserves, and their growing cost, and to a delicate relationship (and dependence) with a restricted number of world regions and nations. A greater diversification of the energy supply is one of the solutions to solve these problems, especially in the field of renewable resources. Despite the fact that biofuels are still more expensive than fossil fuels, they present many advantages: they are not subject to a progressive exhaustion and their impact on the environment is very low, due to their inclusion in natural cycles. Among biofuels a relevant role is played by the biomass, a renewable resource, that is obtained from the fraction of products, waste and residues from agriculture, forestry and related industries, as well as the biodegradable fraction of industrial and municipal waste. This resource allows a relevant control of the greenhouse gas emission, a minor dependence from oil and gas resources and represents a way for the economic valorisation of local resources and a opportune utilisation of agricultural wastes.

In this context, the aims of the chapter are to highlight the opportunities and the bonds that the biomass utilisation involves, to analyse the main public policies adopted to boost a large-scale production and consumption, to measure the different degree of utilisation at national (Italy) and European levels and to draw future scenarios. Particular attention will be paid to the

---

* Corresponding Author: E-mail: zolin@unive.it

European policies starting from the 1992 reform process until nowadays, having in mind the shortage of cereals in the world market. Where possible and opportune, the situation of Italy and of the European union will be compared with that of India, mainly, as regards public decisions/programmes. The chapter concludes with an analysis of the role that the biofuel production is expected to offer in terms of new opportunities to diversify income and employment in rural and marginal areas.

*Keywords: Biomass energy, Rural and marginal areas, Energy public policies.*

## Introduction

The increase of world energy consumption satisfied mainly by fossil fuels, raises a series of issues linked to the scarcity of fossil resources, their growing cost, the dependence of Europe and India on imported energy and commitments to reduce greenhouse emissions. In 2005 the European Union-15 imported about 57 per cent (1995, 47 per cent) of its energy needs and India 27 per cent (1995, 17 per cent) of its consumption. When considering the risks of climate change and guaranteeing energy supplies, the development of renewable energy remains one of the best solutions for solving these problems. Despite the fact that bio fuels are still more expensive than fossil fuels, they present many advantages: they are not subject to progressive depletion and their impact on the environment is very low, as they are the result of a natural cycle. Biomass has an important role within the bio fuel context. It is a renewable resource that is obtained from the fraction of products, waste and residues from agriculture, forestry and related industries, as well as the biodegradable fraction of industrial and municipal waste. This resource allows for significant control of greenhouse gas emissions, a lower dependence on oil and gas resources and represents a means of economic enhancement for local resources and an efficient way of utilising agricultural waste. In fact, biomass production could play an important role in rural areas, protecting jobs and revitalising neglected regions where maintaining productive activities is the only way of eliminating the risk of abandon. With regard to India, since agriculture is the major industry in the villages and the majority of biomass is collected by family workers (especially women) goals related to food security and energy are prominent.

As far as the European union is concerned, the introduction of a carbon credit would offer new opportunities for agricultural producers. It could significantly contribute to higher incomes and support stronger productivity growth in agriculture, with positive implications for renewable energy. However biofuel production poses new food security and environmental risks.

## World Production and Consumption of Energy

In the decade 1995-2005 primary energy production[1] increased by 26.4 per cent and consumption by 26.8 per cent (annual average rate 2.4 per cent). The largest increase in energy demand was for the non-OECD countries (which include China and India) that have higher economic growth rates and a more rapid population growth than the OECD nations. According to the International Information

---

1 Primary production/Total production of primary energy: any kind of extraction of energy products from natural sources and conversion to a usable form is known as primary production. Primary production takes place when the natural sources are exploited, for example in coal mines, crude oil fields, hydro power plants or fabrication of biofuels. Transformation of energy from one form to another, such as electricity or heat generation in thermal power plants, or coke production in coke ovens, is not primary production.

Administration forecast (EIA-International Energy Outlook, 2007) much of the growth in energy demand among the non-OECD nations occurs in the non-OECD Asia region. In this case energy demand has an average growth rate estimated at 3.2 percent per year[2]. In terms of percentage rate, non-OECD nations increased their weight in total world production from 57.3 per cent in 1995 to 64.6 per cent in 2005. Liquids continue to provide the largest quantity of world supply, but their share fell from 39 per cent in 1995 to 36.8 in 2005. The highest growth concerned coal production and consumption, which rank second as far as total supply is concerned. Petroleum and coal are followed by dry natural gas. Related environmental benefits and the efficiency of dry natural gas make it an attractive energy producing fuel choice in many countries, however, higher petroleum and natural gas prices make coal the best choice in those nations where coal resources are available in sizeable quantities.

In 2005 petroleum, dry natural gas and coal represented 86.3 per cent and 86.4 per cent of supply and demand respectively. Net nuclear electric power increased by 18 per cent, in the decade being analysed, and an increase in nuclear power production is expected to increase in the near future in many countries (EIA, 2007)[3].

Renewable resources play a marginal role in the supply and demand of energy.

**Table 33.1: World Production and Consumption of Primary Energy**

| *World* | *Production (Quadrillion (10 15) Btu)* | | | *Consumption (Quadrillion (10 15) Btu)* | | |
|---|---|---|---|---|---|---|
| | *2005* | *1995* | *Δ per cent 2005/1995* | *2005* | *1995* | *Δ per cent 2005/1995* |
| Petroleum | 169.277 | 141.833 | 19.4 | 169.4 | 142.4 | 19.0 |
| Dry Natural Gas | 105.331 | 80.259 | 31.2 | 107.6 | 81.0 | 32.8 |
| Coal | 122.246 | 88.521 | 38.1 | 122.6 | 88.5 | 38.6 |
| Net Hydroelectric Power | 28.997 | 25.340 | 14.4 | 29.0 | 25.3 | 14.4 |
| Net Nuclear Electric Power | 27.473 | 23.260 | 18.1 | 27.5 | 23.3 | 18.1 |
| Net Geothermal, Solar, Wind, and wood and Waste Electric Power | 4.285 | 2.169 | 97.6 | 4.3 | 2.2 | 97.6 |
| Biomass. Geothermal and Solar Energy Not Used for Electricity Generation | 2.529 | 2.551 | -0.9 | 2.5 | 2.4 | 1.5 |
| Total Primary Energy | 460.139 | 363.933 | 26.4 | 462.798 | 365.046 | 26.8 |

*Source*: EIA (Energy Information Administration)–International Energy Annual 2005, our elaboration.

## European Union and Italy

Data related to the supply and demand of energy resources for current EU-27 nations is available for the period from 1990 to 2005. In view of the varying degree with which nations contributed, it was

---

2 Strong increases in energy consumption are predicted for the other non-OECD regions.

3 Cuts in nuclear power generation are expected only for OECD Europe, where several nations have programmed the elimination and non-replacement of some old reactors and others where nuclear power is forbidden.

decided that we should concentrate on consumption trends from the establishment of EU 15, namely 1995[4].

The EU's energy production and consumption are unsustainable due to their limited resources and their impact on the environment. Europe is in fact a net importer of energy, while internal production levels are decreasing.

**Table 33.2: Energy Production and Consumption (per cent) by Region**

| *Energy Production: Per cent in Different Groups of Country* | *Energy Consumption: Per cent in Different Groups of Country* | | | *Energy Production: Per cent in Different Groups of Country* | | |
|---|---|---|---|---|---|---|
| | *2005* | *1995* | *Δ per cent 2005/1995* | *2005* | *1995* | *Δ per cent 2005/1995* |
| **World Total** | 100.0 | 100.0 | 26.4 | 100.0 | 100.0 | 26.8 |
| OECD | 35.4 | 42.7 | 4.8 | 52.4 | 58.4 | 13.6 |
| Non OECD | 64.6 | 57.3 | 42.6 | 47.6 | 41.6 | 45.3 |
| **Other Groups** | | | | | | |
| OPEC | 21.4 | 20.1 | 34.9 | 6.8 | 5.5 | 56.7 |
| EU | 7.8 | 10.6 | -6.4 | 16.9 | 19.3 | 10.6 |

*Source*: EIA (Energy Information Administration)–International Energy Annual 2005, our elaboration.

**Table 33.3: EU-15 Production, Net Imports and Consumption of Primary Energy**

| *EU 15* | *2005 (1.000 toe)* | *1995 (1.000 toe)* | *Δ per cent 2005-1995* |
|---|---|---|---|
| Total production | 711.349 | 738.988 | -3,7 |
| NET imports | 878.330 | 652.393 | 34,6 |
| Gross inland consumption | 1.536.644 | 1.368.697 | 12,3 |

*Source*: EUROSTAT, our elaboration.

Faced with a 12 per cent increase of Gross inland consumption of primary energy[5] (from 1.37 to 1.54 billion toe) in the years 1995 to 2005, net imports (+35 per cent) rose from[6] 47.7 per cent gross inland consumption of primary energy in 1995 to 57.2 per cent in 2005. The production of primary energy showed a decrease of almost 4 per cent, shifting from 739 million toe in 1995 to l 711 million toe in 2005.

Energy supplies for EU15 are in the main nuclear energy and natural gas, both of which progressively increased in the decade 1995-2005. The quota represented by nuclear energy rose from

4 Austria, Belgium, Denmark, Finland, France, Germany, Greece, Republic of Ireland, Italy, Luxembourg, Netherlands, Portugal, Spain, Sweden, United Kingdom.

5 Gross inland consumption represents the quantity of energy necessary to satisfy inland consumption of the geographical entity under consideration. Gross inland consumption is calculated as follows: primary production + recovered products + total imports + variations of stocks - total exports-bunkers. It corresponds to the addition of consumption, distribution losses, transformation losses and statistical differences.

6 Net imports are calculated as total imports minus total exports.

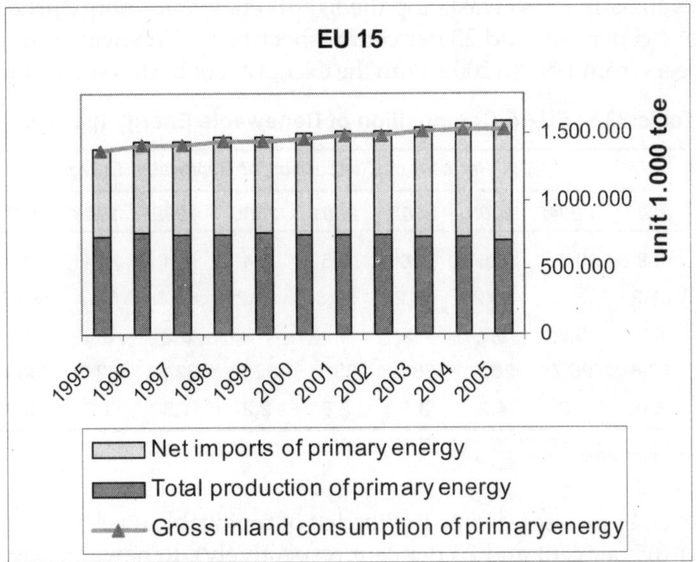


**Figure 33.1: Production, Net Imports and Consumption of Primary Energy (1.000 toe)**
***Source*: EUROSTAT, our elaboration.**

204 to 232 million toe (about +14 per cent). Renewable energy shifted from 73 million to over 100 million toe which represents a 2005/1995 increase of 38 per cent and a 14 per cent share of total primary energy production. The same period showed a sizeable drop in the production of energy derived from coal and lignite (-36 per cent) and crude oil (-25 per cent) and which, in 2005, represented 12 and 17 per cent respectively of total primary energy amounts.

**Table 33.4: EU-15 Primary and Renewable Energy Production by Fuel Type**

| *EU 15* | *2005 (1.000 toe)* | *1995 (1.000 toe)* | *Δ per cent 2005-1995* |
|---|---|---|---|
| **Primary production** | | | |
| Coal and lignite | 86.453 | 135.778 | -36,3 |
| Crude oil | 119.178 | 158.573 | -24,8 |
| Natural gas | 171.437 | 167.045 | 2,6 |
| Nuclear energy | 232.411 | 204.557 | 13,6 |
| Renewable energy | 100.015 | 72.685 | 37,6 |
| **Renewable energy production** | | | |
| Solar | 771 | 242 | 218,6 |
| Biomass and wastes | 65.186 | 43.605 | 49,5 |
| Geothermal | 5.173 | 3.358 | 54,1 |
| Hydro power | 22.849 | 25.130 | -9,1 |
| Wind | 6.037 | 350 | 1.624,9 |

*Source*: EUROSTAT, our elaboration.

At European level, biomasses/waste top the list of renewable energy production followed by hydroelectric power (65 per cent and 23 per cent respectively). Renewable sources show a steady increase in the ten years from 1995 to 2005, with the exception of hydroelectic power (-9 per cent).

**Table 33.5: EU-15 Composition of Renewable Energy (per cent)**

| *EU15* | *Per cent in Composition of Renewable Energy* | | | | | | | | | | |
|---|---|---|---|---|---|---|---|---|---|---|---|
| | *2005* | *2004* | *2003* | *2002* | *2001* | *2000* | *1999* | *1998* | *1997* | *1996* | *1995* |
| Solar | 0,8 | 0,7 | 0,6 | 0,6 | 0,5 | 0,4 | 0,4 | 0,4 | 0,4 | 0,4 | 0,3 |
| Biomass and wastes | 65,2 | 62,9 | 63,2 | 63,2 | 59,8 | 61,2 | 60,6 | 60,7 | 60,9 | 60,8 | 60,0 |
| Geothermal | 5,2 | 5,5 | 5,7 | 4,5 | 4,0 | 3,9 | 5,2 | 5,0 | 4,9 | 4,9 | 4,6 |
| Hydro power | 22,8 | 25,7 | 26,3 | 28,2 | 33,1 | 32,2 | 32,3 | 32,7 | 33,0 | 33,4 | 34,6 |
| Wind | 6,0 | 5,2 | 4,2 | 3,6 | 2,6 | 2,2 | 1,5 | 1,2 | 0,8 | 0,6 | 0,5 |

*Source*: EUROSTAT, our elaboration.

At European level, biomasses/waste top the list of renewable energy production followed by hydroelectric power (65 per cent and 23 per cent respectively). Renewable sources show a steady increase in the ten years from 1995 to 2005, with the exception of hydroelectic power (-9 per cent).

**Table 33.6: Italy Production, Net imports and Consumption**

| *Italy* | *2005 (1.000 toe)* | *1995 (1.000 toe)* | *Δ per cent 2005-1995* |
|---|---|---|---|
| Total production of primary energy | 27.597 | 29.219 | -5,6 |
| NET imports of primary energy | 160.475 | 134.651 | 19,2 |
| Gross inland consumption of primary energy | 186.766 | 161.262 | 15,8 |

*Source*: EUROSTAT, our elaboration.

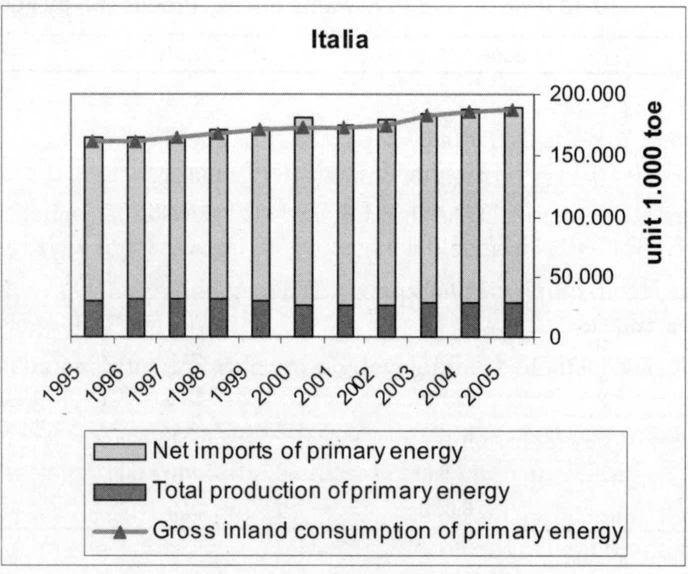


**Figure 33.2: Italy Production, Net imports and Consumption (1995-2005)**
***Source*: EUROSTAT, our elaboration.**

The gross inland consumption of primary energy in Italy increased by 16 per cent between 1995 and 2005 and resulted higher than that of the other EU nations, rising from 161 million to 187 million toe. Contemporarily, the production quota of primary energy decreased by 5,6 per cent, from 29 million toe in 1995 to 27 million toe in 2005.

Furthermore, subsequent to alternating fluctuations in the decade under analysis, Italy's net import quota was noticeably higher than that of the EU15, reaching 86 per cent of gross inland consumption of primary energy in 2005. The dominance of net import quotas is partly attributable to the total absence of nuclear energy production which is not the case for other EU15 nations.

**Table 33.7: Italy Primary and Renewable Energy Production by Fuel Type**

| *Italy* | *2005 (1.000 toe)* | *1995 (1.000 toe)* | *Δ per cent 2005-1995* |
|---|---|---|---|
| **Primary production** | | | |
| Coal and lignite | 60 | 44 | 36,4 |
| Crude oil | 6.146 | 5.289 | 16,2 |
| Natural gas | 9.886 | 16.347 | -39,5 |
| Nuclear energy | 0 | 0 | |
| Renewable energy | 11.504 | 7.540 | 52,6 |
| **Renewable energy production** | | | |
| Solar | 23 | 8 | 187,5 |
| Biomass and wastes | 3.387 | 1.115 | 203,8 |
| Geothermal | 4.791 | 3.167 | 51,3 |
| Hydro power | 3.101 | 3.249 | -4,6 |
| Wind | 202 | 1 | 20.100,0 |

*Source*: EUROSTAT, our elaboration.

Particularly evident is Italy's reluctance to use coal and its total lack of nuclear power both of which have dominated renewable energy supplies in relation to total energy production over recent years. Furthermore, Italy has a much higher quota of energy imports compared with the rest of the EU15 nations.

**Table 33.8: Italy Composition of Renewable Energy (Per cent)**

| *Italy* | *Per cent in Composition of Renewable Energy* | | | | | | | | | | |
|---|---|---|---|---|---|---|---|---|---|---|---|
| | *2005* | *2004* | *2003* | *2002* | *2001* | *2000* | *1999* | *1998* | *1997* | *1996* | *1995* |
| Solar | 0,2 | 0,2 | 0,1 | 0,2 | 0,1 | 0,1 | 0,1 | 0,1 | 0,1 | 0,1 | 0,1 |
| Biomass and wastes | 29,4 | 26,3 | 19,5 | 19,0 | 18,4 | 18,4 | 17,0 | 15,9 | 14,8 | 13,6 | 14,8 |
| Geothermal | 41,6 | 41,2 | 47,7 | 40,1 | 35,5 | 36,3 | 41,8 | 43,5 | 42,5 | 42,1 | 42,0 |
| Hydro power | 27,0 | 31,0 | 31,5 | 39,4 | 44,8 | 44,6 | 40,8 | 40,2 | 42,5 | 44,2 | 43,1 |
| Wind | 1,8 | 1,3 | 1,2 | 1,4 | 1,1 | 0,6 | 0,4 | 0,2 | 0,1 | 0,0 | 0,0 |

*Source*: EUROSTAT, our elaboration.

Renewable energy quotas show a 53 per cent increase in this decade (from 7.5 to 11.5 million toe) which put them in the lead in 2004 with regard to total product. Natural gas, which is in second place with regard to total product as a whole, has decreased by almost 40 per cent shifting downwards from 16 to 9 million between 1995 and 2005.

Geothermic energy prevails in Italy (around 42 per cent of the total). There has been a progressive downturn in the percentage represented by hydroelectric power in relation to the production of biomass which now ranks second as far as total renewable energy is concerned.

## India

In the decade 1995-2005 energy production increased by 23.7 per cent in India and consumption by 41.6 per cent.

Despite the fact that India is the world's eleventh greatest energy producer and the world's sixth greatest energy consumer, it is in fact a net energy importer, mainly because of the large imbalance between oil production and consumption.

Due to the vast availability of coal reserves (about 10 per cent of the world's total) more than half of India's energy needs are met by coal. During the decade 1995-2005, demand exceeded production.

About 30 per cent of India's energy demand is satisfied by petroleum; more than 60 per cent of that oil is imported. At present India is the world's sixth greatest oil consumer. In the period 1995-2005 petroleum production fell by 5.5 per cent while demand increased by 52.0 per cent.

Natural gas has shown the fastest growth rate of any fossil fuel with regard to India's energy. In the decade 1995-2005 consumption increased by 81.7 and production by 51.2 per cent.

Production and consumption of Hydroelectric Power also increased by about 34 per cent, in the last decade.

**Table 33.9: India Production and Consumption of Primary Energy**

| *World* | *Production (Quadrillion (10 15) Btu)* | | | *Consumption (Quadrillion (10 15) Btu)* | | |
|---|---|---|---|---|---|---|
| | *2005* | *1995* | *Δ per cent 2005/1995* | *2005* | *1995* | *Δ per cent 2005/1995* |
| Petroleum | 1.390 | 1.471 | -5.5 | 4.983 | 3.278 | 52.0 |
| Dry Natural Gas | 1.092 | 0.722 | 51.2 | 1.312 | 0.722 | 81.7 |
| Coal | 7.774 | 6.388 | 21.7 | 8.646 | 6.613 | 30.7 |
| Net Hydroelectric Power | 0.990 | 0.741 | 33.6 | 0.990 | 0.741 | 33.6 |
| Net Nuclear Electric Power | 0.190 | 0.078 | 145.8 | 0.190 | 0.078 | 145.8 |
| Net Geothermal, Solar, Wind, and wood and Waste Electric Power | 0.077 | 0.005 | 1.480.5 | 0.077 | 0.005 | 1480.5 |
| World Natural Gas Plant Liquids Production | 0.218 | 0.079 | 176.6 | 0.006 | 0.005 | 9.8 |
| Total Primary Energy | 11.731 | 9.484 | 23.7 | 16.205 | 11.442 | 41.6 |

*Source*: EIA (Energy Information Administration)–International Energy Annual 2005), our elaboration.

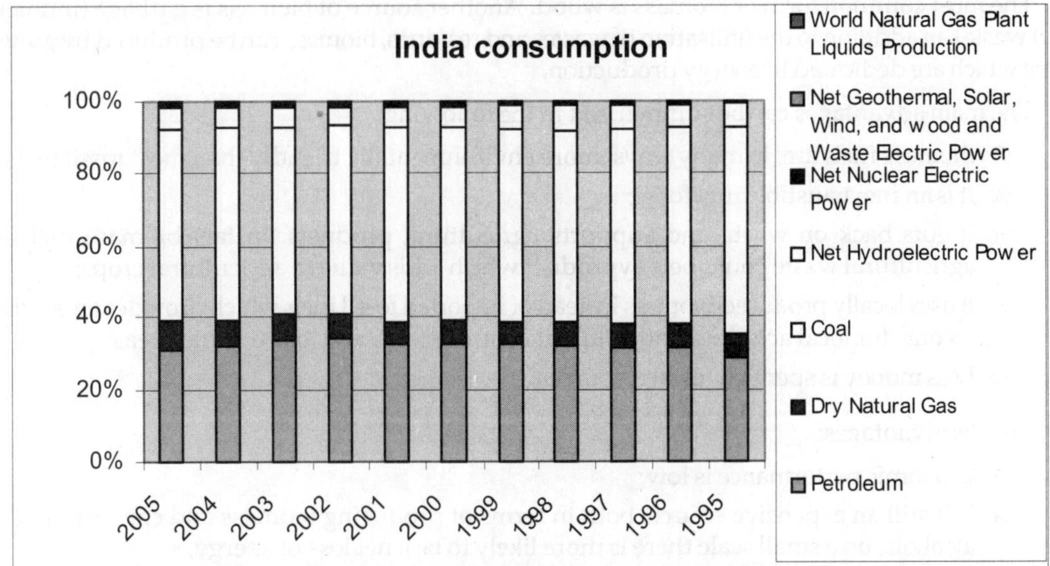


**Figure 33.3: India Consumption by Fuel Type**

***Source*: EIA (Energy Information Administration)–International Energy Annual 2005), our elaboration.**

With regard to India's renewable energy supplies, we can identify three traditional biomass fuel types (wood-fuel, crop residues and dry dung). Generally they are home grown or collected by household labour.

Quantities and incidence have not been officially estimated and have no direct economic value. According to accepted estimates (Shuka, 2007), about a quarter of the primary energy consumed in India in 2004 was obtained from biomass: 58 per cent from fuel wood, 33 per cent from crop residues and 9 per cent from dry dung.

## Biomass Policies: India and European Union Compared

Biomass resources are the total amount of organic materials (wood, agricultural crops, manure or waste and municipal waste) that can be used as a source of fuel or energy[7].

If the vegetable biomass is the result of the photosynthesis process of plants, it can be equally applied both to animal and vegetable derived material[8]. Burning biomass (biopower[9]) is not the only way to release its energy. It can be converted into other typologies such as biofuels[10] and bioproducts[11].

---

7 Biomass excludes organic material which has been transformed by biological processes into substances such as coal or petroleum.

8 Although fossil fuels originate from ancient biomasses, they are not considered to be biomasses because they contain carbon that has been out of the carbon cycle for a very long time.

9 They can satisfy three areas of energy demand : domestic use for the production of heat, industrial use (fire wood and agrifood for the production of electricity and heat) whole local authorities and districts.

10 Production of fuel for means of transport.

11 They belong to the bio products for everyday use such as plastics, glue etc.

The most common form of biomass is wood. Another source of biomass is garbage (municipal solid waste). In addition to the utilisation of wastes and residues, biomass can be produced by growing crops which are dedicated to energy production.

The main advantages can be summarised in the following:

- ☆ Biomass fuels are, in many ways, more environmentally friendly than most fossil fuels.
- ☆ It is an inexhaustible fuel force;
- ☆ It cuts back on waste and supports agricultural products. In fact biomass fuel from agricultural waste could be a byproduct which adds value to agricultural crops;
- ☆ It uses locally produced biomass instead of imported fossil fuels which provides an economic income for local activities and would increase income and jobs in rural areas;
- ☆ Less money is spent on fossil resources.

The disadvantages:

- ☆ Economic performance is low;
- ☆ It is still an expensive source, both in terms of producing biomass and converting it into alcohols, on a small scale there is more likely to be a net loss of energy;
- ☆ It could significantly contribute to global warming and pollution if burned;
- ☆ It is not economically viable to transport biomass fuels over long distances, therefore power plants are usually located near biomass sources;
- ☆ There is a lack of technologies able to ensure high energy performance during the conversion processes; some biomass conversion projects concern animals and are relatively small and therefore limited;
- ☆ Research and public support are needed to reduce production costs;
- ☆ Low levels of public awareness;
- ☆ The land used for energy crops may be in demand for other purposes, such as faming, conservation, housing, resorts or agricultural use (mainly for food).

In this context the support of the public authorities is essential (Menanteau, Finon, Lamy, 2003). Without such aid, market forces would result in the limited diffusion of renewable energy in limited niche markets. Public involvement can be justified as a means for correcting negative external effects, which result from the use of fossil fuels, and an increase in efficiency by stimulating technical changes. Renewable energy contributes to the preservation of a public good. Due to the non-excludable and non competitive characteristics of air and climate stability as a public good, private actors are not interested in investing in something which everyone can acquire free of charge (market failure).

## The European Union Situation

Unmistakeably, the European Union is a group of European countries which participates in the world economy as a single economic unit. The EU's goal is to create a barrier-free-trade-zone and to enhance economic wealth by creating a more efficient market place. The functions and power of the European institutions do not cover all the issues. With regard to energy the role of the European Union is to drive sustainable processes, leaving the single Member States to deal with specific areas of intervention[12]. The European guidelines and support are more complex and articulated, yet at the same time more lenient, than those of a nation where all the functions are centralised. In 2004 the Commission concluded that with the measures and policies currently in force in the EU-15 countries, it was not likely that they would meet the overall target of 22 per cent renewable energy[13]. Thus, there was an urgent need to implement new measures in the EU-15 countries in order to reach the target, and thus contribute towards satisfying the objectives concerning overall renewable energy the climate. The 10 new EU member countries included national renewable electricity targets as part of their accession agreements, but the likelihood that they will reach their targets is limited, particularly due to the fact that most of them have not introduced adequate measures to support renewable energy in the electricity sector. It is crucial that more efforts be made and that effective support systems be introduced at adequate levels, such as feed-in tariffs. The renegotiation of targets should be accompanied by the introduction of efficient support systems to facilitate their development.

Currently, the main EU guidelines and regulations concerning renewable energy and energy saving are to be found in the following documents:

- ☆ Commission of the European Communities, Communication from the Commission to the European Council and the European Parliament: An energy policy for Europe, Brussels 10.1.2007;
- ☆ Commission of the European Communities, Intelligent Energy Programmes;
- ☆ Commission of the European Communities, Communication from the Commission Action Plan for Energy Efficiency: Realising the Potential. Saving 20 per cent by 2020; Brussels October, 2006;
- ☆ Commission of the European Communities, Green Paper on Energy Efficiency or Doing More with Less, 22 June 2005;
- ☆ Directive 2001/77/CE on Electricity Production from Renewable Energy Sources.

With regard to the long-term roadmap concerning renewable energy, the EU is committed to a general objective where 20 per cent of total EU energy consumption must be renewable energy, and where a minimum of 10 per cent of total fuel consumption must be represented by biofuels. Both objectives must be reached by 2020. The Commission has also set out proposals for integrating renewable energy into the following sectors; transport, electricity, heating and cooling.

The proposed chronography package includes legally binding objectives which will, however, allow the individual states the freedom to come up with their own personalised combination of renewable energy solutions. Contemporarily, and in order to implement the general objectives on a national scale, the Member States will need to define both the specific and sectoral objectives for each of the sectors concerned by renewable energy, namely electricity, biofuels, heating and cooling.

The schedule also emphasises the need for the development of biofuels to be coordinated at EU level; by so doing the Commission hopes that the minimum biofuel quotas now fixed at 5.75 per cent of total fuel consumption by 2010, will eventually increase to 10 per cent by 2020. In order to more rapidly satisfy these renewable energy goals, the framework programme for innovation and

---

12 The paper does not consider the national energy policy of each member state.

13 This is part of the communication from the Commission "*The share of renewable energy in the EU*", May 2004, where it stated that there will be only an increase from 14% in 2000 to 18-19% in 2010.

competitiveness (CIP) (2007-2013), the European Union decided to present an "Intelligent Energy-Europe" programme. This programme supports energy enhancement and efficiency, the use of new and renewable energy sources; their more widespread penetration into the market, the diversification of energy and fuels; increased renewable energy quotas (based on the EU's objective these energy sources should represent 12 per cent of the total by 2010) and a reduction of energy consumption, with particular emphasis on the transport sector[14].

At the end of 2006, the European Commission adopted the 2007-2012 plan of action aimed at promoting energy efficiency with a view to reducing consumption by 20 per cent by 2020. The plan of action includes measures aimed at promoting more energy efficient products, buildings and services; making energy more effective and improving its distribution; reducing the impact that transport has on energy consumption; promoting funding and investment in the sector; promoting and strengthening rational behaviour with regard to energy consumption and encouraging the spread of efficient energy initiatives at an international level.

The Green Paper on Energy Efficiency is the Commission's means for promoting energy saving in the European Union; on the basis that increases in current energy consumption damage both the environment and the EU's economic growth, public authorities are being urged to make the general public and industry make more responsible, energy saving decisions. The green paper targets the transport, energy production and the construction sectors as being the main areas to work on with a view to reducing energy consumption. As far as the regulations are concerned, Directive 2001/77/CE of 27 September 2001 established indicative objectives for the production of electricity from renewable energy sources, which by 2010 should amount to 22 per cent of total electricity consumption within the EU, with 25 per cent and 78 per cent of total consumption in Italy and Austria respectively.

With regard to biomasses, the following must not be overlooked:

- ☆ Commission of the European Communities, Biomass Action Plan (December 2005).
- ☆ Commission of the European Communities, An EU Strategy for biofuels (February 2006).

The biomass action plan presents measures aimed at intensifying energy production from wood and agricultural waste by offering incentives based on market mechanisms and the elimination of barriers that hinder the development of the market. These incentives will enable Europe to reduce its dependence on fossil fuels, curb greenhouse gas emissions and stimulate economic growth in rural areas.

The EU's strategy on biofuels has three main aims:

1. To promote the more widespread use of biofuels within EU states and in developing countries, and to guarantee that their use is compatible with the environment, respecting the Lisbon objectives while addressing the issue of competitiveness.
2. Providing conditions for the wide-scale use of biofuels and enhancing their competitiveness with regard to cost, optimising the production of the raw materials, encouraging research into "second generation" biofuels by promoting them on the market and consequently increasing the number of demonstration projects and eliminating non-technical barriers;

---

14 The programme guarantees continuity for the "Intelligent Energy – Europe" programme which expired on 31 December 2006.

3. Examining the opportunities available to developing countries–including those affected by the EU's CAP sugar reform–thanks to the production of raw biofuel materials and the biofuels themselves, as well as defining the EU's potential role in the development of sustainable biofuel production.

As far as European Union policies are concerned, the mid-term reform of the CAP (Common Agricultural Policy) has introduced income aid support for farmers which is no longer linked to the production of crops. This reform also introduced aid for energy crops and within the instrument set aside for this purpose, it maintained the opportunity to grow non food crops. In general, due to the high costs of both land and labour (relatively expensive production factors in Europe) dedicated biomass production is actually more expensive per unit of energy produced than that produced by available residues and wastes.

## India Framework

India's energy policies are influenced by the following factors:

1. Growing economy and limited domestic reserves, with dependence on the supply of energy resources;
2. Rising household incomes, with a need for an adequate supply of electricity and clean cooking fuels;
3. Indoor, urban and regional environmental impact, calling for the adoption of cleaner fuels and cleaner technologies.

Published literature on the matter has established the relationship between energy consumption and economic growth, even if the direction of its causation remains controversial (Antle, Heiderbtink, 1995; Suri, Chapman, 1998; Aqell, Butt, 2001). Unlinked to this relationship, the availability of fossil energy sources in India is limited and may not be sufficient in the long-term to sustain its rapid economic development. In fact, the Indian economy is forecast to undergo high growth levels over the following years. It is predicted that the demand for energy will increase and will have to cope with growing costs because of the limited supply of fossil fuels.

India has a strong tradition of energy planning and intervention programmes. Programmes for promoting renewable energy and for improving energy efficiency date back to the 1940s. Nevertheless, oil imports are currently the major reason for the increasing trade deficit in the balance of payment.

India has significant reserves of coal, but it has scarce oil and gas resources and it is becoming rapidly more dependent on them. With this in mind, India is seeking to diversify its oil suppliers (like other major importing economies outside the Gulf states). The instruments applied are long-term purchase agreements, contracts for exploration, development and production, aid packages and investments. However, the diversification of suppliers has problematic consequences due on the one hand to the social, economic and politic instability of the nations with which India is dealing and on the other hand to direct competition with nearby Asian countries. With a view to reducing at least the potential global fluctuation in energy market, India is building strategic facilities for the storage of crude oil reserves on its east and south coasts.

Despite the fact that coal is the most important energy source, India's fuel basket pursues all options of available energy-fuel types (conventional and non conventional).

Other policies implemented by the Indian government may be summarised as follows:

- ☆ Creation of markets that promote competition through cuts in state subsides in all petroleum products, bearing in mind social objectives (household demand for kerosene and cooking gas);
- ☆ Shift to natural gas and liquefied natural gas. Natural gas, however, is a non-tradable commodity due to the absences of pipelines or liquefied natural gas terminals. The construction of a pipeline from Central Asia would have to cross Pakistan and such a pipeline is not politically feasible. As a consequence, India is seeking new suppliers outside Central Asia (especially Oman and Qatar);
- ☆ Raised domestic oil and gas production and the utilisation of clean coal technology. Due to its long-standing experience in promoting renewable energy technologies (notably biogas and wind energy[15]), India is looking to increase the use of renewable energy sources[16].

With regard to renewable resources, it is appropriate to mention that India has a fully fledged ministry dedicated to the production of energy from non-conventional energy sources. Also worthy of note is the fact that large numbers of poor households have insufficient income to buy commercial fuels (MNES, 2006).

Biomass in India is the most important source of energy for over one hundred million households and traditional industries (it represents a quarter of the primary energy consumed in India). Biomass is the principle provider of energy for domestic use; wood fuels contribute 56 per cent of total biomass energy.

In India, biomass is traditionally used by rural and semi-rural households to satisfy their domestic energy needs. Most biomass is collected on common land by family labourers who are generally women. With the rise in population and greater competition for the use of available land have increased deforestation and the degradation of common land (Shukla, 2007). One of the main issues for Indian policy makers is to develop the market for biomass energy services by removing tariff distortions which favour fossil fuels, and to supply infrastructure with a view to producing and delivering bio-energy services. India has advanced in this field with its important technological innovations concerning the production of Biomass gasifiers.

However, in absence of a developed market in rural areas, most biomass fuels are not traded and the social costs are associated with pollution (mainly indoor pollution).

## Conclusion(s)

Two thirds of the renewable energy in Europe is generated from biomass. Nearly half of the biomass energy used by households is derived from wood. In rural areas, wood is often a suitable alternative for fossil fuels, because it is amply available and most people are not connected to a natural gas or heat grid.

According to International Fund for Agricultural Development (IFAD, 2008) information, about two thirds of India's more than 1 billion people live in rural areas, and almost 170 million of them are poor. Over 80 per cent of the population in rural areas depends on biomass for meeting its cooking

---

15 India is emerging as a growing market for solar, wind and hydroelectric power.

16 India's programme on biomass-based electric power is a recent development.

energy needs. Only 44 per cent of rural households have access to electricity. Agriculture and breeding are important sources of food and income for many poor rural people, but their waste and manure contribute to the production of greenhouse gases. Transforming agricultural waste and manure into biogas will; improve the lives of poor rural people by providing energy for cooking and lighting, increase jobs and substitute the time spent collecting fire wood with economic activities and decrease the release of greenhouse gases, mainly in rural areas (World Bank, 2005).

However, the use of cultivable crop land for fuel remains strongly controversial. Biomass production depends on land, water, fertilisers and/or pesticides and energy. The strengthening of biomass production creates competition for land and water, conflicts based on food security or energy security and feasible environmental improvement, when there is an absence of opportune and dedicated policies (Berndes, Hoogwijk, van den Borek, 2003).

Both agricultural commodities and energy prices have increased significantly in recent years. In the past no significant correlation between them appeared to exist, but many authors and researches state the correlation has now strengthened. We would like to conclude by quoting from the FAO 2007 report: *International cereal prices remained high and volatile, however, reflecting sustained demand, particularly from the biodiesel industry, coupled with historically low levels of stocks and insufficient increases in production. Rises in international prices have translated into higher retail prices of basic food in many countries. Most affected by the food inflation are those developing countries that depend heavily on imports to cover their cereal consumption requirements.*

As far as wood is concerned, it is worth noting that wood has a standing tradition in rural areas due to its availability, sustainability and the fact that it is an environmentally friendly, renewable natural resource. Wood also represents a significant source of income for forest owners in the context of forest restitution processes. In this case, an increase in demand for forest products could see a decrease in forestation. However a recent study (Foster, Rosenzweig, 2003), which examined the connection between economic growth and the growth of forests in India, concluded that as a result of several of India's policies (economic incentives, conservation methods, trade measures that limited importation of forest products) forest owners have been given an incentive to increase the supply of trees. Furthermore, the researchers do not support the assertion that increased agricultural productivity increases forestation by decreasing the need for expansion of agricultural land, and that the growth in rural employment increases tree area by moving labour out of forest-resource extraction.

## References


1. Antle, J.M. and Heiderbrink G. (1995), "Environment and development: theory and international evidence", *Economic Development and Cultural Change*, 43, JSTOR.
2. Aqeel A. and Butt M.S. (2001), "The relationship between energy consumption and economic growth in Pakistan", *Asia-Pacific Journal*, vol. 8.
3. Berndes G., Hoogwijk M., van der Broek (2003), "The contribution of biomass in the future global energy supply: a review of 17 studies" *Biomass and Bioenergy*, vol. 25, issue 1.
4. Commission of the European Communities (2004) *The share of renewable energy in the EU*, Brussels.
5. Commission of the European Communities (2005), *Biomass Action Plan*, Brussels.
6. Commission of the European Communities (2005), *Green Paper on Energy Efficiency or Doing More with Less*, Brussels.
7. Commission of the European Communities (2006), *An EU Strategy for biofuels*, Brussels.

8. Commission of the European Communities (2006), Communication from the Commission Action Plan for Energy Efficiency: Realising the Potential. Saving 20 per ent by 2020; Bruxel.

9. Commission of the European Communities (2007), Communication from the Commission to the European Council and the European Parlament: *An energy policy for europe*, Brussels.

10. Commission of the European Communities (2007), *Intelligent Energy Programmes*, Brussels.

11. European Union and Council of European Union, Directive 2001/77/CE on *Electricity Production from Renewable Energy Source*, Brussels.

12. EIA (Energy Information Admistration) (2007), *International Energy Outlook*, World Energy and Economic Outlook.

13. FAO (2007) *Crop prospects and food situation*, Food and Agriculture Organization of the United Nations, Rome.

14. Foster A.D. and Rosenzweig M.R. (2003) "Economic Growth And The Rise Of Forests", *Quarterly Journal of Economics*, 118:2.

15. International Fund for Agricultural Development (IFAD) (2008), *Rural Poverty Portal*, www.ifad.org.

16. Menanteau O., Finon D., Lamy M. L. (2003), "Prices versus quantities: choosing policies for ptomoting the development of renewable energy", *Energy policy*, vol. 31, issue, 8.

17. MNES (2006), *Annual report 2005-2006*, Ministry of Non Conventional Energy Sources, New Delhi.

18. Rai S.N. and Chakrabarti S.K. (2001) "Demand and supply of fuelwood and timber in India", *Indian Forester*, vol. 127.

19. Shukla P.R. (2007), *Biomass energy strategies for aligning development and climate goals in India*, Netherlands Environment Agency, Indian Institute of Management, Ahmedabas, MNP Report 500101002/2007.

20. Suri V. and Chapman D. (1998), "Economic growth, trade and energy: implications for the environmental Kuznet curve", *Ecological Economics*, Elsevier.

21. World Bank (2005) *World Bamk group progress in Renewable energy and energy efficiency: 1990-2004*, Washington D.C., USA.

# Chapter 34

# Bioremediation of Organochlorine Contaminated Soil by Biostimulation

**K.L. Lapsiya, N.S. Deshmukh, D.V. Savant, P.K. Dhakephalkar, T.Y. Yeole and D.R. Ranade***

**Microbial Sciences Division, Agharkar Research Institute, G.G. Agarkar Road, Pune – 411 004, M.S.**

## ABSTRACT

During conventional bleaching of pulp with chlorine/chlorine based chemicals a number of organochlorines get generated which are collectively termed as AOX (Adsorbable Organic Halides). These organochlorines are highly toxic and recalcitrant. Current wastewater treatment employed does not necessarily ensure removal of AOX. The effluent after treatment is either discharged into a receiving water body or is used for irrigation. There are reports of accumulation of AOX in soil irrigated with such waste-water which adversely affects crops. Therefore, such soil should be remediated from AOX. Bioremediation seems to be an ecofriendly approach. With this aim stimulant such as Diammonium phosphate, Magnesium peroxide was tested to enhance the rate of AOX degradation using native microflora. Addition of nutrients after every 10 days and aeration by Oxygen Releasing Compounds (ORC) resulted in 7 to 8 folds increase of AOX degradation.

*Keywords: Bioremediation, AOX, Biostimulation, ORC.*

---

* Corresponding Author: E-mail: ranadedilip@rediffmail.com

## Introduction

Pulp production from wood not only requires large volume of freshwater, but is also responsible for the generation of huge volume of effluent (~200 $m^3$/metric ton of pulp). Conventional bleaching of pulp is done with chlorine/chlorine based chemicals which gives rise to a number of organochlorines compounds such as chlorophenols, chloroguaiacols, chlorosyringols (Leuenberger *et al.*, 1985) collectively termed as AOX (Adsorbable Organic Halides). These organochlorines are highly toxic and recalcitrant (Deshmukh *et al.*, 2004; Thakur, 2004; Leuenberger *et al.*, 1985; Savant *et al.*, 2006). The current wastewater treatment employed is efficient with respect to removal of COD and BOD. However, it does not necessarily ensure removal of AOX (Deshmukh *et al.*, 2004). The effluent after treatment is either discharged in a receiving water body or is used for irrigation (Haggblom *et al.*, 1989; Rai *et al.*, 2005).

Various studies have shown the toxic effects of this effluent on fishes in the receiving water bodies (Aaltonen *et al.*, 2000; Sepulveda *et al.*, 2004) and accumulation of AOX in sediments and sludges (Andretta *et al.*, 2004; Kimmo Suominen, 1999; Kahkonen *et al.*, 1998; Welker and Schmitt, 1997; Palm and Lami, 1995). There are reports on the toxic effects of Pulp and Paper (PAP) industry effluent on crops. Dutta and Boissya (1996) showed that rice seeds collected from fields irrigated with PAP industry effluent lost their viability to a great extent (germination percentage was lower). Dutta and Boissya (1998) also showed that PAP industry effluent had negative effect on the leaf area of rice plant and therefore resulted in minimum number of grain yield from plants. Malla and Mohanty (2005) showed that there was significant decrease in percentage germination of green gram (*Phaseolus aureus* Roxb.) seeds and reduction in root and shoot length when PAP industry effluent was used for irrigation.

More emphasis is given to develop an efficient microbiological treatment to reduce the AOX from PAP industry wastewater. Attempts are being made to isolate microorganisms which are capable of degrading AOX from wastewater however; there are no efforts to treat soil, irrigated with PAP industry effluent, to make it free from AOX. Soil exposed to AOX is likely to develop certain microbes resistant to AOX, some of which can degrade AOX. Any effort to enhance activity of these microbes in soil will lead to increase in indigenous biodegradation of AOX. In this study we report use of stimulants such as nutrients and Oxygen Releasing Compound (ORC) to stimulate the indigenous microflora of soil thereby increasing the overall rate of degradation of AOX in soil.

## Materials and Methods

### Chemicals

All the chemicals were of HPLC grade.

### Biodegradation Analysis

Biodegradation of AOX from soil and sludge was monitored by AOX Analyzer (multiX 2000, Analytik Jena, Germany) using column adsorption method as per DIN Standard 38414/18 for AOX analysis.

### Construction of Soil Microcosm

Soil microcosms were constructed in porcelain pots using 2 kg soil. The soil was first air dried, homogenized and sieved through 2 mm sieve. Porcelain pots were filled with 2 Kg of soil. BCWW was used as source of AOX. For spiking soil, AOX was extracted from BCWW by USEPA Method 525.2

using SPE C18 cartridges and resuspended in ethyl acetate. One ml of AOX in ethyl acetate was added to approximately 50 g of soil. The soil was mixed thoroughly using mortar and pestle and ethyl acetate was evaporated to dryness. This soil was mixed with rest of the soil in the pot. The final concentration of AOX in the soil was expected to be 10 ppm.

### Biostimulation Using Nutrients (N and P) for AOX Degradation from Soil

Effect of DAP as the biostimulant was tested in the AOX spiked soil. Eighteen porcelain pots with 2 Kg soil spiked with AOX were used for experiment. DAP was added to the test sets to have: 0.25, 0.50, 0.75 and 1.00 (g per cent), respectively. DAP was first dissolved in water which was used to adjust moisture content of the soil (30 per cent) and then added to the soil. In sets 1, 3, 5 and 7 DAP was added after every 10 days and in sets 2, 4, 6, 8 DAP was added only once throughout the experiment. Samples were drawn at 0 day and then after every 10 day. The pots were kept at ambient temperature (30 ± 2 °C). For both the sets common control was kept wherein soil was spiked with AOX and moisture adjusted to 30 per cent.

### Biostimulation Using ORC for AOX Degradation from Soil

Magnesium peroxide ($MgO_2$) has been used as biostimulant for bioremediation of soil contaminated with hydrocarbons, BTEX (Odencrantz *et al.*, 1996, Koenigsberg and Sandefur, 2001). $MgO_2$ when combined with water liberates oxygen atom and becomes $Mg(OH)_2$ which is an inert material. Fifteen porcelain pots with 2 Kg soil spiked with AOX were used for experiment. $MgO_2$ was used at different concentrations *viz.*, 0.25 per cent, 0.50 per cent, 0.75 per cent and 1.00 per cent (w/w). Control pots had soil spiked with AOX. The moisture content of soil in all the pots, including control, was adjusted to 30 per cent by adding appropriate quantity of water. The soil in all the pots was mixed thoroughly and the samples were drawn at '0' day and then after every 10 days for AOX analysis. The pots were kept at ambient temperature (30 ± 2 °C) throughout the experiment.

## Result and Discussions

### Biostimulation Using Nutrients (N and P Source) for AOX Degradation from Soil

DAP in the concentration of 0.5 per cent w/w added after every 10 day was found to be optimum for maximum AOX degradation. This concentration corresponds to 100:10:1 ratio of C:N:P, which is the optimum concentration to bring about maximum activity of microflora from soil.

**Table 34.1: AOX Degradation (per cent) after Addition of DAP Once and After Every 10 Days**

| *DAP Added (g per cent)* | *AOX Degradation (per cent)* | | | | | | | | | | | | | |
|---|---|---|---|---|---|---|---|---|---|---|---|---|---|---|
| | *Addition Once (Days)* | | | | | | | *Addition on Every 10th Day (Days)* | | | | | | |
| | *0* | *10* | *20* | *30* | *40* | *50* | *60* | *0* | *10* | *20* | *30* | *40* | *50* | *60* |
| 0.25 | 0.0 | 7.40 | 11.00 | 13.90 | 15.60 | 17.30 | 19.09 | 0.0 | 8.05 | 13.76 | 17.41 | 21.17 | 23.92 | 26.66 |
| 0.50 | 0.0 | 8.30 | 21.40 | 29.60 | 32.40 | 35.90 | 37.14 | 0.0 | 8.94 | 19.93 | 27.52 | 39.80 | 50.29 | 62.87 |
| 0.75 | 0.0 | 9.00 | 19.30 | 21.02 | 21.90 | 22.70 | 23.33 | 0.0 | 9.00 | 19.30 | 24.38 | 36.40 | 42.14 | 53.77 |
| 1.00 | 0.0 | 10.70 | 11.43 | 13.20 | 13.9ì | 14.02 | 14.14 | 0.0 | 14.00 | 15.02 | 16.98 | 17.00 | 17.57 | 18.75 |
| Control | 0.0 | 0.92 | 1.63 | 3.72 | 5.38 | 6.17 | 7.52 | 0.0 | 0.92 | 1.63 | 3.72 | 5.38 | 6.17 | 7.52 |

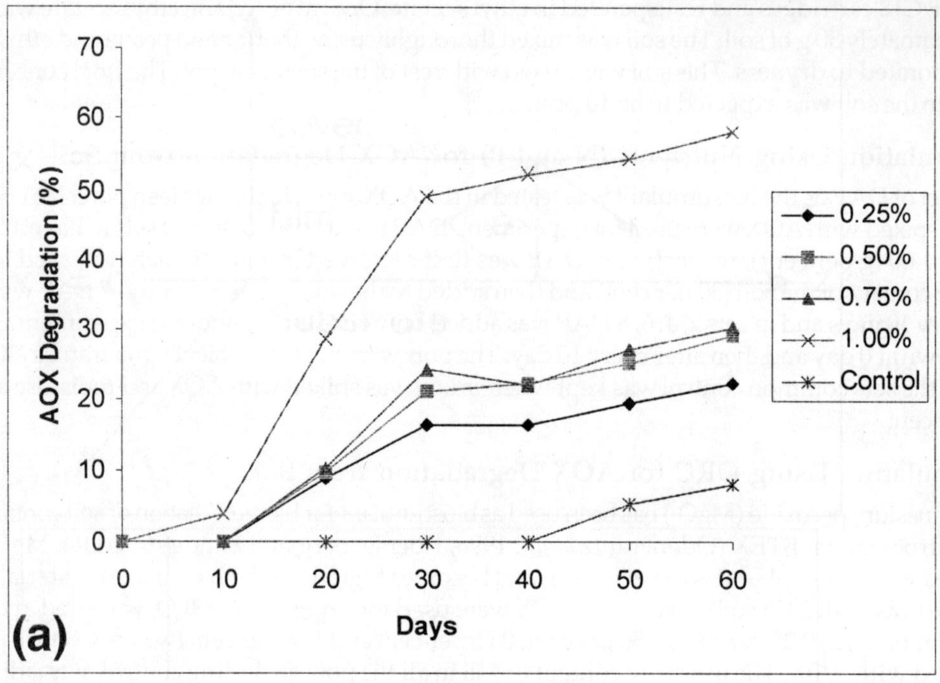


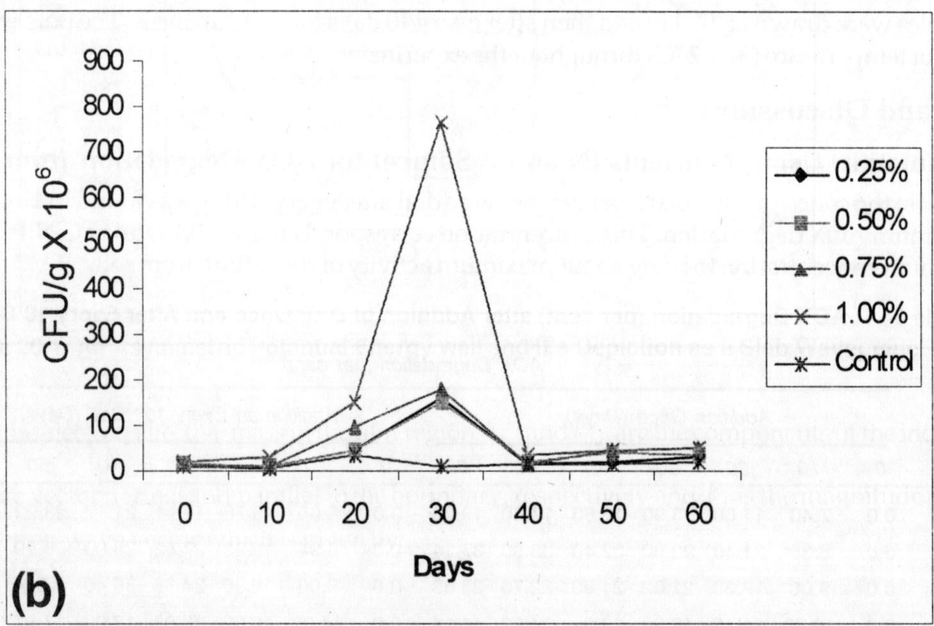


Figure 34.1: Effect of $MgO_2$ on (A) AOX Degradation (per cent) and (B) CFU of Soil

Along with carbon, nitrogen and phosphorous are also required for the growth of bacteria. In contaminated soil the contaminant is mostly used as carbon source by indigenous microflora, but rate of degradation is sometimes slow due to lack of other nutrients. In such cases supply of nutrients can result in stimulation of indigenous microflora and increase in rate of degradation. DAP is used as a source of N and P. For optimum growth of microbes C:N:P ratio should be 100:10:1. Therefore when DAP was added in the concentration of 0.5 per cent maximum AOX degradation was seen.

## Biostimulation Using ORC for AOX Degradation form Soil

The results of effect of biostimulation with ORC on AOX degradation are shown in Figure 34.1. It is seen from the figure that $MgO_2$ added at a concentration of 1 per cent was optimum to bring out maximum degradation of AOX (~58 per cent) from soil (Figure 34.1A) and also for maximum growth of indigenous microflora (Figure 34.1B). However, after 30 days there was a sudden decrease in the number of CFU and rate of AOX degradation was nearly same after 30 days. This could be due to lack of nutrients.

Oxygen plays an important role in activity of aerobic bacteria. It is the final electron acceptor in electron transport chain and thus is necessary for the overall growth of the bacteria. During dechlorination by aerobic bacteria, oxygen atom replaces chlorine atom and brings out degradation of organochlorines. Above 1 g per cent $MgO_2$ there is saturation and rate of oxygen liberation is reduced, therefore 1 g per cent of $MgO_2$ is the optimum concentration.

## Conclusion

Supply of nutrients, in the form of DAP, or ORC showed that rate of AOX degradation can be enhanced by stimulating the native microflora. Combination of DAP and ORC might result in higher rate of degradation when concentration of nutrients, such as N and P, is less in contaminated soil. Otherwise ORC alone can be used to increase rate of AOX degradation.

## Acknowledgements

This work was financially supported by Swedish International Development Agency (SIDA), Sweden through Asian Institute of Technology (AIT), Thailand under the Asian Regional Research Program for Environmental Technology (ARRPET).

## References

1. Aaltonen, T.M., Jokinen, E.I., Lappivaara, J., Markkula, S.E., Salo, H.M., Leppänen, H. and Lammi, R. (2000) Effects of primary- and secondary-treated bleached kraft mill effluents on the immune system and physiological parameters of roach. *Aquatic Toxicology* 51, 55-67.

2. Andretta, C.W.S., Rosa, R.M., Tondo, E.C., Gaylarde, C.C. and Henriques, J.A.P. (2004) Identification and molecular characterization of a *Bacillus subtilis* IS13 strain involved in the biodegradation of 4,5,6-trichloroguaiacol. *Chemosphere* 55, 631-639.

3. Deshmukh, N.S., Lapsiya, K.L., Savant, D.V., Yeole, T.Y. and Ranade, D.R. (2004) Anaerobic degradation of AOX from bleach composite wastewater of pulp and paper industry. In *Biotechnological Approaches for Sustainable Development* ed. Reddy, S.M. and Khanna, S. pp.222-227. New Delhi: Allied Publishers Pvt. Ltd.

4. Dutta, S.K. and Boissya, C.L. (1996) Effect of paper mill effluent on germination of rice seed (*Oryza sativa* L. Var MASURI) and growth behavior of its seedlings. *J Industrial Pollution Control* 12(2), 123-128.

5. Dutta, S.K. and Boissya, C.L. (1998) Effect of paper mill effluent on chlorophylls, leaf area, and grain number in transplanted rice (*Oryza sativa* L. Var MASURI). *J Industrial Pollution Control* 14(2), 141-145.

6. Häggblom, M.M., Janke, D. and Salkinoja-Salonen, M.S. (1989) Transformation of chlorinated phenolic compounds in the genus *Rhodococcus*. *Microbial Ecol* 18, 147-159.

7. Kahkonen, M.A., Suominen, K.P., Manninen, P.K.G. and Salkinoja-Salonen, M.S. (1998) 100 years of sediment accumulation of history of organic halogens and heavy metals in recipient and nonrecipient lakes of pulping industry in Finland. *Environ Sci Technol* 32(12), 1743-1753.

8. Koenigsberg, S. and Sandefur, C. (2001) The efficacy of oxygen release compound: A six year review presented at "Sixth Annual In-Situ and On-Site" Bioremediation Conference, San Diego, CA, June 3-7, 2001.

9. Leuenberger, C., Giger, W., Coney, R., Graydon, J.W. and Molnar-Kubica, E. (1985) Persistent chemicals in pulp mill effluents: Occurrence and behavior in an activated sludge treatment plant. *Wat Res* 19(7), 885-894.

10. Malla, L. and Mohanty, B.K. (2005) Effect of paper mill effluent on germination of green gram (*Phaseolus aureus* Roxb.) and growth behavior of its seedlings. *J Environ Biol* 26(2 suppl), 379-382.

11. Odencrantz, J.E., Johnson, J.G. and Koenigsberg, S.S. (1996) Enhanced intrinsic bioremediation of hydrocarbons using an oxygen-releasing compound. *Remediation*(Autumn), 99-114.

12. Palm, H and Lammi, R. (1995) Fate of pulp mill organochlorines in the Gulf of Bothnia sediments. *Environ Sci Technol* 29, 1722-1727.

13. Savant, D.V., Abdul-Rahman, R. and Ranade, D.R. (2006) Anaerobic degradation of Adsorbable organic halides (AOX) from pulp and paper industry wastewater. *Biores Technol* 97, 1092-1104.

14. Sepúlveda, M.S., Gallagher, E.P., Wieser, C.M. and Gross, T. S. (2004) Reproductive and biochemical biomarkers in largemouth bass sampled downstream of a pulp and paper mill in Florida. *Ecotox and Environ Safety* 57, 431-440.

15. Suominen, K. (1999). *Ecotoxicology and biogeochemical functioning of bleached pulp mill recipient and nonrecipient lake sediments.* Academic Dissertation. Department of Applied Chemistry and Microbiology, Division of Microbiology, University of Helsinki, Finland.

16. Thakur, I.S. (2004) Screening and identification of microbial strains for removal of color and adsorbable organic halogens in pulp and paper mill effluent. *Process Biochemistry* 39, 1693-1699.

17. Welker, A and Schmitt, T.G. (1997) A basic investigation in origins of AOX- substances in paper sludges. *Wat Res* 31(4), 805-815.

# Chapter 35

# Making Construction Industry Green: Some Measures

**V. Kumar[1], D. Mohan[2], T. Kanthi Kiran[3] and S.P. Sharma[4]**

**[1]Professor and Head,**
**[2]Reader, Environmental Engineering Division,**
**[3]Undergraduate Student,**
**Civil Engineering Department, Institute of Technology,**
**Banaras Hindu University, Varanasi – 221 005**
**[4]Graduate Student, Department of Civil Engineering,**
**Indian Institute of Technology, Delhi**

## ABSTRACT

Buildings as the form of shelter have remained as one of the basic requirements for human beings. The model of development adopted after industrial revolution may result in an acute shortage of resources, even for the bare subsistence of future generations. To achieve sustainability, recycling and reuse of the building-materials alongwith durable construction need to be put into practice at the largest possible extent. This may involve environment-friendly planning, design and construction strategies alongwith the recycling and reuse of the construction materials with its durable performance.

The Construction and Demolition (C&D) wastes form a considerable portion of the solid wastes in today's world. The C&D waste management is a major area of concern for the construction industry and the environmentalists alike. The recycling and reuse of C&D wastes,

* Corresponding Author: E-mail: [2]devmohan9@gmail.com; [3]kiranitbhu@gmail.com

especially the aggregates, in the manufacture of concrete, might be a great avenue to explore the possibilities of green and clean construction. This paper elaborates on the details regarding the issues mentioned above, so that the green construction industry may be realized in practice at a large scale.

*Keywords:* *Building construction, Green planning, Durable construction, Construction and demolition wastes, Recycling and reuse, Sustainable development.*

## Introduction

Right from the prehistoric age, buildings and built-up environment have been at the top of the priority of human-beings in their quest for better living standards. The conventional model of development in context of the residential buildings has been primarily homocentric so far. This approach, however, has done much harm than good, to the environment. The results have been diminished comfort levels, unhygienic living conditions and acute shortage of available material and energy resources. Thus, there is an urgent need for considering appropriate planning, design, construction and operation of buildings, and to adhere strictly to an improved model of development in case of all future building projects.

The model of development followed largely so far lays primary emphasis on the progress of human-society at the cost of not only abiotic natural resources, but also it many times disregards even the 'right to life' of many creatures. As a result of large scale application of this type of developmental model, there have been very perceptible and adverse environmental impacts.

The industry related to building construction has not been an exception to this type of development, as in such activities also, due care for preservation of environmental quality has not been taken till very recent past. Since building is one of the basic requirements for all human-beings, environmentally conscious approach in carrying out various activities at pre-construction as well as post-construction stages, will lead to partial amelioration to the damage caused already to the environment by such types of activities earlier. In addition, this will also give a very strong and symbolic message to other industries, catering at present to various acquired needs of the human society.

Traditionally, the scope and functionality of the very important aspect, of planning has been kept quite limited, both in terms of the time and space. Planning is being done within the sheer limitation of the life of building. Afterwards, the construction phase was been taken up, where the buildings have been constructed mostly by following the conventional ways of design and construction.

The Figure 35.1 presents a glimpse (in the form of a flow diagram) of the environmental impact of various activities, carried out during different phases of building construction, operation and demolition (when the building becomes unserviceable). This flow diagram, in its simplest form, clearly highlights the linear and homocentric characteristics of the conventional approach of development towards the planning, construction and operation of any type of building.

## Green Model for Construction Industry

Keeping in view, the above-mentioned facts, it is imperative that one need to look for a better model of development in context of buildings. This desired model should involve more number of loops, so as to make optimal usage of already scarce material and energy-resources. The new model should also be capable of taking full advantage of different environment-friendly and innovative

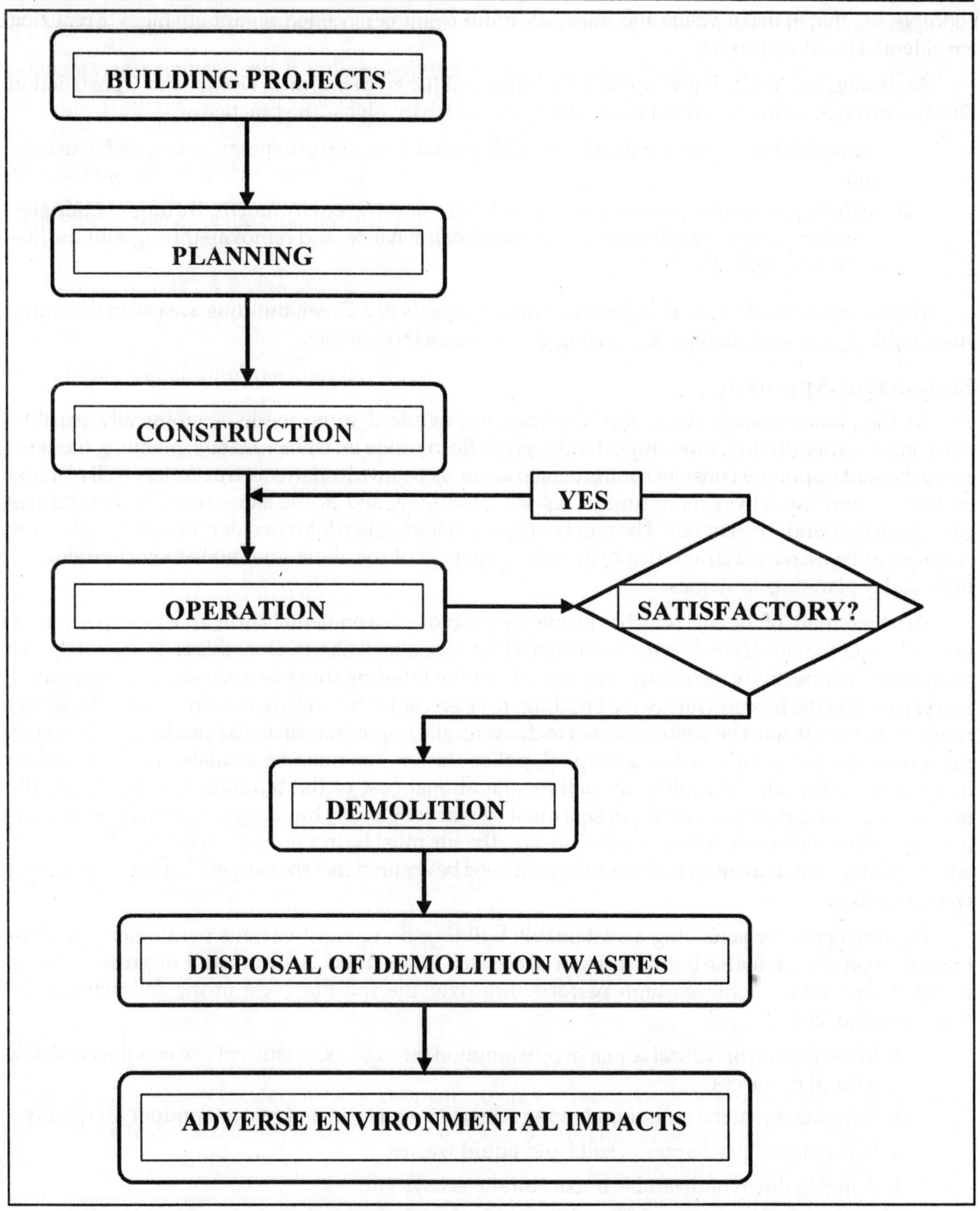


**Figure 35.1: Flow-Diagram Showing the Conventionally Adopted (Homocentric) Model of Development in Building-Construction Industry**

concepts, like that of the recycling and reuse of various resources, needed at various stages of realizing a residential building project.

According to a White Paper on Sustainability, published in Building Design and Construction (2003), a green building is defined as the structure, which involves "the practice of:

- ✰ Increasing the efficiency with which buildings and their sites use energy, water, and materials, and
- ✰ Reducing building impacts on human health and the environment, through better site-selection, design, construction, operation, maintenance, and removal–throughout the life-cycle of the building."

Thus, it becomes obvious that three important aspects of a Green Building are green planning, sustainable design and intelligent recycling and reuse methodologies.

## Sustainable Approach

At the planning stage itself, the buildings are required to be made economically feasible, environment-friendly and, most importantly, aesthetic to reside in. Unfortunately, planning has been given the least important consideration amongst all the steps involved in construction. A well planned building is sure to deliver its optimum service conditions and at the same time, it also ensures environment-friendly behaviour. The number of parameters, taken into consideration, at the planning stage, must be increased dramatically in order to get rid of the above-mentioned shortcomings of present day planning-strategies.

An assessment of the site, which is primarily a sort of environmental audit, must me made prior to purchase of the site, according to South Hams District Council (2001). Once this is done, factors like availability, connectivity, economy and security of the building must be assessed and confirmed, keeping in view the functionality of the buildings to be erected in future. In case of residential building-complex: (i) the site must be well connected to domestically important places like market, bus terminal, *etc.*; (ii) basic services like that of a hospital, police station, *etc.* must be available at a convenient distance from the site; (iii) initial as well as operational cost of the building (including the site improvement and maintenance expenses) must be reasonable; (iv) buildings, especially residential ones, must be safe for satisfactory operation, *viz.*, the site must be in a properly inhabited location. In addition to the requirements mentioned above, it need be ensured that the site gets full benefits of other civic amenities.

Requirements for achieving a sustainable building design and various parameters used for identification of sustainable building material are well understood. A sustainable or green building design is one which achieves high performance over the full life cycle in the following areas (Subramanian, 2007).

- ✰ Minimization of natural resource consumption through more efficient use oı non-renewable natural resources.
- ✰ Bringing down the emissions having adverse impacts on ambient and indoor air quality.
- ✰ Reduction of discharges of solid and liquid wastes.
- ✰ Bringing down undesirable impact on site-ecosystem.
- ✰ Achieving the best possible quality of indoor environment for comfortable and satisfactory physiological conditions and psychological perception.

In context of structural design, according to Sinha (2002), the basic aim is that the structure should be safe, it should fulfill its intended purpose during its life time and it should be economical with regard to its initial costs and also, from the point of view of its operational and maintenance costs. Structural design involves two main steps as mentioned below:

- Selection of appropriate type and form of structure depending upon the operational requirements and
- Detailed design of various components of the chosen structure.

The above-mentioned aspects must be in compliance with the economy, safety and serviceability conditions required for that particular structure for which the design is being prepared. Further, the aesthetic value of any residential structure also depends on the design to a certain extent. For example, visible cracks, which develop quite quickly (without any obvious reason) in a structural element soon after its construction, may be avoided by providing adequate cover and proper design and placement of the reinforcement. Further, design of the structure must take into account the appearance, durability and also, performance of the structure during normal working conditions. Deflection of the structure and cracking of concrete, adversely affect the appearance and therefore, serviceability of the structure.

## Construction and Demolition (C&D) Wastes

Owing to an enormous growth in the population, and also fuelled by the rapid industrialization, the requirement of infrastructure and civil works has multiplied manifold since the world war. It is only logical to assert that the construction industry, in its entire mammothness, only contributes in a much larger way to the total solid-wastes produced globally. According to Huang *et al*. (2002), Construction and demolition (C&D) wastes may also be produced significantly from environmental disasters, such as earthquakes, hurricanes, tornadoes, and floodwater. Construction and demolition wastes frequently constitutes between 15 and 20 per cent of municipal solid wastes (MSW) in Taiwan. Some studies in the United States reported that the rates could be as high as 23 per cent. One study further pointed out that these amounts vary world wide from 13 to 29 per cent.

As mentioned, C&D wastes contribute a major share of the solid waste produced in every part of the world. The term is well defined by splitting it into its logical halves. The composition of C&D wastes normally includes but is not limited to dirt, stones, bricks, blocks, gypsum wallboard, concrete, steel, glass, plaster, lumber, shingles, plumbing, asphalt roofing, heating, and electrical parts. Yet these materials frequently vary constantly due to the changing nature of construction materials over time.

The construction wastes may be typically defined as the wastes emanating from the activities involving construction, remodeling and repair of individual residences, commercial buildings and other civil engineering structures. The demolition wastes are those normally produced from razed buildings. They may also be produced significantly from environmental disasters, such as earthquakes, hurricanes, tornadoes, and floodwater. These also include the debris left after wars.

The most common constituents of C&D wastes are concrete and brick-work remains, wood, glass, various metals and plastics, tiles and ceramics, and paper and cardboard, etc. Concrete rubble is the product of demolition of the structural members like the beams, columns, slabs, stairs *etc*., whereas the brick work remains are obtained from the demolition of walls. Tiles and ceramics are obtained from baths, toilets, and kitchen demolition.

## Recycling and Reuse of C&D Wastes

Recycling of the C&D wastes is not a new concept as such. It had been put to use right since the times of the World War–II, when Japan was left in shambles, from the war debris. The need to recycle the aggregates was realized when the Japanese had to deal with the problems of disposal of the War-debris, simultaneously alongside reconstruction. Also, in Europe, and other developed countries, recycling of building materials also started from the end of World War–II when bricks and other materials, which were recovered from the ruins of war, were utilized for reconstruction. However, other than in Japan, recycling of C&D waste as a means of sustainable use of materials started in Asia only fairly recently.

Until very recently, majority of the Construction and Demolition (C&D) wastes have been primarily disposed off in landfills. According to Peng, Scorpio and Kibert (1997), adverse environmental impact of such an approach is very high as shown in Figure 35.2. Moreover, this practice has created an acute problem of scarcity of suitable space for safe disposal of solid wastes in present era.

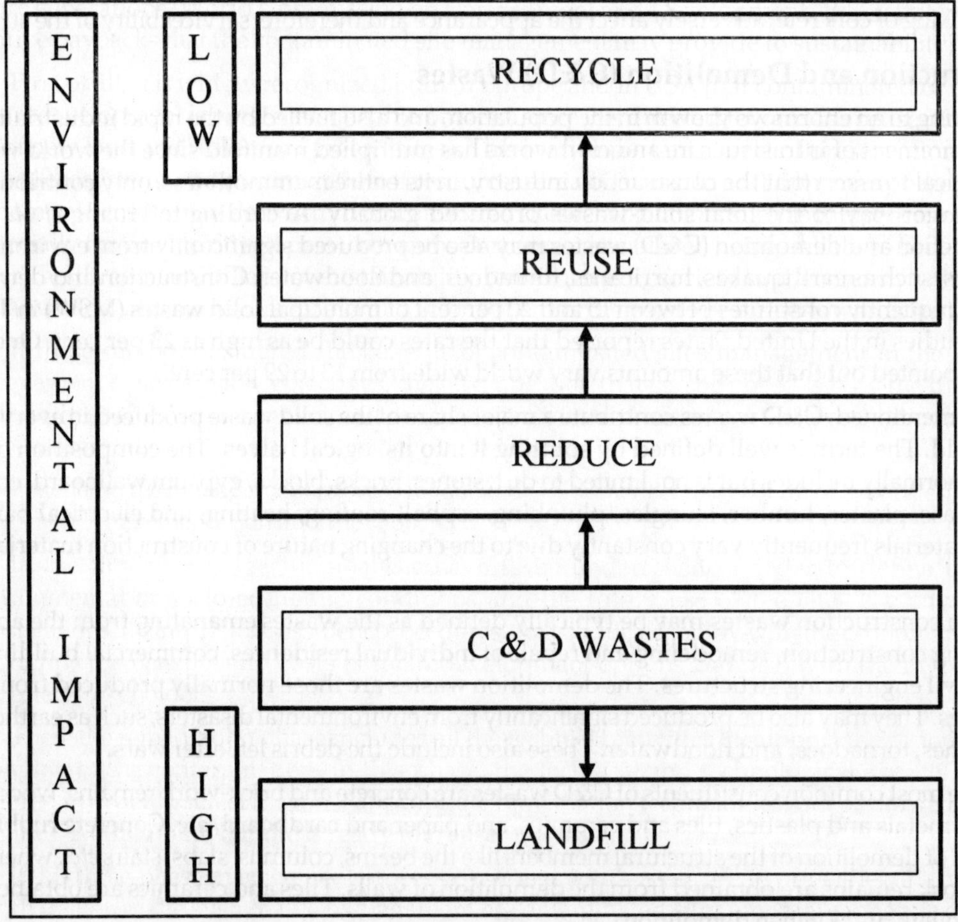


**Figure 35.2: Environmental Impact of C&D Waste Management Strategies**

## Necessity for Recycling and Reuse

Thus far, the C&D wastes have been limited to disposal at landfills, even at an expense of the pre-disposal treatments required. The limited number of landfill sites makes it imperative that their usage is only a short-term based plan. For example, it is estimated that all the landfill sites in Hong Kong would be exhausted in about 8 years of time (Poon and Chan, 2007).

Also, the demand for the disposal sites, and the definite requirement for the pre-disposal treatment of the wastes, help fuel the costs involved shoot up. The growing eco-consciousness only means more stringent regulations regarding the corresponding treatment processes. The C&D wastes are required to be treated to quite a high level, so as to control the hazardous constituents and to curb any further environmental degradation, *e.g.*, the soil degradation or the ground water pollution because a large set of hazardous chemicals may become involved with the C&D wastes (Lawson *et al.*, 2001).

On the other hand, the pressure on the natural aggregates has raised manifold, owing to the rapid industrialization and the ever-increasing population. The natural aggregate deposits, again, being limited in nature, are getting rapidly exhausted. The solution to these two very different problems may be obtained by interlinking the two ends, so as to reach an amicable compromise, for sustainable development. In order to dispose of or at least reduce the accumulation of certain kinds of waste, it has been suggested to reuse some of these waste materials to substitute a percentage of the primary materials used in the concrete (Batayneh *et al.*, 2007). Thus, waste recycling has a great potential in our endeavor to being eco-friendly. Thais practice of recycling and reuse of C&D waste will bring down adverse environmental impacts of building construction.

Therefore, recycling and reuse of C&D wastes, which have a very low environmental impact (Peng *et al.*, 1997), provide an innovative solution to this type of waste-disposal problem. This approach is advantageous in two ways:

- ✰ It solves the problem of disposal of C&D wastes and
- ✰ It simultaneously caters for the material requirements in building construction.

Important factors, which determine the success of a C&D waste management programme, may be (Peng *et al.*, 1997) as mentioned below:

- ✰ Good site and its location
- ✰ Proper equipment
- ✰ Experience in C&D recycling operation
- ✰ Trained supervisors and employees
- ✰ Knowledge of secondary materials' markets
- ✰ Financial capacity
- ✰ Knowledge of environmental and safety regulations

Thus, by adoption of an appropriate strategy for C&D waste management, the process of building construction can be made both economical and environment-friendly at the same time. According to Peng *et al.* (1997), a case study in Florida has yielded the results as shown in Table 35.1 in connection with the C&D waste management approach.

It is evident from the Table 35.1, that necessary steps have been taken in the direction of recycling and reuse of the C&D waste materials. However, certain materials like plastics, dry wall, roofing materials, glass, *etc.* have not been recycled to the required extent. Further steps are required to be taken in the direction of recycling and reuse of all the materials enumerated above to the largest feasible extents in different cases.

**Table 35.1: Typical Composition of Mixed C&D Debris**

| *Material* | *Volume (per cent)* | *Recycled (per cent)* | *Landfilled (per cent)* |
|---|---|---|---|
| Construction wood | 25.0 | 70.0 | 5.0 |
| Trees/Stumps | 5.0 | 100.0 | 0.0 |
| Cardboard | 17.0 | 75.0 | 25.0 |
| Miscellaneous paper | 0.8 | 0.0 | 100.0 |
| Concrete/masonry | 2.5 | 20 | 80 |
| Plastics | 2.0 | 0.0 | 100 |
| Metals | 7.0 | 95.0 | 5.0 |
| Roofing materials | 13.0 | 0.0 | 100 |
| Dirt | 2.0 | 30.0 | 70.0 |
| Dry wall | 15.0 | 0.0 | 100.0 |
| Glass | 0.1 | 0.0 | 100.0 |
| Building insulation | 4.0 | 0.0 | 100.0 |
| Miscellaneous materials | 5.8 | 0.0 | 100.0 |
| Unacceptable materials* | 0.8 | 0.0 | 100.0 |
| Overall | 100 | 55 | 45 |

*: Batteries, paint, metallic cans, *etc.*.

This was the case with a developed country like United States of America. The scenario in a developing nation like India would be entirely different. In all such developing countries, major steps need to be taken in the direction of employing the processes of recycling and reuse of C&D wastes to the maximum feasible extent in various buildings to be constructed in near future.

## Recycled Aggregates: Important Issues

Among the C&D waste, concrete rubble makes up the largest proportion besides excavated soil, and hence its recycling is most important. Although the use of recycled C&D waste as road sub-base is quite common in a few European countries, many laboratory and field studies have shown that the size fraction of the concrete rubble corresponding to coarse aggregate can be satisfactorily used as a substitute for natural aggregates in concrete. Research and experimental efforts, including those published by Waste Management on the use of recycled aggregates for different concrete applications, have been conducted by many researchers and it has been proven that high quality concrete products could also be produced with recycled aggregates. A number of European countries, Japan and a few state agencies in the United States have already modified their specifications to make provision for the use of recycled aggregates in different construction projects.

### Production

The various steps involved in the production of recycled aggregates may be broadly categorized as crushing, separation and screening. In the crushing stage, large-sized building rubbles are crushed to a size of about 300mm. The material not passing the sieves is crushed again, in a cyclic manner. The material passing, on the other hand, is subjected to magnetic separation, in order to remove any ferrous scrap. The material is then subject to vibratory screening where particles of size less than 5mm are separated and used as a recycled fine aggregate.

The particles over 5 mm size are then subject to screening by wind. The remains are then hands picked manually and are crushed again so as to obtain a uniform size of about 20 mm. This material may be used as a recycled coarse aggregate. The details of the mechanism involved in the production of recycled aggregates are mentioned elsewhere (Eguchi *et al.*, 2007)

## Engineering Properties

As with natural aggregate, the quality of recycled aggregates, in terms of size distribution, absorption, abrasion, etc. also needs to be assessed before using the aggregate (Rao *et al.*, 2006)

### Size Distribution

The best particle distribution shape is usually achieved by primary crushing and then secondary crushing, but from an economic point of view, a single crushing process is usually most effective. Primary crushing usually reduces the C&D concrete rubble to about 50 mm pieces and on the way to the second crusher, electromagnets are used to remove any metal impurities in the material. The second crusher is then used to reduce the material further to a particle size of about 14–20 mm (Rao *et al.*, 2006).

### Water Absorption

The water absorption in recycled aggregates ranges from 3 to 12 per cent for the coarse and the fine fractions with the actual value depending upon the type of concrete used for producing the aggregate. It may be noted that this value is much higher than that of the natural aggregates whose absorption is about 0.5–1 per cent (Rao *et al.*, 2006). The high porosity of the recycled aggregates can mainly be attributed to the residue of mortar adhering to the original aggregate.

## Properties of the Concrete

## Freshly Prepared Concrete

### Workability

The workability of recycled aggregate concrete for the same water content in the concrete is lower, especially when the replacement levels exceed 50 per cent (Rao *et al.*, 2006). In order to improve the workability, varying methods of concrete mixes were used in the preparation of recycled coarse aggregate, in terms of saturation. It was found that, extra water corresponding to absorption of the aggregate mixed during concrete preparation produced the most consistent results as far as workability is concerned.

### Bulk Density

The bulk density of fresh concrete made with natural aggregates is in the range of 2400 kg/m$^3$, whereas the concrete made with recycled aggregates is significantly lighter, 2150 kg/m$^3$ (approx.), regardless of the type of cement.

## Hardened Concrete

### Compressive Strength

Though researchers have reported a reduction in strength in RAC, it should be noted that the extent of reduction is related to the parameters such as the type of concrete used for making the RA (high, medium or low strength), replacement ratio, water/cement (w/c) and the moisture condition of the recycled aggregate. For example, it was found that at a high w/c (between 0.6 and 0.75), the strength of RAC is comparable to that of reference concrete even at a replacement level of 75 per cent. It was also found that the strength of RAC and reference concrete are comparable even at 100 per cent replacement, provided that the water–cement ratio was higher than 0.55. However, as the water–cement ratio is reduced to 0.40, the strength of RAC was only about 75 per cent of the reference mix. Apart from the water–cement ratio, the moisture condition of the RA also appears to affect the compressive strength (Rao *et al.*, 2006).

### Flexural and Tensile Strength

The ratio of the flexural and the splitting strengths to the compressive strength is in the range of 16–23 per cent and 9–13 per cent, respectively. These values are about 10–15 per cent lower compared to the recommendations of ACI 363R. Another study showed a reduction in strength of 15–20 per cent compared to reference concrete at 100 per cent replacement. In another study, where the direct tensile strength of concrete was determined, it was found that difference in the tensile strength of RAC and reference concrete at 28 days was less than 10 per cent (Rao *et al.*, 2006). Studies have also shown that use of supplementary cementitious admixtures, such as silica fume, etc. helps improve the properties of RAC. The modulus of elasticity for RAC has been reported to be in the range of 50–70 per cent of the normal concrete depending on the water–cement ratio and the replacement level of RA (Rao *et al.*, 2006).

### Creep and Shrinkage

The use of RA in concrete induces a large shrinkage due to the high absorption of these aggregates. Some studies by show that in RAC at the age of 90 days, the shrinkage could be about 0.55–0.8 mm/m, whereas the comparable value for NAC is only about 0.30 mm/m. However, the test results for creep in normal laboratory conditions are not so clear, though some studies have shown the tendency to be reversed, *i.e.* the creep after 1 year is about 20 per cent lower than concrete with NA. Though more work is needed in the area, it appears that the overall behaviour of RAC and NAC may be comparable when viewing the combined effect of shrinkage and creep (Rao *et al.*, 2006).

### Durability

Durability studies have been done to better understand the effect of using different qualities of RA on the properties of the RAC. Though, some studies have shown that RAC is significantly more permeable than NA concretes, it should also be noted that the durability properties can be improved by using fly ash, condensed silica fume, etc.

## A Case for Illustration

Hong Kong, with a population of about 6.8 million and about 2.1 million household, has a mountainous nature. As land is scare in this territory, the only way to gain additional floor spaces is by demolishing old buildings and replacing them by high-rise buildings. In 2004, there were over 20 million tonnes of construction and demolition (C&D) wastes generated in Hong Kong. In the past, the inert portions of these C&D materials, such as rock, concrete and soil, have been beneficially reused as fill materials in forming land for the fast growing perpetual development. However, the increasing awareness on environmental protection by the public has resulted in having most reclamation projects either been deferred or much reduced in scale. This in turn results in a substantial reduction of the public filling capacity to accommodate the surplus of C&D materials. If these materials are not managed properly, it will also accelerate the depletion of the already limited precious landfill spaces. Hong Kong is now facing a crisis on how to accommodate these surplus materials. Apart from putting more efforts in minimizing its generation and the formation of temporary fill banks for temporary accommodation of these materials, recycling is one of the means to alleviate the demand on public fill capacity (Poon and Chan, 2006).

In order to facilitate the use of recycled aggregate in this region, the Hong Kong government has set up a temporary recycling facility in 2002 and is actively promoting the use of recycled aggregate in concrete. According to the quality control requirements of the recycling facility, the recycled aggregate contained less than 0.5 (wt) per cent of wood and particles less dense than water and less than 1 (wt) per cent of other foreign materials. Therefore, the recycled aggregate used in this territory could be considered as recycled concrete aggregate. In the current specifications, it is allowed to use 100 per cent coarse recycled aggregate in proportioning low-grade concrete (grade 20). In high-grade concrete (*i.e.*, grades 25–35), only 20 per cent coarse recycled aggregate content is permitted. The fine portion (<5 mm), together with the coarse portion of recycled aggregate, is usually prohibited in proportioning concrete mixtures as it is very difficult to control the workability and dimensional stability of the concrete mixtures.

## Conclusion(s)

From the facts mentioned above, it becomes clear that, at planning stage itself, appropriate measures should be taken so that the structure ultimately develops into a green-building. This strategy adopted at the planning stage should bring down the energy consumption by the structure from external sources because the maximum advantage of the natural energy may be taken from heat and light available from the sun, and air-circulation carried out by the flow of wind, throughout the building.

The properties of recycled aggregates are quite different from those of the natural aggregates. But, the properties of concrete prepared from the recycled aggregates, though aberrant from those of the concrete prepared from natural aggregates, under controlled circumstances, are found to be satisfactory enough for usage.

Efforts need to be made to incorporate the application of recycled C&D wastes to relieve terrific stress at present on the naturally available fresh resources. As a desirable consequence, adverse environmental impacts caused on account of indiscriminate exploitation of newer and newer natural resources to yield fresh building materials can also be brought down significantly and thus, it will be possible to achieve environment-friendly characteristics and also, sustainability in the huge building-construction industry.

## References

1. Batayneh, M., Marie, I., and Asi, I. (2007), Use of selected waste materials in concrete mixes, J. Waste Management, Vol. 27, No.–12, pp. 1870–1876.

2. BDC: Building Design and Construction (2003), White Paper on Sustainability, November, 2003. Detailed Report Available at: www.bdcmag.com.

3. Eguchi, K., Teranishi, K., Nakagome, A., Kishimoto, H., Shinozaki, K., and Narikawa, M. (2007), Application of recycled coarse aggregate by mixture to concrete construction, J. Construction and Building Materials, Vol. 21, No.–7, July 2007, pp. 1542–1551.

4. Huang W., Lin D., Chang N., and Lin, K. (2002), Recycling of construction and demolition waste via a mechanical sorting process, J. Resource, Conservation and Recycling, Vol. 37, No.–1, Dec. 2002, pp. 23–37.

5. Lawson, N., Douglas I., Garvin S., McGrath C., Manning D. and Vetterlein J. (2001), Recycling construction and demolition waste–A UK perspective, J. Environmental Management and Health, Vol. 12, No. 2, pp. 149–157.

6. Peng, C.-L., Scorpio, D.E. and Kibert, C.J. (1997), Journal Construction Management and Economics, Volume, 15, pp. 49-58.

7. Poon, C., and Chan, D. (2007), The use of recycled aggregate in concrete in Hong Kong, J. Resource, Conservation and Recycling, Vol. 50, No.–3, May 2007, pp. 293–305.

8. Rao, A., Jha, K. N., and Misra, S. (2007), Use of aggregates from recycled construction and demolition waste in concrete, J. Resource, Conservation and Recycling, Vol. 50, No.–1, March 2007, pp. 71–81.

9. Sinha, S. N. (2002), Reinforced Concrete Design: Indian Edition, Tata McGraw-Hill, New Delhi, India.

10. South Hams District Council Supplementary Planning Guidance (2001), Draft on Sustainable Development of Small Sites, June 2001, South Hams District, U.K.

11. Subramanian, N (2007), Sustainability–Challenges and Solutions, Vol. 81, No.–12, pp. 39–50.

# Chapter 36

# Innovative Green and Clean Technologies for Sustainable Environment

**V.S. Sapkal[1], R.S. Sapkal[2] and G.B. Shinde[3]**

**[1]Professor and Head, UDCT, Amravati, M.S.**

**[2]Reader, Department of Paper and Pulp Technology, UDCT, Amravati, M.S.**

**[3]Lecturer, Department of Chemical Engineering, SVM Engineering College, Chincholi, Nashik, M.S.**

## ABSTRACT

"Green technology" or Clean technology is the application of the environmental sciences to conserve the natural environment and resources, and by curbing the negative impacts of human involvement. Today, at the beginning of the 21st Century, we must step forward to help solve the climate and energy crises and protect our environment.By using our local ingenuity, entrepreneurship, and creative talent, we can create solutions that change the course of global events. By capitalizing on the innovation in our community, we can create thousands of new jobs, grow city revenues, expand our local economy, and demonstrate how deploying clean and green technology is fiscally responsible.Without Clean Technology innovation, communities will become more polluted, the effects of global warming will increase, and our consumption of fossil fuels will continue our dependency on foreign oil. Thus, we can lead the nation in becoming more energy efficient, producing and using electricity from clean renewable sources, creating green buildings, diverting waste from landfills, creating greener street systems, delivering recycled water, and reducing greenhouse gas emissions.

Some environmental technologies that retain sustainable development are; recycling, water purification, sewage treatment, remediation, flue gas treatment, solid waste management, renewable energy, and others. Using the syngas is more efficient than direct combustion of the original biofuel as more of the energy contained in the fuel is extracted.Use of biofuels such as biodiesel for transportation produces no net output of carbon in the form of $CO_2$ and achieves 30-50 per cent greenhouse gas savings.The techno-economic analysis of biodiesel production has indicated that cost of biodeisel will vary from Rs. 25.70 to Rs. 56.70 per liter with cost of oil ranging from Rs. 25/- to 55/- per Kg compared to Rs.36.5/- per liter for diesel. With further availability of cheap oils and taking into consideration the carbon credits earned due to green fuel under CDM mechanism, the biodiesel costs will further reduce and be more competitive with diesel.

In this chapter a case study of an innovative clean technology for production of biodiesel by transesterification using reactive distillation have been illustrated.By using reactive distillation technique, energy conservation is achieved alongwith reduction of carbon footprints.

Thus, it is the need of an hour to promote innovative green and clean technologies for sustainable environment.

*Keywords: Biofuel, Sustainable environment, Biodiesel, Innovative technology, Transesterification, Reactive distillation.*

*Our problems are man-made, therefore they may be solved by man. No problem of human destiny is beyond human beings.*

***– John F. Kennedy***

## Introduction

"Green technology" or Clean technology is the application of the environmental sciences to conserve the natural environment and resources, and by curbing the negative impacts of human involvement. Sustainable development is the core of environmental technologies.Some environmental technologies that retain sustainable development are recycling, water purification, sewage treatment, remediation, flue gas treatment, solid waste management, renewable energy and others. Energy sources such as solar power create less problems for the environment than traditional sources of energy like coal and petroleum. Scientists continue to search for clean energy alternatives to our current power production methods. Some technologies such as anaerobic digestion produce renewable energy from waste materials. Since electric motors consume 60 per cent of all electricity generated, advanced energy efficient electric motor (and electric generator) technology that are cost effective to encourage their application, such as the brushless wound-rotor doubly-fed electric machine and energy saving module, can dramatically cut the amount of carbon dioxide ($CO_2$) and sulphur dioxide ($SO_2$) that would otherwise be introduced to the atmosphere.Thus, some technologies assist directly with energy conservation, while other technologies are emerging that help the environment by reducing the amount of waste produced by human activities.

## Technologies Retaining Sustainable Development

Some environmental technologies that retain sustainable development are; recycling, water purification, sewage treatment, remediation, flue gas treatment, solid waste management, renewable energy, and others.When applying sustainable development as a solution for environmental issues, the solutions need to be socially equitable, economically viable, and environmentally sound.

### Recylcling

Recycling is the reprocessing of materials into new products. Recycling generally prevents the waste of potentially useful materials, reduces the consumption of raw materials and reduces energy usage, and hence greenhouse gas emissions, compared to virgin production. Recycleables include glass, paper, aluminium, asphalt, iron, textiles and plastics. By harvesting and burning landfill gas, its global warming potential is reduced in addition to providing energy for heat and power. Some technologies such as anaerobic digestion can be used as a distinct waste management strategy to reduce the amount of waste sent to landfill and generate methane, or biogas.

Biodegradable waste, such as food waste or garden waste, is also recyclable with the assistance of micro-organisms through composting or anaerobic digestion. Paper recycling began in Britain in 1921, when the British Waste Paper Association was established to encourage trade in waste paper recycling. Resource shortages caused by the world wars, and other such world-changing occurrences greatly encouraged recycling. The next big investment in recycling occurred in the 1970s, due to rising energy costs. Recycling aluminium uses only 5 per cent of the energy required by virgin production; glass, paper and metals have less dramatic but very significant energy savings when recycled feedstock is used so, first, we should start by reducing the quantity of waste we create. We can increase the amount we recycle, and can reuse more products and convert solid waste and biosolids into biodiesel, methanol, biogas, and electricity.

## Water Purification

Water purification is the process of removing contaminants from a raw water source. The goal is to produce water for a specific purpose with a treatment profile designed to limit the inclusion of specific materials. Chemical analysis, while expensive, is the only way to obtain the information necessary for deciding on method of purification.Various methods can be employed for water purification, such as ultrafiltration, reverse osmosis, ion exchange, etc. Along with maximizing water conservation efforts, the quantity of recycled water we produce and distribute should be increased. Recycled water has proven its value for irrigation, cooling and manufacturing, and there are many opportunities for growth in these uses.

## Membrane Technology

Membrane technology is the most innovative technology in which a semipermeable membranes are used to achieve separation or purification. In this, the energy savings as well as environmental protection is achieved. The footprints are reduced to a large extent.In purification methods using membranes, the latent heat is conserved and $CO_2$ emissions are eliminated against applying sterilization by heating or cooling using traditional methods. Thus energy conservation and environmental protection is achieved thereby making it a cleaner technology.

## Sewage Treatment

Sewage treatment, or domestic wastewater treatment, is the process of removing contaminants from wastewater, both runoff (effluents) and domestic. It includes physical, chemical and biological processes to remove physical, chemical and biological contaminants. Its objective is to produce a waste stream (or treated effluent) and a solid waste or sludge suitable for discharge or reuse back into the environment. The final effluent after sewage treatment can be discharged into a stream, river, bay, lagoon or wetland, or it can be used for the irrigation of a golf course, green way or park. If it is sufficiently clean, it can also be used for groundwater recharge.

## Solid Waste Management

The importance of proper solid waste management is one of the prime functions of the civic body, as insanitary management of solid wastes is a cause of much discomfort. Since waste management is the fundamental requirement for public health these wastes must be managed properly.The systemmatic management of solid wastes either in generation of electricity or recycling these wastes to form finished goods can help to protect environment from being polluted and also energy can be obtained, thus employing clean technology.

## Renewable Energy

Renewable energy effectively uses natural resources such as sunlight, wind, rain, tides and geothermal heat, which are naturally replenished. Renewable energy technologies range from solar power, wind power, hydroelectricity/micro hydro, biomass and biofuels for transportation. About 13 per cent of the world primary energy comes from renewables, with most of renewable energy coming from traditional biomass, like wood-burning. Using waste biomass to produce energy can reduce the use of fossil fuels, reduce greenhouse gas emissions and reduce pollution and waste management problems.

Hydropower is the next largest renewable source, providing 2-3 per cent, and modern technologies like geothermal, wind, solar, and marine energy together produce less than 1 per cent of total world energy demand. Brazil has one of the largest renewable energy programs in the world, involving production of ethanol fuel from sugar cane, and ethanol now provides 18 per cent of the country's automotive fuel. The global reduction of greenhouse gases is dependent on the adoption of energy conservation technologies at industrial level as well as this clean energy generation. That includes using unleaded gasoline, solar energy and alternative fuel vehicles, including plug-in hybrid and hybrid electric vehicles.

## Biodiesel Production by Transesterification Using Reactive Distillation: A Case Study

Clean technology is a means to create electricity and fuels with a smaller environmental footprint. Investments in clean technology have grown considerably since coming into the spotlight around 2000. Biofuel is considered as a means of reducing greenhouse gas emissions and increasing energy security by providing an alternative to fossil fuels. Biodiesel the most promosing biofuel is produced from oils or fats using transesterification and is a liquid similar in composition to mineral diesel. Its chemical name is fatty acid methyl (or ethyl) ester (FAME).

Pure biodiesel (B100) needs to meet the requirements of ASTM D-6751 to avoid engine operational problems.

**Table 36.1: ASTM D-6751 Standards for Biodiesel**

| | |
|---|---|
| Flash point (closed cup) | 130°C min. (150°C average) |
| Water and sediment | 0.050 per cent by vol., max. |
| Kinematic viscosity at 40°C | 1.9-6.0 mm2/s |
| Ramsbottom carbon residue, per cent mass | 0.10 |
| Sulfated ash | 0.020 per cent by mass, max. |
| Sulfur | 0.05 per cent by mass, max. |
| Copper strip corrosion | No. 3 max |
| Cetane | 47 min. |
| Carbon residue | 0.050 per cent by mass, max. |

*Contd...*

**Table 36.1–Contd...**

| | |
|---|---|
| Acid number — mg KOH/g | 0.80 max. |
| Free glycerin | 0.020 per cent mass |
| Total glycerine (free glycerine and unconverted glycerides combined) | 0.240 per cent by mass, max. |
| Phosphorus content | 0.001 max. per cent mass |
| Distillation | 90 per cent @ 360°C |

Biodiesel contains only trace amounts of sulfur typically less than the new EPA standards. Biodiesel offers safety benefits over petroleum diesel because it is much less combustible, with a flash point greater than 150°C, compared to 77°C for petroleum diesel. Biodiesel B20 is a proven fuel with over 30 million successful US road miles, and over 20 years of use in Europe. Biodiesel is biodegradable in water, up to 98 per cent in three weeks.

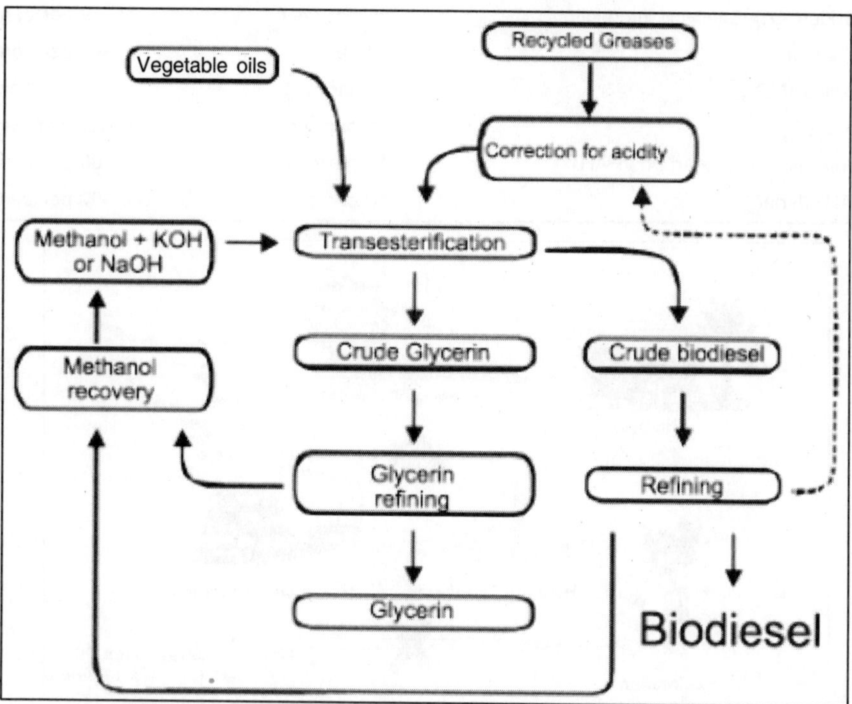


**Figure 36.1: Schematic Process for Biodiesel Production**

Biodiesel production by transesterification using reactive distillation is one of the most promising clean technologies which involves less energy for production of biodiesel fuel from waste/nonedible/edible vegetable oils and it reduces carbon footprints to the large extent. Reactive distillation involves simultaneous reaction and separation in a single piece of process equipment. It is applicable to equilibrium limited reactions where one product is the most volatile component. Process intensification results in lower energy and capital costs. Each reaction system requires a unique approach for successful design and operation. Biodiesel, a mixture of mono-alkyl esters of fatty acids, is currently manufactured by trans-esterification of triglycerides with methanol using NaOH or KOH as liquid base catalyst. This catalyst is corrosive to the equipment, but this is easily overcome with little cost penalty, by constructing the reaction vessels out of stainless steel. However, the main drawback is that the liquid base catalyst has to be neutralized afterwards– typically using HCl–thus producing waste salt streams. Moreover, due to the presence of free fatty acids it reacts to form soap as unwanted by-product hence requiring more expensive separation. Another method to produce these fatty esters is the batch-wise esterification of fatty acids using $H_2SO_4$ as catalyst. However, the problem here is the batch operation mode that is not suitable for very large-scale production and again it involves costly neutralization and separation of the homogeneous catalyst.

**Table 36.2: Average Biodiesel Compositions Compared to Conventional Diesel**

| *Emission Type* | *B20* | *B100* |
|---|---|---|
| Total unburned hydrocarbons | -20 per cent | -67 per cent |
| Carbon monoxide(CO) | -12 per cent | -48 per cent |
| Carbon dioxide($CO_2$)- Lifecycle production | -16 per cent | -79 per cent |
| Particulate matter | -12 per cent | -47 per cent |
| Nitrogen oxides ($NO_X$) | +2 per cent | +10 per cent |
| Sulfur oxides($SO_X$) | -20 per cent | -100 per cent |
| Polycyclic aromatic hydrocarbons(PAH) | -13 per cent | -80 per cent |
| Nitrated PAH's(Npah) | -50 per cent | -90 per cent |

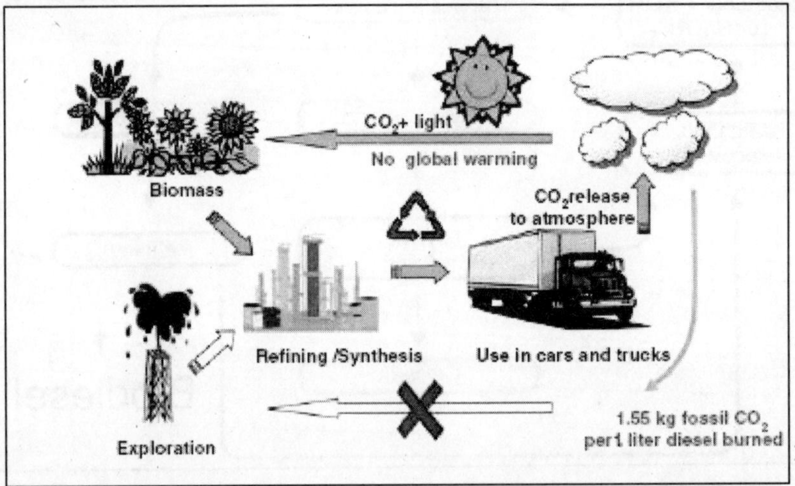


**Figure 36.2: Lifecycle of Biodiesel as Environmentally-Friendly Fuel**

## Biodiesel Production Process

For achieveing more benefits transesterification can be coupled with reactive distillation. The reaction conditions such as reactants ratio, temperature and pressure determine the phase equilibrium. For temperatures below 100 °C, only low conversions can be achieved and an excess of alcohol is required to have only one liquid phase. For stoichiometric reactants ratio, two liquid phases exist at temperatures below 100°C. If temperature exceeds 100°C and the system is closed at over pressure then three phases exist: vapours–liquid–liquid. Liquid separation is undesirable due to catalyst deactivation in the presence of water. The best solution is working at temperatures above 100 °C, in a system with continuous water removal. By removing water as by-product the equilibrium is shifted towards ester formation.

Biodiesel is produced by the transesterification reaction as shown below:

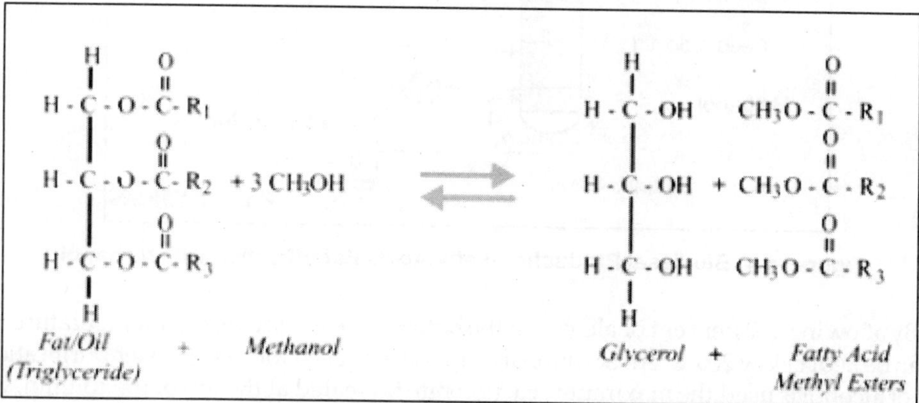


Hence, the ester will always be separated in the bottom of the reactive distillation column (RDC). Water is present as side product, and typically is removed as top product due to its lower boiling point, together with the alcohol if it is volatile (and not completely converted) or if it forms azeotropes with water.

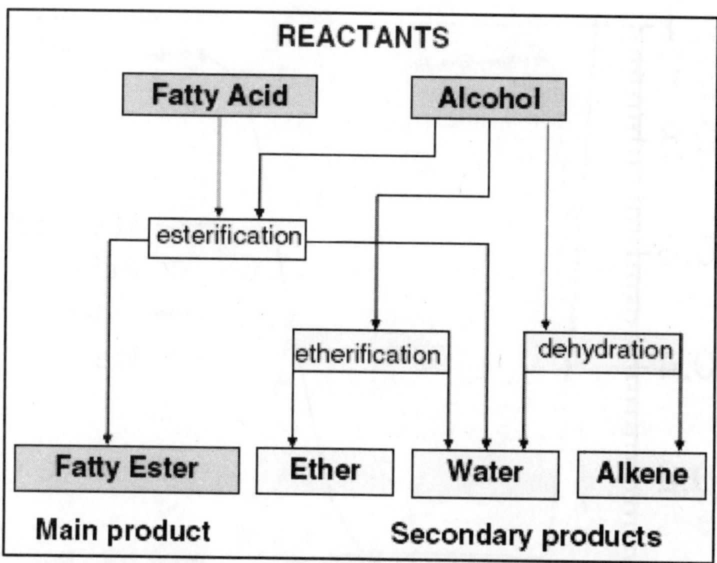


**Figure 36.3: Reaction Pathways and Possible Products**

Figure 36.4 presents the flowsheet for esterification of dodecanoic acid with methanol. High purity final products are feasible. Note however that pure fatty acid esters cannot be the bottom product of the reactive distillation column because of the high boiling points and lower thermo-

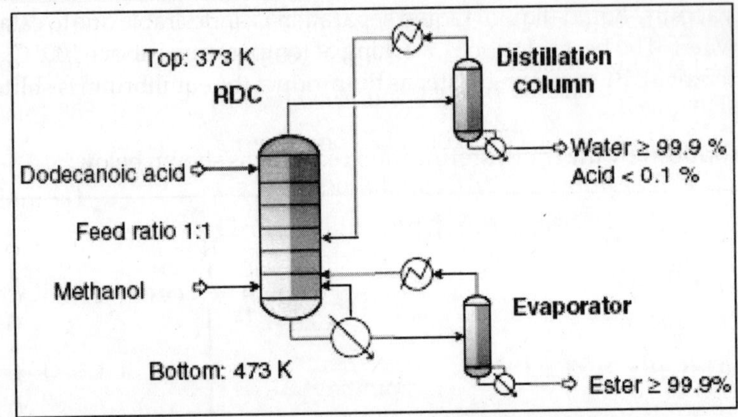


**Figure 36.4: Biodiesel Production-Fatty Acids Esterification with Mehanol**

stability. By allowing 1–2 per cent of alcohol in the bottom stream, the reboiler temperature in the RD column can be kept below 200°C. An additional evaporator is used for further ester purification. When an excess of alcohol is used the maximum reaction rate is located at the top of the column, with total acid conversion in the bottom but partial conversion of alcohol in the top. For the optimal reflux ratio the maximum reaction rate is located in the centre of the column, providing complete conversion of both reactants at the ends of the column. The composition and temperature profiles in the reactive distillation column are shown in Figure 36.5.

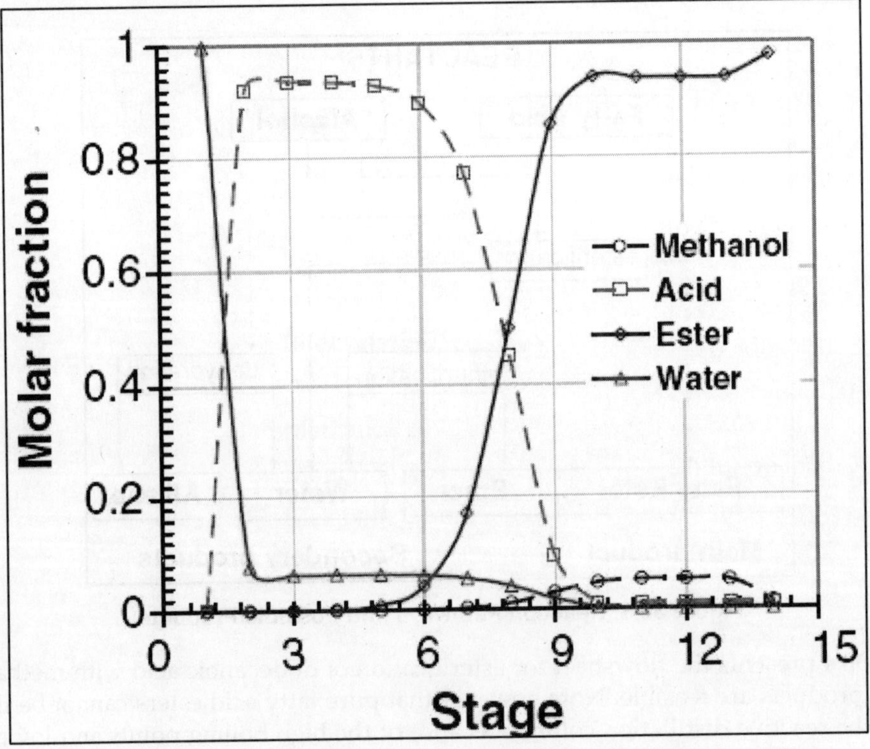


**Figure 36.5: Reactive Distillation Column Profiles–Liquid Composition**

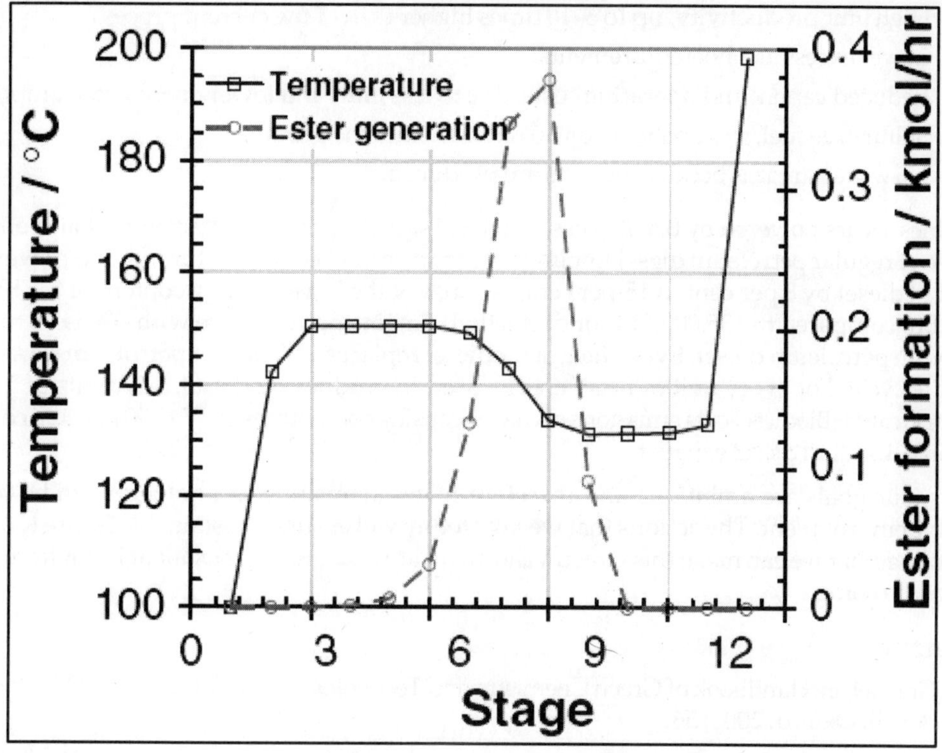


**Figure 36.6: Reactive Distillation Column Profiles–Temperature and Ester Generation**

## Conclusion(s)

Through innovative partnerships with Clean Tech companies and leaders throughout our community, we can help solve the climate crisis while creating a new economic base. It will require unique partnerships between the City, residents, businesses, schools, and universities. It will require changing policies locally and advocating for aggressive change and investment at the State and Federal levels. Most importantly, it will require the enthusiasm and participation of every resident on this earth.

From the above case study,we can conclude that the biodiesel fuel can be produced by a sustainable continuous process based on catalytic reactive distillation. Pure biodiesel (100 per cent biodiesel) reduces carbon dioxide emissions by more than 75 per cent over petroleum diesel. B20 Biodiesel is one the most popular biodiesel blends presently available thoughout much of the U.S., Canada and Europe. Using a blend of 20 per cent biodiesel reduces carbon dioxide emissions by 15 per cent. The integrated design in above case study ensures the removal of water by-product that shifts the chemical equilibrium to completion and preserves the catalyst activity.

Manufacturing of fatty acid esters by reactive distillation can be applied to a variety of alcohols and fatty acids, as a multifunctional reactor, the actual applications depending on the feedstock at hand. The process proposed here can dramatically improve the economics of current biodiesel synthesis and reduce the number of downstream steps. The key benefits are:

1. High unit productivity, up to 6–10 times higher than of the current process.
2. Lower excess alcohol requirements.
3. Reduced capital and operating costs, due to less units and lower energy consumption.
4. Sulfur-free fuel, since solid acids do not leach into the product.
5. No waste streams because no salts are produced.

Engines in cars powered by B20 Biodiesel fuel will significantly fewer harmful exhaust emissions than those of regular petroleum diesel Engine performance (fuel economy, torque, and power) is less than that of diesel by 8 per cent to 15 per cent, because of the lower energy content of the biodiesel (121,000 Btu compared to 135,000 Btu for diesel fuel). Engine performance with B20 is virtually the same as with petroleum diesel. Every liter of biodiesel replaces 0.95 liter of petroleum-based diesel over its life cycle. For every unit of fossil energy used to produce biodiesel, 3.37 units of biodiesel energy are created. Biodiesel cuts emissions of cancer-causing compounds by 75 to 90 per cent compared to those in petroleum diesel exhaust.

Thus, our goals are ambitious and therefore many innovative technologies can be used for sustainable environment. The actions that we take today will ensure a sustainable future tomorrow. However, together we can make this Green Vision a reality and ensure a sustainable environment for generations to come.

## References

1. T.E. Graedel, in: Handbook of Green Chemistry and Technology, ed. J.H. Clark and D.J. Macquarrie (Blackwell, Oxford, 2002) 56.
2. Dennis J. Miller and Carl T. Lira, Biorenewable fuels and chemicals via reactive distillation.
3. J. Sheehan, V. Camobreco, J. Duffield, M. Graboski and H. Shapouri, An Overview of Biodiesel and Petroleum Diesel Life Cycles (National Renewable Energy Laboratory, 1998).
4. Sneesby, M. G., TadeH, M. O., and Smith, T. N. (1997c). Implications of steady-state multiplicity for operation and control of etherifcation columns. *Distillation and absorption '97, Institute of the Chemical Engineers Symposium Series*, 142 (pp. 205}216).
5. Omota, F. Dimian A.C. Bliek A. (2003). Fatty acid esterification by reactive distillation. Part1: Equilibrium-Based Design. Chemical Engineering Science 58, 3159–3174.
6. Omota, F. Dimian A.C. Bliek A. (2003). Fatty acid esterification by reactive distillation: Part 2–kinetics-based design for sulphated zirconia catalysts. Chemical Engineering Science 58, 3175–3185.
7. Darnoko, D. and Cheryan, M. (2000) Kinetics of Palm Oil Tranesterification in a Batch Reactor, *JAOCS*, 77, (12), pp.1263-1267.

## Chapter 37

# Experimental and Theoretical Studies for Blending of Fuel with Tyre in Cement Industry

**S.R. Asthana[1]*, R.H. Niturkar[2], B. Garnaik[1] and A.T. Pise[2]**
**[1]National Chemical Laboratory, Pune, M.S.**
**[2]Government College of Engineering, Karad, M.S.**

### ABSTRACT

Cement industry is highly energy consuming industry and it consumes about 4 GJ per tons of cement produced. The scrap tyre can be obtained at low cost than fossil fuel. High calorific value of tyre makes effective alternative fuel. Environmentally safe disposal of Tyre is an added advantage. Having realized the importance of blending of tyre with petcoke in cement plant, both experimental and theoretical study has been carried out at National Chemical Laboratory, Pune. Study reveals encouraging results.

*Keywords: Tyre, Clinker, Blending, Pyrolysis, Heat value.*

## Introduction

Cement, aluminum, fertilizer, iron and steel, glass and paper are identified as energy intensive industries. Cement industry is highly energy consuming industry and consumes about 4 GJ/t of cement produced. The energy costs accounts about 30–40 per cent of the manufacturing cost of cement.

* Corresponding Author: E-mail: sr.asthana@ncl.res.in

Blending of alternative fuels with fossil fuel will not only help in reducing the energy cost but also reduce production cost. Although there are several waste derived fuels that can be blended use of scrape tyres is found most suitable due to its high calorific value and ready availability. Scrap tyre will definitely be cheaper than fossil fuel. Apart from utilizing heat energy, blending of tyre will also help in safe disposal of tyre as it is not biodegradable.

Tyre is basically composed of rubber materials (mostly polybutadiene, styrene butadiene rubber and polyisoprene or natural rubber), carbon black and some fibrous materials. It has high volatile and fixed carbon contents with a heating value greater than that of coal. The gas phase contains a mixture of light hydrocarbons, carbon dioxide, carbon monoxide and $H_2$. This composition confers a very high heating value. Therefore, these gases can be used to provide the energy to the pyrolysis process in cement production. There are a number of factors that promote the use of alternative fuel in the cement kiln such as high inside temperature, long residence time of the fuel (0.5 to 1.5 hr.), alkaline environment, no ash residue, complete destruction of rubber and cotton, no smoke, no odour, overall reduction in emission.

## Literature Review

Recent work that has been carried out globally on tyre as alternative fuel use in cement plant is reviewed as follows:

A study by Asthana and Patil (2006) has revealed that the use of alternative fuels will help in reducing energy costs and providing a competitive edge for a cement plant. Furthermore, this will reduce the burden of waste disposal considerably. However, in India this practice is not prevalent and needs to be studied and commercialized.

Mokrzycki *et al.* (2003) have found the high incineration temperature, the large area of the furnace, the significant length of the kiln and the alkaline environment inside the kiln are the factors promoting the use of cement kilns. According to the European Cement Association, CEMBUREAU the proportion of alternative fuels used in clinker kiln between 1990 and 1998 in particular countries were as follows: Belgium-18 per cent; Italy-4.1 per cent; Portugal-1.3 per cent; Spain-1 per cent; Sweden-2 per cent; Switzerland-25 per cent; Czech Republic-9.7 per cent; Grate Britain-20 per cent; Germany-15 per cent; Poland-1.4 per cent.

Worrell *et al.* (1999) have carried out an in-depth analysis of the US cement industry and found that the use of blended cement is a key cost-effective strategy for energy efficiency improvement and $CO_2$ emission reductions. They have also estimated total carbon emissions of 160 million t from the cement process, for the world cement industry.

Kantee *et al.* (2004) studied and revealed that the alternative fuels *i.e.*, processes waste materials can replace the conventional fuels in cement manufacturing. This is because of energy cost and environmental standards encouraged cement manufacturers world wide. The study has used a commercial modeling tool to model the four stage preheater kiln system of full-scale cement plant, using petcoke as a fuel. The goal is to optimize process control and alternative fuel consumption, while remaining clinker product quality same.

Aylon *et al.* (2006) have performed experiments and analyzed the emission. The study has revealed that the shredded tyre of 2 mm size when used in pyrolysis process does not produce any pollutant beyond the permissible limits.

Li A.M. *et al.* (1999) performed experiments with wood chips and shredded tyre under simulated operating conditions of cement manufacturing process. The gas chromatograph was done to analyze the components of the gas such as $H_2$, $CO$, $CH_4$, $CO_2$, $O_2$, $N_2$, $C_2H_4$ and $C_2H_6$. The influence of Final Pyrolysis Temperature (FPT) on products was found obvious, raising FPT caused increased yield of gas and decreased yield of semi-coke. FPT also affects the yield and heat value of the gas. The percentages of the raw materials heat content in gas increase with increasing FPT.

Experiments performed by Laregoiti *et al.* (1999) used tyre Pyrolysis was performed at 400, 500, 600, and 700°C and the gases generated are analyzed. These gases were composed of $CO$, $CO_2$, $H_2S$, $CH_4$, $C_2H_4$, $C_3H_6$, $C_4H_8$, which give high heat value. They conclude that these gases evolved at 400°C.

Reto Giere *et al.* (2006) have performed experiment with mixture of 95 wt. per cent coal plus 5 wt. per cent tyre. It was found that the emissions of most other metals and metalloids, and HCL increase when TDF was added. The enhanced metal emissions might be due to formation of gaseous metal chloride species in the stack gases. On the other hand, an emission of CO decreases slightly, whereas those of NOx, $SO_2$, and total particulate matter remained virtually unchanged. These results help in assessing the environmental impact of energy recovery from scrap tires.

Pipilikaki *et al.* (2005) have performed experiment with a mixture of coal and pet-coke and than coal, pet-coke and TDF. The study revealed that 6 per cent of the total fuel used was TDF and the results found were very promising.

Xiang-guo *et al.* (2006) have performed thermogravimetric analysis of co-combustion of coal and tyre blends also comparison of the TG-DTG profile between experimental and calculation result. They conclude that incorporation of waste tyre can improve combustion characteristics of high ash coal, in ignition performance, peak weight loss.

Diez *et al.* (2003) they performed experimental study on pyrolysis of tyre, influence of final pyrolysis temperature on emission and heat value. They have analysis solid, liquid and gases of tyre pyrolysis.

Literature review revealed that blending of waste derived tyres in cement production process are both technically and economically feasible. Therefore, this study is taken up at NCL, Pune for further research.

## Technical Feasibility of Use of Tyre

From literature it is found that at 400–700°C tyre pyrolysis gases composed of $CO$, $CO_2$, $H_2S$, and hydrocarbons such as $CH_4$, $C_2H_4$, $C_3H_6$, $C_3H_8$, etc, having gross calorific values about 30000–40000 kJ/kg. Also, at 600–850°C the non-condensable gas obtained from rubber tyre pyrolysis has a high gross calorific value (around 35000 kJ/kg). In addition, the emissions produced from the combustion of the tyre pyrolysis gas are under the limits, except TOC, $SO_2$ and HCl values. While the TOC emission levels could be avoided by using a proper burner, it would be necessary to incorporate an acid gas cleaning system to reduce HCl and $SO_2$ concentrations. For this sulfuric acid recovery furnace can be used.

Also Steel component of tyre mixed results in increase in strength of cement. And this will reduce burden of waste disposal considerably with utilizing heat content. In this study it is observed that 6 per cent of the total fuel used was tyre and the results were very promising there no adverse effect on cement properties. Therefore, use of alternative fuel in cement industry is ecologically and economically justified.

## Experimental Studies of Fuel and Tyre Blending

Having realized that tyre is a good candidate for blending, experimental vis-à-vis theoretical studies were taken up. Experimental parameters, set-up and results obtained are presented as below:

## Sample Selection and Preparation

Tyre were found most appropriate fuel for blending as compared to other options due to their properties such as high calorific value, high temperature, long residence time, no ash residual, complete destruction of rubber, no smoke, no odour, over all reduction in emission, metal incorporate in cement increase strength, etc. Initially, cycle, two wheeler and four wheeler tyres were considered for the study. It was realized that heat value will be high for four wheeler tyres based on polymer contain and they are readily available in large quantity therefore it is focused for further study.

After selecting four wheeler tyres for blending different brand tyres were obtained such as Appolo, MRF, and CEAT etc. The next step was to shred the tyres. Difficulties arise during shredding as tyres contained steel wires. Grinding of tyres to fine powder was done at re-moulding workshop.

## Experimental Set-up and Results

Experiments (4 nos.) were arranged at NCL with sample weight of 4 g each. The waste tyre samples were prepared by shredding and grinding of tyre to fine powder. Raw petroleum coke is crushed and analyzed for composition. It has fixed carbon 89.80 per cent, ash 0.42 per cent, moisture 1.8 per cent and volatile material 9.78 per cent.

After preparation sample was placed in a U.S. make Nelson reactor. Argon gas flow of 3–5 kg/cm$^2$ was maintained throughout the experiment to create an inert atmosphere inside the reactor. A qualitative analysis has been performed by heating the sample for 30, 60 and 100 minutes at 200, 300, 350 and 400°C. The system is heated from ambient temperature to the desired temperature at a rate of 20°C/min. Results obtained are summarized in Table 37.1 below:

**Table 37.1: Qualitative Analysis of Sample**

| *Sl.No.* | *Temp. in Reactor (°C)* | *Status of Sample* |
|---|---|---|
| 1. | 250 | No change in physical state |
| 2. | 300 | Slight change in state to softness having odour |
| 3. | 350 | Almost softness with strong odour. |
| 4. | 400 | Odourless powder form |

The same procedure is repeated for sample of petcoke and tyre.

The ratio of blending was taken as 4:1 for petcoke and tyre. A weight loss analysis has been carried out to find heat value of the blended fuel. An analysis of sample after heating at 400°C for 100 minutes was carried out and it has been found that total weight loss was 65 per cent in case of tyre while it was 11.12 per cent in petcoke and tyre blend.

## Interesting Result Obtained

At 400°C tyre sample came in odourless powder form. No sludge formation was found during pyrolysis of tyre. Further, it is found interesting observation as it was expected that tyre will become sticky which may result in sludge formation in the plant. Since tyre became powder, introduction of tyre in kiln or preheater is expected to be convenient. It is obvious that mixing of tyre with cement will increase mechanical strength due to steel content present in the tyre.

## Theoretical Studies of Fuel and Tyre Blending

Although experiments were showing encouraging results, it was decided to carry out theoretical modeling to increase confidence level in the study. Initial results are presented.

### Assumptions

Mass of Petcoke ($m_{petcoke}$) = 0.108 kg/kg of clinker (Source: J.K. Laxmi plant)

Calorific Value (CV) of Petcoke = 31411.72 kJ/kg

Calorific Value (CV) of tyre =33440 kJ/kg (Source: Sharma *et al.*, 2000)

### Equations Used

Heat generated ($Q_{p+t}$) due to combustion of petcoke and combustion of tyre =

$$m_p * CV_p + m_t * CV_t$$

*where,*

$m_p$ = mass of petcoke, (kg)

$CV_p$ = calorific value of petcoke, (kJ/kg)

$m_t$ = mass of tyre, (kg) and

$CV_t$ = Calorific value of tyre (kJ/kg)

Table 37.2 presents results obtained.

**Table 37.2: Heat Generation with Different Percentage Blending**

| *Sl.No.* | *Trials* | | *Heat Generated(kJ/kg)* |
|---|---|---|---|
| | *Per cent of Petcoke* | *Per cent of Tyre* | |
| 1. | 100 | 0 | 3392.4600 |
| 2. | 99 | 1 | 3394.6511 |
| 3. | 98 | 2 | 3396.8464 |
| 4. | 97 | 3 | 3398.9619 |
| 5. | 96 | 4 | 3401.2279 |
| 6. | 95 | 5 | 3403.4160 |
| 7. | 94 | 6 | 3405.6090 |

Figure 37.1 presents graph based on Table 37.2.

Findings are in agreement with a study carried out by Pipilikaki (2005) which suggests that upto 6 per cent blending yield good results. Although, heat value of tyre is slight higher than petcoke, it is quite logical that blending with tyre will yield more heat. However, blending percentage can not be increased beyond limit due to generation of hazardous chemicals such as dioxins and furans. Generation of these chemicals is restricted by Central Pollution Control Board (CPCB).

## Thermal Energy Savings due to Blending

Thermal energy saved due to 6 per cent tyre blending is calculated as 13.1412 kJ/kg of clinker. This is due to savings in the amount of petcoke burnt which is $6.48 * 10^{-3}$ kg/kg of clinker. The findings

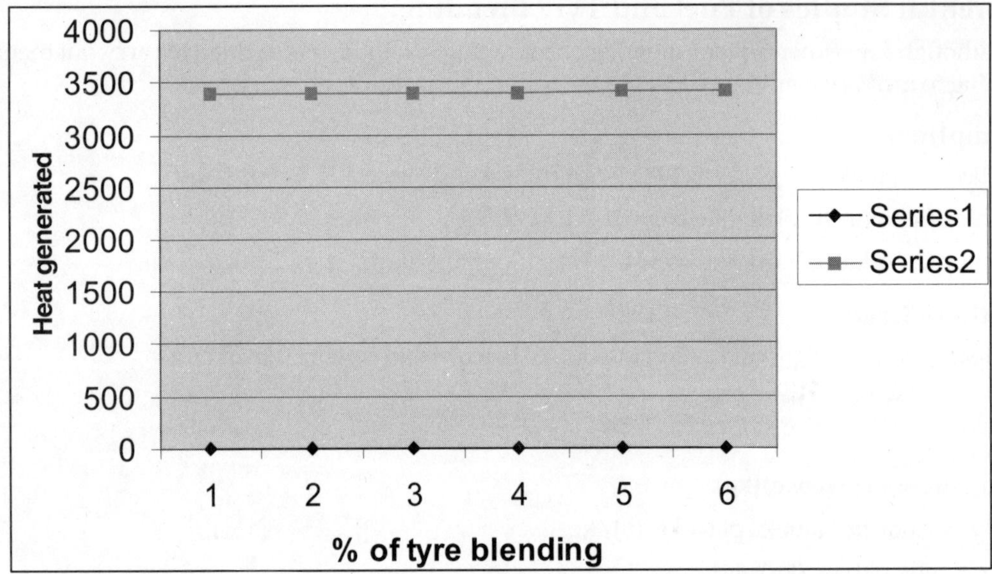


**Figure 37.1: Graph of per cent Tyre Blending vs. Heat Value**

when incorporated for an Indian cement plant J.K.Laxmi Cement Plant producing 4200 TPD clinker revealed that saving of 27.216 TPD petcoke is possible with 6 per cent blending of waste derived tyre. Further, taking cost of petcoke as INR 3,300 (as per plant data), economical savings would be of the order of INR 89,812.8 per day.

## Conclusion

Blending of fuel with alternative fuel is technically and economically viable option. The experiments have revealed an interesting observation that at 400 °C tyre sample becomes odorless powder. 65 per cent weight loss is suggestive of good heat value that tyre can generate. Further, no sludge and sticky mass formation was found during pyrolysis of tyre suggesting that introduction of tyre in kiln or preheater will be convenient. Also, mixing of tyre with cement will increase mechanical strength due to presence of steel content in the tyre.

Use of alternative fuel in cement industry is ecologically and economically justified. This will reduce burden of waste disposal considerably with utilizing heat content. Theoretical study suggests that blending of tyre with petcoke will increase heat generation and there is a significant scope of thermal energy savings. This saving will definitely result in monetary gain. More extensive study with more experiments is underway at National Chemical Laboratory, Pune.

## References

1. Asthana, S.R., Patil, R.K., "Use of Alternative Fuels in Indian Cement Industry". Advances in energy research, Indian Institute of Technology Bombay, 2006.

2. Li *et al.* Pyrolysis of tyres. Influence of the final temperature of the process on emissions and the calorific value of the products recovered. Journal of Analytical and Applied Pyrolysis. 50, 1999, pp. 149–162.

3. Giere Reto *et al.*, Chemical composition of fuels and emissions from a coal + tire combustion experiment in a power station. Fuel. 85, 2006, pp. 2278–2285.

4. Xiang-guo Li. *et al.*, Thermogravimetric analysis of the co-combustion of the blends with high ash coal and waste tyres. Thermochimica Acta, 441,2006, pp.79–83.

5. Laresgoiti *et al.*, Chromatographic analysis of the gases obtained in tyre Pyrolysis. Journal of Analytical and Applied Pyrolysis. 55, 2000, pp. 43–54.

6. Pipilikaki *et al.*, Use of tire derived fuel in clinker burning. Cement and Concrete Composites. 27, 2005, pp. 843–847.

7. Diez *et al.*, Pyrolysis of tyres. Influence of the final temperature of the process onemission and the calorific value of the products recovered. Waste Management, 24, 2004, pp.463–469.

8. Aylon *et al.*, Emission From the Combustion of gas–phase products at tyre pyrolysis. Journal of Analytical and Applied Pyrolysis. 79, 2007, pp. 210–214.

9. E. Mokrzycki, Alicja U, Bochericzyk: Alternative fuels for the cement industry, Applied Energy 74, pp, 95-100, 2003.

10. Kaantee U, Zevenhoven R, Backman R, Hupa M, "Cement Manufacturing Using Alternative Fuels and Advantages of Process Modelling", Fuel Process Technology 85,2004,293-301.

11. Worrell *et al.*, "Potentials for energy efficiency improvement in the US cement industry" Energy 25 (1999) 1189-1214.

12. Sharma *et al.*, "Disposal of waste tyres for energy recovery and safe environment". Applied Energy 65 (2000) 381-394.

13. Log sheet of J.K. Laxmi plant, Sirohi, Rajasthan, India.

# Chapter 38

# RoHS Directive: Impact on Environment, Health and Electronic Industry

**N.P. Deshpande**
**Department of Electronics and Telecommunications,**
**MIT College of Engineering, Pune, M.S.**

## ABSTRACT

Introduction of RoHS directive by European Union in 2006 is a positive step in minimizing impact of hazardous substances on environment and human health. However there are other implications that need to be carefully considered. The directive is recent and its long-term benefits are yet to be established fully. This paper looks into important impacts that this directive has made on environment, human health and electronic industry. Most electronic product manufacturers have taken measures to comply with directives, although sometimes, somewhat unwillingly. An awareness into the ill effects of use of hazardous substances is however absolutely necessary.

*Keywords: RoHS, Lead-free, Reflow soldering, Tin whiskers.*

## Background

In 1998, it was discovered by European Union (EU) that a large amount of hazardous waste was being dumped into landfill sites. The extent of waste was so large that an estimate of volumes that were dumped was 3-5 times faster than average municipal waste. India generated 3.3 lakh tones of e-waste in 2007. E-waste is expected to touch 4.7 lakh tones by 2011. The environment was clearly in danger. In order to address the issue of environmental pollution, member states of EU decided to create the West Electrical and Electronic Equipment (WEEE) directive. The objectives of WEEE were-

1. To improve manufacturer's design, to reduce the creation of waste
2. Make manufacturers responsible for certain phases of waste management
3. Separate collection of electronic waste from other types of waste
4. Create systems to improve treatment, reuse and recycling of WEE.

Subsequent reviews and revisions to original WEEE directive resulted ultimately into a RoHS directive today known by the reference 2002/95/EC.

## Introduction

The European Union restricted the use of six hazardous substances in electrical and electronic equipment placed on market after 1st July 2006. Six banned substances are: (1) Cadmium, (2) Mercury, (3) Hexavalent Chromium, (4) Ploybrominated Biphenyls (PBB), (5) Ploybrominated Diphenyl Ethers (PDBEs) and (6) Lead (Pb). Amongst these six substances, lead is most significant material for electronics industry. The industry has hence adopted a more familiar term- "lead free", although making a product lead-free alone does not imply RoHS compliance. The RoHS directive 2002/95/EC was adopted in February 2006 and became a law to be enforced in all states of European union with effect from 1st July, 2006. The directive is closely related to Waste Electrical and Electronic equipment Directive (WEEE) 2002/96/EC which sets collection, recycling and recovery targets for electrical goods. The main purpose of directive is to minimize the effect of use of hazardous materials on environment and human health.

The directive is applicable to everything that can be identified as a homogeneous material *i.e.* any substance that can be separated mechanically. For example, a radio receiver has case, printed wire board, resistors, capacitors, loudspeakers etc. All these can be mechanically separated.

Other countries across the globe have also introduced their own guidelines that will eventually become law in near future.

The poisonous effects of lead are well known. There are many possible symptoms of lead poisoning. Lead can affect many different parts of the body. Over time, even low levels of lead exposure can harm a child's mental development. The possible health problems get worse as the level of lead in the blood gets higher. Possible complications include:

- ☆ Reduced IQ
- ☆ Slowed body growth
- ☆ Hearing problems
- ☆ Behaviour or attention problems
- ☆ Failure at school
- ☆ Kidney damage

The symptoms of lead poisoning may include:

- ☆ Irritability
- ☆ Aggressive behaviour
- ☆ Low appetite and energy
- ☆ Difficulty sleeping
- ☆ Headaches

- ☆ Reduced sensations
- ☆ Loss of previous developmental skills (in young children)
- ☆ Anemia
- ☆ Constipation
- ☆ Abdominal pain and cramping (usually the first sign of a high, toxic dose of lead poison)
- ☆ Very high levels may cause vomiting, staggering gait, muscle weakness, seizures, or coma.

For example, if a person swallows a lead object, the poison can stay in the body that will cause serious health problems. Unfortunately the problems show up slowly over time. The damaging effects are more serious in children and pregnant women. Lead is more harmful to children since it can affect children's developing nerves and brain. The younger the child, the more harmful the lead can be. Unborn children are even more vulnerable.

Considering the ill effects as above, manufacturers of electronic components and products have taken measures to make their goods RoHS compliant. This has certainly resulted in controlling the pollution. However, other effects on product quality and reliability and commercial implications are also significant. This chapter discusses the impact of introduction of RoHS directive on environment, human health and electronic industry.

Discussed below are the effects of changing over from traditional Tin-Lead (Sn-Pb) solders to a variety of "lead-free" solders. The alternative solders that are considered for evaluation are: (1) Tin-Copper (SnCu); (2) Tin-Silver-Copper (SAC); (3) Bismuth-Tin-Silver (BSA); (4) Tin-Silver-Bismuth-Copper (SBAC).

## Life Cycle Impacts on Environment

The study considers life cycle impact on environment in terms of material consumption, water resources, air resources, landfills, energy/power consumption, human toxicity, ecological toxicity, leachability and recycling.

Eutectic lead based solders were traditionally used in assembling electronic products due to its low melting point, good reflow properties and relative ductility. The primary points of concern in use of these solders, however, remain to be the human toxicity and damage to the environment. Considering the use of huge amounts of lead based solders worldwide, it is really a challenge to enforce any legislation in the interest of protecting the environment. Only a systematic study can convince the stakeholders, particularly electronic industry, to adopt a changeover from lead based to lead-free solders.

There are five stages involved in the life cycle of a solder- Raw material extraction/acquisition, material processing, product manufacture, product use and disposal. These stages also require certain types of resources.

Raw material extraction stage involves activities related to the acquisition of natural resources including mining of non-renewable material and transportation to the processing facility.

Naturally obtained materials are not normally suitable for direct use. This calls for material processing in terms of separation, purification chemical reactions etc. An outcome of material processing is generation of certain amount of waste.

Product manufacture and use stage involves soldering operation- either wave soldering or reflow soldering. The input to this stage is (rather) a large amount of energy required to melt the solder.

Solders are materials that are invariably recycled. This process may involve transportation of products at other location. Product is disassembled for material recovery. Materials that cannot be recovered are disposed off suitably.

## Impact Categories

In analyzing the effect of lead based and lead-free solders on environment, it is useful to categorize the impacts. There are 16 different impact categories that are considered to be significant for such a study. The categories are summarized in Table 38.1.

**Table 38.1: Impact Categories**

| *Impact Category* | *Inventory Type* | | *Description* |
|---|---|---|---|
| | *Input* | *Output* | |
| **A. Impact on natural resources** | | | |
| Non-renewable resource Use/depletion | Material, Fuel | - | Materials like metals and fossil fuels are not regenerated naturally |
| Use of renewable resources | Material, Water | - | Materials that are available in nature and are replaced by nature |
| Energy usage | Electricity, Fuel | - | The amount of energy consumed |
| Landfill space | - | Solid/hazardous or radioactive waste to landfill | The volume of landfill space required for hazardous or radioactive waste |
| **B. Abiotic Ecosystem Impacts** | | | |
| Global warming | - | Air | The mass of emitted greenhouse gases, relative to Carbon dioxide |
| Stratospheric ozone depletion | - | Air | The mass of ozone depleting chemicals emitted, relative to CFC-11 |
| Photochemical smog | - | Air | The mass of smog producing chemicals emitted, relative to ethane. |
| Acidification | - | Air | The mass of acidifying chemicals emitted, relative to sulfur dioxide. |
| Air particulates | - | Air | The mass of particulates emitted that have an effective aerodynamic diameter less than 10 micrometers |
| Water eutrophication | - | Water | The mass of eutrophication chemicals released relative to phosphate |
| **C. Human health and ecotoxicity** | | | |
| Occupational chronic non-carcinogenic health effects | Material | - | Weighted score based on amount and toxicity of non-carcinogenic releases affecting workers |
| Occupational carc-inogenic health effects | Material | - | Weighted score based on amount and toxicity of carcinogenic releases affecting workers |
| Chronic, carc-inogenic health effects–public | - | Air, soil, water | Weighted score based on amount and toxicity of carcinogenic releases affecting general public |
| Carcinogenic health effects–public | - | Air, soil, water | Weighted score based on amount and toxicity of carcinogenic releases affecting general public |
| Aqatic ecotoxicity | - | Water | Weighted score based on the amount and toxicity of releases affecting fish |

## Study Results

Study results of lead based and lead free solders on above impact categories reveal following-

**Among all the Solders**

- ✰ *SnPb* has highest impact score in: (1) Renewable resource use; (2) Water eutrophication; (3) Occupational health-non-cancer; (4) Occupational health-cancer; (5) Public human health-non-cancer.
- ✰ *SAC* has highest impact score in: (1) Renewable resource use; (2) Energy use; (3) Landfill space use; (4) Global warming; (5) Stratospheric ozone depletion; (6) Photochemical smog; (7) Acidification; (8) Air particulates; (9) Water quality; (10) Public human health-non-cancer.
- ✰ *SnPb* has lowest impact score in: (1) Non-renewable resource use; (2) Photochemical smog; (3) Acidification; (4) Water quality.
- ✰ *BSA* has lowest impact score in: (1) Renewable resource use; (2) Energy use; (3) Global warming; (4) Occupational health-non-cancer; (5) Occupational health-cancer; (6) Public human health-non-cancer; (7) Public human health-cancer.

**Among the Lead-Free Solders**

- ✰ *BSA* has lowest impact scores in all categories except non-renewable resource use.
- ✰ *SAC* has highest impact score in all categories except aquatic ecotoxocity and occupational cancer.
- ✰ *SABC* has highest impact score in occupational cancer and aquatic ecotoxicity and the lowest impact score in non-renewable resource use.

## What Can Manufacturers Do to Reduce Impact on Environment and Human Health

Upstream processes result in large impact on environmental and human health factors, particularly for lead-free solders containing silver. In solder manufacturing, virgin material is always mixed with recycled material. Manufacturers can take advantage of this fact. Reclaimed metal derived from either post-industrial or post-consumer recycling produces less environmental impacts. For example, implementing a 25 per cent reduction in virgin material content in SAC solder paste results in corresponding decrease of overall impact on air particulates, landfill space used, photochemical smog and acidification to the extent of 22 per cent.

The generation of energy required for application of solder paste results in the largest contribution to the overall impact score for many of the impact categories, regardless of the solder type. The power consumption can be reduced significantly by replacing older less efficient reflow assembly equipment or by optimizing current equipment to perform at the elevated temperatures required for lead-free solders.

## Impact on Industry and Techno-Commercial Aspects

Switching over to lead-free solder affects many aspects of manufacturing and assembly process. Identifying and procuring lead-free components can be time consuming and expensive. Old components already lying with the manufacturer cannot be used with lead-free solders due to elevated temperatures required (about 260°C, which is higher by 30+C compared to SnPb solders). Particularly susceptible components are connectors, LEDs and capacitors that contain plastic materials.

High soldering temperatures also affect copper clad laminates of PCB. Some commonly used laminates like FR-4 may not be able to withstand 260°C peak temperature required in lead-free soldering operation. New high temperature laminates are being developed for lead-free solders. However, they are expensive compared to FR-4 laminates. High temperatures involved in lead-free soldering will induce additional thermal stress on laminates and may cause unacceptable levels of warping.

Cleaning lead-free flux residues is more difficult due to flux-induced metallurgical reactions and more insoluble tin-salt formation. Cleaning assumes more importance in situation where pure tin components or tin surface finishes are used. These lead to tin whisker formation, which may cause a short circuit if tin whiskers do not get removed during cleaning operation.

The higher temperatures in lead-free soldering lead to shortened bath life for soldering machine owing to increased corrosivity of the high Sn-content solders. Increased erosion of copper from PWB circuitry has been observed during wave soldering, leading to faster copper build-up in SAC and SnCu solders.

Reflow profile defines the time-temperature relationship to promote the formation of reliable solder joints in PWB assembly. Lead-free solder profiles are typically characterized by longer soak times during pre-heat and by higher peak temperatures compared to conventional SnPb solders. A longer soak time is necessary to bring both the board and the components to equilibrium prior to reaching the reflow temperature to limit thermal shock and improper joint formation.

The impact of introduction of RoHS directive on some small-scale manufacturers is so large that some establishments have closed down the operations due to economic viability in meeting RoHS compliance requirements. Large scale manufacturers; on the other hand have adopted the directive. Most large scale MNCs has taken a positive step towards implementing the recommendations of the directive. For example, Advanced Micro Devices (AMD) website states that-

"Although lead containing solder cannot be completely eliminated from all applications today, AMD engineers have developed effective technical solutions to reduce lead content in microprocessors and chipsets to ensure RoHS compliance while minimizing costs and maintaining the product features. There is no change to fit, functional, electrical or performance specifications. Quality and reliability standards for RoHS compliant products are expected to be identical compared to current packages."

## Conclusion(s)

While there is no doubt about need for a legislation or regulatory directive like RoHS in the interest of protecting the environment, as is the case often, there are other equally important techno-commercial issues involved in implementing the guidelines of the directive. An attempt to meet RoHS directive affects product reliability in many instances. This is really a matter of concern. The cost implications are sever in some cases. Adhering to this directive implies an upgrade in manufacturing process and substitution of alternative materials, all of which adds to cost. Hundreds and thousands of products already manufactured may have to be discarded suddenly. Manufacturers of electronic goods are not yet fully aware of all the recommendations. Manufacturers in India are still at awareness stage. Locating and procuring raw materials that are RoHS compliant is a problem, both for dealers/ distributors as well as manufacturers.

On the positive side, many large-scale manufacturers like Wipro and HCL have announced a zero increase in the price of their goods.

In the larger interest of protecting our own environment and ultimately for the well being of human race, however, the producers of electrical and electronic goods must adopt the recommendations of RoHS, irrespective of its techno-commercial implications.

## Acknowledgements

The author acknowledges the inspiration provided by Principal, Dr. S. N. Kaul and support provided by Executive Director, Prof. Mangesh Karad.

## References

1. Ashmore C., "Will Your Printing Process be affected by Pb-Free?" Circuits Assembly, p. 22. March 2005 <http://circuitsassembly.com/cms/images/stories/pdf/0503/0503screenprinting.pdf
2. Analog Device <www.analog.com>
3. Bath, J., C. Handwerker, E. Bradley, "Research Update: Lead-free solder alternatives", Circuit Assembly, May 2000, p. 31-40. <www.circuitassembly.com>
4. Hwang, J., K. Chew, V. Kho, "Lead-free Implementation: Drop-in Manufacturing," A[pex 2004, Technical Conference Proceedings, Anaheim, CA, 2004.
5. Grusd A, C Jorgenson, "Lead-free Alloys: Fitting the square peg in square hole", Circuitree, September 1999, p. 98-102
6. Lang B., J Huckabee, D Romm, "Component Supplier's Qualifications of Lead-free Green Package Solutions." Electronics Goes Green 2004. Berlin, Germany.
7. Morris J., M. O'Keef. "Equipment Impacts of Lead-free Soldering," Apex 2003, Technical Conference Proceedings, Anheim, CA, 2003.
8. McGrath B. "The Effects of Lead-free on PCB Fabrication," Printed Circuit Design and fabrication, p. 44-47, February 2005, http://pcdandm.com/pcdmag/mag/0502/0502mcgrath.pdf
9. Sweatman K, S. Suenaga, M. Yoshimura, T. Nishimura, " Erosion of Copper and Stainless Steel by Lead-free Wave Solder." Proceedings of IPC Printed Circuits Expo, 2004.

# Chapter 39

# Carbon Credits/ Clean Development Mechanism and its Potential in Sugar and Distillery Industries

**Harshvardhan M. Modak* and Nitin Deshpande**

**Trio-Chem Techno-Legal Services P. Ltd.,**
**17/A, "Ishwari" Gokul Society, Pune – 411 030, M.S.**

## ABSTRACT

Global warming experienced over a period of time is due to emissions of various gases (called Green House Gases [GHG]) including carbon dioxide and methane. It was recognized world over that steady warming of the climate in future is likely to adversely affect life on the planet earth. The efforts of United Nations (UN) to reduce global warming and save earth from its adverse impact evolved a protocol called Kyoto protocol. Under its auspices, Clean Development Mechanism (CDM) promotes the investment in pollution control to generate (CERs or Carbon Credits). How these Carbon Credits can be traded for conversion into money is explained. The potential in Sugar and distillery sector in terms of co-generation projects; spent wash disposal etc. for generation of Carbon Credits is illustrated.

Opportunities for Carbon Credits generation in sugar and distillery sector are discussed. The various steps required for the same are explained. The precautions to be exercised for the same are also identified. Overall it's advocated that Carbon Credits constitute an additional avenue for financing future investments in sugar for energy and distillery for pollution control.

---

* Corresponding Author: E-mail: drmodak@triochem-tls.com

## Introduction

The energy incident on earth from the Sun is partially radiated back into cosmos. However, certain gases generated and released into the earth's atmosphere due to human activity have tendency to reflect back the infra red part of it. This infra red part of spectrum increases the atmospheric temperature, called global Warming (Figure 39.1). This is called Green House Effect. The gases, which bring about this effect, are called Green House Gases (GHG). Amongst others they mainly are Carbon Dioxide ($CO_2$) and Methane ($CH_4$). Such gases are said to have Global Warming Potential (GWP).

It is scientifically established that the climactic changes like sudden increase in rain fall, floods or droughts, melting of polar ice crust, changes in hot or cold weather; cyclones etc. are related to the Global Warming. It was a matter of concern for the climatologists in Intergovernmental Panel on Climate Change (IPCC) and it was predicted that uncontrolled Global Warming may severely affect

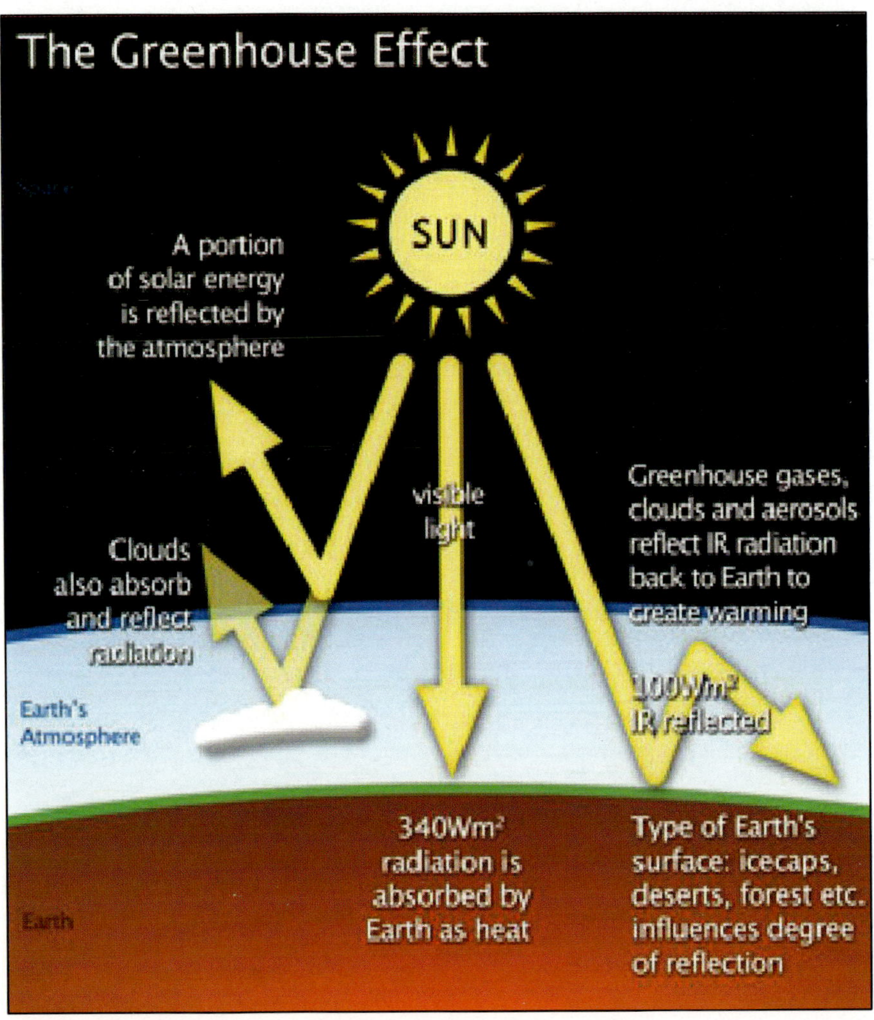


**Figure 39.1**

the human, animal and vegetation well-being and may also threaten the entire life system on the face of the earth.

The United Nations Organization (UNO) formed United Nations Framework Convention for Climate Change (UNFCCC) to deliberate on this issue and evolve measures to protect present and future generations of humankind on the basis of equity and in accordance with their common but differentiated responsibilities and respective capacities. Several rounds of talks later, at a convention in Kyoto, Japan a protocol was adopted, called Kyoto Protocol. It classified the countries in the world into two categories *viz.* Annex I and Non-Annex I, on the basis of the individual country's global warming potential and whether it is a developed or under-developed economy. The Annex I countries were asked to meet specified time bound targets for reduction in emissions of their GHG, failing which they would be slapped with heavy penalties. The Non Annex I countries do not have any such targets. India is a non Annex I country. The emissions reductions are to be brought about under three specific flexible mechanisms *viz.* (a) Joint Implementation (JI) among developed countries. (b) Emissions Trading (ET) among developed countries and (c) Clean Development Mechanism (CDM) between a developed and developing countries.

Since only CDM involves developing country like India, it is of the main interest of us. CDM enables developed country (like those in Europe, Japan, Canada etc.) to finance emissions reduction projects in developing country (like India). Successful emissions reduction from such projects as duly verified in prescribed manner, result into Certified Emissions Reductions (CERs) denominated into so many MT of $CO_2$ equivalent gases ($CO_2$e). These CERs can be sold at carbon trading exchanges (Figure 39.2) like shares to convert them into money at prevalent prices. Alternatively they can be assigned in advance in anticipation of their actual generation at appropriate time to various banks, financial institutions, trading companies, investing companies etc. In return these organizations can lend some part of the money into such emissions reduction projects. Thus CDM can emerge as a potential tool to bring in requisite finance for the particular project, which can bring about emission reduction. Worldwide billions of dollars worth carbon credit markets have emerged and are increasing day by day.

The reduction in use of coal, oil or any fossil fuels ( and $CO_2$ thereby) through use of renewable fuels like bagasse, rice husk is a popular example now known to many. We shall look into other such possible project ideas in sugar and allied products sector and analyze them to identify if they can be classified as potential candidates under CDM or Carbon Credits to bring in investment for their execution.

## Basic Eligibility of Projects for Carbon Credits

In order for a project to qualify under CDM to generate Carbon Credits, it must meet two basic criteria, *viz.* Additionality and Sustainability in Development.

### Additionality

The emission reduction should happen from an additional activity. One should not consider the reduction from routine and normal practices of the business. In other words it should not be from, "business as usual". *E.g.* If prior to consideration for CDM project, there is a present normal practice of electricity being routinely generated from back pressure turbines and using the balance low pressure steam for other process; then it is, "business as usual". Such electricity generation, which is already happening and being practiced routinely, will not qualify for CDM benefits. It is a part and parcel of the electricity baseline in that factory. *It becomes the base line beyond which only extra generation of electricity can be considered for CDM.*

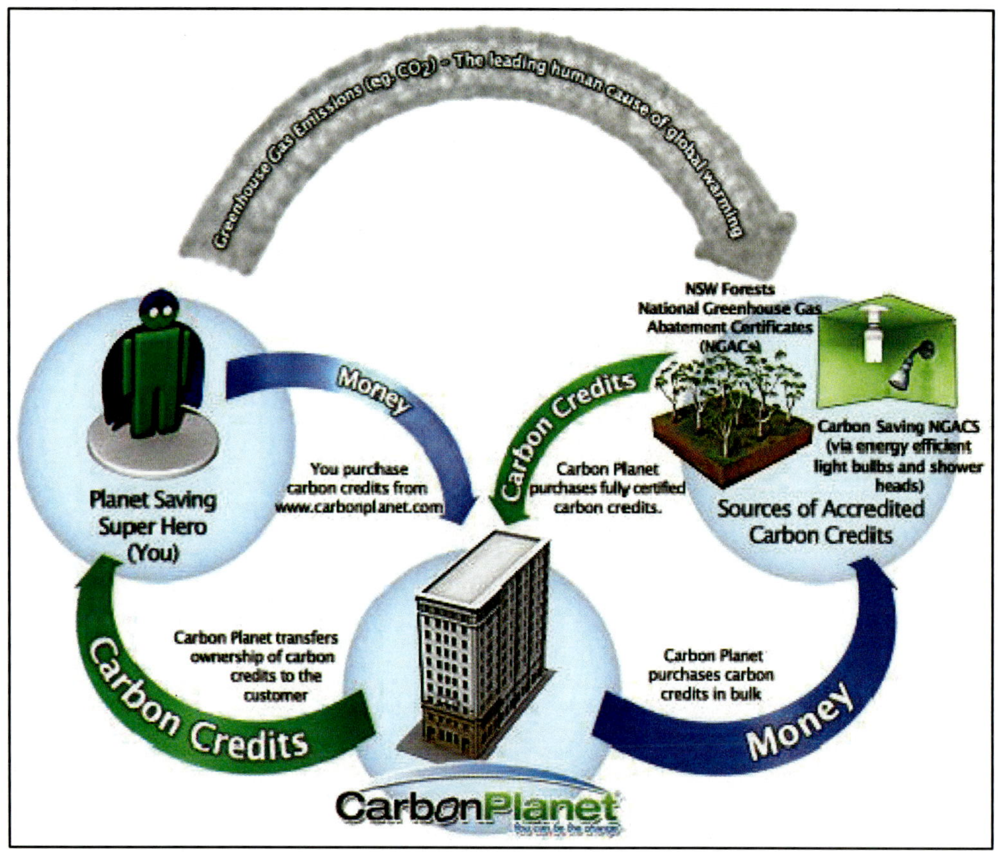


**Figure 39.2**

On the other hand same present boiler is burning coal and if fuel is changed to bagasse, then emission reduction equivalent to coal saved will generate Carbon Credits. Alternatively if the present boiler is already using bagasse and it is modified for co-generation of additional electricity to be sold to electricity board, then emission reduction equivalent to extra electricity will qualify for CDM.

### Sustainability

For every CDM project host country approval (*i.e.* approval from country where project takes place) is essential. This approval is given by judging a project if it meets sustainability criteria. Sustainability consists of economic, social, environmental and technical well- being. In other words, CDM project after implementation should bring additional investment, create employment or largely help community, indeed lead to environmental protection by reducing pollution and should involve technology transfer and it should be technologically replicable. Apart from the above basic criteria, there are number of other pre-conditions, one has to assess the project in question for, to verify if the same can qualify for the CDM benefits.

### Registration of the Projects under CDM

Considerable documentation has to be prepared in prescribed formats before the submission of the project for registration. The diagram (Figure 39.3) will illustrate the requisite steps involved in getting project registration and issuance of Carbon Credits/Certified Emission Reductions (CERs). Figure 39.3.

The Project Identification Note (PIN) and Project Design Document (PDD) are two basic documents, which one has to prepare in the prescribed format before submission for registration. They are scrutinized by Designated National Authority for the sake of Host country Approval. Considerable time is consumed at the each level of CDM activity to ultimately get the project into CERs.

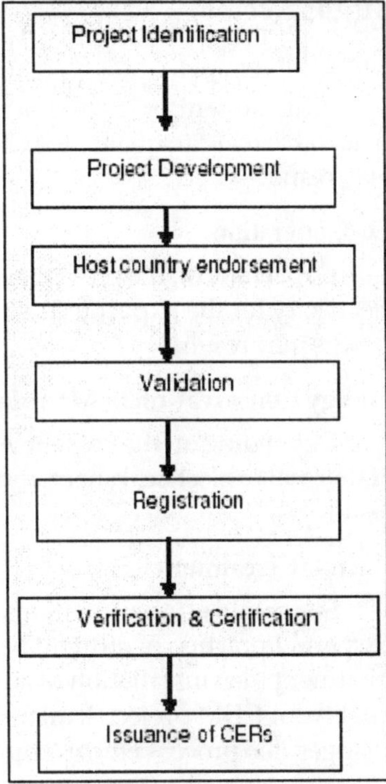


**Figure 39.3**

**Verification and Validation**

It must be emphasized here that the calculations in terms of the emissions reductions are subject to strict verification by Designated Operational Entity (DOE). They are independent organizations well versed in the tricks of the trade and free from any sort of influence. The points of measurement are to be equipped with suitable calibrated tools of measurement. The record keeping must be impeccable, such that it facilitates the verification. The project friendly and professional culture of the management should be apparent to these authorities. Any attempts to falsify the records shall culminate into cancellation of the whole registration. It shall result into permanent loss of considerable efforts, amounts paid for fees etc.

**Expenditure**

Considerable amounts have to be paid for consultancy, project validation, verifications, registration fees etc. before any CERs materialize. Secondly the project cycle is for maximum of ten years and the carbon credits materialize for the emissions reductions for a particular year. About six months may require for the emission data for the year to get validated and getting converted into valid Carbon Credits.

**No Success Scenario**

Depending on various factors, the project may fail to get passed for the CDM. It may happen at Host Country Approval and at CDM Executive Board. Secondly in spite of the project getting registered, validation of resulting data also may fail, if proper precautions are not observed. Thirdly selling the Carbon Credits is also a skillful process, since it may have to be bundled into adequate numbers, attracting potential buyers. Actual sale at a desired price can also be dependent on circumstantial factors. Thus getting Carbon Credits and its conversion into cash is not a child's play.

## CDM Potentials in Sugar and Distillery Sector

It must be borne in mind that the India signed the Kyoto Protocol in 2002. Thus in order for the projects to qualify for any benefits, they must be after 2002. Secondly for CDM registration, the documentation for the same must be before its execution. Thus as of today, CDM projects must be conceived NOW and documentation for the same may be developed subsequently. The CDM projects must be developed under previously approved methodology. Approval of new methodology is a complex affair.

## Sugar Sector

### Fuel Switch

If the present practice is of running the boilers on coal or any fossil fuel, then switching over to renewable fuel can qualify for CDM project. Depending on extent of saving fossil fuel, carbon credits may result.

### Co-Generation

Depending on the present practice of electricity generation, if there is a scope to produce additional electricity for the export to the national/local grid, then to the extent of surplus electricity, carbon credits may result.

### Energy Conservation in Manufacturing Process

Depending on the present level of energy consumption for the production process, if any saving of thermal or electrical energy is brought about, the carbon credits to the extent of energy saving can result.

### Effluent Treatment

Sugar mill effluent contains COD which has tendency to get converted into methane. If there is perpetual practice of effluent being let off into adjoining fields for irrigation, without any aerobic treatment then installation of anaerobic treatment for methane collection and its use in boilers may qualify for CDM project. It all depends on the present practice at ETP, the disposal of effluent and the changes into process etc. resulting into stoppage of methane let off into atmosphere.

### Compost Usage in Agriculture Replacing Chemical Fertilizers

Chemical especially nitrogenous fertilizers are energy intensive in their production process. Saving of such fertilizers in sugar cane management by use of organic compost can ultimately save considerable energy and may qualify for CDM project. Nitrogenous fertilizers have tendency to release Nitrous Oxide, which is very potent GHG.

## Distillery Sector

Methane is very potent GHG. It is 21 times more potent than carbon dioxide. Spentwash is on the other hand very potent for methane generation. Ethanol to the extent of its mixing with petrol saves fossil fuel. Thus above information is important starting point for development of CDM projects.

### Biogas from Spent Wash

A biogas project capturing biogas followed by its usage in boilers or electricity generation can qualify for CDM project. The electricity generated may have additional CDM potential. It must be a new project or an expansion. Such activity must not have earlier been performed.

### Composting of Spent Wash

The methane generation capacity of spentwash is altogether destroyed by composting. Proper full proof composting activity is a potential candidate for CDM project. It must not have been practiced before its consideration for CDM project.

### Ethanol for Petrol Blending

To the extent of its mixing with petrol, ethanol saves the consumption of fossil fuel. Thus ethanol blending can qualify for CDM project. However it may better be at the hands of refineries, which actually carry out such blending. In that case, distilleries may negotiate their pricing with buyers of ethanol, considering carbon credits.

**Energy Saving**

Energy saving activities like replacing spent wash reboilers by RO may qualify for CDM projects. However RO must be properly run by keeping proper verifiable records.

**Preparation of Dry Ice and its Use in Refrigeration**

If the energy calculations and its study indicate any energy saving by use of dry ice in refrigeration plants, the same idea may be developed for a CDM projects.

## Conclusion

Overall it can be said that the sugar and distillery sectors hold large potentials for CDM or in other words Carbon Credit benefits.

1. It must be borne in mind that the projects should be new and processed for CDM methodology before its implementation.
2. Tendency for falsification of records and malpractices in project management must be avoided to prevent getting the same debarred.
3. Equipment installed for project must be properly utilized along with proper and well recorded maintenance, preferably preventive type.
4. Carbon credit projects require extensive and intensive documentation.
5. Carbon credits require considerable time for preparation and expenditure for registration. It has to undergo detail verification by third party auditors, who cannot be influenced.
6. Carbon credits require long time for realization into cash.
7. Carbon credits are billion dollars market and are an opportunity for serious players ready to undertake painstaking efforts to convert the same into great wealth.

## Acknowledgement

The authors thank their CDM collaborators M/s. ICF International, New Delhi for their guidance in this field. The thanks are also due to the STAI, New Delhi for providing opportunity to present this paper at this All India Seminar, Jaisalmer, Rajasthan, India.

# Chapter 40

# Ozone Layer Depletion Modelling

**Akanksha[1], S. Raghunandan Murthy[1] and V.K. Srivastava[2]**
**[1]Department of Chemical Engineering Indian Institute of Technology, Hauz Khas, New Delhi – 110 016**
**[2]ABES Institute of Technology, Ghaziabad**

Widespread use of chlorofluorocarbons has led to about 3 per cent ozone depletion by the end of the 80's. This fact has resulted in sharp decline in subsequent chloroflurocarbon use and manufacture in conformation to Montreal Protocol (1987). Turco and Whitten (1975) have reported dynamic solution to establish quantitative effect of chloroflorocarbons. A one dimensional time dependant model was used and results were arrived at considering the prevalent increase of chloroflorocarbon production and emmision at that time. Miller *et al.* (1978) described a far more detailed model and presented results for a wide variety of species with time and altitude. Taking cognizance of this fact, the ozone layer stability and the projected ultimate equilibrium condition has been studied in this chapter. The steady state solution of the system outlined the fact that the ozone layer can reach a steady state for any pollutant concentration values. The natural limit of the ozone concentration, as well as the upper limits for the pollutant concentration was calculated. The procedure suggested a way of assessing the effect of anthropogenetic activities on ozone layer. The transmission coefficients of the UV radiation were calculated establishing the lower limits of the "permitted" pollution. Ozone layer stability was studied in the framework of irreversible thermodynamics. The following 16 chemical reactions describing the ozone creation and destruction processes in the stratosphere were considered

$O_2 + hv \rightarrow 2O,$

$O + O_2 + M \rightarrow O_3 + M,$

$O_3 + hv \rightarrow O_2 + O^*,$

$O_3 + O^* \rightarrow 2O_2,$

$HO + O_2 \rightarrow HO_2 + O_2,$

$HO_2 + O \rightarrow OH + O_2,$

$NO_2 + hv \rightarrow NO + O,$

$NO + O_3 \rightarrow NO_2 + O_2,$

$NO_2 + O \rightarrow NO + O_2,$

$Cl + O_3 \rightarrow ClO + O_2,$

$ClO + O \rightarrow Cl + O_2,$

$ClO + NO \rightarrow NO_2 + Cl,$

$HO_2 + Cl \rightarrow HCl + O_2,$

$HCl + OH \rightarrow H_2O + Cl,$

$CF_2Cl_2 + hv \rightarrow CF_2Cl + Cl,$

$CFCl_3 + hv \rightarrow CFCl_2 + Cl$

In the above, M stands for a catalyst.

Species continuity equation was applied for 11 relevant species taking into account vertical motion through eddy diffusion term. This gave set of 11 partial and coupled differential equations dependent on time and altitude. For various species, the basic species continuity equation is given as

$$\frac{dX_i}{dt} = J_i C_i + \sum_{n=1}^{l} k_n \prod_{m} C_m^{v_m}$$

*where,*

n: Any single reaction

v: stoichiometric coefficient

m: Number of species in *n*th reaction

l: Number of reactions

The eleven species considered were O, $O_3$, HO, Cl, ClO, NO, $NO_2$, $HO_2$, F-11, F-12, HCl. The model of ozone layer lead to a system of non-linear differential equations. To solve these equations, numerical discretization method was employed using semi-implicit finite difference schemes. The simultaneous equations thus obtained were solved using matrix inversion technique for a tridiagonal matrix wherein the M×M matrix is transformed to M×3 matrix reducing the computer storage memory and time. Steady state solution of the system indicated that ozone layer stability is independent of pollutant concentration viz Freon-11 and Freon-12 and is instead strongly dependent on the relation between $HO_2$ and $NO_2$ concentration in the stratosphere. Hypothetical fluctuations appearing in the system were also studied to get a better measure of the stability of the system. Real and negative eigen values were obtained indicating good damping characteristics. Natural limits of pollutants have been calculated knowing ozone lifetime. Thereafter, putting a condition on change in UV transmission coefficient, which is consonant with the risk of skin cancer, the upper limits of pollutants were also obtained which indicated that current concentrations of Freons were well within the permitted range. Dynamic simulation has been carried out which gave predicted future concentrations of all species with altitude. Despite halt in production of CFC's, the concentration does not show much decline.

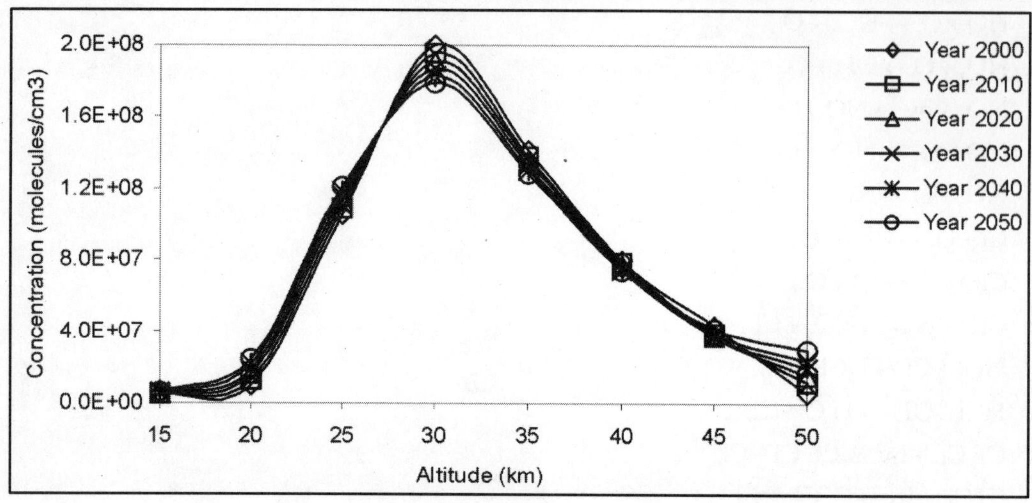


**Figure 40.1: Predicted Future ClO Concentration with Altitude**

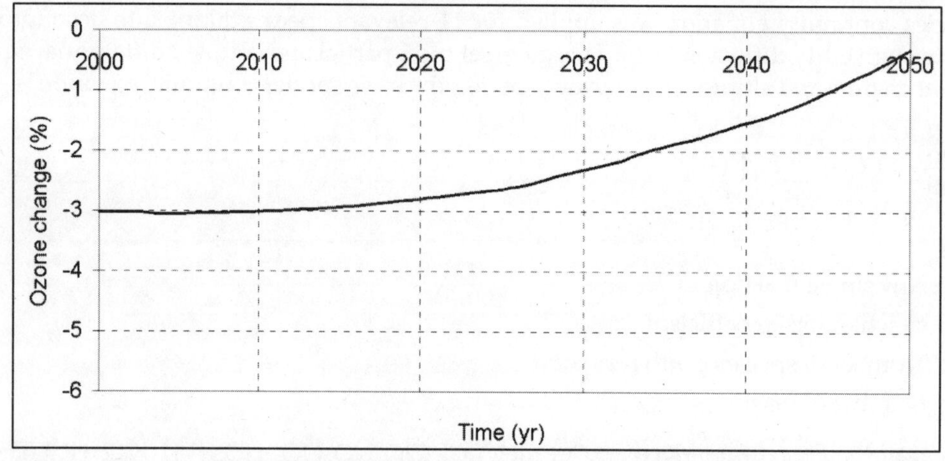


**Figure 40.2: Predicted Total Column Ozone Depletion**

Consequently, because of the photochemical dissociation of the molecules already in the stratosphere, free chlorine concentration keeps on increasing. However, the increasing free chlorine seems to take the HCl route rather than the depletion causing ClO route (Figure 40.1). Solution indicated that ozone will reach normal levels by 2050 (Figure 40.2). Freon concentrations owing to their large lifetimes may take much longer to reach accepatable levels.

## References


Miller C, Meakin P., Franks R.G.E., Jesson, J.P., The fluorocarbon Ozone Theory V One-dimensional modeling of the atmosphere: The base case, Atmospheric Environment (1978), 12, 2481-2500.

Turco R.P., Whiten R.C., Chlorofluoromethanes in the atmosphere and some possible consequences for ozone, Atmospheric Environment (1975),9, 1045-1061.

## Chapter 41

# Theoretical Investigation and Temperature Distribution in a Vortex Tube Using Natural Refrigerants–Air, Argon, Carbon Dioxide and Oxygen

**T.P. Ashok Babu[1] and V.J. Joy[2]**

**[1]Professor and Head of Department of Mechanical Engineering, National Institute of Technology Karnataka, Srinivasnagar Post, Surathkal, DK – 575 025, Karnataka**

**[2]Senior Grade Lecturer, Mechanical Engineering Department, Government Engineering College, Hassan – 573 201, Karnataka**

## ABSTRACT

Vortex tube is a simple device with no moving parts that is capable of separating a high-pressure flow into two lower pressure flows of different temperatures. The device consists of a simple circular tube, a vortex chamber, one or more tangential nozzles, a cold end orifice, and a hot end control valve. It is possible to use the zero ODP and low GWP refrigerants which are already exist in the nature.

In this work an attempt is made to use the environmental substance like air, Oxygen, Carbon di oxide and Argon as refrigerants to produce the cooling effect. This chapter describes the theoretical study of the temperature distribution phenomenon in a vortex tube and performance analysis for different inlet pressure. The COP of four natural refrigerants such as air, argon, carbon dioxide and oxygen are analyzed and performance is compared.

* Corresponding Author: E-mail: [1]tpashok@rediffmail.com; [2]joy.vj73@gmail.com

Applying I law and II law of Thermodynamics to vortex tube the temperature distribution equation has been derived. Program has been written to find the temperature distribution and COP of the system at different inlet pressures.

The COP versus cold mass fraction of vortex tube for different Non dimensional entropy factor $\phi_{ir}$ shows that the COP increases as cold mass fraction increases. The maximum COP is obtained between mass fractions 0.7 to 0.8. From the study it is concluded that when different working fluids are used, the argon gives least COP. Carbon dioxide as working fluid gives highest COP for all cold fractions. The oxygen and air as working fluid gives almost same COP at different pressure and cold fraction.

***Keywords:*** *Natural refrigerants, Vortex tube, Vortex chamber, Cold mass fraction, Nozzle, Orifice, Control valve.*

## Introduction

Vortex tube is a simple device with no moving parts that is capable of separating a high-pressure flow into two lower pressure flows of different temperatures. The device consists of a simple circular tube, a vortex chamber, one or more tangential nozzles, a cold end orifice, and a hot end control valve. The operation of a vortex tube can be best described by referring to Figure 41.1. High pressure gas enters the vortex tube through one or more nozzles. The gas expands through the nozzle and achieves a high angular velocity, causing a vortex-type flow in the vortex chamber. There are two exits to the vortex tube. From vortex chamber, part of the gas swirls to the hot end and remaining gas exits via the cold end. The hot exit is placed near the outer radius of the tube at the end away from the nozzle and the cold exit is placed at the center of the tube at the same end as the nozzle. By adjusting a control valve, it is possible to vary the fraction of the incoming flow that leaves through the cold exit, referred to as the cold fraction. The streams of gas leaving through the hot and cold ends of the tube are at higher and lower temperature, respectively, than the gas entering the nozzle. This effect is first discovered by George J. Ranque, who was granted a French patent for the device in 1932, and a United States patent in 1934. The initial reaction of the scientific and engineering communities to his invention was disbelief and apathy. Since the vortex tube was thermodynamically highly inefficient, it was abandoned for several years.

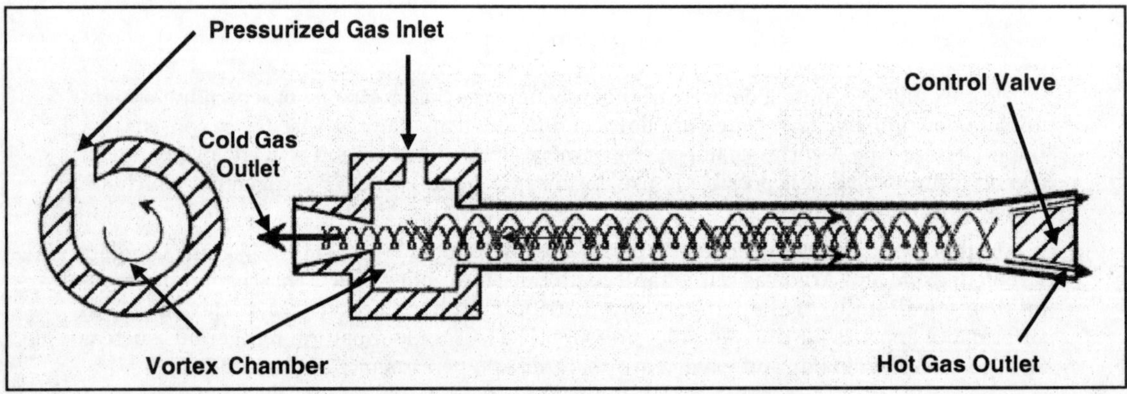


**Figure 41.1: Working Principle of Vortex Tube**

Interest in the device was revived by Hilsch in1947, a German engineer, who reported an account of his own comprehensive experimental and theoretical studies aimed at improving the efficiency of the vortex tube. In memory of their contribution, the vortex tubes are also called Ranque-Hilsch Vortex Tube (RHVT).

A RHVT has many advantages compared to the normal commercial refrigeration device. It is simple, no moving parts, no electricity or chemicals, small and lightweight, low cost, maintenance free, instant cold air, durable, adjustable temperature. But, its low thermal efficiency is a main limiting factor for its application. Also the noise and availability of compressed gas may limit its application. Therefore, when compactness, reliability and lower equipment cost are the main factors and the operating efficiency becomes less important, the RHVT becomes a nice device for heating gas, cooling gas, cleaning gas, drying gas, and separating gas mixtures, liquefying natural gas and other purposes. The underlying physical mechanism processes that determine the cooling of gas in this device have not been resolved completely in the past. The research on vortex tube involved the compressible fluid dynamics of turbulent and unsteady flow, thermodynamics and heat transfer. These aspects make the research complicated and challenging.

## Theoretical Analysis

Figure 41.2 shows the control volume considered in the thermodynamical analysis. For the analysis, only the gas properties at the walls, at the inlet and exhaust are of interest, *i.e.* for the analysis details of the internal process do not need to be considered.

In this system, there are three open boundaries namely $A_{in,}$ $A_h$ and $A_c$. In the steady flow, mass conservation between inlet flow ($m_{in}$), cold ($m_c$) and hot ($m_h$) outlet flow gives the relation as,

$$m_{in} = m_c + m_h \tag{1}$$

With $m_c$ the cold exhaust mass flow, expressed as $ym_{in}$, and $m_h$ the hot exhaust mass flow, expressed as $(1 - y)m_{in}$, then

$$m_{in} = ym_{in} + (1 - y)m_{in} \tag{2}$$

*where*, $y$ is the cold mass fraction and is defined as, $y = \dfrac{m_c}{m_{in}}$

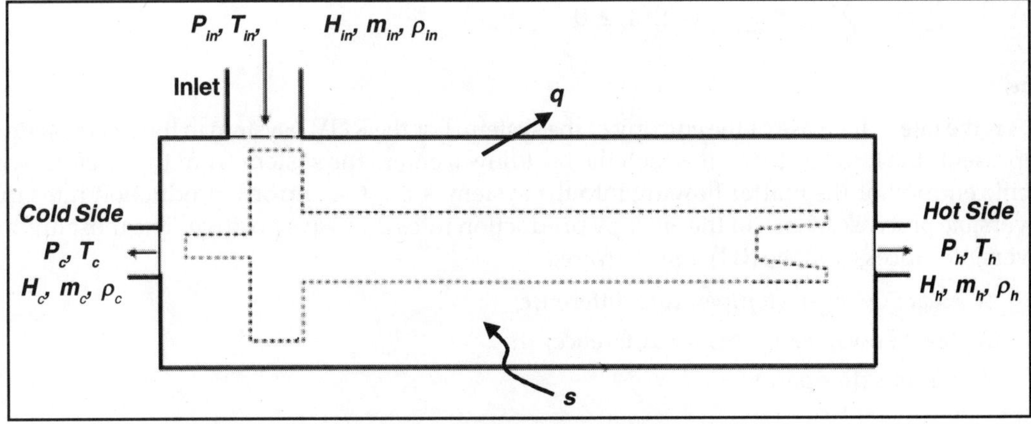


**Figure 41.2: Control Volume**

## The First Law of Thermodynamics Applied to Vortex Tube

The vortex tube satisfies the first law of thermodynamics. We can write,

$$u = \sum q + \sum H - \sum pv \tag{3}$$

*where,*

$u$ is the rate of change of the internal energy. The internal energy is only state dependent. In the analysis, the system is considered as steady, and $u = 0$. $q$ is heat flows at the various regions of the boundary. If heat flows from outside into the system, it is counted as positive. The RHVT system is assumed to be insulated, so $q = 0$. $H$ is the enthalpy flow into the system defined as $H = mh$ where, m is mass and h is specific enthalpy. Because the openings are large, the velocities at these openings are much lower than the sound speed, so the velocity term in these enthalpy flows can be neglected. In that case, for a perfect gas the specific enthalpy depends linearly on the temperature and is defined as $h = c_pT$, here, $c_p$ the specific heat at constant pressure is constant. In Equation 3 $v$ is the rate of change of the volume of the system at various moving boundaries. For the control volume, this term is zero.

Applying the first law on the system shown in Figure 41.2, it follows that

$$0 = m_{in}\,[c_pT_{in} - yc_pT_c - (1 - y)c_pT_h] \tag{4}$$

Taking, $\Delta T_h = T_h - T_{in}$, $\Delta T_c = T_c - T_{in}$ and $\Delta T = T_h - T_c$

The following relationships are derived.

$$\Delta T_h = y\Delta T$$

$$\Delta T_c = (y - 1)\Delta T$$

$$y\Delta T_c = (y - 1)\Delta T_h \tag{5}$$

These relationships express the inlet and exhaust temperatures as functions of the cold fraction $y$.

## The Second Law of Thermodynamics Applied to Vortex Tube

The vortex tube also satisfies the second law of thermodynamics. We can write,

$$s = \sum \frac{q}{T} + \sum s_f + \sum s_i \text{ With } s_i \geq 0 \tag{6}$$

*where,*

$s$ is the rate of increase of the entropy of the system. For the RHVT system in the steady state $s = 0$. $T$ represents the temperatures at which the heat flow $q$ enters the system from the outside. $s_f$ is the specific entropy of the matter flowing into the system. $s_i$ are the entropy production rates due to irreversible processes. Each of the entropy production rates is always positive. The most important irreversible processes in the RHVT system are:

- ✰ Heat flow over a temperature difference;
- ✰ Mass flow over a pressure difference;
- ✰ Viscous dissipation.

Applying the second law of thermodynamics to the RHVT with $q = 0$ and

Assuming $p_c = p_h = p_a$ and substituting $\gamma = \frac{c_p}{c_v}$, $\lambda = \frac{\gamma - 1}{\gamma} = \frac{R}{c_p}$ and

$$\frac{s_i}{m_{in} R} = \left( \frac{1}{\lambda} \ln \frac{T_h^{1-y} T_c^y}{T_{in}} + \ln \frac{p_{in}}{p_a} \right) = \phi_{ir} > 0$$

The factor $\phi_{ir}$ called as Non-dimensional entropy factor.

The second law combined with first law via the factor $\phi_{ir}$. We get

$$T_c^y (T_{in} - yT_c)^{1-y} = (1-y)^{(1-y)} e^{\lambda\phi_{ir}} T_{in} \left( \frac{p_a}{p_{in}} \right)^{\lambda} \quad (7)$$

In this way, $T_h$ and $T_c$ can be written as functions of the other parameters. For $\phi_{ir} = 0$, the process inside the RHVT system is reversible.

When the cold fraction is close to zero, in this extreme case, almost all the inlet gas exhausts from the hot side and the temperature at the cold side is absolute zero. The hot exhaust temperature is very close to the inlet temperature. Practically, it is not possible to generate this extremely low temperature.

When the cold fraction is close to one, almost all the gas will escape from the cold side. According to the adiabatic process assumed, the cold exhaust temperature will depend on the pressure drop over the RHVT system. Again because of the first law, the hot exhaust temperature becomes infinite. In practice, the hot exhaust temperature will never be infinite, since the process inside the RHVT system will never be reversible.

The equation 7 is solved using C program and temperature distribution is analyzed for different pressures and for the different mass fraction. The Table 41.1 shows the sample reading showing temperature distribution for air, argon, carbon dioxide and oxygen at $\phi_{ir}$ =0.25. The temperature distribution graph at 4 bar pressure is shown in Figure 41.3. For all cases with nonzero $\phi_{ir}$, as the cold fraction(y) increases the hot and cold exhaust temperatures increases. For the given cold fraction, if the $\phi_{ir}$ increases between zero to $\ln(p_{in}/p_a)$, the hot exhaust temperature decreases and the cold exhaust temperature increases.

## Coefficient of Performance (COP) of RHVT System

The coefficient of performance (COP) of a cooler is normally defined as the cooling power gained by the system divided by the work power input. So the COP of the cooler, denoted by $COP_c$ is expressed as

$$COP_c = \frac{y(T_{in} - T_c)}{\lambda T_{in} \ln\left( \frac{p_{in}}{p_a} \right)} \quad (8)$$

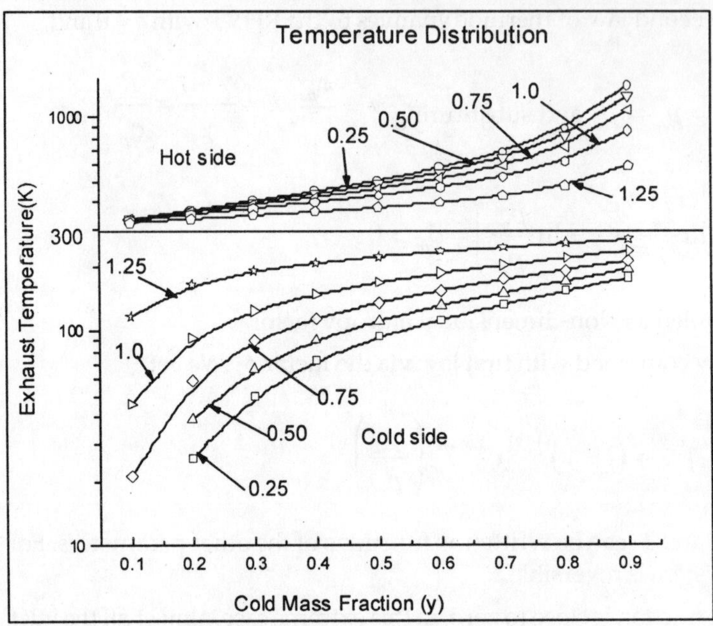


**Figure 41.3: Hot and Cold Side Temperatures versus Cold Fraction for Inlet Pressure 4 Bar Inlet Temperature 300K and $P_c = P_h = 1$ bar. The $\phi_{ir}$ value 0.25, 0.5, 0.75, 1.0, and 1.25 are shown in graph. The limiting value of $\phi_{ir}$ is 1.38. (working fluid air).**

**Table 41.1: Sample Reading Showing Temperature Distribution for Air, Argon, Carbon Dioxide and Oxygen at $\phi_{ir}$ = 0.25**

| *y* | *Air* | | *Argon* | | *Carbon Dioxide* | | *Oxygen* | |
|---|---|---|---|---|---|---|---|---|
| | $T_c$ | $T_h$ | $T_c$ | $T_h$ | $T_c$ | $T_h$ | $T_c$ | $T_h$ |
| | | | | **The Pressure 2.00 bar** | | | | |
| 0.2 | 84 | 354 | 61 | 360 | 100 | 350 | 84 | 354 |
| 0.4 | 140 | 407 | 116 | 423 | 156 | 396 | 141 | 406 |
| 0.6 | 178 | 483 | 157 | 515 | 192 | 462 | 179 | 482 |
| 0.8 | 212 | 652 | 194 | 724 | 223 | 608 | 213 | 648 |
| | | | | **The Pressure 4.00 bar** | | | | |
| 0.2 | 26 | 369 | 13 | 372 | 39 | 365 | 27 | 368 |
| 0.4 | 73 | 451 | 50 | 467 | 90 | 440 | 74 | 451 |
| 0.6 | 113 | 581 | 87 | 620 | 131 | 554 | 114 | 579 |
| 0.8 | 153 | 888 | 127 | 992 | 171 | 816 | 154 | 884 |
| | | | | **The Pressure 6.00 bar** | | | | |
| 0.2 | 14 | 372 | 7 | 375 | 23 | 369 | 15 | 375 |
| 0.4 | 52 | 465 | 32 | 479 | 68 | 455 | 53 | 465 |
| 0.6 | 90 | 615 | 64 | 654 | 109 | 587 | 91 | 614 |
| 0.8 | 129 | 984 | 101 | 1096 | 148 | 908 | 130 | 980 |

The Table 41.2 shows sample reading of variation of COP with varying pressure and cold mass fraction for air, argon, carbon dioxide and oxygen as working fluid. The Figure 41.4 to COP versus pressure for the above fluids. From the study it is concluded that Carbon dioxide as working fluid gives highest COP for all cold fractions. The argon gives least COP.

**Table 41.2: Sample Reading Showing Pressure versus COP for Different Pressure with Air, Argon, Carbon Dioxide and Oxygen ($\phi_{ir}$ = 0.25)**

| *Pressure in Bar* | *COP for Air* | | | | *COP for Argon* | | | |
|---|---|---|---|---|---|---|---|---|
| | *y = 0.2* | *y = 0.4* | *y = 0.6* | *y = 0.8* | *y = 0.2* | *y = 0.4* | *y = 0.6* | *y = 0.8* |
| 2 | 0.727 | 1.077 | 1.232 | 1.185 | 0.575 | 0.885 | 1.031 | 1.019 |
| 4 | 0.461 | 0.764 | 0.944 | 0.990 | 0.345 | 0.601 | 0.768 | 0.832 |
| 6 | 0.372 | 0.646 | 0.820 | 0.891 | 0.279 | 0.498 | 0.658 | 0.740 |
| 8 | 0.327 | 0.581 | 0.751 | 0.830 | 0.240 | 0.442 | 0.596 | 0.686 |
| 10 | 0.304 | 0.537 | 0.705 | 0.790 | 0.216 | 0.407 | 0.556 | 0.645 |
| 12 | 0.277 | 0.507 | 0.670 | 0.759 | 0.201 | 0.382 | 0.525 | 0.617 |
| | *COP for Carbon Dioxide* | | | | *COP for Oxygen* | | | |
| 2 | 0.857 | 1.234 | 1.389 | 1.320 | 0.736 | 1.084 | 1.238 | 1.186 |
| 4 | 0.559 | 0.900 | 1.087 | 1.106 | 0.465 | 0.770 | 0.951 | 0.995 |
| 6 | 0.459 | 0.769 | 0.950 | 1.008 | 0.396 | 0.652 | 0.827 | 0.897 |
| 8 | 0.404 | 0.694 | 0.879 | 0.943 | 0.341 | 0.586 | 0.757 | 0.836 |
| 10 | 0.370 | 0.648 | 0.828 | 0.898 | 0.301 | 0.542 | 0.711 | 0.796 |
| 12 | 0.347 | 0.612 | 0.789 | 0.866 | 0.285 | 0.512 | 0.676 | 0.765 |

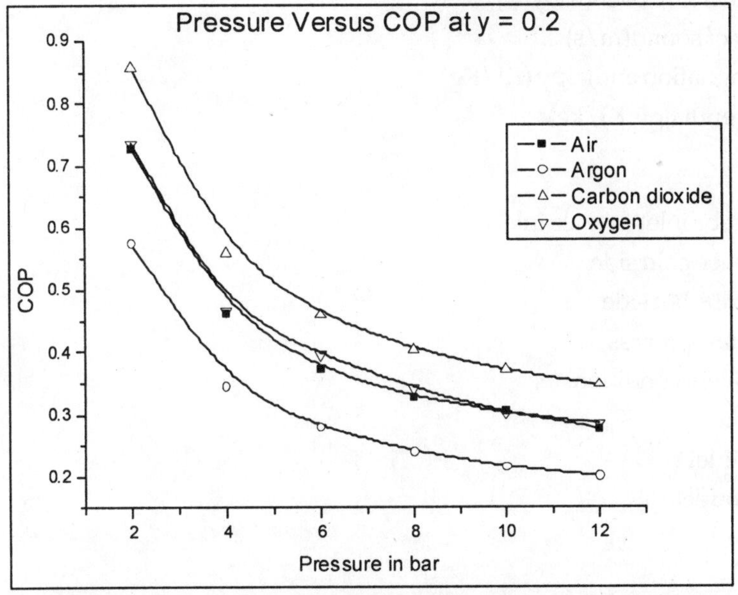


**Figure 41.4: COP versus Pressure for Different Working Fluid at Cold Fraction of 0.2**

## Conclusion(s)

The COP versus cold mass fraction of vortex tube for different Non-dimensional entropy factor $\phi_{ir}$ shows the COP increases as cold mass fraction increases. The maximum COP is obtained between mass fractions 0.7 to 0.8. When different working fluids are used, the argon gives least COP. Carbon dioxide as working fluid gives highest COP for all cold fractions. The oxygen and air as working fluid gives almost same COPs at different pressure and cold fraction.

## Nomenclature

p: Pressure (bar)
T: Temperature(K)
v: Specific volume ($m^3$/kg)
m: Mass flow rate(Kg/s)
y: Cold Mass fraction
M: Molecular weight
R: Gas constant (KJ/kg K)
$M_a$: Mach number
$C_d$: Orifice constant
$C_p$: Specific heat at constant Pressure(KJ/kg K)
$\phi_{ir}$: Non-dimensional entropy generation factor
q: Heat flow rate(KJ/s)
U: Internal energy(KJ/kg)
u: Rate of change of the internal energy(KJ/kg)
s: Rate of increase of entropy(KJ/K-s)
a: Velocity of sound(m/s)
H: Total stagnation enthalpy (KJ/Kg)
h: Specific enthalpy(KJ/kg)

## Subscripts

in: Vortex tube inlet (nozzle inlet)
c: Vortex tube cold side
h: Vortex tube hot side
i: Irreversible process
a: Atmosphere condition
is: Isentropic
n1: Nozzle inlet
n2: Nozzle outlet

## Greek Symbols

γ: Ratio of specific heats
η: Efficiency

$\rho$ Density(kg/m$^3$)

$\mu$: Dynamic viscosity(Ns/m$^2$)

$\nu$ : Kinematic viscosity(m$^2$/s)

$\zeta$: Gas Thermal Conductivity(KW/m-K)

## References

1. J J Van Deemter, "On the theory of the Ranque-Hilsch cooling effect", *Applied Scientific Research (SeriesA)*, Volume 3, Pages 174-196, 1952.
2. Merwin Sibulkin, "Unsteady, viscous, circular flow part 3. Application to the Ranque-Hilsch vortex tube", *Journal of Fluid Mechanics*, Volume 12, Pages 269–293, 1961.
3. J P Holaman and G D Moore, "An experimental study of vortex chamber flow", *Transactions of the ASME*, Pages 632-636, Dec 1961.
4. A J Reynolds, "A note on vortex-tube flow", *Journal of Fluid Mechanics*, Volume 14, Pages18–20, 1962.
5. J L Smith Jr., "An analysis of the vortex flow in the cyclone separator", *Transactions of the ASME, Journal of Basic Engineering*, Volume 84, Pages 609–618, Dec. 1962.
6. C U Linderstrom-Lang, "Gas separation in the Ranque-Hilsch Vortex tube", *International Journal of Heat Mass Transfer*, Volume 7, Pages 1195-1206, Pergamon Press, 1964.
7. H Takahama, Nobuo, "Studies on vortex tubes (Second report)", *Bulletin of the JSME*, Volume 9(33), Pages 121-130, 1966.
8. H H Bruun, "Experimental investigation of the energy separation in vortex tubes", *Journal of Mechanical Engineering Science*, Volume 11(6), Pages 567-582, 1969.
9. A Williams, "The cooling of Methane with vortex tubes", *Journal of Mechanical Engineering Science*, Volume 13(6), Pages 369-375, 1971.
10. Y Soni and W. J. Thomson, "Optimal Design of the Ranque-Hilsch Vortex Tube", *Transactions of the ASME*, Pages 316-317, May 1975.
11. M P Escudier, J Borntein and N Zehnder, "Observation and LDA measurement of confined turbulent vortex flow", *Journal of Fluid Mechanics*, Volume 98(1), Pages 49-63, 1980.
12. H Takahama and H Yokosawa, "Energy separation in vortex tubes with a divergent Chamber"' *Transactions of the ASME: Journal of Heat Transfer*, Volume 103, Pages196–203, May 1981.
13. M Kurosaka. "Acoustic streaming in swirling flow and the Ranque-Hilsch (vortex-tube) effect", *Journal of Fluid Mechanics*, Volume 124, Pages 139–172, 1982.
14. K Stephan, S Lin, M Durst, F Huang, and D Seher, "A Similarity relation for energy separation in a vortex tube", *International Journal of Heat and Mass Transfer*, Volume 27(6), Pages 911-920, 1984.
15. R T Balmer, "Pressure-driven Ranque-Hilsch temperature separation in liquids", *Transactions of the ASME: Journal of Fluids Engineering*, Volume 110, Pages 161–164, June 1988.
16. B Ahlborn, J U Keller, R Staudt, G Treitz and E Rebhan, "Limits of temperature separation in a vortex tube", *Journal of Physics D: Applied Physics*, Volume 27, Pages 480–448, 1994.

17. Rachel W Martin and Kurt W Zilm, "Variable temperature system using vortex tube cooling and fiber optic temperature measurement for low temperature magic angle spinning NMR" *Journal of magnetic Resonance*, Volume 168, Pages 202-209, 2004.

18. J J Derksen, "Simulations of confined turbulent vortex flow", *Computers and Fluids, Elsevier*, Volume 34, Pages 301-318, 2005.

19. Pongjet Promvonge and Smith Eiamsa-ard, "Investigation on the Vortex Thermal Separation in a Vortex Tube Refrigerator", *Science Asia*, Volume 31, Pages 215-223, 2005.

20. C M Gao, K J Bosschaart, J C H. Zeegers, A T A M de Waele, "Experimental study on a simple Ranque-Hilsch vortex tube", *Cryogenics*, Volume 45, Pages173-183, 2005.

21. Upendra Behera, P J Paul, S Kasthurirengan, R Karunanithi, S N Ram, K Dinesh and S Jacob, "CFD analysis and experimental investigations towards optimizing the parameters of Ranque–Hilsch vortex tube", *International Journal of Heat and Mass Transfer*, Volume 48, Pages 1961-1973, 2005.

22. N F Aljuwayhel, G.F. Nellis and S.A. Klein, "Parametric and internal study of the vortex tube using a CFD model", *International Journal of Refrigeration*, Volume 28, Pages 442-450, 2005.

23. Y Fukumoto and V L Okulov, "The Velocity field induced by a helical vortex tube", *Physics of Fluids* Volume 17, Pages 107101.1 -19, 2005.

24. Donald V Nicol and Mark J Lane, "Vortex rube system and Method for processing Natural Gas", US Patent No 6,932,858 B2, August 23, 2005.

25. Smith Eiamsa-ard and Pongjet Promvonge, "Numerical prediction of vortex flow and thermal separation in a subsonic vortex tube", *Journal of Zhejiang University Science A*, Volume 7(8) Pages 1406-1415, 2006.

26. H M Skye, G F Nellisn and S A Klein, "Comparison of CFD analysis to empirical data in a commercial vortex tube", *International Journal of Refrigeration*, Volume 29, Pages 71-80, 2006.

27. J P Holman, Heat Transfer, McGraw-Hill, SI Metric edition 1989.

28. J A Owczarek, "*Fundamentals of gas dynamics*", International Textbook Company, University of Wisconsin, 1964.

29. Exair.com. http://www.exair.com

30. Arizona Vortex Tube Manufacturing Company. http://www.arizonavortex.com

31. AirTX The Air Research Technology Company. http://www.airtxinternational.com

32. ITW Air management. http://www.itw-air.com

33. ITW Vortec. http://www.vortec-nl.com

34. Newman Tools Inc. http://www.newmantools.com

35. Vortexair.biz. http://www.vortexair.biz

# Chapter 42

# Heat Transfer Modeling of Cement Rotary Kiln: A Case Study

**S.R. Asthana[1]*, A.G. Gavit[2] and R.K. Patil[3]**
**[1]CEPD Division, National Chemical Laboratory, Pune, M.S.**
**[2]Government College of Engineering, Pune,**
**[3]Mechanical Engineering Department, M.I.T., Pune,**

## ABSTRACT

Cement industry is high energy intensive. Rotary kiln is the heart of the thermal energy system. Therefore attempt has been made to look into thermal energy optimization possibilities for J.K. Laxmi cement plant. Plant data is collected and collated. This is followed by modeling using simple heat transfer equations. Theoretical energy requirement is calculated based on literature and chemical equations. Actual heat transfer is then established based on plant data. Error analysis has been carried out to validate the model. Findings suggest that there is a scope of at least 12 per cent energy savings in the rotary kiln.

*Keywords: Cement industry, Rotary kiln, Energy, Optimization, Heat transfer modeling.*

## Introduction

Cement industry is considered as one of the large energy consuming industry [1]. It consumes about 4 GJ/t cement produced. Indian cement industry accounts for 10.3 per cent of total fuel consumption in the manufacturing sector [2]. The energy costs account for about 26 per cent of the

* Corresponding Author: E-mail: sr.asthana@ncl.res.in

manufacturing cost of cement. Where, on an average, 25 per cent electrical energy and 75 per cent thermal energy is used as an input energy. The best practice in India for specific energy consumption is reported as 3.06 GJ/t while in some other countries it is even lower than 2.95 GJ/t [2].

Petroleum coke blended with high ash coal (hereafter mentioned as coal) is mainly used in cement production in Indian industries. Out of total coal requirement averagely 60 per cent is used in pyroprocessing basically for combustion purpose in precalciner and 40 per cent is used in kiln. Modeling of precalciner and cooler has been already carried out at National Chemical Laboratory, Pune therefore for this study heat modeling of kiln is taken up.

## Literature Review

Recent work that has been carried out globally on energy conservation in cement plant is reviewed as follows:

Mujumdar *et al.* [3] have evaluated the possible ways of reducing energy consumption in rotary cement kilns. A comprehensive one-dimensional model was developed to simulate complex processes occurring in rotary cement kilns. The model assesses the influence of various design and operating parameters on kiln performance. Several ways of reducing energy consumption in kilns were then computationally investigated. The developed model and the presented results are found useful for enhancing the performance of rotary cement kilns.

A study by Engin and Ari [4] deals with energy audit analyses of dry type rotary kiln system working in cement plant in Turkey. It was found that about 40 per cent of total input energy was being lost through hot fluid gas (19.15 per cent), cooler stack (5.61 per cent) and kiln shell (15.11 per cent) convection and radiation). Some possible ways to recover the heat losses are also introduced and discussed in the chapter. Finding showed that approximately 15.6 per cent of the total input energy could be recovered.

Biswas *et al.* [5] estimated the scope of energy savings in some areas, in and around the rotary kiln in noncoking coal based sponge iron plant. This is done by estimating: the optimum fuel/reductant, combustion air rate for the process requirement in rotary kiln; the size of induced draft fan for exhaust gas treatment; and air requirement for the coal injection and combustion for utilization of unburnt fuel at the post combustion chamber.

Liu *et al.* [6] have studied the cement industry in China to determine the prospects for renovation and for building new facilities during 1990, and, in particular, the prospects for improved energy efficiency. Their findings suggests that state-of-the-art precalciner kilns offer small energy-efficiency advantages, but important environmental and product quality advantages over improved vertical kilns.

Carvalho M. De O. Madivate [7] presents a study that has been carried out with two kiln feed samples used for the production of Portland cement. A method for the direct measurement of the Theoretical Energy Requirement (TER) on a high-temperature calorimeter HT 1500 from the company SETARAM (Lyon, France) is presented. Based on the results obtained, the influence of the composition of the kiln feeds on the TER is discussed.

A study by Li *et al.* [8] has investigated the influence of Final Pyrolysis Temperature (FPT) on the yield and characteristics of pyrolysis products in order to provide base data for engineering purposes.

Rasul *et al.* [9] have presented a simple model to assess the thermal performance of cement industry with integrated view to improve the productivity of the plant. The model developed was on the basis of mass, energy and exergy balance and is applied to an existing Portland cement industry in Indonesia.

Integrated reaction engineering based mathematical model has been presented by Mujumdar *et al.* [10]. The paper focuses on clinker formation in cement industry. Separate models for pre-heater, calciner, rotary kiln and cooler were developed and coupled together to build an integrated simulator. Appropriate models for simulating gas-solid contact and heat transfer in pre-heater were developed. Calciner has been modeled by considering simultaneous combustion of coal particles and calcinations of raw meal. The individual models were coupled with each other via mass and energy communication through common boundaries. The coupled model equations were solved iteratively. The model predictions agree well with the observations and experience from cement industry. The model was used to gain better understanding of influence of operating conditions on energy consumption in cement plant.

## Plant Data

Communication has been taken up with various Indian cement plants whose daily thermal requirement is more than 3 GJ/t of clinker. J.K.Laxmi plant at Sirohi, Rajasthan is selected for further study as they have also attempted blending of coal with waste derived fuel in the past.

As mentioned earlier, main components during pyroprocessing process are cyclone, precalciner, rotary kiln and cooler. Keeping kiln as focus of study we collected information related to kiln *i.e.* its temperature, mass flow rate, dimension, material etc.

The plant uses a dry process with a series of cyclone type pre-heaters and an incline-kiln. The kiln is 4.2 m in diameter and 61.17 m long. The average daily production capacity is 4200 TPD, and the specific energy consumption has been estimated to be 3.2 GJ/t of clinker. The raw material and clinker compositions are presented in Table 42.1.

Figure 42.1 presents simple schematics of the pyroprocessing displaying flow of solid, air and fuel.

**Table 42.1: Raw Material and Clinker Composition [13]**

| *Component* | *Raw Material* | *Clinker* |
|---|---|---|
| $SiO_2$ | 13.21 | 24.29 |
| $Al_2O_3$ | 3.7 | 5.73 |
| $Fe_2O_3$ | 2.33 | 3.61 |
| CaO | 42.8 | 63.30 |
| MgO | 1.47 | 2.28 |
| $SO_3$ | 0.46 | 0.71 |
| $K_2O$ | 0.46 | 0.71 |
| $Na_2O$ | 0.19 | 0.30 |
| $H_2O$ | 0.5 | - |
| Organics | - | - |
| Ignition Loss | 35.5 | - |
| Alkali | - | 1.85 |
| Unaccounted | - | - |

## Mass and Heat Balance

Mass and heat balance for J.K. Laxmi plant is carried out and presented below:

### Mass Balance

The average compositions for dried coal have been presented in Table 42.2. Based on the coal composition, the net heat value has been calculated as 30160 kJ/kg-coal.

The mass balance of the kiln system is summarized in Table 42.3. All gas streams are assumed to be ideal gases at the given temperatures.

**Table 42.2: Percentage Analysis of Coal and Peat Coke [13]**

| *Component* | *Peat Coke* | *Coal* |
|---|---|---|
| Carbon | 87 | 66 |
| Hydrogen | 3.5 | 4.5 |
| Oxygen | 0.6 | 11.5 |
| Sulphur | 6.6 | 0.9 |
| Nitrogen | 0.8 | 0.5 |

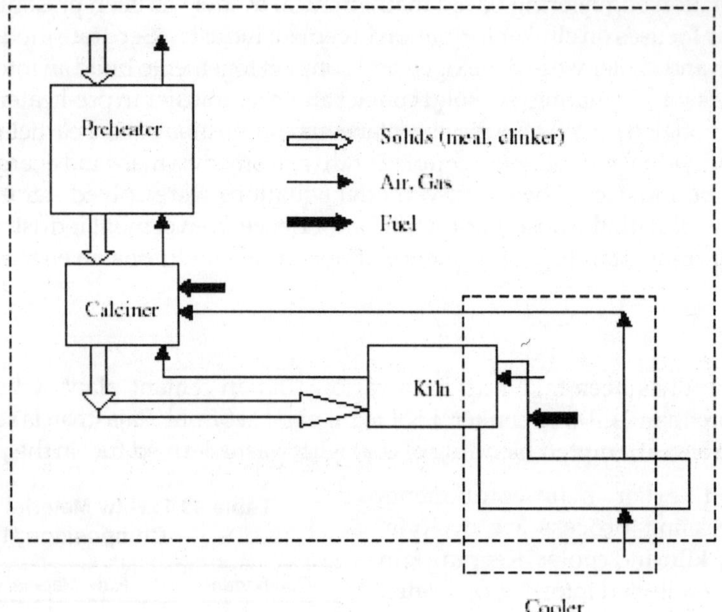


**Figure 42.1: Schematics Diagram of Pyroprocessing Displaying Flow of Solid, Air and Fuel**

**Table 42.3: Mass Balance**

| | *Input (kg/kg clinker)* | *Output (kg/kg clinker)* |
|---|---|---|
| Raw material | 01.81 | - |
| Cooling air | 01.68 | - |
| Coal | 00.108 | - |
| Preheater exhaust | - | 01.8 |
| Clinker | - | 01.00 |
| Cooling dust | - | 00.00683 |
| Cooler–hot air | - | 01.10 |
| Dust from preheater | - | 00.0479 |

## Heat Balance

In order to analyze the kiln system thermodynamically, following assumptions are made:

1. Steady state working conditions.
2. The change in the ambient temperature is neglected.
3. Cold air leakage into the system is negligible.
4. Raw material and coal compositions do not change.
5. Averaged kiln surface temperatures do not change.
6. Feed rate of raw material and coal are considered as constant.

The complete energy balance for the system is calculated and presented in Table 42.4. Since most of the heat loss sources have been considered, there is only a 397 kJ/kg-clinker of energy difference from the input heat. This difference is nearly 12 per cent of the total input energy and can be attributed to the assumptions and nature of data. The distribution of heat losses to the individual components exhibits reasonably good agreement with some other key plants reported in the literature [2,4,9].

**Table 42.4: Heat/Energy Balance**

| *Stream* | *Flow Rate kg/kg of Clinker* | *Specific Heat kJ/kg °C* | *Temperature °C* | *Enthalpy kJ/kg of Clinker* | *Percentage Per cent* |
|---|---|---|---|---|---|
| **Entering the system** | | | | | |
| Combustion of coal | 0.108 | CV=30160*1 | 3016*1 | 93.95 | |
| Raw feed | 1.81 | 0.9 | 80 | 130.32 | 4.05 |
| Ambient air | 1.68 | 1.0 | 35 | 58.8 | 1.83 |
| Coal | 0.108 | 0.9 | 50 | 4.86 | 0.15 |
| Total | | | | 3210 | 100 |
| **Leaving the system** | | | | | |
| Reaction energy | - | - | - | 1657.11*2 | 51.62 |
| Clinker | 1.00 | 0.8 | 220 | 176 | 5.48 |
| Preheater exhaust | 1.80 | 1.0 | 325 | 585 | 18.22 |
| Hot air from cooler | 1.10 | 1.0 | 340 | 374 | 11.65 |
| Heat loss by dust | 0.05473 | 1.09 | 340 | 20.28 | 0.63 |
| Losses | - | - | - | 397.59 | 12.39 |
| Total | | | | 3210 | 100 |

*1: Calorific Value (CV) = 80.8 × C + [ (287 × $H_2$–$O_2$/8) ] + 22.45 × S–6 × $H_2O$ kcal/kg.

*2: Reaction energy = 17.196 × $Al_2O_3$ + 27.112 × MgO + 32 × CaO–21.405 × $SiO_2$ – 2.468 × $Fe_2O_3$ kJ/kg clinker.[14]

## Theoretical Heat Required (TER) to Produce One kg Clinker [14]

$$TER = 4.11 \times H_{Al} + 6.48 \times H_{Mg} + 7.646 \times H_{Ca} - 5.116 \times H_{si} - 0.59\, H_{Fe}$$

$$= 4.11 \times 5.73 + 6.48 \times 2.28 + 7.646 \times 63.3 - 5.116 \times 21.29 - 0.59 \times 3.61$$

$$= 411.27 \text{ kcal/kg}$$

$$= 1719.09 \sim 1720 \text{ kJ/kg of clinker.}$$

## Heat Transfer Modeling of Rotary Cement Kiln

After estimating TER, modeling for actual thermal energy requirement for rotary kiln has been taken up. Heat transfer modeling of rotary kiln are analyzed and discussed here.

### Heat Generation Due to Combustion of Fuel within the Kiln

Heat generated is transferred to various components within the kiln and its transmission takes place to the material at relatively lower temperature. Mode of this heat transfer is very complex, as it not only involves the gas and solid but also the rotating kilns [5]. Further, complexity is caused due to the presence of dust, flames in the gas and cyclic temperature variation at the inside wall as it repeatedly gets covered and exposed by the bed of material. Moreover, heat utilization in the process depends on the heat generated due to combustion, heat gained by incoming flue gas, heat lost through the reactor wall and heat carried away by the outgoing flue gas. The coal and air consumption rate is decided by the constituent energy balance requirement of the process.

## Total Heat Requirement

Combustion of flue gas over bed at selected locations and fuel within the bed of material increases the temperature of flue gas and bed respectively. Table 42.5 indicates the temperature of the flue gas, solid and wall at seven equidistant locations in the kiln.

**Table 42.5: Measured Inside Temperature (°C) at Various Location of Kiln [13]**

| Kiln capacity, TPH, kiln size, D(m), L(m) | $t_g1$, $t_s1$, $t_w1$ | $t_g2$, $t_s2$, $t_w2$ | $t_g3$, $t_s3$, $t_w3$ | $t_g4$, $t_s4$, $t_w4$ | $t_g5$, $t_s5$, $t_w5$ | $t_g6$, $t_s6$, $t_w6$ | $t_g7$, $t_s7$, $t_w7$ |
|---|---|---|---|---|---|---|---|
| 175 | 1050 | 1140 | 1250 | 1380 | 1950 | 2080 | 1350 |
| 4.217 | 940 | 980 | 1030 | 1100 | 1260 | 1370 | 1300 |
| 61.14 | 820 | 828 | 836 | 850 | 863 | 900 | 871 |

## Estimation of Total Heat Requirement at any Stage

The estimation of total heat requirement consists of the following steps.

### Heat Gained by the Bed of Material [15]

$$Q_s = M_s C_s (t_{sm} - t_{s\infty}) \qquad (1)$$

*where,*

$Q_s$ is heat transfer rate kJ/h to solid,$M_s$ is solid material feed rate (TPH),$C_s$ is specific heat of the soild kJ/kg °C [13], $t_{sm}$ is the maximum temperature (°C) of solid inside the kiln and $t_{s\infty}$ is the ambient temperature (°C) of the solid.

### Heat Lost through the Kiln Wall and Fefractory [15]

$$Q_{lk} = (t_{ki} - t_{ko})/[(1/h_{ri}A_{ri}) + (1/2\pi k_r L).\ln(D_{si}/D_{ri}) + 1/(2\pi k_s L).\ln(D_{so}/D_{si}) + (1/h_\infty A_{so}) \qquad (2)$$

*where,*

$Q_{lk}$ is rate of heat loss (kJ/h) through kiln/reactor wall, $t_{ki}$ and $t_{ko}$ are inside and outside temperatures (°C) of the kiln shell respectively, $h_{ri}$ and $h_¥$ are heat transfer coefficients [11,12] (W/m$^{2}$°C) of rotary kiln at inside refractory and at the outer shell respectively. $A_{ri}$ and $A_\infty$ are areas (m$^2$) of rotary kiln at inside the refractory and at the outer shell respectively, L is length (m) of the rotary kiln, $K_r$ and $K_s$ are thermal conductivities (W/m°C) of refractory and kiln shell respectively, $D_{si}$ and $D_{ri}$ are inside diameter (m) of kiln shell and inside diameter (m) of the refractory respectively. $D_{so}$ is outside diameter (m) of the shell.

Actual heat lost through wall includes not only the heat lost through kiln shell but also through shells of inlet hood, outlet hood of rotary kiln, preheater surface and inlet area of cooler. Outer combined areas as mentioned is almost twice then that of the kiln and therefore $Q_{lt}$ is taken as two times the $Q_{lk}$.

**Heat Required for the Calcinations Process to Continue [3] is as Follows**

Major chemical reaction were considered in the present model, are:

$CaCO_3 = CaO + CO_2$ 43740 kcal/kmol (3)

$CaO + SiO_2 = C_2S$ -33240 kcal/kmol (4)

$Cao + C_2S = C_3S$ 470 kcal/kmol (5)

$3CaO + Al_2O_3 = C_3A$ -5800 kcal/kmol (6)

$4CaO + Al_2O_3 + Fe_2O_3 = C_4AF$ -12000 kcal/kmol (7)

Reaction (3) indicates that a great amount of heat is required for desired calcination/transformation to takes place. $Q_p$ is calculated in

kJ/h (8)

Total heat required $Q_t$ for the process at any stage:

$$Q_t = Q_s + Q_{lt} + Q_{p.} \ldots\ldots \text{ kJ/h} \quad (9)$$

The values of $Q_s$, $Q_{lt}$, $Q_p$ and $Q_t$ are estimated for the plant and presented in Table 42.6.

**Table 42.6: Estimated Heat Requirement Against Heating of Bed of Material, Heat Loss from Wall and Heat of Reaction etc.**

| *Capacity (TPH), D (m), L (m)* | *$Q_s$ kJ/h Eq.1 (I)* | *$Q_{lt}$ kJ/h Eq.2 (II)* | *$Q_p$ kJ/h Eq.8 (III)* | *$Q_t$ kJ/h (IV= I+II +III)* |
|---|---|---|---|---|
| 175<br>4.217<br>61.14 | $4399 \times 10^5$ | $7.4 \times 10^5$ | $5942 \times 10^5$ | $10348 \times 10^5$ |

## Mechanism of Heat Transfer

The mechanism of heat transfer is very complex because it involves not only the gas and solid, but also wall of rotating kiln. Apart from the conduction through kiln wall, entire heat transfer involves four following steps.

- ☆ Gas–to–solid by radiation and convection ($Q_{gs}$)
- ☆ Gas–to–wall by radiation and convection ($Q_{gw}$)
- ☆ Exposed wall–to–solid by radiation ($Q_{wes}$)
- ☆ Covered wall–to–solid by radiation and convection ($Q_{wcs}$)

Estimation of above mentioned heat transfer steps are carried out using optimistic operational data on temperature along the various locations of the kiln using following empirical relations. Figure 42.2 present heat flow path in a rotary kiln.

### Gas–to–Solid

The equation governing the heat transfer from gas to solid is

$$Q_{gs}/A_{se} = \{\sigma/[(1/\varepsilon_g + 1/e_s - 1)]\}(T_g^4 - T_s^4) + h_{gs}(T_g - T_s) \quad (10)$$

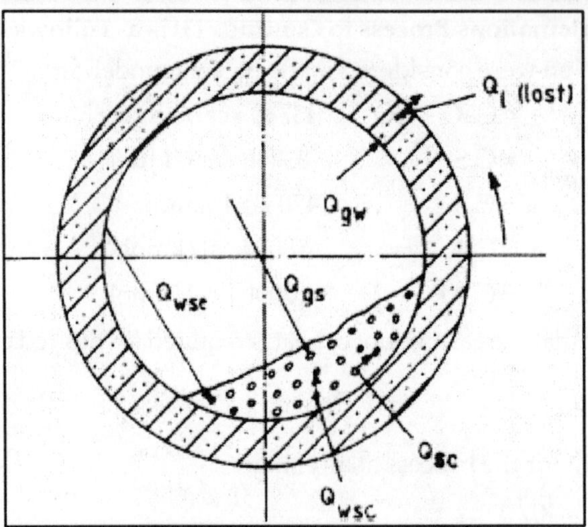


**Figure 42.2: Heat Flow Path in a Rotary Kiln**

*where,*

$A_{se}$ is the exposed solid surface, s is Stefan-Boltzman constant ($kW/m^2K^4$), $\varepsilon_g$ and $\varepsilon_s$ are emissivity of gas and solid respectively. Solid and gas emissivity are taken as 0.9 and 0.4 respectively [3]. $T_g$ and $T_s$ are the fourth power average temperature (K), $T_g$ and $T_s$ are the arithmetic average temperatures of gas and solid respectively, and subscript gs is used for gas to solid, se for solid exposed, g for gas and s for solid. $h_{gs}$ is convective heat transfer coefficient ($W/m^{2o}C$) from gas to solid.

Also, $T_g = \Sigma T_{gi} A_i / \Sigma A_i; T_g^4 = \Sigma T_{gi}^4 A_i / S A_i$ (11)

**Gas–to–Wall**

The equation governing the heat transfer from gas to wall is

$$Q_{gw}/A_{we} = \{\sigma/[1/\varepsilon_g + 1/\varepsilon_w - 1]\}\,(T_g^4 - T_w^4) + h_{gw}\,\{T_g - T_w\} \quad (12)$$

*where,*

$A_{we}$ is the exposed wall surface, subscript gw is gas to wall and we is wall exposed. The wall emissivity may be assumed as 0.9 [3].

**Exposed Wall–to–Solid**

The equation governing the heat transfer from exposed wall to solid is

$$Q_{wes}/A_{se} = \{\sigma/[(1/\varepsilon_w) + (1/\sigma_s) - 1]\}\,\{T_{wes}^4 - T_s^4\} \quad (13)$$

*where,*

$T_{wes}$ is temperature (K) of exposed surface of the kiln wall and subscript wes is exposed wall to solid, se is solid exposed, w is wall and s is solid.

**Covered Wall–to–Solid**

The equation governing the heat transfer from covered wall to solid is

$$Q_{wcs}/A_{sc} = \{\sigma/[1/\varepsilon_w + 1/\sigma_s - 1]\}\,\{T_{wcs}^4 - T_s^4\} + h_{ws}\,(T_{wcs} - T_s) \quad (14)$$

*where,*

wcs is covered wall to solid and $T_{wcs}$ is temperature (K) of covered surface of the kiln wall and $Q_{ws} = Q_{wcs} + Q_{wes}$, taking $T_w \sim T_{we}$, $A_{se}$ is unexposed/covered solid surface, $A_{sc}$ is covered solid surface and $A_s = A_{se} + A_{sc}$.

Plant wise estimation of above mentioned heat transfers is carried out and presented in Table 42.7.

**Table 42.7: Estimated Value of Various Heat Transfer Rate (Eqs. 10 to 14) Based on Temperature Data Presented in Table 42.5**

| *Capacity (TPH), and area, $A_{we}$, $A_{se}$, $A_{sc}$, $A_c$ ($m^2$)* | *$Q_{gw}$ kJ/h Eq.12 (I)* | *$Q_{gs}$ kJ/h Eq.10 (II)* | *$Q_{ws}$ kJ/h Eq.13 and 14 (III)* | *$Q_{cs}$ kJ/h (IV) = (I–II)* | *$Q_s$ kJ/h (V) = (II+III +II)* | *$Q_s$ kJ/h (VI) from Table 42.6* | *Error Per cent $\frac{V-VI}{VI} \times 100$* |
|---|---|---|---|---|---|---|---|
| 175<br>304,212<br>292,504 | $2454 \times 10^5$ | $1300 \times 10^5$ | $11195 \times 10^5$ | $-8741 \times 10^5$ | $3754 \times 10^5$ | $4399 \times 10^5$ | 15.58 |

## Data Analysis

Based on the data of temperature of gas, wall and solid at various locations along the bed of material, heat gained by material ($Q_s$) from Eq.1, heat loss through the reactor system wall ($Q_{lt}$) from Eq.2, heat required ($Q_p$) from Eq.8 and the total heat rate required for the process to continue are calculated and tabulated in Table 42.6. Estimated values of $Q_{gw}$, $Q_{gs}$, $Q_{ws}$ are mentioned in Table 42.7. Moreover $Q_s$, the estimated heat transfer to solid (Table 42.6), is found more than some heat rate estimated in Table 42.7. This difference however, is due to inadequate selection of various constants of the empirical relations. These are found to vary within 10-15 per cent and hence the mentioned procedure may be utilized for design and estimation of the size of the rotary reactor.

## Conclusion

Table 42.1–42.7 presents the various calculations related with kiln. Theoretical energy requirement of this plant is 1720 kJ/kg of clinker where as actual energy supplied is 2286 kJ/kg of clinker. The energy difference is 566 kJ/kg of clinker. Keeping allowance for losses and unaccounted figure still we can reduce actual supply energy up to 2000 kJ/kg of clinker. The study suggests that there is a scope of energy saving up to 12 per cent. This can be achieved through air optimization, coal optimization and reduction of losses through kiln wall. Further, this study can be applied to any rotary kiln based cement plant worldwide.

## Acknowledgements

The authors would like to acknowledge the support of J.K.Laxmi cement plant, and Mr.Rajpal process engineer, J.K. Laxmi cement plant. The information reported in this article is the work of experts of National Chemical Laboratory (NCL), Pune. Authors are also highly thankful to Prof. S.D. Mahajan College of Engg. Pune for his valuable suggestions.

## References


1. Schumacher, K., Sathaye, J.,: "Indian's Pulp and Paper Industry; Productivity and Energy Efficiency", Journal of Economic Research,4,pp.1-46,2001.

2. Banerjee, R., Khurana, S., "Energy Balance and Cogeneration for a Cement Plant", Applied Thermal Engineering, 22, pp. 485-494, 2002.

3. Mujumdar, K.,S., Arora, A., and Ranade, V.,V., "Modeling of Rotary Cement Kilns: Applications to Reduction in Energy Consumption", Ind. Eng. Chem. Res., 45, pp.2315-2330,2006.

4. Engin, T., Ari, V., "Energy Auditing and Recovery for Dry Type Cement Rotary Kiln Systems-A Case Study", Energy Conservation and Management, 46, pp.551-562, 2004.

8. Biswas D, Asthana S, Rau V, "Some Studies on Energy Savings in Sponge Iron Plants", Journal of Energy Resources Technology, Transactions of American Society of Mechanical Engineers, 125 (1), pp.228-237, 2003.

6. Liu, F., Ross, M., and Wang, S.,"Energy Efficiency of Chinas Cement Industry". Energy,20, 7, pp.669-681, 1995.

7. Carvalho, M., De., O., Madivate, "Theoretical Energy Requirement for Burning Clinker", Cement and Concrete Research, 29, pp. 695-698, 1999.

8. Li, A., M., Li, X., D., Li, S., Q., Ren, Y., Chi, Y., Yan, J., H., Cen, K., F., "Pyrolysis of Solid Waste in a Rotary Kiln: Influence of final Pyrolysis Temperature on the Pyrolysis Products", Journal of Analytical and Applied Pyrolysis, 50, pp.149–162, 1999.

9. Rasul, M., G., "Assessment of the thermal performance and energy conservation opportunities of a cement industry in Indonesia", Applied Thermal Engineering, 25, pp. 2950-2965, 2005.

10. Mujumdar, K., S., Ganesh, K., V., Kulkarni, S., B., Ranade, V., V., "Rotary Cement Kiln Simulator (RoCKS): Integrated Modeling of Pre-heater, Calciner, Kiln and Clinker Cooler", Chemical Engineering Science, 62, pp. 2590-2607, 2007.

11. Martins, M., A., Oliveira, L., S., Franca, A.,S., "Modeling and Simulation of Petroleum Coke Calcinations in Rotary Kiln", Fuel, 80, pp. 1611-1622, 2001.

12. F. Marias., H. Roustan., and A. Pichat., "Modelling of a rotary kiln for the pyrolysis of aluminium waste", Chemical Engineering Science, 60, pp. 4609-4622,2005.

13. Log sheet of J.K. Laxmi plant, Sirohi, Rajasthan, India.

14. K.E.Peray, Cement Manufacturers Handbook,Chemical Publishing Company, New York, 1979.

15. Holman J.P., Heat Transfer, McGraw Hill Book Co., New York, 1989.

# Chapter 43

# Climate Change Mitigation and Adaptation Strategies for the City of Bangalore

**Archana Kumta[1], Raghuram Tata[1] and Rakesh Kumar[2]**
**[1]Xavier Labour Relations Institute (XLRI), Jamshedpur**
**[2]Deputy Director and Head, Zonal Lab, National Environmental Engineering Research Institute, Mumbai**

## ABSTRACT

Concerns regarding the deteriorating urban environmental conditions and their long range implications should make our policy makers sit up and take notice. Climate change has compounded these factors even more as cities with large growth rate may face more daunting challenges with changing climate. For the next two decades a warming of about 0.2°C per decade is projected. Even if the concentrations of all greenhouse gases and aerosols had been kept constant at year 2000 levels, a further warming of about 0.1°C per decade would be expected. Earlier IPCC projections of 0.15 to 0.3°C per decade can now be compared with observed values of 0.2°C. The pace and scale of growth of Bangalore has outstripped the capacity to maintain acceptable standards of public health, environmental safety and sustainable economic growth. The ability of the family to earn is compromised. Expenditures on health and health care are excessive. Loss of lives and damage to property due to natural disasters has been frequent in the last few years. The chapter presents the case of Bangalore to understand the climate change impacts as also address the adaptation issues for better quality of life.

## Introduction

An increase in the global temperature due to enhanced greenhouse effect is termed as global warming. Anthropogenic activities are adding these greenhouse gases into the atmosphere at an alarming rate. According to the Intergovernmental Panel on Climate Change (IPCC) water vapor is the most abundant greenhouse gas; it occurs naturally and makes up about two thirds of the natural greenhouse effect. The sources of other important green house gases are fossil fuel combustion, deforestation, rice paddy cultivation, ruminants, landfills, use of nitrogenous fertilizers etc.

Professor Martin Parry (2007)[8], a climate scientist, said destructive changes in temperature, rainfall and agriculture were now forecast to occur several decades earlier than thought. According to the scientists, the effects of climate change would be felt sooner and the people who were affected would have to learn to live with these effects. Parry declared that the world leaders had not yet accepted their countries would have to adapt to the likely consequences. The impacts of global warming according to the report by the UN's Intergovernmental Panel on Climate Change (IPCC) says hundreds of millions of people in developing nations will face natural disasters, water shortages and hunger due to the effects of climate change. He said the international response to the problem had failed to grasp that serious consequences such as reduced crop yields and coastal flooding were now inevitable. "Mitigation has got all the attention but we cannot mitigate out of this problem. We now have a choice between a future with a damaged world or a severely damaged world." Dr Rajendra Pachauri, chair of the IPCC, said, "Wheat production in India is already in decline, for no other reason than climate change." The greenhouse effect is a natural phenomenon in which gasses such as carbon dioxide trap the sun's warmth. Human activity can enhance this effect and destabilize the climate.

The G8 summit in Germany agreed a landmark deal on climate change today with the west's leading economies pledging to make "substantial" cuts in greenhouse gas emissions. Head-to-head talks between Tony Blair and George Bush early this morning paved the way for an agreement under which the major polluters–including the United States–will "consider" reducing emissions by 50 per cent by 2050.

"Extreme weather events" are likely to become more intense and more frequent, while higher global temperatures could affect crops and water supplies and spread disease. The effect on ecosystems could be equally severe, with up to 30 per cent of plant and animal species at risk of extinction if the average rise in global temperatures exceeds 1.5-2.5°C. The experts[8] warn that the consequences of rising temperatures are already being felt on every continent sooner than expected and about 1°C of warming is already in the climate system. If the warming is not kept below 2°C up to 3.2 billion people would face water shortages and up to 600 million would face hunger.

Professor Parry said that he was pessimistic about the chances of keeping the increase in global average temperatures below 2°C. "And it's evident from the work of the IPCC that even with a maximum of 2°C we're not going to avoid some major impacts at the regional level." In February, the IPCC report concluded that global warming was "very likely"–a probability of 90 per cent or greater–to have been caused by human activity. The report further examined how climate change could be addressed and argued that global warming could be avoided without excessive economic cost only if the world acted immediately. The report predicts that while impact of a warmer globe would have mixed effects with increase in crop yields in northern Europe while overall impact would be negative especially in Africa, "mega-deltas of South and East Asia and small islands and in the polar regions".

By 2020, the report warns, shortage of water for almost 250 million Africans along with insufficient food due to climate variability and change. Studies have confirmed that Africa is one of the most

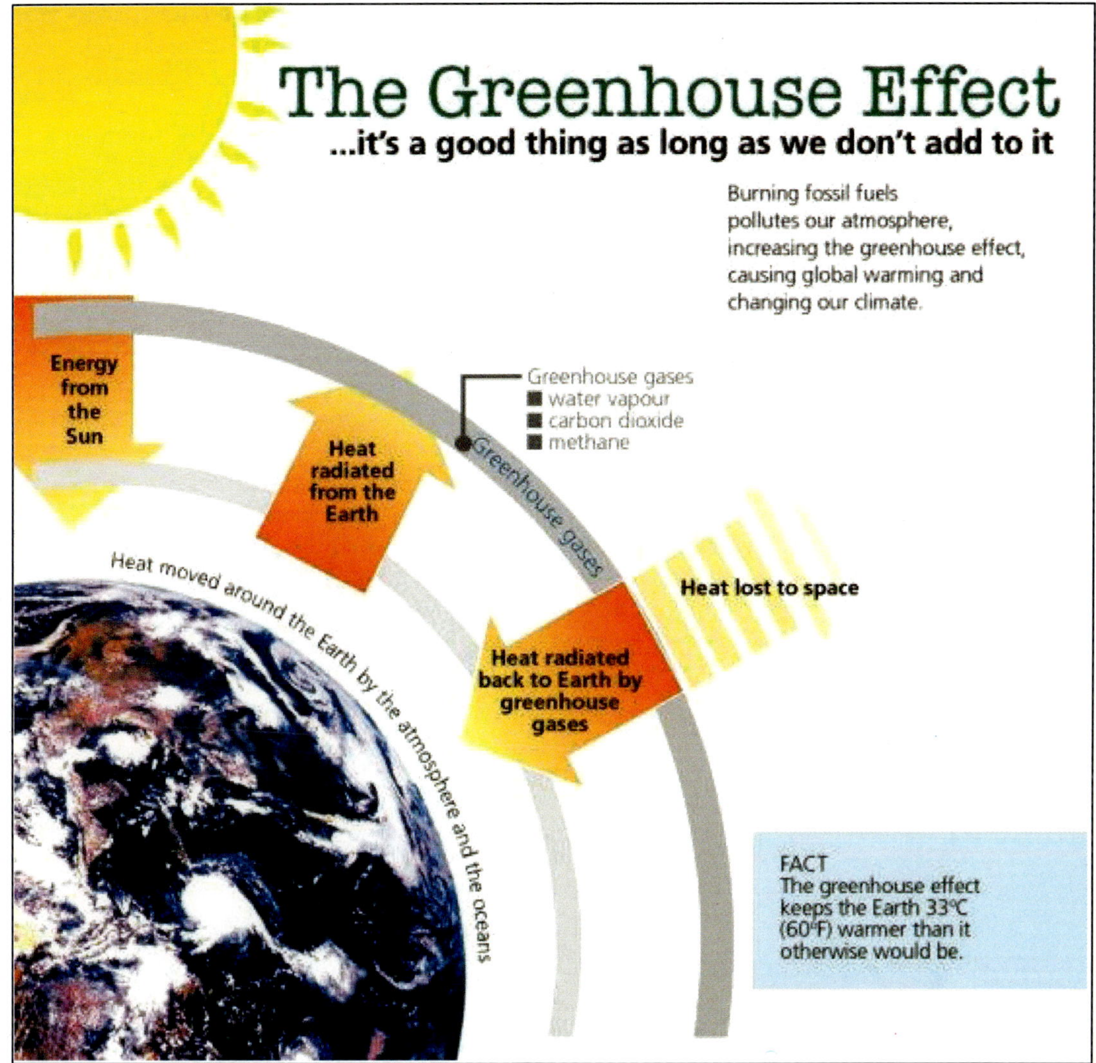


**Figure 43.1**

vulnerable continents to climate variability and change because of multiple stresses and low adaptive capacity. We need to understand why it is important for all the sectors to develop a strategy to address greenhouse gases and global warming issues.

Swedish scientist Svante Arrhenius[1] (1896) described what came to be known as the greenhouse effect. He theorized that doubling the amount of carbon dioxide in the atmosphere might raise the Earth's temperature by 5 or 6 degrees Celsius. Arrhenius suspected that coal and fossil fuels burned by industrialized countries were adding to the levels of carbon dioxide in the atmosphere.

By the end of the century, a small community of scientists[2] was pursuing the question of how solar variability might relate to short-term weather cycles, as well as long-term climate changes. In

1913, attempts to correlate weather patterns with the sunspot cycle were hindered due to inaccurate, unstandardized weather data, and lack of good statistical techniques for analyzing the data. After that series of investigations by various scientists improved upon the knowledge and addressed the climate change issue.[(3-5)]

First World Climate conference sponsored by the World Meteorological Organization was held at Geneva in 1979 was prompted by "extreme weather events of the past decade." In 1982, ice cores from the Greenland ice sheet showed dramatic temperature oscillations in a single century from the past, an extremely short period for climate change. Scientists also called 1981 the warmest year on record. In 1988, the IPCC was formed and in their first report in 1990, emphasized that human activities were causing a rise in the greenhouse gases. UN Framework Convention on Climate Change at Rio de Janeiro formed the treaty which aimed at reducing emissions of greenhouse gas in order to combat global warming[5] and the climate system was recognized as "commons."

In 1997, the Kyoto Protocol[6] was negotiated to reduce greenhouse gas emissions 5.2 per cent below 1990 levels by 2012 in the developed countries including the former communist bloc. Kyoto Protocol came into legal force on February 16, 2005 and was ratified by more than 140 countries. The objective of the protocol was "stabilization of greenhouse gas concentrations in the atmosphere at a level that would prevent dangerous anthropogenic interference with the climate system. By December 2006, a total of 169 countries and other governmental entities had ratified the agreement (with notable exceptions of USA and Australia)[7]. India and China which have ratified the protocol are not required to reduce carbon emissions under the present agreement.

According to the IPCC Fourth Assessment Report by Pachauri and Jallow[10] (2007) warming of the climate system is unequivocal as it is now evident from observations of increases in global average air and ocean temperatures, widespread melting of snow and ice and rising global mean sea levels. Climate change has affected the storm tracks, winds and temperature patterns which may have been likely due to anthropogenic activities.

Most of the observed increase in globally averaged temperatures since the mid-20th century is very likely due to the observed increase in anthropogenic greenhouse gas concentrations. This is an advance since the TAR's (Third Assessment Report) conclusion that "most of the observed warming over the last 50 years is likely to have been due to the increase in greenhouse gas concentrations". Discernible human influences now extend to other aspects of climate, including ocean warming, continental-average temperatures, temperature extremes and wind patterns.

Some projections for future changes in climate:

- ✰ For the next two decades a warming of about 0.2°C per decade is projected for a range of SRES emission scenarios.
- ✰ Even if the concentrations of all greenhouse gases and aerosols had been kept constant at year 2000 levels, a further warming of about 0.1°C per decade would be expected.
- ✰ Earlier IPCC projections of 0.15 to 0.3°C per decade can now be compared with observed values of 0.2°C.

## Important Factors for Climate Change and Bangalore

### Characteristics of Bangalore

The immediate effects of environmental degradation on the local communities are severe. High incidence of respiratory problems like asthma and bronchitis, diseases linked to poor sanitation and water like cholera and typhoid, gastroenteritis and exposure to toxic substances rob families of health, vigor and dignity. The ability of the family to earn is compromised. Expenditures on health and health care are excessive. Loss of lives and damage to property due to natural disasters has been frequent in the last few years. Climate change is likely to compound the factors much more. Bangalore has been selected as one the study region due to various unique features such as:

1. Fastest growing urban center
2. IT sector boom
3. Good place to live
4. Metropolitan
5. Sensitive and eco-friendly city
6. Indicator of good or worse which can happen to cities due to CC and also its role

## Likely Impacts of Global Warming

Some of the impacts of global warming in the city of Bangalore are going to be felt not just in years but over generations to come. Due to the high influx of population from other towns and cities to Bangalore a huge strain is felt on the natural resources like water bodies. Many of the suburbs are already facing problems of water shortage and flooding when there are heavy downpours leading to loss in life and property. This causes a huge burden on the government to ensure rehabilitation and facilities to be provided for the lower and lower-middle income groups which live in unplanned communities in and around the city.

Though Bangalore may not be historical emitters, however its emission in future can have large scale impacts which may also be felt by the city. The growth trend can make Indian cities turning from climate victim into climate perpetrator in future. It will be harder to take the moral high ground if actions accelerate the evacuation of people from poorer and low-lying areas.

## Major Sectors of Greenhouse Gase Emissions

Though there are many sources of GHG, the major sectors which are responsible for not only GHG but also other pollutants are:

- ☆ Vehicular sector
- ☆ Household sector [Domestic energy use]
- ☆ Manufacturing sector–But since Bangalore has a majority of software industries.
- ☆ Municipal refuse sector

## Adaptation Strategies of Bangalore Towards the Global Climatic Change Impact

Policy plans are need of the hour for the city as an example for climate change preparedness and adaptation for the city of Bangalore and how the same can get integrated to the national and international goal of CC mitigation and adaptation.

In the present chapter vehicular and household sectors have been addressed in detail.

# Methodology

## Study Region

Of many cities in India on the world map, Bangalore is emerging as the IT hub of the country. Bangalore has a wonderful climate and is called "The Pensioner's Paradise". With well developed residential areas, broad roads with well grown trees on both sides, good shopping malls, people prefer to move here permanently.

**Table 43.1: City Profile for Bangalore**

| *City Summary*[12, 13, 14] | |
|---|---|
| Urban Population | 9.8 Million |
| Area | 225 Sq Km |
| City Roads | 1500 Kms |
| Num Properties | 6.5 Lakh |
| Total Water Supply | 950 MLD |
| Per Capita Water Supply (BWSSB) | 147 PCLD |

Bangalore is called the Silicon Valley of India but the establishment of the Electronic City, ITPL on the out skirts has converted it into the Silicon City of the world. The city is the playground of many Indian as well as multinational Information Technological (I.T) and Bio Technological (B.T) companies. Bangalore has made a niche for itself in the international arena in terms of this growth and with major players in the IT industry gaining foothold here, Bangalore has come to be best known within India for being the country's un-official high-tech capital.

Table 43.1 gives a broad summary of the different parameters required for the study of the impact of climate change in the city of Bangalore.

The population in the Bangalore Urban Region and Municipalities has increased from 29 lakhs in 1981 to 42 lakhs in 1991 and more than 90 lakhs in 2004, an increase of 45 per cent and 43 per cent respectively during the corresponding period. The extent of developed area has also increased considerably from 174.7 sq.kms in 1971 to about 437 sq.kms in 2006.

The city evolved as the garden city of the country due to its salubrious climate. Today it has grown into radial city with 9 road corridors forming the principal radial arms, with concentric orbital corridors. The total road length of the city is about 4200 kms out of which arterial roads account for 500 kms. Most of the corridors carry traffic volumes much above their capacity, though number of grade separators have been constructed and are being constructed, most of them are located in the developed part of the city and causing a trigger of congestion at adjacent junctions.

The public transport system (bus) is overstressed carrying about 50 lakh commuters in a daily basis. Congested streets and longer route lengths due to urban sprawl have only served to reduce bus frequencies further. Traffic Management measures such as one way systems, parking restrictions, junctions improvements etc., are being implemented to ease the congested street network. The ever increasing traffic is fast deteriorating the limited improvement in level of service these traffic management measures can offer.

Industrial activity in and around Bangalore City, particularly Information Technology and Bio Technology related activities continues to grow and attract a large number of people to the city. The rapid urban population growth with relatively small investments in transport (roads and buses) has resulted in an increasing gap between the transport demand and supply.

## Data and Data Analysis: Sectoral analysis

### Vehicular Sector

The number of motor vehicles registered in Bangalore increased from 2,36,000 in 1983 to 6,84,497 by 1992 and 21 lakhs by December 2004, a phenomenal rate of increase 289 per cent and 278 per cent respectively during the mentioned period. Out of the total 21 lakhs registered vehicles in Bangalore, 15.42 lakh vehicles account for 2 wheelers and 3.3 lakhs vehicles account for cars, which means that 88.17 per cent of total vehicles are personal vehicles. This does not include the floating vehicle population.

The result is a low average speed on the network. Traffic management measures such as one way systems are very much evident, but they too are facing capacity problems. The sprawl of the city further loads the radial routes, as there has been a spate of new layouts developed by Bangalore Development Authority.

BMTC takes on the role of the sole provider for public transport. In Bangalore Omni Bus operations to local factories is quite significant. Factory worker trips to the larger factories are completely taken care off by these services. BMTC carries 28 lakh passenger trips a day at present and has a fleet strength of 3267 buses with 3125 schedules and 45,000 trips.

In order to reduce traffic congestion and ensure smooth Transportation Management major projects such as construction of Fly-overs, widening of existing road, construction of Sub-ways/Grade Separators and Pedestrian Sub-ways needs to be undertaken. Along with this there is also a need to improve Public Transportation by introducing Metro Railway and Mass Rapid Transits facility. Certain short term initiatives such as One-way system, Junction improvement and Restriction on certain type of vehicles is also necessary to make Bangalore City free from traffic jams and to ensure safe environment.

Due to the bad condition of roads and high vehicular traffic, the carbon dioxide emission into the air is very high which is a major source of green house emissions.

GHG contributions from transport sectors could be divided in following ways:

- ☆ High number of personalized vehicles
- ☆ Low occupancy of personalized vehicles
- ☆ Congestion
- ☆ Fuel use and adulteration
- ☆ In-adequacy of Public transport system
- ☆ Roads and infrastructure

Currently, Bangalore has approximately 3.3 lakhs cars. They travel about 26,400,000 km per day. Average occupancy of cars is about 1.1-1.5 PCU. It has been assumed that each car travels on an average about 80 KM/day. Taking the average emissions from each types of vehicles in gms/km emissions from each of them, they emit about 19047 T/day which will increase to 114284 T/day by the year 2025 as shown in Table 43.2.

**Table 43.2: Greenhouse Gas Emissions from Vehicular Sector in Bangalore**

| *Types of Vehicle* | *Vehicle Population, 2006* | *Vehicle km Travelled per Day, 2006* | *Vehicle Population, 2025* | *Vehicle km Travelled per Day 2025* | *$CO_2$ Emission Factor for Normal Fuel* |
|---|---|---|---|---|---|
| 2-3 wheelers | 2098370 | 167869600 | 12590220 | 1007217600 | 25 g/km |
| Cars/LMV | 434428 | 34754240 | 2606568 | 208525440 | 98 g/km |
| LDDV | 75125 | 6010000 | 450750 | 36060000 | 310 g/km |
| HDDV | 171101 | 13688080 | 1026606 | 82128480 | 700 g/km |
| Total GHG 2006 = 19047411 kg/day | | | Total GHG, 2025 =114284466 kg/day | | |

Assuming an average travel of about 80 kms/day, though there will be variations due to location, places to be visited, congestion etc. Emission factors have been taken from data of ARAI (Automotive Research Association of India, Pune 2007 study and averaged over different years/models.

The overall GHG emissions considering just the $CO_2$, in business as usual scenario, the load in atmosphere will be about 114284 tonnes per day by the year 2025. Long term scenario of 50 years shall be atleast 6-8 times more.

If cars are replaced with efficient public transport system (say about 50 per cent), it would lead to one bus replacing at least 30-40 cars. Each day it would lead to significant reduction in GHG emission. For example a bus emits about 700 g of $CO_2$ per km compared to 98 g/km by a car. However, a bus can replace about 30 cars, which is equivalent to saving of $CO_2$ emission by about 2.2 kg/km. [30 cars × 98 g/km$^{-1}$ bus × 700 g/km = 2240 g/km]

In future scenario, it would need to address the existing road infrastructure and the load that it can take of personalized transport system. Modal shift from personal to public mode of transport provides immense benefits:

1. Saving of road space
2. Saving of fuel
3. Saving of direct costs to individuals
4. Support to public transport system
5. Higher speed of the public transport and therefore efficiency
6. Obviating Parking problems
7. Sustainability wrt road usage

And top of all these additional benefits can also be accrued as given hereunder:

8. Green House Gas Emission reduction
9. Health benefits by less exposure to pollutants to public as well as motorists
10. Other pollutants [$SO_2$, CO, NOx, PM] reductions
11. Better quality of life

Other more efficient public transport system (other than buses) will bring much larger benefits to the city. For examples, construction of the Metro Rail in Bangalore will ensure that the traffic congestion, noise levels and green house gases such as $CO_2$ and $N_2O$ besides reducing other pollutants such as Carbon monoxide, sulphur oxide, nitrogen dioxide and ozone to a great extent according to the Environmental Management Action Plan prepared by BMRCL[18] for the Namma Metro.

This would also help in reducing congestion on roads; provide low cost transportation for poor and avoidance of local area pollution. It will result in other eco-benefits such as high fuel efficiency, better well being, time savings and therefore better income and higher standard of living.

## Corporate Sector

In the large IT parks and many BPOs and call centres use of Personal Computers and Laptops is very common. It has been calculated that PC's and related printers, wireless routers use a lot of power. On an average two PC's and related equipment use approximately 270 watts when switched on[16]. During the day if the computers are set to hibernate if they are inactive for 15 minutes will save a total of 1780 KWH per year, 3560 lbs of green house gas, and $178 per year!! The computer uses zero power when hibernated.

According to Innes[11], the sustainable community is promoted where many players in different roles and with differing interests and values are all provided with a flow of meaningful information.

The community needs the opportunity for joint learning and innovative responses to this feedback from the environment and from other changes.

It is this distributed intelligence which allows players in a community to anticipate and constructively address both individually and collectively the systemic problems the community continually faces and to deal with the threats and opportunities of natural and manmade disasters, the shifting global economy, and inequitable distributions of resources.

## Household or Domestic Energy Use Sector

Improving access to energy services has emerged as a cornerstone to development and economic growth. It is proven that the world supply of current sources of energy are depleting. Growing $CO_2$ emissions from fossil-based energy has been found to contribute significantly to climate change. Table 43.3 presents the computations for a household energy use and possible GHG emissions and reductions possible if appropriate steps are taken. Total population of Bangalore is approximately 9.8 million and its average electricity/energy use is 772.4 kWH. Computation of total $CO_2$ production for each tones of coal burnt for electricity production has been carried out.

Domestic sector emissions will need to be looked at from the point of view of availability of energy and later efficiency of its use and savings. As shown above, per household in Bangalore contributes to about 496 tonnes $CO_2$/month. If it makes an attempt to use less energy with a view to save at least a days consumption in a month, it would avoid 16.5 tonnes of $CO_2$.

With an assumption that of the total population of 9.8 million, each household comprises of 5 persons, there would be 19,80,000 household. If we take the average emission savings of 16.5 tonnes/ month by each household, 32670000 tonnes per month $CO_2$ can be avoided. Besides the GHG emissions avoided, it will also benefit through:

1. Energy saving for the country
2. Cost savings to families
3. Less resource use
4. Long term sustainability of fossil fuel
5. Poor people support
6. Better quality of Life.

## Methods for GHG Reduction

1. *Vehicular Sector*: Change in fuel from use of fossil fuels to renewable sources like solar or electric cars (*e.g.* Reva) will help to reduce the greenhouse gases in the atmosphere. By ensuring good public transport system like public buses, trains, and metro rail will help in the demand reduction of individual vehicle use. Ensuring that the roads are in good condition will also help to increase the fuel efficiency and therefore help to reduce the greenhouse gas emissions into the atmosphere.
2. *Household and Domestic energy use Sector*: By use of CFL lamps instead of incandescent lamps can reduce the energy consumption, switching off the mains and also allowing the PC's to hibernate when not in use at homes will ensure that energy is saved. Use of energy efficient equipment with Energy Star is a good investment with good ROI. Use of efficient lighting in the houses allowing natural light, use of solar heaters and cooling systems will ensure reduced consumption of fossil fuels and in turn reduced GHG emissions. By ensuring

regular energy audits are conducted in both households and commercial complexes will ensure reduction in the wastages and plugging of energy losses.

**Table 43.3: Household Energy Use and Equivalent GHG Emissions and Reductions Possibilities**

| *Sl.No.* | *Nos. of Appliances* | *Nos. of Unit* | *Total Usage in Hrs* | *Daily Consumption Factor* | *Total Consumption Per Day (kWH)* | *Nos. of Day Usages in each Month* | *Consumption per Month (KWH)* | *Amount of COAL Required in Tones/ KWH* | *Tones of $CO_2$ Emission per House* |
|---|---|---|---|---|---|---|---|---|---|
| 1. | Bulb | 1 | 4 | 0.06 | 0.24 | 30 | 7.2 | 1.8 | 4.62 |
| 2. | Tube Light | 1 | 8 | 0.04 | 0.32 | 30 | 9.6 | 2.4 | 6.17 |
| 3. | Fan | 1 | 8 | 0.08 | 0.64 | 30 | 19.2 | 4.8 | 12.33 |
| 4. | Room A/C | 1 | 6 | 2.1 | 12.6 | 30 | 378 | 94.5 | 242.77 |
| 5. | Fridge | 1 | 24 | 0.2 | 4.8 | 30 | 144 | 36 | 92.48 |
| 6. | Electric Iron | 1 | 1 | 0.75 | 0.75 | 30 | 22.5 | 5.625 | 14.45 |
| 7. | Geyser | 1 | 1 | 3 | 3 | 30 | 90 | 22.5 | 57.80 |
| 8. | Hotplate | 1 | 0 | 1 | 0 | 30 | 0 | 0 | 0.00 |
| 9. | Oven | 1 | 1 | 0.4 | 0.4 | 30 | 12 | 3 | 7.71 |
| 10. | Mixer | 1 | 0.5 | 0.4 | 0.2 | 30 | 6 | 1.5 | 3.85 |
| 11. | Toaster | 1 | 0.5 | 0.75 | 0.375 | 30 | 11.25 | 2.8125 | 7.23 |
| 12. | Washing Machine | 1 | 1 | 0.32 | 0.32 | 30 | 9.6 | 2.4 | 6.17 |
| 13. | Vacuum Cleaner | 1 | 1 | 0.6 | 0.6 | 30 | 18 | 4.5 | 11.56 |
| 14. | Radio/Tape | 1 | 3 | 0.05 | 0.15 | 30 | 4.5 | 1.125 | 2.89 |
| 15. | T.V. | 1 | 4 | 0.08 | 0.32 | 30 | 9.6 | 2.4 | 6.17 |
| 16. | VCP/VCR/CD/LD | 1 | 1 | 0.03 | 0.03 | 30 | 0.9 | 0.225 | 0.58 |
| 17. | Computer | 1 | 2 | 0.5 | 1 | 30 | 30 | 7.5 | 19.27 |

**Assumption**

1. Rate of $CO_2$ produced per Tones of Coal = 3.67
2. Percentage of carbon in coal = 0.7

Tones of $CO_2$ emission per house = Amount of coal required in Tones per kWH * Percentage of carbon in coal * amount of coal used

**Results**

| | |
|---|---|
| Total Consumption Per Day | = 25.7 kWH |
| Total Consumption per Month | = 772.4 kWH |
| Amount of COAL required | = 193.1 Tones |
| Total $CO_2$ emission per house | = 496.0 Tones |
| Electiricity saved for | = 1.0 Day |
| $CO_2$ Emission saved | = 16.5 Tones |

## GHG Management and Adaptation Issues

Renewable energy resources, energy conservation and energy efficiency are realistic options for a sustainable future, given the shortcomings of conventional energy sources. They are an asset towards energy security and to reduce the energy dependence of local and national authorities.

Local governments are effective in promoting the generation of energy from renewable resources. With their executive power they can implement sustainable energy at a locally appropriate level. Local governments can incentivize and penalize when the domestic and industrial sector do not comply with the standards and laws set by the central bodies. The local authorities can easily monitor the progress of activities at public places and of public interest eg: like solar street lighting instead of using electricity or fossil fuels etc which can ensure that their contribution towards greenhouse effects are minimized even further. Replacing the incandescent lamps in houses/offices and streets by CLF lamps can reduce the carbon emissions into the atmosphere.

Local governments can arrange good medical and paramedical support systems to cater to all emergency health situations. This will help to ensure that loss in productivity due to ill health is minimized to some extent.

They can thus liberate peripheral and developing economic regions from their dependence upon national investment decisions in energy. Providing access to energy services in this decentralized manner can not only improve the economy of these regions but also give control to the people of each region over their energy use and needs.

The city will need to plan for each sector as some of the climate change impacts are irreversible in a time frame of 100-150 years. It would need to start adapting itself to deal with some of these impacts. Besides the attempts made to reduce GHG emissions as stated above, adaptation planning will need different but immediate steps.

## Conclusion(s)

Major cities worldwide have been considered as engine of growth. However, if they are not developed and planned well, many of these cities have also indicated that citizen can experience poorest quality of life in cities. Besides the associated problems of population and infrastructure, global warming due to climate change has brought an unexpected issue to be resolved. Though the process takes long time to manifest itself, its impact once started will also be felt for longer duration.

With a view to understand the climate change impact on cities and how to integrate adaptive measures, various urgent steps are warranted. Sectoral analysis carried out in the study has shown that transport sector will be major impact making sector in terms of its use and likely growth, followed by domestic energy use and manufacturing sector. Transport sector is also a sector which indicates that there are options which could be implemented quickly and it would provide higher returns in terms of overall GHG emission reductions per day.

Similarly, energy use and its optimization in domestic and manufacturing sectors indicate that a lot can be achieved through simple and effective measures. Bangalore city can show a good example of the measures a city can adapt for reducing climate change impact as also adaptive measures

## References

1. http://www.cnn.com/SPECIALS/2005/changing.earth/interactive/timeline/content.1.html
2. Dubois, Eugene (1895). *The Climates of the Geological Past and Their Relation to the Evolution of the Sun*. London: Sonnenschein.

3. http://www.copernicus.org/EGU/egs/milankovic.htm
4. http://www.guardian.co.uk/climatechange/story/0,1517938,00.html
5. http://en.wikipedia.org/wiki/United_Nations_Framework_Convention_on_Climate_Change
6. http://unfccc.int/resource/docs/convkp/kpeng.html
7. http://en.wikipedia.org/wiki/Kyoto_Protocol
8. http://www.guardian.co.uk/environment/2007/sep/18/climatechange.internationalnews
9. http://en.wikipedia.org/wiki/Charles_David_Keeling
10. Pachauri RK and Babu Jallow (2007) "Working Group I Contribution to the IPCC Fourth Assessment Report" at Nairobi on 6th February 2007
11. http://www.sfu.ca/casr/id-newfuel-2.htm
12. http://en.wikipedia.org/wiki/Template_talk:Metropolitan_Cities_of_India
13. http://www.bmponline.org/gis/maps.shtml
14. http://www.bcp.gov.in/english/trafficpolice/aboutus/aboutus.htm
15. http://www.adb.org/Documents/Books/Cities_Data_Book/07chapter7.pdf
16. http://www.builditsolar.com/References/Half/DirtCheap.htm
17. http://climatesci.colorado.edu/2007/03/16/two-papers-on-the-urban-heat-island-effect-on-temperatures/
18. http://www.bmrc.co.in/pdf/draftemp.pdf

# Chapter 44

# Removal of Toluene Vapour from Air Stream Using a Biofilter Packed with Polyurethane Foam

**R.S. Singh*, B.N. Rai, Kiran Singh and S.N. Upadhyay**
**Department of Chemical Engineering and Technology, Centre of Advanced Study, Institute of Technology, BHU, Varanasi – 221 005, U.P.**

## ABSTRACT

Biodegradation of toluene vapor was investigated in a biofilter packed with polyurethane foam pieces supporting a mixed microbial population growth. The biofilter consisted of five modular sections of 0.16 × 0.16 × 0 × 0.20 $m^3$ size. Each section was filled with foam cubes upto a height of 0.15 m. Inlet concentration of toluene was varied from 0.04 to 2.5 g $m^{-3}$ and the volumetric flow rate of toluene loaded air from 0.06 to 0.90 $m^3 h^{-1}$.

Removal efficiency from 68.2 to 99.9 per cent and elimination capacity from 10.85 to 90.48 g $h^{-1}m^{-3}$ were observed during steady state operations at various loading rates. More than 90 per cent removal efficiency was observed upto an inlet loading rate of 76.3 g $h^{-1}m^{-3}$. The dynamic behaviour of the biofilter was evident from its response to change in process conditions. During the entire period of runs lasting over 168 days, the inlet toluene loading was changed many times, but no severe fluctuations in pH and pressure drop were observed. The stability of the biomass was also evident from the fast response of the biofilter to shutdown and restart-up.

*Keywords: Biofiltration, Biofilter media, Removal efficiency, Elimination capacity, Polyurethane foam, VOCs.*

---

* Corresponding Author: E-mail: rssingh.che@itbhu.ac.in, ram_sharans@rediffmail.com

## Introduction

Air emissions from a variety of industries contain non-odorous and/or malodorous volatile organic compounds (VOCs) that have been subjected to increasingly stringent environmental regulations due to their contribution to the green-house effect and photochemical smog. The emission of VOCs is usually at low to intermediate levels but is in large volumes, hence use of conventional methods like adsorption, condensation, incineration, etc. for their control are not economical [1]. Therefore, use of methods based on aerobic biodegradation in biofilters has received the attention in recent decades. Several groups of researchers all over the world have studied the behaviour and performance of such biofilters [2–4]. Biofiltration uses aerobic microorganisms, immobilized on solid particles of a suitable bed media such as peat, compost, wood chips, polyurethane foam, and/or their mixtures packed in a column [3, 5]. Thus biofilter provides a large contact area between the gas stream and the microbial growth supported on the porous granular solid surface. These microorganisms exist both in the pores of the medium as well as in the bio-film [6]. As the polluted gas stream passes through the filter, VOCs get partitioned into the bio-film where biological oxidation to harmless end-products occurs under aerobic conditions [4]. Relatively low cost and excellent operational stability are some other attractive advantages of biofilters [1]. As the biodegradation process is controlled either by mass transfer or biochemical reaction or both, the removal efficiency of biofilters varies with media characteristics (*e.g.* particle size, surface area, porosity), carbon and energy sources, nutrients, pH and temperature [7].

Toluene is a VOC of particular interest due to its use in the synthesis of several compounds such as tri-nitro toluene (TNT), chloramine-T, saccharin and many dyestuffs. Even at low concentrations, toluene is reported to be carcinogenic, causing damage to the liver and kidney and paralyzing the central nervous system [8]. Thus, the widespread use of toluene in various industrial operations, its high vapour pressure (28.6 mmHg at 25 °C) and high polarity characterized by its water solubility (0.53 g $l^{-1}$ at 25 °C) warrants its removal before emission into environment [9]. Hence several workers have used toluene as the model VOC in their experimental studies. Performance of most of the synthetic as well as other packing media has been evaluated using simulated air stream containing toluene.

The effectiveness of a biofilter largely depends on the type of bed medium. The medium is crucial in order to maintain the long-term operational stability of the biofilter. Packing media not only provide support for the growth and sustenance of the microbial population but also contribute to sorption effects, thereby ensuring adequate residence time for metabolic destruction of gaseous pollutants [10, 11, 12]. Natural media such as compost, soil, peat, wood chips, bark, etc are the commonly used packing media in the biofilters. Natural packing media have the advantages of large surface area, high void fraction, high microbial population density with sufficient nutrients, good water retaining capacity, low bulk density and low cost. However, some of their disadvantages like strong aging effect, stability and variation in properties from place to place have encouraged workers to look for alternatives.

Synthetic packing materials including plastic Pall Rings, granular activated carbon (GAC), carbon foam, polyurethane foam, and extruded diatomaceous earth pellets have been used successfully by different workers in biofilters. Pall Ring packings are easy to handle. However, experiments have indicated that startup period is relatively long presumably because of poor biofilm growth on the surface [13]. The relatively low specific surface area is also a disadvantage in achieving high elimination capacity, however, their stability, low cost and high porosity make the attractive. Structured packings made from stainless steel or plastic combine high porosity and large specific surface. Good performance has been observed with these type of packings but they are quite expensive [14, 15].

A summary of the reported experimental work on biodegradation of toluene and other aromatics is given in Table 44.1. It is seen that the benzene, toluene and xylene have been used as the target VOC by several workers. Only a few workers have used polyurethane as the packing media. It has been found to be an effective packing medium. Improved performance were reported for $H_2S$ and toluene removal, in biofilters packed with polyurethane foam cubes over other types of packing, especially at high gas flow rates and low concentrations. This has been attributed to open structure and high porosity, low density, easy inoculation procedure and ability to absorb water, and low pressure drop over the packing even at a relatively high gas velocity [16-18]. Moe and Irvine [16, 17] found polyurethane foam to be an excellent medium for biodegradation operation. They concluded that large specific area, non-biodegradability and physical strengths were its main positive features, and in case of clogging problem, biomass removal was also quite convenient.

Moe and Bing [18] evaluated the biofilter performance over a 3 months period for three loading conditions intended to simulate processes generating contaminated gases only during daytime operation, daytime operation with weekend shutdown periods, and with long term (9-day) shutdown. The biofilter showed stable performance even under discontinuous loading conditions.

Liu *et al.* [20] studied simultaneous degradation of toluene and ethyl acetate into two identical compost based biofilters acclimated with toluene and ethyl acetate respectively. At EBRT of 30 s maximum elimination capacities for toluene and ethyl were found to be 50 and 450 g $m^{-3}h^{-1}$ respectively. Presence of ethyl acetate significantly reduced the elimination capacity of toluene. However, elimination capacity of ethyl acetate was unaffected by presence of toluene. Cox and Deshusses [21] studied co-treatment of odors and VOCs in biotrickling filters operating at acidic and neutral pH. In the acidic condition, toluene degradation started immediately but at a lower rate. However, after several weeks of operation, the toluene elimination capacity (EC) at low pH reached a steady value identical to that found in the neutral-pH filter. At a volumetric load of 100 $m^3 m^{-3} h^{-1}$, maximum elimination capacities of 70 g toluene $m^{-3} h^{-1}$ (at 1.7 $gm^{-3}$ toluene) and 20 g $H_2S$ $m^{-3} h^{-1}$ (at 170 $ppm_v$ $H_2S$, the highest concentration tested) were observed.

Elmrini *et al.* [22] investigated the biofilter response to step and abrupt variations (shock loading effects) in inlet xylene concentrations. EC and removal efficiencies (REs) as high as 61 g $m^{-3}h^{-1}$ and 96 per cent, respectively, were noticed within 24 h of change in loading rate, which revealed potential stability of biofilm to withstand shock loadings. The effect of nitrogen concentration on the removal of VOCs was studied by Delhomenie *et al.* [24] and high removal for toluene (90 per cent) at a nitrogen concentration of 2 g $l^{-1}$ was observed. ECs as high as 95 and 108 g $m^{-3}h^{-1}$ were achieved for toluene and xylene, respectively.

In an earlier work, Singh *et al.* [25,26] used a circular cross-section biofilter packed with agro-wastes such as maze cobs and yellow-gram stalk (*Cajanus cajan*).With maze cobs as the biofilter medium, removal efficiencies in the range of 54.5 to 85.3 per cent and maximum elimination capacity of 174.6 g $h^{-1}m^{-3}$ were obtained. With yellow gram stalk more than 90 per cent removal efficiency was obtained upto an inlet loading of 300 g $h^{-1} m^{-3}$ at flow rate of 0.18 $m^3 h^{-1}$. Maximum elimination capacity of 303 g $h^{-1}m^{-3}$ was observed during the operation. It was also observed that at low loadings the mass transfer and at high loadings the bio-reaction was the controlling mechanism [26]. For long duration operations, deterioration in biofilter media leading to compaction of the bed resulting in the increased pressure drop across the bed was observed in both the cases. Effect of shock loading and dynamic behaviour [22], pH [21], temperature [31,32], presence of other compounds on toluene degradation [20,21], $N_2$ [24], degradation pattern of different VOCs [24,29] and moisture content of bed [33] on biofilter performance have been studied.

**Table 44.1: Summary of Experimental Work on Biodegradation of Organics Biofilters**

| *Reference* | *VOCs* | *Filter Media* | *Operating Condition* | *Remark* |
|---|---|---|---|---|
| Moe and Irvine [16,17] | Toluene | Polyurethane (open pore, high density), Self-manufactured | Concentrations: 50-200 $ppm_v$<br>EBRT: 1-4 min<br>Duration: 300 days | Excellent medium,<br>RE > 99 per cent |
| Moe and Bing [18] | Mixture of n-butyl acetate, MEK, MPK and toluene | Open polyurethane (high density), mixed culture of fungi | Concentrations: 124, 50.5, 174 and 44.6 g $m^{-3}$ respectively,<br>Duration: 94 days<br>EBRT:15 s | Fungi are efficient in degrading VOC, even under fluctuating conditions |
| Rene *et al.* [19] | Toluene | Compost and ceramic beads | | RE: 40-95 per cent<br>EC: 3.5–128 $g/m^3/h$<br>Depended upon initial loading rate |
| Liu *et al.* [20] | Toluene and ethyl acetate | Two identical biofilter packed with compost | Inoculation: 3 months | Presence of ethyl acetate significantly reduced the EC of toluene. |
| Cox & Deshusses [21] | $H_2S$ and toluene | | Two identical biotrickling filters, ne operated at pH 4.5 and the other at pH 7.0 | $H_2S$ did not affect toluene degradation at concentrations up to 170 $ppm_v$ at either pH. |
| Elmrini *et al.* [22] | Xylene | Peat with a mineral additive | Operated for more than two months under shock loading conditions (Abrupt variation in inlet concentration of xylene) | Biofilter regained removal efficiency more than 96 per cent in less than 24 h, stable operation, no effect of shock load |
| Tang and Hwang [23] | Toluene | Three laboratory scale biofilters packed with mixtures of chaff/ compost, diatomaceous earth (DE)/compost and granular activated carbon (GAC)/compost | | Maximum steady-state ECs of 58, 60 and 97 g $m^{-3}h^{-1}$, respectively |
| Delhomenie *et al.* [24] | Toluene, xylene and trimethyl-benzene | Three parallel biofilters with three modules packed with compost and organic binder | More than 6 months of operation, performance evaluated at varying nitrogen concentration | Biodegrability scale followed the degree of substitution around the aromatic ring: toluene >xylene > trimethylbenzene, Effect of $N_2$ was studied. More than 90 per cent toluene removed. |

*Contd...*

**Table 44.1–Contd...**

| *Reference* | *VOCs* | *Filter Media* | *Operating Condition* | *Remark* |
|---|---|---|---|---|
| Sorial *et al.* [28] | BTEX | 6 mm R-635 Celite pellets | Inlet concentration = 45 to 225 ppmv, pH=7.2 ± 0.2; EBRT = 0.67-2.0 min, 175 days operation | RE upto 99 per cent, Toluene degraded more rapidly than others. |
| Lu *et al.* [29] | BTEX | Coal particles of 2.1 cm diameter | Single BTEX exp. B = 6.03-229.6, T = 6.38–270, E = 6.76 -245 and X= 4.77-245 $g/m^3/h$; EBRT = 0.8-6.0 min, mixed BTEX 37.44-408.5 $g/m^3/h$ EBRT=1.50–4.5; Temp. = 25 ± 2 °C | BTEX biodegraded in the following order T > B > X > E. Under similar BTEX load RE and EC of single substrate feed were higher than that of the mixed feed |
| Smith *et al.* [30] | Toluene | 1.12m of 6mm pelletized Celite biological attachment medium | 250 and 500 ppmv of toluene. Loading were 4.1 and 6.2 kg $COD/m^3$ day. | For toluene loading 4.1 $COD/m^3$ day, over 99.9 per cent VOC removal by backwashing removal strategy. |
| Park *et al.* [31] | Toluene | Supported media (wood chips). Microorganism: *Pseudomonas putida* DK-1. | Pressure drop: 0.2-1.0 (cm $H_2O$ $m^{-1}$ packing) Humidity level: 40-60 per cent. Temp: 30°C. Gas inlet concentration: 4.11,8.23, and 12.34 g $m^{-3}$ | EC increased upto 70 per cent. Removal efficiency was more than 95 per cent in the temperature range of 32-35°C. Optimal temperature range was 32-35°C. The increase in gas flow rate caused decrease in EC. Optimal gas flow rate 65ml $min^{-1}$. The biodegradation rate was independent of inlet loading up to 251g $h^{-1}$ $m^{-3}$. The EC increased with inlet loading. |
| Lu *et al.* [32] | Benzene, toluene, ethyl benzene, and o-xylene. (BTEX). | Coal particles coated with micro-organisms. | Feed rate: 2.2 ml $min^{-1}$ EBRT: 4min. Temperature: 25°C. Duration: 2 weeks. | RE of BTEX vapors was up to 90 per cent.BTEX removal efficiency increased as the operating temperature increased from 15 to 30°C. Optimum temperature range is 25-35°C.The temperature effects was more significant for the low BTEX loading. |
| Ranasinghe and Gostomaski [33] | Toluene | Compost | Temperature 30±0.5 °C, Initial concentration 1000 ppmv. Moisture content of bed varied significantly. | Moisture content was most likely the parameter that influenced changes in performance. |

The aim of the present study was to evaluate the performance of a biofilter packed with polyurethane foam cubes inoculated with a mixed culture of microorganisms for treating a simulated toluene vapour loaded air stream under various operating conditions. The removal efficiencies and elimination capacities were monitored by varying the process parameters and operating conditions.

## Materials and Method

### Experimental Setup

Biological degradation of toluene was carried out in a laboratory scale down flow biofilter, continuously for six months. The schematic arrangement of the experimental biofilter system used is shown in Figure 44.1. It consisted of the following main components: (1) toluene bubbler unit (2) pre-humidification unit (3) mixing unit, and (4) biofilter column.

The toluene bubbler unit is used to convert liquid into vapor phase. An aluminum block heater, provided in this unit, was used to keep the temperature of liquid toluene constant (43°C) throughout the experiment. The liquid toluene was fed in the glass bubbler kept in the block heater. Air flow was controlled by a 0-1 lpm rotameter. The evaporation rate of toluene was varied by varying the air flow rate.

For controlling the humidity, a pre-humidification unit was used. It consisted of a square cross-section ($0.10 \times 0.10$ m$^2$) Perspex column of height 0.42 m. The column was packed with porous lava stone. A thick and uniformly punched polyurethane foam sheet was placed at the bottom of column for uniform distribution of air. The column was filled with water and air from compressor was bubbled through it. Humidity of outlet gas stream was periodically measured with the help of a humidity meter.

The mixing unit consisted of a horizontal rectangular column made of Perspex ($0.06 \times 0.16 \times 0.075$ m$^3$). In this unit, the humidified air was mixed with the toluene vapor. Vertical baffles, made of stainless steel sieve were used for proper mixing. Toluene loaded air was supplied to the top of the biofilter.

The biofilter column consisted of a nutrient distribution section, a head space, five modular sections of size ($0.16 \times 0.16 \times 0.20$ m$^3$) and a leachate collection section at the bottom (Figure 44.1). Each section was clamped between square flanges made of Perspex sheet. Rubber packing was provided between flanges to avoid leakage. The column was supported on a mild steel structure.

All modular sections were filled with prehumidified small cubical pieces of identical size (1 cm$^3$) of polyurethane foam up to a height of 0.15 m. The filter material in modular section was supported on a stainless steel sieve plate for avoiding excess compression of media at the bottom. Two sampling ports, one at the top and other at the middle, were provided in each modular section to collect gas and medium samples. To ensure proper distribution of nutrient, a perforated Perspex sheet was provided at the bottom of nutrient distribution section.

The gas samples were collected in 15 ml gas samplers from sampling ports located at the top of each modular section. Biofilter media samples were collected periodically from sampling ports at the middle of each modular section. The percolated liquid was recycled manually at regular intervals of time.

In order to generate air stream of desired toluene concentration, the main air stream was divided into two parts. One part of the stream was allowed to bubble through the bubbler containing pure toluene. Outlet stream from the bubbler was mixed with humidified main air. Flow rates of the main

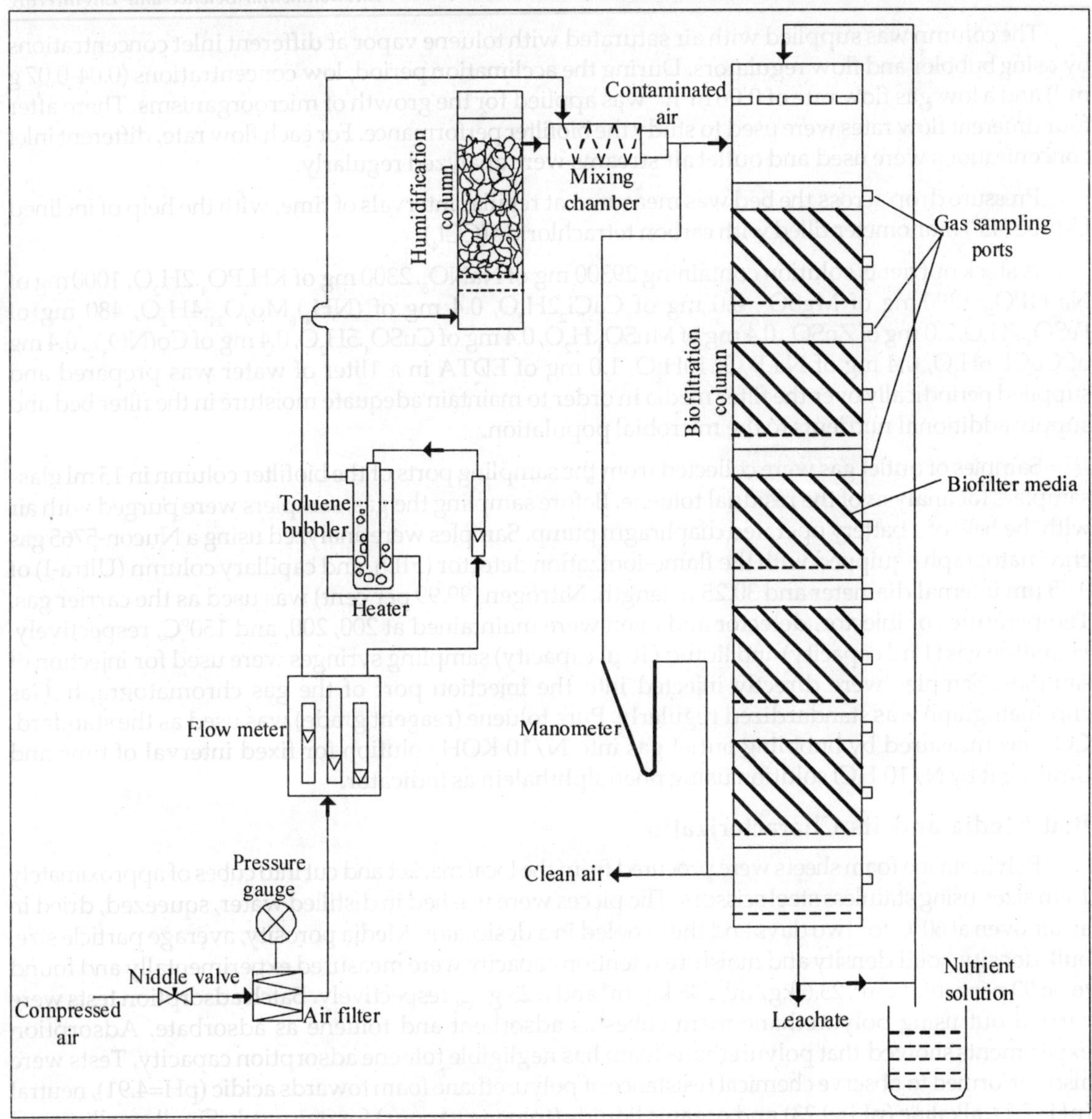


**Figure 44.1: Experimental Setup**

air stream and the air stream to the bubbler were measured by rotameters of 0-15, 0-50 lpm and 0-1 lpm capacities, respectively, coupled with a calibrated orifice meter. Toluene loaded air stream was passed through the packing material (polyurethane foam), covered with biofilm. Gas samples from each section, were collected in 15ml gas samplers for analysis. A nutrient solution was supplied at the rate of 1 l $day^{-1}$ over the filter material in order to maintain adequate moisture in the support medium and also supply essential nutrients to the attached microbial population.

The column was supplied with air saturated with toluene vapor at different inlet concentrations by using bubbler and flow regulators. During the acclimation period, low concentrations (0.04-0.07 g $m^{-3}$) and a low gas flow rate of 0.06 $m^3h^{-1}$ was applied for the growth of microorganisms. There after four different flow rates were used to study the biofilter performance. For each flow rate, different inlet concentrations were used and outlet air streams were analyzed regularly.

Pressure drop across the bed was measured at regular intervals of time, with the help of inclined U-tube (45°) manometer filled with carbon tetrachloride ($CCl_4$).

A stock nutrients solution containing 29300 mg of $NaNO_3$, 2300 mg of $KH_2PO_4.2H_2O$, 1000 mg of $Na_2HPO_4$, 1290 mg of $MgSO_4$, 8.0 mg of $CaCl_22H_2O$, 0.4 mg of $(NH_4)_6Mo_7O_{24}.4H_2O$, 480 mg of $FeSO_4.7H_2O$, 2.0 mg of $ZnSO_4$, 0.4 mg of $MnSO_4.H_2O$, 0.4 mg of $CuSO_4.5H_2O$, 0.4 mg of $Co(NO_3)_2$, 0.4 mg of $CoCl_2.6H_2O$, 0.4 mg of $Na_2B_4O_7.10H_2O$, 1.0 mg of EDTA in a 1liter of water was prepared and supplied periodically over the filter media in order to maintain adequate moisture in the filter bed and supply additional nutrients to the microbial population.

Samples of outlet gas were collected from the sampling ports of the biofilter column in 15 ml glass samplers for analysis of the residual toluene. Before sampling the gas, samplers were purged with air with the help of a battery operated diaphragm pump. Samples were analyzed using a Nucon-5765 gas chromatograph equipped with the flame-ionization detector (FID), and capillary column (Ultra-I) of 0.25 μm internal diameter and 30.25 m length. Nitrogen (99.99 per cent) was used as the carrier gas. Temperatures of injector, detector and oven were maintained at 200, 200, and 150°C, respectively. Hamilton gas (1 ml capacity) and liquid (10 μl capacity) sampling syringes were used for injection of samples. Samples were directly injected into thc injection port of the gas chromatograph. Gas chromatograph was standardized regularly. Pure toluene (reagent grade) was used as the standard. $CO_2$ was measured by bubbling outlet gas into N/10 KOH solution for fixed interval of time and titrating it by N/10 HCl solution using phenolphthalein as indicator.

## Bed Media and its Characterization

Polyurethane foam sheets were procured from the local market and cut into cubes of approximately 1-cm sizes using stainless steel scissors. The pieces were washed in distilled water, squeezed, dried in an air oven at 60°C for two days and then cooled in a desiccator. Media porosity, average particle size, bulk density, solid density and moisture retention capacity were measured experimentally and found to be 90 per cent, 1 $cm^3$, 23.6 $kg/m^3$, 236 $kg/m^3$ and 6.23 g/g, respectively. Batch adsorption tests were carried out using polyurethane foam cubes as adsorbent and toluene as adsorbate. Adsorption experiments showed that polyurethane foam has negligible toluene adsorption capacity. Tests were also performed to observe chemical resistance of polyurethane foam towards acidic (pH=4.91), neutral (pH=7.1), alkaline (pH=9.23) and organic liquids (toluene vapour) for one week. Small swelling and weight loss in the foam pieces were observed in the acidic and basic solutions. In the vapour phase toluene the colour of foam pieces turned light brown from the originally light yellow.

## Microorganisms

Methods of preparation of microbial culture for inoculum and activation was same as used earlier [25]. To confirm the presence of microorganisms in biofilter column, microbial analysis was done at regular intervals of time. The microbial culture was maintained on the slants, which were prepared by dissolving peptone 1.1 per cent, beef extract 1.0 per cent, sodium chloride 0.5 per cent in distilled water with 3.0 per cent agar. The pH of the medium was adjusted to 7.0 and 5.0 ml of it was poured into 20 ml of test tube. It was sterilized at $10.33 \times 10^4$ $Nm^{-2}$ pressure for 20 minutes. Cells from the surface growth of an old slant were taken in the sterile wire loop and used to inoculate the medium. The cultures were incubated at 30°C for one day.

## Results and Discussion

### Continuous Biodegradation Experiments

Operation and performance of a biofilter is generally expressed in terms of removal efficiency and pollutant elimination capacity as a function of pollutant loading at different EBRTs [25]. Experimental data have been used to calculate these parameters.The efficient functioning of any biofilter unit strongly depends on the inlet pollutant load [12]. Initially the biofilter was acclimated for 25 days with very low inlet flow rate and inlet concentration of toluene to achieve sufficient microbial growth in the biofilter. Effects of variation in toluene concentration and gas flow rate were investigated in five distinct phases and during startup period after 10 days shutdown as described in Table 44.2. The results are shown in Figure 44.2. Effects of variation in pH and temperature on the performance of biofilter were observed throughout the test period.

**Table 44.2: Experimental Scheme for Continuous Toluene Degradation Experiments**

| *Nature of Biofilter Operation* | *Flow Rate ($m^3 h^{-1}$)* | *EBRT (min)* | *Inlet Concentration Range ($g\ m^{-3}$)* | *Operating Time (Days)* |
|---|---|---|---|---|
| Acclimatization | 0.06 | 16.875 | 0.04–0.07 | 25 |
| Phase I | 0.36 | 2.813 | 0.51–2.01 | 30 |
| Phase II | 0.54 | 1.875 | 2.0–2.2 | 16 |
| Phase $^{22}$I | 0.72 | 1.406 | 1.2–2.1 | 25 |
| Phase IV | 0.90 | 1.125 | 1.2–2.4 | 23 |
| Phase V | 0.36 | 2.813 | 2.3–2.5 | 24 |
| Shut down period | | - | - | 10 |
| Restart operation | 0.54 | 1.873 | 1.3–1.8 | 13 |

Variation of normalized toluene concentration along the bed height at fixed EBRT was also investigated. The axial concentration profile provided valuable information on the degradation behaviour of pollutant in the bed and that could be utilized as an effective design tool for biodegradation process. Different inlet concentrations were tested for biodegradability in the biofilter.

During Phase I, flow rate was maintained at a low level (0.36 $m^3 h^{-1}$). Inlet concentration was varied between 0.51 to 2.01 $g\ m^{-3}$. In the initial stage of this phase, inlet toluene concentration was also kept low to allow the biofilter to achieve the steady-state quickly. Steady-state was obtained on the 38$^{th}$ day. Maximum removal efficiency of 99.24 per cent was achieved on the 43$^{rd}$ day and on the same day, the inlet toluene concentration was increased from 0.865 to 1.950 $g\ m^{-3}$ but very slight variation in removal efficiency was observed. In Phase II, inlet concentration was nearly kept at the same level as it was in the last stage of Phase I but flow rate was increased and maintained at 0.54 $m^3 h^{-1}$. Due to increase in flow rate, EBRT decreased as compared to Phase I but very slight variation in the removal efficiency was observed. In Phase III, flow rate was increased again and maintained at 0.72 $m^3 h^{-1}$ and inlet concentration was kept at the same level as it was at the end of Phase II. Decrease in the removal efficiency from 99 to 83.6 per cent was observed. No significant variation in removal efficiency was observed with increase in concentration and flow rate in Phases I and II and this showed that bio-

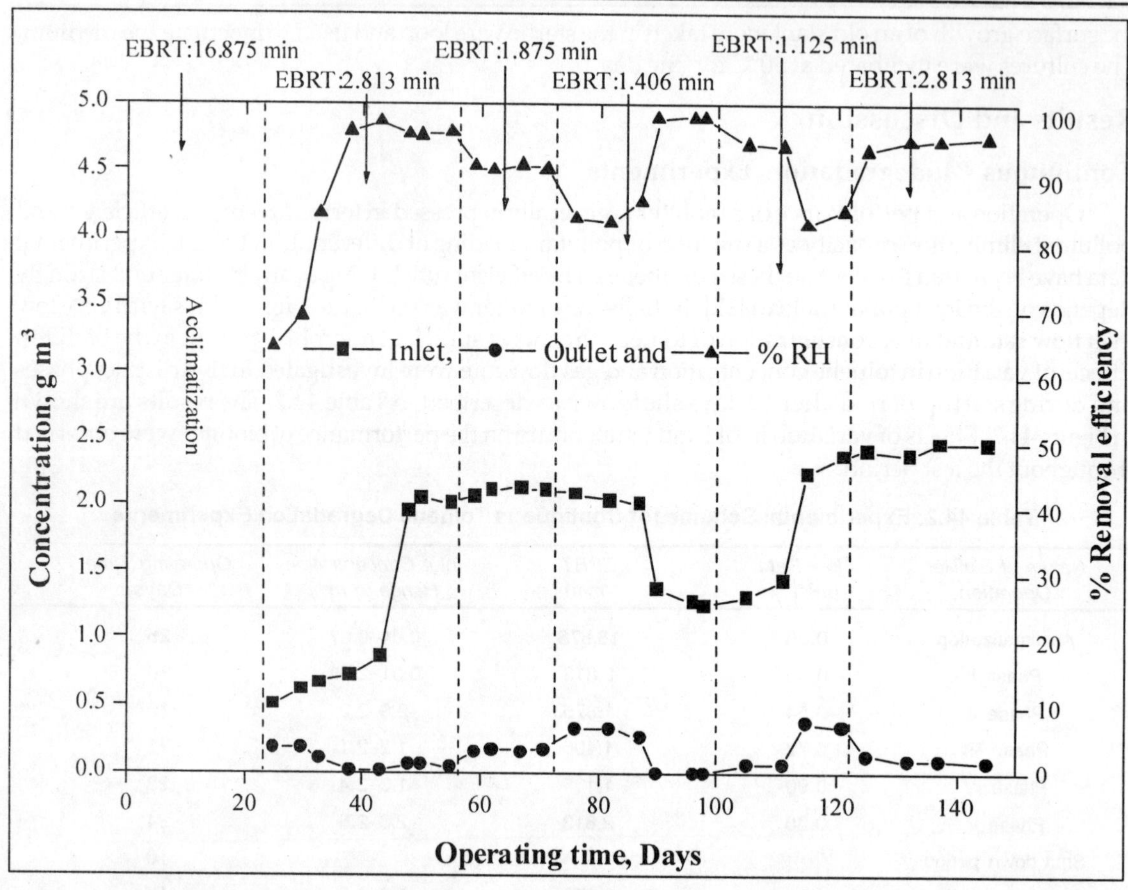


**Figure 44.2: Biofilter Performance with Change in Gas Flow Rate and Inlet Toluene Concentration**

reaction was the controlling mechanism and potential of biofilter was still unutilized. But in Phase III decrease in RE was observed with increase in flow rate, keeping the inlet concentration at the same level as it was at the end of Phase II. This is due to shifting of the mechanism from reaction controlling to diffusion controlling regime. In this phase the potential of biofilter was fully utilized [31]. In Phase IV, flow rate was increased further and maintained at 0.90 $m^3\ h^{-1}$. Inlet concentration was varied between 1.2995 and 2.3676 $g\ m^{-3}$. Removal efficiencies in the range of 83.3 to 95.8 per cent were obtained. In Phases III and IV, biofilter responded very well to changes in inlet concentrations. In Phase V, biofilter was operated again at the same flow rate and at slightly higher inlet concentration as it was in the last stage of Phase I. Biofilter responded very quickly to changes in flow rate and inlet concentrations and achieved nearly the same level of removal efficiency.

Some of the practical problems associated with waste gas treatment are need of intermittent operation and sudden fluctuations in inlet concentration and flow rate. An active biofilter should be able to handle such adverse situations in order to provide maximum removal of the target contaminant. This, however, depends on stability and distribution of the biomass in the biofilter. The transient behaviour of the toluene biofilter was investigated by subjecting it to intermittent operation. The results, shown in Figure 44.3, indicate that the biofilter has been able to maintain its performance even after periodic shutdown operations and sudden fluctuations in the inlet concentration and flow rate. Similar bahaviour was observed by others [10, 19, 23, 25].

## Effect of Inlet Loading Rate on Removal Efficiency

The two most important parameters, which affect the performance of a biofilter, are gas flow rate and inlet pollutant concentration. When these two parameters are combined into a single parameter, it is known as 'pollutant loading'.

An increase in the inlet pollutant concentration enhances the transfer rate of the pollutant to the biofilm. However, higher concentrations of some recalcitrant pollutants may cause inhibition of the metabolic activity of the microbial population [1]. On the other hand, for a specific filter bed volume, the gas flow rate is the key parameter for varying the gas residence time, which should be long enough to permit the transfer of pollutants from the gas phase to the biofilm and their degradation by the immobilized microorganisms.

Figure 44.4 shows the effect of inlet toluene load on the removal efficiency throughout the experimental period. During unsteady state period (initial stage of Phase I), removal efficiency increased

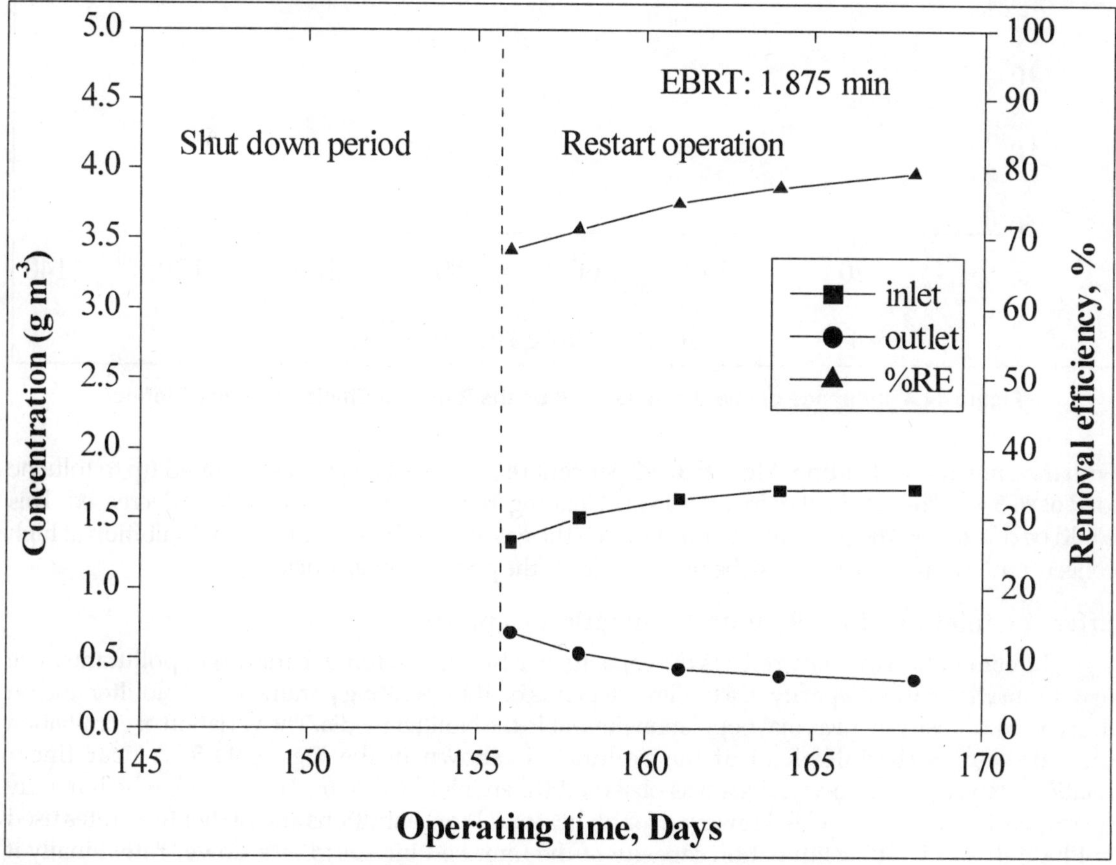


**Figure 44.3: Shutdown and Restart Operation in Biofilter at the Gas Flow Rate of 0.54 $m^3$ $h^{-1}$**

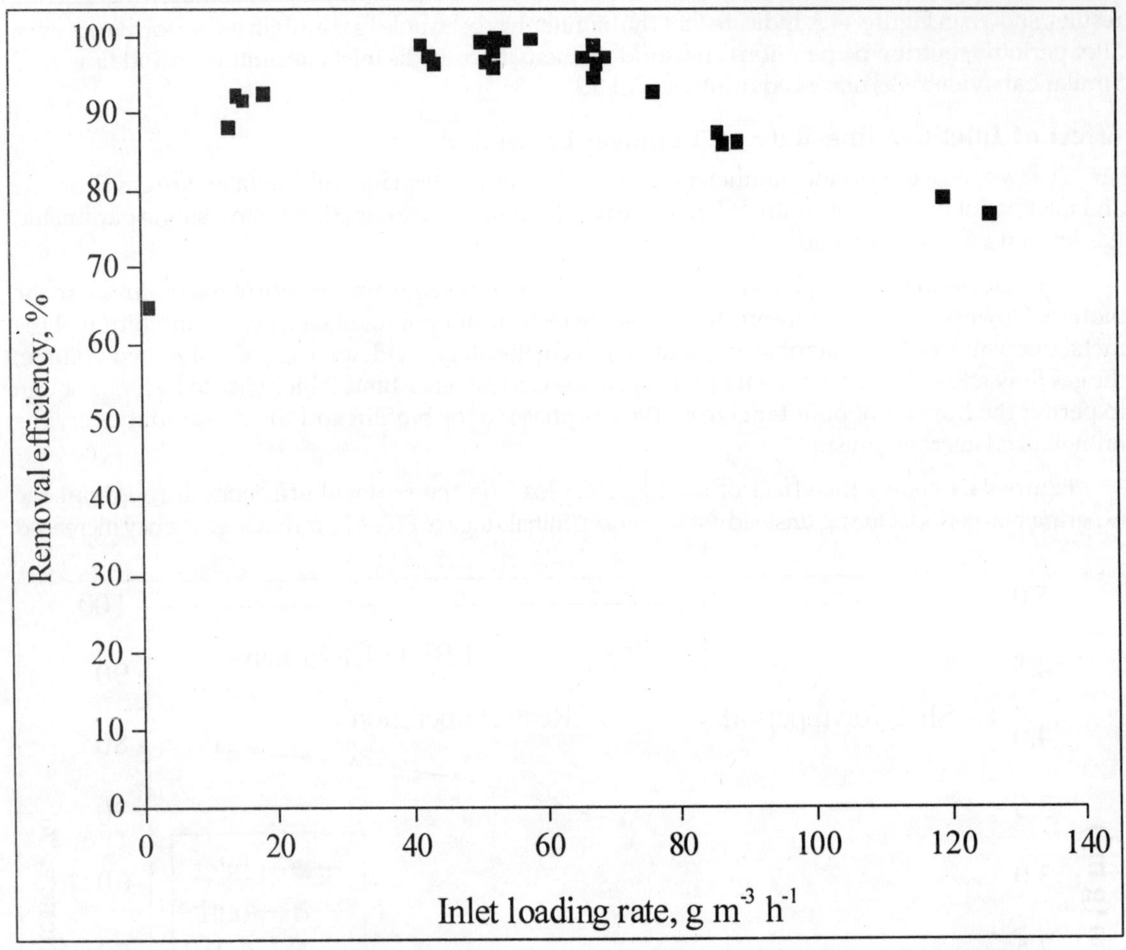


**Figure 44.4: Influence of Inlet Toluene Load on the Removal Efficiency of the Biofilter**

with increase in inlet loading. More than 90 per cent removal efficiency was obtained up to toluene load of 76.3 g $m^{-3}$ $h^{-1}$. On further increase in inlet loading, removal efficiency gradually decreased. This could be due to insufficient contact time between the gas and the biofilm. Substrate inhibition at high concentration of toluene may also be responsible for the poor removal efficiency.

### Effect of Inlet Loading Rate on Elimination Capacity

The elimination capacity reflects the capacity of a biofilter to remove the target pollutants. The maximum elimination capacity of a biofilter depends on the operating parameters of biofilter and on the type and density of microbial population present in the biofilter media. The variation of elimination capacity versus the inlet load of the pollutant is shown in the Figure 44.5. A near linear relation between the two variables was observed till an inlet load of 69.31 g $m^{-3}$ $h^{-1}$, which mainly corresponds to Phases I and II. However, for higher initial concentrations and higher flow rates used in Phase III and IV, the elimination capacity of the filter bed increased at a slower rate. Finally it approached a stage at which it was independent of the inlet load. The maximum elimination capacity

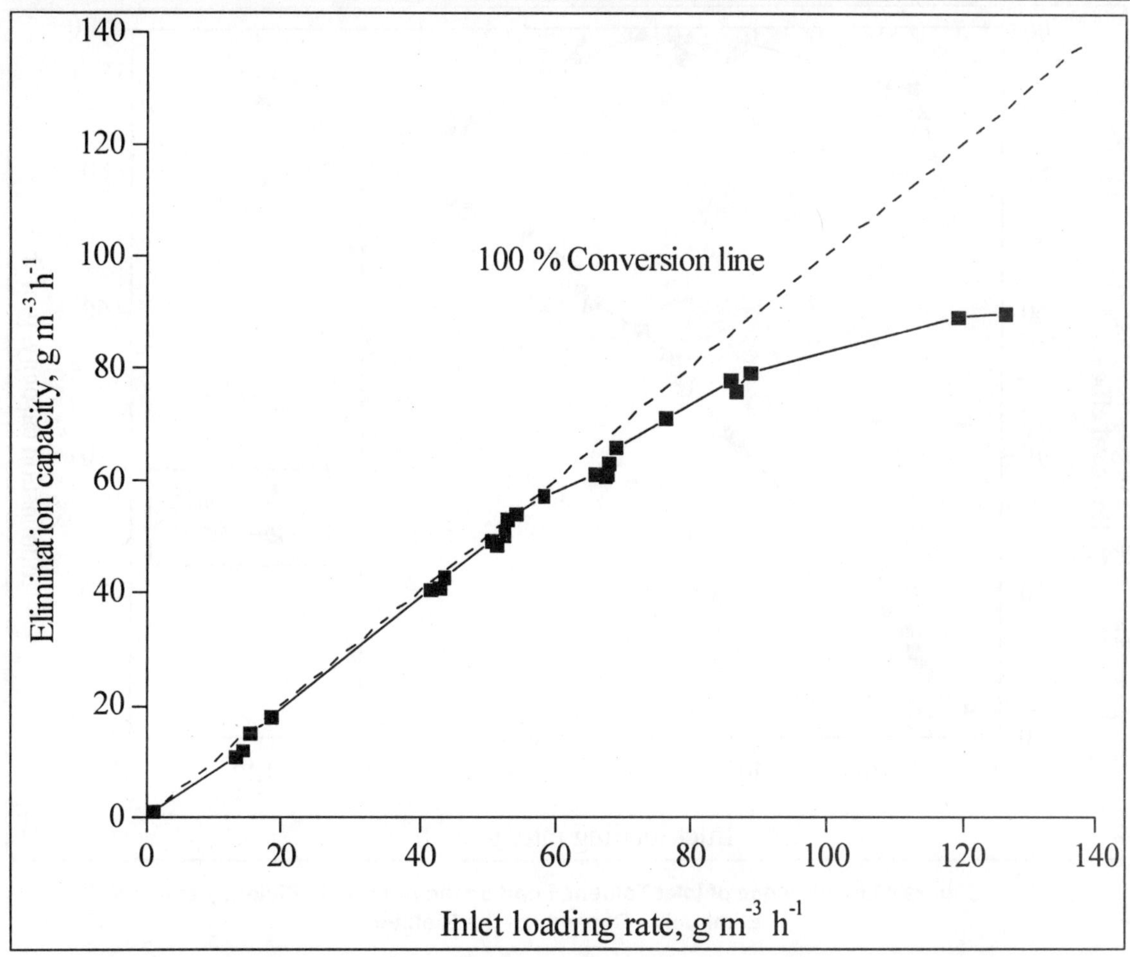


**Figure 44.5: Influence of Inlet Toluene Load on the Elimination Capacity of the Biofilter**

of 90.48 g $m^{-3}$ $h^{-1}$ was observed at an inlet loading of 126.27 g $m^{-3}$ $h^{-1}$. The above behaviour may be explained in the light of the fact that at low loading of pollutant, overall process is controlled by mass transfer and at high loading, bioreaction in the biofilm is the controlling mechanism. In this region biofilm does not become fully active.

Variation of removal efficiency and elimination capacity with respect to inlet load is shown in Figure 44.6. The two performance parameters show opposite trends with respect to inlet loading. Removal efficiency and elimination capacity versus inlet loading plots intersect corresponding to an inlet loading of 104 g $m^{-3}$ $h^{-1}$ and this inlet load may be considered as the optimal operating point for the biofilter.

## Axial Variation of Toluene in Biofilter

The axial variation of toluene concentration in the biofilter was studied and, the normalized toluene concentration profiles at flow rates of 0.36, 0.54, 0.72 and 0.90 $m^3$ $h^{-1}$ were obtained as a function of the biofilter column height. The results for the flow rate of 0.36 $m^3$ $h^{-1}$ are shown in Figure

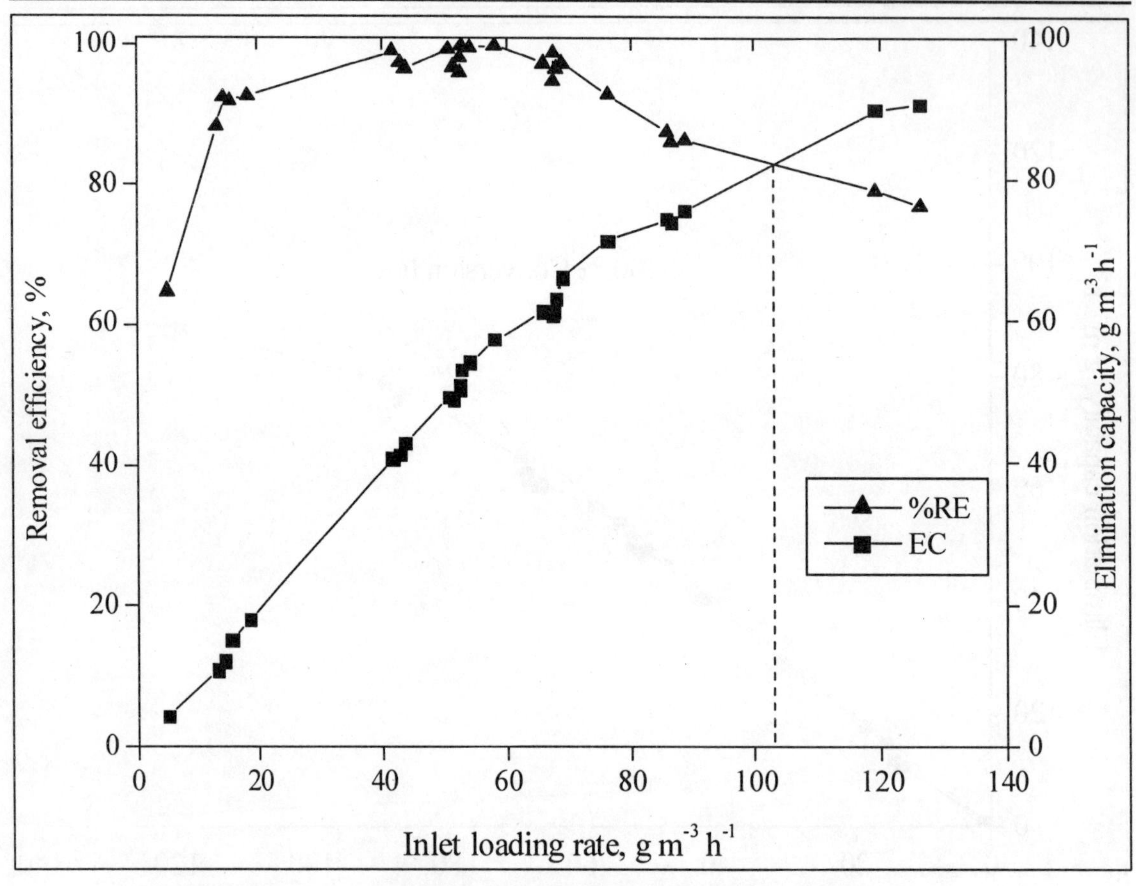


**Figure 44.6: Influence of Inlet Toluene Load on the Removal Efficiency and Elimination Capacity of the Biofilter**

44.7 as a typical example. This figure indicates that the removal is more efficient in the upper part than in the lower part of the biofilter. Nearly 70 per cent of the toluene is removed in the first 0.30 m of the bed height, while the rest of the 0.45 m removed only additional 20-30 per cent. This may be due to higher concentration of toluene, presence of more microbial population and sufficient amount of oxygen for toluene oxidizing microorganisms in the upper section (near the inlet) of the filter bed [Rene *et al.*, 19]. Similar results were observed for other flow rates.

## $CO_2$ Production

Variation of $CO_2$ production with time may provide valuable information about performance of a biofilter. Normally the term 'carbon recovery' is used to assess the amount of carbon utilized by microorganisms and it is defined as

$$\text{Per cent Carbon recovery} = \frac{\text{Mole of } CO_2 \text{ produced}}{\text{Mole of C in pollutant consumed}} \times 100\ (\%)$$

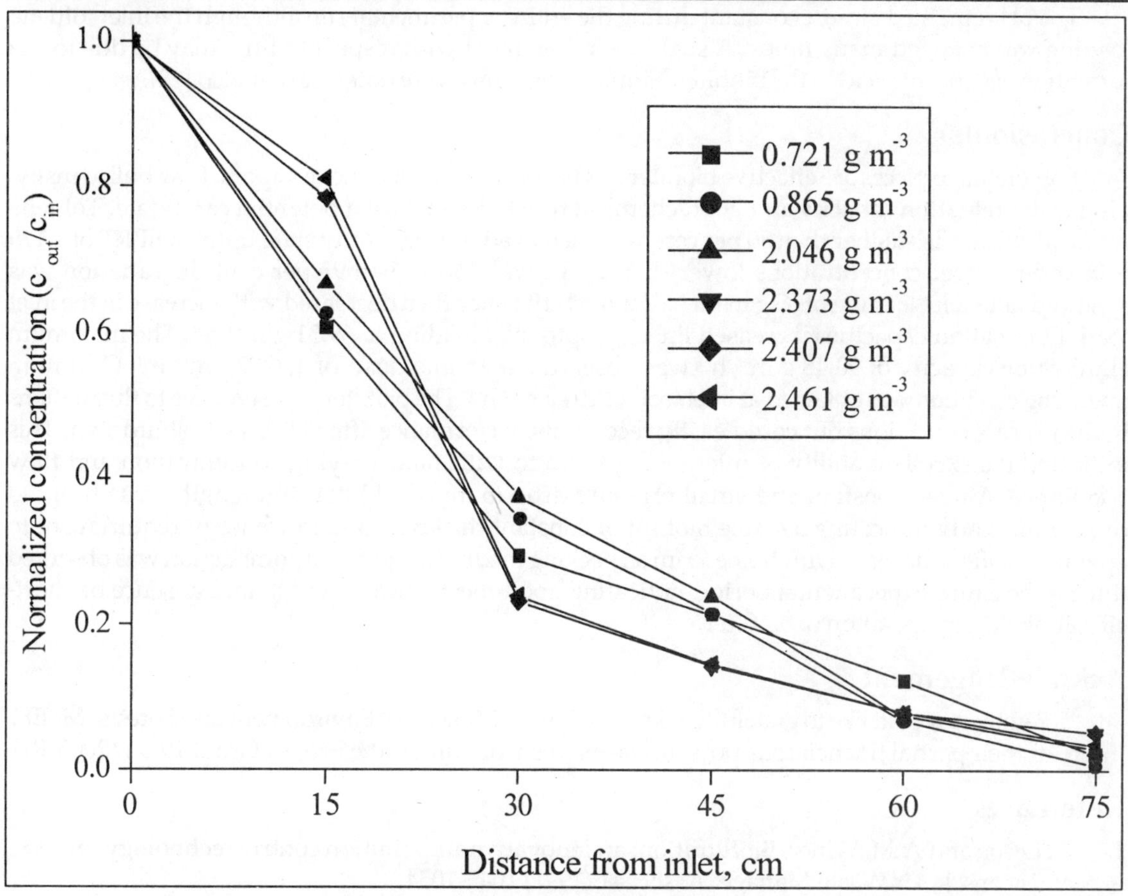


**Figure 44.7: Normalized Toluene Concentration Profiles as a Function of the Biofilter Height at EBRT of 2.813 min**

In the present work $CO_2$ production was monitored by chemical method. More than 90 per cent carbon recovery was observed during entire continuous biodegradation experiment. This is similar to that observed earlier [25].

## Bed Behaviour

The temperature in the biofilter bed is influenced by exothermic biological reactions in the bed as well as temperature of the inlet gas stream. The temperature itself has a significant influence on the performance of the biofilter and microbial activity. Throughout the continuous biodegradation experiment the temperature of biofilter media was measured with the help of digital thermometer and was found to be in the range of 29 to 41°C. The removal efficiency of toluene throughout this range was almost constant and more than 95 per cent. No significant variation in the performance of biofilter was observed with respect to variation in temperature.

As compared to our earlier studies with agro-wastes as biofilter media [25, 26] very little deterioration and compaction in the media was observed in the present study. No significant variation in pressure drop per unit length of the biofilter media was observed during the experiment. At the end of experiment, maximum pressure drop of 5 mm $CCl_4$ per unit length of biofilter was found.

The pH remained almost constant during the entire experimental run although the inlet toluene loading was changed many times. A slight decrease in pH with respect to time may be due to the formation of carbonic acid in the biofilter. Similar variations were noted earlier also [25].

## Conclusion(s)

Polyurethane pieces are effective biofilter media for removal of toluene vapors. Low bulk density, high water retention capacity and good chemical resistance make it a potential candidate. Toluene removal efficiencies higher than 80 per cent were achieved during steady state upto an EBRT of 1.125 min and the inlet concentrations lower than 2.46 g $m^{-3}$. More than 95 per cent degradation was achieved upto inlet loading of 65 g $m^{-3}$ $h^{-1}$. Removal efficiency then decreased with increase in the inlet load. Elimination capacities increased linearly upto inlet loading of 69.31 g $m^{-3}$ $h^{-1}$. The maximum elimination capacity of 90.48 g $m^{-3}$ $h^{-1}$ was observed at an inlet load of 126.27g $m^{-3}$ $h^{-1}$. Optimum operating condition was obtained at inlet load of 104 g $m^{-3}$ $h^{-1}$. The biofilter was sensitive to fluctuations in the process conditions but could easily recover its performance after 11 days of shutdown. This indicated the excellent ability of microbial growth to withstand varying concentration and flow conditions. Almost constant and small pressure drop (5 mm $CCl_4$) per unit length was observed during the study indicating a stable biofilm of constant thickness and low energy requirement to operate a biofilter using polyurethane foam as packing media. Insignificant compaction was observed during the entire experimental period indicating good mechanical strength and absence of short-circuiting during operation of biofilter.

## Acknowledgement

The authors would like to gratefully acknowledge the Ministry of Environment and Forests (MoEF), India for their partial financial supports towards this work under the Project Grant 19-94/2003-RE.

## References

1. G. Leson and A.M. Winer, Biofiltration: an innovative air pollution control technology for VOC emissions, J. Air Waste Manage. Assoc. 41 (1991) 1045-1054.
2. M. Heitz., L.B. Beu, K. Kiared, A. Leroux, R. Brezezinski, G. Viel, Biological purification of exhaust air containing toluene vapour in a filter-bed reactor, The Canadian Journal of Chemical Engineering 25 (1997) 921-929.
3. S. J. Ergas, K. Kinney, M.E. Filler and K.M. Scow, Characterization of a compost biofiltration system degrading dichloromethane, Biotechnol.Bioeng. 44 (1994) 1048-1054.
4. C. Alonso, M.T. Siudan, G.A. Sorial, P. Biswas, P.J. Smith, R.C. Brenner, Gas treatment in tricle-bed biofilter: biomass, how much is enough? Biotechnol. Bioeng. 54 (1997) 583-594.
5. A. Mallakin, O.P. Ward, Degradation of BTEX compounds in liquid media and in peat biofilters, J. Ind. Microbiology 16 (1996) 309-318.
6. J. Huang, K.L. Pinder, Effects of calcium on development of anaerobic acidonenic biofilms, Biotechnol. Bioeng. 45 (1995) 212-218.
7. S.N. Upadhyay, V. Kumar, Biological Control of Air Pollutants, first ed., Concise Encyclopedia of Biresource Technology, The Haworth Reference Press, New York, 2004.
8. M. Murata, M. Tsujikawa, S. Kawanishi, Oxidative DNA damage by minor metabolites of toluene may lead to carcinogenesis and reproductive dysfunction, Biochem Biophys Res Commun 261 (1999) 478-483.

9. C.W. Spicer, S.M. Gordon, M.W. Holdren, T.J. Kelly, R. Mukund, Hazardous air pollutant handbook–measurement, properties and fate in ambient air, Boca Raton. FL: Lewis Publishers, CRC Press LLC, 2002.

10. M. Zilli, E. Palazzi, L. Sene, A. Converti, M.D. Borghi, Toluene and styrene removal from air in biofilters, Process Biochem. 37 (2001) 423–429.

11. D.S. Hodge, V.F. Medina, R.L. Inslander, J.S. Devinny, Treatment of hydrocarbon fuel vapors in biofilters, Envir. Tech. 12 (1991) 655-662.

12. D.S. Hodge, J.S. Devinny, Modelling removal of air contaminants by biofiltration, J. Envir. Engg, 121 (1995) 21-32.

13. N.Y. Fortin, M.A. Deshusses, Treatment of methyl tert-butyl ethers vapors in biotrickling filters: Reactor startup, steady-state performance, and culture characteristic, Environ. Sci. Technol. 33 (1999) 2980-2986.

14. T.S. Webester, H.H.J. Cox, M.A. Deshusses, Resolving operational and performance problems encountered in the use of pilot/full-scale biotrickling filter reactor, Environ.prog. 18 (1999) 162-172.

15. F.J. Weber, M.R.G. Hartman, Z. Lewandowski, Z., W.L Jhnes, Use of activated carbon as a buffer in biofiltration of toluene, Appl. Microbiol. Biotechnol. 43 (1995) 365-369.

16. W. M. Moe, R. L. Irvin, Polyurethane foam medium for biofiltration, I: Characterization, Journal of Environ. Engg. 8 (2000) 817-825.

17. W. M. Moe, R. L. Irvin, Polyurethane foam medium for biofiltration. II: Operation and Performance, Journal of Environ. Engg. 8 (2000) 826-832.

18. W. M. Moe, Qi Bing, Performance of a fungal biofilter treating gas-Phase solvent mixtures during intermittent loading, Water Research 38 (2004) 2259-2268.

19. E.R. Rene, D.V.S. Murty, T. Swaminathan, Performance evaluation of a compost biofilter treating toluene vapour, Process Biochemistry 40 (2005) 2771-2779.

20. Y. Liu, X. Quan, Y. Sun, J. Chen, D. Xue, J. S. Chung, Simultaneous removal of ethyl acetate and toluene in air streams using compost-based biofilters, Journal of Hazardous Materials B95 (2002) 199–213.

21. H. H.J. Cox, M. A. Deshusses, Co-treatment of $H_2S$ and toluene in a biotrickling filter, Chemical Engineering Journal 87 (2002) 101–110.

22. H. Elmrini, N. Bredin, A. Shareefdeen, M. Heitz, Biofiltration of xylene emissions: bioreactor response to variations in the pollutant inlet concentration and gas flow rate, Chem Engg. J. 100 (2004) 149–158.

23. H. M. Tang, S.J. Hwang, Transient behaviour of the biofilters for toluene removal, J Air Waste Manage Assoc. 47 (1997) 1142–1151.

24. M.C. Delhomenie, L. Bibeau, J. Gendron, R. Brzezinski, M. Heitz, Degradation of toluene, xylene and trimethylbenzene vapors by biofiltration: a comparison, J Air Waste Manage Assoc. 53 (2003) 217–226.

25. R.S.Singh, B.N.Rai, S.N.Upadhyay, Performance Evaluation of an Agro-Waste based Biofilter Treating Toluene Vapours, Environmental Technology 27 (2006) 349-357.

26. R.S. Singh, S.S. Agnihotri, S.N.Upadhyay, Removal of Toluene Vapour Using Agro-Waste as Biofilter Media, Bioresource Technology (In Press).

27. G.A. Sorial, F.L. Smith, M.T. Suidan, A. Pandit, P. Biswas, R.C. Brenner, Evaluation of Trickle Bed Biofilter Performance For BTEX Removal, Jl. of Environmental Engg. June (1997) 530-537.

28. C. Lu, W. Chu, M.R Lin, Removal of BTEX Vapor from Waste Gases by a Trickle Bed Biofilter. WasteWater Management Association 50 (2000) 411-417.

29. F.L. Smith, G.A. Sorial, P. Biswas, R.C. Berner, Development and Demonstration of an Explicit Lumped-Parameter Biofilter Model and Design Equation Incorporating Monod Kinetics. Air and Waste Management Association 52 (2002) 208-209.

30. D.W. Park, S.S. Kim, S. Haam, I.S. Ahn, W.S. Kim, Biodegradation of toluene by a laboratory scale biofilter inoculated with pseudomonas pudita DK-1, Environmental Technology 23 (2002) 309-318.

31. C. Lu, M.R. Lin, C. Chu, Temperature Effects of Trickle-Bed Biofilter for Treating BTEX Vapors, Environmental Engineering 125 (1999) 775-779.

32. R.J. Abumaizar, K. Walter, H. S. Edward, Biofiltration of BTEX contaminated air streams using compost-activated carbon filter media, Journal of Hazardous Materials 60 (1998) 111–126.

33. M.A. Ranasinghe, P.A. Gostomaski, A Novel Reactor for Exploring the Effect of Water Content on Biofilter Degradation Rates, Environmental Progress 22 (2003) 103-136.

# Chapter 45

# Mixing Height Studies During ISRO-GBP Land Campaign

**Neha Gera[1], N.C. Gupta[1], B.S. Gera[2], V. Mohanan[2], Joginder Malik[2], G. Singh[2], V.K. Ojha[2] and J.L. Pandey[2]**

**[1]University School of Environment Management, G.G.S. Indraprastha University, Kashmere Gate, Delhi**

**[2]National Physical Laboratory, Dr. K.S. Krishnan Marg, New Delhi – 110 012**

## ABSTRACT

Site specific sodar studies were made as part of nation wide multi-technique and multi-institutional field studies to understand the role of aerosols and trace gases for their radiative forcing over the Indian region. The campaign was organized over Delhi-Hyderabad corridor by the Indian Space Research Organization (ISRO). The sodar data of mixing height pertaining to 8 sites with latitudinal variation has been analyzed. The results reveal a general trend of increase in mixing height with decrease in latitude of the observational site. However, meteorological factors are seen to cause significant departures in the observed trend and thus stress upon the need to use site specific mixing height data for air quality studies.

## Introduction

A national level multi-disciplinary and multi-institutional land campaign of field studies was organized, during 1-29 February 1 to 29, 2004, by the Indian Space Research Organization (ISRO) under Geosphere Biosphere Program (GBP). The basic aim of campaign, called ISRO-GBP land campaign, was to understand the role of aerosols and trace gases in radiative forcing over the Indian sub continent. The National Physical Laboratory (NPL), New Delhi being one of the participating institutes measured several air quality and meteorological parameters viz: aerosol optical depth, column ozone and water vapor, UV-B radiation, green house gases (GHGs), aerosols size and mass concentrations. Besides these parameters, on site continuous measurement of atmospheric mixing height and other meteorological parameters such as temperature, pressure, relative humidity, wind speed and direction were also carried out for better understanding of the observations. The mixing height of the atmospheric boundary layer was measured using a mono-static sodar. The observations analyzed pertain to 8 sites along the Delhi-Hyderabad corridor of nearly 3200 km.

The Latitudinal and Longitudinal coverage of the observational sites varied from 17.3° to 28.6° N and 77.2° to 78.2° E (Figure 45.1). Thus the latitudinal variation of about 9 degrees is available to examine latitudinal variation in mixing height. The studies assume significance as such studies have not been reported so far. The preliminary work is likely to attract the interest of researchers for more focused studies in this area. The measurements of the campaign are first of its kind and have provided extensive data base for understanding the role of aerosols and trace gases for their radiative forcing over the Indian region. The data is useful for satellite data validation and climate modeling. The detailed discussion on the data base is beyond the scope of present work. The work reported here presents the analysis of mixing height data in relation to latitudinal variations of observational sites and the role of meteorological parameters in determining mixing height characteristics.

## Instrumentation and Data Structure

Sodar is acoustic remote sensing technique for continuous monitoring the dynamics of ABL thermal structures in real time and space. A mono-static sodar designed, developed and fabricated at

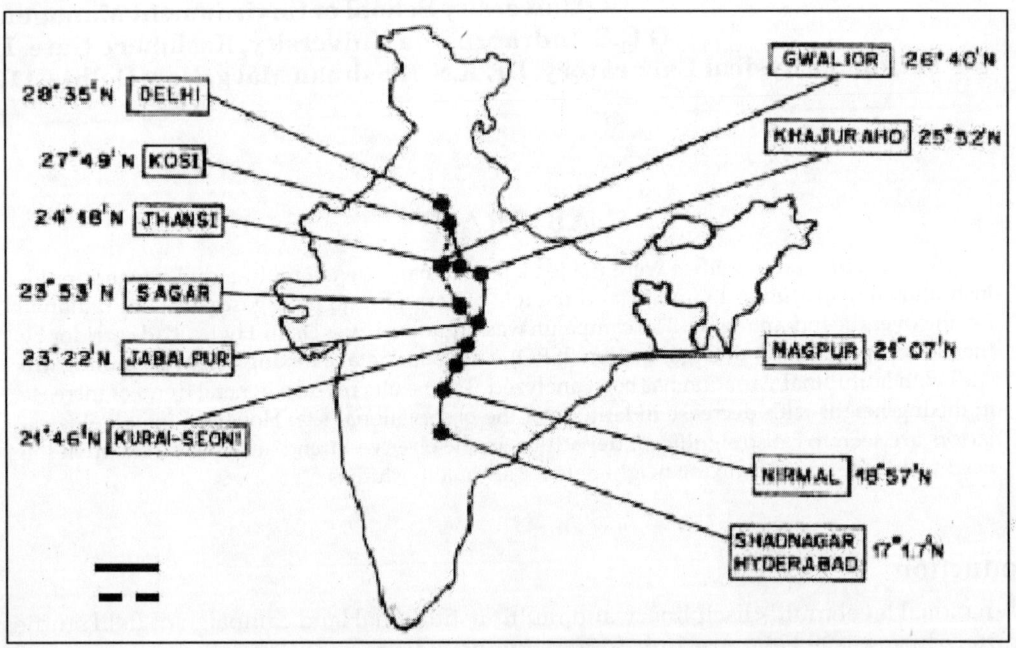


**Figure 45.1: Observational Sites with Latitudes for Delhi-Hyderabad-Delhi Corridor**

**Aerosol optical depth, column ozone and water vapor using Microtops-II sun photometers, UV-B radiation using Biometer and Microtops-II, green house gases (GHGs) *viz.* $CO_2$, $CH_4$, and $N_2O$ by grab sampling and gas chromatography, aerosols size and mass concentrations using QCM, Anderson, PM10-HVS samplers.**

the NPL was operated at all the sites both during the onward and return journey. The sodar system (acoustic power 10W, transmit pulse 100 ms) was capable of monitoring the ABL up to height of 1 km.

High power acoustic burst at 2 KHz is repeatedly transmitted vertically, every 6 sec. The backscattered signal from atmospheric turbulent regions along the propagation path is processed to produce echograms prevailing meteorological phenomenon.

Sodar provides facsimile display of the dynamics of prevailing thermal structures in real space and time. Different sodar echograms (Figure 45.2) portray in real time the relative contribution of different meteorological phenomena playing a part in the background atmosphere for the development of mixing height in real space and time. These phenomenons mainly include thermal plumes or the free thermal convection, spiky layer exhibiting forced convection or the mechanical mixing, rising

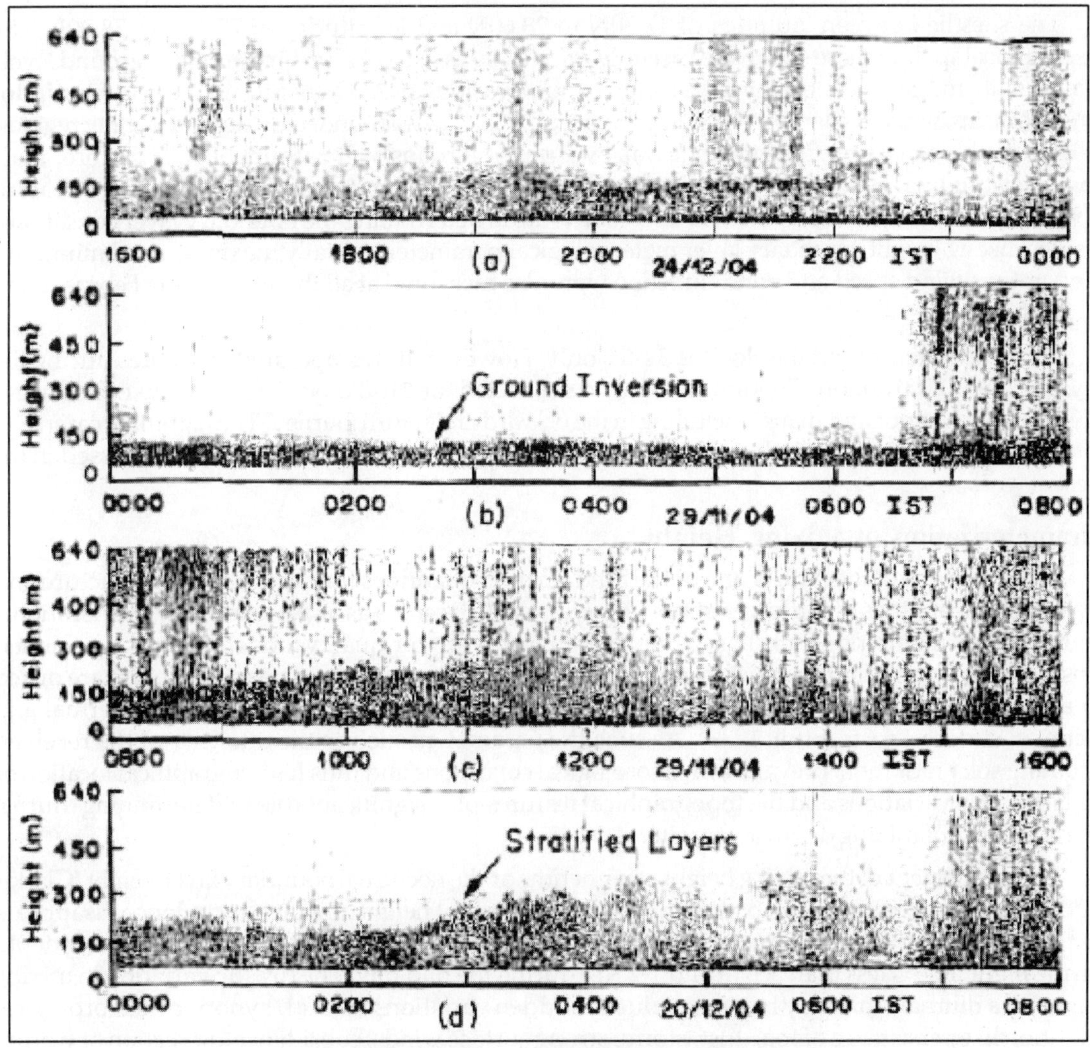


**Figure 45.2: Sodar Records of Various Thermal Structures Under Different Meteorological Conditions of the Atmosphere**

layer or the eroding inversion, elevated layer, stratified layers associated with wind shear. Among these various structures ground based inversion, thermal plumes and the rising layer thermal structures are characteristic features of atmospheric boundary layer in accordance with diurnal cycle of solar heating of the ground.

Multi-disciplinary observations were made at eleven sites along the Delhi-Hyderabad corridor route covering nearly 3200 km distance and spanning through six states of India [*viz.* Delhi, Uttar Pradesh (UP), Rajasthan (Raj.), Madhya Pradesh (M.P.), Maharashtra (Mah.) and Andhra Pradesh (A.P.)]. These sites were Kosi near Mathura (U.P.), Orchha near Jhansi (U.P.) and Tikamgarh (M.P.), Sanuda near Sagar (M.P.), Tikaria near Jabalpur (M.P.), Kurai near Seoni (M.P.), Bussapur near Nirmal (A.P.), Shadnagar near Hyderabad (A.P.), Ramampeth near Hyderabad (A.P.), Khajuraho near Chhatarpur (MP), Dholpur (Raj.) and NPL New Delhi (Figure 45.1).

The sites lie between latitudes of 17.30N to 28.60N and longitudes of 77.20E to 78.20E. The observational spots at these sites were carefully selected so as to have an estimate of background levels of air quality measurements along the route. Among various sites, Delhi falls in the northern Indo-gangatic plains, while Hyderabad is situated in deccan plateau, surrounded by semi-arid, hilly regions. In between these two sites, the other sites fall over mixed terrains over Satpura hills. Therefore, apart from the latitudinal effects related to climatic conditions and the coverage of sunshine, the impact of topographical features at each site will also contribute in determining the boundary layer conditions. Besides mixing height, boundary layer meteorological parameters of daily maximum and minimum temperature, wind speed and wind direction were also measured at all the sites during February 1 to 29, 2004.

Sodar operated round the clock at 3 sites only. However, it was operated at all sites during the daytime (0800-1800 hours). Continuous data was collected for 2 to 3 days at few sites like Shadnagar (Hyderabad), Nagpur and Jhansi, including in the onward and return journey. The nighttime campaign data is not sufficient. As such data of daytime pertaining to unstable ABL has been discussed in the present work.

## Parameterization of Mixing Height

Mixing height of the ABL is one of the fundamental quantities for describing the structure and dispersion capabilities of the lower atmosphere. As such, it serves as a key parameter in operational environmental monitoring for being an input parameter to any air-quality model. The mixing height is considered as the maximum vertical extent up to which pollutants released near the ground are mixed up as a result of turbulent processes generated in the lower atmosphere by way of thermal and frictional surface effects (Stull, 1989). As such the parameters such as the intensity of the received incoming solar radiation, prevailing meteorological conditions and thus his geographical location or the latitudinal variations and the topographical features play significant role in determining diurnal characteristics of mixing depth at any site.

A measurement of the mixing height at night time or the nocturnal boundary layer height (Clarke, 1970) is ideally taken as the top of the stable ABL (inversion) height at which turbulence disappears and above which shear stress and sensible heat flex become negligible. In general, the mixing depth during night time is less than 500 m and 1-2 Km during the mid day. The drastic variation in mixing height on a diurnal scale may be caused due to sudden variations in local/synoptic meteorological condition like persistence of fog, dust, storm, strong surface winds, subsidence, advection of warm/cold air mass etc.

The differences in the ABL mixing height characteristics, on diurnal scale, from day to night appear not only in height but also in dynamic structure. For example, vertical mixing due to turbulence within the nocturnal boundary layer is much lower in the night than that recorded during the daytime, due to the effects of negative buoyancy associated with temperature inversion.

Under clear weather conditions during the day the region located in lower latitudes is relatively more heated as compared to the region located at higher latitudes. Thus unless the local topographical features and the prevailing meteorological factors play a significant role to override the latitudinal effect, the mixing height is expected to decrease with increase in latitude due to differential solar heating of the ground. Under disturbed weather conditions associated with factors such as rain, clouds, dust storm or thunderstorm etc., the latitudinal variations in mixing height may exhibit departure from expected trend.

Parameterization of mixing height is significant in planning air quality management strategies and for applying the similarity theory or the so-called resistance law (Yu, 1978) to the study of atmospheric turbulence in the planetary boundary layer (PBL).

Traditional methods to determine the mixing height are based on the use of characteristics of profiles of wind speed, potential temperature, standard deviation of vertical wind speed, sodar echo intensity, radio sonde, tethered balloon, aircraft measurements etc. Sodar has been increasingly used for mixed layer height (MLH) estimation, and it has shown to provide reliable results under different meteorological and site conditions (Beyrich, 1995). Also the studies based on Inter-comparison of sodar data and the vertical ozone profiles (Beyrich *et al.*, 1995) do support the consideration that the mixing height estimation from sodar measurements represents a height scale characterizing the mixing potential of the boundary layer (Hanna *et al.*, 1982).

Estimation of mixing height under different conditions of sodar structures has been discussed in detail elsewhere (Singal *et al.*, 1994). However, for the sake of continuity of relevance to the present work, this may be mentioned that under inversion or the stable atmospheric conditions, top of the ground-based inversion (Figure 45.2b)/rising layer (Figure 45.2c) gives a fairly good estimate of the mixing height. However, under unstable convective conditions of the atmosphere, represented by thermal plumes on sodar records (Figure 45.2c), the mixing height can be estimated using an empirical relationship (Singal *et al.*, 1997), connecting sodar observed height of thermal plumes and the actual mixing height coming from model (Holzworth, 1992). The empirical relation (Singal *et al.*, 1994) is given below.

Mixing height, $Y = 4.35 * Ph + 95$

*where,* Ph is the sodar observed height of thermal plumes

In this context, it may be mentioned that sodar echograms are manifestations of the resultant turbulent conditions caused due to the contributing local/synoptic meteorological factors *viz.* wind vector profile, vertical transfer of heat, momentum and moisture flux, albedo, advection of cold/warm fronts, land-sea breeze circulation, subsidence etc. As such the influence of topographical factors and the prevailing meteorological conditions are embedded in the site specific sodar structural details. Thus, for a given sodar system, the empirical relationship developed by Singal *et al.* (1994), based sodar derived echo intensity profile and conventional techniques (Holzworth, 1992) can be assumed to be fairly good for different sites. In pursuit of the same, the said relationship has been re–verified for hilly terrains (Singal and Kumar, 1998) as well. In view of the said considerations, we have used the above relationship for estimation of mixing height under convectively unstable atmospheric boundary layer conditions at different observational sites in the present work.

## Results and Discussion

Hourly averaged sodar data of thermal plumes height has been used to compute hourly averaged mixing height of unstable ABL, using the empirical relationship mentioned above. The data is analyzed to examine the trends in diurnal variation of mixing height in relation to latitudinal variations of different sites (Figure 45.3). It is seen that in general mixing height increases with decrease in the latitude of observational site. This is in accordance with the expectations, as the regions at lower latitudes will receive more solar energy due to higher solar zenith angle at these latitudes as compared with higher latitudes.

In the forenoon hours, Kosi site near Mathura (U.P.) with latitude 27° 49′ N has shown minimum mixing height while Bussapur in Nirmal (A.P.) with latitude 18° 57′ N has shown maximum mixing height. In between these latitudes lies Kurai in Seoni (M.P.) with latitude 21° 46′ N. Here the mixing height is also seen to lie in between the said curves (Figure 45.3) for minimum and maximum mixing height. However, similar examination of mixing height around noon hours, when the sun is vertical, surface heating and winds are strong, shows some interesting features. It is seen that sites lying in between 21° and 23° N latitudes (Sagar with 23° 53′ N and Nagpur with 21° 07′ N of latitudes respectively), which were observing mixing heights in between maximum and minimum in the forenoon suddenly start observing highest values (Figure 45.3). This is somewhat in contradiction to the concept of expecting a higher mixing height at comparatively lower latitudes.

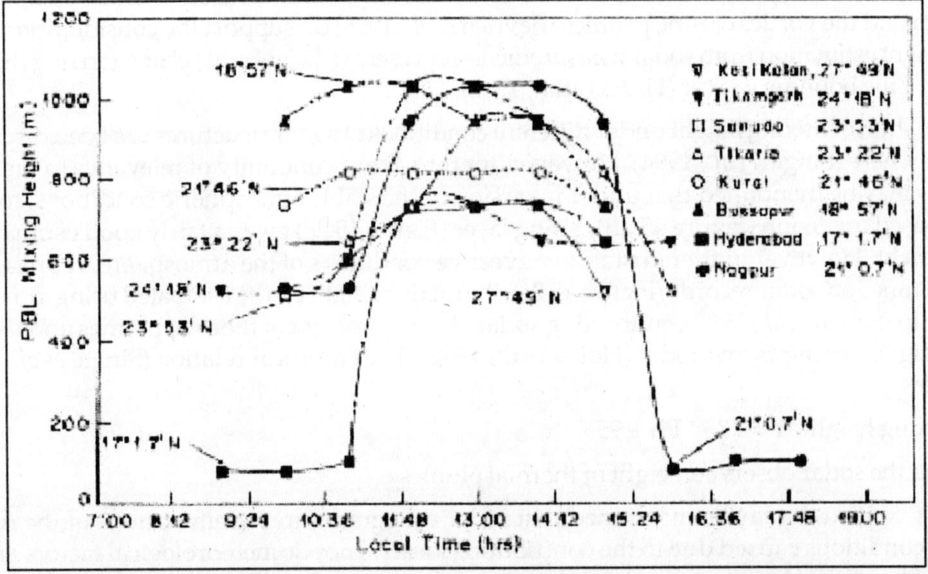


**Figure 45.3: Latitudinal and Diurnal Variation of Mixing Height Over Delhi-Hyderabad Corridor**

In view of this observation synoptic wind flow data from the India Meteorological Department (IMD) was carefully examined. The examination of monthly averaged synoptic wind flow pattern over the Indian subcontinent has revealed that the convergence of winds from the northern and south-eastern sectors taking place at around the trans-boundary region located in between latitudes 21° and 23° N during the month of February (Figure 45.4). The strong wind convergence is considered to impart additional substantial vertical lift to the thermally buoyant air parcels; thereby the mixing height is seen to increase sharply around this zone of wind convergence. This observation clearly

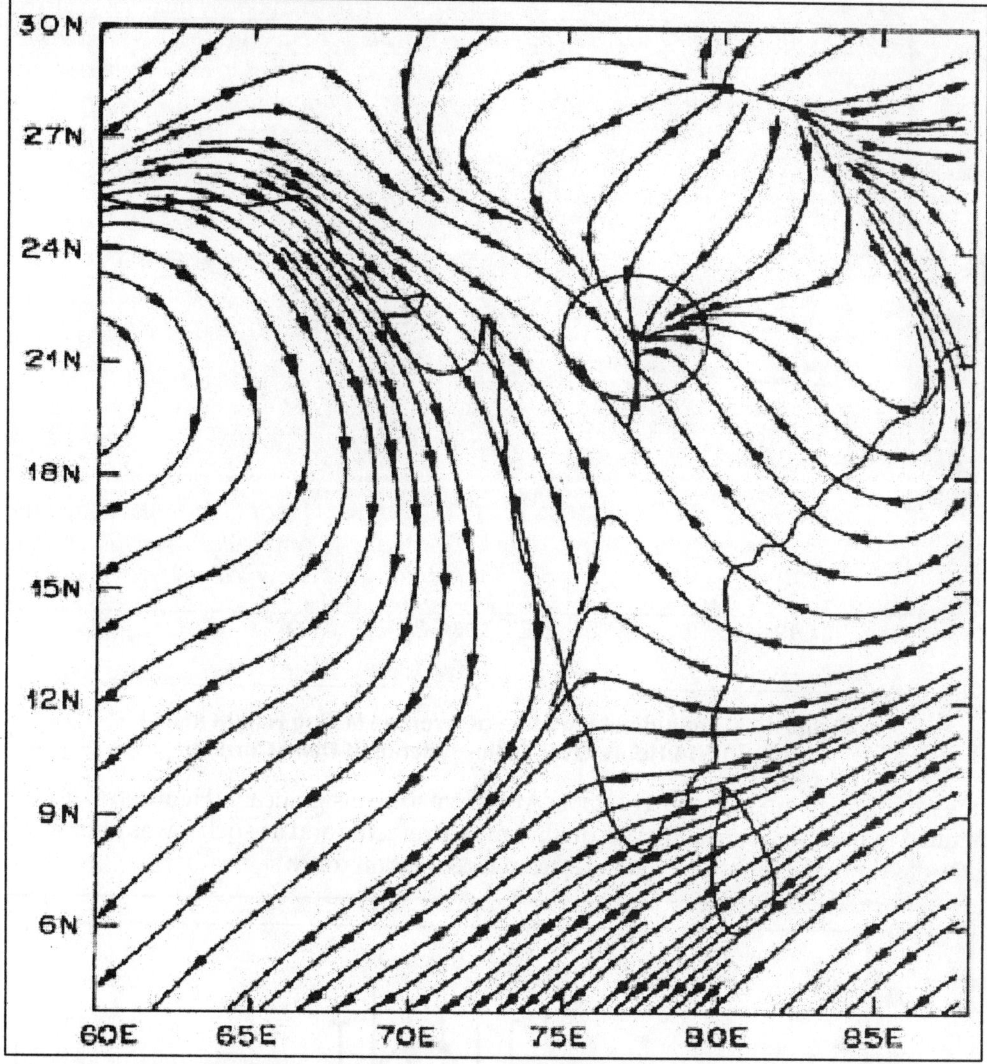


**Figure 45.4: Normal Synoptic Wind Flow Over Indian Sub-continent during February**

shows that besides the latitudinal effects related to solar heating of the ground, meteorological effects on local and synoptic scale play a significant role in determining the mixing depth at any place. The effect is more clearly visible in the plot (Figure 45.5) showing latitudinal variation of averaged (maximum) mixing height during 1200-1400 hours. The sudden increase in mixing height at around 23° N of latitude is associated with the meteorological effects of large-scale synoptic wind convergence.

Further, Figure 45.6 shows the averaged diurnal variation of mixing height integrated over the entire campaign path considering all the sites (Figure 45.1) along the Delhi-Hyderabad corridor. Figure 45.6 shows a commonly observed and expected trend of observing maximum mixing height around noon hours. However, there is a large variability in mixing height during the day over the Delhi-Hyderabad route under study. The large variability is considered to be due to the spatial distribution of topographical and climatologically different characteristics of observational sites. The

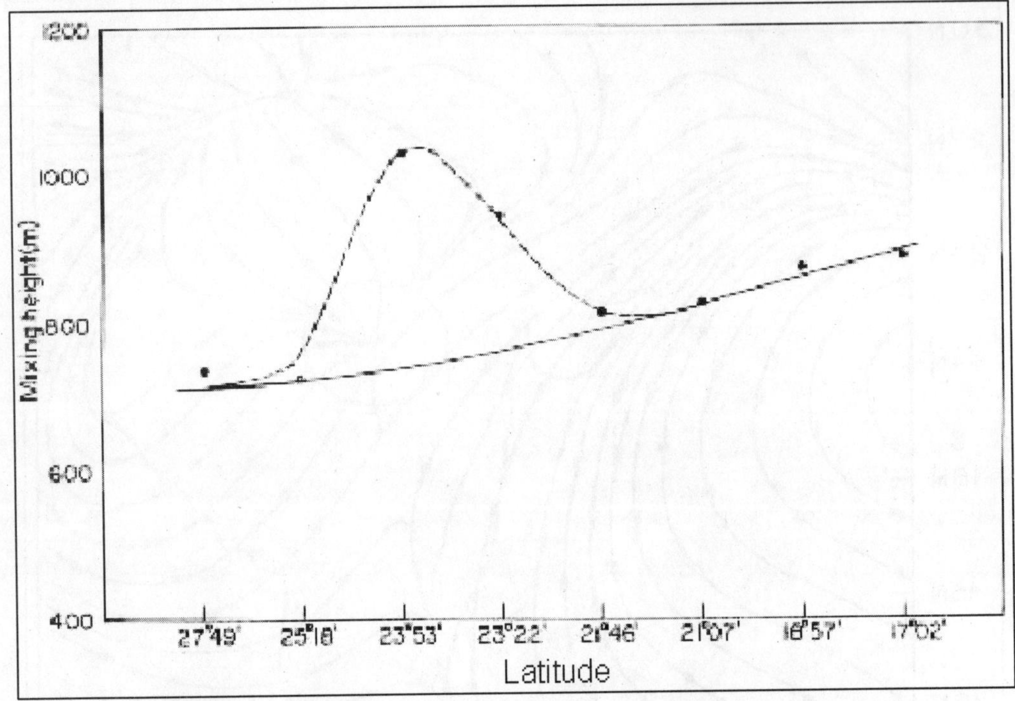


**Figure 45.5: Latitudinal Variation of Average Mixing Height (max.) during 12-14 Hours Over Delhi-Hyderabad-Delhi Corridor**

nighttime variations in mixing height of stable ABL seem to be insignificant (Figure 45.6). In fact sodar was operated in parallel as an aid for air quality measurements and as such it was operated mostly during the daytime. The nighttime data was from three stations only.

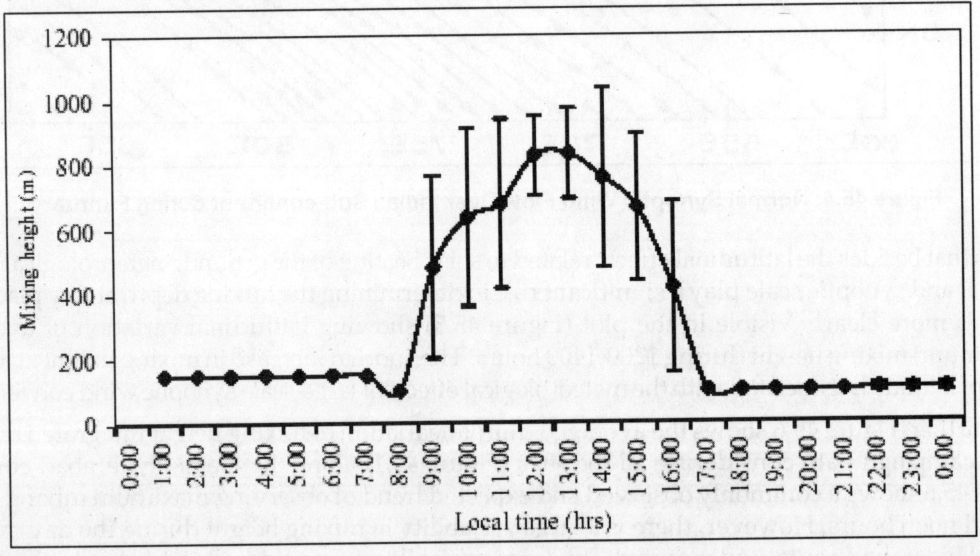


**Figure 45.6: Diurnal Variation of Averaged Mixing-Height Over Delhi-Hyderabad-Delhi Corridor**

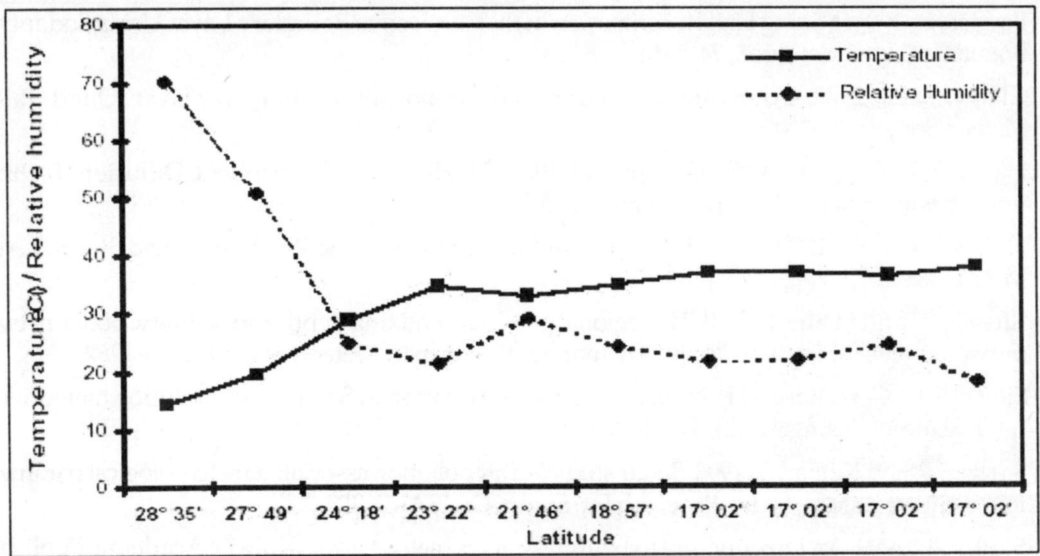


**Figure 45.7: Latitudinal Variation of Averaged Surface Temperature and Humidity (1200-1400 hours) During the Observational Period**

Figure 45.7 shows the variation in maximum of surface temperature and humidity observed during 1200 to 1400 hours as a function of latitudinal variations of observational sites. The data pertains to the actual site observations of temperature and humidity. It is seen that near surface temperature increases with decrease in latitude of the site. This further strengthens our consideration for the convergence of synoptic winds playing a part in the observed sudden increase in the mixing height around the convergence zone.

## Conclusion(s)

Studies of mixing height and meteorological parameters of the ABL boundary layer in relation to latitudinal variations over Delhi-Hyderabad corridor has revealed in general a decrease of mixing height with increase of latitude of observational site in accordance with the differential heating of the ground. However, besides latitudinal variations in mixing height, synoptic scale meteorological factors, like convergence of synoptic surface winds flow can play a significant role in determining the variational trend and override the latitudinal effect in the mixing height variation trend at place. The results stress upon the need to use site specific data for meaningful air quality studies and dispersion modeling in EIA.

## Acknowledgements

The authors are thankful to Dr. Vikram Kumar, Director NPL for rendering his support and encouragements for participating in the ISRO-GBP land campaign. Special thanks are due to Director, ISRO for funding the field studies. Thanks are also due to Dr. R. Sridharan, Dr. C.B.S. Dutt and Sr. Shyam Lal for the organizational support.

## References

1. Beran, D.W. and Hall, F. 1973. Remote sensing applications in the air pollution meteorology, Preprints 2nd Joint Conf. Sensing of Environmental Pollutants, Washington, DC, Instrum. Soc. Amer., 231-246.

2. Beyrich, F., 1995. Mixing Height Estimation in the Convective Boundary Layer Using Sodar Data. Boundary Layer Meteorol., 74, 1-18.

3. Clarke, R. H., 1970. Observational studies in the atmospheric boundary layer. Quart. J. Roy. Meteor. Soc., 96, 91-114.

4. Hanna. S.R., Briggs, G.A., Hosker, Jr. R.P., 1982. Handbook on Atmospheric Diffusion. Technical Information Center, U.S. Department of Energy.

5. Holzworth, G.C., 1992. A study of air pollution potential in the Western United States. J.Appl. Met., 1, 366-382.

6. Russell, P.B. and Uthe, E.E., 1974. Regional patterns of mixing depth and stability: Sodar network measurements for input to air quality models. Bull. Amer. Meteor. Soc.,54, 1275-1287.

7. Singal, S. P., Gera, B.S and Pahwa, D.R., 1994. Application of Sodar to air pollution meteorology. Int. J. Remote sensing., 15(2), 427-441.

8. Singal, S.P. and Kumar, S., 1998. Sodar studies of air pollution associated meteorological parameters in the hills of Assam, India. Proc. 9th Int. Symp. On ISARS, 302-305.

9. Stull, R.B., 1989. An introduction to boundary layer meteorology. Kulwer Academic Publishers, Netherland.

10. Wycoff, R.J., Beran, D.W. and Hall, F., 1973. A comparison of the low level radiosonde and the acoustic echo sounder for monitoring atmospheric stability. J. Appl. Meteor., 12, 1196-1204.

11. Yu, T.W., 1978. Determining height of the nocturnal boundary layer. J. Appl. Meteor., 17(1), 28-33.

# Chapter 46

# Impact of SI Engine Exhaust Emission on Environment

**T.P. Ashok Babu[1] and P.A. Hubballi[2]**
**[1]Professor and Head, [2]Research Scholar,**
**Department of Mechanical Engineering,**
**National Institute of Technology Karnataka,**
**Srinivasnagar – 575 025, D.K. Dist., Karnataka**

## ABSTRACT

Vehicles use SI or CI engines. SI engines are used in aviation vehicles, two wheelers, cars, vans etc. SI engines use gasoline as the main fuel. SI engine emissions are mainly carbon monoxide, carbon di oxide, nitrogen oxides and un-burnt hydrocarbons. Ozone layer depletion is also caused by Nitrous oxides. Motor vehicles emission accounts for 72 per cent of nitrogen oxides. Carbon di oxide is the element causing the global warming. For the past 50 years the carbon dioxide concentration in ppm is increased from 310 to 385 in city areas where vehicles density is high. About twenty percent of carbon dioxide emitted in the atmosphere comes from burning of gasoline in the engines of the vehicles. This is true for most of the developed countries. Vehicles with poor gas mileage contribute the most to global warming. For example, according to the E.P.A's 2000 Fuel Economy Guide, a new Dodge Durango sports utility vehicle (with a 5.9 liter engine) that gets 12 miles per gallon in the city will emit an estimated 800 pounds of carbon dioxide over a distance of 500 city miles. In other words for each gallon of gas a vehicle consumes, 19.6 pounds of carbon dioxide are emitted into the air. A new Honda Insight that gets 61 miles to the gallon will only emit about 161 pounds of carbon dioxide over the same distance of 500 city miles. Sports utility vehicles were built for rough terrain, off road driving in mountains and deserts. When they are used for city driving, they are so much overkill to the environment.

* Corresponding Author: E-mail: tpashok@rediffmail.com

Another burning problem is the exhaust of fossil fuels. Hence it is the need of the hour to find an alternate fuel which is eco-friendly and also renewable source of fuel for engines.

In this work an attempt is made to find an alternative renewable fuel and also to reduce the exhaust emissions. The experimentation is carried out on a four cylindered four stroke S. I. Engine. The engine is made to run in these experiments with initially 100 per cent gasoline and then with DNS100. In all these experiments the operating throttle position is fixed at 50 per cent. Heating values of the fuel is tested by standard methods. The results showed that the heating value of fuel DNS100 is 30.93 MJ/Kg.

The engine performance and pollutant emissions were investigated. Experimental results indicated that the Torque, BP, IP, IMEP and BMEP decrease considerably by 23 per cent, 23.77 per cent, 22.56 per cent, 21.67 per cent and 23 per cent respectively. No major change is found in Mechanical Efficiency, where as the Indicated thermal efficiency decreases by 14.31 per cent, Brake thermal efficiency decreases by 12.36 per cent and Volumetric efficiency decreases by 7.32 per cent. Obviously because of the leaning effect, the Specific fuel consumption increases by 65.34 per cent.

It is found that there is decrease in harmful exhaust gas air pollutants. CO is decreased by 51.41 per cent, $NO_x$ is reduced by 90.75 per cent, $CO_2$ is reduced by 11.36 per cent and HC by 5.57 per cent.

From the study it is concluded that using DNS100 as SI engine fuel, engine not only reduces the green house gases but also the other harmful pollutants with compromise in engine's performance.

*Keywords: P100, DNS100, S.I., Denatured spirit, Ethanol.*

## Introduction

The increasing industrialization and motorization of world led to a steep rise in demand for petroleum products. Petroleum based fuels are stored fuels in the earths crust with limited reserves, these stored fuels are irreplaceable. With the present known reserves and growing rate of consumption, it is feared that they are not going to last long. The uneven distribution of petroleum reserves over the globe has given rise to frequent disruptions and uncertainties in their supply as well as price.

Many countries today solely dependent on imports to meet their fuel oil requirements and many more will be added in future as their limited reserves of petroleum deposits get exhausted. The situation is grave in developing countries like India, which imports 70 per cent of required fuel, spending 30 per cent of its total foreign exchange earnings on oil imports. This situation has created a problem to increase prices of these fuels by more than two-folds during last 10 years. As time passes, this trend will be aggravated further causing greater scarcity and hardships.

Apart from problem of fast vanishing reserves and irreplaceable nature of petroleum fuels, another important aspect of their use is the extent and nature of environmental pollution caused by their combustion in vehicular engines. Petroleum fuelled vehicles discharge significant amount of pollutants like CO, HC, NOx, $CO_2$, Soot, particulates and aldehydes.

The ill effects of automotive pollution on the environment are severe such as Acid Rain, Acidification lakes and water ways, Reduction or disappearance of trout and salmon, Reduction or disappearance of amphibians, Modification of phytoplankton and Zooplankton, Growth of peat moss prevalent in small lakes, Acidification of water leading to dissolution of heavy metals causing Hg, Cd, Pb and Cu to pass in to the food chain, Death of forests, Acidification of soil and Aging of buildings and monuments etc.

In view of these problems of fast dwindling reserves of irreplaceable petroleum fuels and hazards of environmental pollution caused by their combustion led to an intensive search for alternative fuels in transport sector.

After extensive research over the years, it has been realized that future energy needs of transport sector may not be fulfilled by one particular alternate fuel. The ultimate choice of alternatives in a particular region depends on availability and economics, apart from its non polluting characteristics. In this connection the fuel of our interest is alcohol.

## Literature Review

Ethanol-gasoline blends have been tried as the automotive fuels since many years. In this section, following review of papers gives us more details about use of Ethanol-gasoline blends as motor fuel. The review also gives us details about engine performance and exhaust emissions of various experiments with Ethanol-gasoline blend as automotive fuel.

M. Al-Hasan [1] conducted experiments on a four stroke four cylindered SI engine. The fuels used to investigate the performance of the engine are the Gasoline ethanol-blends in different percentages. The variable engine speed is between1000 and 4000 rpm. The experiments were conducted at three fourth throttle opening position. The range of ethanol percentage added to the gasoline is from 5 per cent to the maximum of 30 per cent. In his findings. He mentions that the performance of the engine improves with Gasoline ethanol blends.

Chong-Lin Song *et al.* [2] have investigated the effect of Ethanol and MTBE additives in to the gasoline in different proportions on exhaust emissions and catalytic conversion efficiencies. The measurements were carried out for both the regulated and unregulated exhaust emissions. The results showed that ethanol addition could reduce the regulated emissions than MTBE. The Benzene emission reduction with ethanol addition was much lesser as compared to the addition of MTBE. The formaldehyde emission concentration was in the order MTBE > Base Fuel > Ethanol. It has also been seen that both Ethanol and MTBE have different effect on catalytic conversion efficiencies.

Fikret Yüksel *et al.* [3] have developed a modified carburetor of a four cylindered four stroke SI engine to investigate the effect of 60 per cent ethanol in gasoline. The experiments were conducted for 25 per cent, 50 per cent, 75 per cent and 100 per cent throttle positions. In his findings it is mentioned that, torque and fuel consumption were increased slightly. CO and HC emissions decrease dramatically and increase in $CO_2$ emission.

Hakan Bayraktar [4] has conducted both experimental investigations and model analyses on SI engine using Gasoline ethanol blends. The experiments were carried out with the blends containing 1.5, 3, 4.5, 6, 7.5, 9, 10.5 and 12 vol per cent ethanol and numerical applications have been performed to the extent of 21 vol per cent ethanol. In experimentation engine was operated with each blend at 1500 rpm for two compression ratios of 7.5 and 8.25 at 100 per cent throttle position. In his findings it is shown that performance parameters such as power and efficiency increase with increasing ethanol amount in the blended fuel. While the emissions are considered CO concentrations reduce to the maximum extent. However NOx concentrations are increased.

Hüseyin Serdar Yücesu *et al.* [5] have conducted experiments on a single cylinder variable compression ratio SI engine to study the performance and exhaust emissions using ethanol–gasoline blends. The fuel blends used here are E0, E10, E20, E40 and E60. The experiments were conducted at six compression ratios from 8:1 to 13:1 for each blend. It is found in this experiment that there was increase in the torque at higher compression ratios. The fuels containing high ratios of ethanol; E40 and E60 had important effect on the reduction of emissions. The better decrease was obtained with HC compared with CO.

M. Maheshwari *et al.* [6] have studied few projects sponsored by Ministry of Petroleum and Natural gas. Mainly the study was to use 10 per cent ethanol in gasoline and examine the startability, drivability and mass emission for regulated pollutants without modifying any of the existing components of the engines of the test vehicles. The results show that the addition of 5 per cent to 10 per cent ethanol to gasoline increases the octane number by 1.3 to 3.4 units. Hence the addition of Benzene and MTBE in refineries can replace with ethanol. In startability tests on two wheelers and four wheelers the operations were satisfactory. In mass emission studies two wheelers show slight increase in HC emissions with 5 per cent to 10 per cent addition of Ethanol. But incase of two wheelers fitted with catalytic converters both HC and CO were reduced. In case of four wheelers, reduction of both HC and CO was noticed with marginal increase in NOx.

S.Maji *et al.* [7] have carried out the experiments to evaluate the engine performance and exhaust emissions of CO and HC with Ethanol Gasoline blends in a single cylinder four stroke SI engine. The changing parameters considered in this experiment are engine speed, compression ratio, air fuel ratio, ignition timing and intake air temperature. In his experiments fuel blends E85 and E15 were used as fuels. The findings of this work are, the maximum power with ethanol blends is achieved near stoichiometric ratio. Lower HC emissions with E85. Intake air heating did not improve the performance of the engine with E85.

T.W.Rudolph and J.J.Thomas [8] have conducted experiments on SI engine to evaluate engine performance and emissions of fuel blends 10 per cent methanol, ethanol, MTBE and MTHF. These experiments were conducted at full throttle position at engine speeds 2000, 2400, 2800 and 3200 rpm. Investigations of this work shows that the MTHF-gasoline blend, which has high A/F ratio density products as compared to MeOH-, EtOhH and MTBE-Gasoline blends have the emission and performance profile much similar to gasoline.

Segio M.Correa *et al.* [9] have conducted the experimentations to collect and analyze the data of acetaldehyde and formaldehyde concentrations at the city of Rio de Janeiro, Brazil. The cities of Brazil like Rio de Janeiro are flooded with the vehicles using Gasohol as fuel. The experimental results showed here that the high ambient levels of acetaldehyde and acetaldehyde/formaldehyde ratios measured higher than these in the world. It is also stated here that the acetaldehyde/formaldehyde ratios change depending upon the solar flux, Vehicular emission and temperature.

Wei-Dong Hsieh *et al.* [10] have conducted experiments on a commercial engine GA16DE, of 1600 $cm^3$ multi point injection gasoline engine. The objective of the work was to experimentally investigate engine performance and pollutant emissions using Ethanol Gasoline blended fuels. The experiments were carried out using blend rates of Ethanol with gasoline in the range of 0 per cent, 5 per cent, 10 per cent, 20 per cent, and 30 per cent. It is found in their experiments that the Torque out put values slightly increase with increased blend ratios. Increase in Brake specific fuel consumption with increased blend ratios. And the CO and HC emissions decrease dramatically and $CO_2$ emissions increased. About NOx it is indicated that, the concentrations of these emissions are dependent on engine operating conditions than the ethanol content in the fuel.

## Experimentation

Experimental set up consists of a four cylindered four stroke carburetor SI engine coupled with an eddy current dynamometer. The specifications of engine and dynamometer are given in Tables 46.1 and 46.2. The tests were performed in the speed range of 2500 to 4500 rpm. Totally ten sets of averaged readings were taken. The 50 per cent throttle position is maintained throughout the experimentation. The test fuels used were Gasoline and DNS100. The experiments were conducted at five different speeds between the mentioned range for each fuel. In order to compensate the leaning effect of the used fuel DNS100, the necessary changes to the engine fuel supply system and the ignition systems were made. Before taking any reading with a particular fuel, engine is made to run for at least five minutes to attain the steady state condition. Table 46.3 gives properties of Gasoline and Ethanol.

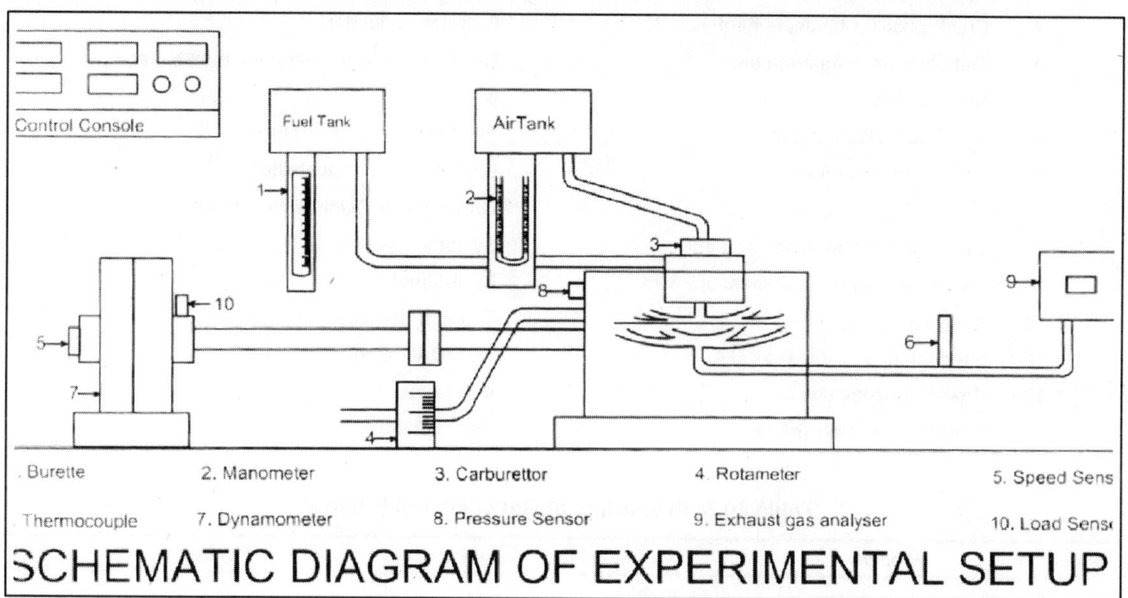


SCHEMATIC DIAGRAM OF EXPERIMENTAL SETUP

**Table 46.1: Test Engine Specifications**

| | |
|---|---|
| Make | Premier |
| BHP | 39 BHP at 5000 rpm |
| Bore | 68.00 mm |
| Stroke | 75.00 mm |
| Cubic capacity | 1.089 liters |
| Compression ratio | 7.8:1 |
| Engine cooling | Water cooled |

## Results and Discussion

All the Engine performance parameters like Mechanical Efficiency, Indicated thermal efficiency, Brake thermal efficiency, Volumetric efficiency, Indicated power, Brake power, Torque, Indicated mean effective pressure, Brake mean effective pressure, Specific fuel consumption and exhaust gas emissions namely CO, $CO_2$, $O_2$, NOx and HC for each tested fuel for different speeds are plotted individually against speed. The values are compared with base fuel gasoline (P100) taking the mean value of each set of values for all the speeds for DNS100.

**Table 46.2. Specifications of the Test Rig**

| | | |
|---|---|---|
| 1. | Dynamometer | |
| | Type | Eddy current. |
| | Model | E-50 |
| | Make | ATE |
| 2. | Load measurement | By load cell |
| 3. | Cylinder pressure measurement | By piezo sensor |
| 4. | Crank position measurement | By rotary encoder |
| 5. | Temperature measurement | By RTDs, with temperature transmitter |
| | No. of points | 6 |
| 7. | Fuel flow measurement | Electronic with transmitter |
| 8. | Air flow measurement | Electronic with transmitter |
| | | Orifice meter Diameter = 30 mm |
| 9. | Jacket water flow measurement | By rotameter |
| 10. | Calorimeter water flow measurement | By rotameter |
| 11. | Analysis | Through computer software |
| 12. | Interfacing | ADC/DAC Card |
| 13. | Power consumption | 4 kW max. |
| 14. | Overall dimension (mm.) | 3800 x 2500 x 1500 Ht. |

**Table 46.3: Properties of Gasoline and Ethanol**

| *Property* | *Gasoline* | *Ethanol* |
|---|---|---|
| Density, kg/m$^3$ | 720 | 794 |
| Boiling point at 1 bar, °K | 316-444 | 351 |
| Oxygen, w per cent | NIL | 35 |
| Carbon, w per cent | 84 | 52 |
| Stoichiometric A : F ratio | 14.8 | 9 |
| Self Ignition Temperature, °K | 330-450 | 478 |
| Flammability limits, per cent by volume in air | 1.0-7.6 | 4.3-19 |
| Lower heating value, MJ/kg | 44.5 | 27 |
| Latent heat of vaporization, KJ/kg | 328 | 842 |
| Ignition Limits (A/F) | 8-19 | 3.5-17 |
| Octane number (Research) | 80-90 | 111 |
| Octane number (Motor) | 85 | 92 |
| Flame Speed, m/sec | 0.43 | 0.52 |

**Brake Mean Effective Pressure**

Figure 46.1 and Table 46.4 indicate that for both the fuels the Brake mean effective pressure decreases with the speed. For DNS00 the value decreases by 23.02 per cent.

**Effect of Brake Power**

Figure 46.2 and Table 46.4 indicate that the brake power increases with increase in the speed for both the fuels. The value decreases by 23.77 per cent for DNS100.

**Effect of Brake Thermal Efficiency**

Figure 46.3 and Table 46.4 indicate that the behavioral pattern of engine is different for both the fuels. For DNS100 the value decreases by 12.36 per cent.

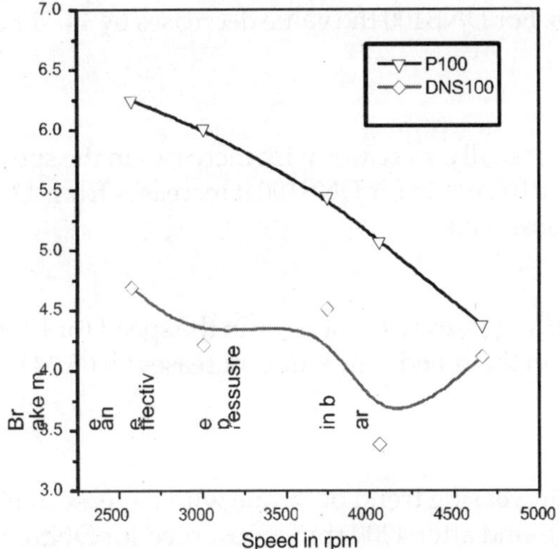


**Figure 44.1: Effect of BMEP Vs. Speed**

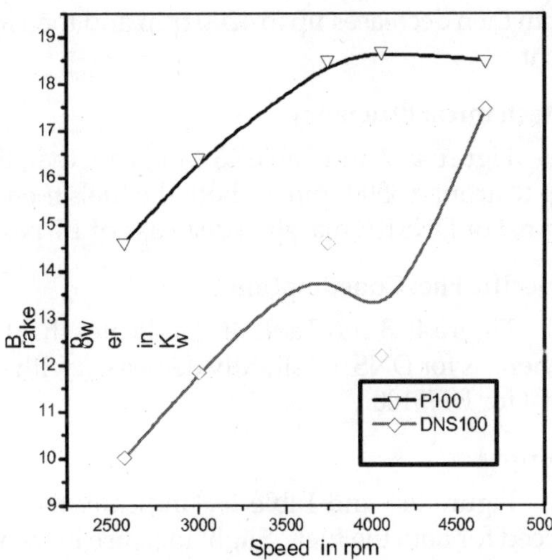


**Figure 44.2: Effect of BP Vs. Speed**

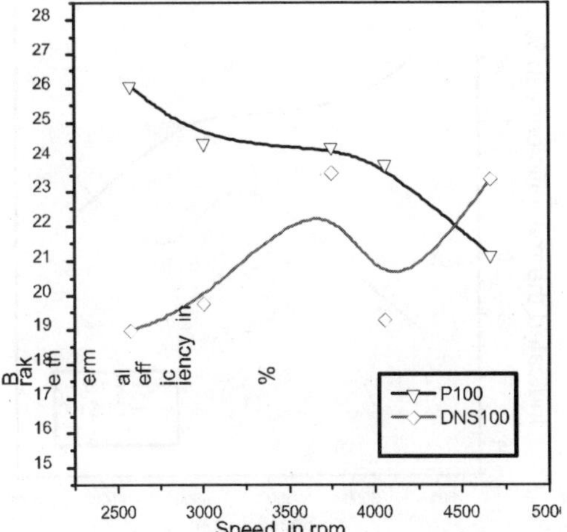


**Figure 44.3: Effect of BThE Vs. Speed**

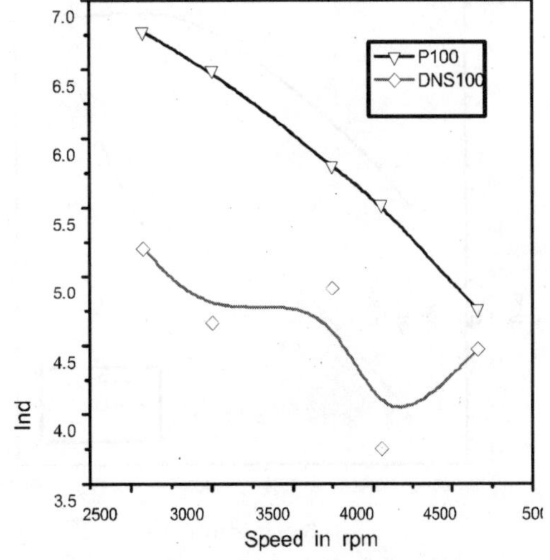


**Figure 44.4: Effect of IMEP Vs. Speed**

**Indicated Mean Effective Pressure**

Figure 46.4 and Table 46.4 indicates that the values decrease with increase in speed for both the fuels except in case of DNS100 it increases after 4000 rpm. For DNS100 the value decreases by 21.67 per cent.

**Indicated Power**

Figure 46.5 and Table 46.4 indicates that the value of indicated power increases with increase in speed for both the fuels. For DNS100 22.56 per cent reduction is observed.

**Indicated Thermal Efficiency**

Figure 46.6 and Table 46.4 indicates that the behavioral pattern of engine is different for both the fuels. For P100 the value decreases with the speed, while for DNS100 the value increases up to 3400 rpm then decreases up to 4200 rpm and then increases. For DNS100 the value decreases by 14.31 per cent.

**Mechanical Efficiency**

Figure 46.7 and Table 46.4 indicate that, the value initially increases with increases in the speed up to around 3500 rpm for both the fuels then falls for P100 while for DNS100 it increases from 4200 rpm. For DNS100 marginal decrease of 1.7 per cent is observed.

**Specific Fuel Consumption**

Figure 46.8 and Table 46.4 indicates that the value increases with increase in the speed for P100, where as for DNS100 slightly decreases with increase in the speed. The value increases by 65.34 per cent for DNS100.

**Torque**

Figure 46.9 and Table 46.4 indicates that there is decreasing trend of torque with increase in the speed for both the fuels. Slight increase in the value is found after 4200 rpm is observed for DNS100. The value decreases by 22.94 per cent for DNS100.

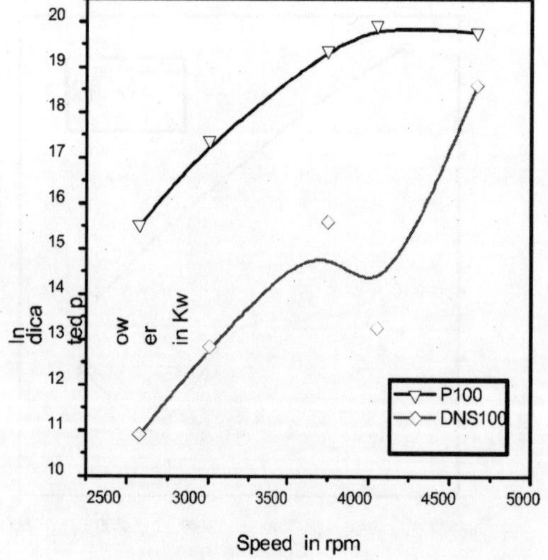


**Figure 44.5: Effect of IP Vs. Speed**

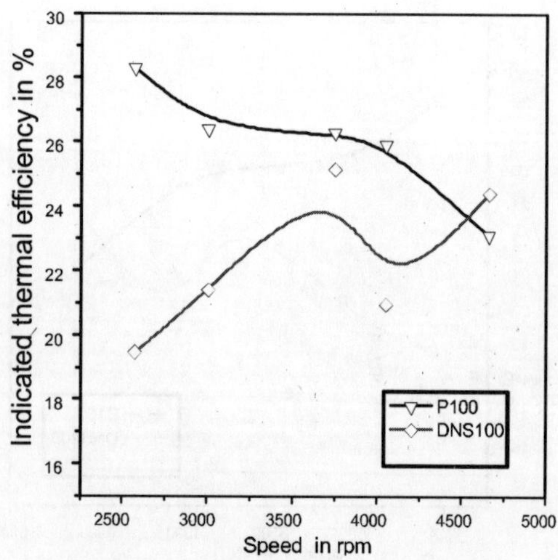


**Figure 44.6: Effect of IThE Vs. Speed**

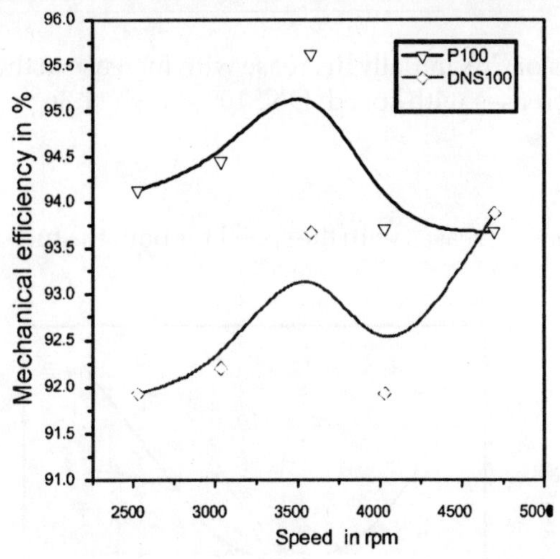


**Figure 44.7: Effect of MechE Vs. Speed**

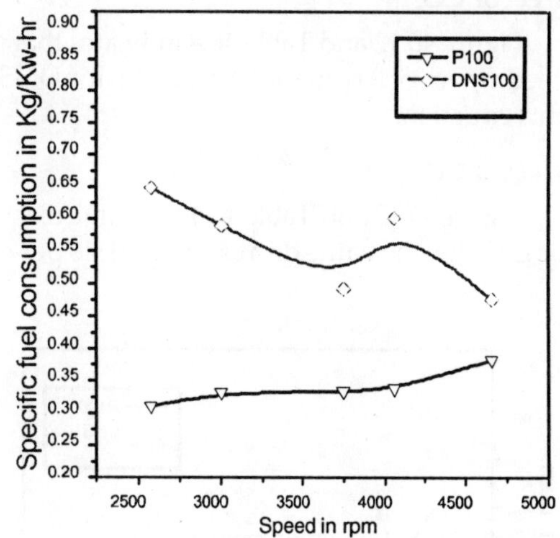


**Figure 44.8: Effect of SFC Vs. Speed**

**Volumetric Efficiency**

Figure 46.10 and Table 46.4 indicates that the value considerably decreases with increase in the speed for both the fuels. For DNS100 7.42 per cent decrease in the value is observed.

**Unburnt Hydrocarbons (HC)**

Figure 46.11 and Table 46.4 indicates that for P100 the HC decreases up to 3700 rpm then increases, while for DNS100 it increases up to 3200 and then decreases. DNS100 shows 5.6 per cent decrease in HC emissions.

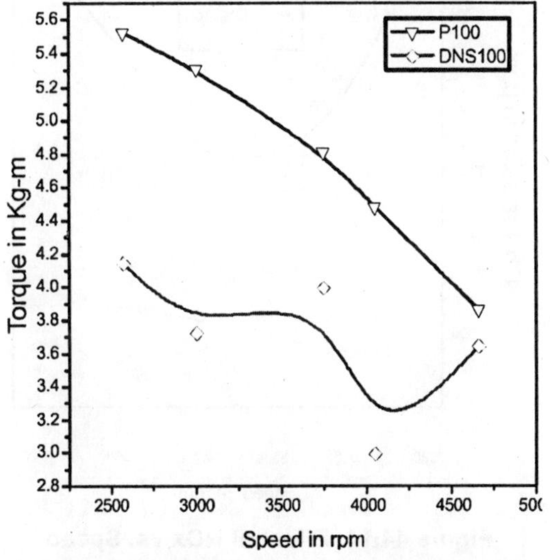


**Figure 44.9: Effect of Torque Vs. Speed**

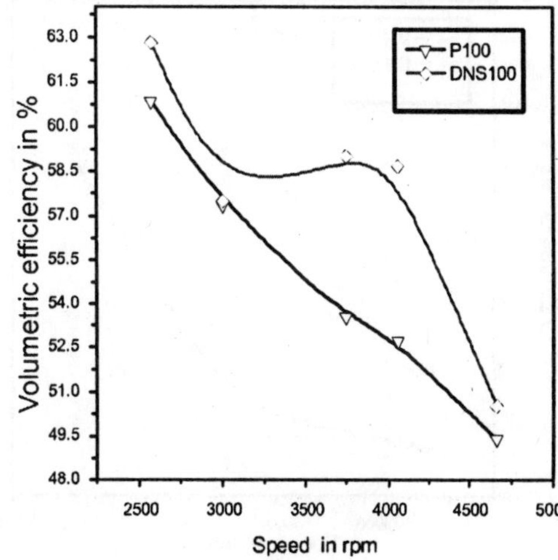


**Figure 44.10: Effect of VolE Vs. Speed**

**Effect of $CO_2$**

Figure 46.12 and Table 46.4 indicates that the values of $CO_2$ initially decrease with increase in the speed up to 3500 rpm for P100, while for DNS100 it increases with speed. DNS100 shows 11.36 per cent reduction.

**Effect of CO**

Figure 46.13 and Table 46.4 the value of CO emissions increases with the speed for both the fuels. For DNS100 the value decreases by 51.40 per cent.

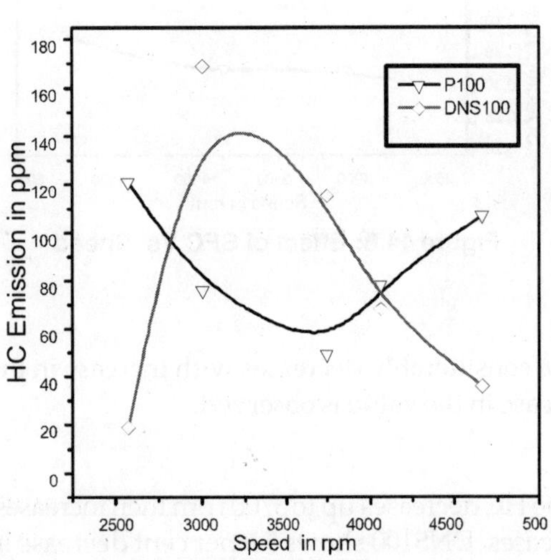


**Figure 44.11: Effect HC Vs. Speed**

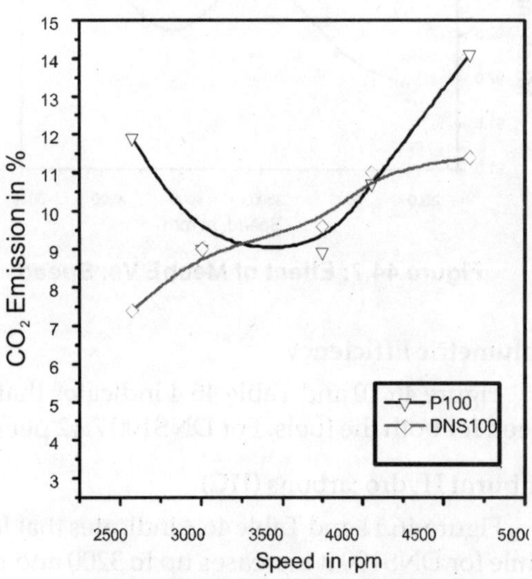


**Figure 44.12: Effect $CO_2$ Vs. Speed**

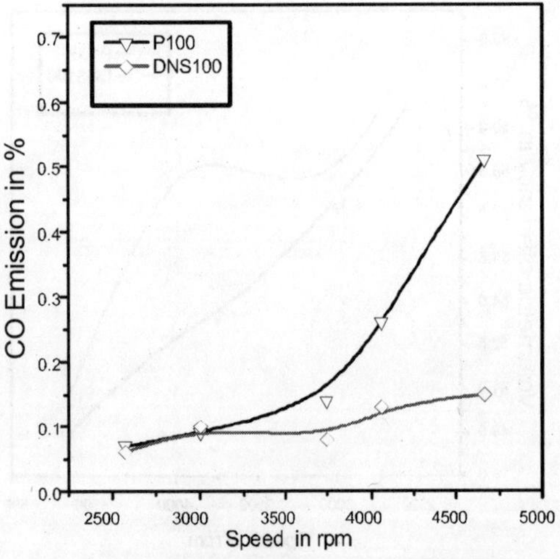


**Figure 44.13: Effect CO Vs. Speed**

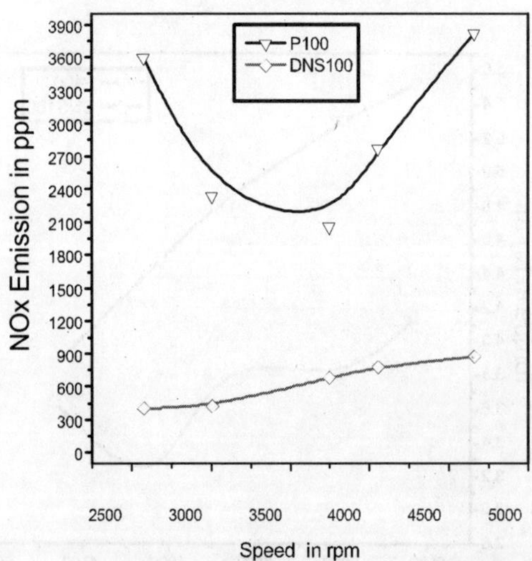


**Figure 44.14: Effect of NOx vs. Speed**

### NOx Emission

Figure 46.14 and Table 46.4 indicates that for P100 the value decreases up to 3500 rpm and then increases with increase in the speed. While for DNS100 the value increases with the speed. The reduction for DNS100 is 90.76 per cent.

### $O_2$ Emission

Figure 46.15 and Table 46.4 indicates that the for P100 the value increases up to 3500 rpm and then falls, while for DNS100 the value continuously decreases with the speed. The value increases by 37.5 per cent for DNS100.

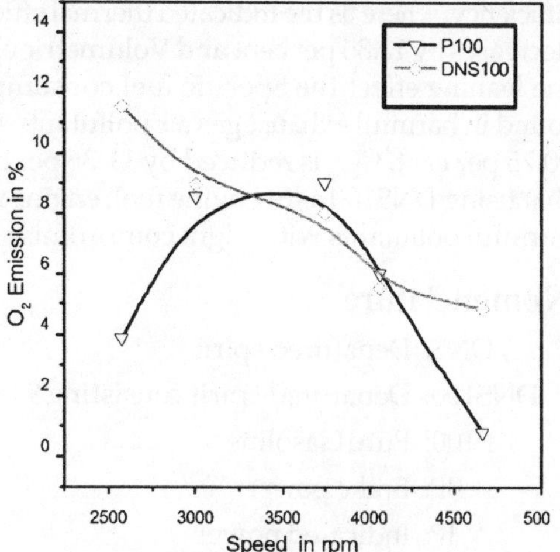


**Figure 44.15: Effect of $O_2$ Vs. Speed**

## Conclusion

The objective of this study was to mainly use the renewable and less polluting fuel for SI engine application. In this direction the study was concentrated using DNS100 as fuel in SI engines. To ease the engine's work few modifications were made to carburetor and ignition system.

**Table 46.4: Comparison of Performance Parameters in Equivalence Index and Per cent**
**Fuels: Gasoline and DNS100 Throttle: 50 per cent**

| *Parameter* | *P100* | | *DNS100* | | |
|---|---|---|---|---|---|
| | | *Eq. Index* | | *Eq. Index* | *In per cent* |
| Mech. E in per cent | 94.32 | 1 | 92.72 | 0.98303647 | -1.6963528 |
| Ithr.E in per cent | 26 | 1 | 22.28 | 0.85692308 | -14.307692 |
| Bthr.E in per cent | 23.94 | 1 | 20.982 | 0.8764411 | -12.35589 |
| Vol.E in per cent | 54.77 | 1 | 50.706 | 0.92579879 | -7.4201205 |
| IP in Kw | 18.392 | 1 | 14.242 | 0.77435842 | -22.564158 |
| BP in Kw | 17.35 | 1 | 13.226 | 0.76230548 | -23.769452 |
| Torque in Kg-m | 4.796 | 1 | 3.696 | 0.7706422 | -22.93578 |
| IMEP in bar | 5.87 | 1 | 4.598 | 0.78330494 | -21.669506 |
| BMEP in bar | 5.43 | 1 4.18 | 0.76979742 | -23.020258 | |
| SFC in Kg/Kw-hr | 0.3394 | 1 | 0.56116 | 1.65338833 | 65.3388332 |
| CO in per cent | 0.214 | 1 | 0.104 | 0.48598131 | -51.401869 |
| $CO_2$ in per cent | 10.92 | 1 | 9.68 | 0.88644689 | -11.355311 |
| $O_2$ in per cent | 5.664 | 1 | 7.788 | 1.375 | 37.5 |
| NOx in ppm | 2915.2 | 1 | 269.4 | 0.09241218 | -90.758782 |
| HC in ppm | 86.2 | 1 | 81.4 | 0.94431555 | -5.5684455 |

The engine performance and pollutant emissions were investigated. Experimental results indicated that the Torque, BP, IP, IMEP and BMEP decrease considerably by 23 per cent, 23.77 per cent, 22.56 per cent, 21.67 per cent and 23 per cent respectively. No major change is found in Mechanical

Efficiency, where as the Indicated thermal efficiency decreases by 14.31 per cent, Brake thermal efficiency decreases by 12.36 per cent and Volumetric efficiency decreases by 7.32 per cent. Obviously because of the leaning effect the Specific fuel consumption increases by 65.34 per cent. Dramatic decrease is found in harmful exhaust gas air pollutants. CO is decreased by 51.41 per cent and NOx is reduced by 90.75 per cent, $CO_2$ is reduced by 11.36 per cent and HC by 5.57 per cent. Finally we emphasize here that using DNS100 as SI engine fuel, engine not only reduces the green house gases but also the other harmful pollutants with slight compromise in the engine's performance.

## Nomenclature

DNS: Denatured spirit

DNS100: Denatured spirit consisting 93.7 per cent Ethanol and 6.3 per cent water

P100: Pure Gasoline

BP: Brake power

IP: Indicated power

FP: Frictional power

IMEP: Indicated mean effective pressure

BMEP: Brake mean effective pressure

BThE: Brake thermal efficiency

MechE: Mechanical efficiency

SFC: Specific fuel consumption

VolE: Volumetric efficiency

HC: Unburnt hydrocarbons

CO: Carbon monoxide

$CO_2$: Carbon dioxide

$O_2$: Oxygen

NOx: Oxides of Nitrogen

A/F: Air fuel ratio

## References

1. Al-Hasan M., Effects of ethanol-unleaded gasoline blends on engine performance and exhaust emission, Energy Conversion and Management 44(2003) 1547-1561.
2. Chong-Lin Song, Wen-Mei Zhang, Yi-Qiang Pei, Guo-Liang Fan and Guan-Peng Xu, Comparative effects of MTBE and ethanol addition into gasoline on exhaust emissions, Atmospheric Environment 40(2006) 1957-1970.
3. Fikret Yüksel, Bedri Yüksel, The use of ethanol–gasoline blend as a fuel in an SI engine, Renewable Energy 29(2004) 1181-1191.
4. Hakan Bayraktar, Experimental and theoretical investigation of using gasoline-ethanol blends in spark-ignition engines, Renewable Energy 30 (2005) 1733-1747.
5. Hüseyin Serdar Yücesu, Tolga Topgül, Can Cinar and Melih Okur, Effects of ethanolgasoline blends in engine performance and exhaust emissions in different compression ratios, Applied Thermal Engineering 26 (2006) 2272-2278.

6. M. Maheshwari, N.K.Acharya and R.K.Malhotra, Indian Experience with the use of Ethanol-Gasoline Blends on Two Wheelers and Passenger Cars, SAE 2004-28-0086.
7. Maji S., A Single Cylinder Engine Study of Power, Fuel consumption and Exhaust emission with ethanol, SAE 2001-28-0029.
8. Rudolph T. W. and J.J.Thomas, NOx, NMHC and CO Emission from Biomass derived Gasoline Extenders, Biomass 16(1988)33-49.
9. Segio M.Correa, Eduardo M. Martins and Graciela Arbilla, Formaldehyde and acetaldehyde in a high traffic street of Rio de Janeiro, Brazil, Atmospheric Environment 37 (2003) 23-29.
10. Wei-Dong Hsieh, Rogan-Hong Chen, Tsung-Lin Wu and Ta-Hui Lin, Engine performance and pollutants emission of an SI engine using ethanol-gasoline blended fuels, Atmospheric Environment 36(2002) 403-410.

# Chapter 47

# Experimental Technique of Calibration of Symmetrical Air Pollution Models

**P. Kumar***
**MAEER's MIT College of Engineering, Pune**

## ABSTRACT

Based on the inherent property of symmetry, of air pollution models a symmetrical Air Pollution Model Index (SAPMI) has been developed to calibrate the accuracy of predictions made by such models, where initial quantity of release at the source is not known. For exact prediction the value of SAPMI should be equal to 1. If the predicted values are overestimating then SAPMI < 1 and if it is underestimating then SAPMI > 1. Specific design for the lay out of receptors has been suggested as a requirement for the calibration experiments. SAPMI is applicable for all variations of symmetrical air pollution dispersion models.

*Keywords: Air pollution model, Validation, SAPMI, Predictions, Receptors layout.*

## Introduction

The transport and dispersion of suspended particulates in air, downwind of their point of discharge, is of great importance to statutory bodies, charged with the responsibility of ensuring that the process is in compliance with air pollution regulations. Most of the theoretical and numerical models are based on the basic premise of symmetry of dispersion of plumes on either side of the central line, downwind of the source. The pollutants may be gases or vapour or they may be particle matter of

---

* Corresponding Author: E-mail: pkumarmbm@yahoo.com

such a small size that their gravimetric settling may be neglected. In all these models quantitative estimate of pollutants at any point away from the point of release, is normally given in fractions of the initially released total quantity at source Q(say). Hence predictions made by these models can be calibrated only when quantity released from the source is correctly known. Due to instrumental errors or observational errors, the computed or observed value of Q at source, may not be correct. This would result in incorrect predictions by the model. In this paper an attempt has been made to develop experimental technique to produce Symmetric Air Pollution Model Index (SAPMI) that could be used to test the accuracy of the predictions made through such models, before they are actually applied for regulatory or purposes, without using the initially released total quantity Q.

## Principle of Computing SAPMI

It is generally assumed that unless horizontal wind shear persists, at or near the source of release of pollutants, the property of horizontal symmetry on either side of the central line (drawn downwind from the source) is maintained–may it be plume or puff model for point, line or area sources. This concept could be used while developing the index SAPMI. This index is based on the experiment described in this chapter, on the concept of dispersion-symmetry, by placing the receivers away from the source point, at and on either side of the central line of the plume, at equal lateral-distances from the line, as shown in Figure 47.1.

Let the total quantity of the pollutant released from the source within finite time interval =

$$Q \text{ (unknown)} \qquad (1)$$

For the purpose of testing our model we assume total weight of dust initially released from source = 100 units.

Based on this assumption let the predicted pollutant amount, by the model, at point DS-2 vide Figure 47.1 be $P_2$.

But predicted dust amount will be different than actual amount of pollutant collected at point- 2 since actual pollutant was Q and not 100 units. Also let the quantity of actual measured pollutant at this point be $A_2$.

When $P_2$ is the predicted pollutant quantity at point DS-2 then the original total pollutant quantity at source is = 100 units.

Hence when $A_2$ is the actually collected quantity of pollutant at point DS-2 then the total pollutant at original source

$$= (100/P_2) \times A_2 = Q'(\text{Say}) \qquad (2)$$

But original quantity of pollutant (Q) will not be same as obtained in equation (2) due to probable prediction-error $e_M$ (say) in the model–if the model is incorrect.

$$\text{Hence, } Q \neq Q' = (100/P_2) \times A_2 \qquad (3)$$

To make Q′ equal to Q equation (2) could be corrected with the prediction error $e_M$.

Hence after adding the correction $e_M$ with the predicted value $P_2$ we get

$$Q = 100 \times A_2/(P_2 + e_M) \qquad (4)$$

Total quantity Q computed by (4) can be used to predict the quantity of pollutant at site DS-1 and site DS-3 as $\{(P_1 \times A_2)/(P_2 + e_M)\}$ and $\{(P_3 \times A_2)/(P_2 + e_M)\}$ respectively. Symmetry in dispersion process

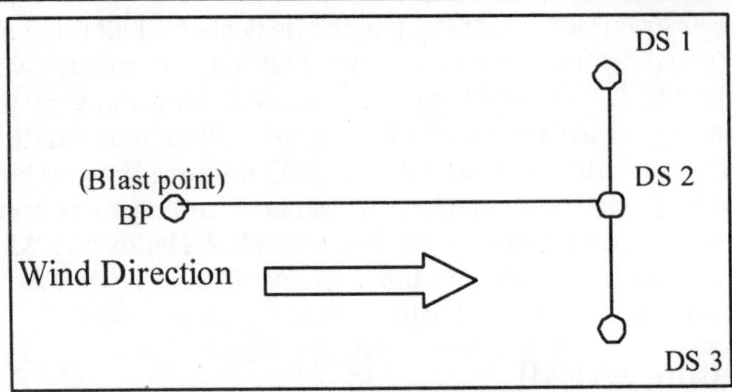


**Figure 47.1: BP is the Blast Point**

**BP to DS-2 is the direction of the wind marked by arrow. Line $DS_1$ and $DS_3$ is at right angle to line joining BP and $DS_3$. Point $DS_2$ biosects the line joining the $DS_1$ and $DS_3$. Distances between BP to $DS_2$, $DS_2$ to $DS_1$ and $DS_2$ to $DS_3$ are 200m, 50m and 50m respectively.**

as per the model however provides that error in the measurement and the prediction at site 1 and 3 should also be equal.

Hence $A_1 - \{(P_1 \times A_2)/(P_2 + e_M)\} = A_3 - \{(P_3 \times A_2)/(P_2 + e_M)\}$ (5)

Solving equation (5) for $e_M$ we get

$e_M = \{(P_1 \times A_2 - P_3 \times A_2)/(A_1 - A_3)\} - P_2$ (6)

Equation (6) gives the quantity of error in the model. Hence for the condition of no error, *i.e.* for $e_M = 0$, equating (6) to zero we obtain the condition for accuracy of the model

*i.e.* $(A_1 - A_3)/A_2 = (P_1 - P_3)/P_2$

$$\frac{(A_1 - A_3)/A_2}{(P_1 - P_3)/P_2} = 1 = SAPMI \quad (7)$$

Equation (7) gives the condition of the accuracy of the prediction made by any model. Thus if,

SAPMI = 1 (prediction is accurate)

< 1 (Model is overestimating)

>1 (Model is underestimating)

## Experiment for Calibration

Calibration experiments were performed at Birla cement factory, Chanderia, which is located in southern region of state of Rajasthan, India. Limestone quarrying is done by blasting the ground crust with detonators. This loosens the crust for mineral collection but it spews sufficient dust in atmosphere.

## Measurement of Dust

High volume air samplers are used for the estimation of the mass(M) of the SPM in a given volume of air (S.τ), where S is the known rate at which air is drawn through the apparatus and τ is the duration of sampling process. If χ (mass of dust per unit volume) is the instantaneous concentration at the sampling position, the average concentration over the duration of sampling is

$$\bar{\chi}=\frac{1}{\tau}\int_0^{\tau}\chi dt=\frac{M}{S\tau}\ \text{(Say)} \tag{8}$$

At the quarry site three Respirable Dust Samplers (RDS) were installed as per lay-out design shown Figure 47.1. Measurements of respirable dust ($PM_{10}$) were taken on Whatman filter papers. Papers were weighed before and after the collection of dust in RDS. Papers were changed at the interval of 15 minutes. All the samplers were simultaneously started/shut-off by single switch point. 20 minutes before the blast, first paper was installed and 15 minutes later second paper was replaced. Thus each paper collected dust for 15 minutes from 18 litre of ambient air, at the rate of 1.2 litre per minute. As the RDS were started 20 minutes before the blast hence after 15 minutes of running of machine (*i.e.* 5 minutes before the blast) filtered dust on each Whatman-paper gave the background dust in the ambient atmosphere. Immediately, therefore, second set of filter papers were replaced in all the three machines. Mine-blast was made 5 minutes later and 10 minutes, thereafter, third set of papers were again replaced. After this 4th and 5th set of papers were replaced at the intervals of 15 minutes each. Each unused filter paper was dried and weighed before use. Total dust accumulated at the filter paper was determined by subtracting the weight of used filter paper from that of used filter paper. Dust per litre of air at 1m a.g.l. at known distances (refer Figure 47.1), from the point of blast was determined by dividing the dust weight on each filter paper by '18.0'.

## Safety Measure

During the blasts stones are thrown out hence to avoid damage to RDS instruments and personnel, instruments at location DS-2 was kept at 200m away from the point of blast and distances between DS-2 to DS-3, and DS-2 to DS-1 were kept as 50 m each.

## Measurement of Surface Wind

The quantity (M) in Equation(8) is a function of displacement in wind direction (along x-axis), cross-wind displacement (along y-axis) and vertical displacement (along z-axis) from central point of dust cloud for the modelling of the dispersion process explicit relationship between cross wind spread and the statistics of wind fluctuation is needed. Cross wind dimension of the dispersing dust cloud is conventionally represented either by width ($Y_0$) between position at which concentration falls 1/10th of the central peak value or by standard deviation($\sigma$) of cross-wind displacement of the material which in practice is defined by

$$\sigma y^2 = Y^2 = \frac{\Sigma Y^2}{\Sigma M} - \left[\frac{\Sigma MY}{\Sigma M}\right] \tag{9}$$

Either definitions are related since $Y_0 = 2.15\sigma$. It is noted that if standard deviation of wind direction is $\sigma_\theta$ and speed is $\sigma_s$ then $\sigma/(\sigma_\theta x)$ or $\sigma/(\sigma_s t)$ do closely approach to unity on an average for time(t) and distance(x). Hence, good correlation between dust cloud spread and the intensity of the turbulence exists irrespective of the condition of the surface reference and stability. This is a significant result, in the theoretical modelling of the dispersing puff of dust cloud. Hence an anemometer was installed close to blast site and wind measurement was made and $\sigma_s$ values were computed.

## Blast Mushroom

Figure 47.2 shows the mushroom of dust cloud immediately after the blast.

L = Horizontal distance of the theodolite. With the help of theodolite angle 'θ' of the top of the dust cloud was determined.

Height of the dust column is given by H = L tan (θ).

**Figure 47.2: Measurement of the Height of the Blast Column with the Help of Horizontal Distance (L) and Angle (θ) of the Inclination of Top of Dust-Cloud Mushroom**

## Modelling of Dispersion of Dust Mushroom

The most recent theory of dispersion modelling is known as gradient transfer theory (or K-theory) which employs the basic differential equation that attempts to describe the advective and turbulent motion of the atmosphere. The other one is a statistical theory. With increasingly better understanding of the convective boundary layer's (CBL) parameterization process, several numerical and analytical solutions have also been developed for dispersion. Gaussian plume diffusion model, which is based on the simple statistical theory, assumes normal distribution as representing dispersion of gas and SPM in atmosphere. This model, although, raises several questions with respect to its assumptions of continuous source geometry, slender plume, neglect of longitudinal diffusion in comparison to advective term by mean flow, but due to its simple form it constitutes the basis of many atmospheric dispersion calculations for regulatory applications (Taylor, 1921, 1959; Batchelor, 1949; Csandy, 1973). For completeness gradient transfer theory and statistical theory are both discussed briefly hereunder.

### Gradient Transfer Theory

For any perturbation parameter $\xi'$ multiplied by the motion term the gradient transport theory assumes

$$\overline{u'\xi'} = -K\frac{\partial \overline{\xi}}{\partial x} \tag{10}$$

*where,*

K implies $\overline{u'\xi'}$ flows down the local gradient of $\xi'$. Although it is the simplest parameterization, known on first order closer relationship, it often fails when larger eddies are present in the flow, as found in convective boundary layer (CBL). Hence its validity is useful for near-neutral or stable atmospheric conditions, only. Batchelor, 1949 and Casandy, 1973, have shown that for the solution of diffusion equation to be consistent with the result of the statistical theory of diffusion, the apparent eddy diffusivities must be related to the plume dispersion parameters as

$$K_x = \frac{\overline{u}}{2}\frac{\partial \sigma_x^2}{\partial x}, K_y = \frac{\overline{u}}{2}\frac{\partial \sigma_y^2}{\partial x}, K_z = \frac{\overline{u}}{2}\frac{\partial \sigma_z^2}{\partial x} \tag{11}$$

If $T_{Lu}$ is the integral time scale of Lagrangian velocity fluctuations in the x-direction and $L_x = u\,T_{Lu}$, then

$$K_x = \sigma_u^2 t = \frac{\sigma_u^2}{\bar{u}}, for \; x \le L_x \tag{12}$$

$$K_x = \sigma_u^2 T_{Lu} = \frac{\sigma_u^2 L_x}{\bar{u}}, for \; x \ge L_x \tag{13}$$

Alternatively following general forms can be used

$$K_x = \frac{\sigma_u^2 x}{\bar{u}}\left[1+\frac{x}{cL_x}\right]^{-1} \tag{14}$$

$$K_y = \frac{\sigma_u^2 x}{\bar{u}}\left[1+\frac{x}{cL_y}\right]^{-1} \tag{15}$$

for all ranges, in which c is a constant of the order of 1.

## Statistical Theory

With its inherent assumptions it is strictly valid only in the region of uniform flow and homogeneous turbulence and is most suitable for convective boundary layer (CBL). For highly convective light wind conditions, plume dispersion parameter should be specified on the basis of the mixed layer similarity theory (Arya, 1995) as below):

$$\sigma_* = \frac{at_*}{(1+ht_*)^{1/2}} \tag{16}$$

$$\text{Here } \sigma_* = \frac{\sigma_x}{h} = \frac{\sigma_y}{h} = \frac{\sigma_z}{h} \text{ and } t_* = \frac{tw_*}{h} \tag{17}$$

*where*, h is the height of convective boundary layer (CBL).

## Model-I

As the order of dispersion range (O≈100) is higher than the order of the thickness of the dust-mushroom (O≈10), hence neglecting thickness of the dust-mushroom, one can obtain the cumulative dispersion effect by integrating effects of all the elevated source elements from surface to height H *i.e.* top of the dust cloud. Near field concentration from an instantaneous blast mushroom source for flat surface (Kumar and Bhandari-2000), therefore, could be computed through the following integral equation:

If $H_* = H/h$

$$\bar{C}(x_*, y_*, z_*, t_*) = \frac{1}{(2\lambda)^{3/2}} \int_0^{H_*} \frac{1}{\sigma_*^3} \exp\left[\frac{(x_* - \bar{u}_* t_*)^2}{2\sigma_*^2}\right] \exp\left[\frac{y^{*2}}{2\sigma_*^2}\right] \exp\left[\frac{(Z_* - H'_*)^2}{2\sigma_*^2}\right.$$

$$\left. + \exp\left[\frac{(Z_* - H'_*)^2}{2\sigma_*^2}\right]\right] dH'_* \tag{18}$$

Ground level concentration (GLC) at the plume centerline (y*=0; z*=0) could be obtained from

$$\bar{C}(x_*,t_*)=\frac{1}{(2\pi)^{3/2}}\int_0^{H_*}\frac{2}{\sigma_*^3}\exp\left[\frac{(x_*-\bar{u}_*t_*)^2}{2\sigma_*^2}\right]\exp\left[\frac{H_*'^2}{2\sigma_*^2}\right]dH' \tag{19}$$

## Model-II

Model–I was improved with the introduction of actual surface profile as the lower boundary condition. All the heights are taken from the blast point as a reference level. The X-axis is taken in the mean wind. Y-axis is at lateral direction at right angle to x-axis. Z-axis is vertical distance from the location of blast. Ground simulation is achieved through MATLAB package which is based on least-square approximation. It was achieved in two steps; preliminary best fit at the first step was linearly added with the best fit of the errors in the second step. Preliminary best-fit sixth order polynomial, *i.e.*

$$P_p(x)=\sum_{i=0}^{6}a_i x^i \tag{20}$$

It has been obtained for the ground profile below the central line of dispersed plume. Following a similar approach, to achieve a better fit to the terrain another correction-polynomial,

$$P_c(x)=\sum_{i=0}^{6}b_i x^i \tag{21}$$

was obtained for departure ($P_p$–$Z_a$) from the actual ($Z_a$) contour heights. Final best fit ($Z_t$) was computed by linear addition of preliminary polynomial $P_p$(x) (equation-20) with the correction polynomial $P_c$(x) (equation-21) as,

$$Z_t=P_p(x)+P_c(x)=\sum_{i=0}^{6}(a_i+b_i)_i x^i \tag{22}$$

If the variation in the topography in height is less than the vertical height of the dust mushroom then reflection theory approach could still be assumed to be valid to reasonable degree of accuracy. Equations (18) and (19) would, therefore, still remain valid.

## Model-III

Model II was better than model–I as it included the ground profile, too, as a lower boundary condition. But since the vertical height of the dust mushroom in surface mine blast goes up to 30 to 50 meters high hence a more appropriate modelling would be with the inclusion of vertical wind shear. Model–II was, therefore, further improvised by inclusion of power law in the observed surface winds, by the in-situ anemometer.

### Power Law

Power law for the vertical variation of mean wind speed is given by

$$\frac{u_2}{u_1}=\left[\frac{z_2}{z_1}\right]^\alpha \tag{23}$$

*where,* $u_1$ and $u_2$ are the mean wind speed at heights $z_1$ and $z_2$ respectively and a is power law index as under,

$$\alpha = \frac{\log u_2 - \log u_1}{\log z_2 - \log z_1} \tag{24}$$

Wind velocity $u_2$, at different levels of the source-elements of dust mushroom, could be accordingly determined if value of $u_1$ (*i.e.* observed surface wind) and height of the source-element above ground level and value of a are known. Power law factor (a) for different stations of India has been documented in Appendix "A" (Anna Mani-1995). a value for Birla Cement Factory, Chittorgarh, for the month of December was obtained as equal to 0.15. This value was included in Model-III.

## Correction Due to Terrain Forcing Factor

Terrain response to dispersion differs with the type of stability conditions in atmosphere. Empirical relation which gives terrain forcing factor ($F_T$) is mentioned below:

$$F_T = (H_A - H + \Delta E)/\Delta E$$

or $$H_A = H - \Delta E + F_T \Delta E$$

*where,*

$H_A$: Adjusted effective height
H: Effective height
$\Delta E$: $E_R - E_S$
ER: Ground level elevation of receptors
Es: Ground level elevation of source.
$F_T$: 0 for stable,
$F_T$: 1 for instable,
$F_T$: 0.1 for Neutral stability.

Figure 47.3 shows vertical dispersion over undulated terrain under unstable stable and neutral stability conditions.

As has been explained in para 6, above, vertical dust column is assumed to be formed of small dust source elements stacked–up together. Each element of dust can be assumed as the elevated point source similar to chimney-source, shown in figure-3. The dispersion results are based on horizontally moving frame of reference; hence as the frame moves along the topography necessary correction in the effective height to the source element has to be applied. In model–III, therefore, $F_T$ correction for dispersion in vertical, has also been applied.

## Data

Preliminary input-data was used for the computations is presented in Table 47.1.

## Discussion

All the three models were used to study the SAPMI values. Table 47.2 presents the results obtained by different models. Gradual improvement in the value of SAPMI was noticed from model I to model III. This indicated that as the error $e_M$ in the model predictions were reduced by applying better physical parameterizations vis-à-vis increasing the prediction accuracy, SAPMI values approached more and more near 1.

**Table 47.1**

| *Description* | *Assumed Value* |
|---|---|
| Distance in meter from blast site | 100.00 m. |
| Angle of the top of the mushroom from the point of observation | 45° |
| Surface pressure in hPa | 980 hPa |
| Surface Wind Speed | 5 m/s |
| Wind variation from mean wind speed | 0.005 m/s |
| Wind speed at 850 hPa | 8 m/s |
| Saturation mixing ratio at surface | 5 g/k |
| Saturation mixing ratio at 850 hPa | 2 g/k |
| Mixing condensation level | 1200 |
| Latitude of blast site | 25° |
| Height (above sea level) of blast point | 596 |
| Height of 850 hPa above sea level | 1500 m |

SAPMI values decreased from 3.9 to 2.39 in Model-II as the undulated topography was incorporated in Model-I. More perfection in the model, however, was achieved in Model-III with the introduction of topography, ground response and vertical wind shear. This brought down the SAPMI value from 2.39 to 1.32. Still, we observe, that its value is not exactly unity. This may be due to inherent limitations in the assumptions that vertical dust cloud has uniform distribution of dust-density and many experimental or observational errors which affect the numerator of the SAPMI computation values. They could be as under.

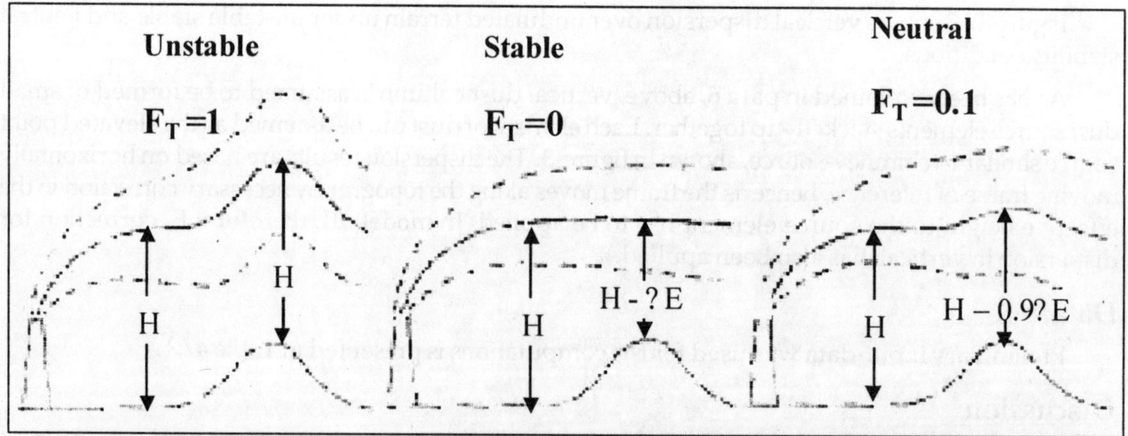


**Figure 47.3: Adjusted Effective Height ($H_A$) of the Central Line of Dispersed Plume at Various Distances from the Elevated Source of Unstable ($F_T$ = 1), Stable ($F_T$ = 0) and Neutral ($F_T$ = 0.1) Atmospheric Conditions.**

**Table 47.2**

| *Type of Model* | *Computed SAPMI Values* |
|---|---|
| Model-I | $SAPMI = \frac{(A_2 - A_4)/A_3}{(P_2 - P_4)/P_3} = \frac{5.20 \times 10^{-3}}{1.33 \times 10^{-3}} = 3.9$ |
| Model-II | $SAPMI = \frac{(A_2 - A_4)/A_3}{(P_2 - P_4)/P_3} = \frac{5.20 \times 10^{-3}}{2.17 \times 10^{-3}} = 2.39$ |
| Model-III | $SAPMI = \frac{(A_2 - A_4)/A_3}{(P_2 - P_4)/P_3} = \frac{5.20 \times 10^{-3}}{3.93 \times 10^{-3}} = 1.32$ |

1. Incorrect identification of central line of plume.
2. Different rate of suction by different dust-samplers due to variation in the manometer readings in the field.
3. Assumption of constant wind with height.
4. Difference in elevation of dust samplers kept at sites 2, 3 and 4 due to variation in ground contour in the field.
5. Fugitive loose dust (sucked in the instrument) may affect the accuracy. Further such a fugitive dust error may not be uniform for all the sites.

Beside experimental causes there could be mathematical approximation errors, too, as follows:-

1. Neglect of thickness of the vertical dust column.
2. Application of reflection property with the assumption that the height of the blast column is of higher order than the order of variation in the ground contour elevation.

Nevertheless it may be clearly observed that SAPMI shows steady approach to unity as the errors of the models are increasingly diluted with better parameterizations, step by step. In other words higher is the value of SAPMI's departure from unity *i.e.* higher is value of (SAPMI–1), the higher is the range of error in the predicted values through the models.

## Conclusions

1. Symmetrical Air Pollution Model Index

   $SAPMI = \{((A_2 - A_4)/A_3)/((P_2 - P_4)/P_3)\}$

   is useful parameter to test the accuracy of predictions made by the models where initial mass of vertical column is unknown.
2. With gradual improvement in physics of the model SAPMI value gradually approaches to unity.
3. Once the extent of model accuracy is established, its inaccuracy range could be accordingly accounted while interpreting the results of the predictions.
4. Even if at the application stage of the model, initial quantity of pollutants released at the source point is accurately known the predictions with the model would not be correct unless SAPMI value is equal to one.

## Acknowledgements

Author acknowledges with thanks the efforts put by Shri Anoop Banerjee, Shri Vikas Mohan, Senior Geologist, in helping to conduct the field experiments at M/s. Birla Cement Works, Chittorgarh, Rajasthan, India. Author is also indebted to Ms. Urvashi Shrivastava and Shri Anoop Banerjee in preparing and typing the manuscript. Author is also extremely grateful to M/s. Birla Cement Works, Chittorgarh, Rajasthan, India, for extending all the facilities and helping hands for the experiments.

## References

1. Anna Mani, 1995. Wind Energy (Resource Survey in India), Allied Publishers Ltd., New Delhi.
2. Arya, S.P., 1995. Modelling and Parametarization of Near-Source diffusion in weak winds, J. Appl. Met, 34, pp 1112-1122.
3. Batchelor, G.K. 1949. Diffusion in a field of homogeneous turbulence, I. Eulerian analysis. Aust J. Sci. Res., 2: 437-450.
4. Berlyand, M.E. 1975. Contemporary Problems of Atmospheric Diffusion and Air Pollution. Hydromet. Press: 448.
5. Csanady, G T, 1973. *Turbulent Diffusion in the Environment*, pp 248 (D. Reidel Publishing,).
6. Kumar, P. and S. Bhandari. 2001. Modelling of Near Source Dust Dispersal after the opencast mine Blasting on flat surface in weak wind in Tropical Countries, Proceedings of EXPLO-2001, Hunter Valley, NSW, Australia, 28th–31st October 2001, pp 289-293.
7. Taylor, G.I. 1921. Diffusion by continuous movements, Proc. London Math. Soc., 2: 196-212.
8. Taylor, G.I. 1959. The present position in the theory of turbulent diffusion. Adv. Geophys., 6: 101-111.

# Chapter 48

# Control of VOCs Using Biofilters

**S.N. Upadhyay**
**Department of Chemical Engineering and Technology,
Centre of Advanced Study, Institute of Technology
Banaras Hindu University, Varanasi – 221 005**

## Introduction

Biofilter is a type of supported film bioreactor used for removal of water soluble biodegradable volatile organic compounds (VOCs), hazardous air pollutants and odorous compounds, present in air emissions at very low to moderate concentrations.

Biofiltration is more economical and efficient as compared to other contemporary control technologies. It not only removes the pollutants from polluted air but destroys them completely. This is one of the attractive features that provide biofiltration an edge over other technologies. Unfortunately, this technology has not received the required attention of Indian researchers. Only a few organizations of India–NEERI, Nagpur; IT-BHU; RRL Thiruvananthapuram; and IIT Roorkee are engaged in the R&D activities in this area.

The biofiltration process occurs in a biofilter, a kind of supported biofilm reactor. These bioreactors are characterized by microorganisms being attached to a porous solid surface in the form of a thin biofilm. Biofilms are dense layers of microorganisms characterized by their ability to adhere to a solid medium and form a fixed layer of biofilm in which the microorganisms are protected against sloughing off. The VOCs consuming microorganisms immobilized on the solid surface of packing media of the biofilter from the biofilm. Peat, compost, volcanic rock, wood pieces, other natural solid biomass and some synthetic packing media, such as polystyrene beads and polyurethane foam, have been used as packing materials in biofilters. Every packing media has some advantages and disadvantages (Upadhyay and Kumar, 2005).

In biofiltration process, polluted air is purified during its passage through the wet biofilter media. The pollutant present in the air diffuses to the biofilm where aerobic microorganisms consume it through their metabolic activities and produce innocuous products like $H_2O$, $CO_2$, etc. The biofilter is fed with regular supply of nutrient solution for the growth of microorganisms. This also maintains proper moisture content inside the filter media (Upadhyay and Kumar, 2005).

Selection of suitable packing media is an important step in the biofilter technology. In view of this, the research work was planned with the objectives of selection and characterization of locally available cheap materials as biofilter media, evaluation of the performance of the biofilters using the selected media under varying process parameters, determination of the optimal operating conditions for the biofilter and study of the reactor stability and durability and mechanical strength of selected media. Here a summary of the results obtained is presented. The detailed results are given elsewhere (Singh, 2007; Singh *et al.*, 2004 a, b; 2006 a,b.)

## Experimental

Two agro-wastes namely yellow gram stalk and maze cobs and one industrial scrap–polyurethane foam, a locally available cost effective polymeric material, were selected as the biofilter media. Benzene and toluene were selected as the target pollutants because these are the most common and extensively used VOCs in various industries as raw material or as solvent. These have been also characterized as one of the 188 toxic pollutants according to United State Environmental Protection Agency (US-EPA). These are easily available and have been widely used by other workers hence extensive data are available for comparison.

To judge the suitability of selected materials as biofilter media, these were characterized and their various physical and chemical properties such as density, porosity, adsorption properties and moisture retention capacity were determined. These materials were also tested against possible hostile conditions such as highly acidic and basic environment, high concentration of pollutants, etc. The results show that the selected media have most of the favourable properties such as high porosity, low bulk density, high water retention capacity and resistant to chemicals.

To know the flow behaviour of packed bed, the residence time distribution (RTD) experiments were conducted and results confirmed plug flow behaviour of the packing. After characterization, continuous biodegradation experiments were performed in three laboratory scale biofilters, at varying operating conditions to evaluate and optimize the biofilter performance. Continuous biodegradation experiments were conducted in four different phases. In the first set of experimental runs, yellow gram stalk, maze cobs and polyurethane foam were used as biofilter media and toluene as the target pollutant. In the last set of runs, polyurethane foam was used as biofilter media and benzene and toluene as the target pollutants.

Pressure drop across the bed, temperature of bed, moisture content of the inlet stream and pH of leachate collected from the bed were also monitored continuously, which provided valuable information about mechanical strength of packing, biomass growth, excessive bio-oxidation, wet or dry conditions in the bed and type of microorganisms present in the biofilter.

Biodegradation of toluene vapor was investigated in a laboratory scale biofilters using cylindrical pieces of yellow-gram (*Cajanus cajan*, a specie leguminosal family) stalk as biofilter media. It consisted of a plexiglas column of diameter 0.15m and effective height 0.70 m with sampling ports at 0.26, 0.49 and 0.70m from the bottom. Inlet concentrations and volumetric flow rates of toluene were varied from 2.56 to 34.73 g/m$^3$ and 0.18 to 0.24 m$^3$/h, respectively.

Using polyurethane foam cubes as biofilter media, biodegradation of toluene and benzene vapors was investigated separately for 168 and 119 days, respectively. For these runs a laboratory scale biofilter consisting of five modular sections of size (0.15 × 0.16 × 0.20m$^3$) and each filled with polyurethane foam cubes upto a height of 0.15 m was used. Inlet concentrations and flow rates of toluene and benzene loaded air were varied from 0.04 to 2.5g/m$^{-3}$, 0.69 to 5.45 g/m$^{-3}$ and 0.06 to 0.90 m$^3$ h$^{-1}$, respectively.

## Results and Discussion

Continuous biodegradation experiments showed that steady state was achieved earlier in the units filled with yellow gram stalk and maze cobs (natural media) than in the unit filled with polyurethane foam (synthetic media) because microorganisms find more favourable conditions to adhere and grow on the surface of natural materials than synthetic. Further, the natural materials generally support of a broad spectrum of microorganisms.

For toluene removal, almost constant removal efficiencies of 90, 80 and 90 per cent were achieved upto inlet loadings of 300, 120 and 77 g h$^{-1}$ m$^3$ using yellow gram stalk, maze cobs and polyurethane foam as biofilter media. For benzene removal, almost constant removal efficiency of 80 per cent was achieved upto inlet loadings of 68 g h$^{-1}$ m$^3$ using polyurethane foam as the biofilter media. Maximum elimination capacities of 342.2, 174.4 and 90.5 g h$^{-1}$ m$^3$ were achieved at inlet loadings of 580, 292.6 and 126.3 g h$^{-1}$ m$^3$ using yellow gram stalk, maze cob and polyurethane foam as biofilter media and toluene as the target pollutant. For benzene, maximum elimination capacity 74.7 g h$^{-1}$ m$^3$ was achieved at the inlet loading of 134.7 g h$^{-1}$m$^3$ using polyurethane foam as biofilter media. Yellow gram stalk was found to be the most effective biofilter media for toluene degradation but experiments were conducted at low flow rates only. Toluene was easily biodegradable than benzene. In all cases, biofilter responded very well to change in operating conditions and could easily recover its performance on re-startup after shutdown. This indicated good stability of the biomass in the biofilters. Axial variation of pollutant concentration was also investigated in the biofilters. Results showed that most of the pollutant got eliminated in the upper section of the biofilter. With the help of this information, optimum height of biofilter column for given operation can be calculated.

Presence of diffusion and bio-degradation controlling zones were observed in the biofilters at low and high inlet loadings. Less deterioration and small compaction of bed resulting in low change in pressure drop was observed with polyurethane media (synthetic) than with yellow gram and maze cob (natural). This indicated good mechanical strength and less power required to operate the biofilter with synthetic media.

Variation of pH in the biofilters with respect to time not only depends on the type of media and pollutant but also on the type of microorganisms in the biofilter. All experiments were carried out in the months of March-September at room temperature (29–41°C) and insignificant variation in the performance of the biofilter was observed. Under steady-state conditions, temperature of biofilter bed was found 3 to 5°C higher than the ambient temperature. Carbon recovery (per cent) was also estimated in the case of toluene removal using maze cob and polyurethane foam as biofilter media. Most of the time more than 70 per cent carbon recovery was observed which indicated the effective biodegradation in the biofilter. More regulated nutrient supply was required in the case of polyurethane foam than yellow gram stalk and maze cobs because the natural media inherently contain most nutrients required for the growth of microorganism.

## Acknowledgement

This chapter is based on the Ph.D. thesis of Dr. R.S. Singh and funding for the research came from AICTE, New Delhi as well as from M.E.F., New Delhi.

## References

1. Singh, R.S., Ph.D. thesis in Chemical Engineering, Banaras Hindu University, Varanasi, India, (2007).
2. Singh, R.S., Agnihotri, S.S., and Upadhyay, S.N., Removal of Toluene Vapour Using Agro-waste as Biofilter Media, Bioresource Technology, Vol. 27, pp.2296-2301 (2006 a).
3. Singh, R.S., Gunjan Kumar and Upadhyay, S.N., Modelling and Simulation of Biofilter for the Removal of Volatile Organic Compound from Polluted Gas Stream, International Conference "BIOCONVERGENCE", TIET Patiala, 18-20 November, (2004 a).
4. Singh R.S., Rai, B.N., and Upadhyay, S.N., Environmental Biotechnology and its Role in Air Pollution Control, International Conference "BIOCONVERGENCE", TIET Patiala, 18-20, November, (2004 b).
5. Singh, R.S., Rai, B.N., and Upadhyay, S.N., Performance Evaluation of an Agro Waste Based Biofilter Treating Toluene Vapours, Environmental Technology, Vol.27, pp.349-357 (2006 b).
6. Upadhyay, S.N., and Kumar, V., Biological Control of Air Pollution in "Concise Encyclopedia of Bioresource Technology" (Pandey, A., Ed.), The Haworth Press Inc., NY (2005).

# Chapter 49

# Energy and Environmental Related Issues with Special Reference to Captive Power Plants in India

**Purvi Pujari and S.V. Dixit**
**CSIR–National Environmental Engineering Research Institute, Nagpur**

*There is now growing awareness on the imperatives for a global energy future which marks a distinct departure from past trends and patterns of energy production and use. These imperatives emerge as much from the need to ensure energy security, as they do from the urgency of controlling local pollution from combustion of different fuels and, of course, the growing challenge of climate change, which requires reduction in emissions of greenhouse gases (GHSs), particularly carbon dioxide–(European Renewable Energy Council).*[1]

The scientific evidence on the need for urgent action on the problem of climate change has now become stronger and convincing. Future solutions would lie in the use of existing renewable energy technologies, greater efforts at energy efficiency and the dissemination of decentralized energy technologies and options. Concerns about environmental degradation, climate change and water shortages and possible solutions to these issues are now in the mainstream. The studies around the world have shown that dealing with climate change is about economic self-interest at every level. Clearly, this is of direct benefit for investors focusing on green issues.

1 GPI REF JN 035. Published by Greenpeace International and EREC - date January 2007-www.energyblueprint.info/

## Current Global Scenario

Scientists say that a rise in global temperatures by more than two degrees may cause catastrophic change. Climate stabilization measures have never been more imperative–a view put forward by German Chancellor Angela Merkel at the June meeting of the G8. Encouragingly, there is almost daily news about advances in environmental technology and shifts in government and corporate policies and consumer behavior that will help tackle climate change and its effects. To provide a sense of the scale of the climate problem, the world currently produces 7 bn tons of carbon per year. This is expected to double in the next 50 years. To negate this, 7 bn tons of emissions must be avoided per year over the next half century. With the current supply and demand imbalances in the world's oil market, the motives for using alternatives to fossil fuels are not just environmental, they are economic. The International Energy Agency says that global energy demand is expected to rise by more than 50 per cent by 2030. It is also thought that crude oil production will peak in the second half of this century. Threats to energy security and political instability only make this picture all the more fragile. Against this, world energy prices are expected to remain high for some time to come.

An overwhelming consensus of scientific opinion now agrees that climate change is happening, is caused in large part by human activities (such as burning fossil fuels), and if left un-checked, will have disastrous consequences. Furthermore, there is solid scientific evidence that we should act now. This is reflected in the conclusions of the Intergovernmental Panel on Climate Change (IPCC), a UN institution of more than 1,000 scientists providing advice to policy makers. In response to this threat, the Kyoto Protocol has committed its signatories to reduce their greenhouse gas emissions by 5.2 per cent from their 1990 level by the target period of 2008-2012.This in turn has resulted in the adoption of a series of regional and national reduction targets. In the European Union, for instance, the commitment is to an overall reduction of 8 per cent. In order to reach this target, the EU has also agreed to increase its proportion of renewable energy from 6 per cent to 12 per cent by 2010. The Kyoto signatories are currently negotiating the second phase of the agreement, covering the period from 2013-2017. Within this timeframe industrialized countries need to reduce their $CO_2$ emissions by 18 per cent from 1990 levels, and then by 30 per cent between 2018 and 2022. Only with these cuts do we stand a reasonable chance of keeping the average increase in global temperatures to less than 2°C, beyond which the effects of climate change will become catastrophic. Alongside global warming, other challenges have become just as pressing. Worldwide energy demand is growing at a staggering rate. Over-reliance on energy imports from a few, often politically unstable countries and volatile oil and gas prices have together pushed security of energy supply to the top of the political agenda, as well as threatening to inflict a massive drain on the global economy. But whilst there is a broad consensus that we need to change the way we produce and consume energy, there is still disagreement about how to do this (Greenpeace International, European Renewable Energy Council (EREC)).

## Rescue Measures

Governments all over the world are responding with increasing urgency. We have seen a raft of policies from governments around the world to encourage the development and use of cleaner energy. And the scale of investment in the sector has been staggering. Some of the bigger drivers behind the sector have been China and the US. In China, some US$184bn each year is expected to be invested to achieve its target of generating 15 per cent of its electricity from renewable energy by 2020. In the US, proposals by the current administration should see a fivefold increase in renewable and alternative fuel use by 2017. Elsewhere, the European Union has committed to source 20 per cent of all energy from renewable sources by 2020. These are positive steps, each causing unprecedented demand in the sector.

In addition, it is worth remembering that the former chief economist of the World Bank, Sir Nicholas Stern, in his report clearly pointed out that the ones who invest in energy saving technologies and renewable energies today will be the economic winners of tomorrow. Inaction will be much more expensive in the long run, than taking action now.

## Upcoming Opportunities

The aggregate economic and environmental outcomes of developing alternative fuels results in clean energy being one of the investment themes with maximum importance for green funds today. This in itself is a big achievement, as only 15 years ago the sector was almost nonexistent. Back then, there were no quoted companies operating in the alternative energy sector. Now there are over 90 listed companies worldwide and many more at an unquoted level, representing great diversification potential. Particularly impressive is the growing profitability of the sector and, if forecast growth in revenues is any indication, future profitability looks bright indeed. Currently at US $55.4 bn, annual revenue from the alternative energy sector is expected to grow to US$ 226.5 bn by 2016. Also impressive is the way that companies have become more efficient, with demand for alternative energy leading to vast technology improvements and economies of scale. This in turn has increased the cost competitiveness of the industry. Wind is nearing cost competitiveness with other power generating sources, for example, and the costs for both wind and solar technologies have halved over the past 15 years. Opportunities in the sector include companies operating in wind, solar, fuel cells, bio-fuels, energy efficiency and air quality among others.

## Energy Efficiency

Till the opportunities in the alternative energy sector are ready to be tapped economically, the best solution in front of us is improving energy efficiency. The idea is to optimally utilize and convert the present energy supply structure into a sustainable system with least dent on the eco-system. The efforts should be aimed at the ambitious exploitation of the industries for the potential for energy efficiency. The focus should be on balancing the current best practice and available technologies for the future. The energy savings should be fairly equally distributed over all the sectors of the economy. A little improvement in the efficiency goes a long way in boosting the energy supply scenario.

With that in mind, governments are prodding businesses to make their products more efficient. A voluntary agreement between the European Union and big carmakers has helped boost fuel economy 12 per cent above its 1995 level, although the target of 25 per cent by 2009 will not be met. The airline industry has promised great strides in jet-engine efficiency to mitigate the environmental damage caused by flying. Best of all, unlike proposals for green taxes or higher electricity prices to pay for expensive windmills and nuclear plants, the prospect of lower power and petrol bills makes efficiency measures attractive to consumers. Given that transport accounts for between one-quarter and one-third of the emissions of most developed countries, a 12 per cent improvement in fuel efficiency sounds impressive. This is only a flavor of the sort of opportunities available in the sector, but they highlight the diversity that the sector affords. It is worth noting, however, that while the advances being made in solar and bio-fuels technologies are impressive, they are yet to achieve the financial strength required for significant inclusion in many green portfolios. Overall, the clean energy sector is of high importance for both the global environment and economy. Heightened concerns about climate change and energy security are expected to underpin rapid growth in the sector for many years to come. For those wanting to make a difference with their capital, it could be argued that this sector provides some very compelling opportunities and confirms that going green is not only good for the environment, but also good for profits.

Improving energy efficiency is the basic philosophy for future energy conservation. At a time when governments around the world are in the process of liberalizing their energy markets, the increasing competitiveness of renewable energy should lead to higher demand. Without political support, however, renewable energy remains at a disadvantage, marginalized by distortions in the world's energy markets created by decades of massive financial, political and structural support to conventional technologies and the failure to internalize environmental and social costs in price of energy. Developing renewable energy will therefore require strong political and economic efforts, especially through laws that guarantee stable tariffs over a period of up to 20 years. At present new renewable energy generators have to compete with old nuclear and fossil fuelled power stations which produce electricity at marginal costs because consumers and taxpayers have already paid the interest and depreciation on the original investments. Political action is needed to overcome these distortions and create a level playing field. In the current background of current political frameworks and barriers that need to be overcome in order to unlock renewable energy's great potential to become a major contributor to global energy supply, a combined global effort is required. In the process, it would also contribute to sustainable economic growth, high quality jobs, technology development, global competitiveness and industrial and research leadership.

Energy Efficiency needs to become the center piece of the national and state energy policy since energy saved is as good as energy produced. The only way to balance economic growth and satisfy domestic comfort without endangering the future of life on the planet, is by gradually raising the level of energy efficiency of domestic and commercial projects.

## Ensuring Energy Security

The World Energy Assessment (UNDP 1999) report defines energy security as: "the continuous availability of energy in varied forms in sufficient quantities at reasonable prices". India's energy security, at its broadest level, is primarily about ensuring the continuous availability of commercial energy at competitive prices to support its economic growth and meet the lifeline energy needs of its households. Reducing energy requirements and increasing efficiency are two very important measures to increase energy security. Investing in and producing renewable energy at sustainable costs will directly affect energy security of the developing nations like India as immediate costs are prohibitive. The only viable solution left for them in this scenario is to improve the efficiency of current technologies to meet their energy demands in the cleanest possible way. This paper looks into the aspect of increasing efficiency to improve status of energy security.

## Solutions?

Fundamental reforms are needed to generate competitive efficiency. Some questions which we should ask ourselves are:

What is the scope for increasing the energy efficiency of the system? What policies can lead to higher efficiency? How do we encourage energy conservation and energy efficiency?

This must be done in a technically efficient, economically viable and environmentally sustainable manner using different fuels and forms of energy, conventional and non-conventional, as well as new and emerging energy sources to ensure supply at all times with a prescribed confidence level considering that shocks and disruption can be reasonably expected. It may be noted that a unit of energy saved by a user is greater than a unit produced as it saves on production, transport, transmission and distribution losses. Thus a "Negawatt" (a negative Megawatt) produced by reducing energy need saves more than a Megawatt generated.

## The Indian Scenario

India faces formidable challenges in meeting its energy needs and in providing adequate energy of desired quality in various forms in a sustainable manner and at competitive prices. India needs to sustain an 8 per cent to 10 per cent economic growth rate, over the next 25 years, if it is to eradicate poverty and meet its human development goals. To deliver a sustained growth rate of 8 per cent through 2031-32 and to meet the lifeline energy needs of all citizens, India needs, at the very least, to increase its primary energy supply by 3 to 4 times and, its electricity generation capacity/supply by 5 to 6 times of their 2003-04 levels. With 2003-04 as the base, India's commercial energy supply would need to grow from 5.2 per cent to 6.1 per cent per annum while its total primary energy supply would need to grow at 4.3 per cent to 5.1 per cent annually. By 2031-32 power generation capacity must increase to nearly 8,00,000 MW from the current capacity of around 1,60,000 MW inclusive of all captive plants. Similarly requirement of coal, the dominant fuel in India's energy mix will need to expand to over 2 billion tones/annum based on domestic quality of coal. Meeting the energy challenge is of fundamental importance to India's economic growth imperatives and its efforts to raise its level of human development. *(Integrated Energy Policy Report of the Expert Committee, Government of India Planning Commission New Delhi August 2006)*[2]

The Indian Government has to achieve a configuration of the various forms of energy which will suit its resources and levels of technological advancement. Which in turn, requires a carefully planned and implementable energy policy and improved efficiencies across the energy chain. India's vast developmental needs require a serious look into the energy supplies and badly require a check on the leakages across the system which automatically puts a dent on energy resources available with the country.

In the recent past the energy production in India has not kept pace with the industrial growth and subsequent energy demands. This has led to serious void in the energy demand supply chain threatening the basic growth momentum. The planning authorities are trying to mitigate this critical risk with multipoint approach. Increasing renewable and fossil fuel free (sic. nuclear and wind) energy production, exploration for onshore and offshore pockets of commercially viable fossil fuels and encouraging private players to participate in energy production are some of the measures that government has undertaken in recent years. Installation of Captive power plants by private players has been one of the popular measures with market now days. Captive power refers to generation from a unit set up by industry for its exclusive consumption. This not only ensures a secured and uninterrupted power supply for the producer but also translates into hefty profits by selling the surplus energy to government transmission bodies for distribution through their network.

The estimates on captive power capacity in the country vary with the Central Electricity Authority putting the figure at about 11600 MW while industry experts feel that it is much higher, close to 20000 MW Industrial sector is one of the largest consumers of electrical energy in India. However, a number of industries are now increasingly relying on their own generation (captive and cogeneration) rather than on grid supply, primarily for the following reasons:

- ☆ Non-availability of adequate grid supply
- ☆ Poor quality and reliability of grid supply
- ☆ High tariff as a result of heavy cross subsidization

---

2 Integrated Energy Policy, Report of the Expert Committee, Government of India, Planning Commission, New Delhi-August 2006.

## Liberalized Policies Regarding Captive Power Plant and its Impact on Environment

Acute power shortage few years back has forced government to liberalize the policy regarding captive power plant to allow putting up in private sector to ensure availability of uninterrupted power supply to industries. Government also promised to purchase surplus power. As a result there was rush to install captive power plants. To hasten the process and to cut capital costs, industrialist started importing closed down units abroad. This is normal scenario in any developing country where power is the most sought commodity.

However, careful planning and proper policy can mitigate the problem likely to crop up. Some of the issues are illustrated below:

1. *Use of obsolete technology*–Many captive power plants in India are essentially, imported ones. Smaller power plants abroad which were closed down due to various reasons *e.g.* economic viability, statutory requirement on environmental pollution, safety and reliability requirements, obsolete technology, very high operation and maintenance cost are imported here.

   These smaller sets were based on design and technology which is outdated and considered unsafe. In many cases virtually scrap, damaged units were imported because of pure economic reasons and attempts were made to assemble the units here. In most of the cases, such imports are not backed by design details, operational history defects/inspection records leading to safety and reliability problems. Such machines are down rated to give operational reliability however, at cost of efficiency. This method though appears profitable in short-run turns out to be the costly one in the long-run.

2. *Wrong economic viability criteria*–Captive power plants are part of industrial setup, hence separate economic evaluation may not have been done for a profitable venture. However recurring loss on account of poor efficiency and higher maintenance cost and operation costs, these sets are economically unviable if judged on basis of lifecycle costs. Being part of industrial setup, penalty on account of unavailability is very high.

3. *System incompatibility*–Each power plant when originally designed, envisages a particular system configuration, whereas these uprooted sets are fitted in different system configuration resulting in lot of compatibility problems causing frequent tripping/shutdown. A very large portion of problems encountered in any power plant are due to system incompatibility. Each startup and shutdown carries a price tag apart from large quantum of heat being dissipated to atmosphere and consumption of machines life.

4. *Large variety, lack of standardization*–There is no guideline regarding output or design for installation of captive power plant. Irrespective whether it is imported or indigenously manufactured one, large varieties of machines, lack of standardization lead to supply problem of quality spares as most of these spares when not backed by designed and drawing details are developed by reverse engineering. Also small quantum does not justify detail testing, inspection or analysis.

   And again, repair and overhaul are done on compromising pattern. Poor quality of spares leads to higher clearance leading of high leakage, bearing instability and malfunction of control system. As a sequel to this higher vibration, higher noise level and larger heat dissipation to atmosphere gives rise to uncomfortable operating condition apart from low availability and higher maintenance costs.

5. *Residual life assessment (RLA) and life extension program (LEP)*–Attempts are made to enhance the working life of these smaller sets by using modern techniques of Residual life assessment and life extension programs. No doubt RLA and LEP are cheapest, easiest and fastest way of meeting power shortage and have been extensively employed all over the world.

   However, absence of proper design, material data and operating history, defect/repair record does not enable proper analysis and reliable results from RLA and LEP. Also this inadvertently leads to:

   (*a*) Extension of obsolete technology which delays adoption of latest technology

   (*b*) Loss of fuel on account of poor efficiency

   (*c*) Environmental damage due to poor emission norms

   (*d*) Risk and safety hazard to community

   (*e*) Inconvenience to operating staff resulting in physical and mental problems

   (*f*) Problems in grid management especially in case of states with surplus power

6. *Who is at receiving end*?–No doubt community at large has to face long term repercussions mainly -

   (*a*) *Environmental pollution*- Higher level of flue gas and fly ash, vibration, noise and heat. Adverse environmental impacts arising from types of fuels used and from higher emissions per unit of production, as compared to large power plants.

   (*b*) *Health Hazard*–on account of uncomfortable working condition due to high level noise vibration and heat causing physical and mental problems.

   (*c*) *Safety considerations*–Explosions, failures, poor reliability and availability.

Social responsibility and accountability are main issues which must be evaluated. Benefits to industrialists vs. loss to community must be thoroughly evaluated while forming such policies. It may not be out of place to mention even industry is not gainer if lifecycle cost is considered in place of capital cost.

## Policy Recommendations

With improving power scenario, it is high time to review this policy in overall national interest.

Policy may include recommendations about:

1. Output slabs–modular design concepts
2. Preferred inlet parameters
3. Evaluation on basis of lifecycle costs- The life cycle costing should be used for all decision-making and alternatives should be compared in terms of expected present discounted values of life cycle cost.
4. Putting up captive power plant on cooperative basis so that unit size can grouped to improve efficiency.
5. Setting up professional houses to supply spares on regional basis.
6. Government's technical control on import of uprooted sets.

The captive power plants have been taken into operation at a time, when the country was struggling through problems like acute power shortage and financial constraints. Now, when the conditions are improving, we must stop and take a look to analyze the complete social and economic costs and benefits of the units. Its time to understand that these discarded units are imported from the countries which had stopped using it and to analyze the reasons behind the action. We should look into the life-cycle costs of these units. Moreover, the people who are operating these units have no operational experience because of large variants right from 0.5 MW to 60 MW. As a result, they are not able to extract the best out of it. There is no standardization in the field as the basic mode of operation and duty is changed with import of uprooted sets. All critical components like rotors and blades should be thoroughly analyzed. Economic viability should be judged.

"Simple and inexpensive improvements in Indian power plants can increase efficiency by more than 1.5 per cent. Replicating these improvements at over 130 small coal power plants could reduce India's $CO_2$ emissions by over 10 million tons per year and reduce fuel costs by over $150 million per year." *(Asia-Pacific Partnership on Clean Development and Climate: Power Generation and Transmission Task Force Summary of Action Plan and Projects)*[3]

Technological choices largely determine the long-term characteristics of industrial society, including impacts on the natural environment. However, the treatment of technology in existing models that are used to project economic and environmental futures can make a huge difference. A detailed study must be made to understand the best practices of REAP (Risk Evaluation and Prioritization) in power plant utilities and acquire information on prior experience associated with execution as it would pertain to Indian utilities. The study should highlight power plant equipment integration and analyze reasons behind failed power generation, which could be due to equipment, location, operational parameters, and constructional features. It should assist in evaluating solutions and prioritizing them based on cost, down time cost, frequency of occurrence of such failure, and reduced force outages. There is a dire need for power plant modernization in India. The correct step towards modernization and efficiency improvement will give all-round results in the future.

One of the unfortunate consequences of technological advance has been the damage we do to the society and environment in the long-term. The onus of giving a better and clean future to the coming generations is on us. The corrective actions should start at every stage to ensure this.

---

3 Asia-Pacific Partnership on Clean Development and Climate - Bureau of Public Affairs Washington, DC, October 31, 2006- http://www.state.gov/r/pa/scp/2006/75384.htm

# Chapter 50

# Non-Edible Vegetable Oil as a Bio-fuel: An Alternative for Sustainable Development of the Rural Community

**P. Lingfa*, P.M.V. Subbarao and M.K.G. Babu**

**Centre for Energy Studies and Department of Mechanical Engineering, Indian Institute of Technology (IIT), New Delhi – 110 01**

## ABSTRACT

The depletion of the world petroleum reserves coupled with the global environmental problems stimulated the search for alternative sources of petroleum-based fossil fuel. In the present work, non-edible vegetable oil (Pongamia Piñata) has been chosen for the study. Biodiesel were produced from this vegetable oil using transesterification process in a small scale biodiesel reactor. The physical and chemical properties were tested and compared with those of the conventional diesel fuel. It was found that they are very close to each other and hence could be used as a substitute for commercial diesel fuel. After testing the emissions and performance characteristics in medium scale diesel engine, it was observed that, biodiesel obtained from non-edible vegetable oil (*Pongamia pinnata*), has clear advantages over the commercial diesel fuel, except a little sacrifice in nitride oxide (NOx). Besides this, after extracting the raw oil from the seeds, remaining oilcake were analyzed and found to be a promising source for hydrogen gas, besides using it for heating purposes. Moreover a brief economic study was also carried out and

* Corresponding Author: E-mail: plingfa@yahoo.com

found that cost of biodiesel and commercial diesel per litre are comparable. Also it was established that the energy consumption for producing unit quantity of biodiesel is less compared to that for conventional diesel fuel, thereby saving a considerable amount of energy resources.

***Keywords:*** *Biodiesel, Transesterification.*

## Introduction

In view of the escalating energy (fuel) crisis and environmental degradation non-edible vegetables oil is one of the sources for replacement of petro-diesel, thereby reducing country's dependence on importing petroleum from overseas country. The concept for producing fuel from non-edible vegetable oil for diesel engine is not radically new concept. When Rudolf Diesel first invented the diesel engine, about century ago, he demonstrated the principle by employing peanut oil and hinted that vegetable oil would be the future fuel in the diesel engine.

## Historical Review

Ever since the advent of the I.C engine, Researchers in various countries tried and carried out many experimental works using vegetable oils as a substitute for diesel in I.C engine and other applications. They concluded that vegetable oils as fuel for I.C engine can used in engine without any modifications to existing design of the engine and can also play a vital role in helping the developing and developed world to reduce the environmental impact of fossil fuels and pollutions.

## Need for Use of Non-edible Vegetable Oil (Biodiesel)

- To provide energy security to the rural people.
- To provide employment and income generation to rural people.
- To reduce expenditures on health problems arising out of vehicular emission.
- Environmental protections, reduced emission, reduce the oil import.
- Rapid increase in price of petroleum and uncertainty about their supply.

## Methods of Production of Biodiesel from Non-edible Vegetable Oil

Biodiesel is a fatty acid or methyl ester made from virgin or used vegetable oils (both edible and non-edible) and animals fats. There are four primary ways to produced vegetable oil as follows:

- Direct use or blending in diesel fuel
- Micro emulsions
- Thermal cracking (pyrolysis)
- Transesterification

Out of thse, transesterification is one of the most popular and best way to produced vegetable oil in the diesel engine. There are three basics route to produced biodiesel by transesterication of vegetable oil as:

- Based catalyst
- Acid catalyst

Based catalyst is the easiest, popular and most commercial used process because of the following reasons:

- ✰ Greater than 90 per cent of all biodiesel is produced by this method.
- ✰ Low temperature (60°C) and pressure (1 atmosphere) are required
- ✰ High conversion (greater than 98 per cent can be obtained.
- ✰ Minimum side reactions.
- ✰ Involves direct conversion and no intermediate steps.
- ✰ Requires ordinary materials of construction of the set up.

### Transesterification Process

| | | |
|---|---|---|
| $CH_2$-OOC-R1 | catalyst | R1-COOR $CH_2OH$ |
| $CH_2$-OOC-R2 + 3ROH | ↔ | R2-COOR + CHOH |
| $CH_2$-OOC-R3 | | R3-COOR $CH_2OH$ |
| Triglyceride Alcohol | | Esters (Biodiesel) Glycerol |

R1, R2, R3, are long chain alkyl group and R, short chain alkyl group

## Steps Involved in Production of Biodiesel Using Medium Capacity Biodiesel Reactor

The production of biodiesel consists transesterification of vegetable oil, which reduces the viscosity of vegetable oils and makes it comparable with that of pure diesel. Any of the vegetable oil can be converted into biodiesel through transesterification using ethanol or methanol.

For the production of biodiesel the following steps are involved:

1. Mixing of alcohol and catalyst
2. Reaction
3. Separation
4. Water washing

### Mixing of Alcohol and Catalyst

The alcohol (methanol) and catalyst (NaOH) were mixed in required proportion to study the effects on transesterification process. The 1 per cent catalyst was use to complete the reaction.

### Reaction

The non-edible vegetable oil (karanja) was charged into the reactor then the required amount of alcohol and catalyst (Molar ratio 12:1) were added to the reactor. Now the mixture was heated for 3 hours at a constant temperature of 60-70° C. It was then stirred vigorously, with electric stirrer until the temperature reached a constant value.

### Separation

After the completion of reaction, the mixture mainly consisted of two products, namely, biodiesel and glycerol. Then, it is separated by using funnel. Since two layers are formed, the upper layer being methyl ester and the rest being glycerol. The two layers are clearly seen. The light layer on the top is the biodiesel while the darker layer is glycerol.

## Biodiesel after Water Washing

After discarding the glycerol layer from the separating funnel the methyl ester was mixed with, distilled water, shaken gently and allowed to settle for 10 minutes. The procedure was repeated for 3 to 4 times. The product obtained is the methyl ester of the Karanja oil.

## Experiment Set-up (Figures 50.1–50.2)

**Table 50.1: Observation for Physical Properties of Biodiesel, Karanja and Diesel**

| *Sl.No.* | *Name of Oils* | *CV Mj/Kg* | *Viscosity at 40°C* | *Specific Gravity (mg/ml)* | *Flash Point (°C)* | *Fire Point (°C)* | *Cloud Point (°C)* | *Pour Point* |
|---|---|---|---|---|---|---|---|---|
| 1. | Karanja oil | 36.00 | 20.18 | 0.921 | 230 | 270 | - | - |
| 2. | Biodiesel | 38.02 | 4.7 | 0.875 | 185 | 225 | 21°C | 8°C |
| 3. | Diesel | 43.47 | 4.2 | 0.825 | 135 | 150 | - | - |

**Figure 50.1: 10-Litre Capacity Biodiesel Reactor**

**Figure 50.2: Karanja Tree with Seeds**

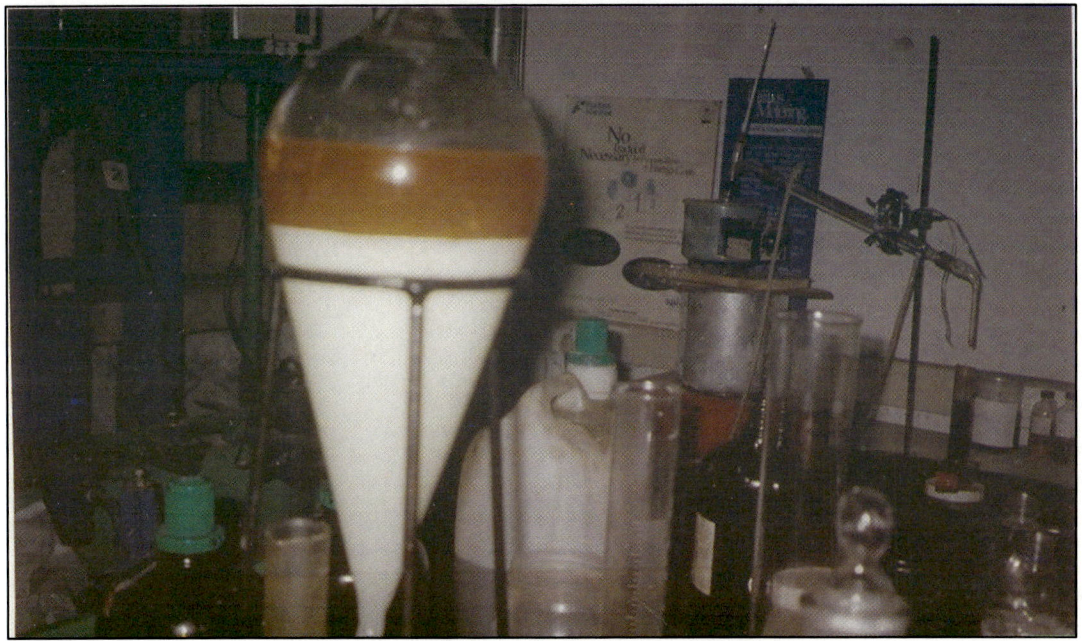

**Figure 50.3: Biodiesel after Water Washing**

**Table 50.2: Observation for Distillation of Commercial Diesel, Biodiesel and Vegetable Oil**

| *Sl.No* | *Volume Recovered (per cent)* | *Temperature (°C) Diesel* | *Temperature (°C) Vegetable Oil* | *Temperature (°C) Biodiesel* |
|---|---|---|---|---|
| 1. | 0 | 140 | 136 | 96 |
| 2. | 10 | 190 | 280 | 260 |
| 3. | 20 | 202 | 290 | 292 |
| 4. | 30 | 248 | 296 | 314 |
| 5. | 40 | 258 | 290 | 318 |
| 6. | 50 | 270 | 288 | 317.8 |
| 7. | 60 | 290 | 276 | 318.9 |
| 8. | 70 | 302 | 272 | 312 |
| 9. | 80 | 320 | 284 | 296 |
| 10. | 90 | 350 | 290 | 306 |

## Engine Test Set-up

The aim of the engine test is to find out feasibility of using biodiesel (karanja oil) produced from small-scale biodiesel reactor as a partial substitute of diesel oil in CI engine and for rural applications.

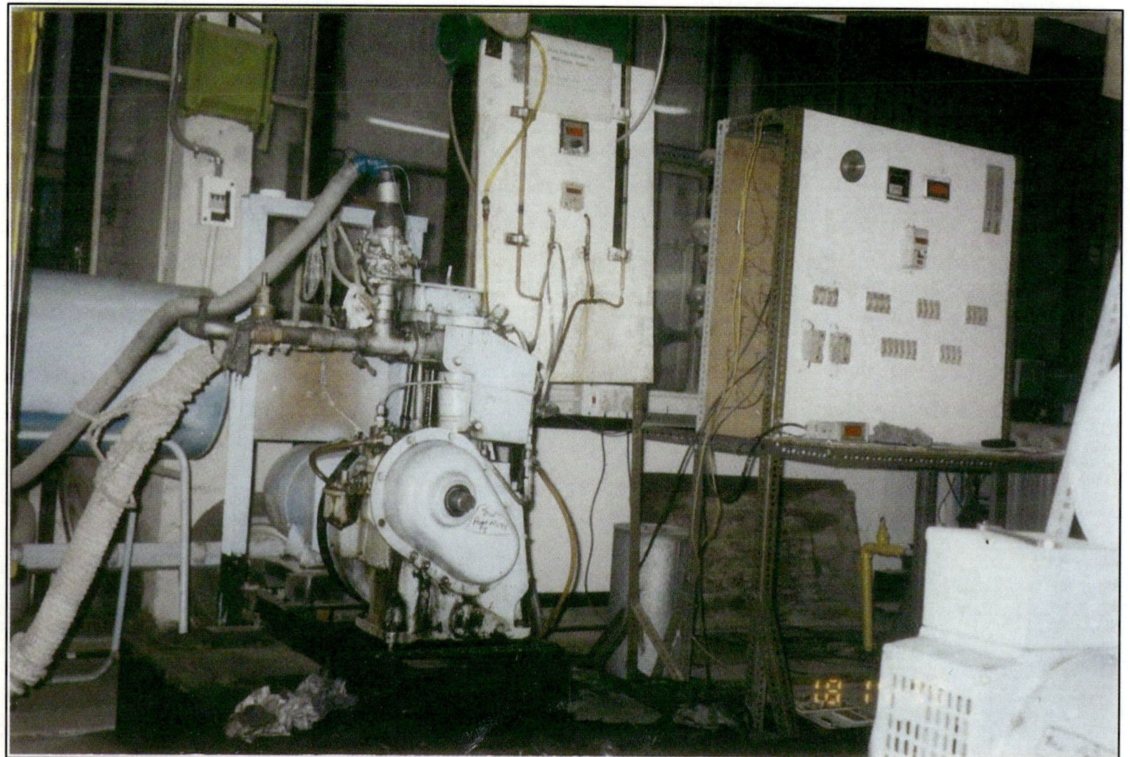

**Figure 50.4: Single Cylinder Four Stroke Diesel Engine**

Hence the performance and emission test were carried out using blends, pure biodiesel and diesel in single cylinder 4-stroke diesel engine at different load conditions.

## Observations

Observations shown in Tables 50.3–50.7.

## Result and Discussion

### Measurement and Comparison of Physical, Chemical Properties of Pure Biodiesel, Raw Karanja and Diesel Oil

The detail measurement of physical, chemical properties of pure biodiesel, raw karanja oil and diesel oil were carried out. From it analysis, it was observed that physical, chemical properties of pure biodiesel and diesel are comparable.

### Engine Test Results

#### Effect on Brake Thermal Efficiency of Brake Power

From the test results it is observed that initially with increasing brake power the brake thermal efficiencies of the diesel fuel and biodiesel and its blends increases. But there was a considerable increase in efficiencies with the blends of biodiesel compared to the efficiency of diesel fuel alone. The lower brake thermal efficiency for B100 could be due to reduction in calorific value and increase in fuel consumption compared to diesel, B10 and B5.

#### Effect on Specific Fuel Consumption of Brake Power

It is observed that specific fuel consumption of all the fuel decrease with increase in brake power. Very little difference can be seen for all the blends of biodiesel at higher load; however as brake power

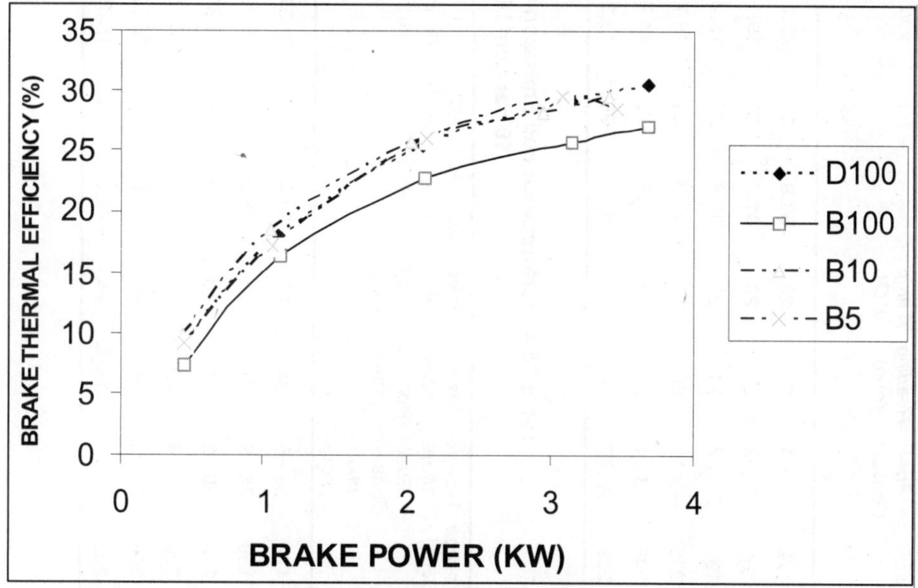


**Figure 50.5: Effect on Brake Thermal Efficiency of Brake Power**

**Table 50.3: Performance and Emission Characteristics of the Engine when Running on Pure Diesel Oil (D100)**

| Speed (rpm) | Voltage (volt) | Time for 10 ml Fuel Consumption (sec) | Manometer Reading (mm) | Inlet Air Temp (°C) | Opacity (%) | HC (ppm) | $O_2$ (% in vol) | $CO_2$ (% in vol) | CO (% in vol) | $NO_x$ (ppm) | BP (V*I) KW | TFC* $10^{-4}$ (kg/sec) | BSFC (kg/kw-hr) | Current (Amp) | Exhaust Gas Temp (°C) | Brake Thermal Efficiency (%) |
|---|---|---|---|---|---|---|---|---|---|---|---|---|---|---|---|---|
| 1308 | 172 | 91.04 | 21 | 20 | 53.9 | 16 | 22.2 | 1.56 | 0.09 | 37 | 0 | 9.06 | 0 | 0 | 175 | 0 |
| 1312 | 190 | 70.93 | 21 | 20 | 81.7 | 21 | 19.1 | 3.01 | 0.15 | 125 | 0.46 | 1.16 | 0.911 | 2.40 | 208 | 9.09 |
| 1313 | 199 | 58.32 | 21 | 20 | 93.8 | 27 | 18.2 | 3.85 | 0.15 | 205 | 1.11 | 1.42 | 0.459 | 5.55 | 256 | 18.03 |
| 1311 | 206 | 43.10 | 20 | 20 | 95.9 | 51 | 16.7 | 5.64 | 0.15 | 382 | 2.11 | 1.92 | 0.326 | 10.25 | 340 | 25.34 |
| 1303 | 209 | 33.24 | 20 | 21 | 98.8 | 72 | 14.7 | 7.76 | 0.14 | 489 | 3.15 | 2.48 | 0.284 | 15.07 | 430 | 29.17 |
| 1306 | 209 | 29.77 | 19 | 21 | 100 | 96 | 13.9 | 8.45 | 0.10 | 491 | 3.68 | 2.77 | 0.272 | 17.64 | 494 | 30.5 |

**Table 50.4: Performance and Emission Characteristics of the Engine when Running on B-10 (10 per cent biodiesel: 90 per cent diesel)**

| Speed (rpm) | Voltage (volt) | Time for 10 ml Fuel Consumption (sec) | Manometer Reading (mm) | Inlet Air Temp (°C) | Opacity (%) | HC (ppm) | $O_2$ (% in vol) | $CO_2$ (% in vol) | CO (% in vol) | $NO_x$ (ppm) | BP (V*I) KW | TFC* $10^{-4}$ (kg/sec) | BSFC (kg/kw-hr) | Current (Amp) | Exhaust Gas Temp (°C) | Brake Thermal Efficiency (%) |
|---|---|---|---|---|---|---|---|---|---|---|---|---|---|---|---|---|
| 1320 | 172 | 96.65 | 21 | 23 | 33.4 | 26 | 21.57 | 0.01 | 0 | 0 | 0 | 8.639 | 0 | 0 | 181 | 0 |
| 1317 | 189 | 78.39 | 21 | 23 | 48.5 | 31 | 19.59 | 2.77 | 0.09 | 108 | 0.43 | 1.065 | 0.892 | 2.28 | 210 | 9.73 |
| 1308 | 200 | 60.52 | 21 | 23 | 53.9 | 48 | 18.80 | 3.75 | 0.13 | 275 | 1.06 | 1.379 | 0.468 | 5.30 | 253 | 18.52 |
| 1305 | 205 | 44.04 | 20 | 24 | 75.4 | 59 | 17.2 | 5.62 | 0.12 | 360 | 2.03 | 1.896 | 0.336 | 9.90 | 336 | 25.79 |
| 1303 | 208 | 33.23 | 19 | 24 | 95.5 | 68 | 16.37 | 5..90 | 0.14 | 427 | 2.95 | 2.513 | 0.306 | 14.20 | 424 | 28.28 |
| 1308 | 208 | 30.06 | 18 | 24 | 98.6 | 112 | 15.08 | 7.1 | 0.25 | 438 | 3.41 | 2.777 | 0.293 | 16.40 | 486 | 29.59 |

**Table 50.5: Performance and Emission Characteristics of the Engine when Running on Pure Biodiesel Oil (B100)**

| Speed (rpm) | Voltage (volt) | Time for 10 ml Fuel Consumption (sec) | Manometer Reading (mm) | Inlet Air Temp (°C) | Opacity (%) | HC (ppm) | $O_2$ (% in vol) | $CO_2$ (% in vol) | CO (% in vol) | $NO_x$ (ppm) | BP (V*I) KW | TFC* $10^{-4}$ (kg/sec) | BSFC (kg/kw-hr) | Current (Amp) | Exhaust Gas Temp (°C) | Brake Thermal Efficiency (%) |
|---|---|---|---|---|---|---|---|---|---|---|---|---|---|---|---|---|
| 1306 | 172 | 75.22 | 22 | 20 | 48 | 19 | 20.02 | 2.25 | 0.16 | 55 | 0 | 1.183 | 0 | 0 | 178 | 0 |
| 1312 | 190 | 60.01 | 21 | 20 | 77.5 | 21 | 19.52 | 3.42 | 0.19 | 145 | 0.45 | 1.483 | 1.186 | 2.37 | 208 | 7.36 |
| 1314 | 201 | 53.69 | 21 | 20 | 94.7 | 24 | 18.68 | 4.08 | 0.19 | 217 | 1.12 | 1.657 | 0.533 | 5.56 | 263 | 16.41 |
| 1314 | 209 | 39.22 | 20 | 20 | 97.9 | 33 | 17.17 | 6.13 | 0.18 | 430 | 2.12 | 2.269 | 0.385 | 10.12 | 340 | 22.67 |
| 1313 | 209 | 29.93 | 20 | 20 | 98.6 | 60 | 14.43 | 8.09 | 0.26 | 515 | 3.15 | 2.974 | 0.339 | 15.10 | 440 | 25.71 |
| 1314 | 212 | 27.03 | 19 | 20 | 99.7 | 80 | 13.25 | 9.36 | 0.41 | 552 | 3.67 | 3.291 | 0.323 | 17.30 | 491 | 27.07 |

**Table 50.6: Performance and Emission Characteristics of the Engine when Running on B-10 (10 per cent biodiesel: 90 per cent diesel)**

| Speed (rpm) | Voltage (volt) | Time for 10 ml Fuel Consumption (sec) | Manometer Reading (mm) | Inlet Air Temp (°C) | Opacity (%) | HC (ppm) | $O_2$ (% in vol) | $CO_2$ (% in vol) | CO (% in vol) | $NO_x$ (ppm) | BP (V*I) KW | TFC* $10^{-4}$ (kg/sec) | BSFC (kg/kw-hr) | Current (Amp) | Exhaust Gas Temp (°C) | Brake Thermal Efficiency (%) |
|---|---|---|---|---|---|---|---|---|---|---|---|---|---|---|---|---|
| 1320 | 172 | 96.65 | 21 | 23 | 33.4 | 26 | 21.57 | 0.01 | 0 | 0 | 0 | 8.639 | 0 | 0 | 181 | 0 |
| 1317 | 189 | 78.39 | 21 | 23 | 48.5 | 31 | 19.59 | 2.77 | 0.09 | 108 | 0.43 | 1.065 | 0.892 | 2.28 | 210 | 9.73 |
| 1308 | 200 | 60.52 | 21 | 23 | 53.9 | 48 | 18.80 | 3.75 | 0.13 | 275 | 1.06 | 1.379 | 0.468 | 5.30 | 253 | 18.52 |
| 1305 | 205 | 44.04 | 20 | 24 | 75.4 | 59 | 17.2 | 5.62 | 0.12 | 360 | 2.03 | 1.896 | 0.336 | 9.90 | 336 | 25.79 |
| 1303 | 208 | 33.23 | 19 | 24 | 95.5 | 68 | 16.37 | 5..90 | 0.14 | 427 | 2.95 | 2.513 | 0.306 | 14.20 | 424 | 28.28 |
| 1308 | 208 | 30.06 | 18 | 24 | 98.6 | 112 | 15.08 | 7.1 | 0.25 | 438 | 3.41 | 2.777 | 0.293 | 16.40 | 486 | 29.59 |

**Table 50.7: Performance and Emission Characteristics of the Engine when Running on B-5 (5 per cent biodiesel: 95 per cent diesel)**

| Speed (rpm) | Voltage (volt) | Time for 10 ml Fuel Consumption (sec) | Manometer Reading (mm) | Inlet Air Temp (°C) | Opacity (%) | HC (ppm) | $O_2$ (% in vol) | $CO_2$ (% in vol) | CO (% in vol) | $NO_X$ (ppm) | BP (V*I) KW | TFC* $10^{-4}$ (kg/sec) | BSFC (kg/kw-hr) | Current (Amp) | Exhaust Gas Temp (°C) | Brake Thermal Efficiency (%) |
|---|---|---|---|---|---|---|---|---|---|---|---|---|---|---|---|---|
| 1313 | 174 | 87.44 | 22 | 23 | 40.1 | 34 | 19.41 | 2.33 | 0.10 | 40 | 0 | 9.49 | 0 | 0 | 170 | 0 |
| 1308 | 190 | 72.45 | 22 | 23 | 64.5 | 42 | 19.04 | 2.96 | 0.11 | 89 | 0.44 | 1.146 | 0.937 | 2.30 | 206 | 9.14 |
| 1314 | 203 | 56.13 | 21 | 23 | 80.8 | 48 | 18.31 | 3.83 | 0.14 | 207 | 1.07 | 1.478 | 0.497 | 5.30 | 257 | 17.23 |
| 1316 | 209 | 42.53 | 21 | 23 | 85.6 | 55 | 16.96 | 5.58 | 0.12 | 368 | 2.17 | 1.952 | 0.328 | 10.26 | 340 | 26.11 |
| 1306 | 298 | 33.36 | 19 | 23 | 99.4 | 67 | 16.01 | 6.04 | 0.22 | 380 | 3.08 | 2.488 | 0.291 | 14.80 | 474 | 29.49 |
| 1306 | 212 | 28.81 | 18 | 23 | 99.6 | 97 | 15.60 | 6.64 | 0.31 | 392 | 3.46 | 2.881 | 0.299 | 16.31 | 504 | 28.59 |

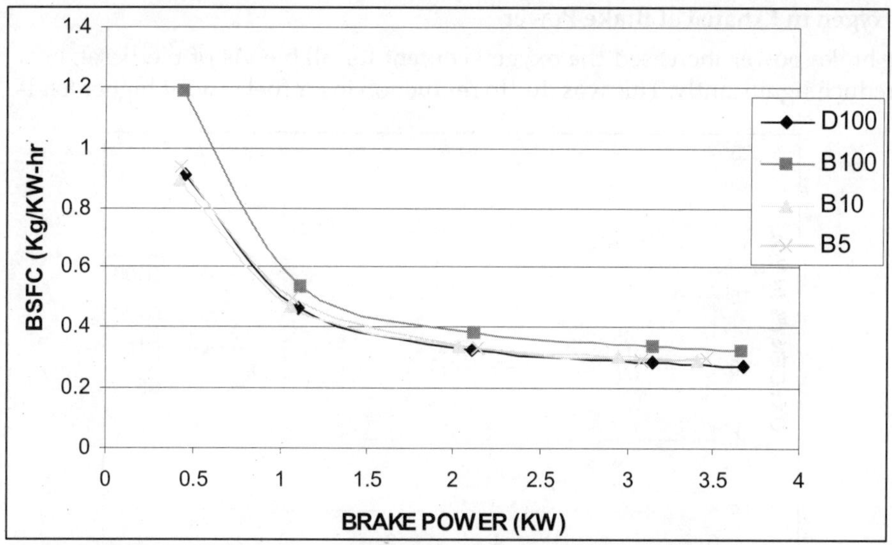


**Figure 50.6: Effect on Specific Fuel Consumption of Brake Power**

increases the specific fuel consumption decreases for all fuels, pure as well as blends. The specific fuel consumption of the B100 is higher than that of diesel; it is because of the lower calorific value.

**Effect on Unburned Hydrocarbon in Exhaust of Brake Power**

There is considerable increase in HC emissions with the increase in brake power; the fuel quantity injected is increase thereby contributing to increased HC emission. Moreover, minimum HC emission level was achieved with neat biodiesel.

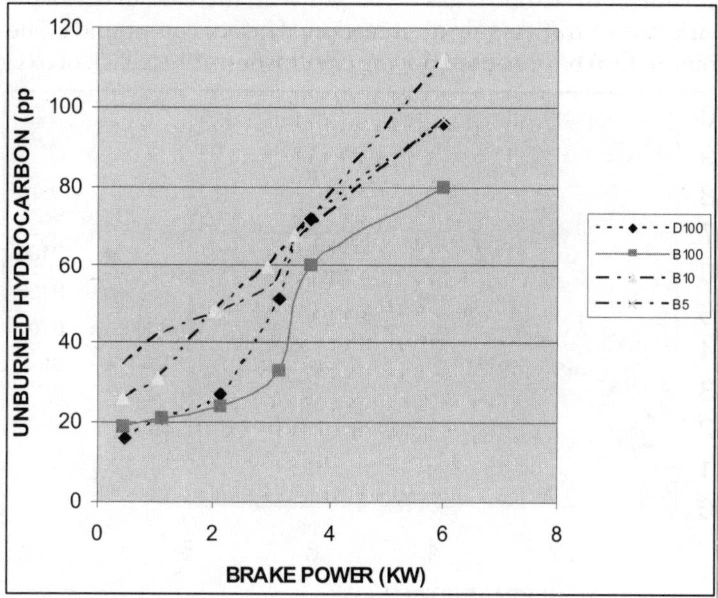


**Figure 50.7: Effect on Unburned Hydrocarbon in Exhaust of Brake Power**

**Effect of Oxygen in Exhaust of Brake Power**

As the brake power increased the oxygen content for all blends of biodiesel, neat diesel fuel tended to reduce significantly. This was due to re-duction in air fuel ratio at higher loads.

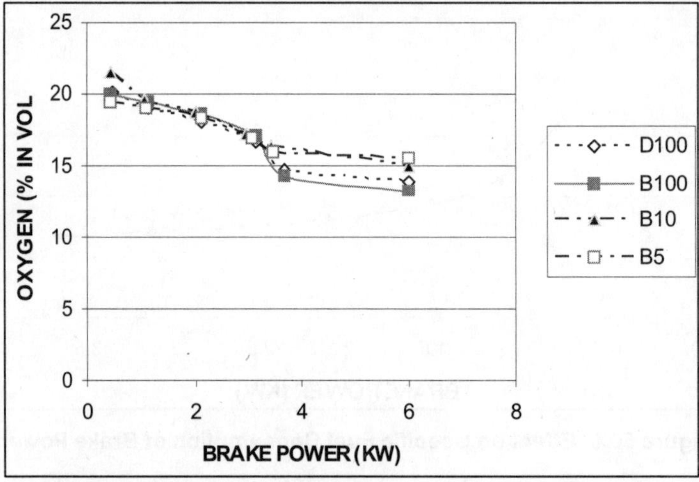


Figure 50.8: Effect of Oxygen in Exhaust of Brake Power

**Effect on Carbon Dioxide on Brake Power**

It was observed that, $CO_2$ content in the exhaust gas increased for all fuels with increasing brake power. Maximum $CO_2$ content at maximum power was also nearly same.

**Effect on CO in Exhaust of Brake Power**

The carbon monoxide in the exhaust gas increased with increase in brake power. The air fuel mixing process was affected by difficulty in atomization of heavy compounds. The resulting locally rich mixture causes more CO to be produced during combustion, due to lack of oxygen.

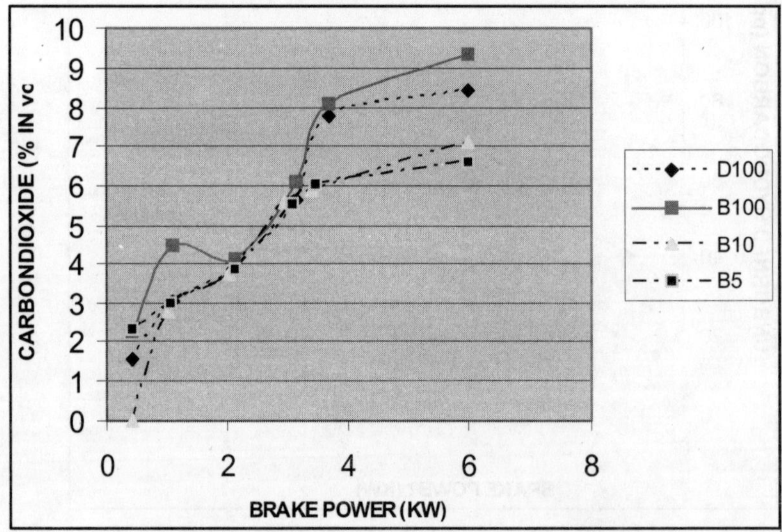


Figure 50.9: Effect on Carbon Dioxide on Brake Power

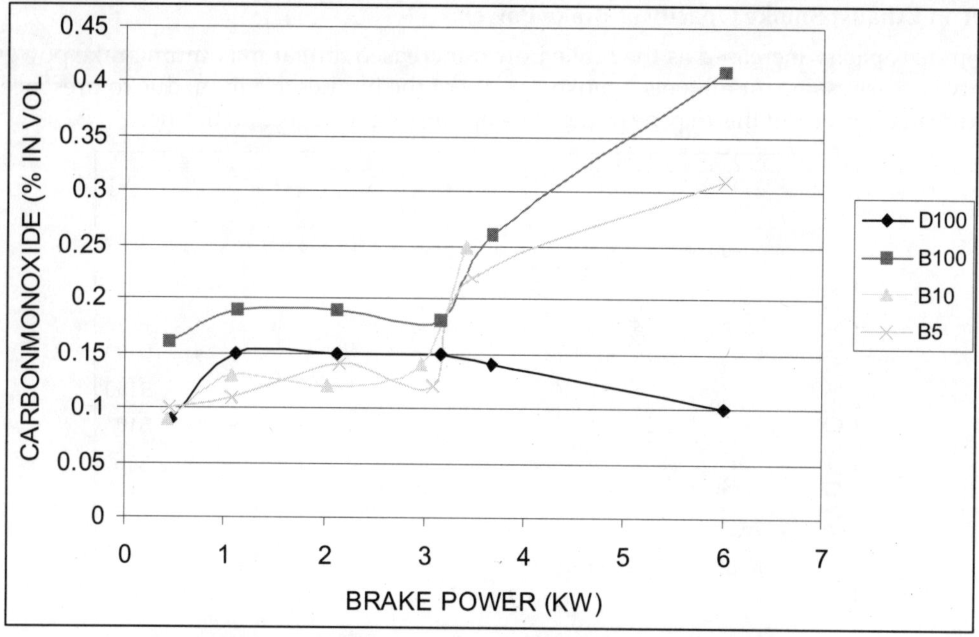


**Figure 50.10: Effect on CO in Exhaust of Brake Power**

**Effect on Exhaust $NO_x$ Emission of Brake Power**

$NO_x$ increase with brake power. This is probably due to formation of higher combustion chamber temperature and presence of oxygen in biodiesel.

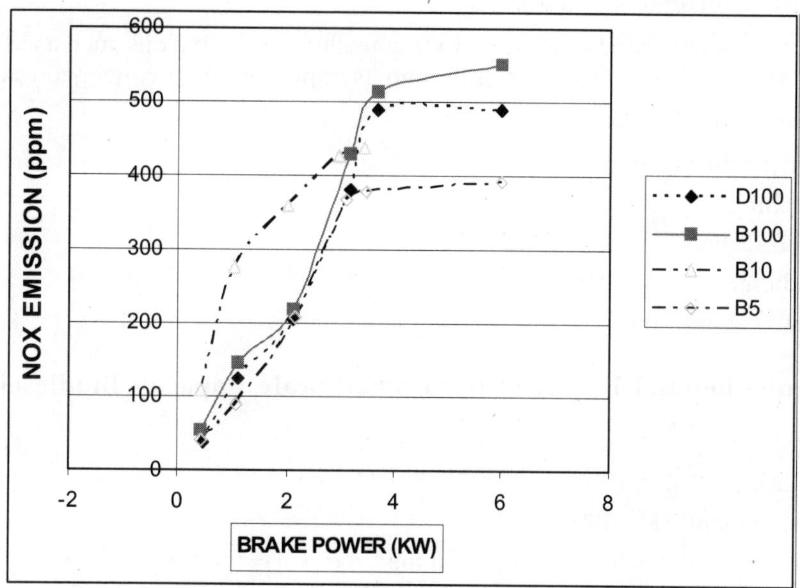


**Figure 50.11: Effect on Exhaust $NO_x$ Emission of Brake Power**

**Effect on Exhaust Smoke Opacity of Brake Power**

Smoke opacity increased as the brake power increased, but at maximum brake power smoke opacity is nearly same for all fuels. Smoke opacity of the biodiesel is high, due to presence of high moisture in the fuel. But this may be reduced by optimizing the injection timing.

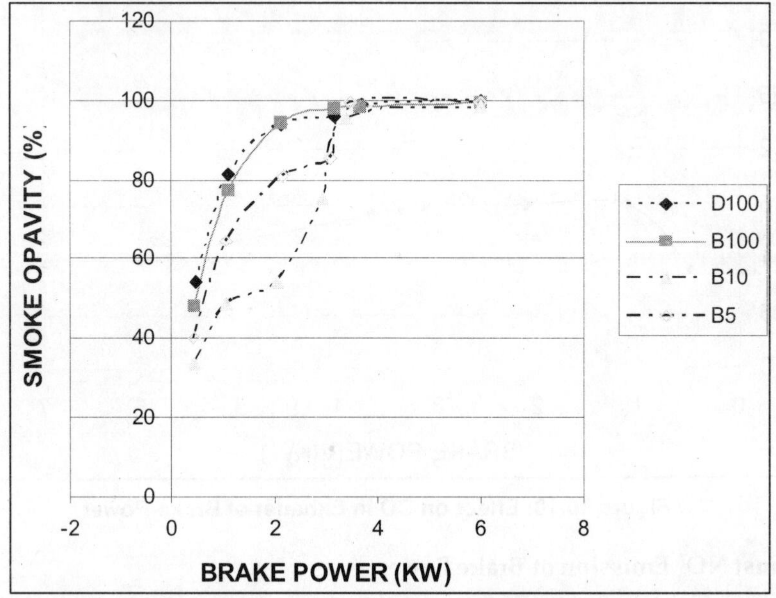


**Figure 50.12: Effect on Exhaust Smoke Opacity of Brake Power**

## Economic of Biodiesel Production

The cost of biodiesel can be reduced if we consider non-edible oils such as karanja, jatropha, neem, mahua etc, which are available in India and its improvement in conversion technology.

For biodiesel, the cost involves are:

- ☆ Seed production from farmers
- ☆ Crushing
- ☆ Esterification
- ☆ Biodiesel

Total cost of biodiesel/litre.

## Economics of Biodiesel Produced from Small-Scale Capacity Biodiesel Reactor

### Calculations

*Yearly Fixed Cost*

(*a*) Cost of plant = Rs 22000/-

(*b*) Care and maintenance (6 per cent of the fixed cost)= Rs 1200/-

(*c*) Life of plant is 10 years

(*d*) Interest is negligible

So the yearly fixed cost:

*Depreciation+maintenance+interest* = 2000+1200 = Rs 3200/-

***Yearly Variable Cost***

(*a*) Seed requirement per day for 10 litre = 32.6 Kgs, so the cost of seed/day @ rate of Rs 5/-Kg = 32.6*5 = Rs 163/-

So cost of seed per year = 300*163 =Rs 48900/-

(*b*) Cost of extraction of oil/year = 32.68*0.90*300 = Rs 8802/-

(*c*) Cost of transesterification per year = 300*10*6.67 = Rs 20,010/-

(*d*) Cost of chemicals:

(*i*) Methanol required/year = 3000*0.2*30 = Rs 18000/-

(*ii*) NaOH required/year = 300*10*10 = 30000gms = 30Kgs = 30*500 = 15000/-

(*e*) Cost of labour for processing like, water washing, studies of properties, storage etc. are neglected.

**Total Cost/Year**

3200+48900+8802+20010+18000+15000 = Rs 113912/-

**Income of Farmer**

(*i*) Quantity of oilcake produced per year @ Rs 5/- per Kg
2.23*10*300* 5 = Rs 33450/-

(*ii*) Glycerol produced for 3000 litre of biodiesel (10 per cent byproduct) per year @ Rs 60/- per litre.

60*300*10*0.1 = Rs 18000/-

**Total Income of Farmer/Year**

33,450+18,000 = Rs 51450/-

So,

Total expenditure – income of the farmer

113912–51450 = Rs 62462/-

Therefore, biodiesel per litre:

= 62462/3000 = Rs 20.8/kg and Rs 18.2/- per litre.

**Other Benefits**

Farmers of good agricultural land can also benefit by planting non-edible oil along the boundary of their fields. Non-edible oil trees are not a large plant that can obstruct air/sun to the main crop. In addition being non-edible it will protect main crop from stray animals. Morover biodiesel cropping will provide income to farmers who are around the degraded land and are facing economic hardship.

## Conclusion

On the basis of present experimental investigations the following conclusion may be drawn:

1. Methyl ester biodiesel was produced from non-edible vegetable and recovery of more than 84 per cent biodiesel was obtained.

2. The viscosity of vegetable oil drastically reduced after esterification.
3. The physical properties of methyl ester (biodiesel) and diesel were found to be comparable.
4. Short term engine performance indicates the suitability for substitute of biodiesel produced from small-scale biodiesel reactor with diesel oil.
5. The cost of biodiesel and diesel per litre are also comparable.

## References

1. A.K Agarwal and L.M Das (2001) "Biodiesel development and characterization for use as a fuel in C.I engine" Transaction of the ASME vol. 123,April 2001,pp 440-447.
2. Y. C. Batt, N.S Murthy, R.K. Dutta "Karanja (Pomgamia Glabra) Oil as a fuel engines". Agricultural Engineering Today, Vol-25 (5-6), 45-57, 2001.
3. H. Raheman and A.G Phadatare (2003) "Karanja esterified oil and alternative renewable fuel for diesel engine in controlling air pollution" Bioresource news vol. 3, pp, 17-23.
4. Parmanik (K. 2003) "Properties and use of Jatropha curcas oil and diesel fuel blends in compression ignition engine" J. Renewable energy V-28, pp, 239-248.
5. OMI Nwafor "Emission characteristics of diesel engine operating on rapeseed methyl ester". Renewable energy vol. 29 (2004) pp, 119-129.
6. Gemma Vicente, Mercedes Martinez (2004) " Integrated biodiesel production and comparison of different homogenous catalyst system" Bioresource Technology vol.92, issue 3,May 2004,pp 297-305.
7. Y. He, Y.D Bao (2004) " Study on cotton seed oil as partial substitute for diesel oil in fuel for single cylinder diesel engine" Renewable energy vol. 30 (2005), pp, 805-813.

# Chapter 51

# Scope of Energy Savings in Integrated Pulp and Paper Mill: A Case Study

**Saroja Asthana[1], Vilas Pharande[2] and S.N. Kaul[2]**
**[1]CEPD Division, National Chemical Laboratory, Pune**
**[2]MIT College of Engineering, Pune**

## ABSTRACT

Energy is most vital resource for human life in modern world. The pulp and paper sector presents one of the energy intensive and highly polluting sector globally and holds sixth position in intensive industries. Indian pulp and paper mill have been found to utilize an extra 30-35 per cent of energy in comparison to mills abroad. A plant study of Star Paper Mills Limited, Saharanpur, U.P. revealed that pulping, paper machine, chemical recovery and waste heat recovery boiler etc. are a major energy intensive process. Therefore, the objective of the study is to identify optimization conditions for a set of controlled parameters in the digester during the pulping operation, evaporation of paper in the paper machine, chemical recovery from liquor and waste heat recovery boiler using pinch technology/exergy analysis. Initial study indicates that there is a scope of at least 20-25 per cent energy conservation in Indian pulp and paper mills.

*Keywords: Digester, Paper machine, Chemical recovery, Waste heat recovery boiler.*

## Introduction

Energy is obtained primarily from the natural resources such as coal, crude oil, natural gas etc. Other forms of energies such as electrical, thermal etc. can be produced from these sources for domestic as well as industrial purposes [1]. In industries, energy is an indicative of the cost of production. So, the conservation of energy in industries is very important [2] and therefore, the study is focused on industrial energy conservation.

The pulp and paper production is highly energy intensive with 75-80 per cent of the energy requirement being used as process heat and 15-25 per cent as electrical power. [2] The combination of these two energy requirements qualifies paper production for the use of co-generation (low pressure steam for process heat and high pressure steam for electricity generation) [2] Environmental Impact of pulp and paper mills is very high in India. The potential pollutants from a pulp and paper mill can be classified into four categories; Liquid effluent, Air pollutants, Solid wastes (Lime Sludge) and Noise pollution. Paper and pulp industry being energy intensive, is considered for optimization study here.

## Literature Review

Asthana, Pharande and Kore (2008) have focused their study on identifying optimization conditions for a set of controlled parameters in the digester of pulp and paper mills using pinch technology. Range of parameters were established based on literature and actual plant data from Indian Pulp and Paper industries. After building a base model, optimized parameters were incorporated for two Indian Integrated Pulp and Paper Mills and were validated to satisfaction. After optimization of percentage of active alkali in white liquor, moisture percentage in wood, concentration of active alkali in white liquor, input temperature of wood in digester and maximum temperature of digester, theoretical energy requirement reduces to 2.79 GJ/t pulp from 3.168 GJ/t pulp in Case 1 and 3.213 GJ/t pulp in Case 2 respectively. This suggests expected percentage energy conservation possibility is 11.74 per cent and 12.97 per cent for mentioned Case 1 and Case 2 respectively. This will result in significant monetary gain. Further, any savings in energy will lead to sustainable development of the nation.

Mathur and Kulkarni (2002) have used pinch technology as a process integration technique and a powerful way to optimize the process design. The yield is superior to the conventional optimization process. The pinch analysis has been found useful in pulp and paper industry for setting energy targets and optimizing the heat recovery system. In a typical case all process heating and cooling duties are reviewed hot effluents being included as potential sources of additional heat. Pinch analysis can also be used to optimize the steam extraction levels from the turbines. In some mills there is a scope to increase power cogeneration by reducing the pressure of medium pressure steam upto 4 bars. The composite curves can also be used to determine the boilers minimum fuel requirement. Pinch technology can help mills to design integrated plants for a wide range of operating conditions. Such as seasonal variations change in raw material or product specification etc. This can help to achieve good flexibility in the design at a lower cost. Pinch analysis has been widely accepted in mills in developed countries resulting in significant energy saving.

Bengtsson and Nordman (2002) have reported newly developed methods and tools based on pinch technology to investigate the potential and economy of excess heat for pre-evaporation of chemo-thermo mechanical pulp effluent and heat pumping in an integrated pulp and paper mill. The matrix method was developed to retrofit the heat exchanger networks. The new curves are also able to provide information about the locations of heaters and coolers in the existing system. The matrix method has been used successfully in order to find an economically feasible heat exchanger network to retrofit for the release of the excess heat. A case study revealed that a pre–evaporation plant can be intergraded with the overall process with some modifications in the existing process. There are also opportunities for heat pumping in the system. Both projects are evaluated as economically viable options.

Bengtsson and Karisson (2007) have attempted to combine the pinch technology and the MIND (method for analysis of industrial energy system) method. It is found possible to identify beneficial and energy efficient measures in a complex industrial energy system.

Sarimveis and Angelou (2003) have reported the utilization of mathematical programming tools for optimum energy management of the power plant in pulp and paper mills.

Masur (1999) have reported a new mathematical model that can be used for checking and controlling the Kraft pulping process. The model is based on the modified relationship between the lignin content and the alkali concentration of cooking liquor.

Wising and Berntsson (2005) have proved and reported that when the mills hot water consumption is decreased, the live steam demand for the mill is also decreased. Also, when decreasing hot water consumption the quantity and temperature of available excess heat increases. This excess heat can be used for evaporation thereby reducing the live steam demand further up to 1.5 GJ/t.

Roger and Thore (2006) have reported that in many pulp mills, hot and warm water systems are over dimensioned and produce more hot and warm water than demand. This over production could be used for other applications if heat could be released at high temperature. In existing mills, warm and hot water is often overproduced and both warm and hot water tanks are commonly overflowed to a cooling tower or drained away. Instead of producing this surplus, the heat could be used for other purposes. Many process integration studies in the pulp and paper industry have been done to investigation the opportunities for energy saving.

Zvolinsci and Johannessen (2006) have reported that the paper drying is an exergy costly operation so a few percentage saving may be of very importance. This means that excess exergy losses should be avoided.

Jeroen and Ernst (1998) have reported that the current paper making processes normally requires energy of the order of 3-9 GJ/t paper mainly for the removal of water that is added initially to fibers and 1.3 to 2.9 GJ/t paper electricity can be manufactured in an integrated pulp and paper mill or in a mill where paper is produced from imported pulp.

Deventer (1997) has reported the feasibility of superheated steam drying of paper and textile. Drying with superheated steam in direct contact with the paper or textile web offers great advantages over conventional ways of drying with respect to energy efficiency.

The literature review suggests that there is an urgent need for energy conservation/optimization in paper and pulp industry globally and more so in India.

## Status of Indian Pulp and Paper Mills

The pulp and paper industry is an important industrial sector having a bearing on the socio-economic development of the country. The Indian paper industry is more than 100 years old, and the first mechanized paper machine was set up sometimes in 1830. This sectors employs more than 0.3 million people directly, and about 1 million indirectly. More than 600 pulp and paper mill units producing nearly 5.5 million tons annually against the installed capacity of around 7.4 millions tons [1]. However, the per capita consumption of paper and board in India is quite low at 5.5 kg when compared with the global average of 54 kg. Table 51.1 present wood based leading paper and pulp industries in India.

**Table 51.1: Wood Based Paper and Pulp Integrated Plants in India**

| *Name of Mill* | *Installed Capacity (tons/annum)* |
|---|---|
| Century Pulp and Paper Limited | 186480 t/a |
| Tamil Nadu Newsprint and Paper Ltd. | 180000 t/a |
| J K Paper Limited | 137000 t/a |
| The West Coast Paper Mills Ltd | 119750 t/a |
| Ballarpur Industris Limited | 89500 t/a |
| Star Paper Mills Limited | 66000 t/a |
| Nepa Limited | 60000 t/a |

Recent estimates indicate energy saving could be as high as 20–25 per cent. [1] Energy consumption in Indian paper mills is [2]: Steam: 08.0–16.0 t/t paper, Fuel: 46.2–73.8 GJ/t paper and Electricity: 1500–1700 KWh/t paper. This accounts to Total Energy of 51.6–80.0 GJ/t paper. The use of waste paper requires about 2.5 times less energy. Further, water use in Indian pulp and paper mill is very high. Annual wastewater discharge is 695.7 million cubic meters and annual consumption is 905.8 million cubic meters, ranking third in water consumption. [4] Average water consumption in Indian wood based pulp and paper mills is 150-200 $m^3$/tonne significantly higher than in developing countries where wood based mills consume 50-75 $m^3$/tonne. [5] The ratio of water consumption and economic value creation in Indian paper mills is poor. One cubic meter of water generates $7.5 U.S. only.

## Energy Consumption in Pulp and Paper Mill

The manufacturing of paper consists of five main stages *viz.* raw material preparation, pulping, bleaching, chemical recovery and paper-making. Out of these, major thermal energy consuming processes are pulping process, chemical recovery and paper-making machine. The raw materials have to be cooked either mechanically or chemically and treated for further use in the production chain. Furthermore, paper making requires considerable amounts of both heat and electricity for forming, pressing and drying of the paper. The comparison of energy consumption in Indian and abroad pulp and paper industry is presented in Table 51.2. The comparative study has revealed that Indian pulp and paper industry is using 30–35 per cent excess energy as compared to industries abroad [3].

### Scope of the Study

Table 51.2 has revealed that the most energy intensive processes are digesters, evaporator, chemical recovery and paper making machine. Therefore, a plant study with an objective to identify optimization conditions for a set of controlled parameters in the digester, paper making machine, chemical recovery and waste heat recovery boiler etc. using pinch technology/exergy analysis is taken up. The study is not suggesting any change in material or design parameters of the equipments in the plants studied. However, it suggests controlling the parameters without changing external environment. Therefore, it is not associated with high investment and Indian industry can adapt these suggestions for their benefit straight away.

## Plant Study

To get a realistic scenario of Indian Integrated Paper and Pulp mill, we wrote to all the industries mentioned in Table 51.1 for seeking their permission to visit and perform the plant study. The positive response were received from J K Paper Limited, The West Coast Paper Mills Ltd, Ballarpur Industris

Limited, Star Paper Mills Limited and Nepa Limited to perform the said study. Star Paper Mills Limited, Saharanpur is chosen for further study as this is one of the largest and oldest integrated plants. Further, Central Paper and Pulp Research Institute (CPPRI) that is a prime research organization for paper and pulp, also located in Saharanput, U.P. This has enabled us to seek expert opinion from both, Star Mills limited as well as researchers at CPPRI.

**Table 51.2: Energy Consumption in India and Abroad [3]**

| *Section/Equipment* | *Fuel (GJ/t of Paper)* | | *Electricity (GJ/t of Paper)* | | *Final Energy (GJ/t of Paper)* | |
|---|---|---|---|---|---|---|
| | *Indian Mill* | *Abroad Mill* | *Indian Mill* | *Abroad Mill* | *Indian Mill* | *Abroad Mill* |
| Chipper | – | – | 0.40–0.46 | 0.33–0.35 | 0.40–0.46 | 0.33–0.35 |
| Digester | 12.5–18.0 | 8.1–9.9 | 0.21–0.22 | 0.15–0.17 | 12.67–18.22 | 8.3–10.02 |
| Evaporator | 11.5–18.5 | 7.7–9.4 | – | | 11.54–18.46 | 7.71–9.43 |
| Washing and Screening | | | 0.52–0.56 | 0.42–0.44 | 0.52–0.56 | 0.42–0.44 |
| Bleaching | 1.6–1.8 | 0.9–1.1 | 0.32–0.33 | 0.24–0.25 | 1.93–2.18 | 1.09–1.32 |
| Chemical Recovery | 2.3–5.1 | 1.3–2.1 | 0.61–0.68 | 0.46–0.49 | 2.92–5.76 | 1.74–2.63 |
| Stock Preparation | – | | 0.99–1.03 | 0.59–0.62 | 0.99–1.03 | 0.59–0.62 |
| Paper Machine | 13.8–18.5 | 7.7–9.2 | 1.67–1.71 | 1.48–1.49 | 15.52–20.17 | 9.19–10.72 |
| Deaeratar | 3.7–5.5 | 1.9–3.0 | - | | 3.69–5.54 | 1.93–3.00 |
| Utilities and Others | | | 0.89–0.91 | 0.58–0.59 | 0.89–0.91 | 0.58–0.59 |
| Total | 46.2–73.8 | 27.9–36.4 | 5.40–6.12 | 4.14–4.50 | 51.55–79.97 | 32.00–40.93 |

Figure 51.1 present the schematics of pulp and paper production process.

Figure 51.2 presents mass balance in digester of star paper mills.

## Scope of Energy Saving in Pulp and Paper Mill

The main areas with significant scope of energy savings are digester, paper making machine, evaporator and waste hear recovery boilers. To being with digester which is heart of pulp making is considered. Figure 51.2 above present mass balance in digester. Further, there are controlled and uncontrolled parameters that govern the energy requirements. Uncontrolled parameters such as Specific heat of wood, white liquor, black liquor, input temperature of white liquor, input temperature of black liquor, H- factor and Kappa Number of pulp. These parameters are difficult to be optimized without changing material and design parameters. However, other controlled parameters such as Percentage of Active Alkali in White Liquor, Moisture Percentage in Wood, Concentration of White Liquor, Input Temperature of Wood in Digester, Maximum Temperature in Digester can be optimized for energy savings.

## Conclusion

As mentioned earlier, Indian pulp and paper mill have been found to utilize an extra 30-35 per cent of energy in comparison to mills abroad. A study of Star Paper Mills Limited, Saharanpur also suggest significant scope of energy savings of the order of 20-25 per cent cumulatively in digester, paper making machine, evaporator and WHRB. The energy conservation will definitely lead to a significant monetary benefit.

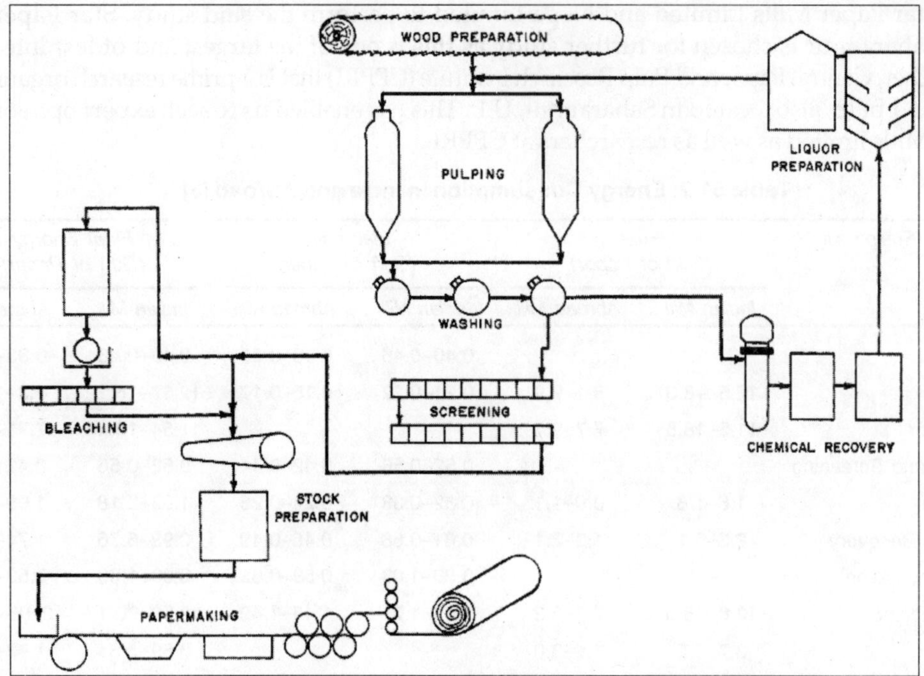


**Figure 51.1: Schematics of Pulp and Paper Production Process**

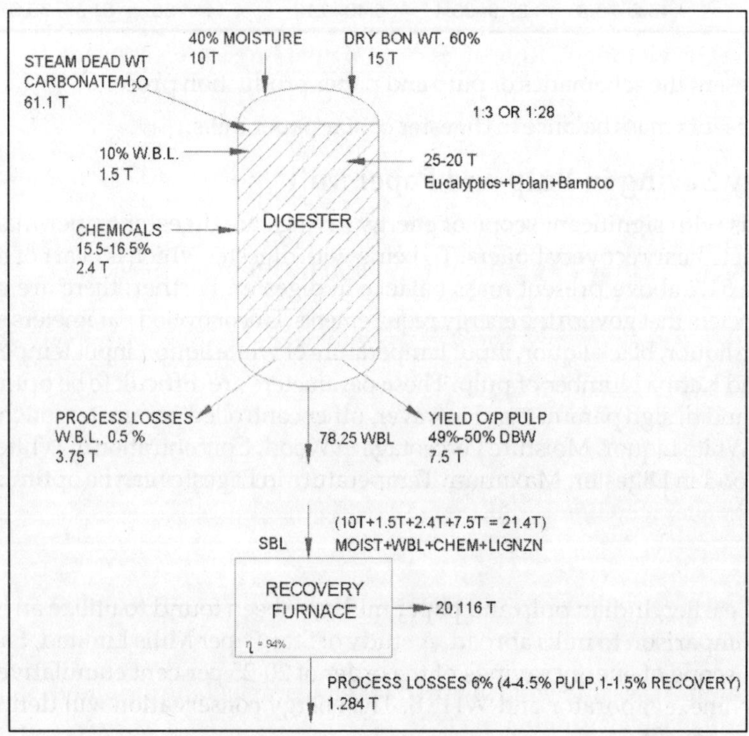


**Figure 51.2: Mass Balance in Digester of Star Paper Mill**

## References

1. Saroja Asthana, Vilas Pharande and Vikramsinh Kore, "Scope of Energy Conservation in Digester in Indian Pulp and Paper Industries for Sustainable Development" International Conference PAPEREX-2007, 7-9 December 2007 at New Delhi.

2. R.M Mathur, B.P.Thapliyal, A.G Kularni, "Process integration through pinch analysis", Central pulp and paper Research Institute Saharanpur, 2002.

3. C.Bengtsson, R.Nordman,T.Berntsson, " Utilization of excess heat in pulp and paper industry a case study of technical and economical opportunities", International journal of Applied Thermal Engineering,22, 2002, pp.1069-1081.

4. C.Bengtsson, M. Karlsson, T.Berntsson, "Co-Ordination of pinch technology and the MIND method applied to a Swedish board mill", International journal of Applied Thermal Engineering, 22, 2007, pp.133-144.

5. H.K Sarimvies, A.S.Angelou, T.R.Resina, "Optimal energy management in pulp and paper mills", International journal of Energy Conversion and Management, 44, 2003, pp.1707-1718.

6. V.Masur, "A mathematical model of kraft pulping related to the alkali concentration in the cooking liquor", International journal of Wood Science and Technology, 33, 1999 pp.381-389.

7. Ulrika Wising, Thore Berntsson, Paul Stuart, "The potential for energy saving when reducing the water consumption in Kraft pulp mill", International journal of Applied Thermal Engineering, 25, 2005, pp.1057-1066.

8. Roger Nordman, Thore Berntsson, "Design of kraft pulp mill hot and warm water systems a new method that maximizes excess heat", International journal of Applied Thermal Engineering, 26, 2006, pp.363-373.

9. Anita Zvolinschi, Eivind Johanessen, Signe Kjelstrup, " The second law optimal operation of a paper drying machine", International journal of Chemical Engineering Science, 61,2006 pp.3653-3662.

10. Jeroen De Beer, Ernst Worrell, Kornelis Blok, "Long term energy efficiency improvements in the paper and board industry", International journal of Energy, 23, 1998, pp.21-42.

11. H.C Van Deventer, "Feasibility of energy efficient steam drying of paper and textile including process integration", International journal of Applied Thermal Engineering, 17, 1997, pp.1035-1041.

# Chapter 52

# Clean Energy Technology Approaches of Mitigating Greenhouse Gases

**S.S. Katoch* and I.K. Bhat**
**National Institute of Technology, Hamirpur – 177 005, H.P.**

## ABSTRACT

The Clean Development Mechanism (CDM) was introduced in Article 12 of the Kyoto Protocol in December 1997 to facilitate the Greenhouse Gases (GHG) emission reduction projects, and lower the cost for emission reductions for industrialized countries, while at the same time assisting developing countries in achieving technology transfer and sustainable development. Stabilizing the atmospheric concentration of greenhouse gases is essential to stabilizing global climate, and this cannot be accomplished without serious reductions in the current rate of use of fossil fuels. Enactments prior to the Electricity Act, 2003, had no specific provisions which would promote renewable or non-conventional sources of energy in India. Renewable energy has real environmental benefits that could provide some insurance for stabilizing our planet's health. World over, the transition to sustainable energy is led by wind power. By the end 2005, wind reached an installed capacity of 60,000 MW globally, and India became the fourth in the world with 5340 MW. Renewables now account for 18 per cent of world investment in power generation according to a UNEP report released in June 2007. The most exciting development is the resurgence of concentrating solar power (CSP) which holds great promise because of its capability to extend power generation to 20-22 hours per day, using various storage modes. An appealing aspect of solar technology is the innumerable variety of applications to which it can be put to use. India's progress in the solar sector has been mostly confined to water heating devices. India which is facing a massive power crunch should urgently put together a comprehensive policy and regulatory framework to fast-forward the growth of this sector. In order to solve the

* Corresponding Author: E-mail: sskatoch@nitham.ac.in

problem of resources for having clean and green energy, the need of the hour is to consider various possible alternatives and solution of reaching them. There is an overall need for research into methods of cost reduction and risk management through technical and economic standardization that can increase the value of renewable energy. Curricula and research programs in academic institutions shall be focused towards application of new lighter material components, cost effective techniques combined with market driven strategies, public involvement, and examining new locations to meet the industrial needs for production of sustainable energy.

***Keywords:*** *Clean development mechanism, Greenhouse gases, Renewable energy.*

## Introduction

Energy is a mainstay of an industrial society. As the population of the world increases and people strive for a higher standard of living, the amount of energy necessary to sustain our society is ever increasing. At the same time, the availability of nonrenewable sources, particularly liquid fuel; is rapidly shrinking. Therefore, there is general agreement that to avoid energy crisis, the amount of energy needed to sustain society will have to be contained and to the extent possible, renewable sources will have to be used. As a consequence, conservation and renewable energy technologies are going to increase in importance and reliable, up-to-date information about their availability, efficiency, and cost is necessary for planning a secure energy future. In order to be successful, renewable technologies will have to become more efficient and cost-effective. Use of conservation and renewable energy can help meet critical national goals for fuel diversity, price stability, economic development, environmental protection, and energy security, and thereby play a vital role in national energy policy.

Renewable energy and energy-efficient technologies are typically characterized by higher up-front costs resulting in significantly reduced fuel and/or operating costs. Many early policy incentives at both central and state levels were intended to reduce the acquisition cost of these technologies, frequently through the use of tax credits proportional to capital investment costs. Improving energy efficiency increases the productivity of basic energy resources by providing the needs of society with less energy. The least expensive and most efficient means is energy conservation, rather than more energy production. Moreover, energy conservation is also the best way to protect the environment and reduce global warming.

Even though there are widely differing views and estimates of the ultimately recoverable resources of fossil fuels, it is fair to say that they may last for around 50-150 years with a peak in production occurring much earlier. However, a big concern is the climatic threat of additional carbon that will be released into the atmosphere. Increased use of nuclear power presents the possibility of additional carbon free energy use and its consequent benefit for the environment. However, there are significant concerns about nuclear waste and other environmental impacts, the security of the fuel and the waste, and the possibility of their diversion for weapon production. According to the estimate, 13.3 per cent of the world's total primary energy supply came from renewable energy in 2003. However, almost 80 per cent of the renewable energy supply was from biomass, and in developing countries it is mostly converted by traditional open combustion, which is very inefficient. In 2003, shares of biomass and hydropower in the total primary energy mix of the world were about 11 per cent and 2 per cent, respectively. All of the other renewables, including solar thermal, solar PV, wind, geothermal, and ocean combined, provided only about 0.5 per cent of the total primary energy. During the same year, biomass combined with hydroelectric resources provided more than 50 per cent of all the primary energy in Africa, 29.2 per cent in Latin America, and 32.7 per cent in Asia (Table 52.1). However, biomass is used very inefficiently for cooking in these countries. Such use has also resulted in significant health problems, especially for women.

**Table 52.1: Share of Renewable Energy in 2003 Total Primary Energy Supply (TPES) on a Regional Basis**

| *Region* | *TPES MTOE* | *Renewables MTOE* | *Share of Renewables in TPES (per cent)* |
|---|---|---|---|
| Africa | 558.9 | 279.9 | 50.1 |
| Latin America | 463.9 | 135.5 | 29.2 |
| Asia | 1224.4 | 400.0 | 32.7 |
| India | 553.4 | 218.0 | 39.4 |
| China | 1425.9 | 243.4 | 17.1 |
| Non-OECD* Europe | 103.5 | 9.7 | 9.4 |
| Former USSR | 961.7 | 27.5 | 2.9 |
| Middle East | 445.7 | 3.2 | 0.7 |
| OECD | 5394.7 | 304.7 | 5.6 |
| USA | 2280.8 | 95.3 | 4.2 |
| World | 10578.7 | 1403.7 | 13.3 |

* OECD–Organization for Economic Cooperation and Development.

*Source*: Kreith and Goswami (2007).

## Greenhouse Gases and Climate Change

Increasing concentrations of $CO_2$ in the atmosphere a major greenhouse gas are also leading to increasing $CO_2$ concentrations in the ocean, increasing acidity and negativity impacting coral reefs and pelagic organisms, including species of phytoplankton and zooplankton. Evidence of global warming comes from analysis of ice core samples, temperature measurements at different levels in several hundred boreholes in the earth's surface, and atmospheric temperature measurements. Historically, pollution has been a relatively local problem, affecting a given river, lake, or bay, or the air in a city. Today, scientists are analyzing pollution on a global scale. The concern about depletion of the stratospheric ozone layer has already led to international action–the Montreal Protocol in 1987–aimed at curbing pollution from the release of chlorofluorocarbon refrigerants into the atmosphere. A more serious problem today is the danger of global climate change due to $CO_2$, an unavoidable byproduct of burning fossil fuels–crude oil, coal, and natural gas. Because of the large amount of fossil fuels currently being burned, $CO_2$ levels in the atmosphere have grown from about 280 parts per million (ppm) in 1900 to over 375 ppm in 2004. Indeed, the level of atmospheric $CO_2$ is increasing by 0.4 per cent per year, and given our dependency on fossil fuels, there is no end in sight. Concern about global climate change led representatives of 166 nations to meet in Kyoto, Japan, in December of 1997 to negotiate a treaty to reduce emissions of carbon dioxide and other greenhouse gases. At that meeting, most of the industrialized nations (including the United States) agreed to reduce emissions to below 1990 levels, a goal that is to be achieved by the year 2010. Stabilizing the atmospheric concentration of

$CO_2$, however, is essential to stabilizing the global climate itself, and this cannot be accomplished without serious reductions in the current rate of use of fossil fuels. This is one of the defining environmental issues of the 21st century (Miller Jr., 2004 and Wright, 2005). The analysis from IPCC (Intergovernmental Panel on Climate Change) suggests that the world emission of greenhouse gases needs to be cut by at least half within 2050 in order to attain the threshold. By current trends, emissions will increase by 50 per cent by 2030.

## Major Greenhouse Gases

Water vapor, methane, nitrous oxide, ozone, and chlorofluorocarbons absorb infrared radiation and add to the insulating effect of $CO_2$ (Table 52.2). Some of these gases also have anthropogenic sources and are increasing in concentration; raising the concern that future warming will extend well beyond the calculated effects of $CO_2$ alone.

**Table 52.2: Anthropogenic Greenhouse Gases in the Atmosphere**

| *Gas* | *Average Concentration 100 Years Ago (ppb)* | *Approximate Current Concentration (ppb)* | *Average Residence Time in Atmosphere (years)* |
|---|---|---|---|
| Carbon dioxide ($CO_2$) | 288,000 | 375,000 | 120 |
| Methane ($CH_4$) | 848 | 1,850 | 12 |
| Nitrous oxide ($N_2O$) | 285 | 316 | 114 |
| Chlorofluorocarbons (CFCs) and halocarbons | 0 | 1.2 | 50-100 |

*Source*: Wright, 2005.

In 1988, the United Nations Environment Program and the world Meteorological Society established the Intergovernmental Panel on climate Change (IPCC) in order to provide accurate and relevant information that would lead to an understanding of human induced climate change. The IPCC has established three working groups: one to assess the scientific issues (Working Group I), another to evaluate the impact of global climate change and the prospects for adapting to it (Working Group II), and a third to investigate ways of mitigating the effects (Working Group III). The IPCC working groups consists of more than two thousand experts in the appropriate fields from over a hundred countries. These stalwarts are unpaid and participate at the cost of their own research and other professional activities. The work of the IPCC has been guided by two basic questions: *risk assessment*–is the climate system changing, and what is the impact on society and ecosystem? *risk management*–how can we manage the system through adaptation and mitigation.

## Climate Recovery through Renewable Energy

Renewable energy is energy derived from resources that are regenerative, for all practical purposes, non-depleting, besides environmentally benign. By these qualities, renewable energy sources are fundamentally different from fossil fuels. While mankind's traditional uses of wind, water, and solar energy are commonplace across all countries, the mass production of energy using renewable energy sources has become more common recently, reflecting the major threats of climate change, depletion of fossil fuels, and the environmental and social risks of fossil fuels. Consequently, many countries promote renewable energies through tax incentives and subsidies.

The role of new and renewable energy has been assuming increasing significance in recent times with the growing concern for the country's energy security. India has the distinction of being the only country in the world to have an exclusive ministry dealing with new and renewable energy sources. During the last two and a half decades, there has been a vigorous pursuit of activities relating to research, development, demonstration, production, and application of a variety of renewable energy technologies for use in different sectors. Ministry of New and renewable Energy (MNRE) has been implementing broad-spectrum programmes covering more or less the entire range of new and renewable energy. These programmes broadly seek to supplement conventional fossil fuel based power through harnessing wind, small hydro, and bio-energy; reach renewable energy to rural to remote rural areas for lighting, cooking, and motive power; use renewable energy in urban, industrial, and commercial applications; and develop alternative fuels and applications for stationary, portable, and transport uses apart from supporting research, design and development of new and renewable energy technologies, products and services. Resource-wise estimated potential is given in Table 52.3.

**Table 52.3: Estimated Potential of Renewable Energy and Cumulative Achievements in India during 2007**

| *Sources/Systems* | *Estimated Potential* | *Cumulative Achievements (as on 30/10/2007)* |
|---|---|---|
| Bio power (agro residues and plantations) | 16,881 MW | 560.30 MW |
| Wind power | 45,195 MW | 7660.20 MW |
| Small hydropower (up to 25 MW) | 15,000 MW | 2014.66 MW |
| Co-generation–bagasse | 5,000 MW | 692.33 MW |
| Urban waste to energy | 2,700 MW | 55.20 MW |
| Solar power | - | 2.12 MW |
| Biomass/co-generation (non-bagasse) | - | 59.00 MW |
| Biomass gasifier | - | 86.53 MW |
| Energy recovery from waste | - | 20.21 MW |
| TOTAL | 84,776 MW | 11,150.55 MW |

*Source*: Singh, 2007.

## Clean and Renewable Energy Options

The relevant key issues and condition which influence any individual country and its specific policy for promoting energy conservation and deployment of renewable energy technologies are determined by promoting energy conservation and deployment of renewable energy technologies are determined by resources, targets, and constraints, the resources of renewable energies and their technical and economic exploitable potentials are one important issue. These conditions and the sources of renewable energies differ in a wide range between continents, countries and even on a regional scale inside individual countries. The limitations can result from the constraints, such as missing infrastructure or lack of financial resources for projects. There are many ways to support renewable energies with special policies depending on technology, resources and policy targets.

### Wind Energy

Wind energy became a significant research area in the 1970s during the energy crisis and the resulting search for potential renewable energy sources. Modern wind turbine technology has made significant advances over the last twenty five years. Today, wind power technology is available as a mature, environmentally sound, and convenient alternative. Many wind farms are megawatt scale, ranging from 1 MW to tens of megawatts. The technology has been vastly improved and capital costs come down to as low as $ 1000 per kW. At this level of capital costs, wind power is already economical at locations with fairly good wind resources. Therefore, the average annual growth in worldwide wind energy capacity from 2000 to 2003 was over 30 per cent and it continued to grow at that rate in 2004 and 2005. The total worldwide installed wind power capacity that was 39 GW in 2003 reached a level of 59 GW in 2005.

Knowledge of wind and its characteristics is essential at all stages of wind power technology– from concept to assessing the wind energy potential of a site, availability of wind vector data at or nearby the site *i.e.*, estimating wind power density for the proposed site or area, is the most important criterion. The approaches practiced widely for wind resources assessment in different parts of the world are as under:

- Ecological indicators
- Onsite measurements
- The measure-correlate-predict (MCP) technique
- Use of global data sets like NCEP (National Centre for Environmental Prediction) and NCAR (National Centre for Atmospheric Research)
- Numerical modeling of wind
- Wind Atlas.

## Concentrating Solar Power

The amount of sunlight striking the Earth's atmosphere continuously is $1.75\times10^5$ TW. Considering a 60 per cent transmittance through the atmosphere cloud cover, $1.05\times10^5$ TW reaches the Earth's surface continuously. If the irradiance on only 1 per cent of the Earth's surface could be converted into electric energy with 10 per cent efficiency, it would provide a resource base of 105 TW, whereas the total global energy needs for 2050 are projected to be about 25-30 TW. The present state of solar energy technologies is such that solar cell efficiencies have reached over 20 per cent and solar thermal systems provide efficiencies of 40 per cent -60 per cent. With the present rate of technological development, these solar technologies will continue improving, thus bringing the costs down, especially with the economies of scale.

Solar power is the technology of obtaining usable energy from sun. Solar energy has been used in many traditional technologies for centuries and has come into widespread use where other power supplies are absent, such as in remote locations and in space. At the equator, the sun provides approximately 1000 watts/$m^2$ of energy on earth's surface. Concentrated solar power (CSP) systems focus direct solar radiation through optical devices onto an area where a receiver is located. The radiation is transformed into heat. CSP plants produce electric power by converting the sun's energy into high temperature heat using various mirror configurations. The heat is then channeled through a conventional generator. Concentrating solar power systems can be sized for village power (10 KW) or grid connected applications (up to 100 MW). Some systems use thermal storage during cloudy periods or at night. Others can be combined with natural gas and the resulting hybrid power plants provide high value power that can be supplied.

## Biofuel

Although theoretically harvestable biomass energy potential is on the order of 90 TW, the technical potential on a sustainable basis is on the order of 8-13 TW. This potential is 3-4 times the present electrical generation capacity of the world. The biggest advantage of biomass as an energy resource is its relatively straightforward transformation into transportation fuels. Biofuels will not require additional infrastructure development. Therefore, development of biofuels is being viewed very favorably by governments around the world. Biofuels, along with other transportation options such as electric vehicles and hydrogen, will help diversify the fuel base for future transportation. Biodiesel production grew almost fourfold, although it started from a much smaller base. In 2005, the world ethanol production had reached about 36 billion liters per year, whereas biodiesel production topped 3.5 billion liters during the same year.

Biofuels are renewable sources of energy derived from biological raw material. Two sources of biofuels, ethanol and biodiesel are gaining worldwide acceptance as one of the solutions for the problems of environmental degradation, energy security, restricting imports, rural employment, agricultural economy, savings in foreign exchange, and reduction in vehicular pollution. Ethyl alcohol (ethanol, commonly termed as alcohol) is close to being a perfect fuel because of low molecular weight, high oxygen content, high combustion efficiency, and non-toxic and non-polluting nature. It is also a renewable raw material from varied saccharine and starchy substances. A motor fuel blended with ethanol does not use tetra ethyl lead as oxygenate and there are also no sulphur dioxide and nitrous oxide particulate emissions. Moreover, the hydrocarbon emissions are not carcinogenic and carbon dioxide emissions are 57 per cent lower than those with petrol. The raw material used for producing ethanol could be sugar (in Brazil), cereals (in USA), sugar beet (in Europe), and Molasses (in India). Brazil uses ethanol as 100 per cent fuel in about 20 per cent of vehicles and 25 per cent blend in gasoline in rest of the vehicles. USA uses 10 per cent ethanol gasoline blends, whereas a 5 per cent blend is used in Sweden. Australia uses 10 per cent ethanol gasoline blend and EU uses 15 per cent ethanol blended gasoline.

Biodiesel is an environment friendly fuel that has almost no sulphur, no aromatics, and about 10 per cent built-in oxygen. It is fatty acid or methyl ester. Biodiesel needs no separate infrastructure for storage, dispensing, and the existing tanks and dispensing stations of conventional diesel can be used. Biodiesel is made from virgin or used vegetable oils (both edible and non-edible) and animal fats. It can be blended in any ratio with petroleum diesel fuels and can also be produced from various sources including tree-borne oilseeds like jatropha, karanja, jojoba, and neem. There has been a remarkable increase in the use of biodiesel in the last one decade. Europe, Germany, France, and Australia have fairly large programmes. However, now there is also a lot of importance being accorded to it in countries like USA, Canada, Malaysia, Indonesia, India, Thailand, the Philippines, Ghana, South Africa, China, and Zambia.

## Hydropower

Hydropower is a renewable energy resource resulting from the stored energy in water that flows from a higher to a lower elevation under the influence of the earth's gravitational field. Ultimately, it is connected to solar energy and the natural hydrologic cycle of evaporating water from lakes and oceans redeposited as rain or snow. In most installations today, hydro energy is converted to electricity by flowing water through turbines to produce rotating shaft work, which turns an electric generator. Technologies for carrying out this conversion are highly developed and efficient, with hydropower installations ranging in scale from a few kWe to over 10,000 MWe. The unexploited potential for hydropower is large. The World Energy Council estimates the gross theoretical potential to be about 40,000 TWh/year, of which 9,000 TWh/year is technically and economically feasible. This is more than three times the current hydropower production level of 2,600 TWh/yr. Without doubt, hydropower is strategically important worldwide.

From country to country and from region to region, large difference is noted in the relative importance of hydropower as a producer of primary electricity. In Norway, Switzerland, Austria, and Brazil, 80-100 per cent of electricity comes from hydropower, while in the US and China, it is about 19 per cent or less (Table 52.4).

**Table 52.4: Potential for Hydropower Development in Selected Countries Based on Technical Potential and Economic Potential in Today's Energy Markets**

| *Country* | *Hydro as per cent of Total Electricity* | *Ratio of Theoretical Potential to Actual* | *Ratio of Economic Potential to Actual* |
|---|---|---|---|
| Norway | 100.00 | 5.77 | 1.80 |
| Brazil | 91.70 | 5.40 | 3.00 |
| Switzerland | 80.00 | - | 1.10 |
| Canada | 63.00 | 3.81 | 1.54 |
| India | 25.00 | 4.20 | 3.00 |
| France | 20.00 | 1.15 | 1.00 |
| China | 17.00 | 10.10 | 6.60 |
| Indonesia | 14.00 | 31.30 | 3.13 |
| USA | 10.00 | 1.82 | 1.30 |
| World total | 19.00 | 18.34 | 2.78 |

*Source*: Tester *et al.*, 2005.

## Clean Development Mechanism

The Clean Development Mechanism (CDM) was one of the three mechanisms established by the Kyoto Protocol in 1997 to meet the climate convention objective of stabilizing greenhouse gas (GHG) concentrations in the atmosphere. Consistent with the objective of the Convention, the CDM has to contribute to sustainable development in developing countries. A market has thus been created for the sale of Certified Emission Reductions (CERs). Sponsors of renewable energy projects can now sell the CERs resulting from their projects to purchasers from developed countries. The sale of CERs has provided an additional source of high quality cash flows to renewable energy projects. The essential steps to be taken by the project participants for completing the CDM project cycle are provided in the Figure 52.1. The baseline for a CDM project activity is a scenario that reasonably represents the anthropogenic emissions of CHGs that would occur in the absence of the proposed project activity. It sets the base from which the amount of total redeuctions in the emissions of GHGs, and thus of the credits, is calculated.

## Emission Trading

With global emissions of carbon dioxide exceeding 25 billion tons per year and growing, it will take radical changes in energy production and use to avoid the dangerous anthropogenic interference

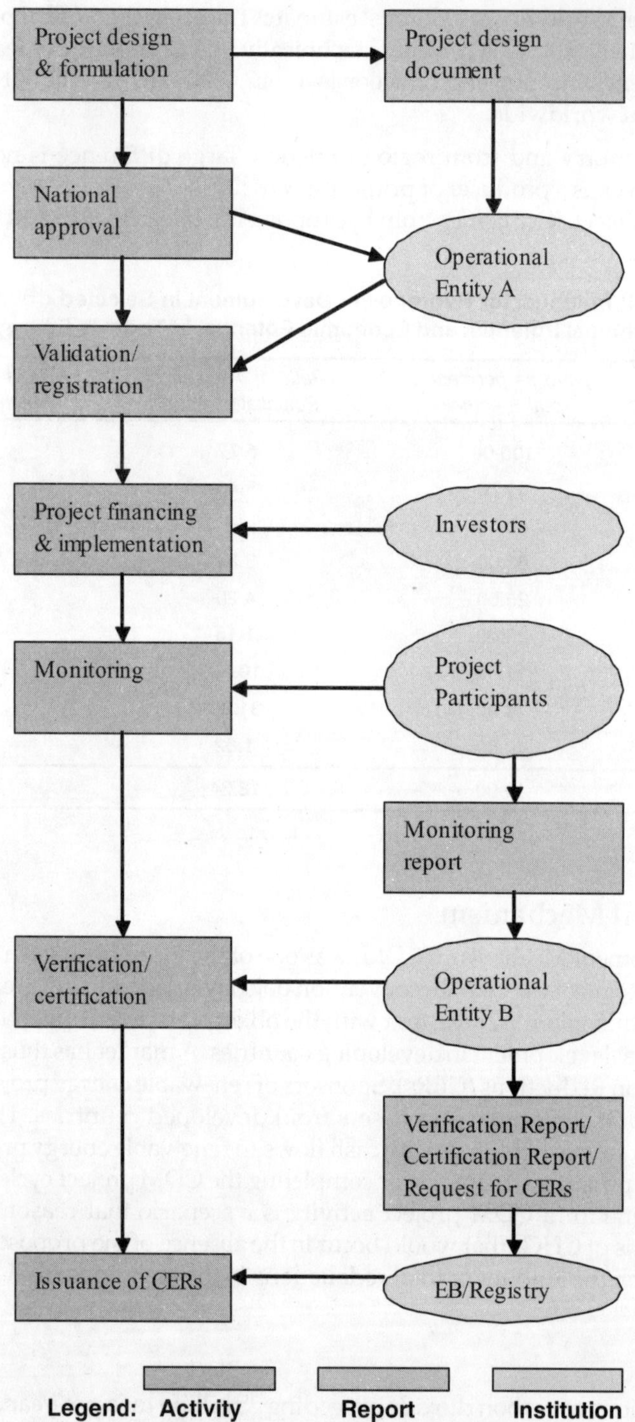


**Figure 52.1: Project Cycle for CDM**

***Source*: Pimparkhedkar, 2006**

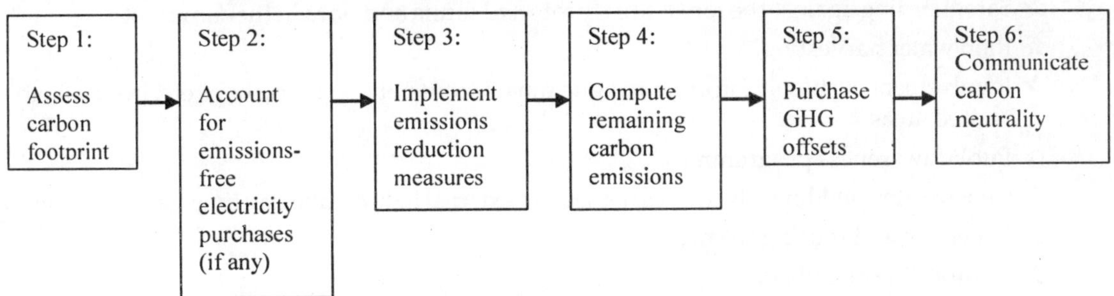


**Figure 52.2: Achieving Carbon Neutrality**
***Source*: Clean Air-Cool Planet, 2006**

with the climate system that the United Nations Framework on Climate Change warns against. National and international policy will be pivotal to any successful long-term effort to reduce GHG emissions and to moderate global warming. At the same time, and partially to the absence of such policy, voluntary greenhouse gas markets are proliferating. Voluntary carbon emissions reductions and offset projects have played a key role in corporate efforts to demonstrate leadership in addressing global warming, and toward branding themselves as sensitive to the climate issue. The retail carbon offsets market especially is emerging as a means of allowing individuals, households, and small companies to be perceived as doing something about global warming by going carbon neutral. Carbon neutrality refers to the idea of reducing and ultimately neutralizing one's carbon footprint (Figure 52.2). The role of carbon offsets is to neutralize that part of a carbon footprint not addressed through direct emissions reductions, the purchase of emissions-free electricity, or other means. Offsets also can be used to render products, services, and events carbon neutral, or to otherwise influence consumer purchasing decisions.

The retail offsets market is growing rapidly. Several dozen companies and organizations now offer the public the ability to purchase offsets in the retail marketplace. These entities generally encourage consumers and small companies to offset their entire personal or business GHG footprints, or to offset specific activity. The price of seeking to go carbon neutral also varies widely. Carbon neutral websites sell offsets that range from $5 to $25 per ton, averaging about $10/ton.

## Essential Contribution from Academic Institutions

For a more sustainable energy future the technology-intensive policy options include increased efficiency of energy production and use, reduced consumption, a new generation of renewable energy technologies, nuclear options that can win and retain public acceptance, and means to use fossil fuels in a climate-friendly way. If fossil fuel prices rise to include costs of carbon management, consumers may also modify their consumption patterns. Environmental and ethical concerns may also contribute to new attitudes about unconstrained economic growth patterns. Sustainability concepts provide a framework to focus the evaluation of energy technology and policy options and their tradeoffs and to guide decision making on energy futures. The key to getting these concepts right is to develop a solid understanding of the multifaceted technologies, geopolitical, sociological, and economic impacts of energy use and abuse. The different initiatives commenced at our institute level are as under:

- ☆ Establishing Energy and Environment Research Center.
- ☆ Increase the forest cover in the campus.

- ☆ Safeguarding against the forest fires (protected fauna and flora in the campus).
- ☆ Rain water harvesting.
- ☆ Involvement of students in organizing seminars/conferences/workshops etc. in sustainable related areas.
- ☆ Public awareness programmes.
- ☆ Constructing building which does not require external light during day time (Smart Building).
- ☆ Recycling and reusing the material.
- ☆ Composting of garbage.
- ☆ Reclamation of sewage.
- ☆ Replacement of ordinary bulbs and tubes with energy efficient compact fluorescent lamps.

However, achieving this vision will require a combination of increased R&D on energy efficiency and policies that encourage conservation and use of high efficiency systems.

## Curricula

The transition to a truly sustainable civilization is hard to picture at present. It requires achieving a stable human population that recognizes the finite limits of Earth's systems to produce resources and absorb wastes and that acts accordingly. If fossil fuel prices rise to include costs of carbon management, consumers may also modify their consumption patterns. Environmental and ethical concerns may also contribute to new attitudes about unconstrained economic growth patterns. Sustainability concepts provide a framework to focus the evaluation of energy technology and policy options and their tradeoffs and to guide decision making on energy futures. The key to getting these concepts right is to develop a solid understanding of the multifaceted technologies, geopolitical, sociological, and economic impacts of energy use and abuse. An alternative, the sustainable and more secure future, is to focus our energy policy on greater efficiency and a move toward renewable energy sources. Therefore, the curricula of different academic institutions and universities must focus on the following inter-disciplinary areas with greater emphasis on sustainable and cleaner energy sources at graduate and post-graduate level:

- ☆ Energy Resources, Economics and Environment
- ☆ Direct Energy Conversion
- ☆ Utilization of Solar Thermal Energy
- ☆ Non-conventional Energy Sources
- ☆ Fuel Cell and Hydrogen Production
- ☆ Wind Energy Conversion Systems
- ☆ Energy Management
- ☆ Waste to Energy
- ☆ Energy Systems Modeling and Analysis
- ☆ Conservation of Energy in Buildings
- ☆ Nuclear Reactor Thermal Hydraulics and Safety
- ☆ Materials and Devices for Energy Applications

## Research

It is clear that oil comprises the largest share of world energy consumption; a reduction in availability of oil will cause a major disruption unless other resources can fill the gap. Natural gas and coal production may be increased to fill the gap, with the natural gas supply increasing more rapidly than coal. Any increase in coal consumption will worsen the global climate change situation. Although $CO_2$ sequestration is feasible, it is doubtful that there will be any large scale application of this technology for existing plants. Presently, there is a resurgence of interest in nuclear power. However, it is doubtful that nuclear power alone will be able to fill the gap. It seems logical that renewable energy technologies of wind, solar, and biomass will not only be essential but will hopefully be able to fill the gap and provide a clean and sustainable energy future.

There are two major pathways for developing non-fossil-fuel energy alternatives: pursuing nuclear power and promoting renewable-energy applications. Nuclear power has been in use for 45 years, and various renewable energy systems have also been developed. Thus, both nuclear and renewable alternatives are at a stage where they could be expanded to provide further energy needs if suitable technological solutions and public acceptance or pressure and government support were provided.

India is endowed with a substantial resource base and non-conventional energy technologies are well suited for both grid-connected power generation and for energy supplies in those remote areas that are not yet connected to the grid. As a result of efforts made during the past quarter century, a number of renewable energy technologies, systems, and devices have been developed and have become commercially available.

## Taking Classroom and Research to Common Man

In understanding energy needs and the technologies for supplying them, it is important to have an idea of the energy requirements of specific activities. Conserving energy is extremely important, but keeps in mind that reducing our use of fossil fuels is not eliminating it. Energy conservation can play an important role in future energy strategy, because it can mitigate impacts on the environment rapidly and economically. Considering the vast potential of energy savings and benefits of energy efficiency, the Government of India enacted the Energy Conservation Act. The Act provides legal framework, institutional arrangement, and a regulatory mechanism at the Central and State levels to embark upon energy efficiency drive in the country. The Energy Conservation Act of 2001 facilitated the creation of the Bureau of Energy Efficiency (BEE) that has served to appropriately elevate the importance of energy efficiency issues.

Many technological advances have occurred in the last two decades, resulting in striking reductions in the energy usage required to operate buildings safely and comfortably. The general procedure for establishing an energy management program in buildings involves five steps:

- ☆ Review historical energy use.
- ☆ Perform energy audits.
- ☆ Identify energy management opportunities.
- ☆ Implement changes to save energy.
- ☆ Monitor the energy management programs, set goals, and review progress.

An overall estimate should be made of how effectively the facility uses its energy resources. This is difficult to do in many cases because so many operations are unique. Equipment that use the greatest amounts of electricity should be examined. Each item should be reviewed and these questions should be asked:

- ✰ Is this actually needed?
- ✰ How can the same equipment be used more efficiently?
- ✰ How can the same purpose be accomplished with less energy?
- ✰ Can the equipment be modified to use less energy?
- ✰ Would new, more efficient equipment be cost-effective?

## Conclusion(s)

Renewable energy can help meet critical national goals for fuel diversity, price stability, economic development, environmental protection, and energy security, and thereby play a vital role in national energy policy.

Clean Development Mechanism (CDM) can provide a tool for technology transfer if investment is channeled into projects that replace old and inefficient fossil fuel technology or create new industries in environmentally sustainable technologies.

The least expensive and most efficient means is energy conservation, rather than more energy production. Moreover, energy conservation is also the best way to protect the environment and reduce global warming. However, achieving this vision will require a combination of increased R&D on energy efficiency and policies that encourage conservation and use of high efficiency systems.

Carbon neutrality offers individuals, businesses, and academic institutions the opportunity to take personal responsibility for the global warming implications of their lifestyles.

## References

1. Bhat I.K., and Katoch, S.S. 2007a. Role of Academia in Combating Global Warming. Proceedings of ISTE Section Annual Convention held at NIT Hamirpur (HP) India during November 18-19, 167-172.
2. Bhat, I.K., and Katoch, S.S. 2007b. Sustainable Energy in Wake of Global Climate Modification. Combat Climate Change Through Market Driven Strategies (Proceedings of Convention on Corporate Response to Climate Change organized by World Environment Foundation at Palampur (HP), India during June 08-10), 19-34.
3. Clean Air-Cool Planet. 2006. A Consumers, Guide to Retail Carbon Offset Providers. Clean Air-Cool Planet, 100 Market Street, Suite 204, Portsmouth, NH 03801.
4. Dhamija, P., Bharadwaj, S.K., and Bist, B.M.S. 2007. Geothermal Energy in India. Renewable Energy 'Akshay Urja', Vol.1, Issue 2, 33-35.
5. Jain, P. and Singh, I.P. 2007. Sustainable Management through Micro and Mini Hydro Power Plants. Renewable Energy 'Akshay Urja', Vol.1, Issue 2, 22-28.
6. Katoch, S.S., Bhat, I.K., and Kumar, A. 2007. Academia Preparedness to Combat Climate Change for Sustainable Development. Proceedings of International Conference on Environmental Management: Scenario and Strategies to 2020 held at UEC Ujjain (MP), India during December 26-27, 147-153.
7. Kelly, T.M., and Winchester, J.W. 2005. Can Hazardous Waste Sites be Breached as a Result of Future Climate Change? Journal of Environmental Engineering, ASCE, Vol.131, No.5, 810-814.

8. Kharul, R. 2006. Wind Resource Assessment for India. Green Energy (World Institute of Sustainable Energy), Vol.2, No.4, 14-15.
9. Kreith, F., and Goswami, D.Y. 2007. Handbook of Energy Efficiency and Renewable Energy. CRC Press, Taylor and Francis Group, Boca Raton.
10. Kureel, R.S. 2007. Biofuel Scenario in India. Renewable Energy 'Akshay Urja', Vol.3, Issue 2, 12-17.
11. Miller Jr., G.T. 2004. Environmental Science–Working with the Earth. Thomson Asia Pte. Ltd., Singapore.
12. Pillai, G.M. 2007. Next Big Thing: Concentrating Solar Thermal Electricity Time for India to Catch Up. Green Energy (World Institute of Sustainable Energy), Vol.3, No.4, 19-25.
13. Pimparkhedkar, S. 2006. Wind Power and CDM. Green Energy (World Institute of Sustainable Energy), Vol.2, No.4, 30-31.
14. Singh, N.P. 2007. 25 Years of Renewable energy in India. Renewable Energy 'Akshay Urja', Vol.1, Issue 2, 16-21.
15. Swain, M. 2007. Wind Energy–An Overview. Renewable Energy 'Akshay Urja', Vol.3, Issue 2, 18-22.
16. Tester, J.W., Drake, E.M., Driscoll, M.J., Golay, M.W., and Peters, W.A. 2005. Sustainable Energy–Choosing Among Options. Prentice-Hall of India Private Limited, New Delhi.
17. Wagh, A. 2007. Solar and CDM: CERs and VERs. Green Energy (World Institute of Sustainable Energy), Vol.3, No.4, 34-35.
18. Wright, T.R. 2005. Environmental Science–Towards a Sustainable Future. Prentice-Hall Inc.

# Chapter 53

# Bioprocesses for Treatment of Flue Gas Containing $SO_2$ and NOx

**Rima Biswas and R.A. Pandey***
**National Environmental Engineering Research Institute,**
**Nehru Marg, Nagpur – 440 020**

## ABSTRACT

Sulfur dioxide ($SO_2$) and oxides of nitrogen (NOx) are of environmental concern as these hazardous air pollutants lead to the formation of acid rain and tropospheric ozone. Concern for environmental and health issues coupled with stringent $SO_2$ and NOx emission standards generates a need for the development of efficient low-cost $SO_2$ and NOx abatement technologies. Under such circumstances, it becomes mandatory for each $SO_2$ and NOx-emitting industry or facility to opt for proper $SO_2$ and NOx control measures.

This article describes major biological and chemobiological processes for the treatment of $SO_2$ and NOx from flue gases based on the catalytic activity of eucaryotes or prokaryotes, the use of micro algae, and a combined physicochemical and biological process (Bio FGD and BioDeNOx). Relatively uncomplicated design and simple operation and maintenance requirements make biological removal a good option for the control of $SO_2$ and NOx emissions in stationary sources.

*Keywords: Sulfur oxide, Oxides of nitrogen, Biological processes, Chemobiochemical processes.*

## Introduction

With the rise in global power consumption, concern for environmental pollutions like sulfur dioxides ($SO_2$) and oxides of nitrogen (NOx) are on rise. Concern for environmental and health issues

* Corresponding Author: E-mail: ra_pandeyy@neeri.res.in

coupled with stringent NOx emission standards generates a need for the development of efficient low-cost $SO_2$ and NOx abatement technologies. Under such circumstances, it becomes mandatory for each $SO_2$ and NOx-emitting industry or facility to opt for proper control measures. Host of commercial processes and technologies are available that can remove these pollutants from the flue gases (Pandey *et al.*, 2005, Jin *et al.*, 2005). But these physico-chemical processes are cost-intensive, as they require chemicals for treatment of these gases and also generate chemical sludge, which warrants treatment either for reuse or disposal. The relatively high usages of the chemicals, power utilization for heat and pressurization, and disposal costs of the waste are the important drawbacks of these physico-chemical processes. Biotechnology is versatile and has been assessed a key technology for a sustainable chemical industry (Lievonen, 1999). Industries that previously never considered biological sciences as impacting their business are exploring ways of using biotechnology to their benefit. Environmental concerns in chemical industries such as those emitting sulfur dioxides ($SO_2$) and oxides of nitrogen (NOx) are developing interest in biotechnological approaches to not only remove pollutants from the environment but prevent pollution in the first place. This is because such biological processes operate at lower temperature; produce less toxic waste, fewer emissions and by-products compared to conventional chemical processes. The present chapter presents an over view of various biological processes for treatment of $SO_2$ and NOx from flue gases.

## Biological Methods of Desulphurization of Flue Gases Containing $SO_2$

In the last two decades the researchers have stressed for developing biological methods for treatment of gaseous streams containing SOx. These methods are either direct or indirect. The direct processes for $SO_2$ removal through biological systems suffer lots of limitations. The major limitation is presence of high concentration of $CO_2$ (10-15 per cent) in the flue gas, which can be inhibitory to the growth of microorganisms in the biofilm. Further, the direct biological processes are slow and it is hard to maintain a consistent supply of the emissions at constant flow rate and at appropriate concentrations of the flue gas constituents into the bioreactor system. This problem can be omitted by absorbing SOx selectively in a suitable solvent and the sulphate/sulfite rich scrubbed solution can then be reduced to elemental sulphur using biotechnological approach [Buisman *et al.*, 1990 and 1994, Cork *et al.*, 1986, Janseen *et al.*, 1995].

In these biological processes the biological entities involved are sulphate reducing bacteria (*Desulfovibrio*, *Desulfotomaculum*, etc.), photosynthetic bacteria (genera of the family of *Chlorobiaceae* and *Chromatiaceae*) and autotrophic *Thiobacillus* sp. The researchers have applied various combinations of anaerobic and aerobic system using these microorganisms in different stages. Some of the investigations regenerate the scrubbing solution that is used in the scrubber for absorption of sulfur dioxide as the first stage of the two stage chemo-biological processes. The brief discussion of these processes are discussed and presented hereunder.

### Direct Biological Method for Flue Gas Desulphurization

Studies have also been carried out by some workers for biofiltration of sulphur dioxide from waste gaseous streams [Clark and Wonorowski, 1992]. The feasibility of the utilization of sulphur dioxide as an electron acceptor from flue gases by *Desulfovibrio desulphuricans* has been studied [Lee and Sublette, 1991] and the economic potential of this approach was further verified [Sublette and Gwozdz, 1991]. *Desulfotomaculum* orient is has been shown to grow on hydrogen, carbon dioxide and sulphur dioxide with production of hydrogen sulphide, while, *D. desulphuricans* transforms $SO_2$ to $H_2S$ using minerals and pretreated sludge as carbon source [Deshmane *et al.*, 1993]. Sulphur dioxide could readily be removed in a laboratory scale experimental biofilter at concentrations as high as 1000 ppm v/v on a simple two component matrix. In spite of the known biocidal properties of sulphur dioxide, the system performed well over a range of concentration and with time.

Studies on various immobilized cell bioreactors to maximize the productivity of the bioreactor for $SO_2$ reduction for effective flue gas desulphurization have been carried out [Selvaraj *et al.*, 1997]. A comparison among a stirred tank reactor containing flocs of mixed cultures of SRB, and a column reactor with cells immobilized in either potassium-carrageenan gell matrix or polymeric porous BIO-SEP beads revealed that mixed SRB cells grown into highly stable BIO-SEP polymeric beads exhibited highest sulphite conversion rates (95 per cent sulphite conversion at $SO_3$ loading rates of 20 M mol/L/h).

## Two Stage Indirect Process for $SO_2$ Removal by Inorganic Iron Oxidation

Dowa Mining Co., Japan, has exploited microbial oxidation of ferrous iron to ferric iron by *Thiobacillus ferrooxidans*, for desulphurization of waste gas containing sulphur dioxide Imaizumi, 1985. This particular process is based on the wet scrubbing of waste gas with ferric sulphate solution of adequate concentration generating ferrous sulphate solution.

$$SO_2 + Fe_2(SO_4)_3 + 2H_2O \rightarrow 2FeSO_4 + 2H_2SO_4 \quad (1)$$

The resultant ferrous sulphate solution is taken to the microbial oxidation tank, wherein *Thiobacillus ferrooxidans* oxidizes the ferrous sulphate to ferric sulphate.

$$2FeSO_4 + 2H_2SO_4 + \tfrac{1}{2}O_2 \xrightarrow{T.\,ferrooxidans} Fe_2(SO_4)_3 + H_2SO_4 + H_2O \quad (2)$$

The ferric sulphate produced through biological oxidation of ferrous sulphate generated in the second stage of the process, is recycled to the wet scrubbing tower to absorb sulphur dioxide contained in the waste gas. The rest of the ferric sulphate solution containing sulphuric acid is neutralized with calcium carbonate powder, producing gypsum and ferric hydroxide precipitates. Both gypsum and ferric hydroxide can be recovered separately through solid-liquid separation. The ferric hydroxide generated in the process has potential market to be used as coagulant for the treatment of water and wastewater. Further, the system operates at low pH (2.0). Therefore, there is no cross contamination with other microorganisms which can affect the process efficiency. Thus, such processes are very selective for treatment of SOx in flue gases.

The environmental problems associated with large-scale production of gypsum have already been discussed in the preceding sections. Sulphur is considered to be the most desirable end product of any FGD process because it can be handled easily, is non-toxic and is non-corrosive solid containing more sulphur per unit molecule (*e.g.*, $S_8$, orthorhombic sulphur) than any other form. Therefore, it can be transported at high value as orthorhombic sulphur than any other sulphur form.

## Microbial Flue Gas Desulphurization with Sorbent Regeneration

Lee and Sublette (1991), proposed to couple the reduction of $SO_2$ to $H_2S$ by mixed culture of sulphate-reducing bacteria containing *D. desulphuricans* to Claus process as a means of by-product recovery from dry generable scrubbing process of desulphurization of flue gas such as the copper oxide process and the NOxSO process. Sorbent regeneration in both the processes produce a concentrated stream of $SO_2$ as in copper oxide process or $SO_2$ and $H_2S$ as in the NOxSO process, which must be disposed of or recovered in a separate process. Two-thirds of $SO_2$ stream obtained from CuO process stream is reduced to $H_2S$ by *D. desulphuricans*. The resulting $H_2S$ may be then combined with remaining one-third $SO_2$ and is fed to a Claus reactor. Alternatively, a stream of $H_2S$ and $SO_2$ obtained from the regeneration of the sorbent from the NOxSO process may be conditioned by mixed cultures of sulphate-reducing bacteria to produce the specific $H_2S/SO_2$ ratio (2:1) required by a Claus reactor.

## Two Step Microbial Reduction of Sulphur Dioxide

Two-step processing of sulphate/sulphite rich solution obtained from scrubbed sulphur dioxide leading to the formation of elemental sulphur have been investigated by Cork *et al.*, 1986. Studies have been carried out on controlled biological reduction of sulphate/sulphite reduction to sulphur utilizing pure cultures of sulphate reducing hetrotrophs and sulphur producing photoautotrophs. In the first step, strictly anaerobic, acetate-utilizing, sulphate-reducing microorganism *Desulfobacter postgateii* converts sulphate into hydrogen sulphide. In the second step of the process, conversion of hydrogen sulphide to elemental sulphur is achieved by a strictly anaerobic photoautotroph *Chlorobium limicola* forma *thiosulfatophilum*. The schematics diagram of the bioprocess is presented in Figure 53.1. The reactions taking place in the bioprocess are represented as follows.

**Step I**

$$CH_3COOH + SO_4^{2-} \rightarrow H_2S + 2HCO_3^- \quad (3)$$

$$2HCO_3^- + 2H^+ \rightarrow 2H_2O + 2CO_2 \quad (4)$$

**Step II**

$$H_2S + 1/2\,CO_2 \rightarrow 1/4\,CH_3COOH + S + 1/2H_2O \quad (5)$$

About 25 per cent of acetate required in the step I is generated in the dark reaction of the step II. This is recycled back to step I (Cork *et al.*, 1986). This means that about 75 per cent of organic reducing power in a steady state bioprocess must be externally provided.

Instead of employing anaerobic photosynthetic bacteria, elemental sulphur can also be reclaimed through partial biological oxidation of sulphide using chemolithotropic colorless sulphur bacteria (*Thiobacillus ferrooxidans*), which oxidize the sulphide to elemental sulphur or sulphate in the presence of oxygen or nitrate. Abma and Buisman (1997), have reported the use of *T. thioparus*, *T. versutus*, *T. neopolitanus*, *T. denitrificans and T. thiooxidans* for the partial oxidation of $H_2S$ to elemental sulphur. The

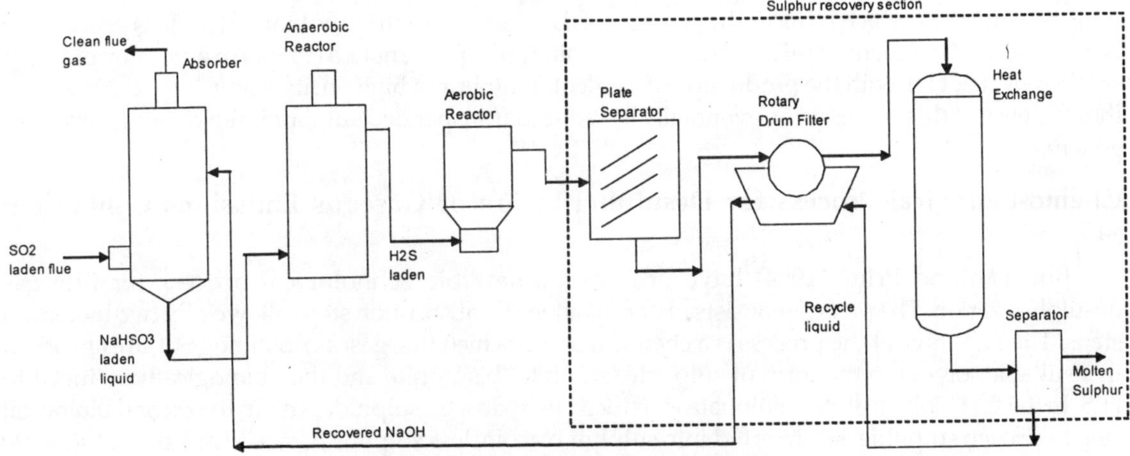


**Figure 53.1: Schematics of Two-Stage Process for Microbial Conversion of Sulphur Dioxide to Elemental Sulphur**

production of elemental sulphur depends upon the dissolved oxygen concentration of the medium (Buisman *et al.*, 1991). As elemental sulphur formation requires four times less oxygen than required for complete oxidation to sulphate, elemental sulphur is the major end product of sulphide oxidation below 0.1 mg/L concentration of dissolved oxygen (Imaizumi, 1985, Janseen *et al.*, 1997). Sublette and Sylvester (1987) have reported better performance of *T.denitrificans* for removing more than 97 per cent of $H_2S$ in an anaerobic stirred tank reactor.

Besides the direct bio-catalytic activity of *Thiobacillus* species for abatement of $H_2S$, an indirect method of combining the physicochemical oxidation of sulfide with ferric sulphate solution to biological ferrous re-oxidation, has been developed. In this process, the $H_2S$ generated from the first biological step of reduction of sorbed solution of $SO_2$ from flue gas, is taken to a chemical absorber unit. In a chemical absorber, hydrogen sulfide is injected into ferric sulphate solution at pH 1.3. Elemental sulphur is formed as per the reaction:

$$H_2S + Fe_2(SO_4)_3 \rightarrow 2FeSO_4 + H_2SO_4 + S^o \tag{6}$$

After recovering the elemental sulphur by solid liquid separation, the ferrous sulphate solution is led to a microbial oxidation tank where *T. ferroxidans* oxidizes the ferrous ion to ferric (Pandey and Malhotra, 1999).

The iron oxidation rate is observed to be five times higher ($3.6 g.L^{-1}.h^{-1}$) in a fixed bed reactor (sintered glass carrier) compared to the suspended cell reactor. Using immobilized *T. ferroxidans* cells, with an estimated bioreactor volume of 60 $m^3$ for a gas stream of 120000 $m^3.h^{-1}$ and with a $H_2S$ concentration of 0.5 ppm, Pagella and Faveri (2000) reported that pH plays a key role both for controlling the growth rate of *Thiobacillus* sp. and the solubility of the materials in the system. The advantages of this process with respect to conventional treatment processes for $H_2S$ abatement (*e.g.* Claus) are mild pressure and temperature conditions (typical of biotechnological processes), lower costs and closed-looped operation without the input of chemicals or output of wastes. Dowa Mining Co. has applied this technology for treating the surplus hydrogen sulphide gas in the plant of Barite Industries Co. at Kosaka, Japan (Imaizumi, 1985), which started its operation in December 1984. It is estimated that the total cost of the process is about one third of the caustic soda absorption process. A modification of this process has been developed at National Environmental Research Institute (NEERI), Nagpur, India. This modification employs a rotating biological rope contactor (RBRC), immobilized with *T. ferrooxidans* (DBT report, 1993), which reduces the problem of jarosite formation in the bioreactor. This process has been tested for treatment of refinery fuel gas containing 4.5 per cent $H_2S$ (v/v). An efficiency of 99 per cent removal of $H_2S$, with the production of elemental sulphur of high purity was achieved. However, the efficiency of this process may eventually be reduced if suspended sulphur is not removed from the system.

## Chemo-biological Process for Desulphurization of Gaseous Emissions Containing $SO_2$

Buisman and Prins (1994) have proposed a new biotechnological process for flue gas desulphurization. The process consists of one alkaline $SO_2$ absorption step followed by two biological steps. The first stage of the process is a chemical one in which flue gases containing $SO_2$ are absorbed in a suitable solvent in the form of sulphate/sulphite/bisulphite and then biologically reduced to $H_2S$. In the first biological step, sulphate is reduced to hydrogen sulphide, and in the second biological step hydrogen sulphide is converted into sulphur by colorless sulphur bacteria (Buisman *et al.*, 1990 and 1991, Janseen *et al.*, 1995). This process is cheaper than the presently used physico-chemical processes and can remove up to 98 per cent of $SO_2$. Moreover, instead of gypsum or waste sorbent, this process produces sulphur that can be potentially reused in the industry

Another similar technology called the Biostar Development has been developed to a pilot-scale, in a joint venture of PAQUES B.V and Hoogovens Technical Services, Netherland (Dean *et al.*, 1996). This technology is also called biological flue gas desulphurization (BIO-FGD), which economically removes the $SO_2$ from flue gas and produces high quality sulphur. The BIO-FGD process, consisting of four units, is represented in a schematic diagram (Figure 53.2). The four units include an absorber, an anaerobic reactor, an aerobic reactor and a sulphur recovery unit.

The flue gas containing the $SO_2$ is washed with small droplets of weakly alkaline water in a counter-current flow. The liquid droplet creates a large surface area for the physical absorption of $SO_2$. Once in the liquid droplet, the $SO_2$ reacts with the added NaOH to form the sodium bisulphite. The optimal efficiency of the absorber unit is governed by pH with optimal efficiency being in the range of 7-8. A low pH decreases the $SO_2$ removal efficiency and increases pH of the aerobic reactor above the required level.

The sodium sulphite solution from the absorber is led to the anaerobic reactor along with the added electron donor (methanol) and is made to flow up through a bacteria-laden fluidized bed reactor. This anaerobic reactor uses the rising biogas produced by the biological system to circulate the medium. The upper chamber of this reactor serves as a polishing chamber through which $H_2S$ gas is

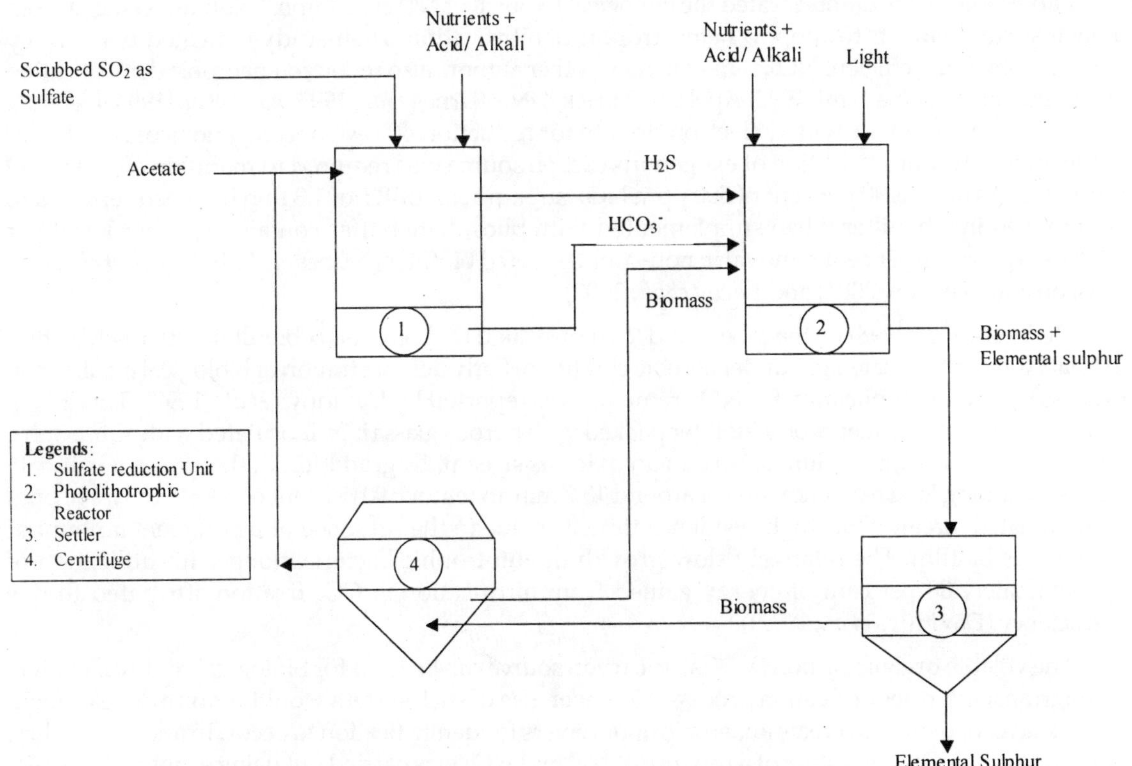


**Figure 53.2: Schematic of BIO-FGD Process**

made to exit, while trapping the remaining biogas. A portion of the biogas is recycled to the bottom of the reactor to aid the liquid flow and the remaining biogas is vented to the atmosphere.

The aerobic unit used in the BIO-FGD process is an airlift loop reactor. Air is injected to mix and circulate the liquid medium and biomass of *Thiobacillus* species, which can partially oxidize the sulphide to elemental sulphur. This circulation ensures that $H_2S$ and bacterial biomass are kept in suspension and have adequate contact with air to form elemental sulphur. The liquid rising through the reactor flows through the separator prior to the exit and biomass is dropped into the slot in the separators. These are then returned back to the unit. The effluent from this unit is then led to sulphur recovery unit, which consists of several steps that purify and recover the molten elemental sulphur.

## Biological Methods for NOx Removal

Biological removal of NOx from contaminated gas streams relies on oxidative route through nitrification leading to generation of nitrate and then followed by denitrification.

### Biofilter System

Biofilter is a type of gas phase bioreactor configuration. This system is effective for contaminants that display high water solubility and high biodegradability (Van Groenestijn and Hesselink, 1993). Biofilters utilize microorganisms immobilized as biofilms on packing materials such as compost, peat and soil (Leson and Winer, 1991). Contaminants in the vapour phase are absorbed by the bed and biologically converted into benign end products, such as water and carbon dioxide.

Barnes *et al.* (1995) demonstrated the biological removal of NOx in compost biofilter, using glucose or lactose supplements to support heterotrophic denitrification. Their study indicated pH as a key factor for obtaining efficient NOx removal. Some other authors also reported anaerobic denitrification in compost biofilter (Sadami, 1993; Apel and Turick, 1993; Barnes *et al.*, 1995; Apel *et al.*, 1995, Lee *et al.*, 1999). Compost it self provided electron donors for reduction of low concentrations of NO. But at higher concentrations, addition of exogenous carbon source was required to maintain efficient NO reduction. As high as 90 per cent of 500 ppm NOx stream at an EBRT of 1.3 min has been reported to be removed in a biofilter when supplemented with phosphate buffer containing either lactate or dextrose against 20 per cent removal in non-supplemented biofilter (Barnes *et al.*, 1995, Lee *et al.*, 2001, Klasson and Davison, 2001, and Lacey *et al.*, 2002).

Du Plesis *et al.* (1998) on the other hand, demonstrated the use of NOx biofilter with thick biofilm to treat NOx contaminated gas under aerobic condition. Early demonstration of biological nitrification process in biofilter application for NOx removal was reported by Davidova *et al.* (1997). This group conducted their experiments on a biofilter packed with porous glass rings inoculated with autotrophic bacteria grown on ammonium source in an oxic gas stream. Degradation in that system was rate limited requiring long residence time of around 13.7 min. to remove 90 per cent of NOx from a 100 ppm contaminated stream. This high residence time was due to the presence of insufficient amount of biomass as biofilm. The relatively slow growth of autotrophic bacteria along with utilization of approximately 80 per cent of energy gained from nitrification in $CO_2$ fixation attributed to this inefficiency (Davidova *et al.*, 1997)).

The volatile organic carbon (VOCs) as carbon source can be used for biological removal of NOx through anaerobic denitrification process. However, use of such system would require development of thick layer of biofilm to create anaerobic underlayers for denitrification to occur. In another studies, biofiltration of NOx, consisting of a mixture of NO and $NO_2$ was carried out using a laboratory scale-biofilter packed with soil (Lukow and Diekmann *et al.*, 1997, Yani *et al.*, 2000, Kim *et al.*, 2000, and

Oukuno *et al.*, 2000). Soil contributed to NOx removal through physical adsorption of NOx on to the soil particles.

Most of the research studies published on biological NOx removal were carried out under mesophilic conditions. Gases emerging from the scrubbers in thermal power plant typically exhibit temperatures between 50°C and 60°C. The bioreactor packing materials should therefore exhibit long-term stability within this temperature range and also must contain suitable concentrations of thermophillic bacteria.

Labedava and co-workers (1998) isolated 82 strains of thermophilic NO consuming bacteria in sediment samples from hot springs in the Kronot reserve (Kamchatka), corroded matter from steel pipes of the heating system in Moscow, and soils of the southern bioclimatic belt in Russia (Chestnut, white alkali soil). The most active strain identified as *Bacillus stearothermophilus* strain 1NM150. Cells of this strain immobilized on a ceramic support demonstrated a high NO uptake in a 4dm$^3$ bioreactor (Labedava *et al.*, 1998). Around 60-90 per cent of NO was removed at 60°C for over 6 months of continuous bioreactor operation with inlet concentrations of 5 ppm and a gas flow rate of 100 dm$^3$h$^{-1}$.

The major drawback of such gas phase bioreactors treating pollutants such as NOx is the presence of oxygen. Flue gases bearing NOx typically contains 8 per cent $O_2$. Development of anaerobicity niches in biofilters depends on formation of thick biofilms creating anaerobic underlayers. This however will pressure drop.

For over a century, denitrification and ammonification had been considered as processes exclusively found in prokaryotes, until denitrification was observed in the filamentous fungus *Fusarium oxysporum* (Shoun and Tanimoto, 1991) and in other fungi (Shoun *et al.*, 1992), which were long thought to be strictly aerobic. The enzymes responsible for fungal denitrification have been isolated in *Fusarium oxysporum*. It uses nitrate dissimilatively but the end product is $N_2O$, probably it lacks $N_2O$ reductase. Studies with *Fusarium solani*, *Cylindrocarpon tonkinase* and *Chactomium* sp. With reported nitrogen as end product of denitrification, but enzymes responsible for $N_2O$ respiration still needs to be determined.

Some other fungal species (such as *Asperigullus* sp. *Alternaria* sp. *F.solani*, *Fusarium verticilliodes*, *Fusarium dimerum*, and *Fusarium javanicum*) have been reported that have exhibited denitrifying activity even under aerobic condition. A toluene degrading fungal bioreactor seeded with *Erophiala lecanii-corni* was shown to aerobically remove NO from a contaminated gas stream, using toluene as a sole source of carbon and energy (Woertz *et al.*, 2001). The fungal bioreactor removed 93 per cent NO at an inlet concentration of 250 ppm and at an EBRT of 1 min, corresponding to an inlet loading rate of 17.22 g NO m$^{-3}$h$^{-1}$ when simultaneously supplied with 90 g toluene m$^{-3}$h$^{-1}$ (Woertz *et al.*, 2001). Interestingly, the fungal mechanisms responsible for aerobic NOx removal in the biofilter may involve nitrogen assimilation, as NOx removal was inhibited by accumulation of ammonium in the biofilter, a known major inhibitor of assimilatory nitrate reduction.

## Biotrickling Filter

Biotrickling filters are similar to biofilters, but contain an inert packing material (such as ceramic or plastic rasching rings) that provide surface area for gas/liquid mass transfer and biofilm growth. A recirculating liquid phase flows through the reactor in order to keep the packing wetted. The major advantage of biotrickling filter is the ability to treat hot air streams and acid producing contaminants (Cox *et al.*, 2001), which makes it possible to handle combustion gas with a minimum of air pre-conditioning.

Majority of the work reported in the literature for biotreatment of NOx using biotrickling filter pertains to nitrification process as nitrate levels in the liquid phase slowly increased with time and that recovery of NO as $NO_3^-$ exceeded 90 per cent (Jin *et al.*, 2005). This obviously necessitates a subsequent denitrifying unit to reduce nitrates to nitrogen gas. Moreover, the biotrickling filters have characteristically high EBRT from 6 min to 13.7 min due to mass transfer limitation of the poorly soluble nitric oxide and oxygen. Use of ultrasonic nebulizer in lieu of a conventional spray nozzle has been demonstrated to reduce the EBRT in biotrickling filter (Davidova *et al.*, 1997).

In another application, Robra *et al.* (1997) demonstrated a pilot scale biotrickling filter packed with light expanded clay aggregates inoculated with nitrifiers isolated from natural substratum and from air and dust of road tunnels. The 55 $dm^3$ trickling filter built to treat polluted air in a road tunnel at Katscherg tunnel in Carinthia, Australia, exhibited good performance over 2 years of experimentation. At gas residence times ranging from 7.5 to 11 s, a maximum removal efficiecies of 90 per cent CO, 15-20 per cent NO, and 95 per cent $NO_2$ was observed.

A bench scale biofilter study conducted by Jinsiriwanit and Deshusses (Paper # 69814) using sulfur oxidizing denitrifying bacteria revealed NO removal of more than 90 per cent at an EBRT of 189s at influent NO concentration of 300 ppm. However, 5 per cent oxygen in the inlet gas inhibited the NO and removal efficiency dropped from 12.4 to 7.7 $g/m^3.h$.

## Hollow Fibre Membrane Bioreactor

Recently, the feasibility of using hollow fiber membrane bioreactor (HFMB) for removal of NO from combustion gas has been evaluated (Min *et al.*, 2002). The HFMB consists of bundles of microporous hydrophobic hollow fiber membrane, a biofilm and a nutrient supply. The HF membranes serve as a support for the microbial populations and provide large surface area for NO and oxygen mass transfer (Ergas, 2005).

Activated sludge from a wastewater treatment plant was acclimated with an $NH_4^+$ and $NO_2^-$ nutrient solution for over two months and was used to inoculate the HFMB. Synthetic ccombustion gas (15 per cent $CO_2$, 5 per cent $O_2$, 77 per cent $N_2$) containing NO passes through the lumen of the microporous hydrophobic hollow fiber membranes. The NO gas diffuses through the membrane pores and partitions in to a nitrifying biofilm where it is oxidized to $NO_3^-$. Over 90 per cent of the nitrogen was converted to $NO_3^-$ by nitrification (Ergas, 2005). The HFMB was initially fed 100 ppm NO at a gas flow rate of 100 $cm^3\ min^{-1}$ for 1 month, and with a liquid recirculation rate of 300 $cm^3\ min^{-1}$. The NO removal rate was 12-14 $g\ m^3\ d^{-1}$, corresponding to efficiencies of 27 $g\ m^3\ d^{-1}$ with 77 per cent removal efficiency at a recirculation rate of 600 $cm^3\ min^{-1}$ at 20°C. The NO elimination capacity of the synthesis combustion gas ranged between 24 and 26 $g\ m^3\ d^{-1}$ at 20°C and remained constant on increase in temperature from 20°C to 55°C.

## Photobioreactor

Photobioreactors are all related to harness the potential of phosynthetic microalgae capable of removing NO and also $CO_2$ from gaseous streams. The first reported use of photobioreactor was made by Yoshihava *et al.* (1996). A marine microalga NOA-113 was isolated from seawater and was inoculated into a long tubular photo bioreactor system for removal of both $CO_2$ and NO from model flue gas. The system achieved only 50 per cent NO removal efficiency (Yoshihara *et al.*, 1996). It was observed that NOA-113 was not suitable for use in an algal NO removal system because this strain could not remove NO at high NO concentration and at higher gas flow-rates (Yoshihara *et al.*, 1996).

To overcome the limitations of NOA-113, another halo-tolerant algal strain *Dunaliella teriolecta* was found which had the capacity of removing NO from waste gases (Nagase *et al.*, 1997). When NO was continuously supplied to unicellular algal culture in a 2 m column bioreactor concentration ranging from 25 to 500 ppm, in presence of light, about 65 per cent of the NO was removed and about 1.6 per cent $O_2$ was continuously evolved by photosynthesis (Nagase *et al.*, 1997). In dark, the rate of NO removal was governed by availability of $O_2$ and NO removal did not occur in absence of $O_2$. The results of the investigations under various cultural conditions indicated that dissolution of NO in the aqueous phase is the rate limiting step in this reactor (Nagase *et al.*, 1997).

In another investigation by the same research group, the solubility of NO in aqueous media was improved by reducing the bubble size of NO in a bubble column photobioreactor (Nagase *et al.*, 1998). The results showed that NO removal was enhanced by increasing the dissolution of NO in water. In addition, the dissolved oxygen can also improve the rate of NO removal. The highest level of NO removal of 96 per cent was achieved with a countercurrent flow type airlift photobioreactor aerated with small bubbles (Nagase *et al.*, 2001).

Investigations into the metabolism of *Dunaliella tertiolecta* for NO removal in bubble column photobioreactor revealed assimilative type of NO uptake. Most of the NO permeated directly into the algal cell by diffusion and was used for its cellular growth. Some of NO was oxidized into the medium by oxygen and $Fe^{3+}$ present in the medium, before its uptake by the algal cells (Nagase *et al.*, 1997). Based on the mass balance calculations and on the change in nitrate and nitrite concentration in the medium in batch culture, it was found that NO taken up in the algal cell was preferentially used as a nitrogen source for cell growth rather than nitrate, which was one of the nitrogen sources present in the algal medium. The members of genus *Dunaliella* are also known to be efficient producers of β-carotene, glycerol and lipids (Ben-Amotz and Avron, 1983 and 1981) and are being mass cultivated as commercial single cell protein (SCP) similar to *Chlorella* and *Spirulina* (Borowitzka and Borowitzka, 1988). Hence Nagase *et al.* (1997) claimed mass production of *D. tertiolecta* in cheap sea water media with NO as nitrogen source.

## Chemobiological Process

One of the major drawback in all the biological processes discussed so far is slow mass transfer of NO and higher oxygen input in the flue gas consuming the major part of organic carbon which is provided for denitrifying bacteria. Absorbtion of NO into aqueous solution of $Fe^{II}$ chelates developed by Demmink *et al.* (1997) is a process in which there is fast mass transfer of NOx in a small scrubber with less $O_2$ input. Regeneration of $Fe^{II}$ chelates was difficult in this process (Shi *et al.*, 1996 and 1997). Economical and technical constraints in this process urged biologists to develop a process where economical regeneration of $Fe^{II}$ chelate was feasible.

In further development of a chemo-biological process (BioDeNOx process) was designed in which NOx is absorbed by a $Fe^{II}EDTA^{2-}$ chelating complex in the scrubber and subsequently NOx of chelated NOx complex was removed by the microbial denitrifying population (Buisman, 1999; Cetinkaya *et al.*, 2000, van Groenestijin, 2001, Abma 2003, van der Mass *et al.*, 2004, and Kumaraswamy *et al.*, 2004).

In BioDeNOx process, the NO removal from the flue gas consists of four steps: two chemical reactions within the scrubber and two microbial reactions within the reactors (Buisman, 1999). In first step of chemical reactor, wet absorption and complexation of NOx by $Fe^{II}EDTA^{2-}$. The rate of reaction is very fast which leads to an increased mass transfer rate of NOx (Kumaraswamy *et al.*, 2004):

$$Fe^{II}EDTA^{2-} + NO \rightarrow Fe^{II}EDTA\text{–}NO^{2-} \quad (7)$$

In the second step of the chemical reaction, the oxidation of $Fe^{II}EDTA^{2-}$ to $Fe^{III}EDTA^{-}$ occurs by oxygen in the flue gas. This reaction is unwanted chemical reaction as it lowers the availability of $Fe^{II}EDTA^{2-}$ for NO absorption. It consumes most of the oxygen input in the scrubber creating microaerophillic environment (Kumaraswamy *et al.*, 2004):

$$2Fe^{II}EDTA^{2-} + \frac{1}{2}O_2 + 2H^+ \rightarrow Fe^{II}EDTA + H_2O \quad (8)$$

In the third step, the biological reduction of NOx to $N_2$ by denitrifying bacteria occurs where ethanol is used as electron donor (Kumaraswamy *et al.*, 2004):

$$6Fe^{II}EDTA\text{–}NO^{2-} + C_2H_5OH \rightarrow 2HCO_3^- + 2H^+ + 3N_2 + H_2O + 6Fe^{II}EDTA^{2-} \quad (9)$$

In the fourth step of the regeneration of $Fe^{II}EDTA^{2-}$ by reduction of $Fe^{II}EDTA^{-}$ with ethanol as electron donor reaction indicates (Kumaraswamy *et al.*, 2004):

$$12Fe(^{II}I)EDTA^- + C_2H_5OH + 5H_2O \rightarrow 2HCO_3^- + 12Fe^{II}EDTA^{2-} + 14H^+ \quad (10)$$

Out of the two microbial reactions mentioned above, NO reduction to $N_2$ via denitrification decides the effluent gas quality where as the microbial $Fe^{III}EDTA^{2-}$ reduction was essential to get back active $Fe^{II}EDTA^{2-}$ complex for NO absorption. Both the microbial processes complete for electron donor.

The core processes of the biological regeneration of $Fe^{II}EDTA$ has been invented (Mass *et al.*, 2004). $Fe^{II}EDTA$ was found to serve as electron donor for biological reduction of nitrate to nitrite and for chemical reduction of nitrite to NO with concomitant formation of $Fe^{II}EDTA$-NO complex. The reduction of NO in $Fe^{II}EDTA$ was catalysed biologically and occurred about three times faster at 55°C than NO reduction at 30°C with ethanol as electron donor.

Nitrate reduction in aqueous solutions of $Fe^{III}EDTA$ is accompanied by the biological reduction of $Fe^{III}$ to $Fe^{II}$, for which ethanol, methanol and also acetate are suitable electron donors.

The continuous reactor experiments demonstration that the iron reduction capacity was the rate limiting step. When treating a flue gas containing 3.3 per cent $O_2$ and 500 ppm NO, approximately 90 per cent of the electron flow was used for reduction of NO. In the competition for electron donor between $Fe^{III}$ reduction and NO reduction, Lovely *et al.* (1991) reported out completion of $Fe^{III}$ reduction by denitrifiers.

NEERI (2005) has developed a chemo-biodenoxing process in which the chemical reaction/ scrubbing of industrial emissions generated from a sodium nitrite plant was carried out using $Fe^{II}EDTA$. The scrubbed solution contains mixture of $Fe^{II}EDTANOx$ and soluble nitrite in the solution. The scrubbed solution could denitrified to $N_2$ using methanol as carbon source and $Fe^{II}EDTA$ present in the solution could be reused for scrubbing of NOx present in the emission of the same plant.

## Discussion

Biological removal of $SO_2$ and NOx from contaminated gaseous stream in air phase bioreactors such as biofilters, biotrickling filters and bioscrubbers have been experimented under mesophilic conditions. High temperature of flue gases in the temperature range of 50°C and 60°C, affects the activity of mesophilic bacteria and also the long term stability of the packing material of the bioreactor in that temperature range. Besides temperature fluctuation on the higher side, presence of oxygen and carbon dioxide in the flue gas also adds to the low efficiency of such airphase bioreactors being operated under pilot or full-scale conditions. Thus such systems require microorganisms that can sustain high temperature and also fluctuations in the concentration of $SO_2$ and NOx. The system should also be attached with a pre-treatment unit for removal of the suspended particulate matters

(SPM) prior to its entry into the air phase bioreactor system. The inadequacy of removal of SPM affects the performance of such biological system.

Biotrickling filter avoids direct contact of the biological species with the fluctuations in the physical and chemical parameters in the flue gas. It could remove 100 per cent of influent $SO_2$ but all the absorbed $SO_2$ was recovered in the biotrickling filter liquid effluent as sulfite (a product of chemical reaction of $SO_2$) and sulfate (product of biological oxidation of sulfite). This needs a second stage biological system to reduce sulfites and sulfates to sulphide. For treatment of NOx containing flue gases, the low solubility of NOx from gaseous phase into aqueous phase forms the rate limiting step and leads to mass transfer limitation of NOx in biotrickling filters. The efforts are being made for increasing the solubility of NOx in aqueous phase through addition of some chelating agents. However, denitrifiers must sustain their growth and their activities in presence of chelating agents. This is an emerging area, which requires more research and development activities. Further, the problem of mass transfer limitations has also been partially taken care of by developing novel bioreactor system which incorporates air-loop lift bioreactors and hollow fibre membrane bioreactors (HFMBs). Such bioreactor systems enhance the solubility of NOx in the aqueous phase. The nitrites and nitrates thus formed as a result of biological nitrification must be stabilized through denitrification step using suitable electron donors and specific denitrifiers prior to discharge of the effluents generated from air phase bioreactors.

Photobioreactors, on the other hand, employs photosynthetic microalgae that are tolerant to high concentration of CO and $CO_2$ in the flue gas. A three-stage biotechnological process for flue gas desulphurization using photosynthetic organisms has been tried but such systems require large volume of reactor with transparent material of construction. Further, in the absence of sunlight as a radiant energy, the source of supply of energy for such system is limited and thus process is not techno-economically viable. Thus, in a nutshell, these bioreactors are able to cater only one or the other factors responsible for affecting the biological $SO_2$ and NOx removal rate.

The BIO-FGD process developed in Netherlands has given a ray of hope for application of biotechnological processes for flue gas desulphurization. It overcomes the drawbacks involved in direct biofiltration of sulphur dioxide and also includes the benefit of a low cost biological approach. The installation cost of this process developed by PAQUES has been estimated to cost approximately 30 per cent lesser than the conventional limestone or gypsum sulphur removal systems. The primary reasons for the lower capital costs are: (1) use of smaller sized scrubbers than the conventional counterparts; (2) reduced need of bulk material handling facilities; (3) does not require purchase of limestone and (4) use of much less power than traditional FGD systems. Other advantages offered by this system, besides efficiently removing the $SO_2$, are the production of high quality reusable sulphur and generation of minimal waste amounts of that requiring disposal. However, the performance of BIO-FGD process with reference to sulphur using plate settler is very poor. Thus it requires development of a suitable separation system for separation and recovery of elemental sulphur produced in the process.

The recent development of the chemobiological process employing high rate chemical absorption of NOx in aqueous solution of ferrous EDTA complex followed by biological denitrification of chelated NOx with regeneration of ferrous EDTA scrubbing solution eliminates most of the problems of biological NOx removal discussed before. Moreover the need of maintaining a suitable reductant such as sulphite, ascorbate, etc. in the ferrous EDTA solution to prevent oxidation of ferrous EDTA to ferric EDTA- not capable of forming ternary complex with NOx, has paved new way for simultaneous removal of SOx and NOx by a single scrubbing solution.

The progressively more stringent regulations for control of pollutants in flue gases from combustion systems are driving engineers and scientists to search for efficient means of limiting escape of these pollutants from the stack. The intrinsic advantages and limitations of the various control strategies are still unknown. Treatment processes based on biotechnological methods are very scanty and almost no work has been duplicated to validate the authenticity of the reported removal effeciencies. The most of the work carried out so far on biological processes are in bench scale level and very few have been elevated to pilot or commercial level. The differences in the composition of the biological flora by various groups of workers is yet another disadvantage making comparisons between any two reported biotechnological methods per say. The biological and chemo-biochemical processes, however, still require extensive research for developing suitable microorganisms and bioreactor systems for large-scale applications. The microbial (biofiltration, chemo-biochemical processes and bio-FGD processes) processes require extensive investigations on occupational health aspects after application to large scale.

## References

1. Abma, W.R. and Buisman, C.J.N., Biological simultaneous removal of $SO_2$ and $NO_x$ from flue gases, in proceedings of *Int. Symp. on Env. Biotechnology, 94th Event of the European Federation of Biotechnology*, Oostende, 1997, 33.
2. Apel, W.A., Turick, C.E., The use of denitrifying bacteria for the removal of nitrogen oxides from combustion gases. Fuel 72, 1715, 2, 1993.
3. Apel, W.A., Barnes, J.M., and Barett, K.B., Biofiltration of nitrogen oxides from fuel combustion gas streams, Proceedings, 88th Annual Meeting, Air and Waste Management Association, San Antonio, TX, June 18–23.
4. Barnes, J.M., Apel W.A., Barrett, K.B., J. of Hazardous Material 41 (2-3), 315, 1995.
5. Ben-Amotz, A., Avron, M. On the factors which determine massive ß-carotene accumulation in the halotolerant alga *Dunaliella bardawil*. *Plant Physiol.* 72. pp. 593. 1983.
6. Borowitzka, M, Borowitzka, L. *Dunaliella*. Algal growth media and sources of algal cultures. In: BOROWITZKA M, BOROWITZKA L (eds) Micro-algal Biotechnology. London: Cambridge University Press. 1988.
7. Buisman, C.J.N., Geraats, B.G., Ijspreert, P. and Lettinga, G., Optimization of sulphur production in a biotechnological sulphide removal reactor, *Biotech. Bioeng.*, 35, 50, 1990.
8. Buisman, C.J.N. and Prins, W.L., New process for biological flue gas desulphurization., *Symposium on Biological Waste Gas Cleaning*, Heidelberg, Germany, 1994.
9. Buisman, C.J.N., Dijkman, H., Verbraak, P.L., Den, Hartog, A.J. Process for purifying flue gas containing nitrogen oxides. United States Patent US5891408. 1999.
10. Cetinkaya, B., Sahlin, R.K., Abma, W.R., Dijkman, H., Meyer, S.F. and Kampeter S.M., Control FCC flue gas emission. *Hydrocarb Process.* 79. pp. 55. 2000.
11. Clark, R.C. and Wonorowski, A.U., Bofilters for sewer pumpstation vents: influence of matrix formulations on the capacity and efficiency of odorant removal by an experimental biofilter, in *"Biotechniques for Air Pollution Abatement and Odour Control Policies"*, A.J. Dragt and J.Van Ham Eds., Elsevier Science, 1992, 183.

12. Cox, H.H.J., Sexton, T., Shareefdeen Z.M. and Deshusses, M.A. Thermophilic biotrickling filtration of ethanol vapors, *Environ. Sci. Technol.* 35, pp. 2612. 2001.

13. Cork, D.J., Jenger, B.E., and Maaka, A., Biocatalytic production of sulphur from process waste streams. *Biotechnol. Bioeng. Symp.*, 16, 149 1986.

14. Davidova Y.B., Schroeder E.D. and Chang D.P.Y. Biofiltration of nitric oxide. *Proceedings, 90th Annual Meeting, Air and Waste Management Association*, Pittsburgh, PA, June 18–22, 1997.

15. Dean, V.A., Eldridge, J.J.and Gruesser, S., Eds., *The Air Pollution Consultant.*, Aspen Law and Business ISSN 1058-6628, Jan/Feb, 6, 1.1,1996.

16. Dean, V.A., Eldridge, J.J. and Gruesser, S., Eds., *The Air Pollution Consultant.*, Aspen Law and Business ISSN 1058-6628,Jan/Feb, 6, 2.21, 1996.

17. Deshmane, V., Lee, C.M., and Sublettc, K.L., Microbial reduction of sulphur dioxide with pretreatment sewage sludge and elemental hydrogen as electron donors, *Appl. Biochem. Biotechnol.*, 39, 739, 1993.

18. Ergas, S.J., Membrane bioreactors, in Bioreactors for Waste Gas Treatment, ed by Kennes C and Veiga MC. Kluwer Academic, Publishers, Dordrecht, The Netherlands, pp. 163. 2001.

19. Groenestijn, Van J.W., Hesselink, P.G., Biotechniques for air pollution control. *Biodegradation*, 4, 283, 1993.

20. Imaizumi, T.T., Some industrial applications of inorganic microbial oxidation in Japan, Presented at workshop on Biotechnology for the Mining, Metal Refining and Fossil Fuel Processing Industry, *Biotechnol. Bioeng. Symposium* No. 16, John Wiley and Sons, New York, 1985, 363.

21. Janseen, A.H., Sleyster, P., Van der Kaa, C., Jo chensen, A., Bonstsema, J. and Lettinga G., Biological sulphide oxidation in a fed-batch reactor, *Biotechnol. Bioeng.*, 47, 327, 1995.

22. Janseen, A.H.H., Ma, S.C., Lens, P., and Lettinga, G. Performance of sulphide oxidizing expanded bed reactor with a spatially separated aeration unit, *Biotechnol. Bioeng.* 1997.

23. Jin, Y, Veiga, M.C, Kennes, C. Autotrophic deodorization of hydrogen sulfide in a biotrickling filter. J Chem Technol Biotechnol 80, 2005.

24. Jin, Y., Viega, M. C. and Kennes, C. Review: Bioprocesses for the removal of nitrogen oxides from polluted air. *J Chem Technol Biotechnol* 80, pp.483, 2005.

25. Jinsiriwanit, S. and Deshusses, M. A. Biotreatment of Combustion Gas. Paper 69814. http://www.engr.ucr.edu/~mdeshuss/pdf per cent 20Files/cp48.pdf.

26. Kim, N.J., Hirai, M. and Shoda, M. Comparison of organic and inorganic packing materials in the removal of ammonia gas in biofilters. *J. Hazard. Mater.* 72, pp. 77, 2000.

27. Klasson K. T. and Davison, B. H. Effect of temperature on biofiltration of nitric oxide. Applied Biochemistry and Biotechnology. 91-93, (1-9), 205, 2001.

28. Kumaraswamy, R., Muyzer, G., Kuenen, J.G. and van Loosdrecht, M.C.M. Biological removal of NOx from flue gas. *Water Sci Technol.* 50. pp. 9. 2004.

29. Lacey, J.A., Lee, B. D, Apel W.A. PCR-DGGE as a method to monitor microbial populations in nitric oxide biofilters. In: Proceedings of the 95th annual meeting and exhibition of the air and waste management association, Baltimore, Md., 23–27, 2002.

30. Lee, K.H. and Sublette, K.L., Simultaneous combined microbial removal of sulphur dioxide and nitric oxide from a gas stream, *Appl. Biochem. Biotehnol.*, 28-29: 623, 1991.

31. Leson, G. and Winer, A.M., Biofiltration: An innovative air pollution control technology for VOC emissions. *J. Air Waste Manag. Assoc.* 41, 8, 1045, 1991.

32. Lee, J.P., Lee, J.S. and Park, S.C. Two-phase methanisation of food wastes in pilot scale. *Applied Biochemistry and Biotechnology*, 77–79, p 585, 1999.

33. Lee, D.H., Lau, A.K. and Pinder, K.L. Development and performance of an alternative biofilter system. *J. Air Waste Manage. Assoc.* 51, 78, 2001.

34. Lievonen, J. Technological opportunities in biotechnology. Espoo, Finland7 VTT, Group for Technological Studies; 1999.

35. Lukow, T. and Diekmann, H. Aerobic denitrifcation by a newly isolated heterotrophic bacterium strain TL1. *Biotechnol. Lett.* 19, 1157, 1997.

36. Min, K.N., Ergas, S.J. and Harrison, J.M, Hollow-fiber membrane bioreactor for nitric oxide removal. Environ Eng Sci 19:575–583 (2002).

37. Nagase, H, Yoshihara, K., Eguchi, K., Yokota, Y., Matsui, R., Hirata, K. and Miyamoto, K., Characteristics of biological NOx removal from flue gas in a Dunaliella tertiolecta culture system. *J Ferment Bioeng.* 83. pp. 461. 1997.

38. Nagase, H., Eguchi, K., Yoshihara, K., Hirata, K. and Miyamoto, K. Improvement of microalgal NOx removal in bubble column and airlift reactors. *J Ferment Bioeng* 86. pp. 421. 1998.

39. Nagase, H., Yoshihara, K., Eguchi, K., Okamoto, Y., Murasaki, S., Yamashita, R., Hirata, K. and Miyamoto, K. Uptake pathway and continuous removal of nitric oxide from flue gas using microalgae. *Biochem Eng.* J 7. pp. 241. 2001.

40. Okuno, K., Hirai, M., Sugiyama, M., Haruta, K. and Shoda, M. Microbial removal of nitrogen monoxide (NO) under aerobic conditions. *Biotechnol Lett.* 22, pp. 77, 2000.

41. Pagella. C., Faveri D. M., $H_2S$ gas treatment by iron bioprocess, *Chem. Eng. Sci.* 55, 2185, 2000.

42. Pandey, R.A. and Malhotra, S., Desulphurization of gaseous fuels with recovery of elemental sulphur: An overview, *Crit. Rev. Environ. Sci. Technol.*, 29, 229, 1999.

43. Pandey, R.A., Biswas, R., Chakrabarti, T. and Devotta, S. Flue Gas Desulfurization: Physicochemical and Biotechnological Approaches. Critical Reviews in Environmental Science and Technology, Environmental Science, 35(6). 2005.

44. du Plessis C.A., Kinney K.A., Schroeder E.D., Chang D.P.Y., Scow K.M. Denitrification and nitric oxide reduction in an aerobic toluene-treating biofilter. *Biotechnol. Bioeng.* 58, 408–415, 1998.

45. Robra, K.H., Wellacher, M.., Kirchmeir, F., Leistentritt, R. and Pucher, K. Abluftreinigung fur straβentunnel und tiefgaragen durch carboxidotrophe mischpopulationen. Biological Waste Gas Cleaning, Proceedings of an International Symposium, Maastricht, The Netherlands, April 28–29. 1997.

46. Shoun, H. and Tanimoto, T. Denitrification by the fungus Fusarium-oxysporum and involvement of cytochrome-P- 450 in the espiratory nitrite reduction. *J Biol Chem.* 266, pp. 11 078, 1991.

47. Shoun H, Kim DH Uchiyama H and Sugiyama J, Denitrification by fungi FEMS Microbiol Lett 94:277–281 (1992).

48. Sublettc, K.L., and Gwozdz, K.J., An economic analysis of microbial removal of sulphur dioxide as a means of by-product recovery from regenerable processes for flue gas desulphurization, *Appl. Biochem. Biotechnol.*, 28, 635, 1991.

49. Selvaraj, P. T., Little, M. H.and Kaufman, E. N., Biodesulphurization of flue gases and other sulphate/sulphite waste streams using immobilized mixed sulfate reducing bacteria, *Biotechnol. Prog.*, 13, 583, 1997.

50. Shi, Y. Littlejohn, D. and Chang, S.G. Integrated tests for removal of nitric oxide with iron thiochelate in wet flue gas desulfurization systems. Environ Sci Technol 30, 3371, 1996.

51. Shi, Y. Wang, H. and Chang, S.G. Kinetics of NO absorption in aqueous iron (II) thiochelate solutions. Environ Prog. 16, 201, 1997.

52. Sublette, K. L. and Sylvester, N. D., Oxidation of hydrogen sulphide by continuous cultures of *Thiobacillus denitrificans*, *Biotechnol. Bioeng.*, 79, 753, 1987.

53. van Groenestijn, J.W., Bioscrubbers, in Bioreactors for Waste Gas Treatment, ed by Kennes, C. and Veiga, M.C. Kluwer Academic Publishers, Dordrecht, The Netherlands, pp 133-162. 2001.

54. van der Maas, P., Klapwijk, B. and Lens, P. BioDeNOx: novel process for NOx removal from flue gases based on chemically enhanced biological NO and iron reduction, in Proceedings of the European Symposium on Environmental Biotechnology, Ostend, Belgium, ed by Verstraete W pp. 341. 2004.

55. Woertz, J.R., Kinney, K.A., McIntosh N.D.P. and Szaniszlo, P.J. Removal of toluene in a vapor-phase bioreactor containing a strain of the dimorphic black yeast *Exophiala lecanii-corni*, *Biotechnol. Bioeng.* 75, pp. 550, 2001.

56. Yani, M., Hirai, M. and Shoda, M. Enhancement of ammonia removal in peat biofilter seeded with enriched nitrifying bacteria. *Environ. Technol.* 21, 1199, 2000.

57. Yoshihara, K.I., Nagase, H., Eguchi, K., Hirata, K. and Miyamoto, K. Biological elimination of nitric oxide and carbon dioxide from flue gas by marine microalga NOA-113 cultivated in a long tubular photobioreactor. *J Ferment Bioeng.* 82. pp. 351. 1996.

# Chapter 54

# Healthy Indoor Air Quality: A Challenge for the Knowledge Economy

**Pravin Karve**
**Thermax Limited**

## Background

Rapid industrialization has given rise to emission of a variety of pollutants like Suspended Particulate Matter (SPM), Sulphur Tri Oxide ($SO_3$), Sulphur Di Oxide ($SO_2$), Carbon Monoxide (CO), Carbon Dioxide ($CO_2$), Oxides of Nitrogen ($NO_X$), Fumes of Hydrochloric (HCl) and other Acid, Mercury, Flouro Carbons, ......... from industrial plants/stacks.

Impact of some of the obvious pollutants was well understood by Humanity and counter measures have been put in place at an early stage. However, the complete impact of these pollutants is just beginning to be understood by Humanity; thanks to the rapid climate change we are all experiencing. However, we are still far away from controlling the fallout of the release of these pollutants to atmosphere!

While we are rapidly turning the focus of our attention, efforts and resources towards addressing this problem, there is a new problem which has started surfacing–the problem of Indoor Air Pollution and the problem of Indoor Air Quality (IAQ). This chapter deals with this new problem of IAQ.

## IAQ and its Significance

Quality of the air in an enclosed space is referred as Indoor Air Quality (IAQ).

If we analyze where we spend our time, we will realize that more than 80 per cent of our time is spent indoors! Therefore, it is obvious that the IAQ has a direct impact on our health and well being. Despite this fact, very little attention is paid towards monitoring the IAQ and maintaining healthy IAQ.

A recent research shows that bad IAQ has 1000 times more impact on Human Health and wellbeing than outdoor air!

Some studies have also indicated that the level of contaminants in indoors air are at least two to five timers higher than outdoors. The World Health Organization estimates that 30 per cent of homes and buildings today contain enough indoor pollutants to cause health affects that range from a sniffle to more serious health problems.

## Why has the IAQ Become so Important Now?

The IAQ has become much more important than before because today's indoor spaces are quite different than what we had 10-15 years back. Here are some of the differences:

- ☆ The offices have become more crowded as compared to offices of the past due to very high cost of real estate. This has reduced the air volume per occupant substantially.
- ☆ Most offices were naturally ventilated with open windows and doors in the past. Now most offices are Air Conditioned with non operable windows. The air is recirculated by the AC system continuously concentrating the contaminants.
- ☆ In the past office indoors were mostly open type with minimum use of particle boards, laminates, etc. Offices used bricks, mortar and glass. Today the offices use a lot of particle boards, laminates, glued aluminum sheets, glued floor tiles/laminates, glued wall papers, etc.
- ☆ We had very few printers, scanners, copiers in the office. Today, almost everyone has a printer cum scanner cum copier on his/her desk.
- ☆ The number of computers inside the office are far more than before.
- ☆ The outdoor air which is used as fresh air make-up air is far more polluted than before.
- ☆ We are spending longer hours at the office than before.
- ☆ We are buying cheaper, toxic chemicals to clean and maintain buildings.
- ☆ People are using much more synthetic perfumes and other personal hygiene products.

All these factors make the job of maintaining good IAQ very demanding.

## Effects of Poor IAQ on Human Body

Poor IAQ with presence of harmful contaminants can have harmful effects on human body. The severity of the effects depends on the concentration of the contaminants, period of exposure and sensitivity of the human body. The physical changes which one's body can undergo due to poor IAQ can include:

- ☆ *Brain and Central Nervous System*: headaches, extreme tiredness, dizziness, fainting, mood swings, confusion, depression, hyperactivity, memory problems, loss of coordination, seizures.
- ☆ *Systemic Reactions*: anaphylactic shock, uticaria, exzema.
- ☆ *Eyes*: infected, itchy, red, watery or puffy, visual problems.
- ☆ *Ears, Nose and Throat*: frequent infections; itching, ringing ears, red earlobes; sneezing, itchy, irritated, blocked, runny or stuffy nose, irritated, hoarse throat, laryngitis.
- ☆ *Mouth*: metallic taste, dryness, cracking, excessive saliva, skin peeling or blistering.

- *Lungs*: infected, coughing, wheezing, tightness, breathing difficulties, asthma.
- *Skin*: cold, itchy, cracked, red, bruised or swollen; hives, rash.
- *Muscles, Bones and Joints*: stiffness, aches, pain, weakness, swelling, muscle cramps, "arthritic" symptoms.
- *Digestive System*: nausea, cramps, bloating, gas, diarrhea or constipation, irritation, food cravings, weight loss or gain.
- *Urinary and Reproductive Systems*: cramps, infections, itching, burning, urinary urgency or frequency.

All these ultimately result in lower productivity and unplanned absenteeism of knowledge workers.

## Poor IAQ also Affects Machines

Besides adversely affecting health of people, bad IAQ also has impact on machines. Multitude of computers operating continuously in BPOs, DCS control rooms get affected due to corrosive gases causing failure of PCBs, CPUs, etc. This causes damage to Process controls and reduces availability of computer networks which ultimately affects productivity.

## What are the Indoor Contaminants?

The contaminants in Indoor Air are mostly invisible and therefore more dangerous. The contaminants include the following:

1. Microbial contaminants like mold, bacteria, viruses, fungus
2. Chemicals/Gases like carbon monoxide, carbon dioxide, nitrogen dioxide, Sulphur dioxide, radon, volatile organic compounds emanating from solvents, paints, varnishes and glues, chemical compounds from pesticides, asbestos, formaldehyde
3. Various types of allergens like pollens
4. Fine dust particles
5. Odors, food smells, Tobacco Smoke, etc.

## Where do the Contaminants come from?

Many sources of contaminants in our indoor environments are found in the workplace as well as our homes. Common contaminants and their sources include:

- Butyric Acid from Road and highways
- Carbon monoxide from Adjacent parking lots and garages
- VOCs from Paints, wood paneling, office equipment (Photocopy, Laminating Computer terminals, printers), air fresheners, cleaning agents, pesticide sprays
- VOCs from scented products (including scent-laden clothing) used by Humans
- Tobacco smoke (including smoke-laden clothing) from Humans
- Mold from Evaporative/desert/swamp coolers, cooling towers, air washers, human beings
- Bactericides, herbicides, fungicides and pesticides sprayed on indoor plants/wood
- Bacteria from Damp corners of Humidification/heating/cooling systems

- ☆ Fungus from Behind insulation
- ☆ Bacteria and Virus from under carpets and from Human beings occupying the indoors
- ☆ Synthetic materials, especially carpeting and under padding, airborne ultra fine dust.

## What Can be Done to Improve the IAQ

There are basically three ways of improving IAQ:

- ☆ *Using ventilation to dilute contaminants*: By venting the indoor air to out doors and taking fresh outdoor air to dilute the contaminants can help in improving the IAQ. However, this approach fails if the outdoor air is also contaminated. This approach requires a lot of power to cool/heat the fresh air to bring it to the Indoor air temperature. It also increases fan power for intake fans. The Utility operators tend to reduce the operation costs by reducing the fresh air intake. This poses grave threat to the IAQ.
- ☆ *Source control*: This method relies on controlling the source of contamination to avoid release of the contaminants into the Indoor Air. By controlling the release of contaminants, it is possible to improve IAQ. However, this method has severe limitations as almost most Indoor sources prove very difficult to control or isolate.
- ☆ *Filtration/Contaminant control*: This method treats the Indoor air to remove/neutralize the various contaminants so that their effect on occupants is minimized. The technological development in this area is in a nascent stage and a lot needs to be done to develop viable and reliable technologies. Some of the currently used technologies include the following:
    1. *Mechanical filters*: These filters separate the suspended particles from the Indoor Air. Various types of filters are being developed to filter out the finest of the particles with lower pressure drops.
    2. *Adsorber filters*: These filters separate specific gases like acid fumes by physical adsorption of gases with filter media like activated carbon. The adsorbed contaminant is then separated from the filter media in a environmentally safe way.
    3. *Mini ESPs*: These are Electro Static Filters which separate the suspended particulate matter by electro charging the particles. However, these filters are not very popular as they generate excessive Ozone.
    4. Photo Catalytic Oxidation (PCO): This technology uses Nano size particles of $TiO_2$ or similar materials to catalyze oxidation of VOCs and CO at room temperature to produce harmless water and $CO_2$. This technology is under development and it would be some more time before this is employed on large scale.
    5. UV Techniques/Laser Technologies: These technologies use Ultra Violate Light/Lasers to kill various airborne microbes. These are very effective methods and are being employed on commercial installations.
    6. Electrophoresis: A process by which harmful elements are separated based on differences in electrical charge. Observed differences between or within species indicate charge divergence over long periods of time, thus facilitating the identification of inter-relationships.
    7. Zeolite Process: A base exchange method of treating effluents in which zeolites can remove oxides etc. The zeolite layer is regenerated by back flushing with brine.

8. Photo voltaic synthesis: This is basically the use and harness of solar energy to synthesize the carbon monoxides and VOCs.
9. Genomics using Plant cells: This is a process of developing counter genes to kill the pollutants. This is still in research stage.

## What Should the Stake Holders Do Now?

Because the work on improving IAQ is at a very nascent stage in India, all stake holders like manufacturers, building owners, Architects/building designers and the occupants need to work closely together to make rapid progress in controlling this problem. Collaborative work done by all these stake holders would help avoid a situation similar to Global Worming happening in the Indoors. I foresee following as some of the initial steps that need to be taken by all:

1. Evolve good standards/recommendations for IAQ
2. Work on making good IAQ a legal requirement through legislations and standards
3. Spread the awareness of importance of IAQ to the occupants
4. Develop multi contaminant solutions
5. Develop systems and equipments and make them commercially available.

# Chapter 55

# Environmental Regulations for Sustainable Construction

**Vijay S. Kulkarni**
**ITD Cementation India Limited, Mumbai, M.S.**

## Introduction

21st century is witnessing indiscriminate urbanization. This has triggered cause-condition-effect chain involving more urban centers, more migrations and more need to augment the infrastructure. Resultantly, development has become synonymous with environmental degradation and progress with pollution.

Construction is a resource intensive activity by itself. Substantial energy is used in the production of materials such as cement, steel, aluminum, wood products, plastics, and paints; movement of materials and components to, and on site, and running and operating of plant and equipment on site. Construction waste, results in the waste of land resources and the contamination of natural resources and ecologies. Unnecessary water usage constitutes a further form of waste, which has an indirect impact on the environment. These impacts are compounded as a result of indiscriminate and unplanned growth.

Tools for environmental management used in 20th century were built mainly around "command and control" paradigm. While this will remain the major drive for environmental protection in 21st century, enhanced public awareness on one hand realization social responsibilities by the business are leading to emergence of new approach built around paradigm of "ecological economics".

With this backdrop, the chapter discusses existing legal framework and obligations of construction sector, identifies environmental issue and strategies, and advocates approach for environmentally sound construction activities.

## Construction Boom in India

Indian economy experiencing boom in all sectors triggered by its economic and investment policies. Metros or the Tier I cities are inundated with burgeoning investments in the industrial and the services sector. Along with large-scale investments has boomed the realty sector creating congestion, arising out of an increasing demand for residential and commercial properties. This congestion in realty structures has forced the respective governments and many investment companies to seek out for alternative smaller cities leading to a demand for Tier II and III cities. The special initiatives taken by various state governments in providing the smaller cities with infrastructural facilities and creation of SEZs, has played a vital role in promoting these small towns into cities of the future.

With these developments the real estate, hospitality and infrastructure sectors, are expanding at 20 per cent per annum. Performance of these sectors is expected to continue to outplace the growth of the overall economy in its foreseeable future. Greater participation and transparency in operation of these sectors, supported by regulatory reforms, has further stimulated this rising trend.

Mr. K P Singh, Chairman, DLF Limited, 42nd AGM on Sept 29, 2007, in his address portrayed the construction boom has follows:

### Homes

Some of the key factors that have spurred demand in housing in India are rising income levels, growth of middle class, increasing incidences of nuclear families, and favourable government policies. According to as RNCOS report, India's estimated demand for housing and real estate will grow at around 14 per cent annually over foreseeable future. Cushman and Wakefield have also highlighted scope for 400 new township projects over the next five years, spread across 30 to 35 cities each with a population of 0.5 million.

### Offices

Recent McKinsey-NASSCOM estimates suggest that the total demand for office space will rise to 500 million sq ft (msf) in the next ten years, with IT/ITES accounting for 60 to 70 per cent. The rest will be from non-IT and manufacturing sectors. India's commercial space (excluding government and public sector) doubled from 50 msf in 2002 to 100 msf in 2005. The rate of growth in India's commercial space is likely to maintained at 20 to 25 per cent per year. India's booming economy and continued growth in services sector, along with increasing demand from other corporate customers, will continue be the key drivers of the demand in the commercial real estate space.

### Retail

India is one of the attractive real markets in the world. The organized retail industry is expected to grow at around 27 per cent annually, growing from USD 6 billion in 2005 to USD 23 billion in 2010. The number of malls is expected to rise manifold in the coming few years.

### Hospitality

India's hospitality sector is growing at 15 per cent annually and the gap between supply and demand is expected to widen as the country liberalizes and economy grows. Leisure and business travel are driving demand for hospitality services, strongly correlated with increasing disposable income and the growing popularity of India as International tourism destination, increasing business activity and corresponding demand for hotel rooms and convention centres.

## Special Economic Zones

Large multi-product SEZ's are seen as growth engines of the Indian economy over the coming years, providing the world class infrastructure, investment opportunities and employment generation avenues. Manufacturing output in India is expected to grow at 14 per cent annually from US$ 300 billion in 2005 to US$ 1100 billion in 2015, with SEZs playing a significant role in making it happen.

## Infrastructure

Infrastructure development in one of the critical focus areas for India as economy sustains an annual growth rate around 9 per cent plus over the next five years. The Government outlay is also expected to increase from 3 per cent to around 8 per cent of GDP, leading to unprecedented opportunities in infrastructure development.

# Environmental Regulations in India for Construction Projects

Major environmental concerns related to construction activities are:

- Generation, storage, and disposal of wastes;
- Noise pollution due to plant, machinery, equipments and vehicle movement;
- Air pollution due to plant, machinery, equipments and vehicle movement;
- Generation and disposal of wastewater;
- Impact on ecology; and
- Consumption of resources such as water, electricity, and diesel.

Environmental Regulations in India arise out of:

- Acts of Central and State Governments
- Rules framed under the Acts
- Notifications issued
- Directives of regulatory bodies
- Decision of High Courts and Supreme Court

Major environmental legislations in India are:

- The Water (Prevention and Control of Pollution) Cess Act, 1977, amended 1992
- The Water (Prevention and Control of Pollution) Cess (Amendment) Act, 2003
- The Water (Prevention and Control of Pollution) Act, 1974, amended 1988 The Water (Prevention and Control of Pollution) Cess Rules, 1978
- The Water (Prevention and Control of Pollution) Rules, 1975
- The Air (Prevention and Control of Pollution) Act 1981, amended 1987 The Air (Prevention and Control of Pollution) (Union Territories) Rules, 1983
- The Air (Prevention and Control of Pollution) Rules, 1982
- The Environment (Protection) Act, 1986, amended 1991
- The Environment (Protection) Rules, 1986 as modified from time to time
- The Batteries (Management and Handling)
- The Municipal Solid Wastes (Management and Handling) Rules, 2000

- ☆ The Recycled Plastics Manufacture and Usage Rules, 1999
- ☆ The Recycled Plastics Manufacture and Usage (Amendment) Rules, 2003
- ☆ The Manufacture, Storage and Import of Hazardous Chemical (Amendment) Rules, 2000
- ☆ The Hazardous Wastes (Management and Handling) Rules, 1989 amended in 2002 and 2003
- ☆ Dumping and disposal of fly ash discharged from coal or lignite based thermal power plants on land 2003
- ☆ Noise Pollution (Regulation and Control) Rules, 2000
- ☆ The Ozone Depleting Substances (Regulation and Control) Rules, 2000
- ☆ The Public Liability Insurance Act, 1991, amended 1992
- ☆ The Public Liability Insurance Rules, 1991, amended 1993
- ☆ Forest (Conservation) Act, 1980, amended 1988
- ☆ Forest (Conservation) Rules, 2003

Even though the list of legal requirements is large, environmental legal requirements relevant to construction activities can be broadly categorized into:

- ☆ Legal requirements resulting in requirement of permits/licenses
- ☆ Legal requirements resulting in requirement of strict adherence to work practices
- ☆ Legal requirements for which we have to ensure that we indemnify ourselves in case of problems as in all criminal activities, participation and abetment is also a crime

Permits (in environmental law these are called "consent to establish" and "consent to operate") are required for setting up any industry/operation/process that will result in effluent *i.e.* wastewater or for establishing and operating industrial plant (that will emit air pollutants). As per law responsibility of taking such permission is with owner. Such permissions when granted stipulate conditions, which need to be fulfilled.

Permits relevant to construction works include:

- ☆ DG sets.
- ☆ Stone crushers
- ☆ Hotmix plant
- ☆ Used lubricating oil

Legal requirements that result into strict adherence to work practices are:

- ☆ Returning used lead-acid batteries to dealers and maintains record of the same.
- ☆ Segregating waste at generation point as per direction of municipal authority
- ☆ Ensuring that noise outside DG enclosure is below 75 dB
- ☆ Obtaining PUC certificate for transport vehicles (as per law this responsibility is with owner of the vehicle)
- ☆ Ensuring that Ozone Depleting Substances which are banned by Government are not used in our activities

- ☆ Noise levels to which workers are exposed are below stipulation in Building and constructions workers Act, 1996
- ☆ Air pollutants to which workers working in tunnels are exposed are below stipulation in Building and constructions workers Act, 199

In case of all criminal activities, participation and abetment is also a crime. Hence, even though many construction works may not need permissions from PCBs it is necessary to ensure that:

- ☆ Wastewater discharged from sites does not violate standards stipulated under Environment Protection Act, 1986
- ☆ Wastewater discharges from sites do not adversely affect water quality in receiving bodies streams, rivers, lakes, land, marine areas
- ☆ Air pollutants emitted from construction activities do not contribute substantially to increase in concentrations of air pollutants that would be in violation of limits specified in Environment Protection Act 1986
- ☆ Noise emitted from construction activities do not contribute substantially to increase in noise levels that would be in violation of limits specified in Environment Protection Act 1986, noise rules.

It may be notes that, organizations are being increasingly pressurized to comply with environmental regulations from regulators, markets, financing institutions and the community. Many companies are struggling to save themselves from the perils of non-compliance. Environmental compliance is complex–dynamic and covers a number of institutions and overlapping legal frameworks. Meeting compliance is not easy as definition of "non-compliance" is rather difficult. Hence, going beyond compliance is the best way to meet compliance! There are benefits of risk reduction, image building and if a right approach is used then even an upside of improving the bottom line.

## Role of EIA in Sustainable Construction

Economic liberalization and improvements in the standard of living is now evident in the changing skyline of urban India marked with high-rise buildings, malls, IT parks, hospitals and townships. Understandably, while exploiting markets for such developmental projects, key factor is commercial exploitation of every inch of land. However, changing land use coupled with redevelopment/ rehabilitation projects compels necessity of looking into stress on meager infrastructure resulting from such developments. Contrary to industrial projects where retrofitting is possible, impacts of construction projects are rarely reversible. With this problem staring at face, since July 2004, MoEF has made prior environmental clearance mandatory for certain categories of construction projects.

Within a short spell of three years, EIA as a preventive environmental intervention tool has trained developers, architects and environmental consultants in "impact identification–quantification–mitigation–management–monitoring" cycle. EIA has made it possible to look at environmental management as "resources management" instead of "regulator/pollution management". Appreciation of property due to implementation of environment friendly measures has demonstrated positive role of environmental economics. Now, project proponents and their consultants dwell upon conservation of non-renewable resources (water, energy, building materials etc), conservation/improvement of landscape, conservation of ecologically sensitive areas including heritage structures, reducing risk to public life because of traffic movement, as integral part of the project.

Impetus provided by EIA has resulted in development and availability of technologies and/or standard practices for use of flyash; solar passive architecture; solar thermal and solar photovoltaic systems; rain water harvesting systems, tertiary treatment of sewage including disinfection; and solid waste composting. These new environmental markets have further triggered growth of environmental professionals who are ready to take responsibility of engineering, procurement, construction and maintenance of environmental systems.

Thus, EIA of construction projects is an example of win-win situation that has helped environment, developers, users, professionals and generations to come to reaffirm that environmental friendly development is feasible.

## Last Words

Tools for internalizing environmental concerns are well established in last 4 decades. However, even today, developers approach is merely compliance driven. Earlier we start looking beyond "compliance" requirements it is better for environment and also for business. There is term "interested party" in ISO 14001. Whereas compliance requirements of regulatory agencies are documented, requirements of people affected due to environmental aspects of organization are not realized so easily. In all these cases, if an organization decides to remain contended with "compliance" then instead of Pollution Control Boards, affected population will take needful action. Our judiciary is with the people. Moral is "stop looking at impacts, start analyzing aspects" and "look beyond compliance".

For ensuring sustainable development environmental management in construction works should follow three mantras "Resource Conservation", "Waste Minimization" and "Continual Improvement". This will bring about "Change in Culture" which is prerequisite for healthy environment.

*"This earth can be a splendor, a magic, a miracle. Our hands have that touch — it is just that we have never tried it. Man has never given a chance to his own potential to grow, to blossom, to bring fulfillment and contentment. It should be a determination every intelligent being that we are not going to allow any vested interest to destroy this planet. To save man is to save the greatest creation of the universe. It has taken four million years for this earth to create man. It is so precious. And the future is much more valuable."*

***Osho in "The Golden Future"***

# Chapter 56

# Environmental Laws in India

**D.T. Devale**
**Advocate and Sr. Law Officer,**
**Maharashtra Pollution Control Board, Mumbai**

Tiwari Committee appointed to evaluate various existing environmental laws in India has observed in its report that "we have in India 500 or more environmental laws in India but we have not undertaken a single systematic study of any environmental law till date. This brings to us the status of the environmental laws in India even today.

While dealing with the environmental laws in India, it will be more appropriate to study the valuation of the environmental laws from ancient time to present period in general and to evaluate specific environmental law in particular.

## Ancient Approach

Protection, conservation and judicious use of environmental resources was the ancient approach. It was Dharma of each individual in the society to protect the nature. People worshiped the objects of the nature. The trees, the water, the land and the animals gained an important position in the ancient time. Govindraja makes a distinction between injury to shed giving plants, flower bearing plants and fruit bearing plants by prescribing the lowest, the middle and the highest punishment for causing injury to the plants. Some of the trees were elevated to the position of the God, the rivers also enjoyed a high position in the life of the society being considered as goddesses having not only self-purifying capacity but also self-purifying ability. Ganga, Jamuna, Saraswati, Kaveri and Godavari are even today considered the goddesses. At ancient time, the knowledge, the motivation and the sanction were the basic postulates which regulated the echo-system. The Dharma of environment was to sustain and to ensure progress and welfare of all.

## British India Period

It was considered a landmark period in the history of legal control in the environmental pollution. An attempt was made for the first time to control specifically water and atmospheric pollution through criminal sanctions.

The Indian Penal Code, 1860 dealt with the water and atmospheric pollution. Section 277 provide that whosoever voluntarily corrupts or fouls reservoir so as to make it less fit for the purpose, for which, it is ordinary used, shall be punished with simple or rigorous imprisonment for a term upto 3 months or fine of Rs.500/- or both.

British India Rule also enacted a number of legislations, which had a direct bearing on one or the other components of the environment. Three specific areas and one general area were selected for preserving the environment.

1. Industrial development required legislations for Regulation of the Factories, Boiler Act, 1983, Explosives Act, 1884, Explosive Sustension Act, 1908, Workmen Compensation Act, 1923.
2. Concern for the protection of forest made enactments such as the Cattle Trespass Act, 1971, the Destructive Insects and Pest Act, 1914.
3. Concern for the protection of animals, wild life and birds including elephants.
4. General areas such as the Smoke Legislations of Bengal, 1905, Bombay, 1912, Madras Town Planning, the Calcutta Improvements, U.P. Municipality, Bihar and Orissa Municipality.

The above enactments had two goals.

(*a*) The care of public health and;

(*b*) The concern for gradual deforestation and the extinction of wild life, birds and animals, but the British India Rule was mainly concerned with the trade and commerce and therefore, aimed at the resources use and the exploitation by the State and industries under State patronage. Every resource was viewed as a commodity that deserves to be exploited. In true sense, the British did not enact a Central Specific Legislation in relation to the environment protection and therefore, the common law principles were brought into play by the courts of law such as remedy for tort, negligence and nuisance etc.

## Environmental Laws in Independent India

At the time of commencement of the constitution, the constituent assembly did not pay any specific attention towards the protection of the environment. The outcome was that the constitution of India at the commencement left out the provision to protect and to improve the existing environment or control pollution. However, the industrial growth resulted in acceleration in the development process, started giving ill effects and gaining public opinion, hence, the attempts were made to minimise them through certain legal controls.

1. First of all, the hazardous activities were found to be regulated through new legal controls, such as Atomic Energy (1962), Inflammable Substance Act, 1952, Explosives (1952), Insecticides, 1968 and Mines (1952).
2. Secondly, there were legislations or the land management such as Requisition and Acquisition of Immovable Property Law (1952), the Land Management, Town Planning and

Slum Clearance Laws in various regions (M.P., 1957, Bengal, 1955, Kerala, 1964 and Jammu and Kashmir, 1953).

3. Thirdly, there were some legislations for preservation of wild and animal.
4. Lastly, there were legislations dealing with the water management such as Interstate Water Disputes, the Water Board Act, 1956 and most comprehensive the Maharashtra Prevention of Water Pollution Act, 1969.

## Specific Legislative Journey

In true sense, specific legislative journey started with the Maharashtra Prevention of Water Pollution Act, 1969 at regional basis but at central level, it can be said to be started by the enactment of the Water (Prevention and Control of Pollution) Act, 1974.

### *Water (Prevention and Control of Pollution) Act, 1974

(*a*) It provide for the prevention and control of water pollution.

(*b*) For maintaining or restoring of wholesomeness of the water.

(*c*) For the establishment of the Boards with the powers and functions relating thereof and for the matters enacted therewith.

Since, the Parliament has no power to make laws for the subject matter, the Parliament has enacted the above legislations in pursuance of Clause-I of Article 252 of the constitution in view of the resolutions passed by all the Houses of the Legislatures of the concerned States. The Act has been enacted to bring uniform approach throughout the country, though the original Act is adopted by all the States, the amendments in 1978 and 1987 are yet to be adopted by some of the States like Orissa, Gujarat etc. Hence, an important power to issue binding directions including closure, prohibition or regulation of the industry can not be exercised by these State Boards and whole concept of uniformity lost.

### Structure and Functioning of the Enforcing Agencies

(a) *Central Government*: The Central Government having overall control with powers to take all such measures necessary or expedient for protection and improving the quality of the environment and for preventing, controlling and abating the environmental pollution under the Environment (Protection) Act, 1986.

(b) *Central Pollution Control Board*:

1. Co-ordinate the activities of the State Boards and to resolve the disputes among them;
2. Provide for technical assistance and guidance to the State Boards;
3. Perform for such functions of any State Board as may be specified in the Order made under Section 18;
4. Lay down, modify or annual in consultation with the State Government the standards for a stream or well;
5. Plan and cause to be executed a nationwise programme for prevention, control or abatement of pollution.

(c) *State Government*:

1. May give such directions in writing to the State Boards in the performance of its functions under the Water and Air Acts;

2. May restrict the applications of Air/Water Acts to certain areas under Section 19 of the Act/s.
3. Borrowing powers, Budget of State Board, Annual report of the State Board, Annual Accounts and Audit of the State Board are subject to the scrutiny of State Government.
4. State Government may establish the Water and Air Laboratory or specify laboratory or institute as a State Water Laboratory and make rules about it, can appoint Government Analyst and approve the appointments of the Board Analysts.
5. State Government: To constitute Appellate Authority for hearing the Appeals under Section 25, 26 and 27 of the Water Act and under Section 21 as well as 31A of the Air Act.
6. State Boards shall furnish such reports, returns, statistics, accounts and other information as may be required by the Central Pollution Control Board or State Government from time to time.
7. State Government may on its own or on an application made to it entertain the revision to satisfy legality or propriety of any order made by the State Board under Section 25, 26 and 27 of the Water Act, but no such provision under Air Act.
8. State Government may make such contribution to the State Board as may it think necessary to enable the Board to perform its functions.

(*d*) State Pollution Control Board:- State Government shall appoint, constitute such Board with its Members including the Chairman, not exceeding 5 officials of State Government, not exceeding 5 officials of State Government, not exceeding 5 members of Local Authority, not exceeding 3 non-officials to represent the interest of the agriculture, fisheries or industry or trade or any other interest, 2 persons to represent Govt. companies or corporations and for full time Member Secretary possessing qualification, knowledge and experience of scientific, engineering or management aspects of pollution control. In case of Air Act, two persons having special knowledge or practical experience in respect of the matters about the improvement of quality of air etc.

It is possible that 15 out of 17 members could be persons not having any experience or qualification in pollution control or environmental protection. The members of the Board should have prescribed qualification and expertise in the environment protection in the related areas. The Board should be free from external influence of any type in exercise of the powers and functions. At present, there is a overdoes of Government control on the above activators.

The Board itself can be superseded by the Government, thus the functions of State Pollution Control Board included advisory, deliberative and investigatory in nature as well as to take adequate steps to control but most of times the actions of the State Pollution Control Board are influenced by the State Government or Members appointed by the State Government.

## How State Pollution Control Board Perform its Functions

(*a*) By grant of consent/s–two types of consent *i.e.*

(i) Consent to Establish.

(ii) Consent to Operate.

(*b*) By imposing prohibition on use of stream or wells.

(*c*) By laying down the standards.

Grant of consent is a basic document to regulate the establishment or operation of any industry, operation or process or any treatment and disposal system or an extension or addition thereof likely to cause pollution.

At the time of the establishment itself, the conditions are imposed to control pollution and after compliance of these conditions only, Consent to Operate issued. Consent to Establish is granted only after fulfilling the locational guidelines and notifications such as Dahanu Taluka Notification dated 20/6/1991.

CRZ Notification dated 19/2/1991 in Coastal Zone, Murud-Janjira Area in Raigad District, Mumbai Metropolitan Regional Policy, Bhatsa River Basin Notification, EIA Notification and other Notifications as well as Environmental Guidelines.

After grant of Consent to Operate, the Board has to monitor the compliance of the consent conditions.

## How State Pollution Control Board Monitors

1. By obtaining necessary information about the pollution prevention etc.
2. By inspection.
3. By taking samples of effluent.
4. By imposing prohibition on the use of stream or well for the disposal of polluting matters.
5. By grant/refusal/review of Consent/s.

## Available Remedies

What remedies available to the Board after monitoring the performance of consent conditions and after watching non-compliance ?

1. The Board may take emergency measures in case of pollution of stream or well by issuing order for restraining or prohibiting the discharge under Water Act.
2. By making applications to the courts for restraining apprehended pollution.
3. By issuing directions of closure, prohibition or regulation of any industry, operation or process or by stoppage or regulation of supply of electricity/water or any other service.

## Penalties and Procedures

The environmental laws in India engaged criminal sanctions for effective enforcement.

The Water and Air Acts prescribed minimum imprisonment unto 3 months or with fine unto Rs.10,000/- or both or an additional fine of Rs.5,000/- per day during the continuance of contravention after conviction for obstructing the functioning of the Board or for contravention of any of the provision of those Acts, for which, no penalty has been provided in the Acts.

Other violations such as contravention of consent conditions, violation of directions dealt sternly with minimum imprisonment of 18 months and fine and maximum 72 months and fine and in case of repetition of non-compliance, the imprisonment minimum of 2 years and maximum upto 7 years and fine. Now cognizance of offences can be taken by the Courts on a complaint made by the Board or the authorised officer of the Board or on a complaint made by any person by giving 60 days notice of the intention to make complaint. Both the Water and Air Acts provides for criminal responsibility of companies and Government Departments.

## Environment (Protection) Act, 1986

The Environment (Protection) Act, 1986 provides for punishment/imprisonment upto 5 years or with fine, which extend to Rs.1 Lakh or with both and in case of continuance of non-compliance after conviction, imprisonment upto 7 years and an additional fine of Rs.5,000/- per day.

In practice, the above compliants/offences, remained undecided for years together ranging from 7 to 10 years or more than that, thereby it imposes several limitations on the preventive nature of legislation, which may give rise for manipulation of these provisions by lingering prosecutions engaging good Advocates and Solicitors by paying huge professional charges, which could have been spent on pollution prevention and control.

All the above Acts provide that a person is held responsible, if he is proved to have committed the prohibited Act knowingly and person could not be liable for the offence, if he proves that the offence was committed without knowledge or he has exercised due diligence to prevent the commissioning of such offence. It becomes very difficult to prove offences against the defaulting persons. Thus, the prosecuting agency is also now a days reluctant in taking penal actions.

In such circumstances, the Board has no alternative than to take recourse to Section 33/22A of the Water and Air Acts respectively for restraining apprehended pollution, the reason is that the prevention is better than cure. However, in practice these applications remained pending for years together and in cases, where the courts issued orders for restraining apprehended pollution, then it becomes very difficult to execute those orders because the remedy for execution is also prosecution for not obeying the orders issued by the courts. In majority of cases, the courts are not issuing restraining orders without hearing, which requires a lot of period and before orders of the courts, the polluters could continue pollution, thereby the provisions of restraining pollution become ineffective in general.

Therefore, there was an urgent need for incorporating a provision enabling the Board to take quick action without waiting for the Court's order, thereby on the lines of the Environment (Protection) Act, 1986, the provisions empowering the powers for issuance of directions under Section 33A and 31a of both the Acts incorporated, which empowered the Board to issue directions to any person officer of the authority, which includes closure, prohibition or regulation of any industry, operation or process or stoppage or regulation of electricity/water or any other service. Thus, by issuance of the above directions, pollution can be prevented/stopped immediately. The effectiveness of these provisions has been sured by way of criminal sanctions with imprisonment minimum of 18 months and maximum upto 6 years.

Taking into consideration the above experience, it was noticed that the penal provisions remained as threats for a while and fade away after some years, the Board has therefore resorted to preventive measures provided under Section 33A/31A for both under the Water and Air Acts respectively. Today, preventive measures are playing very important role in implementation of both the Acts. The Board had taken effective action under Section 33A and 31A of the above Acts by issuing directions including upgradation of existing pollution control devices, adopting cleaner technologies and prescribing most stringent conditions to decrease pollution load by restraining the discharges into the stream or river and direct disposal on land for irrigation/gardening purposes including re-circulation of the water in the process itself. The results of such directions are found to be encouraging. The Board could have secure about 150 crores investment from the industries in the State of Maharashtra for the installation of most modern technologies for the treatment and disposal or emission etc. The Board has also introduced a system of bank guarantee for ensuring the compliance of the above directions and in case of non-compliances, the bank guarantees are forfeited and credited to the accounts of the Board.

As far as the remedies available to the individual, organisations or public at large under the above Acts, they can file prosecution after giving 60 days notice but in practice, nobody has resorted to this remedy taking into consideration very technical nature of offences, non-availability of technical reports and data as well as vexatious nature of the proceedings in the Courts of Law. Therefore, the individual, organisation or public at large have resorted to public interest litigations, wherein our Hon'ble High Court and Supreme Court have played an important role in achieving the desired results by issuing the noteworthy Directions/Judgments/Principles.

The Maharashtra Pollution Control Board, which initially was entrusted with the implementation of the water pollution prevention only, now entrusted with the implementation of the Water (Prevention and Control of Pollution) Cess Act, 1977, the Air (Prevention and Control of Pollution) Act, 1981 and the Environment (Protection) Act, 1986 and Rules made thereunder. The manpower and infrastructure initially provided for the Water Pollution Prevention Board is now entrusted with the above multidisciplinary role, however, the manpower and infrastructure is not so far strengthened to play the above multidisciplinary role effectively. What is necessary is that the Board should become an environment protection agency with suitable manpower and infrastructure and play an important role in the prevention, protection and improvement of environment.

## Maharashtra Pollution Control Board's Role in Implementation of Various Environmental Laws and Notifications

After Bhopal tragedy, the Parliament has enacted the most comprehensive and general legislation of environment protection, namely the Environment (Protection) Act, 1986. It has made drastic changes in the role of the Board. Now, the Central Government shall have overall control on the activities of any authority for taking adequate measures to protect the environment. Now, it can issue directions to any authority with regard to the protection and improvement of environment.

### Hazardous Waste (Management and Handling) Rules, 1989 with Amendment

The Central Government has laid down the procedure and safeguards under the Hazardous Waste (Management and Handling) Rules, 1989, which imposes responsibility on the Board for granting authorization and monitoring the compliance of those rules. In a writ petition filed by the Research Foundation for Science and Technology in the year, 1995 before Hon'ble Supreme Court of India, it has come to know for the first time that there are 3427 industries generating hazardous waste, but only 275 industries obtained authorisation, for which, the Board had to carry out extensive survey of the industries generating hazardous waste within short span of time. It was revealed that though the State Government was under obligation to identify and develop the sites for disposal of hazardous waste, not a single site was identified and fully developed till filing of the said petition, thereby, the authorised industries were disposing of the hazardous was in their own premises. Since, no standards were laid down for the hazardous waste generated in the process, the hazardous waste was disposed off in the nearby area in an unscientific manner. There were instances of import of hazardous waste from the developed countries in the name of the raw material for the manufacturing of products, but in practice, it was being dumped into the Indian Territory. Now, the above rules have been amended, which also make Industries Association/Industrial unit responsible for developing site for hazardous waste treatment and disposal. Earlier NEERI was appointed to do inventorisation of hazardous waste generating units, which had prepared draft report on the basis of consent proceedings showing

8545 units in the State of Maharashtra, but could not submit a list of such units to the Maharashtra Pollution Control Board in spite of repeated requests because of not having proper realistic data.

The Research Foundation for Science, Technology and Natural Resources Policy has filed writ petition No.657 of 1995 before Hon'ble Supreme Court of India, wherein, the Hon'ble Supreme Court of India taking into consideration the precarious condition of the hazardous waste treatment and disposal has constituted a High Powered Committee with Prof. M.G.K.Menon as its Chairman to examine in depth the matters relating to hazardous wastes and to give its report/recommendations at the earliest with a request to submit its first report within a period of 2 months of its constitution. However, a High Powered Committee could not file its report within a time, High Power Committee had filed its report before Hon'ble Supreme Court of India. However, in the meantime, there has been amendment in the Rules, namely 2000, which further makes above rules more complicated by extending 18 categories to 44 hazardous waste categories. Therefore, whatever inventorisation done by the various Boards before Amendment becomes incomplete. Now, whole exercise of inventorisation of hazardous waste will have to be repeated. Even implementation of amended rules has also become tedious. Enforcing agencies without proper infrastructure and manpower are unable to secure compliances.

## Manufacture, Storage and Import of Hazardous Chemical Rules, 1989 with Amendment

The Central Government thereafter made Manufacture, Storage and Import of Hazardous Chemicals Rules, 1989, for which the Board has to ensure compliance of the above rules. Maharashtra Pollution Control Board's role is limited for enforcement of directions and procedure in respect of isolated storage of hazardous chemicals (*i.e.* notification of major accident, notification of sites, safety reports in respect of isolated storages and preparation of Site Emergency Plan etc.)

With regard to import of hazardous chemicals and enforcement of directions and procedure on importing of hazardous chemicals, in respect of other activities like manufacture, storage and import of hazardous chemicals, Ministry of Environment and Forests, Government of India, Chief Controller of Imports and Exports, Chief Inspector of Factories, Dock Safety/Mines, Atomic Energy Regulatory Board are responsible for implementation of the above rules.

## Recycled Plastics Manufacture and Usages Rules, 1999

The Recycled Plastics Manufacture and Usages Rules, 1999 imposes prohibition on the use of carry bags or containers made of recycled plastics and prescribes condition for manufacture of the same *i.e.* carry bags and containers made of virgin plastics shall be in natural shed or white or carry bags and containers made of recycled plastics and used for the purposes other than storing and packaging foodstuff shall be manufactured using pigments and colourants as per IS:9833:1981 standards and the recycling of plastics shall be strictly in accordance with IS standards.

- ☆ Marketing and codification of recycled plastics is necessary and the minimum thickness of carry bags made of virgin plastics or recycled plastics shall not be less than 20 Microns.
- ☆ Maharashtra Pollution Control Board is responsible for monitoring the compliance of provisions of these rules. But, the District Collector/Deputy Commissioner of the concerned District will be responsible for enforcement of the provisions pertaining to se, collection, segregation, transportation and disposal of carry bags and containers made of virgin plastics and recycled plastics.

- ☆ The minimum thickness of carry bags of virgin plastic or recycled plastic shall be not be less than 20 Microns. Now Maharashtra Govt. has accepted more stringent norms for carry bags of virgin plastics/recycled plastic, prescribing thickness not less than 50 microns by passing Maharashtra Non-Biodegradable Garbage (Control) Act, 2006 and also on prescribing weight of 100 bags with registration as well as the size of not less than 8 × 12 inches. The notification issued by Govt. of Maharashtra after taking policy decision to impose prohibition/ restrictions on the manufacturing of and use of plastic carry bags, at the instance of impact of 26th July, 2005 *i.e.* heavy rains in Mumbai and most of parts of the Maharashtra State.

## Bio-Medical Waste (Management and Handling) Rules, 2000

Now, the Board is further entrusted with the implementation of the Bio-Medical Waste (Management and Handling) Rules, 1998 in spite of specific resolution passed by the Board to entrust the implementation of those rules to some other specialised Body/Agency because of not having adequate manpower and infrastructure for the implementation of the same with the Board, which requires monitoring of hospital waste collection, treatment and disposal. The Board has now also asked for the expertise staff for the effective implementation of those rules. Besides, the above, the Board has monitored the environment statement, conduct public hearing without having any manpower and infrastructure for the same. Thus, though the role of the Board has become multidisciplinary, the manpower and infrastructure is not provided accordingly. Even though the Board is self-sufficient as far as financial resources are concerned, for sanction of manpower and infrastructure facility for various Environmental Laws, Rules and Notifications, the Board is required to obtain previous sanction of the State Govt.

### Role of the Maharashtra Pollution Control Board

- ☆ Being the prescribing authority, function under Maharashtra Government, under advice of Advisory Committee appointed by the State Government.
- ☆ Grant or renew or cancel or suspend the authorisation after satisfying itself about possessing of necessary capacity to handle bio-medical waste with the applicant.
- ☆ May refuse to grant or renew authorisation after giving reasonable opportunity of being heard.
- ☆ To inspect and verify all records related to handling of hazardous waste-accident reporting thereof/maintained by the authorised person.

## Municipal Solid Waste (Management and Handling) Rules, 2000

- ☆ Maharashtra Pollution Control Board is responsible to monitor the compliance of the standards regarding ground waster, ambient air quality, leachate quality and the compost quality including incineration standards specified in the Rules.
- ☆ Maharashtra Pollution Control Board grants authorisation for setting up the waste processing and disposal facilities including landfills after examining proposal taking into consideration the views of other agencies such as State Urban Development Deptt., Town and Country Planning Deptt., Air Port or Airbase Authority, Ground Water Board or any other such agency–within 45 days stipulating compliance criteria and standards.
- ☆ To co-ordinate with the Central Pollution Control Board with regard to the implementation and review of standards and compilation of monitoring data.

- ✩ To prepare and submit an Annual Report with regard to the implementation of the rules by 15th September every year.
- ✩ Monitoring of performance of waste processing and disposal facilities once in 6 months as per compliance criteria. (*i.e.* improvement in existing landfill sites by 31/12/2001, identification of landfill sites for future use etc. by 31/12/2002 and setting up of the waste processing and disposal facilities by 31/12/2003 as per compliance criteria)–but Secretary Incharge of Urban Development Deptt. in Metropolitan Cities and Dist.Collector or Deputy Commissioner in other cases will be appropriate authority for enforcement of the rules.

## CRZ Notification

- ✩ Maharashtra Pollution Control Board's role is limited in regulations of setting up and expansion of the industries, operations or processes etc. in the CRZ area. Maharashtra Coastal Zone Management Authority is constituted by the Central Government for implementation of CRZ Notification in 1998 and reconstituted thereafter from time to time.
- ✩ For other activities such as construction, storages in the port etc. Director/Dock Safety, Port Authority/Conservator of Port etc. are the appropriate authorities for the enforcement of CRZ Notification and Dist. Collector is overall incharge for implementation of CRZ Notification.

## Batteries (Management and Handling) Rules, 2000

- ✩ The Batteries (Management and Handling) Rules, 2000 come into force w.e.f. 16/5/2001. It applies to every manufacturer, importer, re-conditioner, assembler, dealer, recycler, auctioneer, consumer and bulk consumer involved in manufacture, processing, sale, purchase and use of batteries or components thereof.
- ✩ Battery means lead acid battery, which is source of electrical energy and contains lead metal.
- ✩ Certain responsibilities are imposed on manufacturer, importer, assembler and re-conditioner such as:
  - (*a*) The batteries are collected back as per the schedule against new batteries sold to original equipment manufacturer and bulk consumer.
  - (*b*) To set up collection centres for collection of used batteries from consumers or dealers and to send it only to the registered recyclers to ensure that no damage to the environment occurs.
- ✩ Similar responsibilities are imposed on dealer, recycler etc. dealing with lead batteries. Registration/Renewal of Registration of recyclers with Ministry of Environment and Forests is made necessary. Some responsibilities are also imposed on consumer or bulk consumer and auctioneer.
- ✩ The State Pollution Control Board is the prescribed authority for ensuring the compliance of the provisions of these rules and it shall file an Annual Compliance Status Report to the Central Pollution Control Board by 30th April of every year
- ✩ Central Pollution Control Board will publish the data received from the State Board and review the compliance of the rules periodically to improve the collection and recycling of used lead batteries and apprise to Ministry of Environment and Forests.

☆ Ministry of Environment and Forests to develop proper record about the distribution, sale etc. of used batteries.

## Hazardous Wastes (Management and Handling) Rules, 1989 as amended in 2000 and guidelines of Hazardous Wastes Rules

The Central Government has issued the notification in exercise of the powers conferred by Sections 6, 8 and 25 of the Environment (Protection) Act, 1986, namely the Hazardous Wastes (Management and Handling) Rules, 1989 on 28/7/1989, which is subsequently amended on 5/2/1990, 31/3/1992, 3/9/1996 and 6/1/2000 respectively.

The rules defined, "hazardous wastes" means categories of wastes specified in the schedule;

"hazardous waste site" means a place for collection, reception, treatment, storage and disposal of hazardous waste duly approved by the competent authority.

Initially there were 18 categories of hazardous waste in the schedule such as Cyanide wastes, Metal finishing wastes, waste containing water soluble chemical compounds of lead etc. Mercury, Arsenic, Thallium and Cadmium bearing wastes, Asbestos, Phenols, Acid/Alkaline/Slurry, Off-specifications and discarded products, containers etc.

(*a*) The rules imposed responsibility on the occupier generating hazardous waste equal to or exceeding the limits given in the schedule to take all practical steps to ensure that such wastes are properly handled and disposed off without any adverse effects, which may result from such wastes and it imposes responsibility on the occupier for proper collection, reception, treatment, storage and disposal of these wastes either himself or through the operator of a facility.

(*b*) The occupier or operator of a facility, to ensure that such hazardous waste is packaged in a manner suitable for storage and transport in accordance with the Rules issued by the Central Govt. under the Motor Vehicles Act, 1989 and other guidelines issued from time to time.

## EIA Notification

☆ Ministry of Environment and Forests, Govt. of India had issued draft EIA Notification dtd.28/1/1993, inviting objections from the public to impose restrictions and prohibitions on the expansion and modernization of any activity or new projects being undertaken in any part of India unless environmental clearance accorded by the Central Govt. or the State Govt.

☆ Central Government issued final Notification, namely, EIA Notification, dtd.27/01/1994 prescribing requirements and procedure for seeking environmental clearance of projects given in the Schedule-I as under:

(*a*) Project proponent undertaking any activity/project under Schedule-I to make an application in proforma including EIA Report, an Environment Management Plan and details of public hearing.

But EIA will not be required for pipeline project and for pipeline and highway projects, public hearing to be conducted in each District through which it passes.

(*b*) In case of site specific projects like Mining, Pit-head Thermal Power stations, Hydropower, major Irrigation, major Ports and Harbours, major minerals above 500 hectors, Green Field Air Ports etc., location of the project site to be informed first and then, after decision regarding site selection, further steps can be taken.

(*c*) Procedure is prescribed for assessment by the Impact Assessment Authority, which is to be completed within 90 days from receipt of requisite documents and data and completion of public hearing.

(*d*) Schedule-I to the Notification dtd.27/01/1994, identified 29 projects requiring environmental clearance from the Central Govt. Procedure for public hearing was also given in the Notification.

(*e*) Experience of the Board in the public hearing–Public hearing adjourned/postponed/ reconvened due to number of reasons such as inadequate information by the project proponent, Public Hearing Panel wants to visit the industry to see the fact first hand.

MPCB communicated its practical difficulties in implementation of earlier EIA Notification of 1994 and comments on the draft EIA Notification issued by the Central Govt. on 15/ 09/2005. The Central Govt. after taking into consideration all objections and suggestions received in response to the draft Notification has finally issued a new Notification bearing S.O.No.1533(E), dtd.14/09/2006 in supersession of earlier Notification dtd.27/ 01/1994, except in respect of things done or omitted to be done before such supersession. Now, the present Notification for the first time transferred a number of matters falling under category 'B' in the schedule under the Notification to the State Environment Impact Assessment Authority before any construction work or preparation of land by the project management except for securing the land is started, for the purposes of grant of environment clearance. Other important matters falling under category 'A' in the schedule are kept under the authority of the Central Govt. in the Ministry of Environment and Forests for grant of environment clearance. The following are the salient features of the new Notification of Environment Impact Assessment issued by Ministry of Environment and Forests, Govt.of India dtd.14/09/2006.

(*i*) The projects/activities requiring prior environment clearance from the Govt. are divided into two parts as stated above. Category 'A' in the schedule for matters falling under MoEF, GoI and Category 'B' falling under State Govt. For the first time, proper distribution of the project/s/activity/s made. Important projects kept with the Central Govt. and other projects falling under the State Authority are transfer to State Govt. for grant of environment clearance.

(*ii*) For the first time, State Level Environment Impact Assessment Authority provision is made in the Notification. Besides that, provision is made for Screening, Scoping and Appraisal Committees at the Central Govt. and State Govt. level are proposed to be constituted by the Central Govt. in consultation with the concerned State Govt. or Union Territory Administration for a period upto 3 years and thereafter to be reconstituted.

(*iii*) An application seeking prior environment clearance shall be made alongwith a copy of pre-feasibility project report except construction projects or activities alongwith necessary forms and conceptual plan in case of construction activities.

(*iv*) At the stage of screening, the concerned State Expert Appraisal Committee, in case of category 'B' projects, to determine whether the project requires environmental studies for preparation of an EIA for its appraisal, prior to grant of environment clearance depending upon the nature and location of the project. If, an Environment Impact Assessment Report requires, the project will be termed as Category 'B1' and other remaining projects will be termed as Category' 'B2' (not requiring EIA).

(*v*) In case of 'B1' projects including applications for expansion and/or modernization and/or change in product, mix of existing project.

(*a*) State Level Expert Appraisal Committee will determine detailed and comprehensive terms of reference (TOR) addressing all relevant environmental concerns for the preparation of an EIA Report, on the basis of the information furnished to it including proposed terms of reference by the Applicant. If necessary a site visit by a group of Expert Appraisal Committee or State Level Expert Appraisal Committee will be carried out.

(*b*) But construction activity (item No.8) will not require any scoping. TOR to be communicated within 60 days after receipt of form-1, if not communicated, the draft TOR submitted by the applicant shall be deemed as final for EIA.

(*c*) Approved TOR to be displayed on the website of MoEF, GoI and concerned State Authority.

(*d*) The concerned State/Central Govt. Regulatory Authority may reject applications for environment clearance on the recommendation of the EAC or SEAC at scoping stage itself alongwith reasons and the decision is to be communicated within 60 days time from the receipt of application.

(*vi*) Public Consultation: It refers to the process by which, the concerns of local affected persons and others, who have plausible stake in the environmental impact of the project or activity are ascertained. All 'A' and 'B1' projects to undertake public consultation except modernization of irrigation projects mentioned in the rules, projects located within industrial estates or parks approved by the concerned authorities, expansion of roads and highways without acquisition of further land, all building/construction projects/area development projects and townships, all 'B2' category projects and projects concerning national defense and security etc.

Public consultation to be done in two components:

1. Public hearing at site or in its close proximity–district wise to ascertain concerns of local affected persons.
2. To obtain responses in writing from other concerned persons having a plausible stake in environmental aspects of the project.
3. Public hearing is to be conducted within 45 days at the request of the applicant by the State Pollution Control Board, in case it is not conducted. The Regulatory Authority shall engage another public agency or authority to complete the process within 45 days. However, if the public agency or authority nominated by the Regulatory Authority/Central Govt. communicate that it is not possible to conduct public hearing due to local situations, then, the Regulatory Authority may decide that public consultation in the case need not include a public hearing.
4. For obtaining responses in writing from the persons having a plausible stake into the environment, the concerned Regulatory Authority and SPCB or Union Territory Pollution Control Committee shall invite response from them by placing on website the summary of EIA Report alongwith a copy of application within 7 days from the request for public hearing, but confidential information shall not be placed on the website.

5. After completion of public consultation, the applicant to make appropriate changes in the draft EIA and EMP and to submit final EIA report to the concerned Regulatory Authority or to submit a supplementary report to draft EIA and EMP.

(*vii*) Appraisal: It is a detail scrutiny by the Expert Appraisal Committee or State Level Expert Appraisal of the application and other documents including outcome of public consultation, where applicant shall be invited for furnishing necessary clarifications and thereafter, to make proper recommendation to the Regulatory Authority either for grant or rejection of the application. Appraisal is to be completed within 60 days from the receipt of final EIA and other documents or the receipt of Form-1 and Form 1A, where public consultation is not necessary.

(*viii*) Besides that, prior environment clearance process for expansion and modernization or change of product mix in existing project, validity of environment clearance, post environment clearance monitoring, transferability of environment clearance and operation of EIA Notification 1994 till disposal of pending cases etc. provisions are made in the new EIA Notification, 2006.

The role of the board in implementing various environmental laws has become multidisciplinary. It has changed from command and control approach to that of facilitator. Earlier board was filing prosecution and the same were remaining pending years together. In 1995, the board has taken review of all its legal actions and decided to resort to the remedy of issuance of directions for compliance. The board granted sufficient time to the polluting industries to comply with environmental norms and where, the defaulting industries not complied, the board had not hesitated to close down such units till effective steps are taken for pollution prevention and bank guarantees were also taken to ensure compliance of directions. The board had also encashed bank guarantees in serious cases of default. The result was that medium and large scale industries provided pollution control devices. In case of SSI units, board promoted common effluent treatment plants, out of financial assistance from the MIDC, State Government and Central Government as well as by the board. After 2001, there was recession and therefore board concentrated on persuasion rather than stringent action. Now board is giving guidance to the industries, local bodies and other institutions. The board has promoted waste processing facilities for 5 local bodies and become facilitator.

# Chapter 57

# Re-Emergence of $CO_2$ as a Viable, Alternative, Natural Refrigerant

**Maneesh Dubey* and S.P.S. Rajput****
**Department of Mechanical Engineering, MANIT, Bhopal, M.P.**

## ABSTRACT

Depletion of stratospheric ozone and climate changes by sustained release of chemicals containing chlorine and bromine led to the enactment of the Montreal and Kyoto Protocols. The HFC's, once thought to be acceptable permanent replacement fluids for refrigeration and air conditioning are now in the list of regulated substances with their production levels being capped and gradually being phased out. There is frenzied activity in the research establishments world-wide to introduce ecologically safe and natural refrigerants. Substantial quantities of refrigerants also find their way in the automobile air conditioning systems. The release of these refrigerants into the atmosphere is still highly unregulated and fraught with uncertainty. $CO_2$ has successfully been introduced in automobile air conditioning systems to ameliorate this problem. Therefore $CO_2$ has been rediscovered as a viable option. The objectives of this paper are to review critically recent developments and sate of art technology for adopting $CO_2$ in various refrigeration, air conditioning and heat pump applications. The focus will be on fundamental issues, design issues, cycle modifications and promising application areas. This review aims to project $CO_2$ as a possible option with some cost effective cycles.

*Keywords:* *$CO_2$(R-744), Refrigerants, Air conditioning, Vapour compression refrigeration, Heat pump.*

---

Author: E-mail: *d.maneesh@gmail.com; **spsrajput@gmail.com

## Introduction

The refrigeration and air conditioning industry is in transition. Environmental concerns with the impact of refrigerant emissions led to reconsider the selection of refrigerants containing chlorine and bromine. HCFC fluids still need to be replaced mostly involving R-22 in air conditioning and heat pump applications. The HFC refrigerants that were once expected to be acceptable permanent replacement fluids are now on the list of regulated substances[1]. The GWP is an index that relates the potency of greenhouse gas to the $CO_2$ emission over 100 year period as shown in Table 57.1.

**Table 57.1: Ozone-Depletion and Global-Warming Potential of Refrigerants (100-yr Values) [2]**

| *Refrigerant* | *ODP* | *GWP* | *Building Applications* |
|---|---|---|---|
| Chlorofluorocarbons | | | |
| CFC-11 | 1.0 | 4,680 | Centrifugal Chillers |
| CFC-12 | 1.0 | 10,720 | Refrigerators, Chillers |
| CFC-114 | 0.94 | 9,800 | Centrifugal Chillers |
| CFC-500 | 0.605 | 7,900 | Centrifugal Chillers, humidifiers |
| CFC-502 | 0.221 | 4,600 | Low temperature Refrigeration |
| Hydrochlorofluorocarbons | | | |
| HCFC-22 | 0.04 | 1,780 | Air-conditioning, Chillers |
| HCFC-123 | 0.02 | 76 | CFC-11 Replacement |
| Hydrofluorocarbons | | | |
| HFC-23 | $<4\times10^{-4}$ | 12,240 | Ultra low temperature Refrigeration |
| HFC-134a | $<1.5\times10^{-5}$ | 1,320 | CFC-12 or HCFC-22 Replacement |
| HFC-245fa | $\sim10^{-5}$ | 1,020 | Insulation agents, Centrifugal chillers |
| HFC-404A | $\sim10^{-5}$ | 3,900 | Low Temperature Refrigeration |
| HFC-407C | $\sim10^{-5}$ | 1,700 | HCFC-22 Replacement |
| HFC-410A | $<2\times10^{-5}$ | 1,890 | Air conditioning |
| HFC-507A | $\sim10^{-5}$ | 3,900 | Low Temperature Refrigeration |
| Natural Refrigerants | | | |
| $CO_2$ | 0 | 1.0 | |
| $NH_3$ | 0 | 0 | |
| Propane | 0 | 3 | |

Due to long atmospheric life, CFCs migrate to the stratosphere and deplete the ozone layer. In the stratosphere they are exposed to powerful ultraviolet rays which break apart the refrigerant chemicals releasing chlorine, because of an important catalytic loss chain reaction involves.

$$CCl_2 + \text{UV rays} = CClF_2 + Cl$$

$$Cl + O_3 = ClO + O_2$$

$$ClO + O_3 = Cl + 2O_2$$

Because of ODP of Freon, International Regulations prescribes that CFCs should no longer be used as refrigerants in industrialized countries. The Montreal Protocol identified the main ozone depleting substances and set specific limits on their production levels in the future. [3].

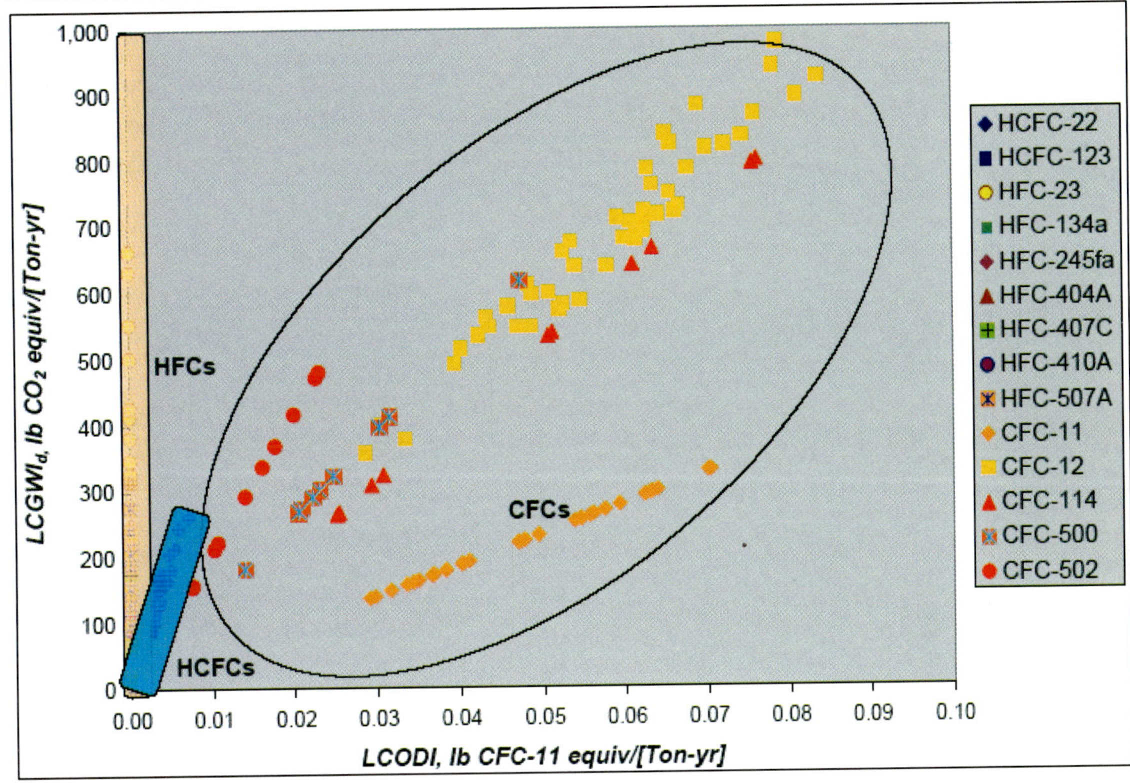


**Figure 57.1: Life-cycle Direct Global Warming and Ozone Depletion for all Refrigerants**

Figure 57.1 provides an instructive depiction of the atmospheric impacts of refrigerants–both ozone depletion and direct global warming–that occurs for different equipment and refrigerants. It also illustrates the inherent trade-off between these two impacts that exists for each possible choice: some deplete less ozone but cause more global warming, and *vice versa*. These variations are a function not only of the refrigerant but also of the performance of the specific refrigeration equipment. [2-4].

There is no way to directly compare the demerits of ozone depletion and global warming since they are distinct phenomena. Increasing the chlorine content in refrigerant molecules generally increases the ODP. Compounds that contain no chlorine, bromine have ODPs that are nearly zero; likewise increasing fluorine count generally increases the GWP. In both the cases, increasing the hydrogen count tends to shorten the atmospheric lifetime. Compounds with very short lives will have low ODPs since most emission decompose before reaching the stratosphere. They also will have low GWP values, since their atmospheric persistence will be comparatively of short duration.

CFCs generally have very high ODPs and GWPs. Most HCFCs have low ODP and GWP. HFCs have ODPs of almost zero, but GWPs that range from very low to very high. Relatively few fluoro chemicals have both very low, or zero, ODP and very low GWP. Among them are R-123 and R-152a, both of which have short atmospheric lifetimes of 1.4 and 1.5 years respectively. R-152a is however flammable. [4].

The influences of chemical compositions on behaviour suggest several future refrigerant candidates. The concern of the industry is limited to impacts on efficiency. The critical point and heat capacity are among the most important parameters driving thermodynamic performance. A review of current refrigerant selections for common applications suggest that increasing the content of

- Carbon increases the molecule size, generally increasing the normal boiling point and molar heat capacities, making large molecules unsuitable.
- Nitrogen makes the resulting compound more reactive, generally increasing toxicity and decreasing stability.
- Oxygen reduces the atmospheric stability, and therefore ODP and GWP, but also may increase flammability, reactivity, and toxicity.
- Sulphur generally increases toxicity and may decrease stability.
- Hydrogen generally decreases atmospheric lifetime, and therefore ODP and GWP, but also increases flammability especially when the number of hydrogen atoms exceeds the number of connected halogen atoms.
- Fluorine attached to carbon increases the GWP, particularly in per fluorinated molecules.
- Chlorine increases lubricant miscibility, but also the ODP and, generally, the toxicity.
- Bromine rapidly increases the ODP but greatly diminishes flammability, and
- Total fraction of attached halogen increases atmospheric lifetime, particularly per halogenated compounds, and therefore ODP and GWP. [4].

The selection of refrigerants for a particular application is governed by the performance and practical efficiencies. No viable refrigerant has at surface that offers higher efficiencies than R-11 or R-123. But environmental concerns again force the researchers to seek alternative refrigerants. [5].

$CO_2$ is an old refrigerant and it is therefore natural to look back why the use of $CO_2$ as a refrigerant declined? The recent revival of $CO_2$ is due to growing concern for environment and development of state of art for $CO_2$ technology that overcome/may find solution of inferior COP with $CO_2$. This chapter describes various properties, cycles, components and application issues related to adopt it as a viable, alternative, natural refrigerant.

## Properties of $CO_2$

The refrigerant properties are important for the design of the heat pump system and its components. The properties of $CO_2$ are well known and they are quite different from all the conventional refrigerants. The critical pressure and temperature of $CO_2$ are 7.38 MPa (73.8 bar) and 31.1 °C, respectively, and it is not possible to transfer heat to the ambient above this critical temperature by condensation as in the conventional vapour compression cycle. This heat transfer process (gas cooling) above the critical point results in the transcritical cycle, *i.e.* with subcritical low-side and supercritical high-side pressure (for a single-stage cycle). [6]. The thermodynamic and transport properties of $CO_2$ compared to other refrigerants are discussed in the following sections.

### Thermodynamic Properties

Figure 57.3 and 57.4 present vapour pressure and slope of the saturation temperature curve of $CO_2$ compared to other fluids. The vapour pressure of $CO_2$ is much higher than other refrigerants, and its higher steepness near the critical point gives a smaller temperature change for a given pressure change. Thus, the temperature change associated with pressure drop in the evaporator will become

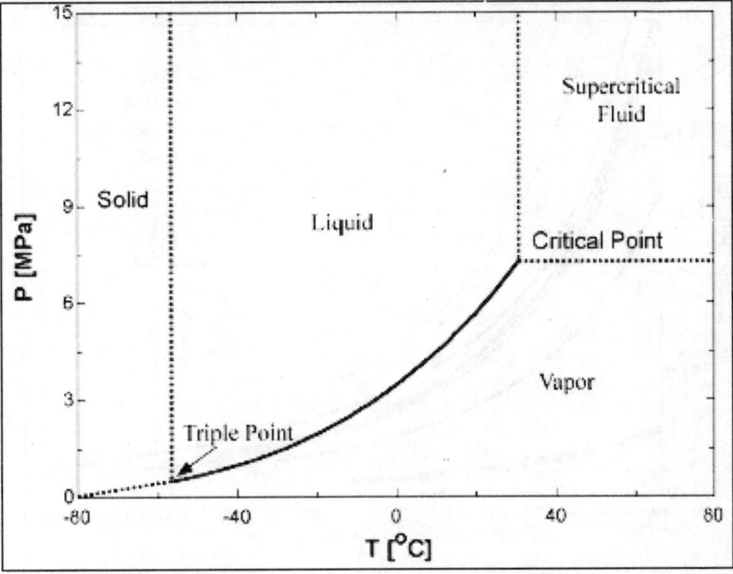


**Figure 57.2: Phase Diagram of $CO_2$**

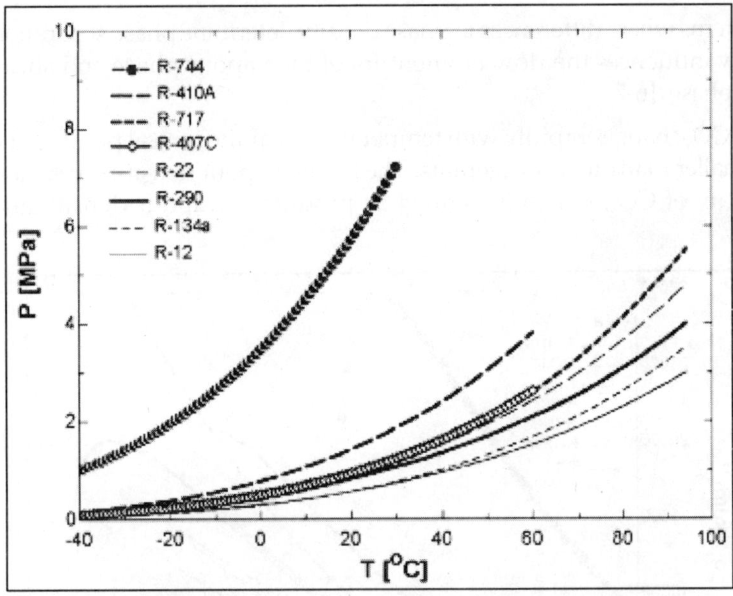


**Figure 57.3: Vapour Pressure for Refrigerants [6]**

smaller. For example, at 0 °C, the temperature change of $CO_2$ for 1 kPa pressure drop is about 0.01 K. On the other hand, the same pressure drop with R-410A and R-134a give the temperature changes of 0.04 and 0.10 K, respectively, *i.e.* about 4–10 times higher, as shown in Figure 57.4. The high vapour pressure and closeness to the critical point result in quite different characteristics of liquid and vapour density of $CO_2$ compared to other refrigerants. The high vapour density may have significant effect on

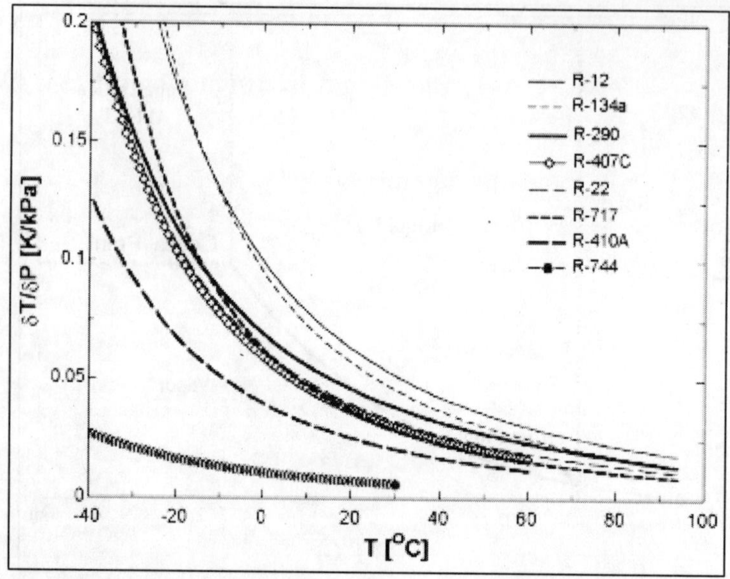


**Figure 57.4: Slope of Saturation Curve dT/dP for Refrigerants [6]**

two-phase flow patterns where differences in phase density determine phase separation characteristics, and vapour density influences the flow momentum of the vapour phase and shear force between vapour and liquid phase. [6-7].

The density of $CO_2$ changes rapidly with temperature near the critical point, and the density ratio of $CO_2$ are much smaller than other refrigerants. The higher vapour density gives the high volumetric refrigeration capacity of $CO_2$, which is defined as product of vapour density and latent heat of

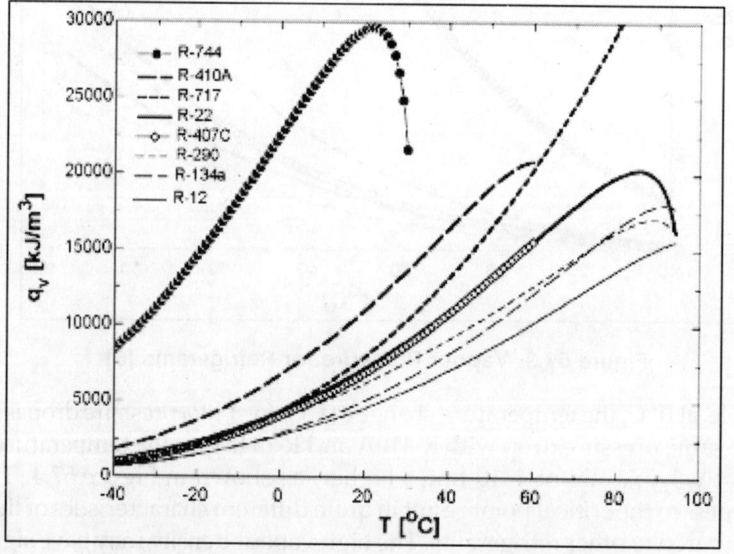


**Figure 57.5: Surface Tension for Refrigerants [6]**

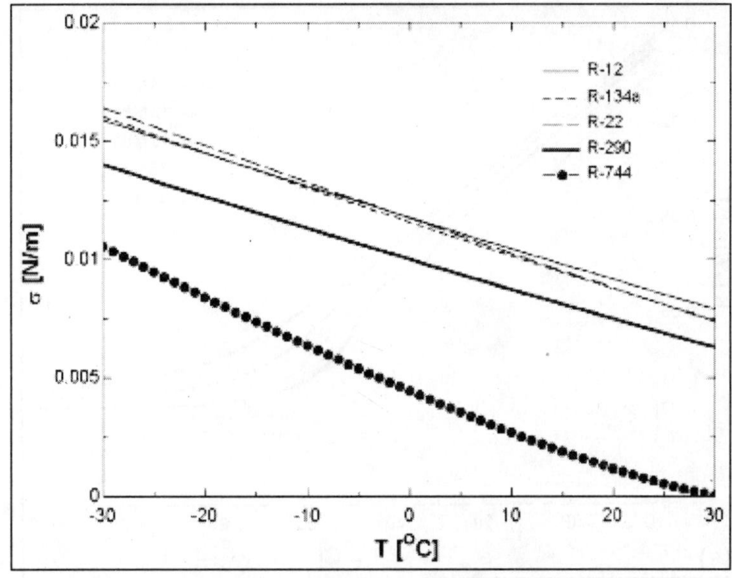


**Figure 57.6: Volumetric Refrigeration Capacity for Refrigerants [6]**

evaporation. The volumetric refrigeration capacity of $CO_2$ increases with temperature, has a maximum at 22°C, and then decreases again. By definition it is zero at the critical point as shown in Figure 57.5. Surface tension of the refrigerants influences nucleate boiling and two-phase flow characteristics. A small surface tension reduces the superheat required for nucleation and growth of vapour bubbles, which may positively affect heat transfer. Figure 57.6 presents surface tension of saturated $CO_2$ liquid at varying temperatures, compared to other fluids. The surface tension of the refrigerants decreases with temperature and becomes zero at the critical point [6].

### Transport Properties

The transport properties of refrigerants play an important role in heat transfer and pressure drop characteristics. Figure 57.7 shows the transport properties of $CO_2$, which are thermal conductivity and viscosity at subcritical and supercritical pressures at varying temperatures.

## Transcritical Vapour Compression Cycle

Compared to conventional refrigerants, the most remarkable property of $CO_2$ is the low critical temperature of 31.1 °C Vapour compression systems with $CO_2$ operating at normal refrigeration, heat pump and air-conditioning temperatures will therefore work close to and even partly above the critical pressure of 7.38 MPa. Heat rejection will in most cases take place at supercritical pressure, causing the pressure levels in the system to be high, and the cycle to be 'transcritical', *i.e.* with subcritical low-side and supercritical high-side pressure (for a single-stage cycle).

### Fundamentals of Transcritical Cycle

During operation at high ambient air temperatures the $CO_2$ system will operate in a transcritical cycle most of the time. Heat rejection then takes place by cooling the compressed fluid at supercritical high-side pressure. The low-side conditions remain subcritical, however, as shown in Figure 57.8. At

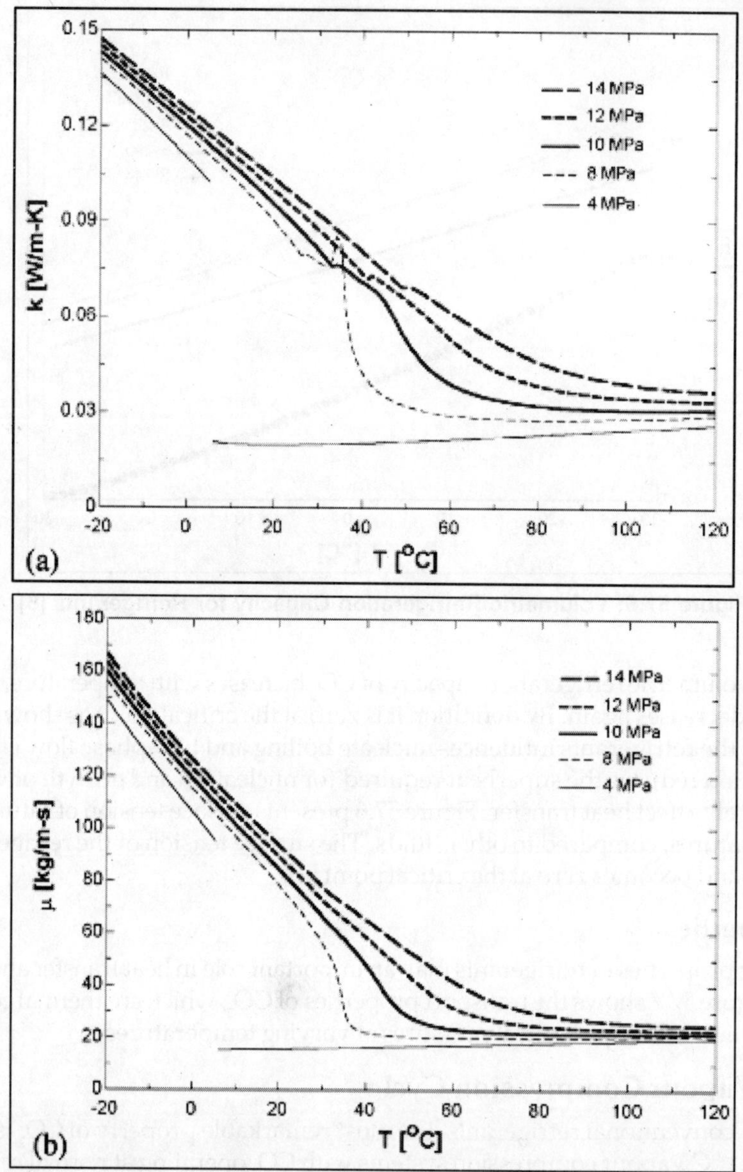


**Figure 57.7: Transport Properties of $CO_2$ (a) Thermal Conductivity (b) Viscosity [6]**

supercritical pressure, no saturation condition exists and the pressure is independent of the temperature. In conventional subcritical cycles, the specific enthalpy in point 3 is mainly a function of temperature, but at supercritical high-side conditions, the pressure also has a marked influence on enthalpy. This effect may be observed as non-vertical or S-shaped isotherms in the supercritical and near-critical region. An important consequence of this is that it is necessary to control the high-side pressure, since the pressure at the throttling valve inlet will determine specific refrigeration capacity. As in conventional systems, the compressor work and thereby also the COP will depend on the

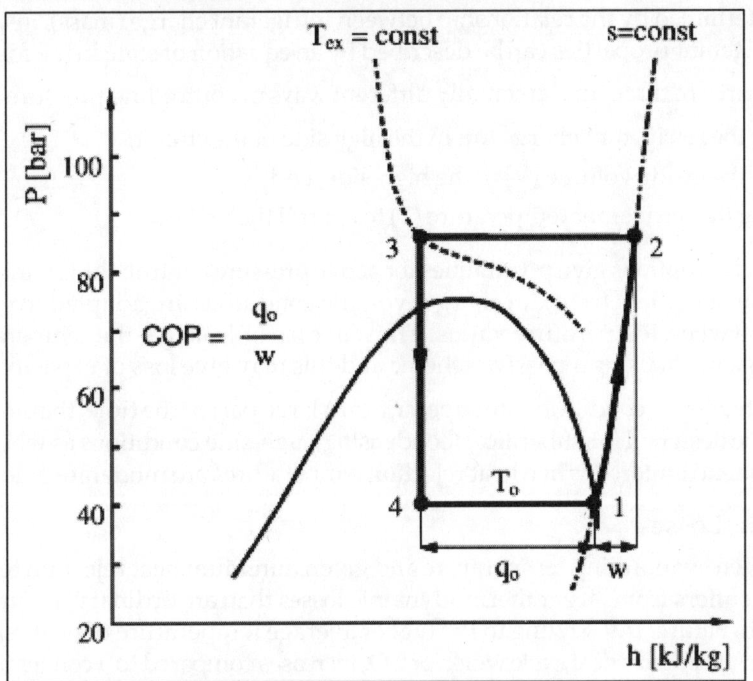


**Figure 57.8: Transcritical Cycle in the $CO_2$ Pressure-Enthalpy Diagram**

discharge pressure. However, while the COP tends to drop with increasing pressure in conventional cycles, the behaviour is quite different in a transcritical cycle. [8].

As the high-side pressure is increased, the COP reaches a maximum above which the added capacity no longer fully compensates for the additional work of compression. Differentiation of cooling $COP = (h_1 - h_3)/(h_2 - h_1)$ with respect to the high-side pressure gives maximum COP for $\frac{\partial COP}{\partial P} = 0$ defined by Inokuty [9].

$$\left(\frac{\partial h_3}{\partial P}\right)_T = -\left(\frac{\partial h_2}{\partial P}\right)_S$$

That is, the 'optimum' pressure is reached when the marginal increase in capacity equals COP times the marginal increase in work. High-side pressure regulation can be applied to maintain the COP at its maximum and/or to regulate the cooling or heating capacity. The optimum pressure increases steadily and almost linearly as Tex is raised, and the influence from varying evaporating temperature is quite small.

## Methods of High Pressure Control

The high-side pressure in a $CO_2$ system may be either subcritical or supercritical. In case of subcritical operation, the system will behave as conventional systems, with high-side pressure determined by condensing temperature. In case of supercritical operation, however, the pressure in the high side is determined by the relationship between refrigerant charge (mass), inside volume and temperature. Refrigerant properties can be described by an equation of state in the following form:

As a result, there are three fundamentally different ways of controlling pressure [10]:

- Varying the refrigerant charge (m) in the high side of the circuit,
- Varying the inside volume (V) of the high-side, and
- Allowing the refrigerant temperature (T) to control the pressure.

While the first two options give possibilities for active pressure control, the last method is actually a passive scheme where the refrigerant charge/volume conditions are adapted to give the desired change in pressure when temperature varies. Thus, in case of leakage, the temperature/pressure relation will change when using a passive scheme and this may give loss of capacity and COP.

Even though high-side conditions are supercritical a large part of the time, the circuit and control system must also be designed for subcritical (condensing) high-side conditions as well, since this type of operation will be encountered when heat rejection temperatures are moderate or low.

## Thermodynamic Losses

Assuming given evaporating temperature and given minimum heat rejection temperature, the transcritical cycle suffers from larger thermodynamic losses than an 'ordinary' Evans–Perkins cycle with condensation, Figure 57.9. Owing to the higher average temperature of heat rejection, and the larger throttling loss, the theoretical cycle work for $CO_2$ increases compared to a conventional refrigerant as R-134a as indicated. The throttling loss in a refrigerating cycle is given by temperatures before and after the throttling device, and by refrigerant properties. With temperatures given, the refrigerant properties become essential. Given the high liquid specific heat and low evaporation enthalpy of $CO_2$ near the critical point, the loss in refrigeration capacity (and the equal increase in compressor power) becomes large.

In reality though, as discussed in subsequent sections, the minimum heat rejection temperature will be lower in the $CO_2$ cycle when heat sink inlet temperature and heat exchanger size is given. In addition, the evaporating temperature tends to be higher for a given duty, heat source temperature, and heat exchanger size. Finally, the compressor losses, which are not shown in Figure 57.9, tend to be lower in $CO_2$ machines.

## Heating Capacity and COP Characteristics

In heat pump operation, the $CO_2$ system obtains a maximum COP at a certain high-side pressure, as explained above. By raising the pressure above this level, the heating capacity may be increased or maintained as the evaporating temperature is reduced. Despite the reduced COP, the overall efficiency of heating may then be increased in bivalent systems due to reduced supplementary heat. Another peculiarity of the $CO_2$ cycle is the smaller influence on heating capacity and COP by varying evaporating temperature, which enables the $CO_2$ system to maintain a high heating capacity at low ambient temperature. [11]. Similar tendencies can also be observed for cooling capacity and cooling COP.

Even at the optimum pressure (opt), the $CO_2$ heat pump output is decreased less than with the other refrigerants as the evaporating temperature is reduced. At -15°C, the capacity ratio $CO_2$/fluorocarbon is about 1.5, and the relative reduction in COP is smaller with $CO_2$ than with the other fluids. By raising the high-side pressure, a further increase in heat pump capacity can be obtained. The actual operation with capacity-boosting will depend on factors like maximum pressure, maximum motor load, and compressor discharge temperature limitations.

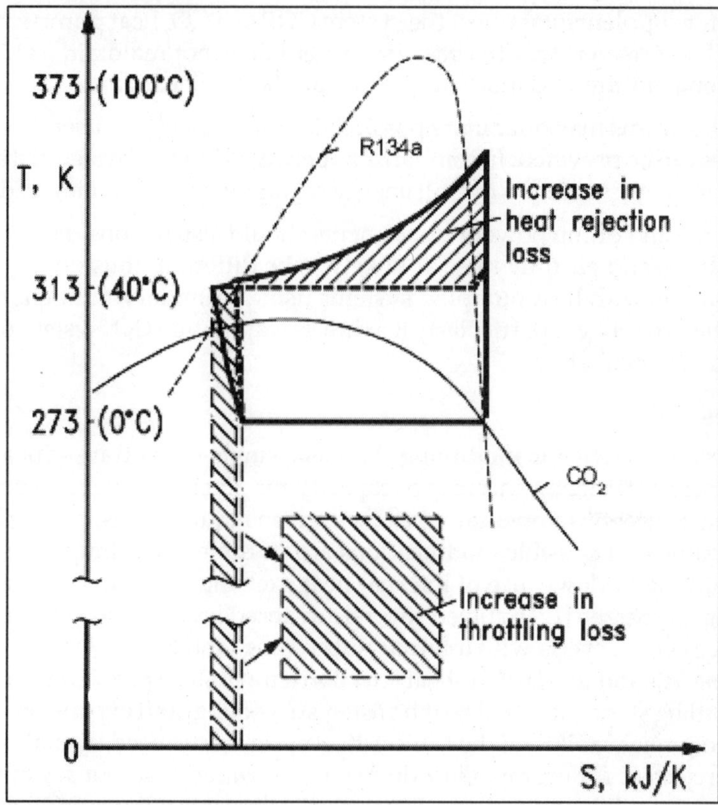


**Figure 57.9: Comparison of Thermodynamic Cycles for R-134a and $CO_2$ in T-S Diagram Showing Additional Thermodynamic Losses for the $CO_2$ Cycle when Assuming Equal Evaporating Temperature and Equal Minimum Heat Rejection Temperature**

## Analysis of Transcritical System Energy Efficiency

Comparisons of energy efficiency and/or TEWI (total equivalent warming impact) between baseline systems and $CO_2$ systems have to account for two important factors: The effect of climate, *i.e.* a seasonal data for comparisons of energy consumption, and a system approach, including the effects of supplementary heat and secondary power requirements for fans or pumps.

Most refrigeration, air-conditioning and heat pump systems are operated in a varying climate. Comparisons based on design point operation apply to conditions that rarely occur–typically at an extreme ambient temperature. In order to obtain a realistic comparison of annual or seasonal energy consumption, realistic climatic data should be applied, *e.g.* temperature occurrence data. Even though the COP of the $CO_2$ system may be somewhat lower at an extreme ambient temperature, the seasonal energy consumption may be reduced compared to a baseline system using conventional refrigerants.

Differences in heating capacity characteristics between $CO_2$ and conventional refrigerants have to be taken into account in the comparison of $CO_2$ to baseline systems, since differences in supplementary heating requirements may significantly affect the system energy efficiency. In this respect, the system energy efficiency is calculated as heating system COP, *i.e.* the ratio of heat pump heat output plus supplementary heat, to heat pump energy use plus supplementary heat input. By reducing the need for supplementary heat the system COP of a $CO_2$ heat pump may often be higher than the baseline. The reason is that the baseline system does not maintain its heating capacity at lower heat source temperature, and more supplementary heat is needed.

In general, $CO_2$ systems may offer more possibilities for efficient and useful heat recovery, since higher temperatures can be provided. In comparisons between different systems, this factor should be taken into account through studies on overall energy requirements for cooling and heating.

Differences in fan and pump power requirements should also be considered, particularly since air-side pressure drops and air flow rates are likely to be different, thus giving differences in fan power. In a comparison with heat pumping systems using secondary fluid circuits, the pumping power should not be overlooked, particularly if a direct-evaporation $CO_2$ system may offer the same environmental and personal safety.

## Modified Cycles

There are several reasons for modifying the basic single-stage transcritical cycle, including improvement of energy efficiency, increase of capacity for given system and component size, and adaptation of the heat rejection temperature profile to given requirements, *e.g.* in a heating system. In principle, a large number of possible modifications are possible, including staging of compression and expansion, splitting of flows, use of internal heat exchange, and work-generating expansion instead of throttling. Lorentzen [12] outlined several advanced heat pump cycles and circuits for $CO_2$, including two-stage cycles, cycles with internal 'subcooling' and cycles with expander. In order to reduce the throttling loss and to adapt the heat rejection temperature profile, cycles with two or more compression/throttling stages, internal heat transfer, subcooling, and expansion work recovery can be applied. The economic viability of the transcritical cycle is enhanced by making use of the high-temperature heat rejected, for example, for domestic hot water in stationary applications and for reheating/defogging in mobile applications. Theoretically the same options are available in subcritical systems, but the relatively small amount of recoverable high-temperature heat has meant that it is usually wasted. The potential payoff is generally greater in $CO_2$ systems. Therefore, in transcritical heat pumps many more options exist for reversing flow between heating and cooling modes and for meeting simultaneous loads. Placement of the reversing valves is further complicated by the existence of the internal heat exchanger, where decisions must be made about preferences for counter vs. parallel flow in heating mode.

## Heat Transfer and Fluid Flow

Most studies on heat transfer and pressure drop in $CO_2$ system have focused on supercritical cooled flow, and flow vaporization, in micro channel tubes and in larger-diameter tubes. The term 'micro channel' is used for flow channels with hydraulic diameter less than 1 mm.

Olson [13] measured heat transfer for cooled supercritical $CO_2$ flow in a 10.9 mm ID (inner diameter) tube. Comparisons were made between the test data and correlations of Gnielinski [14] and a Krasnoshchekov and Protopopov [15] model developed for supercritical fluids. Olson [13] found that the Gnielinski model under predicted the measured coefficients, and more so when using wall-based property data than bulk-based. The special supercritical model worked well as long as the temperature was above the pseudo critical temperature, but gave large scattering and over prediction at lower temperatures.

Pitla *et al.* [16] concluded that for fully developed flow the supercritical-pressure heat transfer coefficient of $CO_2$ increases gradually as the flow is cooled, until a peak is reached at the pseudo critical state (maximum isobaric specific heat).

## Flow Vaporization Heat Transfer

Sun and Groll [17] conducted flow vaporization experiments for pure $CO_2$ on a horizontal 4.6 mm ID (inner diameter) stainless steel tube. Test data were recorded at $CO_2$ mass flux between 500 and 1670 kg/m$^2$-s, heat flux 10–50 kW/m$^2$ and vapor fraction 0–0.95. Evaporating temperatures were maintained between –2 and +10 °C.

In general, the heat transfer coefficient dropped at increasing x: A more or less abrupt drop in heat transfer above a vapor fraction of 0.4–0.6 was observed. The heat transfer was not influenced much by varying mass flux at low vapor fractions, while heat flux variation had significant influence.

## Issues Related to High Operating Pressure

Pressures in $CO_2$ systems are typically 5–10 times higher than with conventional refrigerants, and this gives several effects that influence the design of components and their performance, in particular regarding compression and compressor design, and heat transfer and heat exchanger design. In addition, high pressure may create perceived safety problems unless the underlying issues are taken into consideration properly.

### High Pressure Compression

The compressor for the transcritical $CO_2$ cycle requires thicker walls to contain the high operating pressure, but since the volumetric capacity of the fluid is large, the compressor itself will actually be smaller than refrigeration compressors for the same capacity using conventional refrigerants [18]. With appropriate design it has been shown that leakage losses and piston blow-by may reduce to zero. [19].

### High Pressure Heat Transfer

Tolerable pressure drops in heat exchangers become higher as the pressure level increases, and this gives a possibility of improving heat transfer through higher flow velocities in high-pressure systems. This is of particular importance for single-phase heat transfer in the gas cooler of $CO_2$ systems. High pressure and proximity to the critical point gives increased specific heat, again leading to improved convective heat transfer. Evaporator pressure drop leads to reduced temperature differences due to the corresponding drop in saturation temperature. Pettersen [11] showed how the relation between pressure drop and temperature loss in evaporators. Owing to the higher pressures, optimum compact heat exchanger designs for $CO_2$ generally tend to use small-diameter flow channels. Peterson [20] showed some compact heat exchangers concept for $CO_2$ air conditioning systems having internal tube diameter of 2 mm.

### Compactness of Equipment

Low-side refrigerant line diameters (inner diameter) are typically reduced by 60–70 per cent compared to HFC systems, due to the higher vapor density and flow velocity high-side piping dimensions will also be reduced. Due to small diameter the pressure capability will be sufficient and compressor and heat exchanger size and weight reductions seem possible due to the reduced refrigerant-side volumes and cross-sections.

### High Pressure Safety Issue

Factors discussed by Pettersen *et al.* [21] in relation to $CO_2$ were

- ☆ Flammability. $CO_2$ is non-inflammable. Thus, flammability is not an issue.

- ☆ Inhalation safety. Although $CO_2$ is usually regarded as non-toxic, there are physiological effects from breathing air with a $CO_2$-concentration above a few percent. (2–3 per cent concentration by volume).

The assessment of hazards resulting from accidental leakage of $CO_2$ into occupied space should consider factors like

- ☆ Rate of $CO_2$ outflow at decreasing system pressure,
- ☆ Formation of dry ice inside system at 0.52 MPa system pressure,
- ☆ Amount of $CO_2$ dissolved in lubricant,
- ☆ Room/cabin ventilation rate, and
- ☆ Stratification ($CO_2$ is heavier than air).

Frost burn is probably not a problem with $CO_2$ since the triple point is at 0.52 MPa.

### Explosion or Rupture

High pressure is not a safety issue in itself because the component will be designed for this.

## Component Design Issue

### Compressors

The vapour pressure of $CO_2$ is higher than conventional refrigerants and the transcritical $CO_2$ cycle operates at much higher pressures than the conventional vapor compression systems. Higher pressure gives special requirements regarding the design of suitable components, especially compressors for the $CO_2$ systems. Compressor technology for the $CO_2$ transcritical systems has reached an advanced level after years of development. Fagerli [22] evaluated the possibilities for transcritical compression of $CO_2$ in small hermetic reciprocating compressors. Measured isentropic efficiencies for the $CO_2$ compressor were 9–15 per cent lower than the R-22 machine, while volumetric efficiency was less than 5 per cent lower. Neksa° *et al.* [23] developed a series of semi-hermetic reciprocating $CO_2$ compressors in the range of 1.7–10.7 $m^3/h$ swept volume. The isentropic and volumetric efficiencies were relatively good. The European company Dorin developed the first high pressure semi-hermetic $CO_2$ compressor series in the range of 1.7–10.7 $m^3/h$ swept volume. [24].

### Heat Exchangers

Heat exchangers for mobile and unitary equipment are designed with a finned air-side surface and usually have more than 700 $m^2$ surface area per $m^3$ core volume. This ratio is the loosely defined limit for a compact heat exchanger. The heat exchangers may be mechanically expanded flat-fin/round-tube units or brazed aluminium cores as shown in Figure 57.10. The high working pressure and favorable heat transfer properties of $CO_2$ enable reduced tube diameters and small refrigerant-side surface areas. Pettersen *et al.* [20] reported micro channel heat exchangers (gas cooler and evaporator) could be used for $CO_2$ air-conditioning systems.

### Lubricants

A variety of lubricants can be used for $CO_2$ applications. Heide and Fahl [25] studied the miscibility in $CO_2$ of several candidate lubricants for transcritical systems, including alkyl benzene (AB), alkyl naphthalene (AN), polyalkylene glycol (PAG), polyolester (POE), and polymer ester (PME). The POE lubricants showed good miscibility. PME and aromatic ester derivates have a good potential for applications with high temperatures. [26].

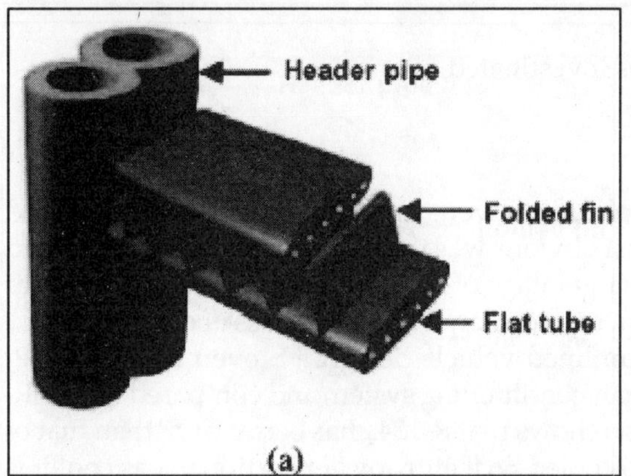


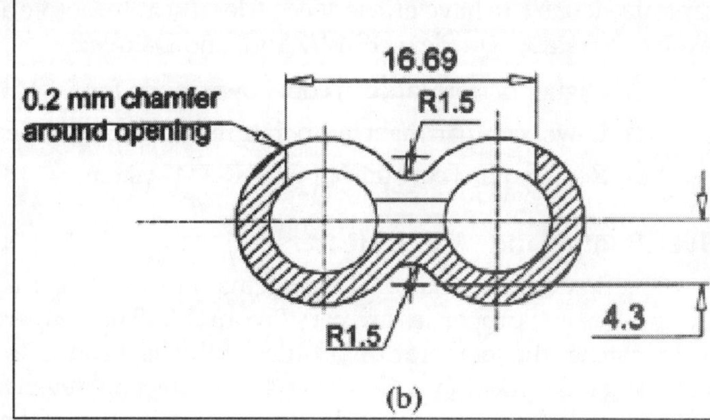


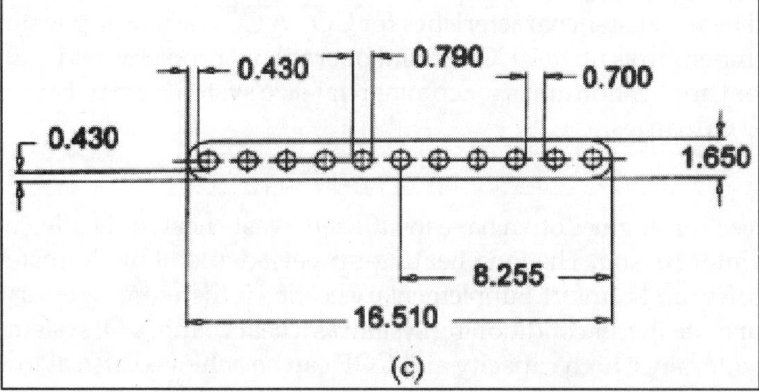


**Figure 57.10: A Prototype Micro Channel $CO_2$ Gas Cooler for a Car Air Conditioning System [76] (a) Geometry of Heat exchanger; (b) Cross Section of Header Pipe; (c) Flat Microchannel tube Channel Tube**

## Valves and Controls

Jain *et al*. [27] has developed components for $CO_2$ air conditioning systems, including electronic expansion valves, accumulators, hoses, o-rings and fittings. In addition, Obrist Engineering in Austria is offering valves, controls and other components for prototyping purposes.

## Application Areas

Recent research on transcritical $CO_2$ systems has investigated a variety of possible applications. Some of the noted applications are as follows:

### Automotive Air Conditioning

For several reasons, mobile air conditioning applications were among the first to be considered for application of the transcritical $CO_2$ cycle. The most obvious was the high leakage rate of R-12 and R-134a. In automobiles, the chief thermodynamic disadvantage of $CO_2$ is its high temperature of heat rejection–makes it possible to design an ultra-compact gas cooler, which increases energy efficiency indirectly by enabling more aerodynamically streamlined vehicle design. J Steven Brown [2002] evaluates the performance merits of $CO_2$automotive air conditioning system and compared it with R-134a automobile air conditioning system. His analysis shows that R-134a has better COP than that of $CO_2$, which is due to large entropy generation in gas cooler. So if entropy generation in gas cooler is minimized, one can have efficient eco friendly automotive air conditioning system. The consistent results from studies made by BMW, Audi and Daimler:

- ✰ Higher performance in cool down mode for R-744 than for R-134a.
- ✰ Lower compartment temperature and faster temperature fall with R-744.
- ✰ Reduced fuel consumption for R-744 system.

### Heat Pumps and Water Heaters

The first application of $CO_2$ systems on the market is heat pump water heaters, where the thermodynamic properties are very favourable. The temperature characteristics of the transcritical cycle matches the temperature profiles of the heat source and heat sink, giving small heat transfer losses and high efficiency. COP above 4 was achievable even for a hot water temperature of 60 °C. The high process efficiency is partly due to good adaptation of the process to the application, but also due to efficient compression and the good heat transfer characteristics for $CO_2$. A $CO_2$ heat pump water heater may produce hot water with temperatures up to 90 °C without operational problems and with only a small loss in efficiency. Important application areas for commercial-size systems are in hotels, apartment houses, hospitals, and food industries.

### Automotive Heat Pumps

Modern cars with efficient fuel-injection engines often have insufficient waste heat for heating of the passenger compartment in the winter season. The long heating-up period and slow defroster action is unacceptable both in terms of safety and comfort. Supplementary heating is therefore necessary, and one attractive solution may be to operate the air conditioning system as a heat pump. $CO_2$ systems have special benefits in heat pump mode, since high capacity and COP can be achieved also at low ambient temperature and with high air supply temperature to the passenger compartment.

### Heat Pump Dryers

Based on theoretical considerations, Steimle [28] reported that energy saving is possible due to better temperature adaptation in the heat exchangers, compared to subcritical processes. It is also possible to achieve higher air temperatures without loss in efficiency, thus increasing the moisture extraction rate.

## Transport Refrigeration

Research interest in $CO_2$ has also been renewed in the area of transport refrigeration for two reasons. The first relates to the relatively high density and capacity of $CO_2$ at low temperatures, compared to alternatives such as hydrocarbons or ammonia; the advent of lightweight compact microchannel heat exchangers presents new opportunities for system optimization. Second, the worldwide availability of $CO_2$ and freedom from HFC-related regulatory uncertainties fits well with the global nature of the transport refrigeration industry. Preliminary test results on a prototype $CO_2$ system for truck refrigeration gave COP data that matched equally sized systems using R-502 and R-507. [29]. Jakobsen and Neksa [30] conducted more detailed simulations including the effects of capacity control and varying compressor efficiency at varying compression ratio. The results showed very similar COP values in freezing mode for $CO_2$ and R-134a over the full range of ambient temperature. In cooling mode, the excess capacity was much greater with R-134a than with $CO_2$ due to differences in refrigerant properties. When the influence on COP by suction throttling or cylinder unloading was included, the estimated COP in freezing mode became slightly (3–10 per cent) higher for the $CO_2$ system than for the R-134a system. One problem with $CO_2$ may be very high compressor discharge temperature for freezing operation at high ambient temperature.

## Commercial Refrigeration

Commercial refrigeration systems for shops, supermarkets, larger kitchens, etc. have large refrigerant emissions, and the energy use is in many cases high. Thus, there is a need for efficient, safe and environmentally friendly refrigeration systems. New concepts based on $CO_2$ have been demonstrated for centralized systems using $CO_2$ as a secondary heat transfer fluid or in a low-temperature cascade stage. The $CO_2$ system was found to reduce the total energy consumption for refrigeration and heating by 32 per cent compared to the R-22 system. Each $CO_2$ unit can also be equipped with a condensing unit in order to reject heat directly to the shopping area when space heating is required. In the warm season with a heat surplus, the waste heat recovery circuit removes the heat. This concept reduces the power demand for the refrigeration units to the same level as for the baseline R-22 system, and the resulting overall energy consumption of the supermarket will then be further reduced.

## Conclusion(s)

The state of the art for the transcritical $CO_2$ cycle technology in various refrigeration, air-conditioning and heat pump applications was critically reviewed. The key conclusions are:

1. After the Montreal and Kyoto Protocols situation began to change dramatically over the last decade. Chlorofluorocarbon refrigerants were phased out, and the prospect of global warming signalled higher energy costs and controls on fluorocarbon refrigerants.
2. Refrigerants will be chosen in the future for their environmental safety and compatibility with various cycle-improving component designs, rather than their simple-cycle ideal efficiency.
3. Re-emergence of the natural refrigerant $CO_2$, its thermodynamic and transport properties, basic $CO_2$ transcritical cycles and some options of cycle modifications (advanced cycles), heat transfer and pressure drop characteristics in $CO_2$ systems, and issues and design characteristics related to high operating pressure were critically reviewed.

4. The re-emergence of the 'old refrigerant' $CO_2$ provides a historical context for making several important observations about the path of technological innovation in air conditioning and refrigeration systems and components. It also explored component design issues and possible applications, and some barriers to be overcome before commercialization.
5. The prospects for $CO_2$ and its transcritical cycle are still uncertain, mainly because of the many simultaneous component and system innovations required, *e.g.* to handle the high pressures, and to develop micro channel evaporators that tolerate frost and condensate while distributing refrigerant uniformly. The greatest obstacle it faces is its high temperature of heat rejection, which explains why its first commercial opportunities are occurring in water heating, and in space heating applications where high delivery temperatures are required. Its inherent advantage–negligible global warming impact–has focused most of the early research on automotive air conditioning applications where direct refrigerant leakage has been a significant contributor to global warming.

## References

1. Lorentzen G, Pettersen J. New possibilities for non-CFC refrigeration. Proceedings of IIR International Symposium on Refrigeration, Energy and Environment, Trondheim, Norway, 1992, pp 147-163.
2. Rubenstein R, didion David, Dozier Jeff. The Treatment by LEED of the Environmental Impact of HVAC Refrigerants. US Green Building Council, Washington DC, September 28, 2004, pp 7-24.
3. The Montreal Protocol on Substances that Deplete the Ozone Layer, Published 2000 by Secretariat for The Vienna Convention for the Protection of the Ozone Layer and The Montreal Protocol on Substances that Deplete the Ozone Layer, United Nations Environment Programme, PO Box 30552, Nairobi, Kenya, pp 1-54.
4. Calm James M, Didion David A. Trade-Offs in Refrigerant Selections: Refrigerants for 21st Century. ASHRAE/NIST Refrigerants Conference, October 6-7, 1997, pp 6-19.
5. Atwood, Hughes HM. Refrigeration and Energy Efficiencies Proceedings of the IIR Conference, Purude, 1990 and ASHRAE CFC Conference, West Lafayette, IN, Purude, July 1990, pp 82-89.
6. Kim MH, Pettersen Jostein, Bullard Clark W. Fundamental Process and System Design Issues in $CO_2$ Vapor Compression Systems. Progress in Energy and Combustion Science, Vol 30, 2004, pp 119-174.
7. Pettersen J. Flow vaporization of $CO_2$ in microchannel tubes. PhD Thesis, Norwegian University of Science and Technology, Norway; 2002.
8. Pettersen J, Skaugen G. Operation of trans-critical $CO_2$ vapour compression systems in vehicle air conditioning. IIR International Conference on New Applications of Natural Working Fluids in Refrigeration and Air Conditioning, Hanover, Germany; 1994. pp. 495–505.
9. Inokuty H. Graphical method of finding compression pressure of $CO_2$ refrigerating machine for maximum coefficient of performance. The Fifth International Congress of Refrigeration, Rome; 1928. pp. 185–92.
10. Pettersen J. Process with high-pressure control. Kohlendioxid: Besonderheiten und Einsatzchancen als Ka¨ltemittel, Statusbericht des DKV, No. 20, German Association of Refrigeration and Air Conditioning; 1998. pp. 64–74.

11. Pettersen J. Experimental results of carbon dioxide in compression systems. ASHRAE/NIST Conference on Refrigerants for the 21st Century, Gaithersburg, MD; 1997. pp. 27–37.

12. Lorentzen G. Revival of carbon dioxide as a refrigerant. International Journal of Refrigeration 1993; 17(5):292–301.

13. Olson D. Heat transfer of supercritical carbon dioxide flowing in a cooled horizontal tube. In: Groll EA, Robinson DM, editors. The Fourth IIR-Gustav Lorentzen Conference on Natural Working Fluids, West Lafayette, IN. 2000. pp. 251–258.

14. Gnielinski V. New equations for heat and mass transfer in turbulent pipe and channel flow. Int Chem Engng 1976;16(2), pp 359–368.

15. Krasnoshchekov E, Protopopov V. A generalized relationship for calculation of heat transfer to carbon dioxide at supercritical pressure. Telofizika Vysokikh Temperatur 1972;9(6):1314.

16. Pitla S, Robinson D, Zingerli A, Groll E, Ramȧdhyani S. Heat transfer and pressure drop characteristics during in-tube gas cooling of supercritical carbon dioxide. ASHRAE 913-RP, report HL 2000-10 No. 3613-1, Herrick Laboratories, Purdue University; 2000.

17. Sun Z, Groll E. $CO_2$ flow boiling in horizontal tubes. Internal report no. HL-2001-8, Ray W. Herrick Laboratories, Purdue University; 2000.

18. Groll E, Cohen R. Review of recent research on the use of $CO_2$ for air conditioning and refrigeration. CLIMA 2000, The Seventh Rehva World Congress, Naples, Italy; 2000.

19. Su¨ss J, Kruse H. Einfluss von Leckage auf die Effizienz von Verdichtern fu¨r ohlendioxid. Ki Luft- und Ka¨ltetechnik 1997; pp. 173–176.

20. Pettersen J, Hafner A, Skaugen G, Rekstad H. Development of compact heat exchangers for $CO_2$ air-conditioning systems. Int J Refrig 1998; 21(3), pp. 180–193.

21. Pettersen J, Bra°na°sM, Hafner A. Some safety aspects of $CO_2$ vapor compression systems. IEA Annex 27 Workshop Proceedings: Selected Issues on $CO_2$ as Working Fluid in Compression Systems, Trondheim, Norway 2000; pp. 61–75.

22. Fagerli B. On the feasibility of compressing $CO_2$ as working fluid in hermetic reciprocating compressors. Dr Ing Thesis. Department of Refrigeration and Air Conditioning, Norwegian University of Science and Technology, Norway; 1997.

23. Neksa° P, Rekstad H, Zakeri G, Schiefloe P, Svensson M. Commercial heat pumps for water heating and heat recovery. IEA/IZWe.V./IIR Workshop on $CO_2$ Technology in M.-H. Kim *et al.*,/ Progress in Energy and Combustion Science 30 (2004) 119–174 171 Refrigeration, Heat Pump and Air Conditioning Systems, Mainz, Germany.

24. Pettersen J, Neksa° P. $CO_2$ refrigeration, air conditioning andheat pump technology development in Europe. Mag Soc Air-Conditioning Refrig Engrs Korea 2002; 31(7), pp 53–64.

25. Heide R, Fahl J. Mischbarkeit von Schmiero¨len mit Kohlendioxid. KI Luft- und Ka¨ltetechnik 2001; No. 10, pp 456–470.

26. Fahl J, Bruns B, Langenberg E, Po¨tke W. Thermoanalyse synthetischer Schmiero¨le mit DSC und TGA. KI Luft- und Ka¨ltetechnik 2001;No. 7, pp 309–313.

27. Jain V, Haramoto C, Shilad I, Pfister J. Components for $CO_2$ A/C systems. The SAE Automotive Alternate Refrigerants Systems Symposium, Scottsdale, AZ; 2000.

28. Steimle F. $CO_2$-drying heat pumps. $CO_2$ Technology in Refrigeration, Heat Pump and Air Conditioning Systems, Trondheim, Norway; 1997.

29. Sonnekalb M, Ko¨hler J. Transport refrigeration with a transcritical refrigeration cycle using carbon dioxide as refrigerant. Proceedings of the International Conference on Ozone Protection Technologies, Washington DC; 1997. pp. 124–33.

30. Jakobsen A, Neksa° P. Carbon dioxide in marine refrigeration applications. Earth Technologies Forum Conference, Washington, DC; 1998. pp. 153–62.

# Chapter 58

# Pesticide Residues in Environment

**H.S. Rathore and Charu Varshney**
**Department of Applied Chemistry**
**Z.H. College of Engineering and Technology**
**Aligarh Muslim University, Aligarh – 202 001, U.P.**

## Pesticides

The term pesticide is derived from Latin word *cida* means *"to kill"*. It is defined as any substance, preparation or organism used to protect plants, wood or plants products from harmful organisms; to regulate the growth of plants; to give protection against the harmful creatures or to render such creatures harmless. Thus the term pesticide embraces large group of class which includes avicides, chemosterilants, defolints, fungicides, growth regulators, herbicides, insecticides, nematicides, piscicides, repellants, rodenticides, surface biocides and wood preservatives etc. Since weeds are not pests like insects, fungi or bacteria so a broader term 'biocide' can be used to include herbicides.

Pesticides are important component of high–input farming. Therefore, successful large–scale agriculture requires that the vegetation should be free from disease and the land reserved for commercial crop should be free from unwanted growth. Thus pesticides are given a share of credit for the green revolution. Hence, there is a strong tendency in our present civilization to resort to the use of pesticides. On the other hand, usage of pesticides in agriculture has imposed many direct and indirect undesirable effects on the environment. Indiscriminate use of pesticides like DDT, dieldrin and aldrin is of great concern because these chemicals are passed on from one organism to another through all the links of food chain and can ultimately be accumulated in the fat of human body.

Pesticides have high stability, low vapour pressure and very low solubility in water, but have substantial solubility in oils and fats. For example, the concentration of DDT in freshwater is about 0.00001 ppm. It tends to become more concentrated from a tropical level to another due to the process of biological magnification. As a result of biological magnification fish and predatory birds are very susceptible to chlorinated hydrocarbons like DDT, dieldrin and aldrin. Dieldrin has been found to affect the calcium metabolism in predatory birds and this has resulted in reproductive failure. Thus, the air we breath, the water we drink, the food we eat and the soil we grow plants on, are all polluted by pesticides. The contaminated environment endangers man, animal and the plant life leading them towards most dangerous future. Therefore there is a genuine need of pesticide residue analysis in our ecosystem.

Men and animals are exposed to these pesticides during manufacture, distribution, and application, consumption of food contaminated with them or through ecological disturbance.

Now the uses of biocides have been increased by several folds. They play an important role in homes, industries, agriculture and in irradiating dangerous diseases in plants and seeds to increase the food production. At present more than 10,000 different pesticides are widely used.

Pliny proposed the use of arsenic to kill insects as early as about 70 A.D. The Chinese used arsenic sulphide in the late sixteenth century. Since the early nineteenth century, certain inorganic compounds like arsenate, Paris green (copper arsenite and acetate), sodium fluorosilicate, Zinc phosphide etc. have been used as insecticide. It was not realized at that time that arsenical pesticides could persist in soil upto forty years.

The era of synthetic organic pesticides began around 1940. The commonly known insecticide DDT was originally synthesized in 1874 and rediscovered as an insecticide in 1939. DDT was soon followed by benzene hexachloride (BHC). Since then thousand of compounds have been synthesized and used in agricultural farms.

## Classification of Organic Pesticides

Pesticides are broadly classified according to their general chemical nature into seven, principal types.

### Organochlorines or Chlorinated Hydrocarbons

These include DDT, BHC, endosulfan, aldrin, heptachlor, chlordane etc. They are not biodegradable and are persistent in the environment for a long period. They are readily soluble in lipids and even a low concentration of these pesticides may accumulate in the body fat of mammals that might pose problems in the long run. Organochlorine pesticides have a special significance in water as they are picked up by micro-organisms mainly unicellular micro-organisms like zooplankton. Fish consume these micro-organisms and in this way pesticides residues residues enter the food chain. This process is known as biomagnification.

Organochlorine pesticide ⟶ Aquatic system ⟶ Zooplankton ⟶ Fish

↓

Mammals and Birds

Recommended dose of chlorinated pesticides ranged from 8-10kg per hectare. Organochlorine pesticides such as DDT, BHC, HCH, aldrin, chlordane heptachlor persist in the soil for many years and even decades.

### Organophosphates (OP) or Phosphorus Containing Pesticides

These include diazinon, dimecron, dimethoate, malathion, parathion etc. More potent and comparatively less persistent organophosphates have an advantage of having a non-lipophilic nature, moderate persistence, systemic nature and more potency than chlorinated pesticides. They are sparingly soluble in water and biological fluids. They are highly toxic since they inhibit the activity of enzymes.

The recommended dose of organophosphates ranged from 4-5kg per hectare. However, because of high toxicity of OP compounds to higher animals, the risk factor was higher because of irreversible nature of their effects.

## Carbamates

These include carbaryl, carbofuran, baygon, dinocap etc. They are relatively soluble in water. They have reversible mode of action, less persistence, systemic nature and adequate potency against crop pests. Normally they do not accumulate in tissues and are rapidly excreted when ingested. Thus these were considered to be less toxic than organophosphates. The widely used carbamate pesticide is isopropyl carbamate (IPC).

## Carboxylic Acid Derivatives

These include TCA, dalapon, 2,4-D, 2,4,5-T, MCPA etc. They are considered as a potent source of environmental hazards, especially when their degradation products remain active in the environment after desired purpose is over. They are used as herbicides both on agricultural lands and in eradicating weeds in non-productive areas such as roadsides and electric transmission lines. Biologically, the phenoxyalkanoic acid compound are not particularly stable. In soils these are fairly readily broken down. In the aquatic environment the anaerobic micro-organisms which often predominate are not always effective in degrading these compounds. In aerobic conditions phenoxyalkanoic acid derivatives do not seem to persist.

## Substituted Ureas

The substituted ureas such as monuron and diuron form another class of biocides. These are primarily herbicides. Their solubility in water is generally low. They do not create problems of persistence and accumulation due to their facile environmental reactions. These are moderately unstable. These fairly readily undergo dealkylation and dearylation. Therefore, they are of local concern only in the area and period of application.

## Triazines

The triazines such as atrazine and simazine are used in large quantities as a pre-emergent herbicide in corn fields. Biologically, dealkylation and substitution are the major routes of metabolic breakdown. In soils, dealkylation appears to be the major route. There is a considerable persistence of the resulting products. In aerobic aqueous systems, including moist substitution of a hydroxyl group for chlorine atom precedes ring cleavage and complete degradation. In anaerobic sediments, the breakdown is likely to be slow. Hence triazines may be of concern in the environment.

## Pyrethroids

This group includes cypermethrin, deltamethrin and fenvalerate which are commonly used in Indian agriculture. They are very stable in sunlight. These are effective against agricultural pests when used at low rate of 0.1 lb per acre. Cypermethrin is effective against biting and_ sucking insects, cowpea mosaic virus. The recommended dose of pyrethroids ranged from 25-500 g active ingredient per hectare.

### Neem Products/Neem Based Formulations

The versatile neem tree is a wonder. The limonoids present in it and its products have made it a harmless and useful insecticide, bactericide, fungicide, pesticide etc. It is likely to provide a solution to many of our incurable diseases. Neem crudes such as its kernel crush and oil which are the potential source of bio-actives were formulated as ready to use dust and water dispersible powder and emulsifiable concentrate respectively.

### Others

These include thiocyanates, dinitrophenols, formamides etc. Thiocyanates have creosote like odours, are relatively safe to use around humans and animals, and give astonishing quick knockdown of flying insects. They interfere with cellular respiration and metabolism.They are used around horses and other farm animals. The nitrophenols have been used as ovicides, insecticides, acaricides, herbicides, fungicides and blossom thinning agents. They are quite toxic and their use resulted in several widely publicized deaths. The formamides comprise a very new, small, but promising group of insecticides. They are used in the control of organophosphates and carbamate resistant pests. Their mode of action is the inhibition of an enzyme, monoamine oxidase. The mode of action of organotins is believed to be the inhibition of oxidative phosphorylation at the site of dinitrophenol uncoupling. This inhibition reduces substantially, and in the case of mites, fatally, the availability of energy in the form of adenosine triphosphate (ATP). The trialkyl tins also inhibit photo-phosphorylation in chloroplasts and can thus serve as algicide

## Sources of Pesticide Residues in Water

Pesticides may enter aquatic ecosystem either directly, indirectly or unintentionally through the following sources (Figure 58.1):

1. Rain Water
2. Spray Drift
3. Run Off from agricultural Fields
4. Industrial Effluents
5. Domestic Sewage

### Rain Water

Pesticides contaminated in vapour phase in air get adsorbed on to the dust particles which ultimately reach soil or water along with the rain water. It has been recorded 187mg/l of total DDT of which 18mg/l was DDE in rain water of Ohio, USA.

### Spray Drift

It was estimated that 30-40 per cent of pesticide used in aerial spray reaches the target while the rest fled into air. Water contaminates heavily with pesticide during accidental spray.Wind may also carry the drift to considerable distances whil the major fall of pesticides takes place near the site of application polluting water system.

### Run Off from Agricultural Fields

Generally water soluble pesticides are transported in water system while the insoluble ones get bound to the particulates and carried by water streams. Caro and Taylor have reported that 0.07 per

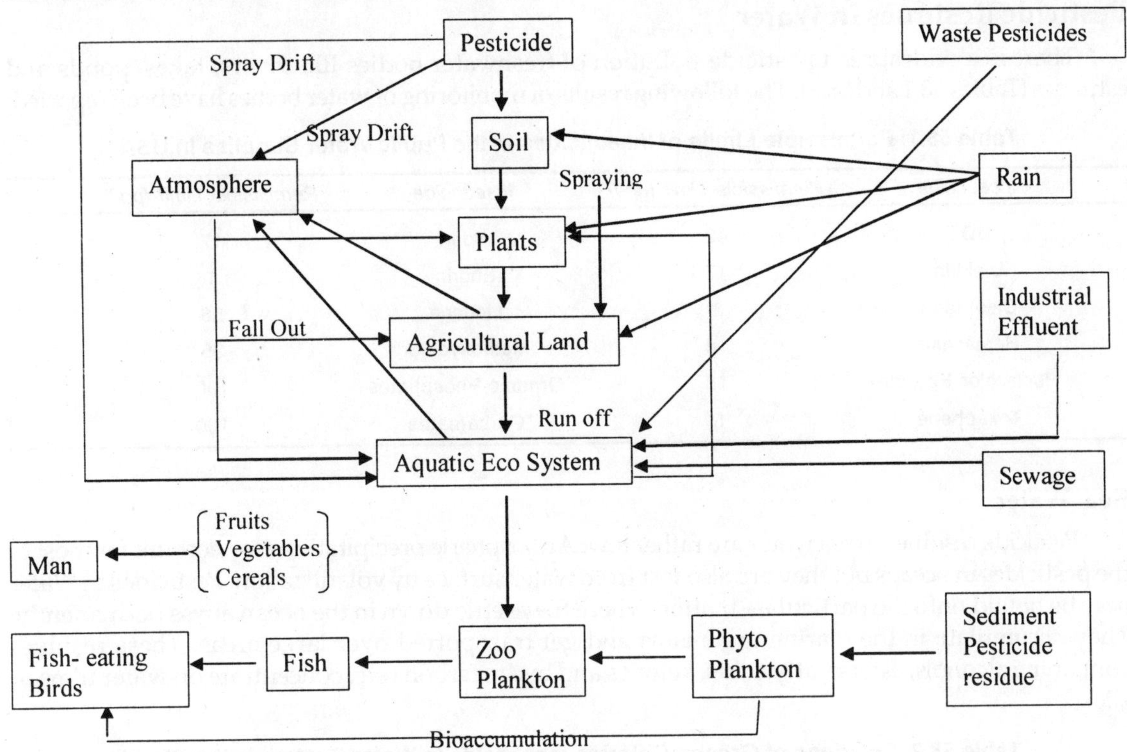


**Figure 58.1: Pesticide Residues in Environment**

cent of dieldrin applied to the soil persists in run off water. Largest quantities of DDT in water were about 70mg/l and 440mg/l in mud.

## Industrial Effluents

A huge amount of pesticide are escaped by pesticide producing industries. It has reported that large quantities of pesticides from industrial in Yorkshire and Lancashire in England. In USA, a wood factory resealed 12 to 70 mg/l of dieldrin in water.

## Domestic Sewage

ICAR have reported that in Jammu and Kashmir, fish mortality was seen in the ponds of Reasi and in Nehru stream where DDT was used in city drains. DDT concentration in Yamuna water has risen from 0.25 ppb to 0.558 ppb. after the Najafgarh Nallah drains sewage into this river at Wazirabad in Delhi.

Besides these sources other factors like the amount of suspended particulates, chemical air crafts, accidental spillage, evaporation from soils and plants, distances between the area treated with pesticide and the receiving water, forest cover, time lapse between treatment and rain fall and residues in horticultural products influence the extent of pesticide pollution in water bodies.

## Pesticide Residues in Water

There is a widespread pesticide pollution of freshwater bodies like rivers, takes, ponds and estuaries (Tables 58.1 and 58.3). The following results of monitoring of water bodies have been reported:

**Table 58.1: Permissible Limits of Insecticides in the Public Water Supplies in USA**

| *Insecticide* | *Permissible Limit (μg/l)* | *Insecticide* | *Permissible Limit (μg/l)* |
|---|---|---|---|
| DDT | 42 | Aldrin | 17 |
| Dieldrin | 17 | Endrin | 1 |
| Chlordane | 3 | Lindane | 56 |
| Heptachlor | 18 | Methoxychlor | 35 |
| Heptachlor Epoxide | 18 | Organic-Phosphates | 100 |
| Toxaphene | 5 | Carbamates | 100 |

## Sea Water

Pesticide residues in sea water are rather low. Atmospheric precipitation may account for most of the pesticides in oceans but they are also lost from water surface by volatilization. Pesticides in Water may be bound onto the particulate matter, where they settle down in the ocean abyss permanently. They accumulate in the marine organisms and get transported over large areas. These residues containing alcohols, esters, fatty acids, ketones and hydrocarbons etc. concentrate on water films or slicks.

**Table 58.2: Residues of Organo Chlorine Insecticids in Water System in the World**

| *Water Body* | *Concentration of Residue ppm.X10$^{6}$(ng/L)* | | | | | | | |
|---|---|---|---|---|---|---|---|---|
| | *DDT* | *DDE* | *Aldrin* | *Dieldrin* | *Lindane* | *Endrin* | *Endo-sulphan* | *Hepta-chlor* |
| River (USA) | 8.2 | - | - | 5.9 | 2.2 | 3.6 | - | 0.1 |
| River (UK) | 8.6 | - | - | 2840 | 53.5 | - | - | - |
| River (W. Germany | 18.5 | - | - | 3.2 | 138.2 | - | 127 | 39.5 |
| South Skunk River (USA) | - | 2 to 1800 | | 2 to 80 | | | | - |
| Mississipi River (USA) | 112.0 | - | 0.6 | 10 | 28 | 540 | - | 2 |
| Streams (USA) | 9.2 | 0.6 | - | 1.1 | 0.5 | 0.3 | - | 1.4 |
| Streams (Ontario Canada) | 64.0 | - | - | 7 | - | - | 19.6 | - |

## River Water

DDT residues account for the largest followed by lindane, dieldrin, endrin, heptachlor and its epoxide. Peak levels of these residues were observed in 1996 and thereafter there has been a decline both in residue levels and their occurrence. According to Edwards (1977) the amount of organochlorine residues is observed one ppb which do not cause any acute toxicity to fish and aquatic biota.

## Estuaries

Estuaries are accumulated with low quantities of pesticides. For example, in Louisiana estuary, less than 1 ppb of dieldrin and endrin were found in surface water and four times in bottom sediment. The DDT residues in esturine molluscs are less than 1 ppm.

## Persistent Pesticides

Pesticide residues are found in various segments of environment. Their abundance on the environment depends on the following factors:

1. Tendency to dissolve in water
2. Tendency to vapourize, and
3. Chemical stability towards degradation processes.

Pesticides in the soil act as reservoirs from where they find their way in water bodies. In India about 97420 tonnes of pesticides are used for agricultural developments every year. An estimate showed that out of 20,000 tonnes of DDT used, about 25 per cent get entered to oceans.

Accumulated pesticides disturb the ecosystem severely. But we are still unable to evaluate and eliminate their permanent effects on living beings. Recently scientific organizations like ICAR, CSIR, Agricultural and non-agricultural universities, Department of Science and Technology and Department of Environment etc. are engaged in developing safer approaches and alternatives to persistent pesticides.

## Exposure to Pesticides

Human beings are exposed to pesticides during their production, formulations, packaging, transportation and application to vegetation. These occasions provide opportunity for contact, inhalation and even ingestion of these toxic agents. In addition to occupational exposures, entry into food channels of man and his economic species may occur through residues in or on raw agricultural commodities or processed or packed food produced there from traces of pesticides may be ingested by every one, for short or long periods at various stages and even throughout the life cycle. This has led to the monitoring of various food stuffs in different parts of the country.

These pesticides get ultimately accumulated in the body tissues. So several samples from FCI godowns and markets of Delhi, Mysore, Hyderabad, Ludhiana, Roorkee and Pant Nagar were analysed. Dr. K. Krishnamurthy of Indian Grain Storage Institute, Hapur had revealed that mixing of DDT/BHC with the grain is a common practice in areas around Hapur and Delhi.

## Bioaccumulation of Pesticides

Pesticides present in water get concentrated in the tissues of plants and animals. The extent of accumulation indicates that how pesticides get transported in the food chain and pose hazardous effects. Experiments have shown that aquatic animals like protozoa-Blepheristna intermediate concentrate DDT by about 3000 fold in 12 hours. It was observed that on exposure to very low concentration of DDT in fish Gambusia affinis accumulates it 25000 folds in a few hours.

### Factors Affecting Bioaccumulation of Pesticides

Bioaccumulation depends on several factors. For example:

1. Lipid and Water Partition Coefficient,

2. Nature of Aquatic Organisms,
3. Environmental Conditions *i.e.* Temperature, Salinity, pH of the Medium etc.
4. Concentration of Pesticide,
5. Duration of Exposure of Pesticide.

Organochlorines accumulate to a greater extent than organo phosphorous and carbamates. They are hydrophobic and concentrate 1000 times over the levels existing in water.

### Measurement of Bio-accumulation

The extent of bio-accumulation can be measured by:

1. Model Ecosystem (Direct Method)
2. By determining octanol per water partition coefficient of the pesticides (Indirect Method).

## Impact of Pesticide Pollution

The use of pesticides is a necessary evil of the modern agro-technology and perhaps their demand would never like to cease. Actually they are directed against economic pests but they also affect non-target organisms. Their total impact is as follows:

### Impact on Non-Target Organisms

1. These organisms accumulate several biocides from the environment which affect their populations with high mortality due to their toxicity.
2. Pesticides result in sterility, death and several dysfunctions.
3. Sometimes pesticide affects the embryonic stage during development rather than the adults. It was observed that when the DDT level in trout eggs was 3 ppm, some fly could survive, but above 5 ppm. Mortality of the young fly was 100 per cent.
4. Sub-lethal doses of DDT bring behavioral changes in non-target organisms. It was observed that on exposing towards DDT, salmon become extremely sensitive to cold water and lay eggs in abnormally warm water. Even 0.1 to 20 ppb levels of DDT cause mosquito fish to prefer saline water than usual for the species.

### Destruction of Insect Predators

Broad spectrum pesticides like endrin, dieldrin and DDT etc. may kill the insect predators. For example, the destruction of vadalia beetle (lady bug predator) which controls the cottony cushion scale *i.e.* a part of citrus trees in California, results in dramatic increase of scale population when DDT was used on citrus trees in 1947.

### Increasing Resistance in Insects

Repeated use of biocide may result in immunity in the target pest. It was first of all reported in house fly in Italy and California. By 1980 DDT resistance was seen in anopheles mosquitoes (60 species), fleas (Plague) and lice carrying typhus. Today 305 species including alfa alfa weevil, cotton ball worm, cabbage maggot, colorado potato beetle and sheep blow fly have developed resistance. DDT in 1945 had effectively controlled mosquito population but within 6 years DDT resistant mosquitoes developed. So a new insecticide ethyl parathion was applied which was replaced by methyl parathion which also became ineffective in 1963. Therefore, Flenthion was used in 1968. Now there is no effective insecticide that can kill pasture mosquito.

### Impact on Environment

Pesticides gradually seep into ground water and contaminate drinking water. These molecules co-distil along with water molecules into the air. Chlorinated hydrocarbons adhered to soil particles may also be carried by dust erosion and eventually return to soil and water systems, thereby affecting the whole biosphere.

## How Pesticides Endanger Our Life

Pesticides acutely affect man, animal, plants, soil as we as the aquatic biota. Toxicity towards the aquatic flora and fauna is an important criterion for a pesticide to be considered as a hazardous pollutant. Pesticides affect in several ways as follows:

### Effects on Man

1. It was recorded higher concentration of DDT in body parts and blood of human beings causing anxiety, tension, cancer, mutations, stress reactions, congenital in uteri malformations and Impotency.
2. Central nervous system is the target of DDT poisoning in man which ultimately leads to death. It dissolves in lipids and accumulates in the body fats (3 to 90 ppm).
3. The most threatening DDT concentrates and accumulates in the food chain. It is continuously recycled in living organisms. The level of DDT estimated in human tissues lies in range of 5 to 10 ppm. According to estimation DDT concentration in man's body fat varies from 3.3 g/$m^3$ in UK to 25 g/$m^3$ in India.
4. It has suggested that low levels of DDT cause cancer, high blood pressure and cirrhosis of the liver. It is well known that human cancer could be caused by 50 per cent of the pesticides used in United States. DDT also interferes with signal transmission at the synoptic gap.
5. Insecticides like BHC, Aldrin, Dieldrin, Chlordane and endo-sulfan are extremely toxic to living beings. These are considered to affect the vital organs, heart, brain, kidneys and liver producing chronic disturbances.
6. Pesticide 2,4,5-T consists of highly toxic component called dioxin or TCDD which is teratogen and liable to cause congenital malformations.
7. Organophosphates like Ethion, Fenthion, Dimethoate, Trithion, Monocrotophos, Diazinon, Dursban, Phosdrin, Thionazin and Metasystox inhibit the production of cholinesterase at the junction between adjoining nerve cells with the result that cholinesterase breaks down acetyl choline secreted by nerve cell axons. Excessive accumulation of acetylcholine interferes with the nerve impulse transmission.
8. The organophosphorus insecticides such as parathion, malathion and TEPP may be absorbed by the lungs, eye membrane and skin to toxic amounts. This excessive absorption leads to larger accumulation of acetyl choline in the body disturbing the normal functioning of blood.
9. Chronic accumulation of pesticides plays a major role in liver and kidney malfunctioning, secretion of excess of amino acids in human blood and urine, blood abnormalities as well as electroencephalo-gram deformations of brain tissues.

10. The danger of long term consumption of pesticide residue in food is far more serious than acute poisoning from the view point of national health.
11. Today children born have to start their life with a body burden of pesticides which increases with age. Pesticides are also reported to cause chronic impotency in man.
12. It had reported that 2,4.5 tri chloro phenoxy acetic acid is teratogenic and fetocidal in lower mammals.
13. Pesticide reported deaths in developing countries are estimated as 10,000 per year while two million persons suffer from acute pesticide poisoning.
14. Pesticide poisoning in Kerala (1951) resulted in death of 112 people while in Indore (1968) 35 cases of malathion poisoning were reported. Bhopal industrial disaster was the worst ever pesticidal MIC accident in history, taking an unprecedented and still unaccounted death toll leaving no fewer than 50,000 effected, quickly dwarfed by the tidal wave of man's suffering that spread across the central city like the poison cloud. This carbaryl (carbamate pesticide) leakage caused an increased risk of sleeping, digestive problems, vision problems, sterility, kidney and liver infections as well as the brain damage.

### Effects on Animals

1. Insecticides such as aldrin, dieldrin, chlordane, endosulphane, heptachlor and gammaxane are reported to affect the wild life by changing their metabolic activities and body chemistry.
2. Recent reports have indicated that several farm animals died by drinking water stagnating in sprayed fields and contaminated water channels.
3. 

   Animals which have become weak and moribund as a result of exposure to pesticides may easily be destroyed by predators.
4. Chlorophenoxyl acid compounds such as 2,4-D affect embryos while triazones may cause mutagenic effects in animals.
5. Organophosphate pesticides cause extreme muscular weakness, tremors and dizziness in poisoned animals.
6. In higher animals, pesticide poisoning results in concentration of these chemicals in their passage through food chains.

### Effects on Birds

Pesticides such as parathion, malathion, aldrin, dieldrin, heptachlor and several other organochlorines are reported to affect severely the metabolism in birds. Many species of hunting birds, particularly those having high levels of DDT are threatened with extinction.

DDT has been implicated as the main villain and the cause of thin and fragile egg shells because of inhibition of enzymes and interference with the hormones that control calcium metabolism.

#### Pesticide Pollution Reduces Egg Laying in Birds

Due to pollution of surface waters with substances such as poly chlorobiphenyls (PCBs) and pesticides, birds laid fewer eggs and fewer youngs are hatched.

Researchers of the Applied Marine Research Laboratory, Netherlands conducted six years experiments with tufted ducks. These ducks feed on zebra mussels found in the water of Netherlands. In polluted under water beds, the mussels accumulate the pollutants in their tissues and such polluted mussels are also eaten by some ducks.

The researchers fed one group of tufted ducks an mussels from polluted waters and other group on mussels from clean water.

The duck fed on mussels from polluted waters showed an increased concentration of pollutants in their tissues. They also have smaller kidneys, sexual organs and exhibited nervous behaviour patterns. Many of the females died at the beginning of breeding season and those that survived spent less time brooding their eggs.

Tufted ducks fed on heavily polluted mussels lay fewer eggs than the ducks feeding in clean water. Only 39 per cent of the eggs hatched while 979 per cent of the eggs from ducks feed on clean mussels produced ducklings. It had reported that percentage of eggs that hatch depends on the PCBs and other pesticide content of the egg yolk.

## Effects on Aquatic Biota

Pesticides affect aquatic organisms as follows:

1. Aquatic animals are extra ordinarily sensitive even to low concentrations of DDT. Even lone part of DDT per trillion seems to cause death in brine shrimps.
2. Acute danger arises from DDT which results in changing the behaviour and metabolic activities in sea animals.
3. DDT is extremely fatal to fish, crustaceans and shell fish.
4. It is repoıted that marine invertebrates can concentrate pesticides by a concentration factor up to 70,000 or more.
5. The $LD_{50}$ values of mosquito fish to DDT were 60 ppb in water.
6. Experiments have shown that DDT causes depression of photosynthesis in planktons. Its very low concentration as 1 ppb. has immobilizing effects on Simocephalus species.
7. Endrin is extremely toxic to fishes like blue gills and rainbow trout with $LD_{50}$ values being 0.006 and 0.007 ppm. respectively, while DDT is much less toxic to these fishes with $LD_{50}$ values being 0.016 to 0.018 ppm. respectively.
8. Pesticides badly accumulate in tissues of aquatic organisms. They make fish unable to tolerate low temperature in water.
9. Even the raptorial hawks and falcons are severely affected by DDT which causes breeding failure and death in them.
10. Recent studies have proved that extremely low quantities of pesticides which enter the aquatic environment can affect productivity of organisms, kill eggs and larvae of clams and oysters, influence the behaviour of fishes such as schooling and feeding and deteriorate water quality.
11. Pesticides induce changes in blood chemistry and enzymatic functions of marine invertebrates. They reduce their backbone collagen content and indirectly interface with food chain.
12. Since slight concentrations of organochlorine pesticides affects reproduction in fishes, there is every possibility that these pesticidal pollutants may adversely affect local fishery.
13. Varieties of crustacea that make up the most valuable marine harvest are of course representatives of some groups of animals that pesticides are designed to kill, particularly sedentary animals which are unable to move away from pollutants. The oysters might be

susceptible because of their tendency to concentrate and store trace chemicals from the surrounding environment.

14. Pesticides-induced mortality pattern of marine molluscs, crustaceans and teleosts are also measurably related to various physico-chemical environmental parameters like concentration of pesticide and duration of its exposure.
15. PCBs are reported to be concentrated in the food chain of marine ecosystems.
16. In a survey conducted by US Public Health Service it was observed that pesticides were the cause of 32 per cent fish kill in 1960, 21 per cent in 1961, 18 per cent in 1962, 30 per cent in 1970 and these pesticidal pollutants ranked second to all other industrial wastes.

### Effects on Grains

Many vegetables, fruits, rice, cereals and grains such as wheat, maize, grams and barley have been found to be contaminated with significant amounts of DDT and BHC. Edwards *et al.*, reported that pesticides have severely interfered with the desired yield of crops, during "All India Symposium on Pesticide Residues in Environment."

### Effects of Pesticides on Soil-Pesticides Contaminate Soil

Pesticides not only pose potential hazards to man, animal, livestock, wild life and fish but they seriously interfere with the desired use of soil and water as follows:

1. The most dangerous characteristic of pesticides, particularly organochlorines, is their long persistence in the soil which causes numerous adverse effects on grain quality.
2. Even the accepted dosages of pesticides create deleterious effects in the long run in the soil.
3. Pesticide retained in the soil also concentrate in some crops, vegetables, cereals and fruits which taint them to such an extent that they are not usable.
4. American Scientists observed that radish and carrots grown on a loamy soil treated with aldrin at one lb/acre contained 0.03 and 0.05 ppm. of it respectively.
5. Retention of pesticides in bottom soil of ponds renders them unsuitable for fish culture.
6. Soil contaminated by pesticides also causes pollution of water sources, water supply and ground or surface water by rain.
7. Pesticides may adhere to the soil particles, causing harm to vegetation.

### Pesticides Contaminate Air Too

Pesticidal introduction into air may occur through spraying operations, volatilizations, evaporation of chemicals from soil, plants or burning of plant products. Large quantities of DDT in human fat are attributed to the practice of spraying it inside the home in India.

In a nutshell, if we continue to rely upon broad spectrum of pesticides, the recovery of natural forms of control will become almost impossible.

## Central Facility for Safety Evaluation of Pesticides

Industrial Toxicology Research Centre (ITRC), Lucknow has established a Central Facility for Safety Evaluation of Pesticides with the following objectives:

1. To generate toxicological data on different pesticides synthesized, manufactured or formulated by national laboratories and private sector industries within the country.

2. To conduct carcinogenic, teratogenic and multigenerational studies on selected strains of experimental animals, in vivo tests using bacterial mutants, human cell lines and lymphocytes to map mutagenic and carcinogenic effects and interference with immune mechanism.
3. To study cytogenic effects of pesticides using bone marrow cells of intact animals.
4. To develop new techniques for monitoring and predicting harmful effects of pesticides on target and non-target organisms and their diffusion in the environment.
5. To provide expertise for diagnosing toxic hazards symptoms and developing methods of their prevention and control.

## Methods of Minimize Pesticidal Pollution

Fortunately, it is possible to treat persistent pesticides which contaminate in water bodies by the following techniques:

1. Less persistent and harmless pesticides can be used instead of highly persistent ones.
2. Proper utilization of persistent pesticides would minimize drifts etc.
3. Pest control may be achieved biologically by the release of sterile males and entocones.
4. Chemical methods like coagulation, volatilization, irradiation, sedimentation and filtration can be adopted which can remove up to 90 per cent DDT.
5. Industrial wastes and effluents containing pesticides can be treated before they are released into water systems.

## Biodegradation of Pesticides

Biodegradation of pesticides is extremely essential to maintain the environmental quality. It can be achieved by the following methods:

1. Pesticides can be metabolized and degraded by the living organisms. They can be detoxified and removed from the cycle.
2. By transferring the aquatic animals containing higher concentration of pesticide to a clear water where they may unload the pesticide residues. For example, freshwater snail can lost 75 per cent DDT within 24 hours in clean water.
3. Microbial degradation of pesticides can be achieved by chemical methods *viz.* hydrolysis, oxidation, dehalogenation and epoxide formation. For example, aldrin can be oxidized to dieldrin by epoxied formation. Generally oxidation may occur by introduction of -OH group into aromatic ring while reduction of pesticide occurs by converting $-NO_2$ group into $-NH_2$ group.
4. Biodegradation of pesticides is strongly affected by soil temperature, moisture and texture etc. Enzyme system existing in soil also plays a significant role in their degradation. However, biodegradation of pesticides does not always result in harmless products. Degraded intermediates may be more toxic to aquatic life than are the original pesticides; For example, biodegraded product DDD is more toxic than DDT.

## Mode of Poisoning of Pesticides

Pesticides are classified mainly into four categories depending upon their mode of poisoning action. These groups are as follows:

1. Physical Poisons
2. Nerve Poisons
3. Protoplasmic Poisons
4. Respiratory Poisons

### Physical Poisons

These pesticides kill the living organism by some physical action. For example, endrin penetrates percutaneously through the epidermis of the skin and produce lethal effects.

Silica and charcoal dusts also interfere with the inhalation of air through nasal passage, thereby accumulating in the lungs.

### Nerve Poisons

DDT, MIC, Malathion, Parathion, Diazinon, Systox etc. act as nerve poisons. DDT initiates extreme nervous excitation and causes release of excessive neuroactive substances, disrupting nerve activity. An estimate indicates that 5 to 25 ppm. of DDT gets contaminated in body tissues causing cancer, leukemia, sterility, tumour and hypertension etc.

### Protoplasmic Poisons

Insecticides and chemicals such as endrin, ziram, lead arsenate and sodium arsenite etc. cause precipitation of protein in the body resulting in liver damage and ultimately death.

### Respiratory Poisons

Pesticides included in this class inactivate the respiratory enzymes like oxidases, peroxidases and reductases. For example, fumigants such as hydrocyanic acid gas, methyl bromide, ethylene dichloride and dibromide cause acute suffocation and block the respiratory tract.

## Interaction of Pesticides

The interaction of pesticide has been reported by Secretary's Commission on Pesticides, USA. The mechanism of such interaction by organophosphorus pesticides has been ascribed to detoxifying enzymes, called aliesterase. Its accumulation in body tissues makes the animal and insects resistant to pesticides. Until now, more than 150 species of insects have been reported to gain immunity towards different pesticides.

## References

1. JT Nyangababo, L Henry, E Omutange. Organochlorine pesticide contamination in surface water, sediment and air precipitation of lake Victoria Basin, East Africa. Bulletin Environ Contam Toxicol 75 (5): 960-967, 2005.
2. GC Hose. Assessing the need for groundwater quality guidelines for pesticides using the species sensitivity distribution approach. Human Ecol Risk Assess 11 (5): 951-966, 2005.
3. YF Tang, YX Wang, HS Cai. Transport and fate of pesticide in surface water environment. Wuhan Huag Xue Xuebao 27 (4): 13-15, 2005.
4. M Jankovska, V Hammer, L Pasek. Problems of water contamination with organochlorine pesticides. Vodni Hospo 55 (3): 51-55, 2006.

5. ASY Chau, BK Afghan, JW Robinson, eds. Analysis of Pesticide in water. Vol I, II, III. Boca Raton, FL: CRC, 1981.
6. Leo ML Nollet, ed. Handbook of Water Analysis. New York: Marcel Dekker, 2000, pp 487-654.
7. WV Valkenburg, B Sugavanan, SK Khetan. Pesticide Formulation: Recent Development and their Application in Developing Countries. New Delhi: New age International (P) Limited, 1998.
8. Y Ming-HO. Environmental Toxicology: Biological and Health Effects of Pollutants. 2nd Ed. Boca Raton, FL: CRC, 2005.
9. AC Edward. Persistent Pesticides in the environment. CRC Press, Cleve land Ohio, 1973.
10. AC Edward, GK Veeresh and HR Krueger. Pesticide Residues in the environment in India. USA Tech. Series No. 32, 1978, pp-524.
11. AC Edward. Environmental Pollution by Pesticides. Plenum Publishing Co. Ltd., New York, 1973.

5. [illegible]

6. [illegible]

7. [illegible] New Delhi [illegible]

8. [illegible]

9. [illegible]

10. [illegible]

11. [illegible]

# Author Index

# Subject Index